To Papa, Prachi, and Manya

Ashish Tewari

Atmospheric and Space Flight Dynamics

Modeling and Simulation with MATLAB® and Simulink®

Springer

Ashish Tewari
Department of Aerospace Engineering
Indian Institute of Technology, Kanpur
IIT-Kanpur 208016
India

Mathematics Subject Classification: 00A71, 00A72, 65L06, 70B05, 70B10, 70E05, 70E15, 70E20, 70E50, 70F05, 70F07, 70F10, 70F15, 70G60, 70H03, 70J10, 70J25, 70J35, 70K05, 70M20, 70P05, 70Q05, 74F10, 76G25, 76H05, 76J20, 76K05, 76L05, 76N15, 76N20, 93A30, 93B52, 93B55, 93B60, 93C05

Library of Congress Control number: 2006935811

ISBN-10: 0-8176-4437-7 e-ISBN-10: 0-8176-4438-5
ISBN-13: 978-0-8176-4437-6 e-ISBN-13: 978-0-8176-4438-3

Tewari: Atmospheric and Space Flight Dynamics

First Indian Reprint 2009
Second Indian Reprint 2012

ISBN 978-81-8489-333-5

This edition is manufactured in India for sale only in India, Pakistan, Bangladesh, Nepal and Sri Lanka and any other country as authorized by the publisher.

This edition is published by Springer (India) Private Limited, (a part of Springer Science+Business Media), Registered Office: 7th Floor, Vijaya Building, 17 Barakhamba Road, New Delhi – 110 001, India

Printed in India by Rashtriya Printers, Delhi.

Preface

It is fitting in this beginning of the second century of powered flight to be writing a book on flight dynamics—in particular, the modeling and simulation of flight dynamics. No one had heard of flight dynamics, or indeed, modeling and simulation, in the early days of aviation. Pioneers such as the Wright brothers, Langley, Curtiss, and Bleriot preferred experimentation with flying models to working with the equations of motion. Until the 1940s, flight dynamics was more of an art than an engineering discipline, with books such as *Stick and Rudder: An Explanation of the Art of Flying* by Langewiesche (McGraw-Hill, 1944) imparting a fundamental, but purely qualitative understanding of airplane flight, which was useful to both budding aeronautical engineers and pilots. After the Second World War, large strides made in mathematical modeling of airplane flight dynamics were first documented in classical texts such as *Airplane Performance, Stability and Control* by Perkins and Hage (Wiley, 1949), which covered linearized, time-invariant flight dynamics, and inspired a multitude of textbooks in the field, such as those by Etkin, McCormick, Miele, etc. With the advent of supersonic, hypersonic, and space flight in the 1950s and 1960s, there was a great need for modeling and simulation of high-performance flight dynamics, which essentially involves a set of coupled, time-varying, nonlinear, ordinary differential equations. While a sprinkling of mathematical models for high-speed flight was available in the introductory texts of the 1960s and 1970s, a thorough analytical treatment of these topics was confined to specialized technical reports. Yet, there was little mention of numerical modeling and simulation. Research articles first began to appear in the 1960s, which presented numerical calculations for special examples with analog—and later—digital computers, concerning high-performance flight dynamics. As numerical science evolved and computing power grew, the sophistication of flight dynamic modeling and simulation progressed to missiles, launch vehicles, re-entry vehicles, and spacecraft. Guidance and navigation for the manned lunar missions (1968–72) essentially utilized such numerical capability. Numerical modeling and simulation of flight dynamics has now emerged as a major discipline.

The chief motivation for writing this book is to present a unified approach to both aircraft and spacecraft flight dynamics. Modern aerospace vehicles, such as the space shuttle, other launch vehicles, and long-range ballistic missiles, do not discriminate between atmospheric and space flight. Unfortunately, nearly all textbooks on flight dynamics do so, and seldom do we find aircraft and spacecraft co-existing within the covers of the same book. Many excellent textbooks are available on modern aircraft dynamics (such as those by Zipfel, Pamadi, Stengel, Etkin, and Schmidt), but they stop short of hypersonic aircraft and sub-orbital trajectories. Similarly, the available textbooks on space dynamics (Hale, Brown, Curtis, etc.) do not go below hypersonic speeds of re-entry vehicles. While it is easy to understand the separate evolution of aircraft and spacecraft in the past, the future of flight lies in integrating the two vehicles into a single unit. The single-stage-to-orbit (SSTO) reusable launch vehicle, which takes off and lands like an aircraft and delivers payload to an orbiting space station, exemplifies the vision of aerospace engineering for the future. Therefore, it is imperative that this new generation of engineers is taught to remove the artificial distinction between atmospheric and space flight. Many aerospace engineering departments realize this need, offering courses that integrate atmospheric and space flight. Examples of such courses are AE-520 *Flight Vehicle Dynamics*, AE-580 *Analytical Methods in Aeronautical and Astronautical Engineering*, and AE-621 *Aircraft and Spacecraft Automatic Control Systems*, offered by the Department of Aerospace Engineering at Ohio State University.

This book is an attempt to bridge the gap between aircraft and spacecraft dynamics, by demonstrating that the two evolve logically from the same set of physical principles. The breadth of topics covered is unparalleled by any other book on the subject. Beginning with kinematics and translational dynamics over a rotating planet, nonspherical gravity model, leading to two-body orbits, orbital maneuvers, rendezvous in space, and lunar and interplanetary travel, atmospheric flight follows logically after chapters on atmospheric modeling, aerodynamics, and propulsion. The attitude dynamics and control of aircraft and spacecraft are presented in an integrated and continuous fashion. Modeling of nonlinear flight dynamics is covered in numerous examples, such as simulation of long-range airplane flight, supermaneuvers, rocket ascent, sub-orbital flight and atmospheric entry, multi-axis rotations of spacecraft, and inertia coupled, open- and closed-loop airplane dynamics, which are not found in other textbooks on flight dynamics. The book culminates with a final chapter covering advanced concepts with six-degree-of-freedom simulation examples, and modeling of structural dynamics, unsteady aerodynamics, aeroelasticity, and propellant slosh dynamics. From the solved examples, the reader can easily build his/her own simulations as independent semester projects. The choice of gravitational models, coordinate frames, attitude control systems, propulsion systems, and flow models to use is left up to the reader, in order to provide an almost unlimited capability to build various simulations.

This book is primarily designed as a textbook for junior and senior undergraduates, as well as graduate students in mechanical, aerospace engineering/aeronautics, and astronautics departments. The book may also be used as a reference for practicing engineers and researchers in aerospace engineering, aeronautics, and astronautics, whose primary interest lies in modeling and simulation of flight dynamics. The contents have evolved from the lecture notes of several 3rd–4th year undergraduate, and graduate-level courses I have taught over the past 14 years. The material in the book has been especially selected to be useful in a modern course on flight dynamics, where the artificial distinction between atmospheric and space flight is removed. At the same time, the material offers the choice of being adopted in separate traditional courses on space dynamics and atmospheric flight mechanics. In this respect, the text is quite flexible and can be utilized by even those instructors who do not necessarily agree with the comprehensive approach adopted in the book. It is, however, suggested that the mix of atmospheric and space dynamics be retained in each course. A detailed discussion of the usage of material by course instructors is given below. The chapters are designed to follow in a sequence such that their concepts evolve logically and fit into each other like a glove. The concepts are introduced in an easy-to-read manner, while retaining mathematical rigor. The theory behind flight dynamic modeling is highlighted and fundamental results are derived analytically. Examples and problems have been carefully chosen to emphasize the understanding of underlying physical principles. Each chapter begins with a list of clearly defined aims and objectives. At the end of each chapter, short summaries and a number of exercises are provided in order to help readers consolidate their grasp of the material presented. Answers to selected problems are included at the back of the book so that a reader can verify his/her own solutions. Full step-by-step solutions to all of the exercises will be available upon request to the publisher in a separate solutions manual designed for course instructors to use with their students. The manual may also be made available to researchers and professionals (nonstudents) who are using the book for self-study purposes. Requests for the solutions manual should be sent to the publisher on an official letterhead with full particulars, including the course name and number for which the book is being adopted.

Perhaps the greatest distinguishing feature of the book is the ready and extensive use of MATLAB® and Simulink®,[1] as practical computational tools to solve problems across the spectrum of modern flight dynamics. The MATLAB/Simulink codes are integrated within the text in order to readily illustrate modeling and simulation of aerospace dynamics. MATLAB/Simulink is standard, easy-to-use software that most engineering students learn in the first year of their curriculum. Without such a software package, the numerical examples and problems in a text of this kind are difficult to understand and

[1] MATLAB/Simulink® are registered products of The MathWorks, Inc., 3 Apple Hill Drive, Natick, MA 01760-2098, U.S.A. http://www.mathworks.com.

solve. In giving the reader a hands-on experience with MATLAB/Simulink as applied to practical problems, the book is useful for a practicing engineer, apart from being an introductory text for the beginner. The book uses the software only as an instructional tool, discouraging the "black-box" approach found in many textbooks that carry "canned" software. The reader is required to write his/her own codes for solving many of the problems contained as exercises. An appendix contains a brief review of some important methods of numerically integrating ordinary differential equations that are commonly encountered in flight dynamics. In summary, the primary features of this book are a unified approach to aircraft and spacecraft flight, a wide range of topics, nontrivial simulations, logical and seamless presentation of material, rigorous analytical treatment that is also easy to follow, and a ready use of MATLAB/Simulink software as an instructional tool. All the codes used in the book are available for downloading at the following website: http://home.iitk.ac.in/~ashtew/page10.html.

The text focuses on the modeling and simulation aspects of flight mechanics in a wide range of aerospace applications. This treatment is more general than that found in many textbooks on atmospheric flight dynamics, which only cover the approximate equations of motion offering analytical closed-form solutions (considered trivial from a modeling and simulation viewpoint). However, it is recognized that the analytical solutions impart an insight into the science of flight dynamics, especially in a junior-level course. For this reason, the discussion of approximate, analytical solutions to special flight situations is offered in the form of exercises at the end of the chapters. The reader is referred to traditional flight mechanics texts for details on approximate, analytical treatment wherever necessary. A course instructor has the freedom to begin with the general derivations of the equations of motion presented in the book, proceeding to the special approximate flight situations (planar, quasi-steady, constant mass, flat nonrotating earth, etc.) for which the traditional, analytical solution is available.

A reader is assumed to have taken basic undergraduate courses in mathematics and physics—particularly calculus, linear algebra, and dynamics—and is encouraged to review these fundamental concepts at several places in the text. I will now briefly discuss the organization and highlights of the topics covered in each chapter in order to provide a ready guide to the reader and the classroom instructor. This will help readers and instructors select what parts of the book will be relevant either in a particular course, or for specific professional study and reference.

It is sometimes felt that a "logical" sequence of topics should begin with atmospheric flight and end with space flight. While such an "earth-to-space" arrangement may appear natural in a documentary on flight, it is not suitable for a textbook. As pointed out above, the material in the text has been ordered such that the physical and mathematical concepts evolve logically and sequentially. Chapters 2–4, which cover kinematics and analytical dynamics, are equally relevant to both space and atmospheric flight. The next three chapters (5–7) on orbital mechanics logically follow from this foundation, as they do not require a model of the atmosphere and aerodynamics. Chapter 8, on rocket propulsion, follows for the same reason. However, before beginning the treatment on atmospheric flight, it is necessary to introduce an atmospheric model, aerodynamic concepts, and air-breathing propulsion, which are carried out in Chapters 9–11. I would add that for an undergraduate student in flight dynamics, the introductory chapters on aerodynamics and propulsion (Chapters 8–11) are especially relevant. Chapters 12–14—the "meat" of the book—put together the concepts of the foregoing chapters in order to present a comprehensive modeling, simulation, control, and analysis of atmospheric, trans-atmospheric, and space trajectories. Chapter 15 culminates the treatment with advanced modeling and simulation concepts applicable to aerospace vehicles. Hence, the first four and the last four chapters pertain to both atmospheric and space flight, whereas the intervening chapters present specialized treatment of either of the two aspects of flight. The following is a detailed overview of each chapter:

Chapter 1 offers a basic introduction and motivation for studying flight dynamics in a comprehensive manner and includes the classification of flight vehicles, as well as the important assumptions made in their modeling and simulation.

Chapter 2 presents the kinematic modeling and coordinate transformations useful in all aspects of flight dynamic derivations. The rigorous vector analysis of rotational kinematics is presented with many numerical examples. Basic identities—such as the time derivative of a vector, its rotation, and representation in various reference frames—are derived in a manner that can be easily utilized for derivation of both translational and rotational equations of motion in subsequent chapters. Several alternative kinematic representations [Euler angles, Euler-axis/principal rotation, rotation matrix, Euler symmetric parameters (quaternion), Rodrigues and modified Rodrigues parameters] are introduced and their time-evolution derived. A reader can cover the first two sections in a first reading, proceed to Chapters 3–12, and then return to the other sections of Chapter 2 before beginning Chapter 13.

Chapter 3 discusses planetary shape and gravity. While a spherical gravity model serves most atmospheric flight applications reasonably well, it is necessary to model the spherical harmonics of a nonspherical mass distribution (Sections 3.2–3.4) for accurate space-flight, rocket-ascent, and entry-flight trajectories.

Chapter 4 is an introduction to analytical dynamics. While presenting the analytical tools for deriving a general model for translational motion, the chapter also discusses the relationship between translational and rotational dynamics of a flight vehicle. This chapter is essentially the starting point for deriving the basic kinetic equations for aerospace flight, and includes dynamics in moving frames, variable mass bodies, the N-body gravitational problem in space dynamics, and its specialization to two-body trajectories with analytical and numerical solutions. The problems at the end of the chapter test the reader's understanding of the important concepts in analytical (Newtonian) dynamics. After reading Chapter 4, a reader can either proceed to space flight dynamics (Chapters 5–8) or go to Chapters 9–12 on atmospheric flight.

Chapter 5 covers orbital mechanics concepts, orbital maneuvers, relative motion in orbit, and orbit determination for three-dimensional guidance, with examples. A basic course can take advantage of the special coordinate frames (celestial, local horizon, planetary) discussed in Sections 5.1–5.3, which are useful not only in space flight, but also in long-range atmospheric trajectories. The later sections of the chapter are useful in designing trajectories for interplanetary missions, where the associated Lambert's problem is solved numerically. An interesting example of Lambert's problem in the chapter is a nonplanar orbital rendezvous between two spacecraft, which is not found in other textbooks on orbital mechanics.

Chapter 6 discusses orbital perturbations caused by gravitational asymmetry (oblateness, and presence of a third body) as well as atmospheric drag. Oblateness effects lead to sun-synchronous and Molniya orbits, whereas a simple atmospheric model is used to predict the life of a satellite in low orbits. Third-body perturbations result in the sphere of influence and the patched conic approach for the design and analysis of lunar and interplanetary missions.

Chapter 7 is devoted to the restricted three-body problem, its solvability, equilibrium points, and numerical solutions, with examples of the earth-moon-spacecraft trajectories. This chapter can be skipped in a basic-level course on flight dynamics.

Chapter 8 introduces the elements of rocket propulsion. The first two sections of this chapter are strongly recommended in all basic courses on flight dynamics. The design of optimal multistage rockets—a crucial problem in sub-orbital and space flight—is covered in Section 8.3 with examples of two- and three-stage rockets.

Chapter 9 begins the modeling of atmospheric flight with a detailed standard atmosphere model, including nondimensional aerodynamic parameters. For example, a 21 layer *U.S. Standard Atmosphere* is considered, ranging from sea level to a geometric altitude of 700 km. This model is utilized in all simulations of atmospheric and trans-atmospheric trajectories in the book.

Chapter 10 introduces aerodynamics, ranging from elementary concepts to models of viscous hypersonic and rarefied flows using computational fluid dynamics. The discussion is aimed at building an appropriate model of aerodynamic force and moment vectors for each flow regime for the purpose of flight dynamic calculations. This chapter is a must for a beginning course on flight dynamics. Those interested in the details of aerodynamic and thermodynamic models may consider the multitude of specialized texts cited in the chapter.

Chapter 11 covers the elements of air-breathing propulsion from the point of view of flight dynamic modeling of the thrust vector and the rate of fuel consumption. The discussion of characteristics and operational limitations of piston–propeller, turbine, and ramjet engines is comprehensive and self-contained. A numerical model of the thrust and specific fuel consumption with altitude and Mach number of a low-bypass turbofan engine is presented and utilized in Chapter 12 for the simulation of fighter airplane trajectories. This chapter also must be a part of the basic undergraduate course in flight dynamics.

Chapter 12 is the heart of the book with its three-degree-of-freedom flight models, including planetary form, rotation, aerodynamics, and propulsion. General equations of motion in the planet-fixed frame are derived from first principles in a systematic fashion. These equations govern the translational flight of all aerospace vehicles (airplanes, rockets, spacecraft, entry vehicles). The chapter contains nontrivial examples of atmospheric and trans-atmospheric trajectories and provides detailed analytical insight into airplane-flight, rocket-ascent, and entry trajectories.

Chapter 13 presents the universal rotational dynamics model applicable to all aerospace vehicles, emphasizing the commonality between the stability and control characteristics of aircraft and spacecraft. The chapter derives several attitude dynamics models based on various useful kinematic parameters introduced in Chapter 2. Single-axis, open-loop, time-optimal impulsive maneuvers are an important part of this chapter. After exhaustively covering spacecraft dynamics with many examples, the chapter culminates with a rigorous derivation, modeling, and simulation of attitude motion in the atmosphere. Examples in the chapter range from spin-stabilized, rotor and thruster controlled spacecraft, to gravity-gradient satellites, thrust-vectored rockets, and six-degree-of-freedom, inertia-coupled, fighter airplanes. The rotational dynamic models can be easily added to the three-degree-of-freedom translational models of Chapter 12 in order to simulate the complete six-degree-of-freedom motion of rigid craft.

Chapter 14 offers the modeling and simulation of closed-loop control systems for a large variety of aerospace applications based upon modern control concepts. The first part of the chapter presents an introduction of linear systems theory, while the later sections cover multivariable control systems applied to aircraft, spacecraft, and rockets with a multitude of interesting examples.

Chapter 15 introduces advanced concepts, such as six-degree-of-freedom and nonlinear modeling and simulation, flexible vehicle dynamics, unsteady aerodynamics and aeroelasticity, and propellant slosh dynamics. The importance of these topics to flight dynamics, and their inclusion in advanced models, are discussed along with several important references for further study.

The following is a suggested coverage of material by course instructors. It is not envisaged that the entire contents can be followed in a single course. The basic undergraduate curriculum traditionally includes flight dynamics as a pair of courses: one on translational flight (airplane performance/space flight dynamics), and the other on attitude motion (airplane stability and control/spacecraft dynamics and control). I suggest that the first course (called

Flight Dynamics-I)—typically offered at the third-year level—should address the translational aspects of both atmospheric and space flight dynamics. Ideally, this course would cover Chapter 1, Sections 2.1–2.4, 3.1, Chapter 4, Sections 5.1–5.2, 8.1–8.2, 9.1–9.3, 10.1, 10.2.1–10.2.4, and 10.3, Chapter 11, and Sections 12.1 and 12.2. The second undergraduate course, *Flight Dynamics-II*, (taken in the fourth year) would focus on the attitude motion of aerospace vehicles, and would cover Sections 2.5–2.8, 13.1–13.6, 13.7.1–13.7.6, 14.1–14.5, and 15.1–15.3. An advanced elective undergraduate- and graduate-level course can be designed to cover rocket and entry trajectories, such as AE-644 *Hypersonic and Trans-atmospheric Flight* offered at the Indian Institute of Technology, Kanpur, having the two basic undergraduate courses discussed above as its prerequisites. Such a course may cover Sections 3.2–3.4, 5.3–5.5, 6.1, 6.2, 8.3, 9.4, 10.2.5, 12.3, 12.4, 13.7.7, 13.7.8, 14.6, and 15.4. Another advanced senior undergraduate and graduate course on interplanetary flight would consist of material covered in Sections 3.2–3.4, 5.3–5.7, 6.1–6.4, Chapter 7, Sections 8.3 and 12.4, having *Flight Dynamics-I* as its prerequisite. Such a course would rely heavily on modeling and simulation, with trajectory design semester projects based upon numerical solutions to Lambert's problem. The following flowchart highlights the suggested coverage of material:

My motivation to study flight dynamics began early in life. I recall, as a five-year-old, being inspired by my late father—a doctor and medical officer—to listen at late hours to a scratchy, live commentary of the *Apollo-11* mission over a radio set in the remote town of Chunar in India. My incessant curiosity about flight in the growing years was sought to be satisfied by him through numerous illustrated books, magazines, and newspaper articles (of which I kept a careful catalog in an old scrapbook). His encouragement was partly responsible for my taking up aeronautical engineering in college at IIT–Kanpur, and later on when he was no longer with us, completing my doctorate in aerospace engineering at the University of Missouri–Rolla, as well as obtaining a private-pilot certificate. I still remember his ideas about hydrogen-fueled airplanes, and his suggestion that I also study astronomy in graduate school. My family, friends, students, and colleagues have continually supported my enthusiasm for all forms of flight to the present day.

I would like to thank the editorial and production staff of Birkhäuser Boston, especially Tom Grasso, for their constructive suggestions and valuable insights during the preparation of the manuscript. I am also grateful to The MathWorks, Inc., for providing the latest MATLAB/Simulink version, utilized in the examples throughout the book.

Ashish Tewari
May 2006

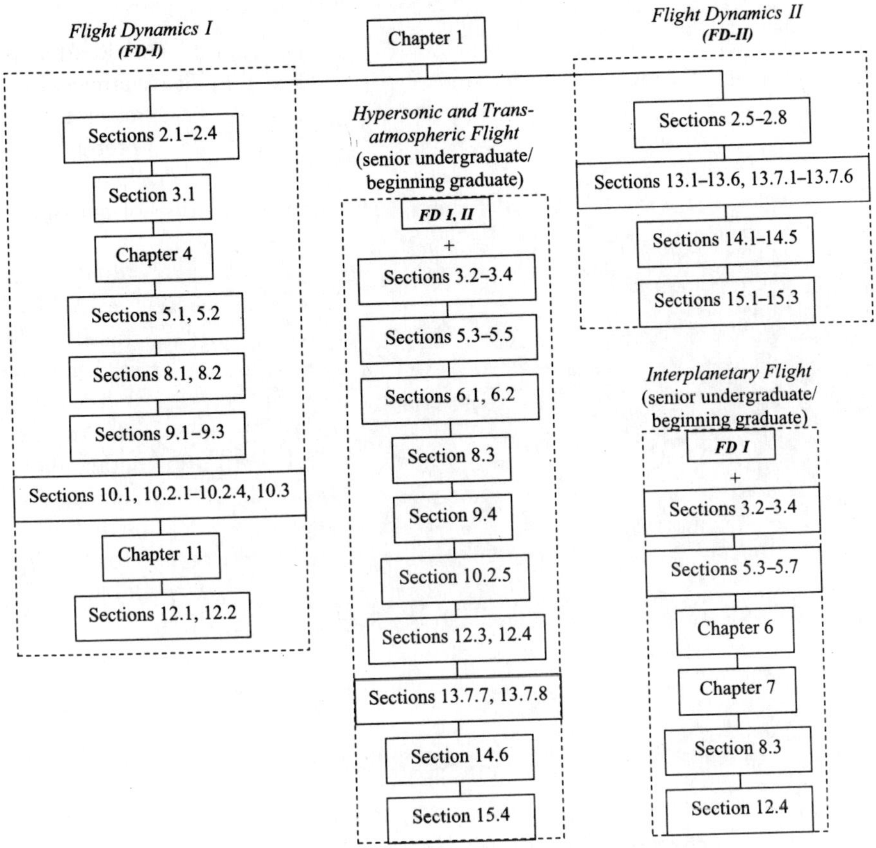
Chapter 1
Flight Dynamics I
(FD-I)
Sections 2.1–2.4
Section 3.1
Chapter 4
Sections 5.1, 5.2
Sections 8.1, 8.2
Sections 9.1–9.3
Sections 10.1, 10.2.1–10.2.4, 10.3
Chapter 11
Sections 12.1, 12.2
Hypersonic and Trans-atmospheric Flight
(senior undergraduate/ beginning graduate)
FD I, II
+
Sections 3.2–3.4
Sections 5.3–5.5
Sections 6.1, 6.2
Section 8.3
Section 9.4
Section 10.2.5
Sections 12.3, 12.4
Sections 13.7.7, 13.7.8
Section 14.6
Section 15.4
Flight Dynamics II
(FD-II)
Sections 2.5–2.8
Sections 13.1–13.6, 13.7.1–13.7.6
Sections 14.1–14.5
Sections 15.1–15.3
Interplanetary Flight
(senior undergraduate/ beginning graduate)
FD I
+
Sections 3.2–3.4
Sections 5.3–5.7
Chapter 6
Chapter 7
Section 8.3
Section 12.4

Contents

1

Introduction

1.1 Aims and Objectives

- To provide a basic introduction to flight.
- To motivate the study of atmospheric and space flight in a comprehensive manner.
- To classify and categorize flight vehicles.
- To introduce the fundamental assumptions in flight modeling and simulation.

1.2 Atmospheric and Space Flight

The study of flight is traditionally divided into two categories: atmospheric and space flight mechanics. The two have evolved separately over the last century. The advent of sustained, powered flight through the air began in 1903 with the *Wright Flyer*, whose main purpose was to fight gravity through the thrust of its engine and the lift produced by its wings—both aerodynamic in nature—in a controllable fashion. As atmospheric flight progressed over the decades, a new methodology was developed for its analysis, largely based on the study of aerodynamic forces and moments. In contrast, space flight, which required neither lift, nor aerodynamic thrust, was contemplated using the theories of astronomy and ballistics. Visionaries such as Jules Verne, Tsiolkovsky and Walter Hohmann, and rocket pioneers, like Robert Goddard, Wernher von Braun, and Sergei Korolev, produced a terminology for space flight which borrowed heavily from Kepler, Galileo, and Newton.

By the time *Sputnik-I* was launched into a low earth orbit in 1957, the dichotomy in the science of flight mechanics was well established, where aerodynamicists would have little to interest them in space exploration, other than the design of launch vehicles, for which the common enemy—aerodynamic drag—was to be minimized. Similarly, space mechanicians were least concerned about airplane flight, which was perhaps considered to be a lowly form of flight, not at all in the same league as celestial mechanics. The revolutionary growth of aeronautical engineering in the first five decades of the twentieth century had given place to an incremental — rather than phenomenal— progress in the latter half of century. With the shattering of the supersonic and hypersonic flight *barriers* in the 1940s and 1950s, it might have been a school of thought that further development in aeronautical engineering would be less challenging, compared to orbital, and lunar flight. Separate programs of aeronautics and astronautics had already been instituted in all major engineering universities, with specialized faculty and courses for each. Research organisations, such as NASA, had different centers for aeronautics and space flight. It was several decades later, in the 1980s, that the term *aerospace engineering* came into vogue. This was a novel concept, dramatized by the development of the NASA's *space shuttle*, which took off vertically like a rocket, went into orbit for several days, then re-entered the atmosphere, and landed horizontally as a glider.

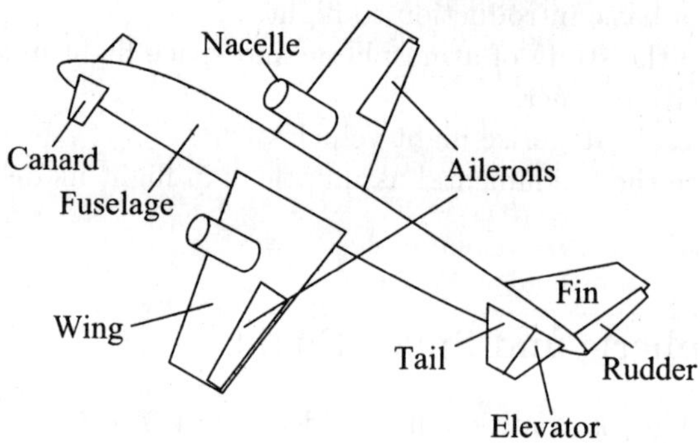

Fig. 1.1. Elements of airplane configuration.

The *atmospheric flight vehicles* are especially adapted for low aerodynamic drag and can be classified into *lifting* vehicles (or *aircraft*) and *non-lifting* (or *ballistic*) vehicles. Lifting vehicles derive their support (*lift*) in air using either static, or dynamic interaction with the atmosphere. In the former the *aerostatic* category lie the hot-air balloons, blimps, and dirigibles, while in the *aerodynamic* lift category we have the *airplanes*, *gliders*, and *rotorcraft* (or *helicopters*). The airplane is a versatile atmospheric vehicle, consisting of fixed *wings*, *fuselage*, *nacelles*, and *empennage* (or *stabilizing and contol surfaces* such as *tail*, *canards*, and *fins*), *elevator*, *ailerons*, and *rudder*, as depicted in Fig. 1.1. While the wings produce the aerodynamic lift, the payload, crew, powerplants, and fuel are housed in the fuselage and nacelles, and the stabilizing surfaces maintain the vehicle in a stable equilibrium, and provide control for maneuvering. An airplane possesses all the features that are found piecemeal in other atmospheric flight vehicles. For example, a glider is an airplane without a powerplant, while a helicopter has rotating—rather than fixed—wings.

The ballistic category of atmospheric vehicles includes *missiles*, *launch vehicles*, and *entry capsules*. Some missiles and launch vehicles incorporate fins as aerodynamic stabilizing and control surfaces. The *spacecraft* are also categorized according to their missions, such as *low-earth orbit*, *medium-earth orbit*, *geosynchronous orbit*, *lunar*, and *interplanetary* spacecraft. Each mission is defined by the payload, and orbital elements of the final orbit. Of course, a reusable launch vehicle, such as the space shuttle, is also a spacecraft with a unique mission.

Traditionally, a flight vehicle is considered as being either atmospheric, or space vehicle depending upon the instantaneous location of the craft. For example, flight above an altitude of 100 km over the earth is generally regarded as space flight. However, when modeling *trans-atmospheric* flight (such as the ascent of a rocket into space, and an atmospheric entry), it is necessary that the artificial distinction betweeen space and atmospheric flight be removed, such that a smooth, continuous trajectory is generated from the governing equations of motion. Consequently, the same set of equations can be used across the atmosphere, and into the space (as we shall demonstrate in Chapter 12), provided due consideration is given to the variation of atmospheric density with altitude by an appropriate atmospheric model (Chapter 9), and an accurate aerodynamic modeling is carried out according to the prevailing flow regime (Chapter 10). When it is certain that the trajectory lies either completely within the atmosphere, or ouside it, one can take advantage of the simplification afforded in the equations of motion, and in certain cases, enjoy closed-form, analytical solutions. For example, space flight is often rendered by exact, analytical solutions for trajectories (Chapters 4 and 5), or by numerical approximations (Chapters 6 and 7).

1.3 Modeling and Simulation

Modeling of flight dynamics consists of *idealization*, selection of a reference *coordinate frame*, and derivation of governing *equations of motion* consistent with the idealization. Idealization is the process whereby necessary simplifying assumptions are made for studying the relevant dynamics. For example, in modeling the translational motion, it is often sufficient to ignore the size and mass distribution, and consider the vehicle as a point mass (or, particle). This is called the *particle* idealization of the vehicle. In this process, the distinct ways in which the vehicle can move, i.e., its *degrees of freedom*, are reduced to only three. Similarly, it is a common practice to treat the vehicle as a *rigid body* when considering its rotational motion, thereby reducing the degrees of freedom, from infinite (for a flexible body) to only six. The idealization must be carefully carried out so that the essential characteristics of the motion under study are not lost. When the degrees of freedom are known, the next step is to select a set of motion variables (two for each degree of freedom), and a reference frame for expressing the equations of motion. For example, when studying high-altitude trajectories, such as those of spacecraft and atmospheric entry vehicles, the reference frame is usually fixed to the planet at its center, and the motion variables are spherical coordinates of position and velocity. On the other hand, low-altitude flight, such as that of an airplane, usually employs a flat, nonrotating planet idealization, with the reference frame fixed to the planet's surface, and motion variables are expressed in Cartesian coordinates. The rotational motion is generally described in reference to a coordinate frame fixed to the vehicle at its center of mass. The equations of motion can be divided into two categories: (a) kinematic equations, which only consider the geometric relationships among the motion variables, and (b) dynamic (or kinetic) equations, that are derived by taking into account the physical laws of motion. The fundamental physical laws pertinent to flight dynamics are Newton's laws of motion and gravitation, as well as the aerothermodyamic principles by which the aerodynamic and propulsive force and moment vectors are derived. Figure 1.2 depicts the various idealizations and reference frames employed in aerospace flight dynamics.

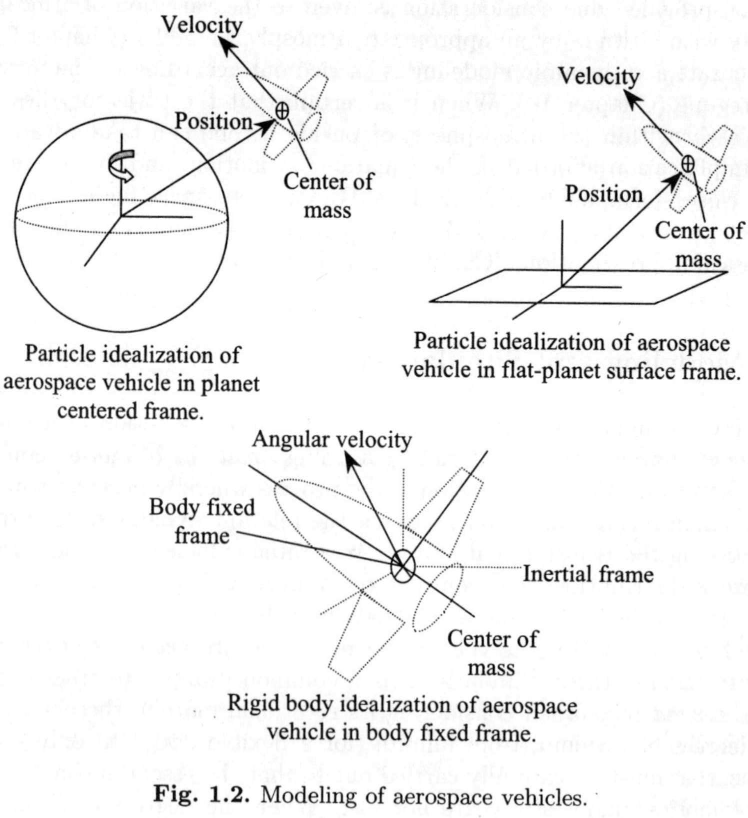

Fig. 1.2. Modeling of aerospace vehicles.

Simulation is the task of solving the governing equations of motion in such a manner that a good approximation of the actual vehicle's motion is attained. Since the governing equations are generally nonlinear, coupled, ordinary differential equations, their solution in a closed form is seldom possible, and a numerical integration subject to appropriate initial condition is often the only alternative. Therefore, simulation of flight dynamics essentially consists of numerical integration of a set of nonlinear, ordinary differential equations. The accuracy attained in the solution depends primarily upon the numerical procedure, and to some extent on the latter's implementation in a computer algorithm. All numerical schemes employ varying degrees of approximation, wherein the derivatives are evaluated by Taylor series expansion. The number of terms retained in such a series is a rough indicator of the scheme's accuracy. The neglected terms of the series are grouped into the *truncation error* of the numerical scheme. Since the neglected higher-order terms must be relatively smaller in size, it is necessary that the numerical integration be performed over steps of small intervals. Therefore, truncation error accumulates as the number of steps required in the integration increases. Generally, a fine balance must be struck between the reduction of the total truncation error, and the number of terms that must be retained in the memory for each computational step. Since the nonlinear numerical integration procedure has to be iterative in nature, one must also look at its *stability* and *convergence* properties. Stability of a numerical scheme allows the truncation error to remain bounded, while convergence implies that the numerical solution reaches essentially a steady state, and does not keep oscillating forever. We will see in an appendix how an accurate numerical solution of governing equations of motion can be carried out. The process of flight dynamic modeling and simulation is graphically depicted by a flow chart in Fig. 1.3.

With the availability of ready-made modeling tools, such as specialized *MATLAB/Simulink* toolboxes, the task of modeling complex system dynamics has become quite simple. However, great caution must be exercised when using such software as learning tools. Often, students have a tendency to employ tailor-made software to solve rather complex problems, without really understanding the inherent modeling assumptions and other limitations of the software. Such a tendency must be definitely curbed in a successful course on flight dynamics. The MathWork's *Simulink* software comes with a toolbox called *Aerospace Block-Set*, which has many useful features for building airplane flight dynamic models in a modular fashion, such as six-degree-of-freedom equations of motion, linearized aerodynamics, a turbofan engine block, second- and third-order actuators, several standard atmosphere and earth gravity models, as well as statistical wind disturbance models. However, these must be properly understood in the context of equations of motion and coordinate systems of modeled dynamics before applying them to solve a given problem. For example, the *3-DOF Animation* and *3-DOF Equations of Motion* blocks pertain to a very specific three-degree-of-freedom system—namely translation in the vertical plane and rotation about a body axis—and

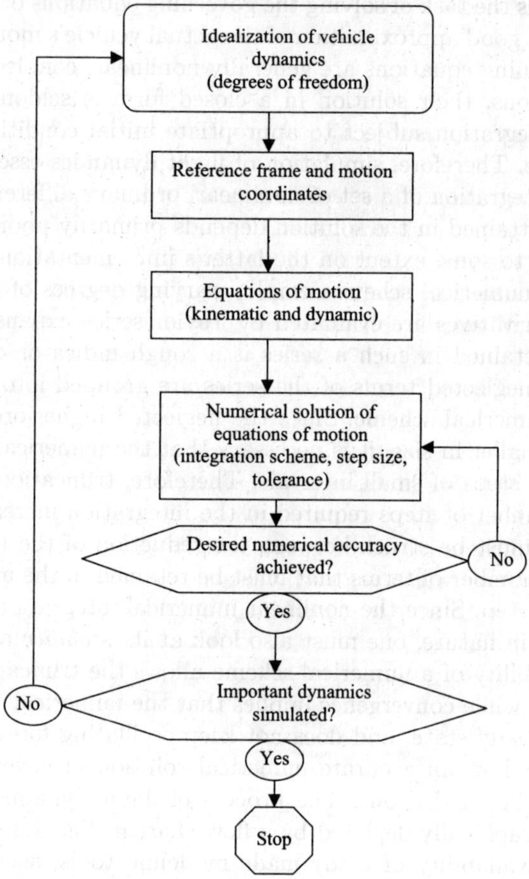

Fig. 1.3. Flight dynamic modeling and simulation procedure.

should not be applied to model the three-dimensional translation of a flight vehicle. The use of specialized software as a *black box* is detrimental to learning the basic concepts underlying flight dynamics. It is a responsibility of the teacher to appropriately admonish an undergraduate class about the limitations of using custom-built software for learning. Ideally, the students must be encouraged to write their own programs from scratch, using only the basic functions and operations of a mathematical software, such as *MATLAB*, or the numerical algorithm libraries of basic programming languages such as *FORTRAN*, *C*, and *Java*. Such an approach is adopted everywhere in this book. Formulation—rather than programming—is the core of flight mechanics. It matters little which programming language is used to solve a given set of equations, provided the latter are correct. While we have utilized *MATLAB* to write programs in this book, the same can be done in any other programming language. The emphasis in exercises is placed on testing whether the reader has understood the correct procedure for the derivation of the flight models.

1.4 Summary

Atmospheric and space flight vehicles, although having evolved separately, obey the same physical principles and share the same modeling and simulation concepts. The categorization of flight vehicles into spacecraft (satellites, lunar, and interplanetary craft) and aircraft (balloons, airplanes, gliders, missiles, launch, and entry vehicles) is by mission rather than by physical distinction, and includes vehicles such as the space shuttle. Modeling of any flight vehicle involves idealization, selection of appropriate coordinate frames, and derivation of the governing equations of motion. Simulation refers to the task of accurately integrating the governing differential equations of motion in time, while including appropriate environmental and control effects. The ready-made modeling and simulation tools, such as the *MATLAB/Simulink* software, enable the analysis of most flight situations with ease. However, a successful simulation with even such versatile tools requires a correct problem formulation and a suitable mathematical model.

2

Attitude and Kinematics of Coordinate Frames

2.1 Aims and Objectives

- To present the kinematic modeling useful in all flight dynamic derivations and analyses.
- To offer a rigorous vector analysis of rotational kinematics.
- To derive the basic identities of coordinate transformations in a manner that can be easily utilized for both translational and rotational equations of motion in subsequent chapters.
- To introduce several alternative kinematic representations (Euler angles, Euler axis/principal angle, rotation matrix, quaternion, Rodrigues and modified Rodrigues parameters) and their evolution in time.

2.2 Basic Definitions and Vector Operations

The basic entity in dynamics is a *vector*. We shall denote vectors by bold-face symbols, and draw them as arrows. A vector has both magnitude and direction and is represented by magnitudes along any three mutually perpendicular axes, called a *coordinate frame* (or, a *reference frame*). Each axis of a coordinate frame is represented by a *unit vector*, defined as a vector of unit magnitude. Let us consider a vector, **A**, written in terms of its magnitudes along a coordinate frame consisting of a triad formed by the unit vectors, $\mathbf{i}, \mathbf{j}, \mathbf{k}$, as follows:

$$\mathbf{A} = A_x\mathbf{i} + A_y\mathbf{j} + A_z\mathbf{k} \,, \tag{2.1}$$

or, simply as

$$\mathbf{A} = \left\{ \begin{array}{c} A_x \\ A_y \\ A_z \end{array} \right\} , \tag{2.2}$$

where it is understood that the *components*, A_x, A_y, A_z multiply $\mathbf{i}, \mathbf{j}, \mathbf{k}$, respectively. A component is the magnitude of the vector projected on to a particular axis. As we shall see below, it is possible to transfer the components of a vector to another coordinate frame by using a *coordinate transformation*. Two (or more) vectors can be added, or subtracted by adding, or subtracting their respective components referred to the same coordinate frame. There are two distinct ways in which vectors can be multiplied: the *scalar product*, and the *vector product*. As the names suggest, the scalar product of two vectors is a *scalar* (a quantity with only magnitude), whereas a vector product of two vectors is a vector. The scalar product (also called *dot product*) of two vectors, **A** and **B**, is defined by

$$\mathbf{A} \cdot \mathbf{B} \doteq AB \cos\theta \,, \tag{2.3}$$

where A, B are the respective magnitudes, and θ is the angle between the two vectors. From this definition, it is clear that the dot product of any two vectors with the same direction is the product of their respective magnitudes, while two mutually perpendicular (*orthogonal*) vectors have a zero dot product. Therefore, it follows that

$$\mathbf{A} \cdot \mathbf{i} = A_x \; ; \mathbf{A} \cdot \mathbf{j} = A_y \; ; \mathbf{A} \cdot \mathbf{k} = A_z \,, \tag{2.4}$$

and we can write the dot product of two vectors, $\mathbf{A} = A_x\mathbf{i} + A_y\mathbf{j} + A_z\mathbf{k}$ and $\mathbf{B} = B_x\mathbf{i} + B_y\mathbf{j} + B_z\mathbf{k}$, as

$$\mathbf{A} \cdot \mathbf{B} = A_xB_x + A_yB_y + A_zB_z = \{A_x \; A_y \; A_z\} \begin{Bmatrix} B_x \\ B_y \\ B_z \end{Bmatrix} = \mathbf{A}^T\mathbf{B} \,. \tag{2.5}$$

In Eq. (2.5) we have denoted the *transpose* of a vector with a superscript T, which is obtained by arranging the components in a *row* [rather than the *column* form of Eq. (2.2)]. Hence, the scalar product is the sum of products of the respective components of the two vectors. For a refresher on vectors and matrices, please refer to Kreyszig [4].

The vector product (also called the *cross product*), $\mathbf{A} \times \mathbf{B}$, is defined as follows:

(a) The magnitude is given by $\mid \mathbf{A} \times \mathbf{B} \mid = AB\sin\theta$, where θ is the angle between the two vectors.

(b) The direction of $\mathbf{A} \times \mathbf{B}$ is normal to the plane formed by the two vectors, and is given by the *right-hand rule*, i.e., when the curled fingers of the right hand point from **A** to **B**, the thumb points in the direction of $\mathbf{A} \times \mathbf{B}$ (Fig. 2.1).

From this definition, it is clear that the vector product of any two vectors with the same (or opposite) direction is zero, while two mutually perpendicular vectors have a vector product with magnitude equal to the product of their respective magnitudes. Furthermore, it follows that $\mathbf{A} \times \mathbf{B} = -\mathbf{B} \times \mathbf{A}$. A coordinate frame $\mathbf{i}, \mathbf{j}, \mathbf{k}$ is said to be *right-handed* if its triad of unit vectors is arranged such that $\mathbf{i} \times \mathbf{j} = \mathbf{k}$. Therefore, with reference to a right-handed frame, we can write the vector product of two vectors, $\mathbf{A} = A_x\mathbf{i} + A_y\mathbf{j} + A_z\mathbf{k}$ and $\mathbf{B} = B_x\mathbf{i} + B_y\mathbf{j} + B_z\mathbf{k}$, as follows:

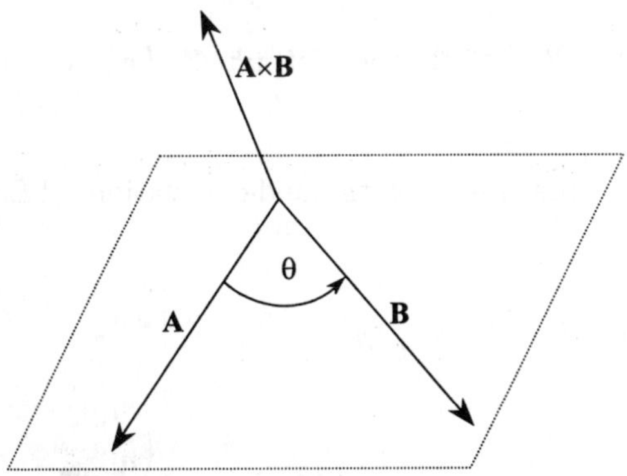

Fig. 2.1. Vector product of two vectors, **A** and **B**.

$$\mathbf{A} \times \mathbf{B} = (A_y B_z - A_z B_y)\mathbf{i} + (A_z B_x - A_x B_z)\mathbf{j} + (A_x B_y - A_y B_x)\mathbf{k} \ . \quad (2.6)$$

By introducing the short-hand notation of the *determinant* of a square matrix, we can write Eq. (2.6) as

$$\mathbf{A} \times \mathbf{B} = \begin{vmatrix} \mathbf{i} & \mathbf{j} & \mathbf{k} \\ A_x & A_y & A_z \\ B_x & B_y & B_z \end{vmatrix} \ . \quad (2.7)$$

The foregoing definitions allow us to derive the following identities involving the vector product:

$$\mathbf{A} \times (\mathbf{B} \times \mathbf{C}) = \mathbf{B}(\mathbf{A} \cdot \mathbf{C}) - \mathbf{C}(\mathbf{A} \cdot \mathbf{B}) \ . \quad (2.8)$$

$$(\mathbf{A} \times \mathbf{B}) \times \mathbf{C} = \mathbf{B}(\mathbf{A} \cdot \mathbf{C}) - \mathbf{A}(\mathbf{B} \cdot \mathbf{C}) \ . \quad (2.9)$$

$$(\mathbf{A} \times \mathbf{B}) \cdot (\mathbf{C} \times \mathbf{D}) = (\mathbf{A} \cdot \mathbf{C})(\mathbf{B} \cdot \mathbf{D}) - (\mathbf{A} \cdot \mathbf{D})(\mathbf{B} \cdot \mathbf{C}) \ . \quad (2.10)$$

$$(\mathbf{A} \times \mathbf{B}) \times (\mathbf{C} \times \mathbf{D}) = (\mathbf{A} \cdot (\mathbf{B} \times \mathbf{D}))\mathbf{C} - (\mathbf{A} \cdot (\mathbf{B} \times \mathbf{C}))\mathbf{D} \ . \quad (2.11)$$

By using MATLAB, one can easily compute scalar and vector products of vectors by using the in-built commands *dot* and *cross*, respectively.

Example 2.1. Let us find scalar and vector products of $\mathbf{A} = -2\mathbf{i} + 5\mathbf{j} - \mathbf{k}$ and $\mathbf{B} = \mathbf{i} + 3\mathbf{k}$ with the following MATLAB commands:

```
>>A=[-2;5;-1];B=[1;0;3];
>>s=dot(A,B) %scalar product
s = -5

>>C=cross(A,B)  %vector product C=AxB
C = 15
      5
    -5

>>D=cross(B,A)  %vector product D=BxA
D = -15
    -5
    5
```

```
>>E=cross(A,cross(A,B)) % vector triple product E=Ax(AxB)
E = -20
    -25
    -85
```

You may verify these results with hand calculations of Eqs. (2.5) and (2.7).

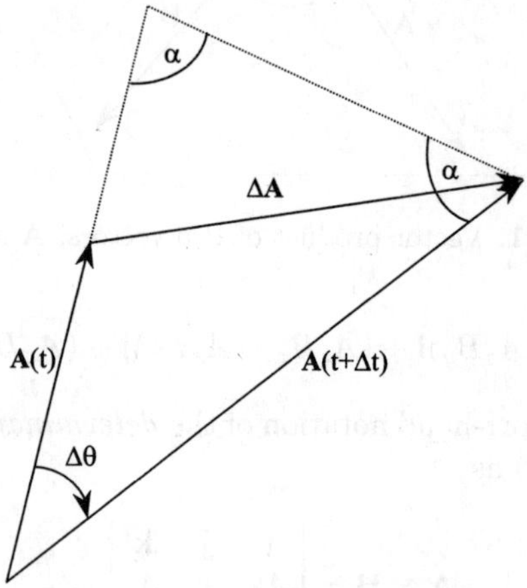

Fig. 2.2. A changing vector.

The time derivative of a vector, **A**, is defined by the following:

$$\frac{\mathrm{d}\mathbf{A}}{\mathrm{d}t} \doteq \lim_{\Delta t \to 0} \frac{\mathbf{A}(t+\Delta t) - \mathbf{A}(t)}{\Delta t} = \lim_{\Delta t \to 0} \frac{\Delta \mathbf{A}}{\Delta t} , \tag{2.12}$$

where $\Delta \mathbf{A}$ denotes the total change caused by changes in both magnitude and direction, as depicted in Fig. 2.2. It can be seen from Fig. 2.2 that an isosceles triangle is formed by extending the vector $\mathbf{A}(t)$ until its magnitude becomes equal to $\mathbf{A}(t+\Delta t)$. In the limit $\Delta t \to 0$, we have $\Delta\theta \to 0$, and thus $\alpha \to \frac{\pi}{2}$. Hence, the two dashed lines in Fig. 2.2 represent the extension (or contraction) and rotation of **A**, respectively, and we have

$$\lim_{\Delta t \to 0} \Delta \mathbf{A} = [A(t+\Delta t) - A(t)]\frac{\mathbf{A}(t)}{A(t)} + \boldsymbol{\omega} \times \mathbf{A}(t)\Delta t , \tag{2.13}$$

where $\boldsymbol{\omega}$ is the *angular velocity* of **A**, directed into the plane of Fig. 2.2 with magnitude given by

$$\omega = \frac{\mathrm{d}\theta}{\mathrm{d}t} \doteq \lim_{\Delta t \to 0} \frac{\Delta\theta}{\Delta t} . \tag{2.14}$$

Upon substitution of Eq. (2.13) into Eq. (2.12), we have

$$\frac{\mathrm{d}\mathbf{A}}{\mathrm{d}t} = \frac{\mathrm{d}A}{\mathrm{d}t}\frac{\mathbf{A}}{A} + \boldsymbol{\omega} \times \mathbf{A} , \tag{2.15}$$

where the first term on the right-hand side represents time derivative due to a change in the magnitude, and the second, that due to rotation. An observer rotating with the same angular velocity as **A** would notice only the first term, while a stationary observer would notice both the terms.

2.3 Coordinate Systems and Rotation Matrix

As noted above, a vector can be expressed in a variety of coordinate frames, each of which is represented by a triad of unit vectors. We will confine our discussion to right-handed coordinate frames. Any number of different coordinate frames can be derived by rotating the orthogonal axes about the origin, O. Sometimes, it may be necessary to use a coordinate frame with a different origin, O'. Hence, the general transformation from one coordinate frame to another consists of a translation of the origin, and a rotation of the axes about the new origin. However, the translation of the origin is handled quite easily by specifying the *displacement* vector, $\mathbf{R}$ from O to O' (Fig. 2.3), and its time derivatives, such that the *displacement*, *velocity*, and *acceleration*, $\mathbf{r}, \mathbf{v}, \mathbf{a}$, respectively, in the original frame are related to their counterparts $\mathbf{r}', \mathbf{v}', \mathbf{a}'$, in the translated frame by

$$\begin{aligned} \mathbf{r} &= \mathbf{R} + \mathbf{r}' , \\ \mathbf{v} &= \frac{\mathrm{d}\mathbf{R}}{\mathrm{d}t} + \mathbf{v}' , \\ \mathbf{a} &= \frac{\mathrm{d}^2\mathbf{R}}{\mathrm{d}t^2} + \mathbf{a}' . \end{aligned} \tag{2.16}$$

Since the translation of coordinate frames is trivial, we shall focus instead on

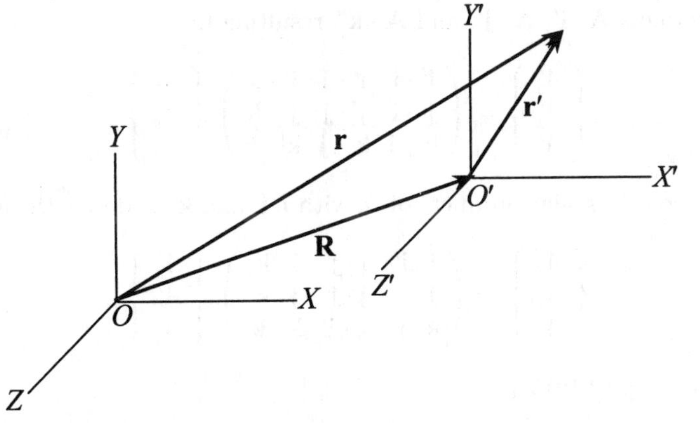

Fig. 2.3. Translation of a coordinate frame.

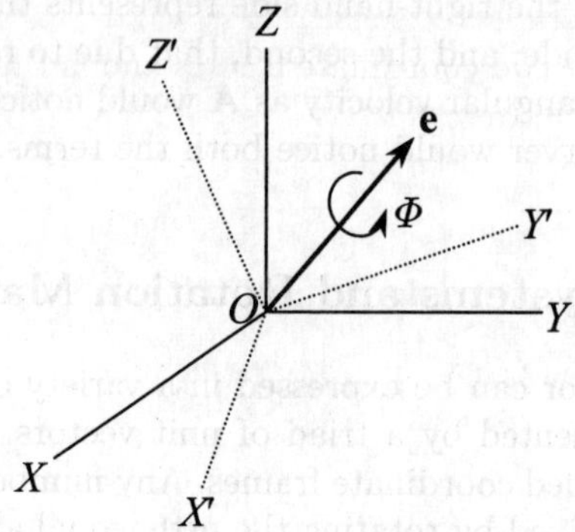

Fig. 2.4. Rotation of a coordinate frame.

coordinate transformations involving rotation. Consider a frame ($OXYZ$) with axes OX, OY, and OZ, denoted by unit vectors $\mathbf{i}, \mathbf{j}, \mathbf{k}$, respectively. The frame is rotated about the origin, O, to produce a new frame, ($OX'Y'Z'$) denoted by $\mathbf{i}', \mathbf{j}', \mathbf{k}'$ (Fig. 2.4). Now, consider a vector, $\mathbf{A}$, alternately expressed in terms of its components in the original and rotated frames, as follows:

$$\mathbf{A} = A_x\mathbf{i} + A_y\mathbf{j} + A_z\mathbf{k} = A'_x\mathbf{i}' + A'_y\mathbf{j}' + A'_z\mathbf{k}' , \tag{2.17}$$

which can be written in a matrix form as

$$\mathbf{A} = (\mathbf{i}\;\mathbf{j}\;\mathbf{k}) \begin{Bmatrix} A_x \\ A_y \\ A_z \end{Bmatrix} = (\mathbf{i}'\;\mathbf{j}'\;\mathbf{k}') \begin{Bmatrix} A'_x \\ A'_y \\ A'_z \end{Bmatrix} . \tag{2.18}$$

In order to find the relationship between the two sets of components, we take scalar products $\mathbf{A}\cdot\mathbf{i}'$, $\mathbf{A}\cdot\mathbf{j}'$, and $\mathbf{A}\cdot\mathbf{k}'$, resulting in

$$\begin{Bmatrix} A'_x \\ A'_y \\ A'_z \end{Bmatrix} = \begin{pmatrix} \mathbf{i}'\cdot\mathbf{i} & \mathbf{i}'\cdot\mathbf{j} & \mathbf{i}'\cdot\mathbf{k} \\ \mathbf{j}'\cdot\mathbf{i} & \mathbf{j}'\cdot\mathbf{j} & \mathbf{j}'\cdot\mathbf{k} \\ \mathbf{k}'\cdot\mathbf{i} & \mathbf{k}'\cdot\mathbf{j} & \mathbf{k}'\cdot\mathbf{k} \end{pmatrix} \begin{Bmatrix} A_x \\ A_y \\ A_z \end{Bmatrix} . \tag{2.19}$$

We can also take scalar products of $\mathbf{A}$ with $\mathbf{i}, \mathbf{j}$, and $\mathbf{k}$, to derive the following:

$$\begin{Bmatrix} A_x \\ A_y \\ A_z \end{Bmatrix} = \begin{pmatrix} \mathbf{i}\cdot\mathbf{i}' & \mathbf{i}\cdot\mathbf{j}' & \mathbf{i}\cdot\mathbf{k}' \\ \mathbf{j}\cdot\mathbf{i}' & \mathbf{j}\cdot\mathbf{j}' & \mathbf{j}\cdot\mathbf{k}' \\ \mathbf{k}\cdot\mathbf{i}' & \mathbf{k}\cdot\mathbf{j}' & \mathbf{k}\cdot\mathbf{k}' \end{pmatrix} \begin{Bmatrix} A'_x \\ A'_y \\ A'_z \end{Bmatrix} . \tag{2.20}$$

We rewrite Eq. (2.19) as

$$\begin{Bmatrix} A'_x \\ A'_y \\ A'_z \end{Bmatrix} = \mathsf{C} \begin{Bmatrix} A_x \\ A_y \\ A_z \end{Bmatrix} , \tag{2.21}$$

where C is the following matrix consisting of the cosines of angles (Eq. (2.3)) between the axes of the two coordinate frames and is thus called the *direction cosine matrix*, or the *rotation matrix*:

$$\mathsf{C} \doteq \begin{pmatrix} \mathbf{i}'\cdot\mathbf{i} & \mathbf{i}'\cdot\mathbf{j} & \mathbf{i}'\cdot\mathbf{k} \\ \mathbf{j}'\cdot\mathbf{i} & \mathbf{j}'\cdot\mathbf{j} & \mathbf{j}'\cdot\mathbf{k} \\ \mathbf{k}'\cdot\mathbf{i} & \mathbf{k}'\cdot\mathbf{j} & \mathbf{k}'\cdot\mathbf{k} \end{pmatrix} . \tag{2.22}$$

The coordinate transformation, Eqs. (2.19), can also be expressed as follows:

$$\begin{Bmatrix} \mathbf{i}' \\ \mathbf{j}' \\ \mathbf{k}' \end{Bmatrix} = \mathsf{C} \begin{Bmatrix} \mathbf{i} \\ \mathbf{j} \\ \mathbf{k} \end{Bmatrix} . \tag{2.23}$$

From Eqs. (2.19) and (2.20), it is clear that the rotation matrix has the following property:

$$\mathsf{C}^T\mathsf{C} = \mathsf{C}\mathsf{C}^T = \mathsf{I} , \tag{2.24}$$

from which it follows that $\mathsf{C}^{-1} = \mathsf{C}^T$. A matrix with this property is said to be *orthogonal*, since the vectors formed out of the columns of the matrix are orthogonal. From Eq. (2.24), we can also deduce the fact that the determinant, $\mid \mathsf{C} \mid = \pm 1$. A rotation for which $\mid \mathsf{C} \mid = 1$ is called a *proper* rotation. It is easy to see from the definition of the rotation matrix, Eq. (2.21), that two successive rotations of a coordinate frame can be represented simply by multiplying the rotation matrices of individual rotations as follows:

$$\mathsf{C}'' = \mathsf{C}'\mathsf{C} , \tag{2.25}$$

where the orientation C'' is obtained by first undergoing a rotation C, followed by a rotation C'. Other properties of the rotation matrix are discussed in the following section.

Example 2.2. Find the rotation matrix that produces the right-handed coordinate frame $\mathbf{i}', \mathbf{j}', \mathbf{k}'$ given by $\mathbf{i}' = 0.1\mathbf{i} + 0.2\mathbf{j} + \sqrt{1-0.01-0.04}\mathbf{k}$, and $\mathbf{j}' = -0.1\mathbf{i} - 0.9726095077\mathbf{j} + \sqrt{1-0.01-0.9726095077^2}\mathbf{k}$.

We check the orthogonality of the given vectors, calculate the third axes of the right-handed frame, and then find C according to Eq. (2.22) as follows:

```
>> iprime=[0.1;0.2;sqrt(1-0.01-0.04)],jprime=[-0.1;-0.9726095077;
       sqrt(1-0.01-0.9726095077^2)]
iprime =     0.10000000000000      jprime =     -0.10000000000000
         0.20000000000000                   -0.97260950770000
         0.97467943448090                    0.20983504362133

>>dot(iprime,jprime)
ans =     1.111077896354118e-010

>> kprime=cross(iprime,jprime)      %k'=i'xj'
kprime =     0.98994949366004
         -0.11845144781022
         -0.07726095077000

>>i=[1;0;0];j=[0;1;0];k=[0;0;1];
```

```
>> C=[dot(iprime,i),dot(iprime,j),dot(iprime,k);...%rotation matrix
      dot(jprime,i),dot(jprime,j),dot(jprime,k);...
      dot(kprime,i),dot(kprime,j),dot(kprime,k)]

C = 0.10000000000000      0.20000000000000      0.97467943448090
   -0.10000000000000     -0.97260950770000      0.20983504362133
    0.98994949366004     -0.11845144781022     -0.07726095077000

>> det(C)   %determinant of C
ans =     1.0000
```

Hence, the rotation matrix consists of the axes of the rotated frame (expressed in the unit vectors of the original axes) as its columns. You may verify that the rotation considered in this example is proper.

2.4 Euler Axis and Principal Angle

From the discussion given above, it is clear that a rotation of a coordinate frame can be represented by a rotation matrix, C. A rotation can also be described by specifying the *axis* of rotation, as well as the angle by which the frame has been rotated. This common experience is formalized in *Euler's theorem*, which states that the relative orientation of any pair of coordinate frames is uniquely determined by a rotation by angle, Φ, about a fixed axis through the common origin, called the *Euler axis*. This unique rotation is termed the *principal angle*. A graphical depiction of the principal rotation is shown in Fig. 2.4, where a *counter-clockwise* rotation is considered positive by the right-hand rule. Euler's theorem thus provides an alternative description of rotation using unit vector, $\mathbf{e}$, representing the direction of Euler axis, and the principal rotation angle, Φ. Before using the new representation, we must know how these two quantities can be derived. An insight into the Euler axis can be obtained by analyzing the *eigenvalues* and *eigenvectors* [4] of the rotation matrix. Let $\mathbf{c}$ be an eigenvector associated with the eigenvalue, λ of C:

$$\mathsf{C}\mathbf{c} = \lambda \mathbf{c} \,. \tag{2.26}$$

By premultiplying Eq. (2.26) by the *Hermitian conjugate* [4] of each side, we have

$$(\mathsf{C}\mathbf{c})^H(\mathsf{C}\mathbf{c}) = \overline{\lambda}\lambda \mathbf{c}^H \mathbf{c} \,, \tag{2.27}$$

or, since C is real and satisfies Eq. (2.24),

$$(\overline{\lambda}\lambda - 1)\mathbf{c}^H \mathbf{c} = 0 \,, \tag{2.28}$$

which implies that

$$\overline{\lambda}\lambda = 1 \,, \tag{2.29}$$

because $\mathbf{c}$ is nonzero. Equation (2.29) states the fact that all eigenvalues of C have unit magnitudes. Now, C, being a (3×3) matrix, has three eigenvalues. Since complex eigenvalues occur in conjugate pairs, it follows that one of the eigenvalues of C must be real, for which Eq. (2.26) becomes

$$\mathsf{C}\mathbf{c_1} = \mathbf{c_1} , \tag{2.30}$$

and which implies that the eigenvector, $\mathbf{c_1}$, associated with $\lambda = 1$ is unchanged by the rotation. It also follows from Eqs. (2.24) and (2.30) that $\mathbf{c_1}^T\mathbf{c_1} = 1$, i.e., $\mathbf{c_1}$ is a unit vector. Therefore, it is clear that Euler axis—being invariant under coordinate frame rotation—is represented by $\mathbf{e} \doteq \mathbf{c_1}$.

The other two eigenvalues of C, being complex conjugates with a unit magnitude, can be written as $\lambda_{2,3} = e^{\pm i\beta} \doteq \cos\beta \pm i\sin\beta$. Thus, from Eq. (2.26), we have

$$\mathsf{C}\mathbf{c_{2,3}} = e^{\pm i\beta}\mathbf{c_{2,3}} . \tag{2.31}$$

From the complex plane representation of a vector (also due to Euler), the factor $e^{i\beta}$ multiplying a vector, implies a rotation by angle β. Thus, the eigenvectors, $\mathbf{c_{2,3}}$ (complex conjugates), undergo a rotation by angle β when the coordinate frame is rotated about the axis $\mathbf{c_1}$. A consequence of C being orthogonal [Eq. (2.24)] is that its eigenvectors are mutually perpendicular. Since $\mathbf{c_{2,3}}$ are perpendicular to $\mathbf{c_1}$, their rotation must be equal to the angle of coordinate frame rotation. Therefore, $\Phi \doteq \beta$. A simple method of obtaining the principal rotation angle is through the *trace* of C [4]:

$$\text{trace}\mathsf{C} = \lambda_1 + \lambda_2 + \lambda_3 = 1 + e^{i\Phi} + e^{-i\Phi} = 1 + 2\cos\Phi , \tag{2.32}$$

or,

$$\cos\Phi = \frac{1}{2}(\text{trace}\mathsf{C} - 1) . \tag{2.33}$$

There are two values of Φ, differing only in sign, that satisfy Eq. (2.33), each having Euler axis, $\mathbf{e}$, in opposite directions. This does not cause any ambiguity, because a rotation by Φ about $\mathbf{e}$ is the same as a rotation by $-\Phi$ about $-\mathbf{e}$.

Example 2.3. Let us find the principal rotation angle and Euler axis for the transformation given in Example 2.2, with the use of the built-in MATLAB program *eig.m*, that enables eigenvalue analysis of a square matrix, as follows:

```
>> [c,D]=eig(C)  %eigenvectors, c, and diagonal matrix of eigenvalues, D

c =    -0.7378    -0.0179 - 0.4770i    -0.0179 + 0.4770i
       -0.0343    0.7067               0.7067
       -0.6742    -0.0164 + 0.5220i    -0.0164 - 0.5220i

D =    1.0000      0                    0
       0           -0.9749 + 0.2225i    0
       0           0                    -0.9749 - 0.2225i

>> phi=acos(0.5*(trace(C)-1))  %rotation angle (rad.)
phi =          2.9172

>> phi=acos(real(D(2,2)))    %confirm rotation angle (rad.)
phi =    2.9172
```

Thus, the rotation angle is $\Phi = \pm 2.9172$ rad. ($\pm 167.1447°$), and Euler axis is given by the eigenvector of C corresponding to the real eigenvalue, $\mathbf{e} = \pm(-0.7378\mathbf{i}-0.0343\mathbf{j}-0.6742\mathbf{k})$. It is also clear that the complex conjugate eigenvectors satisfy Eq. (2.31).

The rotation matrix can be derived from the Euler axis/principal angle using *Euler's formula* (Exercise 2.5).

The simplest coordinate transformations are rotations about the axes of a coordinate frame, called *elementary rotations*. A positive rotation of $(OXYZ)$ about OX by angle Φ is represented by the rotation matrix

$$\mathsf{C}_1 \doteq \begin{pmatrix} 1 & 0 & 0 \\ 0 & \cos\Phi & \sin\Phi \\ 0 & -\sin\Phi & \cos\Phi \end{pmatrix} , \tag{2.34}$$

whereas the rotation matrix for a positive rotation about OY by the same angle is

$$\mathsf{C}_2 \doteq \begin{pmatrix} \cos\Phi & 0 & -\sin\Phi \\ 0 & 1 & 0 \\ \sin\Phi & 0 & \cos\Phi \end{pmatrix} . \tag{2.35}$$

Similarly, a positive rotation about OZ by Φ is given by

$$\mathsf{C}_3 \doteq \begin{pmatrix} \cos\Phi & \sin\Phi & 0 \\ -\sin\Phi & \cos\Phi & 0 \\ 0 & 0 & 1 \end{pmatrix} . \tag{2.36}$$

A more complicated coordinate transformation can be derived by multiple single-axis rotations in a given sequence, using the elementary rotation matrices, as demonstrated below.

Example 2.4. Find the rotation matrix for a coordinate transformation obtained by a 20° rotation about OZ, followed by a $-65°$ rotation about OX'.

We begin by representing the first rotation as follows:

$$\begin{Bmatrix} \mathbf{i}' \\ \mathbf{j}' \\ \mathbf{k}' \end{Bmatrix} = \mathsf{C}_3(20°) \begin{Bmatrix} \mathbf{i} \\ \mathbf{j} \\ \mathbf{k} \end{Bmatrix} . \tag{2.37}$$

The next rotation is by an angle $-65°$ about $\mathbf{i}'$. Thus, we write the final orientation as

$$\begin{Bmatrix} \mathbf{i}'' \\ \mathbf{j}'' \\ \mathbf{k}'' \end{Bmatrix} = \mathsf{C}_1(-65°) \begin{Bmatrix} \mathbf{i}' \\ \mathbf{j}' \\ \mathbf{k}' \end{Bmatrix} , \tag{2.38}$$

or, by substituting Eq. (2.37) into Eq. (2.38), we have

$$\begin{Bmatrix} \mathbf{i}'' \\ \mathbf{j}'' \\ \mathbf{k}'' \end{Bmatrix} = \mathsf{C}_1(-65°)\mathsf{C}_3(20°) \begin{Bmatrix} \mathbf{i} \\ \mathbf{j} \\ \mathbf{k} \end{Bmatrix} . \tag{2.39}$$

Therefore, the required rotation matrix is the following:

$$
\begin{aligned}
\mathsf{C} &= \mathsf{C}_1(-65^\circ)\mathsf{C}_3(20^\circ) \\
&= \begin{pmatrix} \cos(20^\circ) & \sin(20^\circ) & 0 \\ -\sin(20^\circ)\cos(-65^\circ) & \cos(20^\circ)\cos(-65^\circ) & \sin(-65^\circ) \\ \sin(20^\circ)\sin(-65^\circ) & -\cos(20^\circ)\sin(-65^\circ) & \cos(-65^\circ) \end{pmatrix} \\
&= \begin{pmatrix} 0.9397 & 0.3420 & 0 \\ -0.1445 & 0.3971 & -0.9063 \\ -0.3100 & 0.8517 & 0.4226 \end{pmatrix} .
\end{aligned}
$$

The Euler axis and principal angle for this rotation are obtained in the same manner as in Example 2.3 to be $\mathbf{e} = \pm(0.9501\mathbf{i} + 0.1675\mathbf{j} - 0.2630\mathbf{k})$ and $\Phi = \pm 67.6836^\circ$, respectively.

In terms of the elements (i, j) of C—denoted by c_{ij} Eqs. (2.30) and (2.33) lead to the following explicit expressions for the components of Euler axis, $\mathbf{e}$:

$$
\begin{aligned}
e_1 &= \frac{c_{23} - c_{32}}{2\sin\Phi}, \\
e_2 &= \frac{c_{31} - c_{13}}{2\sin\Phi}, \\
e_3 &= \frac{c_{12} - c_{21}}{2\sin\Phi}.
\end{aligned}
\tag{2.40}
$$

It is clear from Eqs. (2.40) that Euler axis is defined only for nonzero rotations.

2.5 Euler Angles

We are now in a position to consider the *orientation*, or *attitude*, of a coordinate frame, relative to another frame. Such a description is quite useful in representing the attitude of a flight vehicles (or any other object), to which a coordinate frame is rigidly attached, in reference to a second coordinate frame. Two previously discussed attitude representations are via the rotation matrix and the Euler-axis/principal angle combination. From the foregoing discussion, it is clear that a general orientation can also be obtained by using successive rotations about the axes of the reference frame. The largest number of such rotations needed to uniquely specify a given orientation, called rotational *degrees of freedom*, is three. Hence, we can employ three angles, each about a particular cordinate axis, to describe a given orientation. Such a representation of the attitude by three angles is called an *Euler angle* representation, and the concerned angles are known as *Euler angles*. The sequence of axial rotations is of utmost importance in the Euler angle representation. You can convince yourself about this fact by rotating a book by three elementary rotations of 90°, each about a different axis, and then carrying out the rotations about the

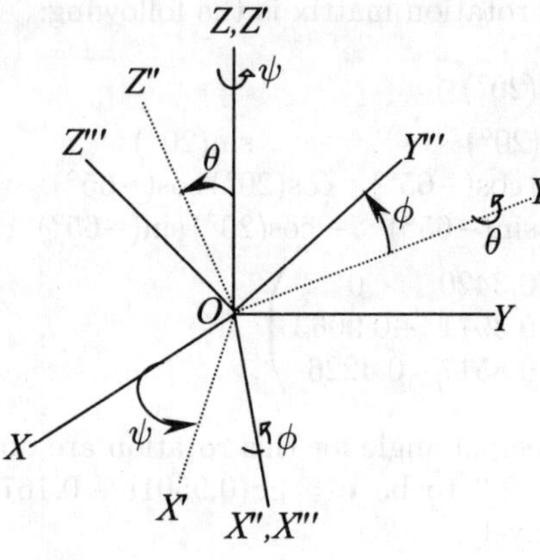

Fig. 2.5. The Euler angle orientation, $(\psi)_3, (\theta)_2, (\phi)_1$.

same axes in a different sequence. The final attitude of the book will be different in each set of rotations. This important property of elementary rotations can also be seen by simply reversing the sequence of rotations in Example 2.4.

We can specify the Euler angles and the axes of sequential rotations using notation such as $(\psi)_3, (\theta)_2, (\phi)_1$, which denotes a rotation of $(OXYZ)$ by angle ψ about OZ, resulting in the intermediate orientation, $(OX'Y'Z')$, followed by a rotation by angle θ about OY', resulting in $(OX''Y''Z'')$, and then a final rotation by angle ϕ about OX'', to produce the new orientation, $(OX'''Y'''Z''')$. This Euler angle orientation, which is a description of the attitude of an aircraft relative to a local horizon using the Euler angles, yaw (ψ), pitch (θ), and roll (ϕ), is depicted in Fig. 2.5. The rotation matrix for the orientation $(\psi)_3, (\theta)_2, (\phi)_1$, is the following:

$$\mathsf{C} = \mathsf{C}_1(\phi)\mathsf{C}_2(\theta)\mathsf{C}_3(\psi) = \tag{2.41}$$

$$\begin{pmatrix} \cos\theta\cos\psi & \cos\theta\sin\psi & -\sin\theta \\ (\sin\phi\sin\theta\cos\psi - \cos\phi\sin\psi) & (\sin\phi\sin\theta\sin\psi + \cos\phi\cos\psi) & \sin\phi\cos\theta \\ (\cos\phi\sin\theta\cos\psi + \sin\phi\sin\psi) & (\cos\phi\sin\theta\sin\psi - \sin\phi\cos\psi) & \cos\phi\cos\theta \end{pmatrix}.$$

However, it is not necessary that all the three coordinate axes should be involved in describing a particular orientation. For example, astronomers and physicists traditionally employ the classical Euler angles, $(\omega)_3, (i)_1, (\Omega)_3$, to represent the orientation of an orbital plane in reference to a celestial coordinate frame (Chapter 5). A set of Euler angles that begins and ends with the same axis—such as $(\omega)_3, (i)_1, (\Omega)_3$—is said to be *symmetric*. The symmetric and asymmetric sets of Euler angles are qualitatively different, and it is generally much easier to handle the former.

Example 2.5. Find the rotation matrix, Euler axis, and principal angle for the orientation $(\alpha)_1, (\beta)_2, (\alpha)_3$, where $\alpha = -45°, \beta = 45°$.

We shall employ MATLAB for this calculation, by first evaluating the elementary rotation matrices, and then multiplying them in the proper sequence to produce the final orientation:

```
>> dtr=pi/180;alfa=-45*dtr;beta=45*dtr;
>> C1=[1 0 0;0 cos(alfa) sin(alfa);0 -sin(alfa) cos(alfa)]
C1 =     1.0000            0            0
         0            0.7071      -0.7071
       0            0.7071        0.7071

>> C2=[cos(beta) 0 -sin(beta);0  1  0;sin(beta) 0 cos(beta)]
C2 =  0.7071            0            -0.7071
      0               1.0000          0
      0.7071            0             0.7071

>> C3=[cos(alfa) sin(alfa) 0;-sin(alfa) cos(alfa) 0;0 0 1]
C3 =  0.7071          -0.7071          0
      0.7071           0.7071          0
      0                0               1.0000

>> C=C3*C2*C1  %rotation matrix for the final orientation
C =   0.5000          -0.8536          0.1464
      0.5000           0.1464         -0.8536
      0.7071           0.5000          0.5000

>> [c,D]=eig(C)  %eigenvectors and eigenvalues of C
c =   0.6786           0.1405+0.5000i  0.1405-0.5000i
     -0.2811           0.6786          0.6786
      0.6786           0.1405-0.5000i  0.1405+0.5000i

D =   1.0000           0               0
      0                0.0732+0.9973i  0
      0                0               0.0732-0.9973i

>> Phi=acos(0.5*(trace(C)-1))/dtr  %principal angle (deg.)
Phi =   85.8009
```

Hence, the rotation matrix for the given orientation is

$$\mathsf{C} = \begin{pmatrix} 0.5 & -0.8536 & 0.1464 \\ 0.5 & 0.1464 & -0.8536 \\ 0.7071 & 0.5 & 0.5 \end{pmatrix},$$

and the Euler axis and principal angle are $\mathbf{e} = \pm(0.6786\mathbf{i} - 0.2811\mathbf{j} + 0.6786\mathbf{k})$, and $\Phi = \pm 85.8009°$, respectively.

In order to specify the attitude by Euler angles, we must be able to determine them uniquely from the rotation matrix. It is clear from Eq. (2.41) that the Euler angles for the representation, $(\psi)_3, (\theta)_2, (\phi)_1$, can be determined according to the following inverse transformation:

$$\begin{aligned} \phi &= \tan^{-1}\frac{c_{23}}{c_{33}}, \\ \theta &= -\sin^{-1} c_{13}, \\ \psi &= \tan^{-1}\frac{c_{12}}{c_{11}}, \end{aligned} \tag{2.42}$$

where c_{ij} represents the element (i, j) of C. Of course, neither c_{11} nor c_{33} must vanish; otherwise the angles ϕ and ψ cannot be determined.

Example 2.6. Let us find the Euler angles for the attitude given in Example 2.2, using the representation $(\psi)_3, (\theta)_2, (\phi)_1$:

$$\begin{aligned}
\phi &= \tan^{-1}\frac{c_{23}}{c_{33}} = \tan^{-1}\frac{0.20983504362133}{-0.07726095077} = -69.7864^\circ \text{ or } 110.2136^\circ, \\
\theta &= \sin^{-1} c_{-13} = \sin^{-1}(-0.9746794344809) = -77.079^\circ \text{ or } -102.921^\circ, \\
\psi &= \tan^{-1}\frac{c_{12}}{c_{11}} = \tan^{-1}\frac{0.2}{0.1} = 63.4349^\circ \text{ or } -116.5651^\circ .
\end{aligned}$$

You may verify the accuracy of the calculated Euler angles by forming the rotation matrix according to Eq. (2.41) and comparing with the result of Example 2.2.

It is clear from Example 2.6 that the Euler angles are not unique. Furthermore, there are certain orientations for which the Euler angles cannot be determined at all from the rotation matrix, C. In such a case, the Euler angle representation is said to be *singular*, and becomes useless. An example of a singular orientation is $(\psi)_3, (\pm 90^\circ)_2, (\phi)_1$, for which $c_{11} = c_{12} = c_{23} = c_{33} = 0$, and the angles ϕ and ψ become indeterminate. For most aircraft, usage of the Euler angle representation $(\psi)_3, (\theta)_2, (\phi)_1$ does not cause a problem, because $(\psi)_3, (\pm 90^\circ)_2, (\phi)_1$ is rarely encountered. However, the same cannot be said of a fighter aircraft, a missile, or a spacecraft, where the vertical attitude is a possibility. Of course, in such a case one can switch to a different set of Euler angles (e.g., $(\psi)_3, (\theta)_1, (\phi)_3$), for which a particular singularity is avoided, but the new representation would have singularity at some other orientation. Thus, a single Euler angle representation cannot be utilized where an arbitrary orientation is possible. This deficiency in attitude representation by three angles (or any three parameters) leads one to search for nonsingular representations that must necessarily involve more than three parameters. Two obvious choices of nonsingular attitude representations are the set of nine elements of the rotation matrix, and the set of four orientation parameters arising out of the Euler axis/principal angle combination. Equally obvious is the fact that any four (or more) orientation parameters are not mutually independent [the elements of the rotation matrix obey Eq. (2.24), and the components of the Euler axis must produce a unit vector]. The Euler angles—like any other three parameter set—have the advantage of being mutually independent, and thus form a *minimal* set for attitude representation. However, their use is limited to those applications where the principal rotation is restricted to nonsingular orientations.

Example 2.7. Find the rotation matrix and the principal rotation for the attitude $(\psi)_3, (90^\circ)_2, (\phi)_1$, which is singular in terms of Euler angles.

By substituting $\theta = 90^\circ$ into Eq. (2.41), we get the following rotation matrix:

$$\mathsf{C} = \begin{pmatrix} 0 & 0 & -1 \\ \sin(\phi-\psi) & \cos(\phi-\psi) & 0 \\ \cos(\phi-\psi) & -\sin(\phi-\psi) & 0 \end{pmatrix},$$

We next determine the Euler axis, $\mathbf{e} = e_1\mathbf{i} + e_2\mathbf{j} + e_3\mathbf{k}$, according to Eq. (2.30):

$$\begin{pmatrix} 0 & 0 & -1 \\ \sin(\phi-\psi) & \cos(\phi-\psi) & 0 \\ \cos(\phi-\psi) & -\sin(\phi-\psi) & 0 \end{pmatrix} \begin{Bmatrix} e_1 \\ e_2 \\ e_3 \end{Bmatrix} = \begin{Bmatrix} e_1 \\ e_2 \\ e_3 \end{Bmatrix},$$

which, along with ${e_1}^2 + {e_2}^2 + {e_3}^2 = 1$, yields

$$\begin{aligned} e_1 &= \pm\frac{1-\cos(\phi-\psi)}{\sqrt{3+\cos(\phi-\psi)^2 - 4\cos(\phi-\psi)}}, \\ e_2 &= \frac{\sin(\phi-\psi)}{1-\cos(\phi-\psi)} e_1, \\ e_3 &= -e_1 \, . \end{aligned}$$

The appropriate sign of e_1 is obtained from that of the principal angle, Φ, which is derived from Eq. (2.33) as follows:

$$\Phi = \cos^{-1}\frac{1}{2}(\mathrm{trace}\mathrm{C} - 1) = \cos^{-1}\frac{1}{2}\{\cos(\phi-\psi)\}$$

or

$$\cos\frac{\Phi}{2} = \frac{1}{\sqrt{2}}\cos\frac{\phi-\psi}{2}\, .$$

Thus, there is no difficulty in determining the rotation matrix and principal rotation for this case for which the Euler angle representation is singular.

2.6 Euler Symmetric Parameters (Quaternion)

Since Euler axis/principal angle representation is free from singularities, a very useful representation can be derived from it, called *Euler symmetric parameters*, or the *quaternion*. A quaternion is a special set composed of four mutually dependent scalar parameters, q_1, q_2, q_3, q_4, such that the first three form a vector, called the *vector part*,

$$\mathbf{q} \doteq \begin{Bmatrix} q_1 \\ q_2 \\ q_3 \end{Bmatrix}, \qquad (2.43)$$

and the fourth, q_4, represents the *scalar part*. The quaternion for attitude representation can be derived from the Euler axis, $\mathbf{e}$, and principal rotation angle, Φ, as follows:

$$\begin{aligned} q_i &\doteq e_i \sin\frac{\Phi}{2} \quad (i = 1, 2, 3), \\ q_4 &\doteq \cos\frac{\Phi}{2}\, . \end{aligned} \qquad (2.44)$$

It is clear from Eq. (2.44) that q_1, q_2, q_3, q_4, must satisfy the constraint equation

$$q_1^2 + q_2^2 + q_3^2 + q_4^2 = 1 . \tag{2.45}$$

This constraint implies that the quaternion yields only three independent, scalar parameters, as in the principal angle/Euler axis, or the Euler angle attitude representations. Since the four elements of the quaternion satisfy the constraint equation, Eq. (2.45), it can be said that attitude orientations vary along the surface of a *four-dimensional* unit sphere without any singularity. This fact is also evident from the principal angle, Φ, and the elements of the unit vector, $\mathbf{e}$, representing the Euler axis. The chief advantage of the quaternion over the principal angle/Euler axis combination (which is also a four-parameter, nonsingular representation) lies in that the former does not require computationally intensive trigonometric function evaluations when derived from the rotation matrix. The rotation matrix, C, can be written in terms of the quaternion by substituting the definitions of Eq. (2.44) into *Euler's formula* (to be derived in Exercise 2.5), leading to

$$\mathsf{C} = (q_4^2 - \mathbf{q}^T\mathbf{q})\mathsf{I} + 2\mathbf{q}\mathbf{q}^T - 2q_4\mathsf{S}(\mathbf{q}) , \tag{2.46}$$

where $\mathsf{S}(\mathbf{q})$ is the following *skew-symmetric* matrix function formed out of the elements of vector $\mathbf{q}$:

$$\mathsf{S}(\mathbf{q}) = \begin{pmatrix} 0 & -q_3 & q_2 \\ q_3 & 0 & -q_1 \\ -q_2 & q_1 & 0 \end{pmatrix} . \tag{2.47}$$

We can write Eq. (2.46) in terms of the individual quaternion elements as follows:

$$\mathsf{C} = \begin{pmatrix} q_1^2 - q_2^2 - q_3^2 + q_4^2 & 2(q_1q_2 + q_3q_4) & 2(q_1q_3 - q_2q_4) \\ 2(q_1q_2 - q_3q_4) & -q_1^2 + q_2^2 - q_3^2 + q_4^2 & 2(q_2q_3 + q_1q_4) \\ 2(q_1q_3 + q_2q_4) & 2(q_2q_3 - q_1q_4) & -q_1^2 - q_2^2 + q_3^2 + q_4^2 \end{pmatrix} , \tag{2.48}$$

which yields the following expressions for calculating the quaternion elements from the elements of the rotation matrix, c_{ij}:

$$\begin{aligned} q_1 &= \frac{c_{23} - c_{32}}{4q_4} , \\ q_2 &= \frac{c_{31} - c_{13}}{4q_4} , \\ q_3 &= \frac{c_{12} - c_{21}}{4q_4} , \end{aligned} \tag{2.49}$$

where

$$q_4 = \pm\frac{1}{2}\sqrt{1 + c_{11} + c_{22} + c_{33}} = \pm\frac{1}{2}\sqrt{1 + \text{trace}\mathsf{C}} . \tag{2.50}$$

Note that two signs are possible in deriving the quaternion from C. However, just as in the case of principal angle/Euler axis derivation from the rotation matrix [Eq. (2.33)] this does not cause any ambiguity, because a rotation by Φ about $\mathbf{e}$ is the same as a rotation by $-\Phi$ about $-\mathbf{e}$. Thus, there is no loss of generality in taking the positive sign in Eq. (2.50). Of course, the derivation given above is valid only if $q_4 \neq 0$. If q_4 is close to zero, one can employ an alternative derivation, such as the following:

$$\begin{aligned} q_2 &= \frac{c_{12} + c_{21}}{4q_1}, \\ q_3 &= \frac{c_{31} + c_{13}}{4q_1}, \\ q_4 &= \frac{c_{23} - c_{32}}{4q_1}, \end{aligned} \tag{2.51}$$

where

$$q_1 = \pm\frac{1}{2}\sqrt{1 + c_{11} - c_{22} - c_{33}}\,. \tag{2.52}$$

Similarly, the two remaining alternative derivations of the quaternion from the rotation matrix involve division by q_2 and q_3, respectively. Among all the four possible derivations, the greatest numerical accuracy is obtained for the one that has the largest denominator term, which implies the largest argument in the square root. There are efficient algorithms that employ such a procedure for the computation of the quaternion, such as the one encoded in the MATLAB file, *quaternion.m*, given in Table 2.1.

The quaternion is a compact, nonsingular representation of attitude that results in algebraic (rather than trigonometric) expressions for the elements of the rotation matrix. Another benefit in using the quaternion over other attitude representations is in its easy combination for successive rotations. When the orientation $\mathbf{q}'', q_4''$ is obtained by first undergoing a rotation $\mathbf{q}, q_4$ followed by a rotation $\mathbf{q}', q_4'$, we can substitute Eq. (2.48) on both sides of the relationship

$$\mathsf{C}(\mathbf{q}'', q_4'') = \mathsf{C}(\mathbf{q}', q_4')\mathsf{C}(\mathbf{q}, q_4)\,, \tag{2.53}$$

in order to obtain the following simple product, called the *composition rule*:

$$\begin{Bmatrix} q_1'' \\ q_2'' \\ q_3'' \\ q_4'' \end{Bmatrix} = \begin{pmatrix} q_4' & q_3' & -q_2' & q_1' \\ -q_3' & q_4' & q_1' & q_2' \\ q_2' & -q_1' & q_4' & q_3' \\ -q_1' & -q_2' & -q_3' & q_4' \end{pmatrix} \begin{Bmatrix} q_1 \\ q_2 \\ q_3 \\ q_4 \end{Bmatrix}. \tag{2.54}$$

The composition rule of Eq. (2.54) is the defining property of the quaternion [1], [2]. Any set $(\mathbf{q}, q_4)$ that satisfies this rule is called a quaternion.[1] The

[1] The quaternion predates vectors and matrices. It was discovered by Hamilton (1805–1865), who spent many years deriving the rather abstract quaternion algebra. However, the relationships for symmetric Euler parameters (quaternion for rotation) that are employed in attitude representation can be derived without utilizing the quaternion algebra. The symmetric Euler parameters form a special quaternion that obeys the constraint given by Eq. (2.45).

Table 2.1. M-file *quaternion.m* for Deriving the Quaternion from Rotation Matrix

```
function q=quaternion(C)
%(c) 2006 Ashish Tewari
T=trace(C);
qsq=[1+2*C(1,1)-T;1+2*C(2,2)-T;1+2*C(3,3)-T;1+T]/4;
[x,i]=max(qsq);
if i==4
    q(4)=sqrt(x);
    q(1)=(C(2,3)-C(3,2))/(4*q(4));
    q(2)=(C(3,1)-C(1,3))/(4*q(4));
    q(3)=(C(1,2)-C(2,1))/(4*q(4));
end
if i==3
    q(3)=sqrt(x);
    q(1)=(C(1,3)+C(3,1))/(4*q(3));
    q(2)=(C(3,2)+C(2,3))/(4*q(3));
    q(4)=(C(1,2)-C(2,1))/(4*q(3));
end
if i==2
    q(2)=sqrt(x);
    q(1)=(C(1,2)+C(2,1))/(4*q(2));
    q(3)=(C(3,2)+C(2,3))/(4*q(2));
    q(4)=(C(3,1)-C(1,3))/(4*q(2));
end
if i==1
    q(1)=sqrt(x);
    q(2)=(C(1,2)+C(2,1))/(4*q(1));
    q(3)=(C(1,3)+C(3,1))/(4*q(1));
    q(4)=(C(2,3)-C(3,2))/(4*q(1));
end
```

efficiency of the quaternion relationship, Eq. (2.54), for the combined rotation is evident from the fact that it involves only 16 multiplications, whereas the rotation matrix representation of the same combination [Eq. (2.53)] requires 27. A MATLAB code called *quatrot.m*, which implements Eq. (2.54), is given in Table 2.2. Note that the code considers the quaternion as a row vector.

Table 2.2. M-file *quatrot.m* for Combining Two Rotations in Terms of Quaternion

```
function qpp=quatrot(q,qp)
% Program for combining two successive rotations, given by quaternions
% 'q' and 'qp', to produce the final orientation given by the quaternion
% 'qpp'. All the quaternions are stored here as row vectors.
% (c) 2006 Ashish Tewari
qpp=q*[qp(4) qp(3) -qp(2) qp(1);
     -qp(3) qp(4) qp(1) qp(2);
     qp(2) -qp(1) qp(4) qp(3);
     -qp(1) -qp(2) -qp(3) qp(4)]';
```

Example 2.8. Derive the quaternion representation for the orientation given in Example 2.5, where we had utilized Eq. (2.53) to obtain the composite rotation matrix formed out of three successive, elementary rotations.

We first derive the quaternion for each elementary rotation, and then employ the relationship of Eq. (2.54) twice to produce the quaternion for the final representation. In order to do so, we use the following MATLAB statements to call *quaternion.m* (Table 2.1), and *quatrot.m* (Table 2.2). It is presumed that the three elementary rotation matrices have been calculated and stored in the MATLAB workspace (Example 2.5).

```
>> q1=quaternion(C1) %quaternion for the first elementary rotation

q1 =      -0.3827 0          0          0.9239

>> q2=quaternion(C2) %quaternion for the second elementary rotation

q2 =     0          0.3827  0          0.9239

>> q3=quaternion(C3) %quaternion for the third elementary rotation

q3 =     0          0          -0.3827 0.9239

>> qp1=quatrot(q1,q2) %quaternion for intermediate rotation

qp1 =     -0.3536 0.3536  -0.1464 0.8536

>> qp2=quatrot(qp1,q3) %quaternion for final rotation

qp2 =     -0.4619        0.1913        -0.4619       0.7325

>> C=C3*C2*C1 % check: final rotation matrix

C = 0.5000          -0.8536       0.1464
    0.5000          0.1464        -0.8536
    0.7071          0.5000        0.5000

>> q=quaternion(C) %confirm quaternion for final rotation

q = -0.4619        0.1913        -0.4619       0.7325
```

Hence, the final orientation is represented by $q_1 = -0.4619$, $q_2 = 0.1913$, $q_3 = -0.4619$, and $q_4 = 0.7325$, which is compatible with the final rotation matrix. The accuracy of the calculated quaternion can also be confirmed by checking that it yields the correct principal rotation as follows:

$$\Phi = 2\cos^{-1} q_4 = \pm 85.8009^\circ$$

and

$$\mathbf{e} = \frac{\mathbf{q}}{sin\frac{\Phi}{2}} = \pm(0.6786\mathbf{i} - 0.2811\mathbf{j} + 0.6786\mathbf{k}) .$$

Another set of four parameters related to the quaternion is the complex *Cayley–Klein matrix* [3], which is sometimes used in attitude representations. However, we shall not consider its derivation here.

2.7 Rodrigues Parameters (Gibbs Vector)

A set of three attitude parameters, $\boldsymbol{\rho} = (\rho_1, \rho_2, \rho_3)^T$, called *Rodrigues parameters*, or the *Gibbs vector*, can be directly derived from the quaternion $(\mathbf{q}, q_4)$ as follows:

$$\boldsymbol{\rho} \doteq \frac{\mathbf{q}}{q_4}, \tag{2.55}$$

which, when substituted into Eq. (2.44), yields

$$\boldsymbol{\rho} = \mathbf{e} \tan \frac{\Phi}{2}. \tag{2.56}$$

The composition rule for Rodrigues parameters can be derived from that for the quaternion [Eq. (2.54)] to be the following:

$$\boldsymbol{\rho}'' = \frac{\boldsymbol{\rho} + \boldsymbol{\rho}' - \boldsymbol{\rho}' \times \boldsymbol{\rho}}{1 - \boldsymbol{\rho} \cdot \boldsymbol{\rho}'}, \tag{2.57}$$

where $\boldsymbol{\rho}''$ represents the final orientation obtained by combining $\boldsymbol{\rho}$ and $\boldsymbol{\rho}'$. By using Euler's formula (Exercise 2.5), one can derive the following expression for the rotation matrix in terms of Rodrigues parameters:

$$\mathsf{C} = (\mathsf{I} - S(\boldsymbol{\rho}))(\mathsf{I} + S(\boldsymbol{\rho}))^{-1}, \tag{2.58}$$

where $\mathsf{S}(\boldsymbol{\rho})$ is the following skew-symmetric matrix formed out of the elements of $\boldsymbol{\rho}$:

$$\mathsf{S}(\boldsymbol{\rho}) = \begin{pmatrix} 0 & -\rho_3 & \rho_2 \\ \rho_3 & 0 & -\rho_1 \\ -\rho_2 & \rho_1 & 0 \end{pmatrix}. \tag{2.59}$$

Clearly, Rodrigues parameters can be derived from the rotation matrix elements, c_{ij}, as follows:

$$\begin{aligned} \rho_1 &= \frac{c_{23} - c_{32}}{1 + \text{trace}\mathsf{C}}, \\ \rho_2 &= \frac{c_{31} - c_{13}}{1 + \text{trace}\mathsf{C}}, \\ \rho_3 &= \frac{c_{12} - c_{21}}{1 + \text{trace}\mathsf{C}}. \end{aligned} \tag{2.60}$$

It is evident from Eq. (2.56) [as well as from Eq. (2.60)] that Rodrigues parameters have a singularity at $\Phi = n\pi$ $(n = 1, 3, 5, \ldots)$, hence their use is limited to principal rotations of $\Phi < 180°$. The three-parameter set is thus similar to the Euler angles in being incapable of representing an arbitrary orientation.

2.8 Modified Rodrigues Parameters

In order to extend the applicability of Rodrigues parameters for principal rotations greater than 180°, a modified three-parameter set is defined as follows:

$$\mathbf{p} \doteq \frac{\mathbf{q}}{1 + q_4}, \tag{2.61}$$

which, when substituted into Eq. (2.44), yields

$$\mathbf{p} = \mathbf{e} \tan \frac{\Phi}{4} \,. \tag{2.62}$$

Clearly, the new set $\mathbf{p} = (p_1, p_2, p_3)^T$—called the *modified Rodrigues parameters*—is nonsingular for principal rotations of $\Phi < 360°$. However, there is a singularity at $\Phi = 360°$. Most nonspinning aerospace vehicles have orientations with $\Phi < 360°$ and are thus represented by the set of modified Rodrigues parameters without singularity. Since $\mathbf{p}$ is a minimal representation, it is advantageous over the four-parameter quaternion in reducing the number of kinematic equations to be solved for the attitude.

The rotation matrix can be expressed in terms of the modified Rodrigues parameters with the use of Eq. (2.46) as follows:

$$\mathsf{C} = \mathsf{I} + \frac{4(\mathbf{p}^T\mathbf{p} - 1)}{(1 + \mathbf{p}^T\mathbf{p})^2}\mathsf{S}(\mathbf{p}) + \frac{8}{(1 + \mathbf{p}^T\mathbf{p})^2}\mathsf{S}^2(\mathbf{p}) \,, \tag{2.63}$$

where $\mathsf{S}(\mathbf{p})$ is the following skew-symmetric matrix formed out of the elements of $\mathbf{p}$:

$$\mathsf{S}(\mathbf{p}) = \begin{pmatrix} 0 & -p_3 & p_2 \\ p_3 & 0 & -p_1 \\ -p_2 & p_1 & 0 \end{pmatrix} . \tag{2.64}$$

By substituting the relationship between the Rodrigues and modified Rodrigues parameters,

$$\rho = \frac{2\mathbf{p}}{1 - \mathbf{p}^T\mathbf{p}} \,, \tag{2.65}$$

into Eq. (2.57), we can derive the following rather complicated composition rule for the modified Rodrigues parameters:

$$\mathbf{p}'' = \frac{(1 - \mathbf{p}^T\mathbf{p})\mathbf{p}' + (1 - \mathbf{p}'^T\mathbf{p}')\mathbf{p} - 2\mathbf{p}' \times \mathbf{p}}{1 + (\mathbf{p}^T\mathbf{p})(\mathbf{p}'^T\mathbf{p}') - 2\mathbf{p} \cdot \mathbf{p}'} \,. \tag{2.66}$$

Example 2.9. Derive the modified Rodrigues parameters for the orientation given in Example 2.5, for which the quaternion was calculated in Example 2.8.

We begin by applying Eq. (2.61) to the calculated quaternion:

$$\mathbf{p} = \frac{1}{1 + 0.7325} \begin{Bmatrix} -0.4619 \\ 0.1913 \\ -0.4619 \end{Bmatrix} = \begin{Bmatrix} -0.2666 \\ 0.1104 \\ -0.2666 \end{Bmatrix} ,$$

leading to the skew-symmetric matrix [Eq. (2.64)],

$$\mathsf{S}(\mathbf{p}) = \begin{pmatrix} 0 & 0.2666 & 0.1104 \\ -0.2666 & 0 & 0.2666 \\ -0.1104 & -0.2666 & 0 \end{pmatrix} ,$$

which corresponds to the following rotation matrix according to Eq. (2.63):

$$\mathsf{C} = \begin{pmatrix} 0.5 & -0.8536 & 0.1464 \\ 0.5 & 0.1464 & -0.8536 \\ 0.7071 & 0.5 & 0.5 \end{pmatrix} .$$

This is the same result as that of Example 2.5.

2.9 Attitude Kinematics

We are now in a position to address the evolution of a coordinate frame's attitude with time. We can adopt any one of the various attitude representations considered above to describe the changing attitude. Let us first take the rotation matrix representation. Consider a rotating coordinate frame, *(oxyz)*, with unit vectors, $\mathbf{i}, \mathbf{j}, \mathbf{k}$, representing the axes *ox, oy, oz*, respectively, whose changing attitude relative to an arbitrary, fixed frame is of interest. When the attitude is changing with time, the rotation matrix representing the orientation of the rotating frame relative to a fixed frame is a function of time, $\mathsf{C}(t)$. In order to find this function, consider an infinitesimal principal rotation, $\Delta\Phi$, in a small time interval, Δt, measured after a given time, t. The rotation is small enough for us to approximate $\cos\Delta\Phi \approx 1$, $\sin\Delta\Phi \approx \Delta\Phi$. Furthermore, we assume that that Δt is so small that the axis of rotation, $\mathbf{e}(t)$, remains essentially unchanged. These assumptions allow us to write the rotation matrix representing attitude change in the time interval Δt with the use of Euler's formula (Exercise 2.5) as follows:

$$\mathsf{C}(\Delta t) \approx \mathsf{I} - \Delta\Phi \mathsf{S}(\mathbf{e}) , \tag{2.67}$$

or,

$$\mathsf{C}(\Delta t) \approx \mathsf{I} - \mathsf{S}(\Delta\Phi\mathbf{e}) , \tag{2.68}$$

where

$$\mathsf{S}(\Delta\Phi\mathbf{e}) = \begin{pmatrix} 0 & -\Delta\Phi e_z & \Delta\Phi e_y \\ \Delta\Phi e_z & 0 & -\Delta\Phi e_x \\ -\Delta\Phi e_y & \Delta\Phi e_x & 0 \end{pmatrix} . \tag{2.69}$$

We chose to resolve the instantaneous axis of rotation in the rotating frame, because $\mathsf{C}(\Delta t)$ describes the frame's orientation at time $t + \Delta t$, relative to its own previous orientation at time t, rather than that relative to the fixed frame. Thus, $\mathbf{e}(t)$ has its components, e_x, e_y, e_z, resolved along the instantaneous axes of the rotating frame, $\mathbf{i}, \mathbf{j}, \mathbf{k}$, at time t. Let the rotation matrices $\mathsf{C}(t)$ and $\mathsf{C}(t + \Delta t)$ denote the attitudes of the rotating frame relative to a fixed frame at times t and $t + \Delta t$, respectively. The rotation matrix, $\mathsf{C}(\Delta t)$, describes the evolution of the attitude in the time interval, Δt, caused by a principal rotation of $\Delta\Phi$ about $\mathbf{e}$. Thus, $\mathsf{C}(\Delta t)$ denotes the rotation required to produce $\mathsf{C}(t + \Delta t)$ from $\mathsf{C}(t)$. Now, these rotation matrices must obey the composition rule for successive rotations, Eq. (2.25), which implies that the final orientation at time $t + \Delta t$ is related to the initial attitude at time t by

$$\mathsf{C}(t + \Delta t) = \mathsf{C}(\Delta t)\mathsf{C}(t) \,. \tag{2.70}$$

By substituting Eq. (2.68) into Eq. (2.70), we can write the time derivative of the rotation matrix as follows:

$$\frac{\mathrm{d}\mathsf{C}}{\mathrm{d}t} \doteq \lim_{\Delta t \to 0} \frac{\mathsf{C}(t + \Delta t) - \mathsf{C}(t)}{\Delta t} = -\mathsf{S}(\boldsymbol{\omega})\mathsf{C}(t) \,, \tag{2.71}$$

where

$$\boldsymbol{\omega}(t) \doteq \lim_{\Delta t \to 0} \frac{\Delta \Phi \mathbf{e}}{\Delta t} \tag{2.72}$$

is called the *angular velocity* vector of the coordinate frame. It is important to emphasize that the angular velocity, $\boldsymbol{\omega}(t)$, has its components, $\omega_x, \omega_y, \omega_z$, at time t resolved along the axes of the rotating coordinate frame (rather than the fixed frame). The skew-symmetric matrix of angular velocity components is thus the following:

$$\mathsf{S}(\boldsymbol{\omega}) = \begin{pmatrix} 0 & -\omega_z & \omega_y \\ \omega_z & 0 & -\omega_x \\ -\omega_y & \omega_x & 0 \end{pmatrix} \,. \tag{2.73}$$

Equation (2.71) can also be derived by simply differentiating with time both sides of the following defining equation of the rotation matrix:

$$\begin{Bmatrix} \mathbf{i} \\ \mathbf{j} \\ \mathbf{k} \end{Bmatrix} = \mathsf{C} \begin{Bmatrix} \mathbf{I} \\ \mathbf{J} \\ \mathbf{K} \end{Bmatrix} \,, \tag{2.74}$$

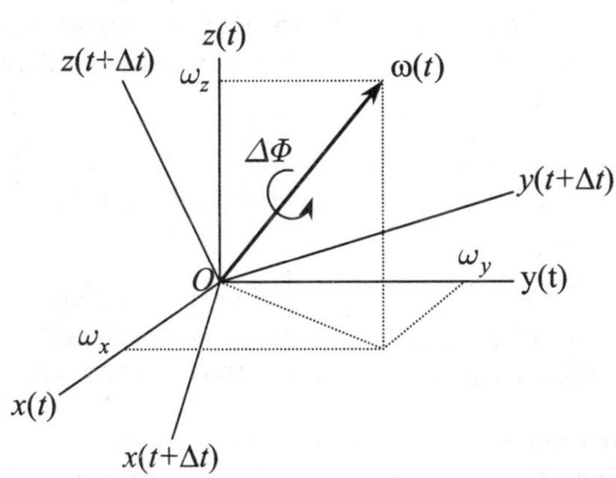

Fig. 2.6. A coordinate frame rotating with angular velocity, $\boldsymbol{\omega}(t)$.

resulting in

$$\begin{Bmatrix} \boldsymbol{\omega} \times \mathbf{i} \\ \boldsymbol{\omega} \times \mathbf{j} \\ \boldsymbol{\omega} \times \mathbf{k} \end{Bmatrix} = \frac{\mathrm{d}\mathsf{C}}{dt} \begin{Bmatrix} \mathbf{I} \\ \mathbf{J} \\ \mathbf{K} \end{Bmatrix} . \tag{2.75}$$

By applying the result of Exercise 2.1, and substituting (2.74) into Eq. (2.75), we have

$$-\mathsf{S}(\boldsymbol{\omega}) \begin{Bmatrix} \mathbf{i} \\ \mathbf{j} \\ \mathbf{k} \end{Bmatrix} = \frac{\mathrm{d}\mathsf{C}}{dt} \mathsf{C}^T \begin{Bmatrix} \mathbf{i} \\ \mathbf{j} \\ \mathbf{k} \end{Bmatrix} , \tag{2.76}$$

from which Eq. (2.71) follows through the orthogonality property, Eq. (2.24).

Recall that the angular velocity, $\boldsymbol{\omega}(t)$, is resolved along the axes of the rotating coordinate frame *(oxyz)*, as shown in Fig. 2.6. Such a choice of axes is very convenient when depicting a rotating *rigid body* (Chapter 13). In such a case, the axes of the rotating coordinate frame are fixed to the rigid body and can be used to represent the attitude and angular velocity of the body relative to an *inertial* (or space-fixed) frame. On rare occasions, it is desired to alternatively express the angular velocity with components along the axes of an inertial frame. In such a case the angular velocity is termed *space-referenced*, or inertial angular velocity, and is related to the body referenced angular velocity by $\boldsymbol{\omega}_I = \mathsf{C}^T(t)\boldsymbol{\omega}(t)$ according to the coordinate transformation of Eq. (2.21).

The differential equation for the rotation matrix is Eq. (2.71), which, with a given function $\boldsymbol{\omega}(t)$ and initial condition, $\mathsf{C}(0)$, should be solved for $\mathsf{C}(t)$ in order to describe the evolution of the attitude. The solution, $\mathsf{C}(t)$, must satisfy the orthogonality condition Eq. (2.24); by differentiating both sides of which, we have

$$\frac{\mathrm{d}}{\mathrm{d}t}\mathsf{C}\mathsf{C}^T = \frac{\mathrm{d}}{\mathrm{d}t}\mathsf{C}^T\mathsf{C} = 0 . \tag{2.77}$$

This implies that $\mathsf{C}\mathsf{C}^T$ and $\mathsf{C}^T\mathsf{C}$ are equal and constant matrices. Thus, if $\mathsf{C}(t)$ satisfies Eq. (2.24) at some initial time $t = 0$, it does so at all other times. One can easily verify (Exercise 2.13) that the matrix differential equation, Eq. (2.71), satisfies Eq. (2.77).

Example 2.10. Consider a coordinate frame, $\mathbf{i}, \mathbf{j}, \mathbf{k}$, rotating with an angular velocity $\boldsymbol{\omega}(t) = 0.1\mathbf{i} - 0.5\mathbf{j} - \mathbf{k}$ rad/s. If the original attitude of the frame is given by

$$\mathsf{C} = \begin{pmatrix} 0.1399200225 & -0.9857942023 & 0.0929095147 \\ -0.9432515656 & -0.1612425105 & -0.2903055921 \\ 0.3011625330 & -0.0470174803 & -0.9524129804 \end{pmatrix} ,$$

determine the attitude of the frame after one second.

Note that since the given angular velocity components resolved in the rotating frame are constant, the matrix differential equation, Eq. (2.71), is now a set of linear algebraic equations with constant coefficients, $\mathsf{S}(\boldsymbol{\omega})$, and can be integrated easily using the evolution rule of Eq. (2.70). It is to be noted that Eq. (2.70) is valid only for an infinitesimal time interval, Δt. Since the given time of 1 s is not small enough given the assigned angular speed, we must break it into several smaller intervals of equal length, and apply Eq. (2.70) to each interval, beginning with the given attitude at $t = 0$. Such an approach is called *time-marching* and is frequently applied to the solution of linear, vector (or matrix) differential equations.

We begin the solution by writing a MATLAB code called *rotevolve.m* (Table 2.3), which obtains the rotation matrix by the time-marching approach of Eq. (2.70). During each time step, the rotation angle is given by $\Delta\Phi = \mid \boldsymbol{\omega} \mid \Delta t$, while the axis of rotation is approximated by $\mathbf{e} = \frac{\boldsymbol{\omega}}{|\boldsymbol{\omega}|}$. Needless to say, these approximations are accurate only if a reasonably small Δt is employed. The size of a time step should be much smaller than the period of rotation and is generally dictated by the desired tolerance in satisfying the orthogonality condition Eq. (2.24). Since the computation error accumulates over the steps, a reasonably small time step must be used for a given time interval. In *rotevolve.m*, the time step is obtained by $10^6 \Delta t = \frac{2\pi}{|\boldsymbol{\omega}|}$. In order to invoke *rotevolve.m* for the present example, the following MATLAB statements are used:

```
>> C= [0.1399200225    -0.9857942023    0.0929095147;
      -0.9432515656  -0.1612425105    -0.2903055921;
      0.3011625330    -0.0470174803    -0.9524129804];% rotation matrix (t=0)
>>C*C' %test for orthogonality of given rotation matrix

ans =    1.0000         0.0000         0.0000
        0.0000         1.0000         0.0000
        0.0000         0.0000         1.0000

>> Cf=rotevolve(C,[0.1;-0.5;-1],1) %call rotevolve for final rot. matrix

Cf =   0.94711618664267          -0.315392238061307         -0.0592086317903142
      -0.313796220025831         -0.871634028765343         -0.376552394359275
      0.0671544766324447         0.375218475325714          -0.924501393941303
```

Therefore, the attitude at $t = 1$ s is given by the rotation matrix

$$\mathsf{C} = \begin{pmatrix} 0.94711618664267 & -0.315392238061307 & -0.0592086317903142 \\ -0.313796220025831 & -0.871634028765343 & -0.376552394359275 \\ 0.0671544766324447 & 0.375218475325714 & -0.924501393941303 \end{pmatrix}$$

We can verify the orthogonality of this matrix, within a tolerance of 10^{-5}. However, since the time-step size is only $\Delta t = 5.5975 \times 10^{-6}$ s, the computation requires 178,651 steps, which is rather inefficient.

The angular velocity, $\boldsymbol{\omega}(t)$, is generally time-varying and is obtained from the *attitude dynamics* (Chapter 13) (rather than kinematics). As seen above, even when $\boldsymbol{\omega}(t)$ has constant elements, a large number of inefficient matrix products must be computed for obtaining the attitude using the rotation matrix. When the angular velocity, $\boldsymbol{\omega}(t)$, does not have constant elements, it becomes rather cumbersome to solve the differential equation, Eq. (2.71), with a time-varying coefficient matrix $\mathsf{S}(\boldsymbol{\omega})$, for the nine elements of $\mathsf{C}(t)$. Instead, it is much easier

Table 2.3. M-file *rotevolve.m* for the Time Evolution of the Rotation Matrix

```
function c=rotevolve(c0,w,T)
%function for evolving the rotation matrix with a constant
%body-referenced angular velocity
%c0=rotation matrix at t=0
%w=angular velocity vector (3x1) (rad/s)
%T=final time (s)
%(c)2006 Ashish Tewari
S=[0 -w(3,1) w(2,1);w(3,1) 0 -w(1,1);-w(2,1) w(1,1) 0];
dt=2*pi/(10^6*norm(w))
cdt=eye(3)-S*dt;
t=dt;
c=cdt*c0;
while t<=T
     c=cdt*c;
     t=t+dt;
end
```

to employ other attitude representations, such as Euler angles, quaternion, or the modified Rodrigues parameters.

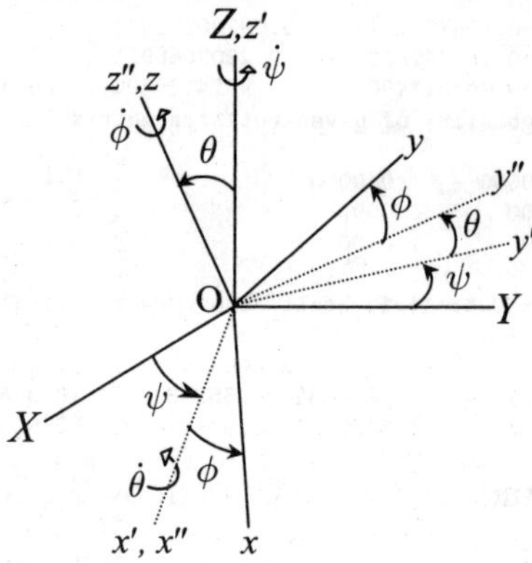

Fig. 2.7. The Euler angle orientation, $(\psi)_3, (\theta)_1, (\phi)_3$.

Let us consider the Euler angle representation, $(\psi)_3, (\theta)_1, (\phi)_3$, which is often employed for spacecraft (as well as orbital plane) attitude. Figure 2.7 depicts the three successive, elementary rotations required in this representation from a stationary frame, *(OXYZ)*, with axes along **I**, **J**, **K**, respectively, to the rotating frame *(Oxyz)* with axes **i**, **j**, **k**, respectively. The rotation matrix relating the final orientation of *(Oxyz)* to *(OXYZ)* can be obtained by

$$\mathsf{C} = \mathsf{C}_3(\phi)\mathsf{C}_1(\theta)\mathsf{C}_3(\psi) = \tag{2.78}$$

$$\begin{pmatrix} (\cos\psi\cos\phi - \sin\psi\sin\phi\cos\theta) & (\sin\psi\cos\phi + \cos\psi\sin\phi\cos\theta) & \sin\phi\sin\theta \\ -(\cos\psi\sin\phi + \sin\psi\cos\phi\cos\theta) & (-\sin\psi\sin\phi + \cos\psi\cos\phi\cos\theta) & \cos\phi\sin\theta \\ \sin\psi\sin\theta & -\cos\psi\sin\theta & \cos\theta \end{pmatrix}.$$

In terms of the local angular rates, $\dot\phi, \dot\theta, \dot\psi$, the angular velocity, $\boldsymbol{\omega}(t)$, can be expressed as follows:

$$\boldsymbol{\omega}(t) = \dot\phi\mathbf{k} + \dot\theta\mathbf{i}' + \dot\psi\mathbf{K}\,, \tag{2.79}$$

where

$$\mathbf{i}' = \mathbf{i}\cos\phi - \mathbf{j}\sin\phi \tag{2.80}$$

and

$$\mathbf{K} = \mathbf{i}\sin\phi\sin\theta + \mathbf{j}\cos\phi\sin\theta + \mathbf{k}\cos\theta\,. \tag{2.81}$$

Upon substituting Eqs. (2.80) and (2.81) into Eq. (2.79), we have

$$\boldsymbol{\omega}(t) = \begin{Bmatrix} \omega_x \\ \omega_y \\ \omega_z \end{Bmatrix} = \begin{Bmatrix} \dot\psi\sin\phi\sin\theta + \dot\theta\cos\phi \\ \dot\psi\cos\phi\sin\theta - \dot\theta\sin\phi \\ \dot\psi\cos\theta + \dot\phi \end{Bmatrix}, \tag{2.82}$$

or,

$$\begin{Bmatrix} \dot\psi \\ \dot\theta \\ \dot\phi \end{Bmatrix} = \frac{1}{\sin\theta}\begin{pmatrix} \sin\phi & \cos\phi & 0 \\ \cos\phi\sin\theta & -\sin\phi\sin\theta & 0 \\ -\sin\phi\cos\theta & -\cos\phi\cos\theta & \sin\theta \end{pmatrix}\begin{Bmatrix} \omega_x \\ \omega_y \\ \omega_z \end{Bmatrix}. \tag{2.83}$$

Equation (2.83) is the required kinematic relationship between the Euler angles, ψ, θ, ϕ, and the angular velocity in the form of three, coupled, nonlinear first-order ordinary differential equations. Note that this relationship has singularities at $\theta = n\pi$ $(n = 0, 1, 3, \ldots)$, which are the inherent singularities of the Euler angle representation, $(\psi)_3, (\theta)_1, (\phi)_3$. Hence, this representation is limited to rotations of $0 < \theta < \pi$, which are applicable to spacecraft attitudes and the orientation of orbital planes. A different Euler angle representation, $(\psi)_3, (\theta)_2, (\phi)_1$ (Fig. 2.5), is commonly employed in aircraft applications and has singularities at $\theta = n\frac{\pi}{2}$ $(n = 1, 3, \ldots)$, which limits its application to $-\frac{\pi}{2} < \theta < \frac{\pi}{2}$. The kinematic relationship for $(\psi)_3, (\theta)_2, (\phi)_1$ will be discussed in Chapter 4.

Example 2.11. Solve the problem given in Example 2.10 using the Euler angle representation, $(\psi)_3, (\theta)_1, (\phi)_3$.

We begin by deriving the $(\psi)_3, (\theta)_1, (\phi)_3$ Euler angles for the initial attitude prescribed in Example 2.10. By using the third column of the given rotation matrix and Eq. (2.78), we have

$$\theta = \cos^{-1} c_{33} = \cos^{-1}(-0.9524129804) = 162.2532°,$$
$$\sin\phi = \frac{c_{13}}{\sin\theta} = \frac{0.0929095147}{0.3048106211} = 0.3048106211,$$
$$\cos\phi = \frac{c_{23}}{\sin\theta} = \frac{-0.2903055921}{0.3048106211} = -0.9524129804\,,$$

which yield $\phi = \theta = 162.2532°$. The third column of the given rotation matrix and Eq. (2.78) produce

$$\sin\psi = \frac{c_{31}}{\sin\theta} = \frac{0.301162533}{0.3048106211} = 0.98803162409,$$
$$\cos\phi = \frac{c_{32}}{\sin\theta} = \frac{-0.0470174803}{0.3048106211} = 0.1542514499,$$

or, $\psi = 81.1266°$. Now, we write a code called *euler313evolve.m*, tabulated in Table 2.4, which provides the kinematic equations for integration according to the intrinsic *Runge–Kutta*, nonlinear differential equation solver (Appendix A) of MATLAB (*ode45.m*). The program is invoked as follows:

```
>> dtr=pi/180; psi=81.1266*dtr; theta=162.2532*dtr; phi=theta;% Euler angles
>> [t,x]=ode45(@euler313evolve,[0 1],[psi theta phi]'); %Runge-Kutta solver
>> plot(t,x/dtr) %plot of Euler angles' time history
```

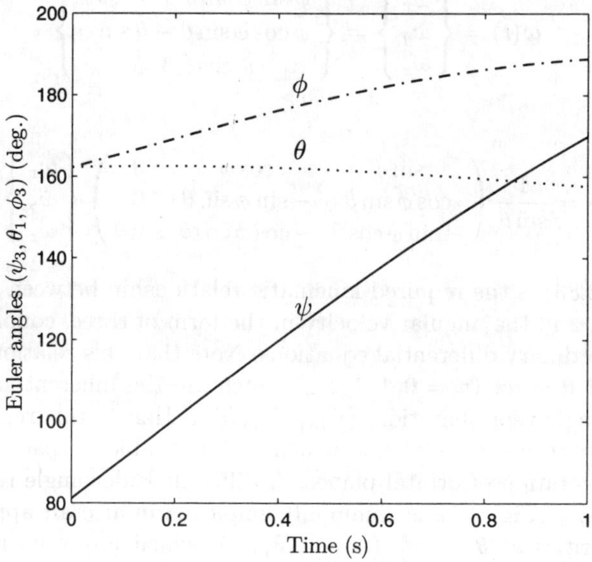

Fig. 2.8. The time history of the Euler angles, $(\psi)_3, (\theta)_1, (\phi)_3$ with $\omega = (0.1, -0.5, -1)^T$.

The resulting time history of the Euler angles, obtained in only 41 time steps, is shown in Fig. 2.8. Note that θ does not cross the singularity at $\theta = 180°$. This is merely due to the specific value of the constant angular velocity and cannot be ensured in a general case where angular velocity components could be time-varying. The final values obtained at $t = 1$ s are $\psi = 169.8530°, \theta = 157.5933°$, and $\phi = 188.9358°$. We can easily verify that a substitution of these angles into Eq. (2.78) results in the same final rotation matrix as that derived in Example 2.10.

There is no qualitative modification required in the above implemented Runge–Kutta approach (Appendix A) for solving the nonlinear differential equations when the angular velocity components are varying with time. Thus, the Euler angle representation for attitude kinematics is much more compact

Table 2.4. M-file *euler313evolve.m* for the Time Evolution of the 3-1-3 Euler Angles

```
function xdot=euler313evolve(t,x)
%x(1)=psi, x(2)=theta, x(3)=phi
% (c) 2006 Ashish Tewari
w=[0.1,-0.5,-1];% angular velocity in rad/s
xdot(1,1)=(sin(x(3))*w(1)+cos(x(3))*w(2))/sin(x(2));
xdot(2,1)=cos(x(3))*w(1)-sin(x(3))*w(2);
xdot(3,1)=w(3)-(sin(x(3))*cos(x(2))*w(1)+cos(x(3))*cos(x(2))*w(2))/sin(x(2));
```

than the rotation matrix. However, it suffers from singularities and cannot be applied when principal rotations are larger than $\theta = 180°$. For a compact kinematical calculation that has a larger range of validity, we must look toward the attitude representations that are based upon the Euler-axis/principal angle combination, such as the quaternion and the modified Rodrigues parameters.

The quaternion has an advantage in being a compact (albeit nonminimal) as well as nonsingular representation for attitude kinematics. Thus, modern flight dynamic applications generally employ the quaternion. The kinematical equations of the rotating frame, $(oxyz)$, in terms of the quaternion can be obtained from its composition rule, Eq. (2.54), as follows. As in the foregoing discussion, we consider an infinitesimal principal rotation, $\Delta\Phi$, in a small time interval, Δt, such that the axis of rotation, $\mathbf{e}(t) = e_x\mathbf{i} + e_y\mathbf{j} + e_z\mathbf{k}$, remains essentially unchanged. In Eq. (2.54), we substitute the orientation at $t + \Delta t$, $\{\mathbf{q}'', q_4''\} = \{\mathbf{q}(t+\Delta t), q_4(t+\Delta t)\}$, which is obtained from the initial attitude, $\mathbf{q}(t), q_4(t)$, by undergoing a rotation $[\mathbf{q}', q_4'] = [\mathbf{q}(\Delta t), q_4(\Delta t)]$. Thus, from the definition of quaternion we have

$$
\begin{aligned}
q_1' &= e_x \sin\frac{\Delta\Phi}{2}, \\
q_2' &= e_y \sin\frac{\Delta\Phi}{2}, \\
q_3' &= e_z \sin\frac{\Delta\Phi}{2}, \\
q_4' &= \cos\frac{\Delta\Phi}{2}\,.
\end{aligned}
\tag{2.84}
$$

By substituting these into Eq. (2.54), approximating $\cos\Delta\Phi \approx 1$, $\sin\Delta\Phi \approx \Delta\Phi$, and utilizing Eq. (2.72), we can write the following equation for the time evolution of the quaternion:

$$\{\mathbf{q}(t+\Delta t), q_4(t+\Delta t)\}^T \approx [\mathsf{I} + \frac{1}{2}\Omega\Delta t]\{\mathbf{q}(t), q_4(t)\}^T\,, \tag{2.85}$$

where Ω is the following skew-symmetric matrix of the angular velocity components:

$$\Omega = \begin{pmatrix} 0 & \omega_z & -\omega_y & \omega_x \\ -\omega_z & 0 & \omega_x & \omega_y \\ \omega_y & -\omega_x & 0 & \omega_z \\ -\omega_x & -\omega_y & -\omega_z & 0 \end{pmatrix}\,. \tag{2.86}$$

Therefore, the time derivative of the quaternion is the following:

$$\begin{aligned}\frac{\mathrm{d}\{\mathbf{q}, q_4\}^T}{\mathrm{d}t} &\doteq \lim_{\Delta t \to 0} \frac{\{\mathbf{q}(t+\Delta t), q_4(t+\Delta t)\}^T - \{\mathbf{q}(t), q_4(t)\}^T}{\Delta t} \\ &= \frac{1}{2}\Omega\{\mathbf{q}(t), q_4(t)\}^T . \end{aligned} \tag{2.87}$$

The linear, algebraic form of the matrix differential equation, Eq. (2.87), is an obvious advantage of the quaternion representation. Unlike the Euler angles, it is not required to evaluate trigonometric functions in the process of solving the kinematic equations. The matrix Ω is either constant, or time-varying, depending upon whether the components of the angular velocity are changing with time. In either case, we can adopt a numerical scheme (Appendix A) to integrate this linear, ordinary differential equation for the quaternion, $\mathbf{q}(t), q_4(t)$. We can break the final time into smaller intervals and utilize Eq. (2.87) in each interval to produce a time history of the changing quaternion. The splitting of time into smaller intervals is necessary when the angular velocity components (Ω) are time-varying, wherein Eq. (2.87) is employed with a different Ω in each time interval. The size of the time interval in such a case would be determined by the rate of change of angular velocity components. Such a *quasi-steady* approximation of the time-varying coefficient matrix is commonly employed.

Example 2.12. Solve the problem given in Examples 2.10 and 2.11 using the quaternion.

Since Ω is a constant matrix here, we can write the solution to the resulting linear, time-invariant state equations involving the quaternion using the *matrix exponential* (Chapter 14) as follows:

$$\{\mathbf{q}(t), q_4(t)\}^T = e^{\frac{1}{2}\Omega t}\{\mathbf{q}(0), q_4(0)\}^T , \tag{2.88}$$

where $\mathbf{q}(0), q_4(0)$ represents the initial attitude at $t = 0$. The matrix exponential, e^{A}, of a square matrix, A, can be computed using an efficient algorithm (Chapter 14), such as that implemented in the intrinsic MATLAB function *expm.m*, which we will employ here. Equation (2.88) is implemented in the program, *quatevolve.m*, which is tabulated in Table 2.5. The computation of the final quaternion is carried out by the following MATLAB statements, assuming the initial rotation matrix, C, is already stored in the workspace:

```
>> q0=quaternion(C) %quaternion at t=0 using "quaternion.m"

q0 = 0.750595682341698  -0.64250495082845  0.131253102359508  -0.0810316784134013

>> qf=quatevolve(q0',[0.1;-0.5;-1],1)' % final quaternion

qf = 0.967373385090801  -0.162601824240507 0.002053412854426  -0.194281146902704

>> Cf=rotquat(qf') %rotation matrix

Cf =     0.94711286        -0.31539123       -0.0592081039
     -0.31379548      -0.87163097      -0.3765525988
     0.067153772      0.375217044      -0.9245012389
```

The final rotation matrix is computed above from the final quaternion according to Eq. (2.46), which is implemented in the M-file *rotquat.m* (Table 2.6). Note that the final attitude agrees with that calculated in Example 2.10 up to the fifth decimal place, which was the tolerance specified in Example 2.10. However, since only one matrix multiplication is required (apart from the steps used in calculating the matrix exponential) in the specified time interval (compared to 178,651 in Example 2.10), this computation is more efficient and suffers from a smaller cumulative truncation error. Of course, the quaternion computation does not have any singularities and thus can be carried out for any angular velocity. The same cannot be said for the Euler angle computation in Example 2.11, where a singularity was avoided only by the given value of the angular velocity and the initial attitude.

Table 2.5. M-file *quatevolve.m* for the Time Evolution of the Quaternion

```
function q=quatevolve(q0,w,T)
%function for evolving the quaternion with a constant
%body-referenced angular velocity
%q0=quaternion at t=0 (4x1) vector
%w=angular velocity vector (3x1) (rad/s)
%T=final time (s)
%(c) 2006 Ashish Tewari
S=[0 w(3,1) -w(2,1) w(1,1);-w(3,1) 0 w(1,1) w(2,1);
   w(2,1) -w(1,1) 0 w(3,1);-w(1,1) -w(2,1) -w(3,1) 0];
q=expm(0.5*S*T)*q0;
```

Table 2.6. M-file *rotquat* for the Computation of Rotation Matrix from the Quaternion

```
function C=rotquat(q)
% rotation matrix from the quaternion
% (c) 2006 Ashish Tewari
S=[0 -q(3,1) q(2,1);q(3,1) 0 -q(1,1);-q(2,1) q(1,1) 0];
C=(q(4,1)^2-q(1:3,1)'*q(1:3,1))*eye(3)+2*q(1:3,1)*q(1:3,1)'-2*q(4,1)*S;
```

Finally, we consider the kinematical representation possible through the modified Rodrigues parameters, $\mathbf{p} = (p_1, p_2, p_3)^T$. Being a minimal representation based upon the quaternion, $\mathbf{p}$, reduces the number of kinematic, first-order differential equations required by the latter. However, its singularity at $\Phi = 360°$ limits its application to principal rotations of $\Phi < 360°$, which gives a larger range of validity than that of the other minimal representations (Euler angles and Rodrigues parameters). By substituting Eq. (2.61) into Eq. (2.87), we have

$$\frac{\mathrm{d}\mathbf{p}}{\mathrm{d}t} = \frac{1}{2}\left[\mathsf{S}(\mathbf{p}) + \mathbf{p}\mathbf{p}^T + \left(\frac{1-\mathbf{p}^T\mathbf{p}}{2}\right)\mathsf{I}\right]\boldsymbol{\omega}\,, \tag{2.89}$$

where $\mathsf{S}(\mathbf{p})$ is given by Eq. (2.64). Being a nonlinear, ordinary differential equation, Eq. (2.89) can be integrated in time using a Runge–Kutta algorithm (Appendix A) to produce the instantaneous value of $\mathbf{p}(t)$, given an initial condition, $\mathbf{p}(0)$, and the time-varying angular velocity, $\boldsymbol{\omega}(t)$.

Example 2.13. Solve the problem given in Examples 2.10–2.12 using the modified Rodrigues parameters.

We begin by writing a code called *mrpevolve.m*, tabulated in Table 2.7, which provides the scalar kinematic equations, Eq. (2.89), for integration according to the intrinsic Runge–Kutta nonlinear differential equation solver (Appendix A) of MATLAB (*ode45.m*). The final rotation matrix is calculated from the final modified Rodrigues vector through Eq. (2.63), which is implemented in the program *rotmrp.m* (Table 2.8). The computations are given by the following statements, beginning with the initial quaternion from Example 2.12:

```
>> p0=q0(1:3)/(q0(4)+1) %MRP at t=0 from initial quaternion (Example 2.12)

p0 = 0.816780801590412  -0.699158976143123   0.142826579846517

>> [t,p]=ode45(@mrpevolve,[0 1],p0'); %Runge-Kutta solver
>> pf=p(size(t,1),:) % final MRP vector at t=1 second

pf = 1.2006339199379   -0.201809630219251   0.0025485478218541

>> Cf=rotmrp(pf') %final rotation matrix using "rotmrp.m"

Cf =  0.947112860753656   -0.315391232427005   -0.0592081033575685
     -0.313795474675653   -0.871630965841634   -0.376552598529159
      0.0671537718084939   0.375217043714908   -0.92450123906771З

>> plot(t,p) %plot of MRP time history
>> for i=1:size(t,1);phi(i)=4*atan(norm(p(i,:)))*180/pi;end
>> plot(t,phi) %plot of principal angle time history
```

The final rotation matrix agrees with that computed in Example 2.12 up to nine decimal places. The resulting plots for the time evolution of the modified Rodrigues parameters, and the principal angle, $\Phi = 4\tan^{-1} \mid \mathbf{p} \mid$ are shown in Figs. 2.9 and 2.10, respectively. As in the case of Euler angles (Example 2.12), the Runge–Kutta iteration is carried out with 41 time steps.

Table 2.7. M-file *mrpevolve.m* for the Time Derivatives of the MRP

```
function pdot=mrpevolve(t,p)
% program for calculating the time derivatives of the modified Rodrigues
% parameters, with a constant angular velocity, w (3x1)
% (c) 2006 Ashish Tewari
w=[0.1;-0.5;-1]; % angular velocity in rad/s
S=[0 -p(3,1) p(2,1);p(3,1) 0 -p(1,1);-p(2,1) p(1,1) 0];
G=0.5*(eye(3)+S+p*p'-0.5*(1+p'*p)*eye(3));
pdot=G*w;
```

Table 2.8. M-file *rotmrp.m* for the Rotation Matrix from the MRP

```
function C=rotmrp(p)
% rotation matrix from the modified Rodrigues parameters
% (c) 2006 Ashish Tewari
S=[0 -p(3,1) p(2,1);p(3,1) 0 -p(1,1);-p(2,1) p(1,1) 0];
C=eye(3)+4*(p'*p-1)*S/(1+p'*p)^2+8*S*S/(1+p'*p)^2;
```

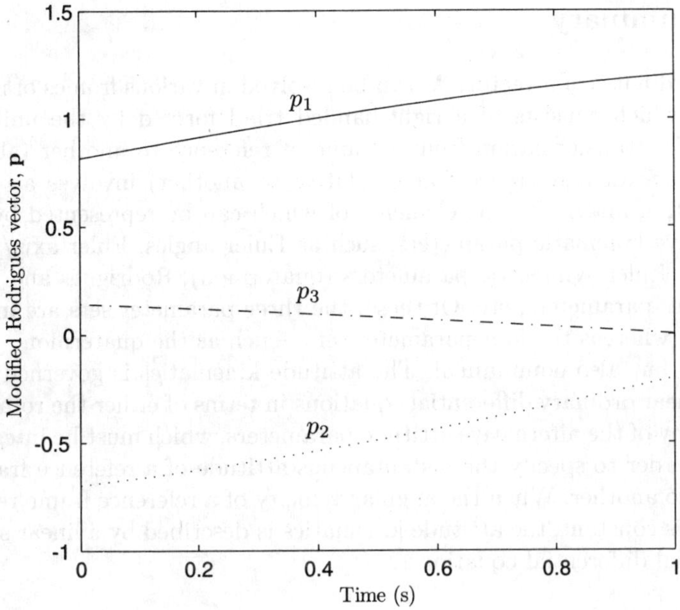

Fig. 2.9. The time history of the modified Rodrigues vector, $\mathbf{p}$, with $\omega = (0.1, -0.5, -1)^T$.

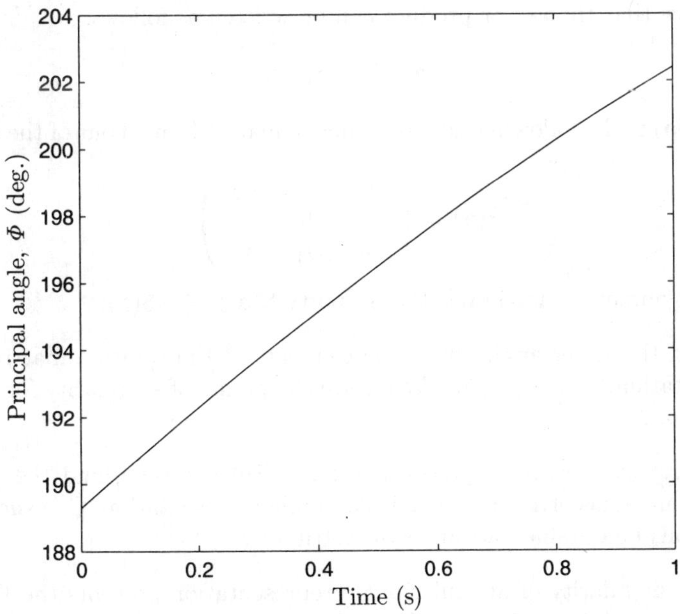

Fig. 2.10. The time history of the principal angle, $\Phi = 4\tan^{-1} | \mathbf{p} |$.

2.10 Summary

The coordinates of a vector, **A**, can be resolved in various frames of reference, each of which consists of a right-handed triad formed by the unit vectors **i**, **j**, **k**. The transformation from a frame of reference to another (also called the *attitude* of a reference frame relative to another) involves an orthogonal rotation matrix, C, the elements of which can be represented by various alternative kinematic parameters, such as Euler angles, Euler axis/principal rotation, Euler symmetric parameters (quaternion), Rodrigues and modified Rodrigues parameters, etc. Of these, the three-parameter sets are inherently singular, whereas the four-parameter sets—such as the quaternion—are nonsingular, but also nonminimal. The attitude kinematics is governed by a set of nonlinear ordinary differential equations in terms of either the rotation matrix or any of the alternative attitude parameters, which must be integrated in time in order to specify the instantaneous attitude of a reference frame with respect to another. When the angular velocity of a reference frame relative to another is constant, the attitude kinematics is described by a linear system of differential differential equations.

Exercises

2.1. Show that the vector product can be written as follows:

$$\mathbf{a} \times \mathbf{b} = \mathsf{S}(\mathbf{a})\mathbf{b} \,,$$

where $\mathsf{S}(\mathbf{a})$ is the following skew-symmetric matrix formed out of the elements of $\mathbf{a}$:

$$\mathsf{S}(\mathbf{a}) = \begin{pmatrix} 0 & -a_3 & a_2 \\ a_3 & 0 & -a_1 \\ -a_2 & a_1 & 0 \end{pmatrix} .$$

A skew-symmetric matrix has the property $\mathsf{S}(\mathbf{a})^T = -\mathsf{S}(\mathbf{a})$.

2.2. Find the Euler angles from the elements of the rotation matrix for the representation $(\psi)_3, (\theta)_1, (\phi)_3$. What are the points of singularity for this representation?

2.3. Derive the general expressions for the Euler axis, **e**, and the principal angle, Φ, in terms of a symmetric Euler angle representation. Can such simple expressions be obtained for an asymmetric set?

2.4. The singularity of an Euler angle representation prevents the Euler angles from describing infinitesimal rotations about singular orientations. Using the representation $(\psi)_3, (\theta)_1, (\phi)_3$, show that if an infinitesimal rotation is performed about OY from the initial orientation of $\phi = \theta = \psi = 0$, the Euler angles change instantaneously to finite values. [In a *gyroscope* (Chapter 14), this peculiarity of Euler angles manifests itself as the phenomenon of *gimbal lock*, in which a jamming of the gimbals occurs at a singular attitude, since they are physically incapable of rotating instantaneously by finite angles.]

2.5. For a general orientation, show that the rotation matrix can be expressed in terms of the Euler axis and principal angle as follows:

$$\mathsf{C} = \mathsf{I}\cos\Phi + (1-\cos\Phi)\mathbf{e}\mathbf{e}^T - \sin\Phi\mathsf{S}(\mathbf{e}) ,$$

where $\mathsf{S}(\mathbf{e})$ is the following skew-symmetric matrix (Exercise 2.1) formed out of the elements of $\mathbf{e}$:

$$\mathsf{S}(\mathbf{e}) = \begin{pmatrix} 0 & -e_3 & e_2 \\ e_3 & 0 & -e_1 \\ -e_2 & e_1 & 0 \end{pmatrix} .$$

This relationship between the rotation matrix and principal rotation angle/Euler axis is called *Euler's formula.*

2.6. Derive Eq. (2.46) using the result of Exercise 2.5 and the definition of the quaternion given in Eq. (2.44). Note the similarity in the form of E and Q.

2.7. Obtain the expressions for the quaternion elements in terms of the elements of the rotation matrix and employing (a) q_2, and (b) q_3, respectively, in the denominator.

2.8. Derive the composition rule for the quaternion, Eq. (2.54).

2.9. Derive the composition rule for the Rodrigues parameters, Eq. (2.57).

2.10. Show that the derivation of the Rodrigues parameters from the rotation matrix, Eq. (2.60), is true.

2.11. Derive the composition rule for the modified Rodrigues parameters, Eq. (2.66).

2.12. Derive an expression for the modified Rodrigues vector, $\mathbf{p}$, in terms of the elements of the rotation matrix, c_{ij}.

2.13. By differentiating $\mathsf{C}\mathsf{C}^T$ (or $\mathsf{C}^T\mathsf{C}$) with time according to the chain rule, and substituting into the result the differential equation for the rotation matrix, Eq. (2.71), show that Eq. (2.77) is satisfied.

2.14. Using the space-referenced angular velocity, $\boldsymbol{\omega}_I = \mathsf{C}^T\boldsymbol{\omega}$, derive an expression for the time derivative of the rotation matrix.

2.15. Derive the kinematic relationship between the time derivatives of the Euler angles, $(\psi)_3, (\theta)_2, (\phi)_1$, and the body-referenced angular velocity components.

2.16. Write a MATLAB program for calculating the time history of quaternion, given a constant body-referenced angular velocity. Use this program to plot the variation of $\mathbf{q}(t), q_4(t)$ for Example 2.12.

2.17. Write a MATLAB program for calculating the time history of the modified Rodrigues parameters, given a varying body-referenced angular velocity, $\boldsymbol{\omega} = (-0.01, 0.5\sin 3t, -0.5\cos 3t)^T$. Use this program to compute and plot the variation of $\mathbf{p}(t)$, given $\mathbf{p}(0) = (0.1, -0.2, 0.3)^T$. Also, plot the variation of the principal angle with time.

3

Planetary Form and Gravity

3.1 Aims and Objectives

- To present an accurate gravity model for flight dynamic applications for use in subsequent chapters.
- To derive the gravitational effects of a nonspherical planetary shape.

3.2 Newton's Law of Gravitation

The well-known *Newton's law of gravitation* expresses the gravitational force between two bodies, regarded as *point masses* (particles, or spheres of negligible volume), m_1 and m_2, as follows:

$$\mathbf{f_g} = m_1\mathbf{g} = -Gm_1m_2\frac{\mathbf{r_{12}}}{r_{12}^3}\,, \tag{3.1}$$

where $\mathbf{r_{12}}$ is the position vector of the center of m_1 relative to that of m_2, and G is the *universal gravitational constant*. Clearly, the force of Newtonian gravity acts along the line joining the two bodies and diminishes in inverse proportion to the square of the distance, r_{12}. Newton's law of gravitation, Eq. (3.1), is also valid if the two bodies have perfectly spherical mass distributions about their respective centers. In such a case, $\mathbf{r_{12}}$ is the position of the center of mass, m_1, relative to that of m_2. By substituting Eq. (3.1) into the equations of motion of the two bodies under mutual gravitation, it can be shown (Chapter 4) that the acceleration of mass, m_1, relative to the center of mass, m_2, is given by

$$\mathbf{g} = -G(m_1+m_2)\frac{\mathbf{r_{12}}}{r_{12}^3}\,. \tag{3.2}$$

Thus, the acceleration due to gravity, $\mathbf{g}$, relative to the center of m_2 also obeys the inverse-square rule.

The *universal gravitational constant*, $G = 6.67259(\pm 0.0003) \times 10^{-11}$ m^3/kg/s^2, has been the object of extensive measurement over the past two centuries and indicates the weakness of the gravitational force in comparison with other "action-at-distance" (such as electromagnetic) forces.[1] However, the product $G(m_1 + m_2)$ is obtained much more simply (and accurately) for a pair of planetary bodies through Kepler's laws (Chapter 4). For example, for the earth of mass M_e and a negligible, spherical test mass, we have $GM_e = 398600.4418 \pm 0.0008$ km^3/s^2. Hence, a flight vehicle near a planet of mass M would experience a gravitational acceleration

$$\mathbf{g} = -GM\frac{\mathbf{r}}{r^3}\,, \tag{3.3}$$

where $\mathbf{r}$ denotes the vehicle's position relative to the planet's center. The dimensions and mass of the vehicle are generally negligible in comparison with the planetary mass, M, and radius, R, respectively.

In the actual world of nonspherical masses, the Newtonian gravity model is accurate only when the distance separating the two bodies is much larger than the individual body dimensions, *i.e.*, when the bodies are reasonably approximated as point masses. However, Newton's law of gravitation is quite useful in deriving a more accurate gravity model.

3.3 Gravity of an Axisymmetric Planet

All heavenly bodies (referred to here by the generic term *planets*) depart from the perfect symmetry of spherical shape, because of their rotation about an axis. One such departure is due to the centrifugal mass displacement caused by rotation, which yields an axisymmetric body bulging at the *equator* (largest circle on the planet's surface, normal to the axis of rotation) and flattened at the *poles* (points of intersection of planetary surface with the axis of rotation). A planet of such a shape is said to be *oblate*. Apart from oblateness, the shape of an axisymmetric planet may depart from a sphere in many other ways. The total departure from the spherical shape is a superposition of all such effects, called *spherical harmonics*. As pointed out above, Newton's law of gravitation is invalid for a test mass close to a nonspherical planet. We shall adopt the energy approach in deriving the nonspherical gravity model.

Gravity is a *conservative* force, as it has no influence on the *total energy* of any two-mass system (Chapter 4). As we will find in Chapter 4, any force that depends only upon the position (as gravity does) is a conservative force and can be expressed as the *gradient* of a scalar function.[2] Consider the gravitational attraction between two point masses, m_1, and m_2, with the mutual

[1] The first accurate estimate of G was made by Henry Cavendish in 1798 with his torsional balance. Although measurements of the gravitational constant continue in the present day, there has been only a modest increase in the accuracy, mainly due to the extreme difficulty of separating an experimental apparatus from other influences.

[2] The gradient of a scalar, Q, with respect to a column vector, $\mathbf{x} = (x_1, x_2, x_3)^T$, is defined as the derivative of the scalar with respect to the given vector,

force given by Eq. (3.1). Let us define a *gravitational potential*, Φ_1, for the point mass, m_1, by

$$\Phi_1 \doteq \frac{Gm_1}{r_{12}} \tag{3.4}$$

and Φ_2, for the point mass, m_2, by

$$\Phi_2 \doteq \frac{Gm_2}{r_{12}} . \tag{3.5}$$

The gradient of Φ_2 with respect to $\mathbf{r_{12}}$ is

$$\frac{\partial \Phi_2}{\partial \mathbf{r_{12}}} = -\frac{Gm_2}{r_{12}^2}\frac{\partial r_{12}}{\partial \mathbf{r_{12}}} = -\frac{Gm_2}{r_{12}^3}\mathbf{r_{12}}^T , \tag{3.6}$$

or, substituting Eq. (3.1), we have

$$\mathbf{f_g}^T = m_1 \frac{\partial \Phi_2}{\partial \mathbf{r_{12}}} . \tag{3.7}$$

Similarly, we can write

$$\mathbf{f_g}^T = m_2 \frac{\partial \Phi_1}{\partial \mathbf{r_{21}}} = -m_2 \frac{\partial \Phi_1}{\partial \mathbf{r_{12}}} . \tag{3.8}$$

The acceleration due to gravity is thus given by

$$\mathbf{g}^T = \frac{\partial \Phi_2}{\partial \mathbf{r_{12}}} = -\frac{\partial \Phi_1}{\partial \mathbf{r_{12}}} . \tag{3.9}$$

This derivation can be extended to more than two bodies. The net gravitational acceleration of $N-1$ point masses, m_i, $(i = 1 \dots N-1)$, on the Nth body of mass m is the vector sum of the gravitational accelerations caused by the individual masses,

$$\mathbf{g}^T = \sum_{i=1}^{N-1} \frac{\partial \Phi_i}{\partial \mathbf{r_i}} , \tag{3.10}$$

where $\mathbf{r_i}$ is the position of the test mass, m, from the ith point mass, m_i. The gravitational potential, Φ_i, due to the point mass, m_i, is given by

$$\Phi_i = \frac{Gm_i}{r_i} . \tag{3.11}$$

This forms the basis of the *N-body problem* discussed in Chapter 4. Such an approach can be extended to the determination of the gravitational acceleration due to a nonspherical body, whose mass is distributed over a large

---

$$\frac{\partial Q}{\partial \mathbf{x}} \doteq \left(\frac{\partial Q}{\partial x_1}, \frac{\partial Q}{\partial x_2}, \frac{\partial Q}{\partial x_3}\right) .$$

Thus, the gradient of a scalar with respect to a column vector is a row vector.

number of elemental masses, m_i. Assuming that the test mass m is negligible in comparison with the sum of remaining $N-1$ masses, which are densely clustered together, away from the test particle, m, the partial derivative can be moved outside the summation sign in Eq. (3.10), leading to

$$\mathbf{g}^T = \frac{\partial}{\partial \mathbf{r_i}} \sum_{i=1}^{N-1} \frac{Gm_i}{r_i} . \tag{3.12}$$

For all the $N-1$ particles constituting the planetary mass, M, we take the limit of infinitesimal elemental mass, $m_i \rightarrow \mathrm{d}M$ and $N \rightarrow \infty$, whereby the summation in Eq. (3.12) is replaced by the following integral:

$$\mathbf{g}^T = \frac{\partial}{\partial \mathbf{r}} \int \frac{G}{s} \mathrm{d}M , \tag{3.13}$$

where s is the distance of the test mass, m, from the elemental mass, $\mathrm{d}M$, as depicted in Fig. 3.1, and can be written as

$$s = \sqrt{r^2 + \rho^2 - 2r\rho\cos\gamma} . \tag{3.14}$$

Here, $\mathbf{r}$, and $\boldsymbol{\rho}$ are the position vectors of the test mass, m, and elemental mass, $\mathrm{d}M$, respectively, from the center of mass of the planet, and γ, is the angle between $\mathbf{r}$, and $\boldsymbol{\rho}$ (Fig. 3.1). It is clear that

$$\mathbf{r} = \mathbf{s} + \boldsymbol{\rho} , \tag{3.15}$$

and ρ is a constant (the planet is a *rigid body*). Therefore, the gravitational potential of the planetary mass distribution is given by

$$\Phi = \int \frac{G\mathrm{d}M}{\sqrt{r^2 + \rho^2 - 2r\rho\cos\gamma}} . \tag{3.16}$$

In order to carry out the integration in Eq. (3.16), it is convenient to expand the integrand in a series (with the assumption $r > \rho$) as follows:

$$\frac{1}{\sqrt{r^2 + \rho^2 - 2r\rho\cos\gamma}} = \frac{1}{r}\Big\{1 + \frac{\rho}{r}\cos\gamma + \frac{1}{2}(\frac{\rho}{r})^2(3\cos^2\gamma - 1) + \frac{1}{2}(\frac{\rho}{r})^3\cos\gamma(5\cos^2\gamma - 3) + \ldots\Big\} , \tag{3.17}$$

or,

$$\frac{1}{\sqrt{r^2 + \rho^2 - 2r\rho\cos\gamma}} = \frac{1}{r}\Big\{P_0(\cos\gamma) + \frac{\rho}{r}P_1(\cos\gamma) + (\frac{\rho}{r})^2 P_2(\cos\gamma) + \ldots + (\frac{\rho}{r})^n P_n(\cos\gamma) + \ldots\Big\} , \tag{3.18}$$

where $P_n(\nu)$ is the *Legendre polynomial* of degree n. The first few Legendre polynomials are the following:

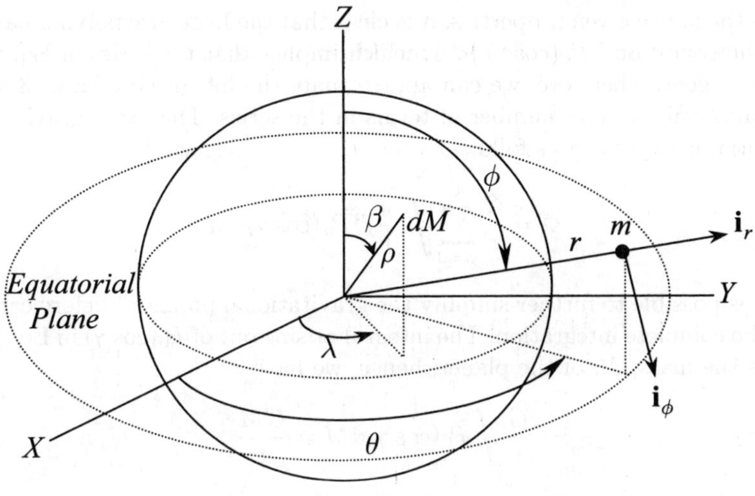

Fig. 3.1. A test mass, m, and an elemental planetary mass, dM, in a planet-centered coordinate frame.

$$\begin{aligned}
P_0(\nu) &= 1, \\
P_1(\nu) &= \nu, \\
P_2(\nu) &= \frac{1}{2}(3\nu^2 - 1), \\
P_3(\nu) &= \frac{1}{2}(5\nu^3 - 3\nu), \\
P_4(\nu) &= \frac{1}{8}(35\nu^4 - 30\nu^2 + 3), \\
P_5(\nu) &= \frac{1}{8}(63\nu^5 - 70\nu^3 + 15\nu) .
\end{aligned} \tag{3.19}$$

Some important properties of Legendre polynomials are as follows:

$$\begin{aligned}
P_n(1) &= 1, \\
P_n(-1) &= (-1)^n, \\
P_n(-\nu) &= (-1)^n P_n(\nu) .
\end{aligned} \tag{3.20}$$

A Legendre polynomial can be generated from those of lower degree with the help of *recurrence formulas*, such as

$$P_n(\nu) = \frac{(2n-1)\nu P_{n-1}(\nu) - (n-1)P_{n-2}(\nu)}{n} . \tag{3.21}$$

Other formulas, called *generating functions*, are also useful in determining the Legendre polynomials, such as the following [5]:

$$P_n(\nu) = \frac{1}{2^n n!}\frac{\mathrm{d}^n}{\mathrm{d}\nu^n}(\nu^2 - 1) . \tag{3.22}$$

From the above given properties, it is clear that the Legendre polynomials satisfy the condition $| P_n(\cos\gamma) |< 1$, which implies that the series in Eq. (3.18) is convergent. Therefore, we can approximate the integrand of Eq. (3.16) by retaining only a finite number of terms in the series. The exact gravitational potential is expressed as follows:

$$\Phi = \frac{G}{r}\sum_{n=0}^{\infty}\int(\frac{\rho}{r})^n P_n(\cos\gamma)\mathrm{d}M \, . \tag{3.23}$$

It is possible to further simplify the gravitational potential before carrying out the complete integration. The integral arising out of $P_0(\cos\gamma)$ in Eq. (3.23) yields the mass, M, of the planet; hence, we have

$$\frac{G}{r}\int P_0(\cos\gamma)\mathrm{d}M = \frac{GM}{r} \, . \tag{3.24}$$

Furthermore, the integral containing $P_1(\cos\gamma)$ vanishes, because the axis $\rho = 0$ is the axis of symmetry of the planet.[3] Therefore, the gravitational potential can be expressed in terms of the Legendre polynomials of the second and higher degrees as

$$\Phi = \frac{GM}{r}\left[1+\sum_{n=2}^{\infty}\int(\frac{\rho}{r})^n P_n(\cos\gamma)\frac{\mathrm{d}M}{M}\right] \, . \tag{3.25}$$

A further simplification of the gravitational potential requires the planetary mass distribution. Let the planet's mass density be given by $D(\rho,\beta,\lambda)$ in terms of the spherical coordinates of Fig. 3.1. Hence, the elemental mass can be expressed as

$$\mathrm{d}M = D(\rho,\beta,\lambda)\rho^2 \sin\beta \mathrm{d}\rho \mathrm{d}\beta \mathrm{d}\lambda \, , \tag{3.26}$$

where β and λ are the *co-latitude* and *longitude*, respectively, of the elemental mass. The assumption of symmetry about the polar axis translates into neglecting any longitudinal dependence of the mass distribution; thus, we have

$$\mathrm{d}M = D(\rho,\beta)\rho^2 \sin\beta \mathrm{d}\rho \mathrm{d}\beta \mathrm{d}\lambda \, . \tag{3.27}$$

Using the spherical trigonometry [2] of Fig. 3.1, we can write

$$\cos\gamma = \cos\beta\cos\phi + \sin\beta\sin\phi\cos(\theta-\lambda) \, , \tag{3.28}$$

where ϕ and θ are the co-latitude and longitude, respectively, of the test mass, m, as shown in Fig. 3.1. Upon neglecting longitudinal variations in the mass distribution, we have from Eq. (3.28)

[3] Axial symmetry is common to all heavenly bodies, with rare exceptions, such as the asteroids, and some moons. However, the mass distribution of even an axisymmetric planet is never exactly axisymmetric and displays a slight longitudinal dependence (which we shall ignore here for our purposes).

$$P_n(\cos\gamma) = P_n(\cos\beta)P_n(\cos\phi) , \tag{3.29}$$

resulting in

$$\Phi(r,\phi) = \frac{GM}{r} + \sum_{n=2}^{\infty} \frac{A_n}{r^{n+1}} P_n(\cos\phi) , \tag{3.30}$$

where

$$A_n \doteq G \int D(\rho,\beta)\rho^{n+2} P_n(\cos\beta) \sin\beta \mathrm{d}\rho \mathrm{d}\beta \mathrm{d}\lambda . \tag{3.31}$$

A more useful expression for the gravitational potential can be obtained as follows in terms of the nondimensional distance, $\frac{r}{R_e}$ (where R_e is the equatorial radius of the planet):

$$\Phi(r,\phi) = \frac{GM}{r}\left\{1 - \sum_{n=2}^{\infty}\left(\frac{R_e}{r}\right)^n J_n P_n(\cos\phi)\right\} , \tag{3.32}$$

where

$$J_n \doteq -\frac{A_n}{GM{R_e}^n} \tag{3.33}$$

are called *Jeffery's constants* and are unique for a planet. Jeffery's constants represent spherical harmonics of the planetary mass distribution, and diminish in magnitude as the order, n, increases. The largest of these constants, J_2, denotes a nondimensional difference between the moments of inertia about the polar axis, SZ, and an axis in the equatorial plane (SX or SY in Fig. 3.1), and is a measure of *ellipticity*, or oblateness, of the planet. In Chapter 6, we shall study the profound effect of J_2 upon the the orbital plane of a satellite. The higher-order term, J_3, indicates the *pear-shaped*, or triangular, harmonic, whereas J_4 and J_5 are the measures of *square* and *pentagonal* shaped harmonics, respectively. The approximate spherical harmonics corresponding to the Jeffrey's constants are indicated in Fig. 3.2 (the actual harmonics do not have the sharp corners shown in the figure). It is seldom necessary to include more than the first four Jeffery's constants in a gravity model. For example, the earth's spherical harmonics are given by $J_2 = 0.00108263, J_3 = -0.00000254$, and $J_4 = -0.00000161$.

The acceleration due to gravity of a nonspherical, axisymmetric planet can be obtained according to Eq. (3.9) by taking the gradient of the gravitational potential, Eq. (3.32), with respect to the position vector, $\mathbf{r} = r\mathbf{i_r} + r\phi\mathbf{i}_\phi$, as follows:

$$\mathbf{g} = -\left(\frac{\partial\Phi}{\partial\mathbf{r}}\right)^T = -\frac{\partial\Phi}{\partial r}\mathbf{i_r} - \frac{\partial\Phi}{r\partial\phi}\mathbf{i}_\phi , \tag{3.34}$$

or,

$$\mathbf{g} = g_r\mathbf{i_r} + g_\phi\mathbf{i}_\phi , \tag{3.35}$$

where

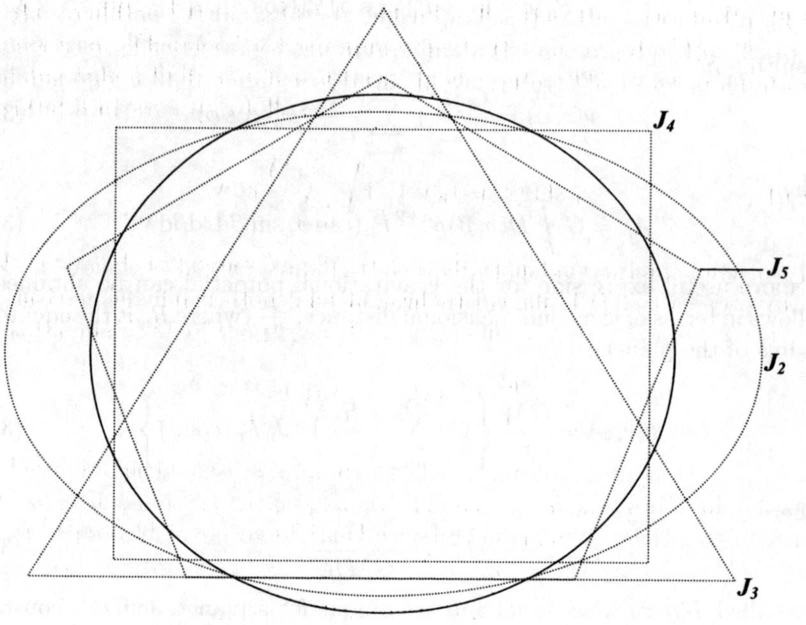

Fig. 3.2. Spherical harmonics of an axisymmteric planet.

$$g_r = -\frac{GM}{r^2}\left[1 - 3J_2(\frac{R_e}{r})^2 P_2(\cos\phi) - 4J_3(\frac{R_e}{r})^3 P_3(\cos\phi) - 5J_4(\frac{R_e}{r})^4 P_4(\cos\phi)\right] , \tag{3.36}$$

and

$$g_\phi = \frac{3GM}{r^2}(\frac{R_e}{r})^2 \sin\phi\cos\phi\left[J_2 + \frac{1}{2}J_3(\frac{R_e}{r})\sec\phi(5\cos^2\phi - 1) + \frac{5}{6}J_4(\frac{R_e}{r})^2(7\cos^2\phi - 1)\right] . \tag{3.37}$$

The unit vectors $\mathbf{i_r}$ and $\mathbf{i}_\phi$ denote the radial and southward directions in the *local horizon frame* attached to the test mass (Fig. 3.1). Due to a nonzero transverse gravity component, g_ϕ, the direction of $\mathbf{g}$ differs from the radial direction, while its radial component, g_r, is smaller in magnitude compared to that predicted by a spherical gravity model. These deviations are quite important in applications such as the flight of an atmospheric entry vehicle, and the long-range navigation of airplanes and missiles. For example, by ignoring the nonspherical gravity, one may commit an error of several hundred kilometers in an entry trajectory from a low earth orbit.

Example 3.1. Construct a model of the earth's gravity using the first four Jeffery's constants in the series expansion of gravitational potential. Compare the acceleration due to gravity with that of the spherical earth model ($R = R_e = 6378.14$ km) for a trajectory in which the latitude (in degrees) varies with altitude, $h = r - R_e$ (in kilometers), as follows:

$$\delta = h - 100, \quad (0 \leq h \leq 200 \text{ km}) .$$

A MATLAB program called *gravity.m*, tabulated in Table 3.1, calculates the acceleration due to gravity according to Eqs. (3.36) and (3.37), but reverses the direction of the radial and transverse components, i.e., $\mathbf{g} \doteq -g_c\mathbf{i_r} - g_\delta\mathbf{i}_\phi$, such that the components, g_c, g_δ, are along unit vectors toward the planetary center and north, respectively. This arrangement is useful when the equations of motion are expressed a north–east–down local horizon frame.

Table 3.1. M-file *gravity.m* for Acceleration due to Gravity in Radially Upward and North Directions

```
function [gc,gnorth]=gravity(r,lat);
% (c) 2006 Ashish Tewari
phi=pi/2-lat;
mu=3.986004e14;%mu=GMe
Re=6378.135e3;
J2=1.08263e-3;
J3=2.532153e-7;
J4=1.6109876e-7;
gc=mu*(1-1.5*J2*(3*cos(phi)^2-1)*(Re/r)^2-2*J3*cos(phi)...
    *(5*cos(phi)^2-3)*(Re/r)^3-(5/8)*J4*(35*cos(phi)^4...
    -30*cos(phi)^2+3)*(Re/r)^4)/r^2;
gnorth=-3*mu*sin(phi)*cos(phi)*(Re/r)*(Re/r)...
        *(J2+0.5*J3*(5*cos(phi)^2-1)...
        *(Re/r)/cos(phi)...
        +(5/6)*J4*(7*cos(phi)^2-1)*(Re/r)^2)/r^2;
```

The program is executed, and its results compared with the spherical model, for the given trajectory as follows:

```
dtr=pi/180;
lat=100*dtr;r=6578.14e3;
for i=1:200
[GC(:,i),GN(:,i)]=gravity(r,lat);
GS(:,i)=3.986004e14/r^2;
R(:,i)=r;
LAT(:,i)=lat;
r=r-1000;
lat=lat-dtr;
end
plot(LAT/dtr,GC,LAT/dtr,GS,':'),xlabel('Latitude,\delta (deg.)'),...
ylabel('-g_r (m/s^2)'),legend('Nonspherical model','Spherical model')
```

```
figure
plot(LAT/dtr,-GN,LAT/dtr,zeros(size(GS)),':'),xlabel('Latitude,\delta (deg.)'),...
ylabel('g_{\phi} (m/s^2)'),legend('Nonspherical model','Spherical model')
```

The resulting plots of $-g_r$ and g_ϕ are shown in Figs. 3.3 and 3.4, respectively. In these figures, it must be remembered that $h = 200$ km when $\delta = 100°$, and $h = 0$ when $\delta = -100°$. Clearly, both the components display an oscillatory dependence on the latitude. The difference with the spherical model vanishes for g_r at $\delta = \pm 45°$, and for g_ϕ at the equator and the poles. The maximum departure from the spherical model is observed for g_r at the poles, and for g_ϕ at $\delta = \pm 45°$. Evidently, while no trajectory can completely escape gravitational variations due to the nonspherical planet, they are the least in magnitude for a flight confined entirely to the equatorial plane.

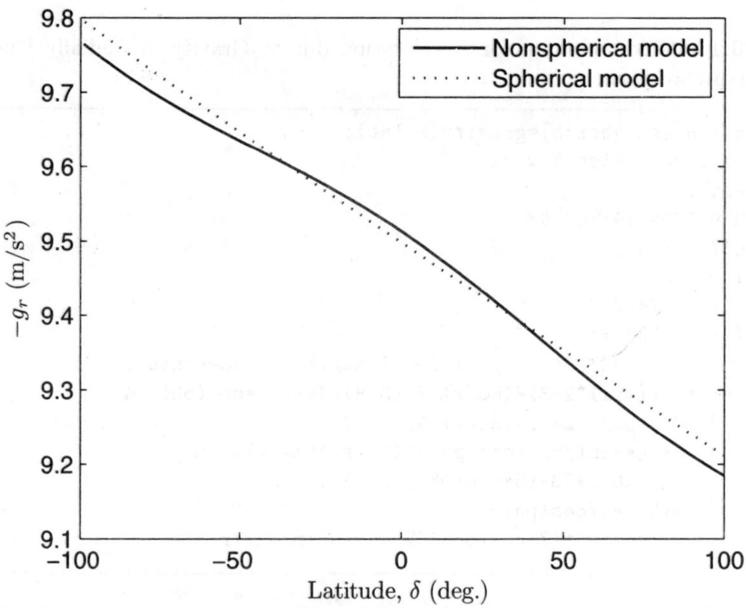

Fig. 3.3. Radial component of acceleration due to earth's gravity for a reference trajectory.

3.4 Radius of a Nonspherical Planet

For a nonspherical planet, the equatorial radius R_e—being larger than the polar radius R_p—cannot be taken to be a reference surface radius. In order to define a datum sea level, which is useful in accounting for atmospheric variations (Chapter 9), a surface radius, R, is required as a function of the local,

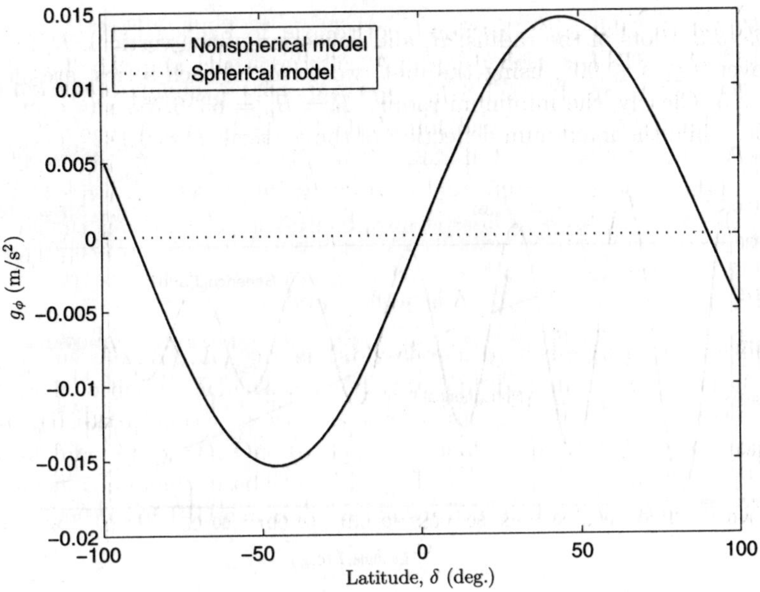

Fig. 3.4. Transverse component of acceleration due to earth's gravity for a reference trajectory.

planetcentric latitude, δ. Britting [7] derived the following series expansion for R:

$$R = R_e[1 - \frac{\epsilon}{2}(1 - \cos 2\delta) + \frac{5\epsilon^2}{16}(1 - \cos 4\delta) - \ldots] , \tag{3.38}$$

where

$$\epsilon = 1 - \frac{R_p}{R_e} \tag{3.39}$$

is called the *ellipticity* of the planet. Generally, ϵ is a small number (for the earth, $\epsilon = \frac{1}{298.257}$), and thus the series in Eq. (3.38) is absolutely convergent. Furthermore, in most cases it is sufficiently accurate to retain only the first two terms of the series, leading to the approximation

$$R \approx R_e(1 - \epsilon \sin^2 \delta) . \tag{3.40}$$

It is important to note that the vertical direction is indicated by the local normal to the nonspherical planetary surface, and departs from the radial direction. This deflection of the vertical at the surface can be expressed by the angle, D, between the local vertical and radial directions, given by Britting [7] as the following series:

$$D = \epsilon \sin 2\delta - \frac{\epsilon^2}{4} \sin 2\delta + 2\epsilon^2 \sin 2\delta \sin^2 \delta + \ldots . \tag{3.41}$$

It is seldom necessary to employ more than the first term in this series.

Example 3.2. Plots of the radius, R, and deflection of the vertical, D, for the earth over $0 \leq \delta \leq 90°$, using the first two terms of each series, are shown in Fig. 3.5. Clearly, the minimum radius, $R = R_p = 6356.755$ km, occurs at the pole, while the maximum deflection of the vertical, $D = 0.192°$, occurs at $\delta = 45°$.

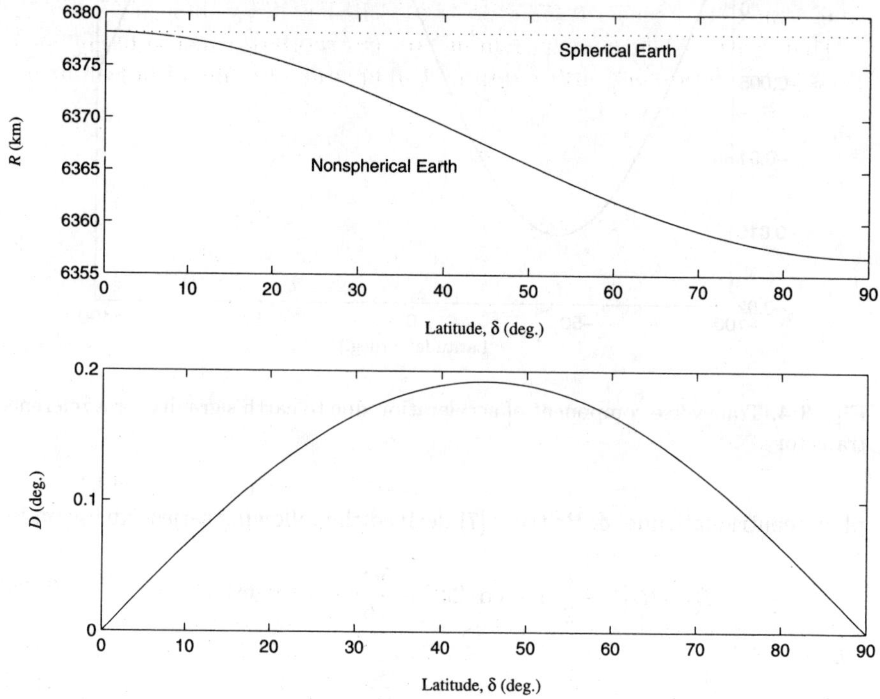

Fig. 3.5. Earth's radius and deflection of the vertical in the northern hemisphere.

3.5 Gravitational Anomalies

The nonspherical gravity model presented above essentially treats the planetary surface to be an axisymmetric *ellipsoid* generated by the surface contour of a constant gravitational potential (called an *equipotential surface*). The actual mass distribution of the planet, however, slightly differs from this model, leading to small gravitational anomalies. One has to consider longitudinal variations in the mass distribution (called *sectorial and tesseral effects*), as well as local variation in mass due to the presence of surface and sub-surface features (mountains, valleys, plains, tectonic zones, etc.), in order to arrive at a more accurate gravity model. The modified equipotential surface obtained with the inclusion of gravitational anomalies is called a *geoid*. Rather than theoretically accounting for density anomalies, it is much simpler to actually measure the acceleration due to gravity over various points on the planetary surface and then construct either a geoid, or a detailed map, of **g**. Such a mapping is usually carried out through specially instrumented spacecraft in circular orbits, such as NASA's *Lageos*, *TOPEX/Poseidon*, and *GRACE*—as well as the French *SPOT-2/Doris*—missions for the earth. Although the longitudinal and local gravitational anomalies are negligible in most flight dynamic applications, there are certain cases, such as that of a geostationary satellite, where even small changes in **g** may cause a significant trajectory change over a long period of time (months and years). For such special applications, the spherical harmonics are carefully modeled [2], using *Legendre functions* [5] (also called *associated Legendre polynomials*), P_n^m. Legendre functions are related to Legendre polynomials by $P_n = P_n^0$. In a general gravity model, the spherical harmonic coefficients are denoted C_n^m, where m is the order, and n the degree, of harmonics. Jeffery's constants are the negative of the zeroth-order harmonic coefficients of the same degree, $J_n = -C_n^0$, which are referred to as *zonal harmonic coefficients*, while the coefficients with $m \neq 0$ are due to a nonaxisymmetric mass distribution, and display longitudinal dependence of gravity. These are clubbed into the *sectorial harmonic coefficients*, C_n^n, and the *tesseral harmonic coefficients* with $n \neq m$. The *Mars Global Surveyor* (MGS) spacecraft of NASA has carried out extensive gravitational acceleration measurements beginning in 1997, which provide data for spherical harmonics of order, $m = 80$, and degree $n = 80$. These data [6] include gravitational anomalies due to local surface features, such as the gorge *Valles Marineris*, the mountain *Olympus Mons*, and the plateau *Tharsis*.[4]

3.6 Summary

While a spherical gravity model serves most atmospheric flight applications reasonably well, it is necessary to model the spherical harmonics of a nonspherical mass distribution for accurate space-flight, rocket-ascent, and entry-flight trajectories. Both the magnitude and direction of the gravity force are modified by the spherical harmonics and can be expanded in infinite series, the largest term of which is due to oblateness (J_2). In addition to spherical harmonics, there are gravitational anomalies caused by nonaxisymmetric

[4] The spherical harmonic coefficients are often *normalized* by $\bar{C}_n^m = C_n^m N$, where

$$N \doteq \sqrt{\frac{(n+m)!}{k(2n+1)(n-m)!}}\,,$$

with $k = 1$ if $m = 0$, and $k = 2$ if $m \neq 0$ [8].

sectorial and tesseral effects and local features, whose geoid model requires accurate experimental data generated by satellite mapping missions.

Exercises

3.1. Using an appropriate rotation matrix (Chapter 2), transform the gravitational acceleration [Eq. (3.35)] to a stationary coordinate frame, XYZ, with origin at planet's center (Fig. 3.1). Express your answer in terms of the co-latitude, ϕ, and longitude, λ.

3.2. Model the variation of Mars' acceleration due to gravity with latitude at an altitude of $h = 100$ km, using the following data provided by the *Mars Global Surveyor* (MGS) mission of NASA [6]:

$$\begin{aligned}
GM &= 42828.371901284\,\text{km}^3/\text{s}^2, \\
R_e &= 3397\,\text{km}, \\
J_2 &= 0.0019554536794 4545, \\
J_3 &= 3.14498094262035 \times 10^{-5}, \\
J_4 &= -1.53773961526397 \times 10^{-5}, \\
J_5 &= 5.71854718418134 \times 10^{-6}.
\end{aligned}$$

3.3. *Gravity gradient*, $\mathsf{G}_\mathbf{g}$, at a point located at $\mathbf{r}$ from the center of a planet is defined by the gradient of the gravitational acceleration, $\mathbf{g}$, with respect to $\mathbf{r}$:

$$\mathsf{G}_\mathbf{g} \doteq \frac{\partial \mathbf{g}}{\partial \mathbf{r}}\,.$$

Derive an expression for the square matrix $\mathsf{G}_\mathbf{g}$[5] for a spherical planet of mass M.

[5] Since the gradient of a scalar with respect to a column vector is a row vector, the gradient of a column vector with respect to another is a square matrix.

4

Translational Motion of Aerospace Vehicles

4.1 Aims and Objectives

- To introduce analytical (Newtonian) dynamics useful in flight dynamic derivations and analyses.
- To offer the tools for a rigorous derivation of the translational motion model.
- To establish a relationship between translational and rotational dynamics of a flight vehicle.
- To present the basic kinetic equations of aerospace flight, covering dynamics in moving frames, variable mass bodies, the N-body gravitational problem in space dynamics, and its specialization to two-body trajectories with analytical and numerical solutions.

4.2 Particle and Body

The complete description of aerospace vehicle dynamics consists of the *translational* motion of a point on the vehicle, and the *rotational* motion of the vehicle *about* that point. When we are merely interested in the *trajectory* (or *flight path*) of a vehicle, we can disregard the rotational motion of the vehicle and confine our attention to translation. Reducing the vehicle dynamics to the motion of a specific point on the vehicle is tantamount to approximating the aerospace vehicle by a point mass, or a *particle*. A *particle* can be defined as an object of infinitesimal dimensions, which occupies a point in space. Consequently, the position, velocity, and acceleration of a particle are each determined by only *three* scalar quantities; thus a particle has three degrees of freedom. No physical object fulfills the precise definition of a particle. The particle is thus a mathematical abstraction, which is used whenever we are interested in studying the path of a physical object, ignoring its size and rotational (or angular) motion. A baseball can be regarded as a particle, if we want to study its trajectory from the time it leaves the hand of the pitcher, and prior to its reaching the bat, provided we ignore the dimensions of the ball and its spin. However, the particle approximation of an object gives an incomplete description of its motion. For example, ignoring the size of the baseball will prevent us from studying how closely the bat misses the ball, or whether the bat hits the ball squarely in its middle. Furthermore, the ignored spin of the baseball would be quite important not only in its interaction with the bat, but also in the deviation of its trajectory caused by aerodynamic forces. Hence, a particle approximation will be inadequate if the baseball's motion is to be accurately simulated.

The particle approximation implies that we can separate the translational and rotational motions of an aerospace vehicle. The translational and rotational motions can be accurately separated for spacecraft in most circumstances, but for an aircraft the two motions are intertwined, because the aerodynamic forces influencing the translational motion depend on the angular orientation (*attitude*) of the aircraft with respect to the instantaneous flight path, and the attitude is governed by the rotational dynamics of the aircraft. However, since attitudinal rotations usually involve a smaller time scale compared to that of translational dynamics, a good first approximation of the aircraft trajectory can be obtained by considering only the particle dynamics and treating the aerodynamic forces as functions of the angles defining the attitude relative to the instantaneous flight path. In such an approximation, the angles relative to the flight path are either specified functions of time, or treated as instantaneously variable control inputs for the trajectory, completely disregarding the rotational dynamics of the aircraft.

A *body* is a collection of particles. If the distance between any two particles of a body remains fixed, then it is referred to as a *rigid body*. In Chapter 2 we saw how each of the position, velocity, and acceleration of a coordinate frame (rigid body) require *six* (rather than three) scalar quantities (six degrees of freedom). An aerospace vehicle is treated as a rigid body whenever we are interested in simulating its rotational motion. Sometimes, the rigid-body assumption is inadequate. Most aerospace vehicles have lightweight structures, which are quite flexible. If the structural vibration of a vehicle is to be studied, then the structural deformation becomes important, and the rigid-body approximation cannot be applied. If a flight vehicle has moving internal parts (such as engines or gyroscopes), then it too cannot be treated as a single rigid body; in such a case, we may take the vehicle as consisting of several rigid bodies, moving with respect to each other. The presence of liquid fuel (or propellant) inside a vehicle also causes a departure from the rigid-body assumption. A nonrigid (or *flexible*) body requires thrice as many scalar quantities as the total number of individual particles to specify its position, velocity, or acceleration. Since a continuous structure consists of an *infinite* number of particles, it possesses infinite degrees of freedom. However, the structural deformations and sloshing of liquids are small in magnitude compared to the translation of a vehicle's center of mass, and generally occur at smaller time scales compared to the rotational dynamics of the vehicle as a rigid body. The effects of nonrigidity become important only when there is a significant coupling between themselves and the motion of the vehicle as a rigid body, or with the aerodynamic loads. Under certain circumstances, such a coupling may cause the structural deformation (or liquid sloshing) to grow in an unbounded manner, which can be catastrophic. Coupling between rigid and flexible motions is important whenever the associated time scales are comparable, or when aerodynamic loads begin feeding the flexible motion. For a spacecraft, the dissipation of energy caused by nonrigid internal motions can be important in studying rotational stability.

The purpose of this chapter is to develop the equations of translational motion for aerospace vehicles employing the particle idealization, and to apply them to certain interesting trajectories for which analytical solutions are possible. Later in the book, we will address the numerical simulation of the more general trajectories.

4.2.1 Particle Kinematics in a Moving Frame

The velocity and acceleration of a particle can be expressed in any coordinate frame. However, only a *stationary* frame can measure the true (or *total*) velocity and acceleration. Consider a particle, p, instantaneously located by the position vector, $\mathbf{R}$, relative to a stationary frame of reference, $(SXYZ)$ (Fig. 4.1). We would like to derive the velocity and acceleration of the particle relative to another frame, $(oxyz)$, which is rotating with an angular velocity, ω, and whose origin, o, is instantaneously located at $\mathbf{R_0}$ with respect to the stationary frame.

From the vector triangle in Fig. 4.1 we can write

$$\mathbf{R} = \mathbf{R_0} + \mathbf{r} \,. \tag{4.1}$$

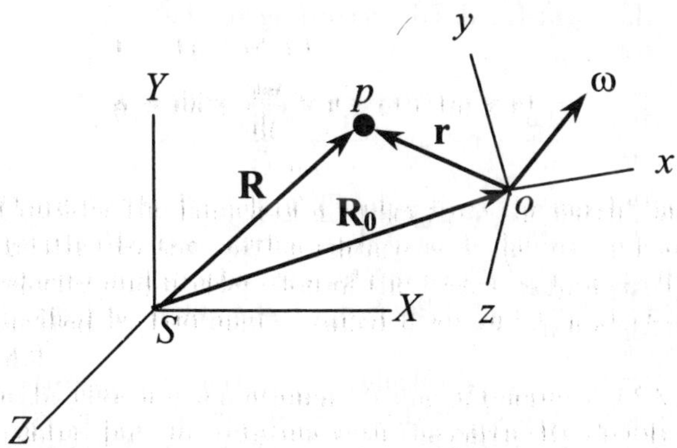

Fig. 4.1. A particle in a moving frame.

Equation (4.1) defines the position, $\mathbf{r}$, of the particle relative to the moving frame. Differentiating both sides of Eq. (4.1) with time, we get

$$\frac{\mathrm{d}\mathbf{R}}{\mathrm{d}t} = \frac{\mathrm{d}\mathbf{R_o}}{\mathrm{d}t} + \frac{\mathrm{d}\mathbf{r}}{\mathrm{d}t} . \tag{4.2}$$

Note that the total derivative of the relative position can be expressed as follows, in accordance with Section 2.2 concerning the derivative of a vector in a rotating reference frame:

$$\frac{\mathrm{d}\mathbf{r}}{\mathrm{d}t} = \frac{\partial \mathbf{r}}{\partial t} + \boldsymbol{\omega} \times \mathbf{r} . \tag{4.3}$$

The *partial derivative* in Eq. (4.3) represents the rate of change of relative position, $\mathbf{r}$, as seen by an observer located in the moving frame, *(oxyz)*.[1] Introducing the velocity vector of the particle, $\mathbf{v} \doteq \frac{\mathrm{d}\mathbf{R}}{\mathrm{d}t}$, and defining the velocity of the point o by $\mathbf{v_o} \doteq \frac{\mathrm{d}\mathbf{R_o}}{\mathrm{d}t}$, we can write Eq. (4.2), after substituting Eq. (4.3), as

$$\mathbf{v} = \mathbf{v_o} + \frac{\partial \mathbf{r}}{\partial t} + \boldsymbol{\omega} \times \mathbf{r} . \tag{4.4}$$

Note that Eq. (4.4) relates the total velocity of a particle (left-hand side), with the velocity measured in a coordinate frame rotating with angular velocity, $\boldsymbol{\omega}$, and whose origin has a linear velocity, $\mathbf{v_o}$ (right-hand side). While here $\boldsymbol{\omega}$ and $\mathbf{v_o}$ are measured in a stationary frame, Eq. (4.4) represents a more general relationship between for any two coordinate frames that have a relative motion described by $\boldsymbol{\omega}$ and $\mathbf{v_o}$. Hence, Eq. (4.4) represents a coordinate transformation for the velocity of a particle from one frame to another, where the two frames have a relative angular velocity, $\boldsymbol{\omega}$, and a relative linear velocity, $\mathbf{v_o}$.

In order to find an expression for acceleration, we differentiate Eq. (4.4) with time using the chain rule of differentiation, and decompose the total derivatives on the right-hand side according to the vector derivative rule of Section 2.2, as follows:

$$\frac{\mathrm{d}\mathbf{v}}{\mathrm{d}t} = \frac{\mathrm{d}\mathbf{v_o}}{\mathrm{d}t} + \frac{\mathrm{d}\boldsymbol{\omega}}{\mathrm{d}t} \times \mathbf{r} + \boldsymbol{\omega} \times \frac{\mathrm{d}\mathbf{r}}{\mathrm{d}t} + \frac{\partial^2 \mathbf{r}}{\partial t^2} + \boldsymbol{\omega} \times \frac{\partial \mathbf{r}}{\partial t} , \tag{4.5}$$

$$\frac{\mathrm{d}\mathbf{v}}{\mathrm{d}t} = \frac{\mathrm{d}\mathbf{v_o}}{\mathrm{d}t} + \frac{\mathrm{d}\boldsymbol{\omega}}{\mathrm{d}t} \times \mathbf{r} + \boldsymbol{\omega} \times \left(\frac{\partial \mathbf{r}}{\partial t} + \boldsymbol{\omega} \times \mathbf{r} \right) + \frac{\partial^2 \mathbf{r}}{\partial t^2} + \boldsymbol{\omega} \times \frac{\partial \mathbf{r}}{\partial t} , \tag{4.6}$$

$$\frac{\mathrm{d}\mathbf{v}}{\mathrm{d}t} = \frac{\mathrm{d}\mathbf{v_o}}{\mathrm{d}t} + \frac{\mathrm{d}\boldsymbol{\omega}}{\mathrm{d}t} \times \mathbf{r} + 2\boldsymbol{\omega} \times \frac{\partial \mathbf{r}}{\partial t} + \boldsymbol{\omega} \times (\boldsymbol{\omega} \times \mathbf{r}) + \frac{\partial^2 \mathbf{r}}{\partial t^2} . \tag{4.7}$$

[1] The partial derivative in the calculus of variations refers to the differentiation of a multivariable functional with respect to one of the independent variables. However, as we do not normally concern ourselves with variational calculus in this book, the partial derivative stands for the time derivative of a vector quantity with reference to a rotating frame, unless stated otherwise.

In Eq. (4.7), we can identify the left-hand side as the *total acceleration* of the particle in the stationary frame, $\mathbf{a} \doteq \frac{d\mathbf{v}}{dt}$, and the acceleration of the origin of the moving frame, o, as $\mathbf{a_o} \doteq \frac{d\mathbf{v_o}}{dt}$. Hence, we write the total acceleration of the particle, p, as follows:

$$\mathbf{a} = \mathbf{a_o} + \frac{\partial^2 \mathbf{r}}{\partial t^2} + \frac{d\boldsymbol{\omega}}{dt} \times \mathbf{r} + 2\boldsymbol{\omega} \times \frac{\partial \mathbf{r}}{\partial t} + \boldsymbol{\omega} \times (\boldsymbol{\omega} \times \mathbf{r}). \qquad (4.8)$$

Equation (4.8) is an expression of coordinate transformation for the acceleration of a particle from one frame to another, where the two frames have a relative angular velocity, $\boldsymbol{\omega}$, and a relative linear acceleration, $\mathbf{a_o}$. The second term on the right-hand side of Eq. (4.8) represents the *linear acceleration* of the particle, p, relative to the moving frame. This is the acceleration of a sandwich thrown from a car traveling in a straight line. The third term on the right-hand side is the contribution to the total acceleration due to the relative *angular acceleration* of the moving frame, $\frac{d\boldsymbol{\omega}}{dt}$, whereas the fourth term is called the *Coriolis acceleration*, caused by the linear velocity of the particle relative to the moving frame, $\frac{\partial \mathbf{r}}{\partial t}$. The last term in Eq. (4.8) is called the *centripetal acceleration* and acts toward the origin, o, of the moving frame. An insect crawling along the wheel-spoke of a moving bicycle experiences all the relative acceleration terms on the right-hand side of Eq. (4.8). However, the relative acceleration *experienced* by the insect would be *opposite* in direction to that seen by an observer located in the stationary frame. We are all familiar with the *centrifugal acceleration*, which tends to throw us in our car seat away from the direction of a turn; but an observer standing on the road sees us turning with the car in the same direction as that of the turn (centripetal acceleration). In order to avoid confusion, we must remember that all the acceleration terms in Eq. (4.8) are measured by a stationary observer.

If the points o and p are located on a rigid body, then the time derivatives of the relative position, $\mathbf{r}$, vanish from Eqs. (4.4) and (4.8); thus we can write

$$\mathbf{v} = \mathbf{v_o} + \boldsymbol{\omega} \times \mathbf{r}, \qquad (4.9)$$

$$\mathbf{a} = \mathbf{a_o} + \frac{d\boldsymbol{\omega}}{dt} \times \mathbf{r} + \boldsymbol{\omega} \times (\boldsymbol{\omega} \times \mathbf{r}). \qquad (4.10)$$

Example 4.1. Consider the launch of a rocket from the earth's surface. If the launch speed relative to the earth's surface is V, let us find an expression for the total velocity and acceleration of the rocket at launch. The direction of launch is specified by two angles, called *azimuth,* A, and *elevation,* E, as shown in Fig. 4.2.

Let us begin by choosing a stationary frame of reference, $(SXYZ)$, located at the earth's center, but *not* rotating with the earth. By denoting the frame $(SXYZ)$ stationary, we are effectively ignoring the orbital motion of the earth around the sun, which is acceptable for the limited purpose of studying the rocket's flight relative to the earth. Hence, we can assume the frame $(SXYZ)$

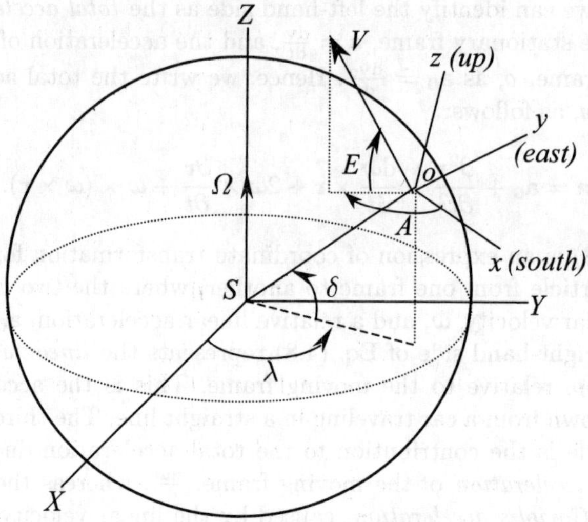

Fig. 4.2. Launch of a rocket from the surface of a rotating earth.

to be fixed with respect to the distant stars; such a coordinate frame is called a *celestial* frame. The launch point is located in the stationary frame by two angles, namely the *celestial longitude*, λ, and the *latitude*, δ (Fig. 4.2). Note that earth's rotation about the axis SZ causes the celestial longitude of the launch point to change with time. Let the unit vectors $\mathbf{I}, \mathbf{J}, \mathbf{K}$ be along the axes SX, SY, and SZ, respectively. The earth's angular velocity relative to the stationary frame is then $\boldsymbol{\omega} = \Omega\mathbf{K}$, whereas the location of the launch point, o, is given by

$$\mathbf{R_o} = R_0 \cos\delta \cos\lambda \mathbf{I} + R_0 \cos\delta \sin\lambda \mathbf{J} + R_0 \sin\delta \mathbf{K} \,. \tag{4.11}$$

The velocity of point, o, is merely due to the rotation of the vector $\mathbf{R_o}$; thus, applying Eq. (4.3), we can write

$$\mathbf{v_o} \doteq \frac{\mathrm{d}\mathbf{R_o}}{\mathrm{d}t} = \boldsymbol{\omega} \times \mathbf{R_o} = \Omega R_0 \cos\delta \cos\lambda \mathbf{J} - \Omega R_0 \cos\delta \sin\lambda \mathbf{I}. \tag{4.12}$$

Similarly, the acceleration of point, o, is obtained by applying Eq. (4.3) to the derivative of $\mathbf{v_o}$ as follows:

$$\mathbf{a_o} \doteq \frac{\mathrm{d}\mathbf{v_o}}{\mathrm{d}t} = \boldsymbol{\omega} \times \mathbf{v_o} = -\Omega^2 R_0 \cos\delta \cos\lambda \mathbf{I} - \Omega^2 R_0 \cos\delta \sin\lambda \mathbf{J}. \tag{4.13}$$

Next, let us choose the orientation of the moving frame ($oxyz$) fixed to the earth's surface at the launch point, such that ox is toward south, oy is toward the *east*, and oz is vertically *upward*; the unit vectors $\mathbf{i}, \mathbf{j}, \mathbf{k}$ specifiy the directions of the axes ox, oy, and oz, respectively. In the moving frame, the rocket's relative velocity is written as

$$\frac{\partial \mathbf{r}}{\partial t} = V \cos E \cos A\mathbf{i} + V \cos E \sin A\mathbf{j} + V \sin E\mathbf{k}. \tag{4.14}$$

At the time of launch, $\mathbf{r} = \mathbf{0}$. Substituting Eqs. (4.12) and (4.14) into Eq. (4.4), the total velocity of the rocket is given by

$$\mathbf{v} = \Omega R_0 \cos\delta(\cos\lambda\mathbf{J} - \sin\lambda\mathbf{I}) + V(\cos E \cos A\mathbf{i} + \cos E \sin A\mathbf{j} + \sin E\mathbf{k}). \tag{4.15}$$

Equation (4.15) is a jumble of unit vectors of two different frames, and we would like to express $\mathbf{i}, \mathbf{j}, \mathbf{k}$ in the stationary frame. According to Section 2.3, a rotation matrix describes the transformation between any two frames.

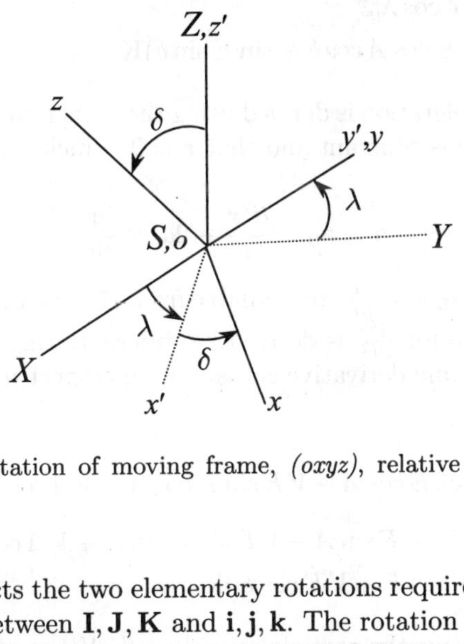

Fig. 4.3. The orientation of moving frame, *(oxyz)*, relative to stationary frame, *(SXYZ)*.

Figure 4.3 depicts the two elementary rotations required to obtain the relative orientation between $\mathbf{I}, \mathbf{J}, \mathbf{K}$ and $\mathbf{i}, \mathbf{j}, \mathbf{k}$. The rotation matrix representing this orientation is thus derived as a product of the two individual rotation matrices—a rotation by angle λ about the axis, *SZ*, and a rotation by angle $\frac{\pi}{2} - \delta$ about the intermediate axis, Sy', as follows:

$$\begin{Bmatrix} \mathbf{i}' \\ \mathbf{j}' \\ \mathbf{k}' \end{Bmatrix} = \begin{pmatrix} \cos\lambda & \sin\lambda & 0 \\ -\sin\lambda & \cos\lambda & 0 \\ 0 & 0 & 1 \end{pmatrix} \begin{Bmatrix} \mathbf{I} \\ \mathbf{J} \\ \mathbf{K} \end{Bmatrix}, \tag{4.16}$$

$$\begin{Bmatrix} \mathbf{i} \\ \mathbf{j} \\ \mathbf{k} \end{Bmatrix} = \begin{pmatrix} \sin\delta & 0 & -\cos\delta \\ 0 & 1 & 0 \\ \cos\delta & 0 & \sin\delta \end{pmatrix} \begin{Bmatrix} \mathbf{i}' \\ \mathbf{j}' \\ \mathbf{k}' \end{Bmatrix}, \tag{4.17}$$

$$\begin{Bmatrix} \mathbf{i} \\ \mathbf{j} \\ \mathbf{k} \end{Bmatrix} = \begin{pmatrix} \sin\delta & 0 & -\cos\delta \\ 0 & 1 & 0 \\ \cos\delta & 0 & \sin\delta \end{pmatrix} \begin{pmatrix} \cos\lambda & \sin\lambda & 0 \\ -\sin\lambda & \cos\lambda & 0 \\ 0 & 0 & 1 \end{pmatrix} \begin{Bmatrix} \mathbf{I} \\ \mathbf{J} \\ \mathbf{K} \end{Bmatrix}, \tag{4.18}$$

or,

$$\begin{Bmatrix} \mathbf{i} \\ \mathbf{j} \\ \mathbf{k} \end{Bmatrix} = \begin{pmatrix} \sin\delta\cos\lambda & \sin\delta\sin\lambda & -\cos\delta \\ -\sin\lambda & \cos\lambda & 0 \\ \cos\delta\cos\lambda & \cos\delta\sin\lambda & \sin\delta \end{pmatrix} \begin{Bmatrix} \mathbf{I} \\ \mathbf{J} \\ \mathbf{K} \end{Bmatrix}, \tag{4.19}$$

where $\mathbf{i}', \mathbf{j}', \mathbf{k}'$ are the unit vectors denoting the intermediate axes, Sx, Sy, and Sz, respectively. Substituting Eq. (4.19) into Eq. (4.15) and collecting terms, we can finally express the total velocity of the rocket as

$$\begin{aligned} \mathbf{v} = {} & [V(\cos E\cos A\sin\delta\cos\lambda - \cos E\sin A\sin\lambda + \sin E\cos\delta\cos\lambda) \\ & - \Omega R_0 \sin\delta\sin\lambda]\mathbf{I} \\ & + [V(\cos E\cos A\sin\delta\sin\lambda + \cos E\sin A\cos\lambda + \sin E\cos\delta\sin\lambda) \\ & - \Omega R_0\sin\delta\cos\lambda]\mathbf{J} \\ & + V(-\cos E\cos A\cos\delta + \sin E\sin\delta)\mathbf{K}. \end{aligned} \tag{4.20}$$

The rocket's acceleration is derived using the facts that the angular velocity of the moving frame is constant and that $\mathbf{r} = \mathbf{0}$, which, when substituted into Eq. (4.8), yield

$$\mathbf{a} = \mathbf{a_0} + \frac{\partial^2 \mathbf{r}}{\partial t^2} + 2\omega \times \frac{\partial \mathbf{r}}{\partial t}. \tag{4.21}$$

The expressions for $\mathbf{a_0}$ and $\frac{\partial \mathbf{r}}{\partial t}$ are available from Eqs. (4.13) and (4.14), respectively. An expression for $\frac{\partial^2 \mathbf{r}}{\partial t^2}$ is derived by differentiating Eq. (4.14) partially with time (i.e., the time derivative is taken with respect to the moving frame) as

$$\begin{aligned} \frac{\partial^2 \mathbf{r}}{\partial t^2} = {} & (\dot{V}\cos E\cos A - V\dot{E}\sin E\cos A - V\dot{A}\cos E\sin A)\mathbf{i} \\ & + (\dot{V}\cos E\sin A - V\dot{E}\sin E\sin A + V\dot{A}\cos E\cos A)\mathbf{j} \\ & + (\dot{V}\sin E + V\dot{E}\cos E)\mathbf{k}, \end{aligned} \tag{4.22}$$

where the dot represents the partial derivative, $\frac{\partial}{\partial t}$. We can transform Eq. (4.22) to the stationary frame using Eq. (4.19), and write

$$\begin{aligned} \frac{\partial^2 \mathbf{r}}{\partial t^2} = {} & \begin{Bmatrix} (\dot{V}\cos E\cos A - V\dot{E}\sin E\cos A - V\dot{A}\cos E\sin A) \\ (\dot{V}\cos E\sin A - V\dot{E}\sin E\sin A + V\dot{A}\cos E\cos A) \\ (\dot{V}\sin E + V\dot{E}\cos E) \end{Bmatrix}^T \\ & \times \begin{pmatrix} \sin\delta\cos\lambda & \sin\delta\sin\lambda & -\cos\delta \\ -\sin\lambda & \cos\lambda & 0 \\ \cos\delta\cos\lambda & \cos\delta\sin\lambda & \sin\delta \end{pmatrix} \begin{Bmatrix} \mathbf{I} \\ \mathbf{J} \\ \mathbf{K} \end{Bmatrix}. \end{aligned} \tag{4.23}$$

The last term on the right-hand side of Eq. (4.21) is

$$2\omega \times \frac{\partial \mathbf{r}}{\partial t} = 2\Omega V[\cos E\cos A(\mathbf{K}\times\mathbf{i}) + \cos E\sin A(\mathbf{K}\times\mathbf{j}) + \sin E(\mathbf{K}\times\mathbf{k})], \tag{4.24}$$

which, using the rotation matrix of Eq. (4.19), becomes

$$
\begin{aligned}
2\boldsymbol{\omega} \times \frac{\partial \mathbf{r}}{\partial t} = 2\Omega V[&\cos E \cos A(\sin\delta \cos\lambda \mathbf{J} - \sin\delta \sin\lambda \mathbf{I}) \\
+ \ &\cos E \sin A(-\sin\lambda \mathbf{J}) - \cos\lambda \mathbf{I}) \\
+ \ &\sin E(\cos\delta \cos\lambda \mathbf{J} - \cos\delta \sin\lambda \mathbf{I})].
\end{aligned} \tag{4.25}
$$

When we substitute Eqs. (4.13), (4.23), and (4.25) into Eq. (4.21) and collect terms, the total acceleration is expressed as follows:

$$
\begin{aligned}
\mathbf{a} = [&(\dot{V}\cos E\cos A - V\dot{E}\sin E\cos A - V\dot{A}\cos E\sin A)\sin\delta\cos\lambda \\
&- (\dot{V}\cos E\sin A - V\dot{E}\sin E\sin A + V\dot{A}\cos E\cos A)\sin\lambda \\
&+ (\dot{V}\sin E + V\dot{E}\cos E)\cos\delta\cos\lambda - \Omega^2 R_0 \sin\delta\cos\lambda \\
&- 2\Omega V(\cos E\cos A\sin\delta\sin\lambda + \cos E\sin A\cos\lambda + \sin E\cos\delta\sin\lambda)]\mathbf{I} \\
&+ [(\dot{V}\cos E\cos A - V\dot{E}\sin E\cos A - V\dot{A}\cos E\sin A)\sin\delta\sin\lambda \\
&+ (\dot{V}\cos E\sin A - V\dot{E}\sin E\sin A + V\dot{A}\cos E\cos A)\cos\lambda \\
&+ (\dot{V}\sin E + V\dot{E}\cos E)\cos\delta\sin\lambda - \Omega^2 R_0 \sin\delta\sin\lambda \\
&+ 2\Omega V(\cos E\cos A\sin\delta\cos\lambda - \cos E\sin A\sin\lambda + \sin E\cos\delta\cos\lambda)]\mathbf{J} \\
&\quad [-(\dot{V}\cos E\cos A - V\dot{E}\sin E\cos A - V\dot{A}\cos E\sin A)\cos\delta \\
&+ (\dot{V}\sin E + V\dot{E}\cos E)\sin\delta]\mathbf{K}.
\end{aligned} \tag{4.26}
$$

Note that the total acceleration could also be obtained by taking the vector derivative of the total velocity derived in Eq. (4.20), according to

$$
\mathbf{a} \doteq \frac{\mathrm{d}\mathbf{v}}{\mathrm{d}t} = \frac{\partial \mathbf{v}}{\partial t} + \boldsymbol{\omega} \times \mathbf{v}. \tag{4.27}
$$

This can be verified by the reader.

Example 4.2. Consider an aircraft whose position and angular velocity relative to a ground station, $(SXYZ)$, with unit vectors $\mathbf{I}, \mathbf{J}, \mathbf{K}$, are $R_0 = -0.2t^2\mathbf{I} + 0.5t^2\mathbf{J} + 30t\mathbf{K}$ m and $\boldsymbol{\omega} = 0.02\mathbf{J} - 0.01\mathbf{K}$ rad/s, respectively. The orientation of a coordinate frame $(oxyz)$ fixed to the aircraft with unit vectors $\mathbf{i}, \mathbf{j}, \mathbf{k}$ is defined by Euler angles, θ (*pitch angle*), ϕ (*roll angle*), and ψ (*yaw angle*), as shown in Fig. 4.4. A restless passenger is walking up and down the aisle such that her position relative to the point, o, in the aircraft is $\mathbf{r} = \cos(\frac{t}{10})\mathbf{i} - \sin(\frac{t}{10})\mathbf{j}$ m. Find the total velocity and acceleration of the passenger at time $t = 100$ s if the initial orientation of the aircraft at $t = 0$ is given by $\theta = 0$, $\phi = 0$, and $\psi = 0$.

From Section 2.5, we know the rotation from $(SXYZ)$ to $(oxyz)$ in terms of the Euler angles of Fig. 4.4 is the following:

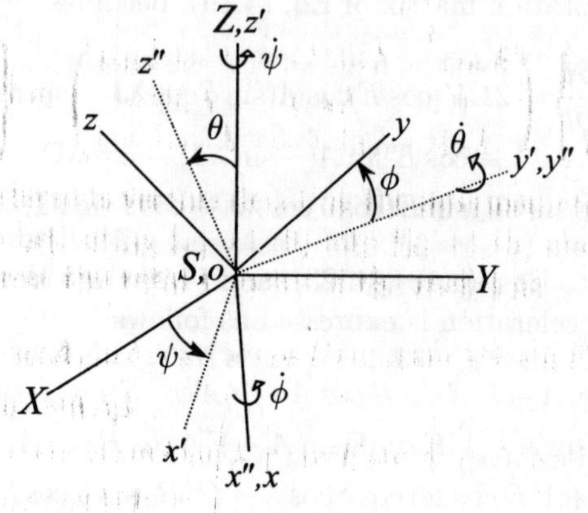

Fig. 4.4. The orientation of aircraft axes, $(oxyz)$, relative to groundstation, $(SXYZ)$.

$$\begin{Bmatrix} \mathbf{i} \\ \mathbf{j} \\ \mathbf{k} \end{Bmatrix} =$$

$$\begin{pmatrix} \cos\theta\cos\psi & \cos\theta\sin\psi & -\sin\theta \\ (-\cos\phi\sin\psi + \sin\theta\sin\phi\cos\psi) & (\cos\phi\cos\psi + \sin\theta\sin\phi\sin\psi) & \cos\theta\sin\phi \\ (\sin\phi\sin\psi + \sin\theta\cos\phi\cos\psi) & (-\sin\phi\cos\psi + \sin\theta\cos\phi\sin\psi) & \cos\theta\cos\phi \end{pmatrix}$$

$$\times \begin{Bmatrix} \mathbf{I} \\ \mathbf{J} \\ \mathbf{K} \end{Bmatrix}. \tag{4.28}$$

The evolution of Euler angles is related to the aircraft's angular velocity as follows (Fig. 4.4):

$$\begin{aligned} \omega &= \dot{\phi}\mathbf{i} + \dot{\theta}\mathbf{j}' + \dot{\psi}\mathbf{K} \\ &= \dot{\phi}(\cos\theta\cos\psi\mathbf{I} + \cos\theta\sin\psi\mathbf{J} - \sin\theta\mathbf{K}) + \dot{\theta}(-\sin\psi\mathbf{I} + \cos\psi\mathbf{J}) + \dot{\psi}\mathbf{K} \\ &= (\dot{\phi}\cos\theta\cos\psi - \dot{\theta}\sin\psi)\mathbf{I} + (\dot{\phi}\cos\theta\sin\psi + \dot{\theta}\cos\psi)\mathbf{J} + (\dot{\psi} - \dot{\phi}\sin\theta)\mathbf{K} \\ &= 0.02\mathbf{J} - 0.01\mathbf{K}. \end{aligned} \tag{4.29}$$

Equation (4.29) results in the following differential equations for the Euler angles:

$$\begin{aligned} \dot{\phi}\cos\theta\cos\psi - \dot{\theta}\sin\psi &= 0, \\ \dot{\phi}\cos\theta\sin\psi + \dot{\theta}\cos\psi &= 0.02, \\ \dot{\psi} - \dot{\phi}\sin\theta &= -0.01. \end{aligned} \tag{4.30}$$

In order to numerically integrate the nonlinear differential equations of Eq. (4.30), we resort to a fourth-order, variable time step, Runge–Kutta algorithm (see Appendix A), as programmed in the MATLAB routine *ode45*. For this purpose, an M-file named *eulerevolve.m* (Table 4.1) is written to calculate the Euler angle derivatives, at each time step, using Eq. (4.30). Of course, we must recall (Chapter 2) that the point of singularity of Euler angle representation ($\theta = 90°$) would render this procedure useless (in such a case, a different representation, e.g., quaternion, should be used). From the resulting plot of the Euler angles (Fig. 4.5), it is evident that the singularity has not been encountered for the given angular velocity and initial condition. The following MATLAB commands are used to solve this example:

```
>> [t,x]=ode45(@eulerevolve, [0 100], [0 0 0]');% Integration of Eq. (4.30)
>> plot(t,x*180/pi),xlabel('Time (s)'),ylabel('Euler Angles (deg.)') %(Fig 4.5)
>> size(x) % size of solution vector
ans =
   61    3
>> phi=x(61,1),thet=x(61,2),psi=x(61,3) %Euler angles at t=100 s
phi =
   -1.9971
thet =
    0.7806
psi =
   -2.6235
>> C=[cos(thet)*cos(psi) cos(thet)*sin(psi) -sin(thet);
    -cos(thet)*sin(psi)+sin(thet)*sin(phi)*cos(psi) cos(phi)*cos(psi)+sin(phi)*sin(thet)
    *sin(psi)...cos(thet)*sin(phi);
    sin(phi)*sin(psi)+sin(thet)*cos(phi)*cos(psi)  -sin(phi)*cos(psi)+sin(thet)*cos(phi)
      *sin(psi)...cos(thet)*cos(phi)] % Rotation matrix at t=100 s

C =
   -0.6173   -0.3518   -0.7037
    0.9085    0.6765   -0.6469
    0.7037   -0.6469   -0.2938
>> RO=[-0.2*100^2;0.5*100^2;30*100];omega=[0;0.02;-0.01]; % Origin; angular velocity
>> v0=[-40;100;30];a0=[-0.4;1;0]; %Velocity and acceleration of point o
>> r=C'*[cos(10);-sin(10);0] % Relative position
r =
    1.0122
    0.6633
    0.2385
>> rdot= C'*[-sin(10)/10;-cos(10)/10;0]%Relative velocity
rdot =
    0.0426
    0.0376
   -0.0926
>> v=v0+rdot+cross(omega,r) %Total velocity of passenger
v =
  -39.9460
  100.0275
   29.8872
>> rdotdot= C'*[-cos(10)/100;sin(10)/100;0]%Relative acceleration
rdotdot =
   -0.0101
   -0.0066
   -0.0024
>> a=a0+rdotdot+2*cross(omega,rdot)+cross(omega,cross(omega,r))%Total acceleration
a =
   -0.4136
    0.9924
   -0.0043
```

Hence, the total velocity and acceleration vectors of the passenger at $t = 100$ s in the chosen stationary frame are $\mathbf{v} = -39.946\mathbf{I}+100.0275\mathbf{J}+29.8872\mathbf{K}$ m/s and $\mathbf{a} = -0.4136\mathbf{I} + 0.9924\mathbf{J} - 0.0043\mathbf{K}$ m/s^2.

Table 4.1. M-file *eulerevolve.m* for Calculating the Euler Angle Derivatives

```
function xdot=eulerevolve(t,x)
%x(1)=phi, x(2)=theta, x(3)=psi
%(c) 2006 Ashish Tewari
omega=[0,0.02,-0.01]';% angular velocity in rad/s
xdot(2,1)=omega(2)/(tan(x(3))*sin(x(3))+cos(x(3)));
xdot(1,1)=(xdot(2,1)*sin(x(3))+omega(1))/(cos(x(3))*cos(x(2)));
xdot(3,1)=xdot(1,1)*sin(x(2))+omega(3);
```

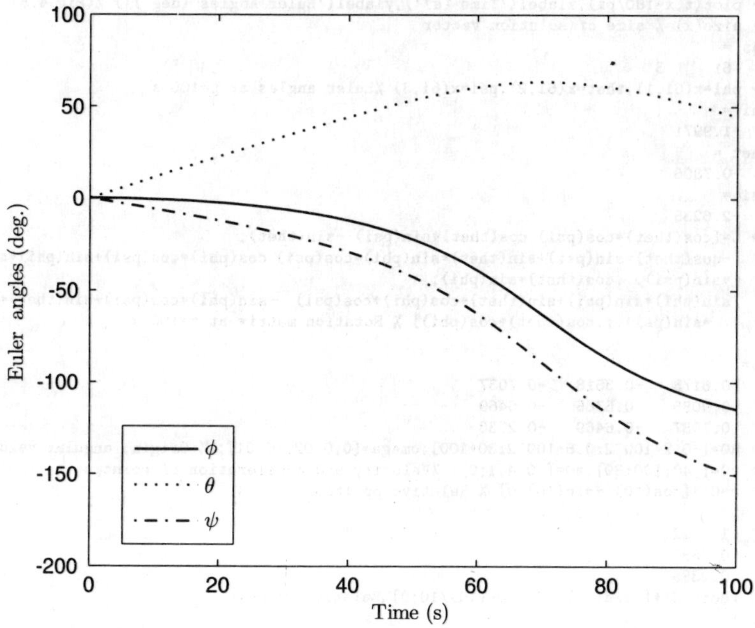

Fig. 4.5. Evolution of the aircraft Euler angles with time.

4.3 Newton's Laws of Motion

In 1687, Newton published the three laws of universal motion of particles in his *Philosophiae Naturalis Principia Mathematica*. Stated briefly, they are

1. A particle at rest remains at rest, and a particle in uniform rectilinear (straight-line) motion remains in uniform rectilinear motion unless acted upon by a force.
2. The force applied on a particle equals the mass of the particle multiplied by its total acceleration.
3. Every applied force on a particle is opposed by an equal reaction from the particle.

The three laws of motion form the fundamental basis of classical mechanics. Note that in the previous section we had defined total velocity and acceleration as being measured in a stationary frame. However, since the laws of motion do not distinguish between a stationary frame and a frame moving uniformly in a straight line, we will relax our requirement of total acceleration, and redefine the latter as being measured in an *inertial* frame; an *inertial* frame is a frame that has zero acceleration relative to a stationary frame. Hence, a stationary frame, and a frame moving uniformly in a straight line, are both inertial. An application of Eq. (4.8) to an inertial frame reveals that no modification is required in the expression for total acceleration. However, Eq. (4.4) implies that the total velocity (which is unimportant according to Newton's laws) should still be defined as being measured relative to a stationary frame.

Since Newton's laws are valid only in an inertial frame, we must make a suitable choice for our frame of reference. The previous section illustrated how the velocity and acceleration relative to a stationary frame can be obtained, using the quantities measured relative to a moving frame, provided the velocity and acceleration of the moving frame itself are known. Hence, we are suitably equipped to apply Newton's laws of motion by measuring acceleration in a moving, rather than an inertial, frame. This is quite useful, as an inertial frame is non-existent in practice. However, we usually need not reduce the acceleration to an absolutely inertial frame before applying Newton's laws. Most often, an approximately inertial frame is good enough, since the acceleration of the moving frame might be negligible in comparison with the relative acceleration of the particle. Recall that in Examples 4.1 and 4.2, we had approximated two different frames as being stationary: a frame at the earth's center (but not rotating with the earth), and a frame attached to the earth's surface (and rotating with the earth). In each case, we had ignored the acceleration of the "inertial" frame, with a differing degree of approximation. For instance, the frame chosen as being stationary in Example 4.2 ignored the earth's rotation on its axis, the orbital motion of the earth around the sun, the orbital motion of the sun around the center of our galaxy, and so on, with the higher-order effects having a progressively diminishing order of magnitude of acceleration. The largest ignored acceleration term of the stationary frame (i.e., the earth's rotation) in Example 4.2 would affect the total acceleration by less than 0.1%. In some applications, such as naval gunnery, or flight of ballistic missiles, the error caused by ignoring the earth's rotation could be significant in trajectory calculations. Hence, one must carefully examine the magnitude of the largest ignored acceleration terms in choosing an adequate inertial frame.

A mathematical statement of Newton's second law applied to a particle of constant mass, m, is as follows:

$$\mathbf{f} = m\mathbf{a} = m\frac{d\mathbf{v}}{dt} = \frac{d(m\mathbf{v})}{dt} = m\frac{d^2\mathbf{R}}{dt^2}, \tag{4.31}$$

where $\mathbf{a}$ is the total acceleration of the particle, and $\mathbf{f}$ is the total force applied on the particle. Recall that the total acceleration is the acceleration of the particle relative to an inertial frame. If the applied force is zero, then Eq. (4.31) implies $\mathbf{a} = \mathbf{0}$, which is the mathematical statement of Newton's first law, and represents the principle of conservation of *linear momentum*, $m\mathbf{v}$. If the applied force is known as a function of time, or as a functional of the position and velocity vectors, then Eq. (4.31) can be successively integrated to find the velocity $\mathbf{v} = \frac{\mathrm{d}\mathbf{R}}{\mathrm{dt}}$ and the position $\mathbf{R}$ of the particle as functions of time, subject to known initial conditions. The force acting on a flight vehicle can be represented as a vector sum of *gravitational, aerodynamic*, and *propulsive* forces, each of which can be prescribed in terms of position and velocity vectors.

Note that since a particle has three degrees of freedom, Eq. (4.31) represents three second-order, scalar differential equations, called *translational equations of motion*. Generally, the motion of a particle is governed by Eq. (4.31) and a set of additional relationships called *constraints*. The constraints arise due to *kinematic* restrictions on the position and velocity vectors. For example, a constraint on the position of a flight vehicle is that its *altitude* (height above the earth's surface) must be greater than, or equal to, zero. The constraints imply that the coordinates of a particle are no longer independent variables, and hence the degrees of freedom are reduced from three. When the constraints are expressed as equations relating the particle's coordinates (and time) in the form

$$\mathbf{F}[\mathbf{R}(X, Y, Z), t] = 0, \tag{4.32}$$

where $\mathbf{F}[.]$ denotes a vector function, they are called *holonomic* constraints. An example of a holonomic constraint is the equation of a curve along which a particle is constrained to move (such as a simple pendulum constrained to move in a planar arc). However, many constraints are either inequalities, or equations that cannot be expressed as Eq. (4.32), and are said to be *nonholonomic*. We have to solve Eq. (4.31) subject to the constraints, in order to find the trajectory of a flight vehicle. An analytical—or numerical—integration of Eq. (4.31) subject to constraints involves solving nonlinear, ordinary, coupled, scalar, differential equations, subject to constraint equations, or inequalities. Hence, Eq. (4.31), along with the constraints, governs the motion of a particle. However, it is seldom that the solution to the equations can be obtained analytically in a closed form.

When Newton's laws are applied to a body, we can obtain equations for both the translation and rotation of the body. We begin by considering a body of constant mass m acted upon by force $\mathbf{f}$ as a collection of a large number of particles of elemental mass ∂m, such that $m = \sum \partial m$. Applying Newton's law of motion, Eq. (4.31), to individual particles, and summing over all particles, we may write

$$\sum \partial\mathbf{f} = \sum \partial m\mathbf{a} = \frac{\mathrm{d}}{dt} \sum \mathbf{v}\partial m, \tag{4.33}$$

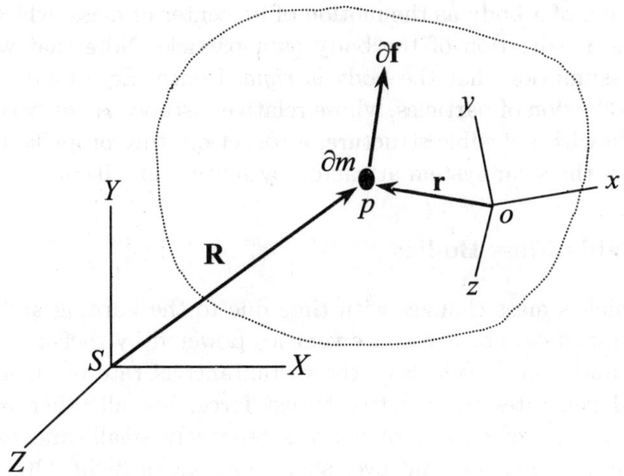

Fig. 4.6. A body of particles.

since the elemental mass is invariant. Due to Newton's third law, the internal forces acting between any two particles are equal and opposite and thus add to zero; hence, we can write $\mathbf{f} = \sum \partial\mathbf{f}$, the net external force. Now, let us choose a point, o, on the body at which a frame ($oxyz$) is fixed to the body (Fig. 4.6). It follows from Eq. (4.2) that the inertial velocity of an elemental particle is given by

$$\mathbf{v} = \mathbf{v_o} + \frac{\mathrm{d}\mathbf{r}}{\mathrm{d}t}, \tag{4.34}$$

which, when substituted into Eq. (4.33), yields

$$\mathbf{f} = \frac{\mathrm{d}(m\mathbf{v_o})}{\mathrm{d}t} + \frac{\mathrm{d}^2}{\mathrm{d}t^2}\sum \mathbf{r}\partial m. \tag{4.35}$$

There is a special point associated with every body, called the *center of mass*, such that the *mass moment*, $\sum \mathbf{r}\partial m$, vanishes about that point. It is not necessary that the center of mass be one of the particles constituting the body. If we choose the point o to be the center of mass, then we can write

$$\mathbf{f} = \frac{\mathrm{d}(m\mathbf{v_o})}{\mathrm{d}t} = m\frac{\mathrm{d}\mathbf{v_o}}{\mathrm{d}t}, \tag{4.36}$$

which governs the translation of the body. Equation (4.36) states that the rate of change of *linear momentum* of the body, defined by $m\mathbf{v_o}$, is proportional to the net external force. If there is no net force acting on the body, its linear momentum is conserved. It is clear from a comparison of Eqs. (4.31) and (4.36) that for the limited purpose of studying the translational motion of a body, we can regard the body as if all its mass were concentrated at the center of mass, and the net external force were applied at the center of mass. Hence, we regard the translation of a body as the motion of its center of mass, which forms the basis for the idealization of the body as a particle. Note that we have not made the assumption that the body is *rigid*. Hence, Eq. (4.36) is applicable to a loose collection of particles, whose relative distance is not fixed, such as a flight vehicle with a flexible structure, a rocket ejecting propellant gases, and the bodies in the solar system attracted by mutual gravitation.

4.3.1 Variable Mass Bodies

A flight vehicle's mass changes with time due to the burning and ejection of the propellant mass. For aerospace vehicles powered by rocket engines (such as missiles and launch vehicles), the instantaneous rate of change of mass is large and generates the reactive thrust force. For all other vehicles, the instantaneous rate of change of mass is relatively small, and the vehicle's mass can be assumed constant over short portions of flight. Other examples of variable mass systems includes the expenditure of munitions from a vehicle, and a docking of two spacecraft. When the instantaneous rate of change of mass is not negligible, it is rather difficult to apply Eq. (4.36), where m denotes the total mass of the system of particles constituting the body, including the mass of ejected particles. In order to write the equation for the translational motion of a vehicle with variable mass, consider Fig. 4.7, which depicts a vehicle of mass $m - \Delta m$, where Δm is the mass of ejected particles. Let the point o' denote the center of mass of the vehicle, while o is the center of mass of the entire body of particles. Let $\mathbf{v_o}'$ denote the instantaneous velocity of the vehicle's center of mass, o'. The center of mass of the ejected body has a relative velocity, $\mathbf{v_R}$, with respect to o'. Of course, the net external force, $\mathbf{f}$, acting at o, does not include the internal reactive force on both bodies caused by the ejection of mass.

Since there is no change in the net linear momentum of the system due to mass ejection (or accretion), we can write

$$m\mathbf{v_o} = (m - \Delta m)\mathbf{v_o}' + \Delta m(\mathbf{v_o}' + \mathbf{v_R}), \tag{4.37}$$

which yields

$$\mathbf{v_o} = \mathbf{v_o}' + \frac{\Delta m}{m}\mathbf{v_R}. \tag{4.38}$$

Substituting Eq. (4.38) into Eq. (4.36), we get

$$\mathbf{f} = m\frac{\mathrm{d}\mathbf{v_o}}{\mathrm{d}t} = m\frac{\mathrm{d}\mathbf{v_o}'}{\mathrm{d}t} + \Delta m\frac{\mathrm{d}\mathbf{v_R}}{\mathrm{d}t} + \mathbf{v_R}\frac{\mathrm{d}\Delta m}{\mathrm{d}t}, \tag{4.39}$$

or, substituting Eq. (4.37) into Eq. (4.39),

$$\mathbf{f} = (m - \Delta m)\frac{\mathrm{d}\mathbf{v_o}'}{\mathrm{d}t} + \Delta m\frac{\mathrm{d}(\mathbf{v_o}' + \mathbf{v_R})}{\mathrm{d}t} + \mathbf{v_R}\frac{\mathrm{d}\Delta m}{\mathrm{d}t}. \tag{4.40}$$

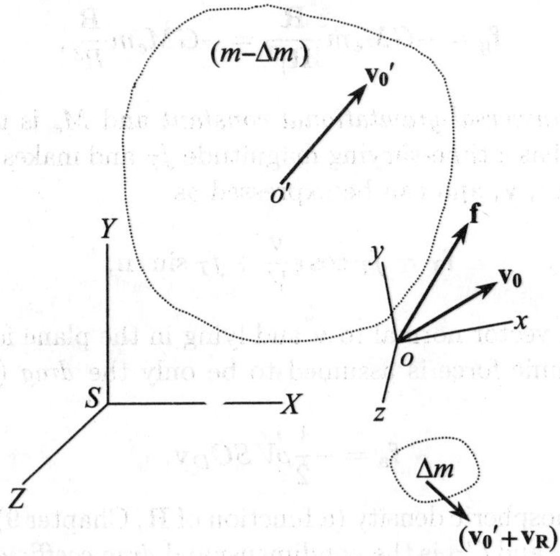

Fig. 4.7. A body ejecting (or accruing) mass, Δm.

Equation (4.40) describes the motion of the primary body of mass $m - \Delta m$ in terms of the acceleration of its center of mass, $\frac{\mathrm{d}\mathbf{v_o}'}{\mathrm{d}t}$. The last two terms on the right-hand side of (4.40) are the opposite of the reactive thrust caused on the primary body due to mass ejection, which is expressed as follows:

$$\mathbf{f_T} = -\Delta m \frac{\mathrm{d}(\mathbf{v_o}' + \mathbf{v_R})}{\mathrm{d}t} - \mathbf{v_R}\frac{\mathrm{d}\Delta m}{\mathrm{d}t}. \tag{4.41}$$

Thus, Eq. (4.40) is rewritten as

$$\mathbf{f} + \mathbf{f_T} = (m - \Delta m)\frac{\mathrm{d}\mathbf{v_o}'}{\mathrm{d}t}, \tag{4.42}$$

Note that $\mathbf{f} + \mathbf{f_T}$ is the net external force acting on the primary body of variable mass $m - \Delta m$. Also, note the similarity between Eq. (4.42) and the particle idealization of Eq. (4.31), where the mass is constant. If we let the mass m in Eq. (4.31) vary with time, we can apply it to represent the motion of a vehicle with a changing mass, with the understanding that $\mathbf{f}$ includes the reactive force $\mathbf{f_T}$.

Example 4.3. Let us derive the governing equations of translational motion for the rocket of Example 4.1. Continuing with the particle idealization, let us write the net external force on the rocket as consisting of gravity, $\mathbf{f_g}$, the reactive thrust, $\mathbf{f_T}$, and the aerodynamic force, $\mathbf{f_a}$:

$$\mathbf{f} = \mathbf{f_g} + \mathbf{f_T} + \mathbf{f_a}. \tag{4.43}$$

The gravitational force—assuming a uniform, spherical earth—is expressed according to *Newton's law of gravitation* (Chapter 3) as follows:

$$\mathbf{f_g} = -GM_e m\frac{\mathbf{R}}{|\mathbf{R}|^3} = -GM_e m\frac{\mathbf{R}}{R^3}, \tag{4.44}$$

where G is the *universal gravitational constant* and M_e is the earth's mass. The thrust force has a time-varying magnitude f_T and makes an angle, ϵ, with the velocity vector, $\mathbf{v}$, and can be expressed as

$$\mathbf{f_T} = f_T \cos\epsilon\frac{\mathbf{v}}{V} + f_T \sin\epsilon\mathbf{n}, \tag{4.45}$$

where $\mathbf{n}$ is a unit vector normal to $\mathbf{v}$ and lying in the plane formed by $\mathbf{f_T}$ and $\mathbf{v}$. The aerodynamic force is assumed to be only the *drag* (see Chapter 10) given by

$$\mathbf{f_a} = -\frac{1}{2}\rho VSC_D\mathbf{v}. \tag{4.46}$$

Here, ρ is the atmospheric density (a function of $\mathbf{R}$, Chapter 9), S is a reference area of the rocket, and C_D is the nondimensional *drag coefficient* (a function of both $\mathbf{R}$ and $\mathbf{v}$). The mass, m, is an explicit function of time due to propellant consumption. Substituting Eqs. (4.43)–(4.46) into Eq. (4.31), we can write

$$-GM_e m\frac{\mathbf{R}}{R^3} + f_T \cos\epsilon\frac{\mathbf{v}}{V} + f_T \sin\epsilon\mathbf{n} - \frac{1}{2}\rho VSC_D\mathbf{v} = m\mathbf{a}. \tag{4.47}$$

A constraint on the rocket's flight is that its altitude must not be negative, which in terms of the coordinates of Example (4.1), can be written as

$$z \geq 0\ (R \geq R_0). \tag{4.48}$$

The rocket's instantaneous position, $\mathbf{R}$, is to be solved for by integrating Eq. (4.47), subject to the nonholonomic constraint of inequality Eq. (4.48), given the initial condition at launch (when $\mathbf{R} = \mathbf{R_0}$) and the input variables ϵ, f_T, and $\mathbf{n}$. The velocity and acceleration vectors required in Eq. (4.47) have already been derived in Eqs. (4.20) and (4.26), respectively, at the time of launch. They need to be evolved in time using Eqs. (4.4) and (4.8), respectively, with $\mathbf{r} = \mathbf{R} - \mathbf{R_0}$ at each time step. The input variables ϵ, f_T, and $\mathbf{n}$ have to be specified at each time instant in obtaining a solution. Due to the nonlinear nature of Eq. (4.47), where the forcing terms are complicated functions of position and velocity, an analytical solution cannot be obtained. This is usually true for any atmospheric flight problem. Instead, a numerical scheme must be employed, such as those implemented in Chapter 12.

4.3.2 Rotation and Translation of a Body

By taking the cross product of both sides of Eq. (4.31) applied to a constituent particle of mass ∂m (Fig. 4.6), with $\mathbf{r}$, we write

$$\mathbf{r} \times \partial\mathbf{f} = \mathbf{r} \times \partial m\mathbf{a} = \mathbf{r} \times \partial m\frac{d\mathbf{v}}{dt}. \tag{4.49}$$

Summing over all the particles, and noting that all internal torques cancel each other by virtue of Newton's third law, the equation of rotational motion of a body is written as follows:

$$\mathbf{M} = \sum \left(\mathbf{r} \times \partial m \frac{d\mathbf{v}}{dt} \right) , \tag{4.50}$$

where $\mathbf{M} \doteq \sum(\mathbf{r} \times \partial \mathbf{f})$ is the net *external torque* about o. Substituting Eq. (4.34) into Eq. (4.50), we can write

$$\mathbf{M} = \frac{d}{dt} \sum (\mathbf{r} \times \partial m \mathbf{v}) - \left(\sum \frac{d\mathbf{r}}{dt} \partial m \right) \times \mathbf{v_o} . \tag{4.51}$$

If the point o is fixed in space, $\mathbf{v_o} = \mathbf{0}$ and the second term on the right-hand side of Eq. (4.51) vanishes. Alternatively, if the point o is chosen to be the center of mass, then it follows from the definition of the center of mass that

$$\sum \frac{d\mathbf{r}}{dt} \partial m = \frac{d}{dt} \sum \mathbf{r} \partial m = \mathbf{0} , \tag{4.52}$$

and the second term on the right-hand side of Eq. (4.51) again vanishes. Therefore, if the point o is either a stationary point, or the center of mass of the body, we have

$$\mathbf{M} = \frac{d}{dt} \sum (\mathbf{r} \times \partial m \mathbf{v}) . \tag{4.53}$$

Equation (4.53) is the equation for the rotational motion of the body about either a fixed point or its center of mass, which we will later discuss in detail. Since fixed points in space are difficult to find, it is a common practice to choose o as the center of mass. Considering Eqs. (4.36) and (4.53), we note that the general motion of a body can be described by the translation of the center of mass and the rotation of the body about its center of mass. In order to solve Eq. (4.53), one must know $\mathbf{r}(t)$ for all the N particles constituting the body, which requires that additional $(3N - 6)$ scalar quantities, describing the relative motion of the particles with respect to the center of mass, be obtained by integrating additional differential equations in time. Such equations of relative motion are obtained by taking into account the internal forces acting on the individual particles. However, if the body is rigid, the relative distances of all particles with respect to the center of mass are fixed, and Eqs. (4.36) and (4.53) are sufficient for describing its general motion. The translational and rotational motions represented by Eqs. (4.36) and (4.53), respectively, can be studied separately, provided the net force and moment vectors do not depend upon the rotational and translational motions, respectively. For an atmospheric flight vehicle, the aerodynamic force depends upon the vehicle's attitude (rotational variables), and aerodynamic and thrust moments depend upon the speed and altitude (translational variables); thus, the two motions are inherently coupled. However, when the time scale of rotation is much smaller than that of translation, then, the two can effectively be decoupled. In such a case, the instantaneous rotational parameters are treated as inputs to the translatory motion, and the position and velocity are treated as almost constant parameters for the rotational motion. The decoupling of rotational and translational motions may not be a good approximation for a high-performance aircraft, or a missile, which would require a simultaneous solution of the two sets of equations [Eqs. (4.36) and (4.53)].

Returning to the translation of a flight vehicle, the aerodynamic and thrust forces are usually expressed in a frame fixed to the body (called *body-fixed* frame), $(oxyz)$, at the center of mass, o. Therefore, it becomes imperative to write Eq. (4.36) in a moving, body-fixed frame, employing the convention of Subsection 4.1.1 as follows:

$$\frac{\mathrm{d}\mathbf{v_o}}{\mathrm{d}t} = \frac{\partial \mathbf{v_o}}{\partial t} + \boldsymbol{\omega} \times \mathbf{v_o}, \tag{4.54}$$

where $\boldsymbol{\omega}$ is the angular velocity of the body about its center of mass. Substituting Eq. (4.54) into Eq. (4.36), we get

$$\mathbf{f} = m\frac{\partial \mathbf{v_o}}{\partial t} + m(\boldsymbol{\omega} \times \mathbf{v_o}). \tag{4.55}$$

The coupling of rotational and translational motions is evident in Eq. (4.55), where $\boldsymbol{\omega}$ is obtained from the solution to the rotational equation, Eq. (4.53).

Example 4.4. Reconsider Example 4.2, where the total velocity and acceleration of a passenger walking inside a maneuvering aircraft were determined relative to a ground station, $(SXYZ)$, at $t = 100$ s to be $\mathbf{v} = -39.946\mathbf{I} + 100.0275\mathbf{J} + 29.8872\mathbf{K}$ m/s and $\mathbf{a} = -0.4136\mathbf{I} + 0.9924\mathbf{J} - 0.0043\mathbf{K}$ m/s^2. Let us apply Newton's laws of motion to find the instantaneous net external force experienced by the passenger and the aircraft, relative to a body-fixed frame, $(oxyz)$, if o is the aircraft's center of mass. We can take the masses of the passenger and the aircraft as 75 kg and 10,000 kg, respectively.

Since the rotation matrix, C, from $(SXYZ)$ to $(oxyz)$ in terms of the instantaneous Euler angles has already been computed in Example 4.2 using Eq. (4.2), we can directly calculate the inertial acceleration of the passenger in the body-fixed coordinates as follows:

```
>> fp=75*C*a  %force in body-fixed coordinates on the passenger (N)
fp =
    -6.8131
    22.3856
   -69.8815
```

Hence, the total force the passenger experiences is $\mathbf{f_p} = -6.8131\mathbf{i} + 22.3856\mathbf{j} - 69.8815\mathbf{k}$ N, or $\mathbf{f_p} = 75(-0.4136\mathbf{I} + 0.9924\mathbf{J} - 0.0043\mathbf{K})$ N. Now, let us calculate the net force on the aircraft as follows:

$$\mathbf{f_a} = m\frac{\mathrm{d}\mathbf{v_o}}{\mathrm{d}t} = 10^4\mathsf{C}\frac{\mathrm{d}^2}{\mathrm{d}t^2}(-0.2t^2\mathbf{I} + 0.5t^2\mathbf{J} + 30t\mathbf{K})\,|_{t=100}, \tag{4.56}$$

which is computed using MATLAB as follows:

```
>> fa=10^4*C*a0
fa =
  1.0e+003 *
        -1.0494
              3.1316
        -9.2838
```

Hence, the net force the aircraft experiences is $\mathbf{f_a} = (-1.0494\mathbf{i} - 3.1316\mathbf{j} - 9.2838\mathbf{k}) \times 10^3$ N. However, since the aircraft is rotating with an angular velocity $\boldsymbol{\omega} = 0.02\mathbf{J} - 0.01\mathbf{K}$ rad/s, it may also experience a net torque by virtue of Eq. (4.53), which we will further explore in Chapter 13.

In atmospheric flight dynamics it is often very useful to employ a particular orientation of the moving frame located at the vehicle's center of mass, such that one of the axes is along the velocity vector relative to the atmosphere. Although the origin of the axes is instantaneously made to coincide with the vehicle's center of mass, its velocity (being that of the wind) is different from the velocity of the center of mass. Such a moving frame is referred to as the *wind axes*. The aerodynamic (Chapter 10) and air-breathing propulsion (Chapter 11) forces are generally resolved along the wind axes, so that the thrust is along, the drag is opposite to, and the lift is perpendicular to the aircraft's velocity relative to the atmosphere.

Example 4.5. Consider an aircraft of mass, $m = 5000$ kg, flying at a constant altitude with acceleration due to gravity, $g = 9.8\ m/s^2$, airspeed, $v_\infty = 100\ m/s$, bank angle, μ, lift, L, drag, D, side force, f_Y, and air-breathing thrust, f_T. If a steady wind of 10 m/s is blowing from the northwest, what should be the constraints for the lift and the bank angle, such that the aircraft makes a steady, coordinated, horizontal turn (Chapter 12) of rate 10 deg/s in the clockwise direction with reference to a groundstation?

Let us choose a stationary frame attached to the groundstation, $(SXYZ)$, with unit vectors $\mathbf{I}, \mathbf{J}, \mathbf{K}$, and wind axes, $(oxyz)$, with unit vectors $\mathbf{i}, \mathbf{j}, \mathbf{k}$ whose origin o coincides with the aircraft's center of mass and has velocity $\mathbf{v_o}$ equal to the wind velocity, such that ox is along the velocity vector of the aircraft relative to the atmosphere, $\mathbf{v}_\infty$ (Fig. 4.8). The vector $\mathbf{v}$ is the aircraft's total velocity. As the aircraft turns relative to the groundstation, both $\mathbf{v}_\infty$ and $\mathbf{v}$ rotate with an angular velocity $\boldsymbol{\omega} = -\omega\mathbf{K} = -\frac{\pi}{18}\mathbf{K}$ rad/s, but the wind velocity remains constant at $\mathbf{v_o} = v_o(\mathbf{I}+\mathbf{J}) = \frac{10}{\sqrt{2}}(\mathbf{I}+\mathbf{J})$ m/s. The magnitudes v and v_∞ are called the *groundspeed* and *airspeed*, respectively. The bank angle, μ, is positive when oy is below the *horizon* (a plane parallel to SXY). For a clockwise horizontal turn, the aircraft must bank to the right as shown in Fig. 4.8, so that a component of the lift points toward SZ. The net force resolved in the wind axes is $\mathbf{f} = (f_T - D)\mathbf{i} + (f_Y + mg\sin\mu)\mathbf{j} + (mg\cos\mu - L)\mathbf{k}$. For a steady, coordinated turn, $f_T = D$, and $f_Y = 0$ (see Chapter 12); hence Eq. (4.36) yields

$$(mg\cos\phi - L)\mathbf{k} + mg\sin\mu\mathbf{j} = m\frac{d\mathbf{v}}{dt} = m(\boldsymbol{\omega} \times \mathbf{v})\,. \tag{4.57}$$

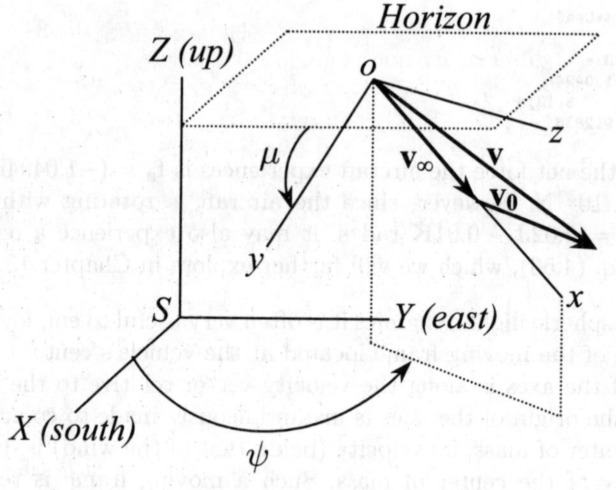

Fig. 4.8. The wind axes, $(oxyz)$, relative to a groundstation, $(SXYZ)$.

Note that $\mathbf{v} = \mathbf{v_0} + \mathbf{v_\infty} = v_0(\mathbf{I}+\mathbf{J}) + v_\infty \mathbf{i}$ m/s, by virtue of Eq. (4.4). In Example 4.2 we employed the rotation matrix, C, for coordinate transformation from *(SXYZ)* to *(oxyz)* in terms of the instantaneous Euler angles [Eq. (4.2)]. Comparing Figs. 4.4 and 4.8, we note that the pitch angle is zero, the roll angle is given by $\phi = \pi + \mu$, and the yaw angle, ψ, varies linearly with time according to $\psi = -\omega t = -\frac{\pi}{18}t$ rad if we begin measuring time when $\psi = 0$. Substituting the Euler angles into (Eq. (4.2), we can write

$$\begin{Bmatrix} \mathbf{i} \\ \mathbf{j} \\ \mathbf{k} \end{Bmatrix} = \begin{pmatrix} \cos\psi & \sin\psi & 0 \\ \cos\mu\sin\psi & -\cos\mu\cos\psi & -\sin\mu \\ -\sin\mu\sin\psi & \sin\mu\cos\psi & -\cos\mu \end{pmatrix} \begin{Bmatrix} \mathbf{I} \\ \mathbf{J} \\ \mathbf{K} \end{Bmatrix} . \tag{4.58}$$

We note from Eq. (4.58) that $\mathbf{K}\times\mathbf{i} = \cos\psi\mathbf{J} - \sin\psi\mathbf{I}$, and that while the airspeed, v_∞, is a constant, the groundspeed is a variable given by $v = \sqrt{2v_0^2 + v_\infty^2 + 2v_0v_\infty(\cos\psi + \sin\psi)}$. Let us substitute the expression $\mathbf{v} = v_0(\mathbf{I}+\mathbf{J}) + v_\infty(\cos\psi\mathbf{I} + \sin\psi\mathbf{J})$ m/s into Eq. (4.57) to obtain

$$\begin{aligned} &(mg\cos\mu - L)(-\sin\mu\sin\psi\mathbf{I} + \sin\mu\cos\psi\mathbf{J} - \cos\mu\mathbf{K}) \\ &\quad + mg\sin\mu(\cos\mu\sin\psi\mathbf{I} - \cos\mu\cos\psi\mathbf{J} - \sin\mu\mathbf{K}) \\ &\quad = -m\omega[v_0(\mathbf{J}-\mathbf{I}) + v_\infty(\cos\psi\mathbf{J} - \sin\psi\mathbf{I})] . \end{aligned} \tag{4.59}$$

Equating the vector components on both sides of Eq. (4.59) and simplifying, we can write

$$\begin{aligned} L\cos\mu &= mg, \\ (L\sin\mu - m\omega v_\infty)\sin\psi &= m\omega v_0, \\ (L\sin\mu - m\omega v_\infty)\cos\psi &= m\omega v_0 . \end{aligned} \tag{4.60}$$

We can combine the last two of Eqs. (4.60) into a single equation; thus, the restriction on the aircraft motion is given by

$$\begin{aligned} L \cos \mu &= mg , \\ L \sin \mu &= m\omega [v_\infty + v_0(\cos\psi + \sin\psi)] . \end{aligned} \tag{4.61}$$

Equations (4.61) represent the conditions to be satisfied by the lift, L, and the bankangle, μ, for executing a horizontal, coordinated, ground-referenced circular turn in the presence of a steady wind from the northwest. These conditions result from the kinematic constraint of having to move in a horizontal circle. They imply that the required lift must be greater than the weight, mg, and the horizontal component of lift (caused by banking) must be sufficiently large to create the desired turnrate, while overcoming the effect of the wind. For a constant airspeed, v_∞, the magnitude of lift, L, is controlled by adjusting the attitude of the aircraft relative to the wind axes (which is also how the sideforce, f_Y, is maintained at zero), while its direction is controlled by changing μ. Hence, the lift vector (determined by L and μ) is a force arising out of the constrained motion and is called the *force of constraint*, rather analogous to the tension in the simple pendulum, which is constrained to move in an arc. By modulating the lift vector according to Eq. (4.61), the pilot (or an automatic flight control system) can fly a horizontal turn about a fixed point. Let us numerically determine the lift vector of the present example, using MATLAB as follows:

```
>> omeg=pi/18;v0=10/sqrt(2);vinf=100;m=5000;g=9.8;% constant parameters
>> t=0:2*pi/omeg;psi=-omeg*t; % time points and yaw angle profile for one full turn
>> mu=atan(omeg*(vinf+v0*(cos(psi)+sin(psi)))/g); % required bank angle profile
>> L=m*g./cos(mu); % lift profile
>> subplot(211),plot(t,180*mu/pi),hold on,subplot(212),plot(t,L)% plots
```

The resulting plots of L and μ for a complete turn are shown in Fig. 4.9. Note that the lift magnitudes are nearly double the weight for the given turn rate and wind velocity. Also, the maximum bank angle and lift are required when the aircraft is banking into the wind, which happens in this case for $\psi = 45°$ (i.e., *southwest* orientation of ox) at $t = 31.5$ s.

4.4 Energy and Angular Momentum

We have seen that the equations of motion for the translation of a particle result in the conservation of linear momentum when there is no net force acting on the particle. In such a case, Newton's laws imply that the particle continues to be in a state of uniform rectilinear motion. The same can be said about a body, by considering the motion of its center of mass. Even when the net force is nonzero, the magnitude and direction of the force can be such that certain quantities, called the *energy* and *angular momentum* of the particle, are conserved. These quantities are quite useful in describing the motion of a particle or a system of particles (body).

Let us define the *work done*, δW, by a force, $\mathbf{f}$, acting on the particle as the dot product between the force and the resulting *small* displacement of the particle, $\delta \mathbf{R}$, given by

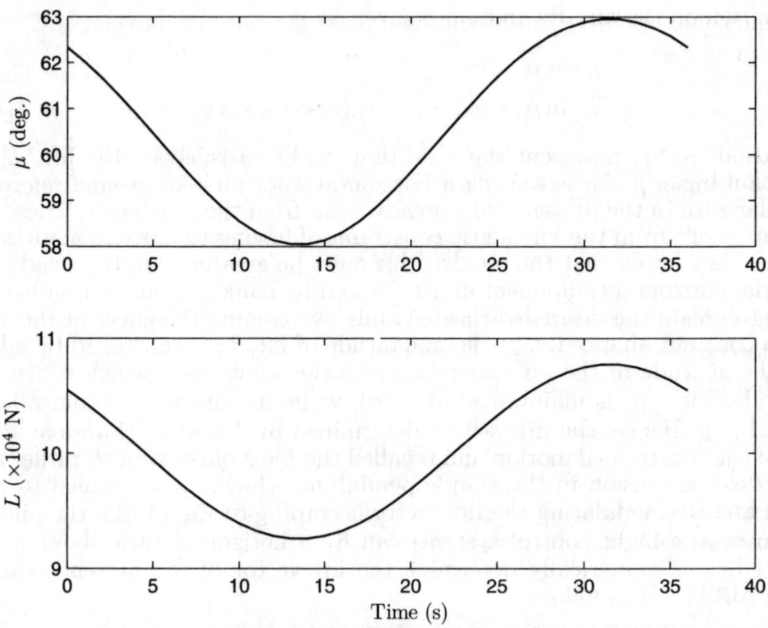

Fig. 4.9. The lift vector modulation for a horizontal, circular turn in the presence of wind.

$$\delta W = \mathbf{f} \cdot \delta \mathbf{R}. \tag{4.62}$$

In this definition, the assumption of small displacement is employed to ensure that the force, **f**, remains unchanged while the particle is displaced. This is necessary, because **f** generally depends upon the instantaneous position **R**. However, we can calculate the net work a force does when the particle traverses a known path joining two arbitrary positions, by employing a summation of the work done over a number of small path segments, during each of which Eq. (4.62) is valid. Hence, the total work done W_{12} by **f** when the particle is displaced from $\mathbf{R_1}$ to $\mathbf{R_2}$ is given by

$$W_{12} = \int_{R_1}^{R_2} \mathbf{f} \cdot \mathrm{d}\mathbf{R} \,, \tag{4.63}$$

where the path between $\mathbf{R_1}$ and $\mathbf{R_2}$ is specified by a curve, $\mathbf{R}(t)$. Note that W_{12} depends upon the initial and final particle positions, as well as the path $\mathbf{R}(t)$ joining the two positions. Since there can be several possible paths joining any two points, the work done along each can be different. The restrictions on the possible paths are that they must be the solution to Eq. (4.31) and obey the kinematic constraints on the motion of the particle.

There are certain forces whose work done is *path-independent*, i.e., W_{12} depends only upon the initial and final particle positions, but not on the path $\mathbf{R}(t)$ joining the two positions. Such a force is said to be *conservative*, because the net work done by returning the particle to its original position, $\mathbf{R_1}$, by a different path would always be zero (i.e., $W_{12} = -W_{21}$, irrespective of the path). Since a closed curve is formed if we return the particle to the original position by a different path, we write the following for a conservative force:

$$\oint \mathbf{f} \cdot \mathrm{d}\mathbf{R} = 0, \tag{4.64}$$

where $\oint$ denotes the integral carried out along a closed curve. Very few physical forces are conservative. From our everyday experience, we can say that friction and other dissipative forces are not conservative. However, the force of gravity is conservative. There must be a specific functional form of the conservative force, $\mathbf{f}(\mathbf{R})$, that meets the requirement of Eq. (4.64). It can be shown by vector calculus [4] that the necessary and sufficient condition for Eq. (4.64) to be satisfied is that $\mathbf{f}$ should be determined by the *gradient* of a scalar function of position, $\Phi(\mathbf{R})$, as follows:

$$\mathbf{f}(\mathbf{R}) = m\boldsymbol{\nabla}\Phi(\mathbf{R})^T \doteq m\left(\frac{\partial \Phi}{\partial \mathbf{R}}\right)^T . \tag{4.65}$$

The scalar function, $\Phi(\mathbf{R})$, whose gradient gives the acceleration caused by the conservative force, is called a *potential function*. Note that the gradient $\boldsymbol{\nabla}\Phi(\mathbf{R})$ is the derivative of a scalar with respect to a vector and is represented by a *row* vector, whose elements are partial derivatives of Φ with respect to corresponding elements of $\mathbf{R}^T$. An important caveat for a force to be conservative is that it should not *explicitly* depend upon time, i.e., a force $\mathbf{f}(\mathbf{R}, t)$ satisfying Eq. (4.65) is *not* conservative.

Substituting Eq. (4.31) into Eq. (4.63), and noting that $\mathrm{d}\mathbf{R} = \mathbf{v}\mathrm{d}t$, we have

$$W_{12} = \int_{t_1}^{t_2} m\frac{\mathrm{d}\mathbf{v}}{\mathrm{d}t} \cdot \mathbf{v}\mathrm{d}t = \frac{1}{2}m\int_{t_1}^{t_2} \frac{\mathrm{d}}{\mathrm{d}t}\mathbf{v} \cdot \mathbf{v}\mathrm{d}t , \tag{4.66}$$

or,

$$W_{12} = \frac{1}{2}m(v_2^2 - v_1^2) \doteq T_2 - T_1 , \tag{4.67}$$

where $v^2 = \mathbf{v}\cdot\mathbf{v}$, and $T \doteq \frac{1}{2}mv^2$ is defined as the *kinetic energy* of the particle. Hence, the work done by a force upon a particle is equal to the change in its kinetic energy. The kinetic energy, being proportional to the square of the speed, is a measure of the energy of motion.

If the force applied to a particle is conservative, we can substitute Eq. (4.65) into Eq. (4.63) to write

$$W_{12} = m\int_{R_1}^{R_2} \left(\frac{\partial \Phi}{\partial \mathbf{R}}\right)^T \cdot \mathrm{d}\mathbf{R} = m[\Phi(\mathbf{R_2}) - \Phi(\mathbf{R_1})] , \tag{4.68}$$

or,

$$W_{12} = V_2 - V_1\,, \tag{4.69}$$

where $-V \doteq -m\Phi(\mathbf{R})$ is defined as the *potential energy* of the particle. Comparing Eq. (4.69) with Eq. (4.67), we can eliminate the work done by a conservative force to write $T_2 - T_1 = V_2 - V_1$, or

$$E \doteq T - V = \text{constant}. \tag{4.70}$$

Equation (4.70) states the important principle that the sum of kinetic and potential energies (called *total energy*, E) of a particle acted upon by a conservative force is conserved. The potential energy is thus seen to be *stored*, which can be exchanged for kinetic energy, and vice versa. This principle can be easily extended to a body of particles, by focusing on the translation of the center of mass. For a spacecraft acted upon only by gravity, which is a conservative force, the law of energy conservation is very useful in determining orbits and in designing orbital maneuvers. An aircraft flying with the thrust equal to the drag has a zero net nonconservative force along the velocity vector, and the resulting trajectory can be analyzed by applying the principle of energy conservation.

For a general, nonconservative force, $\mathbf{f}$, the total energy, E, is not conserved, and we can write

$$\frac{\mathrm{d}W}{\mathrm{d}t} \doteq \frac{\mathrm{d}T}{\mathrm{d}t} = mv\frac{\mathrm{d}v}{\mathrm{d}t}\,, \tag{4.71}$$

which implies that the rate of work done (called *power*) by $\mathbf{f}$ is equal to the rate of change of kinetic energy.

We can obtain the net potential and kinetic energies of a system of particles by summing up Eqs. (4.69) and (4.67), respectively, for all the particles. Let us consider a system of N particles constituting a body. The total kinetic energy of the system is

$$T = \frac{1}{2}\sum_{i=1}^{N} m_i v_i^2\,, \tag{4.72}$$

where $v_i^2 = \mathbf{v_i}\cdot\mathbf{v_i}$. We can express the velocity of particle i as $\mathbf{v_i} = \mathbf{v_o} + \mathbf{u_i}$, where $\mathbf{v_o}$ is the velocity of the center of mass, and $\mathbf{u_i}$ is the velocity of the particle relative to the center of mass. Therefore, the kinetic energy of the system becomes

$$\begin{aligned} T &= \frac{1}{2}\sum_{i=1}^{N} m_i(\mathbf{v_o} + \mathbf{u_i})\cdot(\mathbf{v_o} + \mathbf{u_i}) \\ &= \frac{1}{2}\sum_{i=1}^{N} m_i v_0^2 + \frac{1}{2}\sum_{i=1}^{N} m_i u_i^2 + \mathbf{v_o}\cdot\frac{\mathrm{d}}{\mathrm{d}t}\left(\sum_{i=1}^{N} m_i\mathbf{r_i}\right) \\ &= \frac{1}{2}mv_0^2 + \frac{1}{2}\sum_{i=1}^{N} m_i u_i^2\,, \end{aligned} \tag{4.73}$$

where $\mathbf{r_i}$ is the particle's location relative to the center of mass. The last step of Eq. (4.73) follows from the definition of the center of mass. Hence, the total kinetic energy of a system of particles is sum of the kinetic energy of the center of mass and the kinetic energy of the motion of particles about the center of mass. [If the body is rigid, there is no translation of the particles relative to the center of mass, and the last term in Eq. (4.73) is due to the rotation of the body about the center of mass.] The work the net external force $\mathbf{f}$ does in moving the body from an initial configuration of particles to a final configuration is equal to the change in the total kinetic energy, given by Eq. (4.67).

If the forces acting on all the particles are conservative, then the net work done is equal to the change in potential energy of the system, and the total energy, $T-V$, is conserved [Eq. (4.69)]. The potential energy of the conservative system is then written as follows:

$$-V = -\sum_{i=1}^{N} V_i^{(e)} + \frac{1}{2}\sum_{i=1}^{N}\sum_{j\neq i}^{N} V_{ij} \,, \tag{4.74}$$

where $-V_i^{(e)}$ is the potential energy associated with the external force, $\mathbf{f_i^{(e)}}$, and $-V_{ij}$ denotes the potential energy due to the internal force $\mathbf{f_{ij}}$ between particles i and j. In the remainder of the chapter, we will consider the modeling of interesting conservative systems.

A system of particles (body) may be simultaneously acted upon by both conservative and nonconservative forces, e.g., the rocket of Example 4.3. In such a case, the total energy is not conserved, but we can still talk of a potential energy by limiting the summations in Eq. (4.74) to those particles and forces that constitute a conservative subsystem. The utility of the potential energy in a partially conservative (thus complex) system lies in deriving the equations of motion by an approach based upon *variational principles* (also known as the *Lagrangian method*) rather than Newton's laws of motion. Such an approach becomes valuable when the equations of relative motion among the particles of a nonrigid body are to be derived. We shall discuss the Lagrangian method for analyzing flexible aerospace structures in Chapter 15.

Let us now define another quantity of particle motion, namely *angular momentum*, $\mathbf{H}$, as the moment of linear momentum, $\mathbf{H} \doteq \mathbf{r} \times m\mathbf{v}$, where $\mathbf{r}$ is the relative position of the particle, p, with respect to a point o, which need not be stationary (Fig. 4.1). From Eq. (4.49)—which can be rewritten for the particle replacing ∂m by m and $\partial \mathbf{f}$ by $\mathbf{f}$—as

$$\mathbf{r} \times \mathbf{f} = \frac{\mathbf{dH}}{dt} - \frac{\mathbf{dr}}{dt} \times (m\mathbf{v}) \,, \tag{4.75}$$

it is clear that the angular momentum is related to the rotational motion of the particle about the point o, because the left-hand side of Eq. (4.75) is the torque about o, denoted by $\mathbf{M} \doteq \mathbf{r} \times \mathbf{f}$. A zero angular momentum implies parallel position and velocity vectors, which is the case of *rectilinear* (straight-line) motion. If o is stationary, then $\mathbf{v} = \frac{d\mathbf{r}}{dt}$. Consequently, the second term on the right-hand side of Eq. (4.75) vanishes; thus, it follows that

$$\mathbf{M} = \frac{d\mathbf{H}}{dt}. \tag{4.76}$$

Hence, if the applied torque about a fixed point is zero, then $\frac{d\mathbf{H}}{dt} = \mathbf{0}$, which implies that $\mathbf{H}$ is a constant vector. This is the principle of the conservation of angular momentum for a particle.

Example 4.6. Let us calculate the angular momentum of the aircraft of Example 4.5, about the vertical axis, SZ (Fig. 4.8), as follows:

$$\begin{aligned}\mathbf{H} &= \mathbf{R} \times (m\mathbf{v}) \\ &= R(\cos\psi\mathbf{I} + \sin\psi\mathbf{J}) \times [v_0(\mathbf{I}+\mathbf{J}) + v_\infty\mathbf{i}] \\ &= R[v_0(\cos\psi - \sin\psi)\mathbf{K} + v_\infty(\sin\psi\cos\psi - \sin\psi\cos\psi)\mathbf{K}] \\ &= Rv_0(\cos\psi - \sin\psi)\mathbf{K}\,, \end{aligned} \tag{4.77}$$

where R is the constant radius of turn. Taking the time derivative of $\mathbf{H}$, we find that

$$\frac{d\mathbf{H}}{dt} = -Rv_0(\sin\psi + \cos\psi)\dot{\psi}\mathbf{K} = Rv_0\omega(\sin\psi + \cos\psi)\mathbf{K}. \tag{4.78}$$

Hence, $\frac{d\mathbf{H}}{dt} \neq \mathbf{0}$, due to the steady wind of speed v_0. Had there been no wind, the angular momentum would have been conserved. We will get the same conditions on lift and bank angle as obtained in Eq. (4.61) for executing a steady, horizontal turn in the blowing wind by applying Eq. (4.76) to the present example, with $\mathbf{M} = \mathbf{R} \times [(mg\cos\mu - L)\mathbf{k} + mg\sin\mu\mathbf{j}]$. This is left as an exercise at the end of the chapter.

Angular momentum can also be defined for a body by summing the angular momenta of all the constituent particles, as $\mathbf{H} \doteq \sum(\mathbf{r} \times \partial m\mathbf{v})$, which, when substituted into Eq. (4.53), leads to Eq. (4.76) for a body. Recall that in Eq. (4.53), o is the center of mass of the body. Therefore, if the net torque about the center of mass is zero, the angular momentum of the body about its center of mass is conserved. We again note the similarity between a particle and the center of mass of a body, even where the rotational motion is concerned. We have seen above that the rotational motion of a body about its center of mass is similar to that of a particle about a fixed point. There are several important instances—often in spacecraft dynamics, where the net force passes through the center of mass—when we can apply the principle of angular momentum conservation for a body. We reiterate that in the above analysis we have *not* assumed the body to be rigid. If a body is nonrigid, we must take into account the relative motion among the particles constituting the body.

4.4.1 The N-Body Problem

The problem of determining the motion of N bodies under the influence of mutual gravitational attraction is one of the fundamental problems of astronomy. More generally, the N-body problem is applicable to any system where the mutual forces between any two bodies act along the line joining their centroids (such as electrostatic, gravitational, and impact forces). In all such cases, the net external force and torque about any point are both equal to zero by virtue of Newton's third law, and consequently, the linear and angular momenta of the system are conserved. Our solar system is the example of the N-body problem, where the sun, planets, comets, and asteroids are all gravitationally bound. In flight dynamics, an N-body problem may result from studying the motion of a spacecraft under the influence of one or more celestial bodies (such as the earth, the moon, and the sun). Since the distances between the bodies are generally large in comparison with their dimensions, it is a standard assumption to treat the N-body problem as a collection of N particles, mutually attracted by the Newton's inverse-square law of gravitation (Chapter 3). As the net external force on the system of total mass, m, is zero, we can write the equation of motion of the system's center of mass according to Eq. (4.36) as follows:

$$m\frac{\mathrm{d}\mathbf{v_o}}{\mathrm{d}t} = m\frac{\mathrm{d}^2\mathbf{R_o}}{\mathrm{d}t^2} = \mathbf{0}\,, \tag{4.79}$$

which can be successively integrated in time to yield

$$\mathbf{R_o} = \mathbf{v_o}t + \mathbf{c}\,, \tag{4.80}$$

where $\mathbf{v_o}$ and $\mathbf{c}$ are constant vectors, denoting the velocity and initial position of the center of mass. These two vectors lead to six known scalar quantities, called the *integrals of motion*. However, the N-body problem has a total of $3N$ degrees of freedom, which require the solution of a system of $3N$ second-order, ordinary differential equations for the positions of all particles, resulting in $6N$ motion variables. If we conveniently choose the origin of the inertial frame to be at the center of mass, then $\mathbf{v_o} = \mathbf{c} = \mathbf{0}$, and $\mathbf{R_i}$ denotes the position of the particle i.

Let us see whether we can determine the remaining $(6N - 6)$ variables in order to solve the N-body problem. Since the net external torque on the system is zero, we can write the rotational equations of motion, Eq. (4.53), in the following form:

$$\frac{\mathrm{d}}{\mathrm{d}t}\sum_{i=1}^{N}(\mathbf{R_i} \times m_i\mathbf{v_i}) = \mathbf{0}, \tag{4.81}$$

where $m_i, \mathbf{v_i}$ represent the mass and velocity of particle i. It is clear that a constant angular momentum vector, $\mathbf{H}$, whose time derivative is equal to the left-hand side of Eq. (4.81), can be specified for the system. Hence, we can express the three new integrals of motion (components of $\mathbf{H}$) as

$$\mathbf{H} \doteq \sum_{i=1}^{N} (\mathbf{R_i} \times m_i \mathbf{v_i}) = \sum_{i=1}^{N} \left(\mathbf{R_i} \times m_i \frac{\mathrm{d}\mathbf{R_i}}{\mathrm{d}t} \right) . \tag{4.82}$$

The conservation of total angular momentum is responsible for the near planar shape of the solar system.

In order to derive another integral of motion, the equations of motion for the bodies—idealized as the translation of their respective centers of mass—can be written as follows:

$$m_i \frac{\mathrm{d}^2\mathbf{R_i}}{\mathrm{d}t^2} = \mathbf{f_i} \quad (i = 1, 2, ..., N), \tag{4.83}$$

where $\mathbf{f_i}$ is the net force acting on the particle i, and can be expressed by Newton's law of gravitation as follows:

$$\mathbf{f_i} = G \sum_{j \neq i}^{N} \frac{m_i m_j}{R_{ij}^3} (\mathbf{R_j} - \mathbf{R_i}) , \tag{4.84}$$

where G is the universal gravitational constant, and $R_{ij} \doteq | \mathbf{R_j} - \mathbf{R_i} |$. Substituting Eq. (4.84) into Eq. (4.83), we get

$$\frac{\mathrm{d}^2\mathbf{R_i}}{\mathrm{d}t^2} = G \sum_{j \neq i}^{N} \frac{m_j}{R_{ij}^3} (\mathbf{R_j} - \mathbf{R_i}) ; \quad (i = 1, 2, ..., N). \tag{4.85}$$

Since the gravitational force $\mathbf{f_i}$ is conservative, we can define a gravitational potential, Φ_i such that its *gradient* (Chapter 3) with respect to $\mathbf{R_i}$ determines the force as follows:

$$\mathbf{f_i}^T = m_i \frac{\partial \Phi_i}{\partial \mathbf{R_i}} . \tag{4.86}$$

When we compare Eq. (4.86) with Eq. (4.84), it is clear that

$$\Phi_i \doteq G \sum_{j \neq i}^{N} \frac{m_j}{R_{ij}} . \tag{4.87}$$

The total potential energy of the system is thus given by $-V$, where, according to Eq. (4.74),

$$V \doteq \frac{1}{2} \sum_{i=1}^{N} m_i \Phi_i . \tag{4.88}$$

Note that the potential energy represents the work done in bringing the N bodies from infinite mutual separation to their present positions. It is also clear that

$$\mathbf{f_i}^T = \frac{\partial V}{\partial \mathbf{R_i}} . \tag{4.89}$$

From Eq. (4.70) we expect the total energy of the system to be conserved. In order to show this, we write the time derivative of V as follows:

$$\begin{aligned}\frac{dV}{dt} &= \sum_{i=1}^{N} \frac{\partial V}{\partial \mathbf{R_i}} \frac{d\mathbf{R_i}}{dt} \\ &= \sum_{i=1}^{N} \mathbf{f_i} \cdot \mathbf{v_i} \,. \end{aligned} \tag{4.90}$$

Substituting Eq. (4.31) for the particle i into Eq. (4.90), we have

$$\begin{aligned}\frac{dV}{dt} &= \sum_{i=1}^{N} m_i \frac{d\mathbf{v_i}}{dt} \cdot \mathbf{v_i} \\ &= \frac{dT}{dt} \,, \end{aligned} \tag{4.91}$$

where $T \doteq \frac{1}{2}\sum_{i=1}^{N} m_i \mathbf{v_i} \cdot \mathbf{v_i} = \frac{1}{2}\sum_{i=1}^{N} m_i v_i^2$ is the total kinetic energy of the system. Therefore, it follows from Eq. (4.91) that the total energy, $E = T - V$, is conserved, which gives us another integral of motion. Thus, we have obtained a total of *10* integrals of motion ($\mathbf{v_0}$, $\mathbf{c}$, $\mathbf{H}$, and E), whereas *6N* integrals are required for the complete solution. In the next section, it will be shown that two additional integrals can be obtained when $N = 2$ from the considerations of relative motion of the two bodies. Hence, a two-body problem is analytically solvable. However, with $N > 2$, the number of unknown motion variables exceeds the total number of integrals; thus, no analytical solution exists for the N-body problem when $N > 2$. Due to this reason, we cannot mathematically prove certain observed facts (such as the stability of the solar system) concerning N-body motion. The best we can do is to approximate the solution to the N-body problem either by a set of two-body solutions or by numerical solutions.

4.5 The Two-Body Problem

The problem of two spherical bodies in mutual gravitational attraction is the fundamental problem of translational motion in space dynamics, and has a complete, analytical solution. The motion of a spacecraft under the influence of a celestial body in the solar system is usually approximated as a two-body problem by ignoring the gravitation caused by the other objects, as well as the actual, nonspherical shapes of the two bodies. Such an approximation is valid when the gravitational attraction of the remaining $N - 2$ bodies and gravitational asymmetry resulting from nonspherical shapes are negligible in magnitude. Compare with the inverse-square gravitational attraction between the two primary bodies (called *primaries*). With some exceptions, such as a lunar trajectory, such an assumption can generally be applied. When a spacecraft is one of the primaries, we can get an accurate representation of its trajectory by including the small perturbations caused by the departure from the two-body solution, using an appropriate numerical scheme. However, the starting point of such a scheme is usually the two-body solution, for which a total of 12 integrals are necessary. We have already obtained the 10 known integrals of the N-body problem in the previous section. Rather than trying to find the remaining two integrals in terms of translation of and rotation about the center of mass, it is simpler to separate the two-body motion into two parts: the motion of the center of mass [for which six integrals of motion are available from Eq. (4.80)], and the relative motion between the two bodies. There is a historical basis for such a separation, beginning with the earliest astronomical observations of celestial bodies from the earth. Also, it is more convenient to study the motion of a spacecraft relative to a coordinate frame fixed at the center of the other primary body. In order to get the remaining six integrals of motion, we write Eq. (4.85) for $i = 1, 2$, and subtract the two equations of motion from one another:

$$\frac{\mathrm{d}^2\mathbf{R_{12}}}{\mathrm{d}t^2} + G\frac{(m_1 + m_2)}{R_{12}^3}\mathbf{R_{12}} = \mathbf{0}, \tag{4.92}$$

where $\mathbf{R_{12}} \doteq \mathbf{R_2} - \mathbf{R_1}$ is the position of m_2 relative to m_1. For simplicity of notation, let us replace $\mathbf{R_{12}}$ by $\mathbf{r}$, and denote $\mu \doteq G(m_1 + m_2)$, resulting in the following equation of relative motion:

$$\frac{\mathrm{d}^2\mathbf{r}}{\mathrm{d}t^2} + \frac{\mu}{r^3}\mathbf{r} = \mathbf{0}\,. \tag{4.93}$$

Since Eq. (4.93) is a nonlinear vector differential equation, we do not expect it to have a *closed-form* solution, i.e., an explicit expression for the relative position as a function of time, $\mathbf{r}(t)$. However, we can analytically determine the necessary integrals of motion governing the problem, which allow us to represent the trajectory either as an infinite series expansion, or by related numerical approximations.

Let us begin by taking the vector product of Eq. (4.93) with $\mathbf{r}$:

$$\mathbf{r} \times \frac{\mathrm{d}^2\mathbf{r}}{\mathrm{d}t^2} + \frac{\mu}{r^3}(\mathbf{r} \times \mathbf{r}) = \mathbf{0}\,, \tag{4.94}$$

or,

$$\frac{\mathrm{d}}{\mathrm{d}t}\left(\mathbf{r} \times \frac{\mathrm{d}\mathbf{r}}{\mathrm{d}t}\right) - \frac{\mathrm{d}\mathbf{r}}{\mathrm{d}t} \times \frac{\mathrm{d}\mathbf{r}}{\mathrm{d}t} = \mathbf{0}\,, \tag{4.95}$$

which implies that

$$\frac{\mathrm{d}}{\mathrm{d}t}\left(\mathbf{r} \times \frac{\mathrm{d}\mathbf{r}}{\mathrm{d}t}\right) = \mathbf{0}\,. \tag{4.96}$$

Therefore, it follows that the *specific angular momentum* of m_2 relative to m_1, defined by

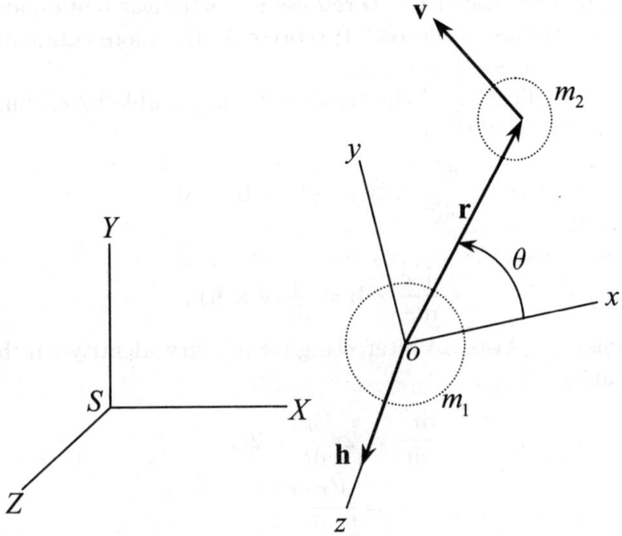

Fig. 4.10. The two-body motion.

$$\mathbf{h} \doteq \mathbf{r} \times \frac{\mathrm{d}\mathbf{r}}{\mathrm{d}t} = \mathbf{r} \times \mathbf{v} \;, \tag{4.97}$$

is conserved, where $\mathbf{v} \doteq \frac{\mathrm{d}\mathbf{r}}{\mathrm{d}t}$ is the relative velocity. We emphasize that $\mathbf{h}$ is the angular momentum per unit mass of m_2 about m_1 and is *not* the total angular momentum of the two-body system about the common center of mass, $\mathbf{H}$ [Eq. (4.82)]. Since $\mathbf{h}$ is a constant vector, it represents three scalar integrals of motion. There are two important consequences of a constant $\mathbf{h}$:

(a) The direction of $\mathbf{h}$ is a constant, which implies that the vectors $\mathbf{r}$ and $\mathbf{v}$ are always in the same plane, and $\mathbf{h}$ is normal to that plane. (You may verify this fact by showing $\mathbf{r} \cdot \mathbf{h} = 0$ and $\mathbf{v} \cdot \mathbf{h} = 0$.) Hence, we can choose a coordinate frame (Fig. 4.10) attached to m_1 such that the z-axis is along $\mathbf{h}$ and the relative motion occurs in the xy-plane.
(b) The magnitude of $\mathbf{h}$ is constant. In terms of polar coordinates, (r, θ), the magnitude of $\mathbf{h}$ is written as

$$h = \mid \mathbf{r} \times \mathbf{v} \mid = r^2 \frac{\mathrm{d}\theta}{\mathrm{d}t} \;, \tag{4.98}$$

which implies that the radius vector, $\mathbf{r}$, sweeps out area at a constant rate $\frac{1}{2} r^2 \frac{\mathrm{d}\theta}{\mathrm{d}t}$, called *areal velocity*. This is the general form of *Kepler's second law of planetary motion*,[2] which was restricted to elliptical trajectories. The two-body trajectories can be classified according to the magnitude and direction

[2] Johannes Kepler (1571–1630) discovered the three laws of planetary motion after a careful analysis of the observational data of the planet Mars, painstakingly gathered by the Danish astronomer Tycho Brahe (1546–1601):

of a constant $\mathbf{h}$. The case of $\mathbf{h} = \mathbf{0}$ represents rectilinear motion along the line joining the two bodies, while $\mathbf{h} \neq \mathbf{0}$ represents the more common trajectory involving rotation of m_2 about m_1.

A further classification of the trajectories is possible by taking the vector product of Eq. (4.93) with $\mathbf{h}$:

$$\frac{\mathrm{d}^2\mathbf{r}}{\mathrm{d}t^2} \times \mathbf{h} + \frac{\mu}{r^3}(\mathbf{r} \times \mathbf{h}) = \mathbf{0} \,. \tag{4.99}$$

We note that

$$\frac{\mathrm{d}^2\mathbf{r}}{\mathrm{d}t^2} \times \mathbf{h} = \frac{\mathrm{d}}{\mathrm{d}t}(\mathbf{v} \times \mathbf{h}) \,, \tag{4.100}$$

since $\mathbf{h}$ is constant. Also, an interesting elementary identity can be obtained by differentiating r^2:

$$\begin{aligned} \frac{\mathrm{d}r^2}{\mathrm{d}t} &= 2r\frac{\mathrm{d}r}{\mathrm{d}t} = 2r\dot{r} \\ &= \frac{\mathrm{d}(\mathbf{r}\cdot\mathbf{r})}{\mathrm{d}t} \\ &= \dot{\mathbf{r}}\cdot\mathbf{r} + \mathbf{r}\cdot\dot{\mathbf{r}} \\ &= 2\mathbf{r}\cdot\dot{\mathbf{r}} = 2\mathbf{r}\cdot\mathbf{v} \,, \end{aligned} \tag{4.101}$$

from which it follows that $\mathbf{r} \cdot \mathbf{v} = r\dot{r}$ (dot represents time derivative). We use this and another vector identity to evaluate the second term in Eq. (4.99) as follows:

$$\begin{aligned} \frac{\mu}{r^3}(\mathbf{r} \times \mathbf{h}) &= \frac{\mu}{r^3}(\mathbf{r} \times (\mathbf{r} \times \mathbf{v})) \\ &= \frac{\mu}{r^3}[(\mathbf{r}\cdot\mathbf{v})\mathbf{r} - (\mathbf{r}\cdot\mathbf{r})\mathbf{v}] \\ &= \frac{\mu\dot{r}}{r^2}\mathbf{r} - \frac{\mu}{r}\mathbf{v} \\ &= -\mu\left(\frac{\mathbf{v}}{r} - \frac{\dot{r}}{r^2}\mathbf{r}\right) \\ &= -\mu\frac{\mathrm{d}}{\mathrm{d}t}\left(\frac{\mathbf{r}}{r}\right) . \end{aligned} \tag{4.102}$$

Substituting Eqs. (4.100) and (4.102) into Eq. (4.99), we have

$$\frac{\mathrm{d}}{\mathrm{d}t}\left(\mathbf{v} \times \mathbf{h} - \frac{\mu\mathbf{r}}{r}\right) = \mathbf{0}. \tag{4.103}$$

Hence, we can define a constant vector $\mathbf{e}$, called the *eccentricity vector*, such that $\mu\mathbf{e} \doteq \mathbf{v} \times \mathbf{h} - \frac{\mu\mathbf{r}}{r}$. Since $\mathbf{e}$ is a constant vector, we expect it to provide

1. The planets orbit the sun in elliptical orbits, with the sun at one focus.
2. The straight line joining a planet and the sun sweeps out equal areas in equal times.
3. The square of the orbital period of a planet is proportional to the cube of the ellipse's semi-major axis.

three more scalar integrals of motion; however, since $\mathbf{e}$ and $\mathbf{h}$ are related by $\mathbf{e} \cdot \mathbf{h} = 0$, we get a total of only five scalar integrals from the two constant vectors. It is also clear that $\mathbf{e}$, being perpendicular to $\mathbf{h}$, lies in the plane of motion formed by $\mathbf{r}$ and $\mathbf{v}$.

An important insight into the two-body motion is obtained by writing the magnitude of the eccentricity vector (called *eccentricity*) as follows:

$$e^2 = \mathbf{e} \cdot \mathbf{e} = \frac{1}{\mu^2}(\mathbf{v} \times \mathbf{h}) \cdot (\mathbf{v} \times \mathbf{h}) - \frac{2}{\mu r}\mathbf{r} \cdot (\mathbf{v} \times \mathbf{h}) + 1 . \tag{4.104}$$

Since $\mathbf{v}$ and $\mathbf{h}$ are mutually perpendicular, it follows that $(\mathbf{v} \times \mathbf{h}) \cdot (\mathbf{v} \times \mathbf{h}) = v^2h^2$. Furthermore, it is true that $\mathbf{r} \cdot (\mathbf{v} \times \mathbf{h}) = (\mathbf{r} \times \mathbf{v}) \cdot \mathbf{h} = h^2$. Therefore, Eq. (4.104) yields the following:

$$1 - e^2 = \frac{v^2h^2}{\mu^2} - \frac{2h^2}{\mu r} + 1 = \frac{h^2}{\mu}\left(\frac{2}{r} - \frac{v^2}{\mu}\right) . \tag{4.105}$$

Let us define *parameter*, p, and another constant, a, both having units of length, by

$$p \doteq \frac{h^2}{\mu} ; \qquad \frac{1}{a} = \frac{2}{r} - \frac{v^2}{\mu} , \tag{4.106}$$

which are related by $p = a(1 - e^2)$. If we use the definition of a, another integral of motion is defined as

$$\varepsilon \doteq -\frac{\mu}{2a} = \frac{v^2}{2} - \frac{\mu}{r}. \tag{4.107}$$

Note that $\frac{v^2}{2}$ is the specific kinetic energy and $-\frac{\mu}{r}$ is the specific potential energy of relative motion of m_2 about m_1. Hence, we call ε the *energy integral* of the relative motion.[3] However, ε is not the total energy, E, of the two-body system, because it does not include the kinetic energy of m_1. The energy integral can be used to describe the various kinds of trajectories of the two-body motion. For a bound orbit of m_2 about m_1, the magnitude of potential energy must be greater than the kinetic energy, which implies $\varepsilon < 0$ and $\frac{1}{a} > 0$. For m_2 to escape the gravity of m_1, the relative kinetic energy must be greater than or equal to the magnitude of potential energy, implying $\varepsilon \geq 0$ and $\frac{1}{a} \leq 0$. Note that since, by definition, $p \geq 0$, the eccentricity, e, can also be used to determine if the orbit is closed ($e < 1$) or open ($e \geq 1$). All practical orbits have $p > 0$.

From everyday observation of celestial bodies, we suspect that a possible two-body trajectory is the *circular* orbit of m_2 about m_1, implying a constant radius, r. Let us see whether a circular trajectory satisfies Eq. (4.93) by writing $\mathbf{r} = r\mathbf{i_r}$, where $\mathbf{i_r}$ is the unit vector along $\mathbf{r}$, and by noting that $\mathbf{v} = \frac{\mathrm{d}\mathbf{r}}{\mathrm{d}t} =$

[3] The name *vis viva*, or *living force*, was historically used for the energy integral, as it is responsible for causing motion.

$\boldsymbol{\omega} \times \mathbf{r} = r\omega \times \mathbf{i}_r$, and $\frac{d^2\mathbf{r}}{dt^2} = \boldsymbol{\omega} \times (\boldsymbol{\omega} \times \mathbf{r}) = -\omega^2 \mathbf{r}$, where $\boldsymbol{\omega} = \omega \mathbf{i_h}$ is the angular velocity in the direction of $\mathbf{h}$ (denoted by the unit vector $\mathbf{i_h}$). Upon substitution into Eq. (4.93), we have

$$-\omega^2 \mathbf{r} + \frac{\mu}{r^3}\mathbf{r} = \mathbf{0}\,, \tag{4.108}$$

which would be satisfied if and only if $\omega^2 = \frac{\mu}{r^3}$. This is Kepler's third law for a circular orbit. From the $\mathbf{v}$ expressed above, we have $v = r\omega$, which implies $v^2 = r^2\omega^2 = \frac{\mu}{r}$, and substituted into the energy integral, Eq. (4.107), leads to

$$-\frac{\mu}{2a} = \frac{v^2}{2} - \frac{\mu}{r} = \frac{\mu}{2r} - \frac{\mu}{r} = -\frac{\mu}{2r}\,, \tag{4.109}$$

or, $r = a$. Also, since $\mathbf{h} = \mathbf{r} \times \mathbf{v} = rv\mathbf{i_h}$, it follows that $p = \frac{h^2}{\mu} = \frac{r^2v^2}{\mu} = \frac{r^4\omega^2}{\mu} = r$. Thus, for circular orbit, $r = a = p$, which implies $e = 0$.

Example 4.7. Determine the total change in velocity required to launch a space shuttle from the surface of the earth to achieve a circular orbit of 200 km altitude. What is the period of the satellite in its orbit? For earth, $Gm_1 = 398,600.4\ \text{km}^3/\text{s}^2$ and surface radius $R_0 = 6378.14$ km.

We begin by noting that the shuttle's mass is negligible in comparison with that of the earth. Hence, we can approximate $\mu = G(m_1 + m_2) \approx Gm_1$. Such an assumption is commonly applied in a two-body problem, because usually there is a large difference between the two masses. Thus, the velocity of the shuttle in a circular orbit of radius $r = 6378.14 + 200 = 6578.14$ km relative to the earth's center is calculated as follows:

$$v = \sqrt{\frac{\mu}{r}} = \sqrt{\frac{\mu}{R_0 + h}} = \sqrt{\frac{398600.4}{6578.14}} = 7.7842596\ \text{km/s}.$$

This is the total change in the velocity required for launching the space shuttle into the given circular orbit. A small part of this velocity comes from the earth's rotation about its axis, which we saw in Example 4.1 translating into the velocity of the launch point according to Eq. (4.12). The remaining velocity must be imparted by the engines. The shuttle's orbital period, P, can be calculated from the orbital rotational rate, ω, by

$$P = \frac{2\pi}{\omega} = \frac{2\pi r}{v} = \frac{2\pi(6578.14)}{7.7842596} = 5309.65\ \text{s}\ (88.49\ \text{min}).$$

4.5.1 Geometry of Two-Body Trajectories

In the previous subsection, we were able to derive a special trajectory of the two-body motion, namely a circular orbit where $e = 0$. However, in order to determine the shape of a general trajectory where $e \neq 0$, let us take the scalar product of $\mu\mathbf{e}$ with the position, $\mathbf{r}$, to write

$$r + \mathbf{e} \cdot \mathbf{r} = \frac{1}{\mu}\mathbf{r} \cdot (\mathbf{v} \times \mathbf{h}) = p\,. \tag{4.110}$$

Furthermore, let us define the angle between the vectors, $\mathbf{r}$ and $\mathbf{e}$, as the *true anomaly*, θ, which, when substituted into Eq. (4.110), leads to

$$r = \frac{p}{1 + e\cos\theta}\,, \tag{4.111}$$

defining the shape of the trajectory in polar coordinates, r and θ. There is no change in r if the sign of θ is changed, which imples that the trajectory is symmetrical about $\mathbf{e}$. Furthermore, the minimum separation of the two bodies (called *periapsis*[4]), $r = \frac{p}{1+e}$, occurs when $\theta = 0$, which indicates that $\mathbf{e}$ points toward the periapsis. [The maximum separation of the two bodies (called *apoapsis*), $r = \frac{p}{1-e}$, occurs when $\theta = \pi$ for $e \leq 1$.] Equation (4.111) is called the *orbit equation* because it specifies the shape of the orbit. By plotting Eq. (4.111) (or converting it to Cartesian coordinates), it is clear that the general orbit is a *conic section*, i.e., the shape we get by cutting a right-circular cone in a particular way. In each case, the *focus* of the orbit is at the center of m_1, and the constant a is called the *semi-major axis*, which has a special geometrical significance. For $e < 1$, the orbit is an *ellipse*, with $a > 0$ and $p = a(1 - e^2) > 0$. The circle is a special ellipse with $e = 0$ and $r = a = p$. For $e = 1$, the orbit is a *parabola*, with $\frac{1}{a} = 0$ (or $a = \infty$), and $p \geq 0$. The *rectilinear* trajectory is a special parabola with $p = 0$. For $e > 1$, the orbit is a *hyperbola*, with $a < 0$ and $p = a(1 - e^2) > 0$. Figure 4.11 depicts the three general orbits. In Fig. 4.11, the periapsis location relative to the focus is denoted by $q\mathbf{e}$, where $q = a(\frac{1}{e} - 1)$ for the ellipse and the hyperbola, and $q = \frac{p}{2}$ for the parabola. In each case, the parameter, p, is the length of the *chord* (also called *semi-latus rectum*) drawn from the focus to the trajectory. Since the chord is perpendicular to the major axis, the frame having $\mathbf{p}$ and $q\mathbf{e}$ as the two Cartesian axes is a convenient choice of a coordinate system for the two-body motion in the orbital plane. Such a coordinate system is called a *perifocal frame*. Another possible coordinate system is a moving frame with origin at m_2, having mutually perpendicular axes along the radial direction, $\mathbf{i_r}$, and the tangential direction, $\mathbf{i_\theta}$, with the velocity vector making an angle above the local horizon, ϕ, called the *flight path angle* (Fig. 4.11):

$$\mathbf{v} = v\cos\phi\,\mathbf{i_\theta} + v\sin\phi\,\mathbf{i_r}\,. \tag{4.112}$$

The moving polar frame, $(\mathbf{i_r},\ \mathbf{i_\theta})$, is used to represent $\mathbf{v}$ in terms of $\mathbf{r}$. By taking the vector product of the angular momentum vector with $\mu\mathbf{e}$, we have

$$\mu\mathbf{h} \times \mathbf{e} = \mathbf{h} \times (\mathbf{v} \times \mathbf{h}) - \frac{\mu}{r}\mathbf{h} \times \mathbf{r}, \tag{4.113}$$

which leads to

[4] The general *apsis* in periapsis (and apoapsis) is replaced by a more specific *gee* when m_1 is the earth, and *helion* when m_1 is the sun.

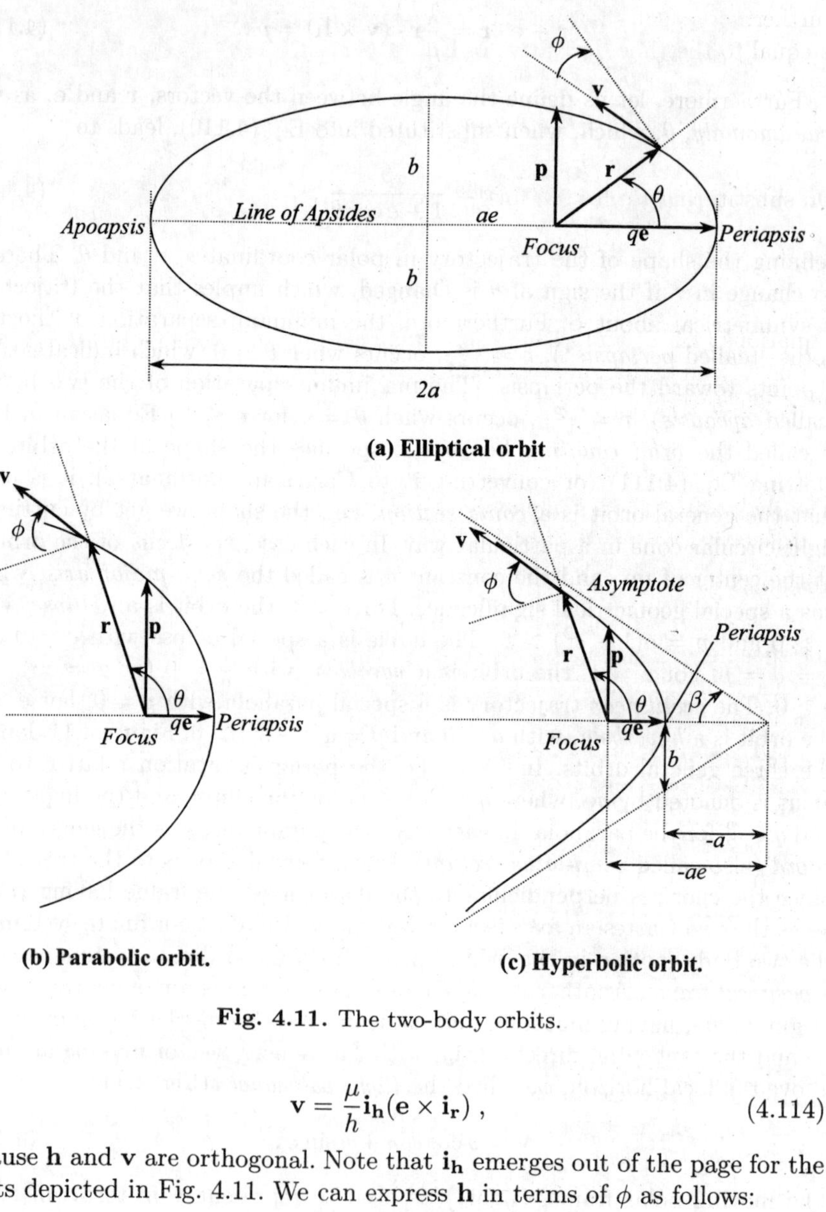

Fig. 4.11. The two-body orbits.

$$\mathbf{v} = \frac{\mu}{h}\mathbf{i_h}(\mathbf{e} \times \mathbf{i_r}) , \tag{4.114}$$

because $\mathbf{h}$ and $\mathbf{v}$ are orthogonal. Note that $\mathbf{i_h}$ emerges out of the page for the orbits depicted in Fig. 4.11. We can express $\mathbf{h}$ in terms of ϕ as follows:

$$\mathbf{h} = \mathbf{r} \times \mathbf{v} = rv\cos\phi\mathbf{i_h} . \tag{4.115}$$

On substituting Eq. (4.111) into Eq. (4.115), with the definition of p given in Eq. (4.106), we have

$$\cos\phi = \frac{\mu(1 + e\cos\theta)}{hv} . \tag{4.116}$$

Furthermore, the radial velocity magnitude is identified from Eq. (4.112) and is equal to the time derivative of Eq. (4.111):

$$\dot{r} = v\sin\phi = \frac{ep\sin\theta}{(1+e\cos\theta)^2}\frac{d\theta}{dt}\,. \tag{4.117}$$

On substituting Eq. (4.98) for $\frac{d\theta}{dt}$, we have

$$\sin\phi = \frac{\mu e\sin\theta}{hv}. \tag{4.118}$$

Equations (4.116) and (4.118) determine ϕ from θ without quadrant ambiguity, or we can divide one by the other, obtaining

$$\tan\phi = \frac{e\sin\theta}{1+e\cos\theta}\,. \tag{4.119}$$

Clearly, positive values of the flight-path angle occur in the range $0 < \theta < \pi$, and negative for $\pi < \theta < 2\pi$.

Example 4.8. A space shuttle in a circular earth orbit of 200 km altitude is *de-boosted* by firing rocket thrusters along the velocity vector, such that the speed instantaneously decreases by 500 m/s. Neglecting the effects of the earth's atmosphere, calculate the speed and flight-path angle when the shuttle reaches an altitude of 100 km.

In Example 4.7, we calculated the circular orbital speed to be 7.7842596 km/s. The speed immediately after de-boost is $v = 7.2842596$ km/s. The resulting trajectory is no longer circular, but an ellipse with the point of de-boost being an *apogee*, because it has $\phi = 0$ and a decreasing radius thereafter. The trajectory is defined by a constant angular momentum, $h = rv = (6578.14)(7.2842596) = 47916.8795\ \text{km}^2/\text{s}$, which implies $p = \frac{h^2}{\mu} = \frac{47916.8795^2}{398600.4} = 5760.223$ km, $e = 1 - \frac{p}{a(1+e)} = 1 - \frac{5760.223}{6578.14} = 0.124338588$, and $a = \frac{6578.14}{1+e} = 5850.6753$ km. The speed at 100 km altitude is calculated using the energy integral, Eq. (4.107), as follows:

$$v = \sqrt{\frac{2\mu}{r} - \frac{\mu}{a}} = \sqrt{398600.4\left(\frac{2}{6378.14+100} - \frac{1}{5850.6753}\right)}$$
$$= 7.4115573\ \text{km/s},$$

from which it follows that $\cos\phi = \frac{h}{rv} = \frac{47916.8795}{(6478.14)(7.4115573)} = 0.99799583$. Due to quadrant ambiguity involved in cosine inverse, we also require $\sin\phi$ from Eq. (4.118). The true anomaly at 100 km altitude is calculated from the orbit equation as $\cos\theta = \frac{p}{er} - \frac{1}{e} = -0.89128738$, which leads to $\theta = \pm 153.0355°$. Both values of θ are theoretically possible. However, $\theta = 153.0355°$ implies a return from the *perigee*, which is impractical, because the perigee radius is $r = a(1-e) = 5123.211$ km, a value less than the earth's radius, implying a violation of the constraint that the altitude is always nonnegative. Therefore, this earth intersecting trajectory has $\theta = -153.0355° = 206.96453°$ at 100 km altitude. From Eq. (4.118), we get $\sin\phi = \frac{\mu e \sin\theta}{hv} = -0.06332201$, which fixes the flight-path angle at $\phi = -3.628084°$.

Example 4.9. A spacecraft is tracked at an altitude of 5000 km and flight-path angle 10°, to be moving at speed 10 km/s relative to a nonrotating earth. Determine the trajectory of the spacecraft.

We begin by calculating the semi-major axis from the energy integral as follows:

$$a = -\frac{\mu}{2\varepsilon} = \left(\frac{2}{r} - \frac{v^2}{\mu}\right)^{-1}$$
$$= \left(\frac{2}{6378.14 + 5000} - \frac{100}{398600.4}\right)^{-1} = -13,315.1949\,\text{km}.$$

The negative semi-major axis indicates a hyperbolic orbit. The angular momentum is calculated next as $h = rv\cos\phi = (6378.14 + 5000)(10)\cos 10° = 112,052.8049\,\text{km}^2/\text{s}$, from which the parameter follows, $p = \frac{h^2}{\mu} = 31,499.7955$ km, leading to $e = \sqrt{1 - \frac{p}{a}} = 1.8345852$. The orbit's shape in its plane is completely determined by any suitable pair of constants, (a, e), (a, p), (p, e), (h, a), etc. An interesting orbital parameter is the angle of the asymptote, β. From Fig. 4.11(c) and the orbit equation, it is evident that $\beta = \pi - \lim_{r\to\infty}\theta = \pi - \cos^{-1}\frac{-1}{e} = \cos^{-1}\frac{1}{e} = 56.9697°$. Hence, the object will depart the earth's gravity ($\lim_{r\to\infty}$) along the asymptote inclined at 56.9697° to the line of apsides. The speed of departure at infinity, called *hyperbolic excess speed* v_∞, can also be calculated by $v_\infty = \lim_{r\to\infty}\sqrt{\frac{2\mu}{r} - \frac{\mu}{a}} = \sqrt{-\frac{\mu}{a}} = 5.4713576\,\text{km/s}$. The hyperbolic excess velocity is vectorially added to the earth's velocity relative to the sun, in order to determine the speed at which a spacecraft is launched on an interplanetary voyage. The lowest velocity of departure at a radius, r, is obtained for the parabolic trajectory ($e = 1$) and is called the *parabolic escape velocity*, $v_{\text{esc}} = \sqrt{\frac{2\mu}{r}}$.

4.5.2 Lagrange's Coefficients

The two-body trajectory expressed in the perifocal frame with Cartesian axes along **e**, **p**, and **h** (represented by the unit vectors $\mathbf{i_e}$, $\mathbf{i_p}$, and $\mathbf{i_h}$, respectively) is the following:

$$\mathbf{r} = r\cos\theta\mathbf{i_e} + r\sin\theta\mathbf{i_p},$$
$$\mathbf{v} = v(\sin\phi\cos\theta - \cos\phi\sin\theta)\mathbf{i_e} + v(\sin\phi\sin\theta + \cos\phi\cos\theta)\mathbf{i_p}. \quad (4.120)$$

The second of Eqs. (4.120) is directly obtained from Eq. (4.114). On eliminating the flight-path angle with the use of Eqs. (4.116) and (4.118), we have

$$\mathbf{v} = -\frac{\mu}{h}\sin\theta \mathbf{i_e} + \frac{\mu}{h}(e+\cos\theta)\mathbf{i_p}\,. \tag{4.121}$$

Given the position and velocity at some time, t_0, we would like to determine the position and velocity at some other time, t. In order to do so, we write the known position and velocity as

$$\begin{aligned}\mathbf{r_0} &= r_0\cos\theta_0\mathbf{i_e} + r_0\sin\theta_0\mathbf{i_p},\\ \mathbf{v_0} &= -\frac{\mu}{h}\sin\theta_0\mathbf{i_e} + \frac{\mu}{h}(e+\cos\theta_0)\mathbf{i_p},\end{aligned} \tag{4.122}$$

or, in the matrix equation form as

$$\begin{Bmatrix}\mathbf{r_0}\\ \mathbf{v_0}\end{Bmatrix} = \begin{pmatrix} r_0\cos\theta_0 & r_0\sin\theta_0\\ -\frac{\mu}{h}\sin\theta_0 & \frac{\mu}{h}(e+\cos\theta_0)\end{pmatrix}\begin{Bmatrix}\mathbf{i_e}\\ \mathbf{i_p}\end{Bmatrix}. \tag{4.123}$$

Since the square matrix in Eq. (4.123) is nonsingular (its determinant is equal to h), we can invert the matrix to obtain $\mathbf{i_e}$ and $\mathbf{i_p}$ as follows:

$$\begin{Bmatrix}\mathbf{i_e}\\ \mathbf{i_p}\end{Bmatrix} = \begin{pmatrix} \frac{1}{p}(e+\cos\theta_0) & -\frac{r_0}{h}\sin\theta_0\\ \frac{1}{p}\sin\theta_0 & \frac{r_0}{h}(e+\cos\theta_0)\end{pmatrix}\begin{Bmatrix}\mathbf{r_0}\\ \mathbf{v_0}\end{Bmatrix}. \tag{4.124}$$

On substituting Eq. (4.124) into Eqs. (4.120) and (4.121), we have

$$\begin{Bmatrix}\mathbf{r}\\ \mathbf{v}\end{Bmatrix} = \begin{pmatrix} r\cos\theta & r\sin\theta\\ -\frac{\mu}{h}\sin\theta & \frac{\mu}{h}(e+\cos\theta)\end{pmatrix}\begin{pmatrix} \frac{1}{p}(e+\cos\theta_0) & -\frac{r_0}{h}\sin\theta_0\\ \frac{1}{p}\sin\theta_0 & \frac{r_0}{h}(e+\cos\theta_0)\end{pmatrix}\begin{Bmatrix}\mathbf{r_0}\\ \mathbf{v_0}\end{Bmatrix}, \tag{4.125}$$

or,

$$\begin{Bmatrix}\mathbf{r}\\ \mathbf{v}\end{Bmatrix} = \begin{pmatrix} f & g\\ \dot{f} & \dot{g}\end{pmatrix}\begin{Bmatrix}\mathbf{r_0}\\ \mathbf{v_0}\end{Bmatrix}, \tag{4.126}$$

where

$$\begin{aligned} f &\doteq 1 + \frac{r}{p}[\cos(\theta-\theta_0) - 1],\\ g &\doteq \frac{rr_0}{h}\sin(\theta-\theta_0),\\ \dot{f} &\doteq \frac{\mathrm{d}f}{\mathrm{d}t} = -\frac{h}{p^2}[\sin(\theta-\theta_0) + e(\sin\theta - \sin\theta_0)],\\ \dot{g} &\doteq \frac{\mathrm{d}g}{\mathrm{d}t} = 1 + \frac{r_0}{p}[\cos(\theta-\theta_0) - 1].\end{aligned} \tag{4.127}$$

The functions f and g were first derived by Lagrange and are thus called *Lagrange's coefficients*. They are very useful in determining the trajectory from a known location and velocity. The matrix

$$\Phi(t,t_0) \doteq \begin{pmatrix} f & g\\ \dot{f} & \dot{g}\end{pmatrix}, \tag{4.128}$$

called the *state transition matrix*, has a special significance because it uniquely determines the *current state*, $(\mathbf{r}, \mathbf{v})$, from the *initial state*, $(\mathbf{r_0}, \mathbf{v_0})$, according to Eq. (4.126). Such a relationship between the initial and final states is rarely possible for the solution to a nonlinear differential equation and is thus a valuable property of the two-body problem. The state-transition matrix is seen to have the following properties:

1. From the conservation of angular momentum,

$$\mathbf{h} = \mathbf{r} \times \mathbf{v} = (f\dot{g} - g\dot{f})\mathbf{r_0} \times \mathbf{v_0} = \mathbf{r_0} \times \mathbf{v_0}, \tag{4.129}$$

it follows that

$$\mid \mathbf{\Phi} \mid = f\dot{g} - g\dot{f} = 1. \tag{4.130}$$

A consequence of the unity determinant is that the inverse of the state-transition matrix is given by

$$\mathbf{\Phi}^{-1} \doteq \begin{pmatrix} \dot{g} & -g \\ -\dot{f} & f \end{pmatrix}. \tag{4.131}$$

Such a matrix is said to be *symplectic*.

2. Given any three points $(t_0,\ t_1,\ t_2)$ along the trajectory, it is true that

$$\mathbf{\Phi}(t_2, t_0) = \mathbf{\Phi}(t_2, t_1)\mathbf{\Phi}(t_1, t_0). \tag{4.132}$$

Example 4.10. Let us calculate the perifocal position and velocity of the space shuttle of Example 4.8 at 100 km altitude, given the position and velocity at the de-boost point. We begin by computing Lagrange's coefficients for $\theta = 206.96453°$, $\theta_0 = 180°$, $r_0 = 6578.14$ km, and $r = 6478.14$ km in a trajectory defined by $e = 0.124338588$ and $p = 5760.223$ km:

$$\begin{aligned} f &= 1 + \frac{r}{p}[\cos(\theta - \theta_0) - 1] = 0.8777381487, \\ g &= \frac{rr_0}{h}\sin(\theta - \theta_0) = 403.258618476, \\ \dot{f} &= -\frac{h}{p^2}[\sin(\theta - \theta_0) + e(\sin\theta - \sin\theta_0)] = -0.000573409435, \\ \dot{g} &= 1 + \frac{r_0}{p}[\cos(\theta - \theta_0) - 1] = 0.875850850. \end{aligned} \tag{4.133}$$

These computations are verified by ensuring that $f\dot{g} - g\dot{f} = 1$. Since the point of de-boost is the apogee, $\mathbf{r_0} = -6578.14\mathbf{i_e}$ km and $\mathbf{v_0} = -7.2842596\mathbf{i_p}$ km/s. Therefore, we have

$$\begin{aligned} \mathbf{r} &= f\mathbf{r_0} + g\mathbf{v_0} = -5773.88443\mathbf{i_e} - 2937.44045\mathbf{i_p} \text{ km}, \\ \mathbf{v} &= \dot{f}\mathbf{r_0} + \dot{g}\mathbf{v_0} = 3.7719675388\mathbf{i_e} - 6.3799249319\mathbf{i_p} \text{ km/s}. \end{aligned}$$

Using MATLAB, we can verify our calculations of $\mathbf{r}$ and $\mathbf{v}$ as follows:

```
>> R=[-f*r0 -g*v0 0]',V=[-fd*r0 -gd*v0 0]'% r, v vectors from Lagrangian coeffs.
>> r=norm(R)
   r = 6.478139991426316e+003
>> v=norm(V)
   v = 7.41155727563704
>> costheta=dot(R,[1 0 0]')/r %cosine of true anomaly
   costheta =  -0.8912873796l147
>> H=cross(R,V) %angular momentum
   H =  1.0e+004 *
                    0
                    0
        4.79168792196855
>> cosphi=norm(H)/(norm(R)*norm(V)) %cosine of flight-path angle
   cosphi =    0.99799583123837
>> sinphi=dot(R,V)/(norm(R)*norm(V)) %sine of flight-path angle
   sinphi =  -0.06327970315068
```

These values agree with those calculated in Example 4.8.

As seen above, the vectors $\mathbf{h}$ and $\mathbf{e}$ (or $\mathbf{r_0}$ and $\mathbf{v_0}$) completely determine the shape and orientation of a two-body trajectory but provide only five of the six required scalar integrals of motion. The missing information is the location of m_2 along the trajectory at a particular *time*. Hence, the sixth integral of motion can be regarded as the value of the true anomaly θ (thus r) at a given time. A convenient choice of the remaining constant is the *time of periapsis*, τ, which fixes the time when $\theta = 0$. On substituting the orbit equation, Eq. (4.111), into Eq. (4.98), we have

$$\frac{\mathrm{d}\theta}{(1+e\cos\theta)^2} = \sqrt{\frac{\mu}{p^3}}\mathrm{d}t\,. \tag{4.134}$$

The integration of Eq. (4.134) provides the integration constant, τ, thereby determining the variation of θ with time, and completing the solution to the two-body problem. Attempts to carry out this integration in a closed form have occupied the minds of the greatest mathematicians (Euler, Lagrange, Laplace, Bessel, Fourier, Gauss, Cauchy, Leibnitz, and Newton) for more than the past three centuries, and have led to significant developments in all branches of mathematics. The integrated form of Eq. (4.134) for an elliptical orbit is called *Kepler's equation.*[5]

4.5.3 Kepler's Equation for Elliptical Orbit

When the two-body trajectory is elliptical, Eq. (4.134) can be written as follows:

$$\frac{(1-e^2)^{\frac{3}{2}}\mathrm{d}\theta}{(1+e\cos\theta)^2} = n\mathrm{d}t, \tag{4.135}$$

where

$$n \doteq \sqrt{\frac{\mu}{a^3}}, \tag{4.136}$$

[5] An interesting historical perspective on the solution to Kepler's equation is found in Battin [11].

is referred to as the *mean motion*. It is clear from Eq. (4.135) that the true anomaly, θ, does *not* vary uniformly with time. However, as seen above, when $e = 0$, the motion is uniformly circular, with the frequency given by n. It remains to be seen whether the mean motion, n, represents the frequency of the periodic motion when $e \neq 0$. Let us introduce an angle called the *eccentric anomaly*, E, defined by

$$\cos E \doteq \frac{e + \cos\theta}{1 + e\cos\theta}. \tag{4.137}$$

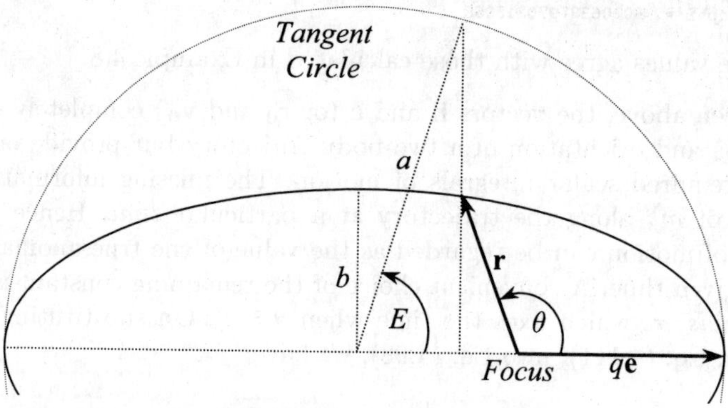

Fig. 4.12. The eccentric anomaly, E.

The geometric significance of the eccentric anomaly is shown in Fig. 4.12, where E describes a corresponding motion in a tangent circle of radius a, sharing the ellipse's center (*not the focus*). Note that $E = 0$ when $\theta = 0$, and $E = \pi$ when $\theta = \pi$. Also, $r = a(1 - e\cos E)$. On differentiating Eq. (4.137) with time, we have

$$\sin E \mathrm{d}E = \frac{(1 - e^2)\sin\theta \mathrm{d}\theta}{(1 + e\cos\theta)^2}. \tag{4.138}$$

But

$$\sin E = \sqrt{1 - \cos^2 E} = \frac{\sqrt{1 - e^2}\sin\theta}{1 + e\cos\theta}. \tag{4.139}$$

Furthermore, from Eq. (4.137) it is clear that

$$1 + e\cos\theta = \frac{1 - e^2}{1 - e\cos E}. \tag{4.140}$$

Therefore, on substituting Eqs. (4.139) and (4.140) into Eq. (4.138), we have

$$(1 - e\cos E)\mathrm{d}E = n\mathrm{d}t. \tag{4.141}$$

Upon integrating from the periapsis ($t = \tau$, $E = 0$) to the present location (t, E), we derive the famous *Kepler's equation*, written as

$$E - e \sin E = n(t - \tau) . \tag{4.142}$$

Note that τ is an integral of motion, which, along with a and e (or any other pair of scalar constants), completely determines the trajectory in the plane of motion. It is clear from Kepler's equation, Eq. (4.142), that the time taken to complete one revolution (the period of motion, P) involves $E = 2\pi$ and is given by

$$P = \frac{2\pi}{n} \doteq 2\pi \sqrt{\frac{a^3}{\mu}} , \tag{4.143}$$

which confirms Kepler's third law of motion ($P^2 \propto a^3$). Hence, the mean motion, n, is indeed the frequency of the elliptical orbit.

The eccentric anomaly is related to the true anomaly (without quadrant ambiguity) using the half-angle trigonometric identities of Eqs. (4.137) and (4.139), written as follows:

$$\tan \frac{\theta}{2} = \sqrt{\frac{1+e}{1-e}} \tan \frac{E}{2} . \tag{4.144}$$

Note that $\frac{E}{2}$ and $\frac{\theta}{2}$ are always in the same quadrant.

Example 4.11. Calculate the time taken by the space shuttle in Example 4.8 to reach the 100 km altitude after de-boost from the 200 km altitude circular orbit.

In Example 4.8 we had calculated the true anomaly at 100 km altitude to be $\theta = 206.96453°$, which, according to Eq. (4.144), implies $E = 3.67212805663$ rad (210.39744°). With a semi-major axis, $a = 5850.6753$ km, the mean motion is calculated to be $n = \sqrt{\frac{398600.4}{5850.6753^3}} = 0.001410782$ rad/s. Using Kepler's equation, the time since perigee is obtained as $t - \tau = \frac{E - e \sin E}{n} = 2647.4984$ s, and the time since de-boost at apogee is thus $t - \tau - \frac{P}{2} = 2647.4984 - \frac{\pi}{0.001410782} = 420.6534$ s.

Usually, the right-hand side of Kepler's equation is expressed as another angle, called the *mean anomaly*, $M = n(t - \tau)$, which results in

$$E - e \sin E = M . \tag{4.145}$$

Kepler's equation is a transcendental equation in E, whose solution in a closed form has beaten the greatest minds in the last three centuries. Given time, t (or mean anomaly, M), obtaining the location, E, from the Kepler's equation solution is thus a nontrivial task, requiring a numerical approximation. Newton was the first to present such an approximation, employing a Taylor series expansion of the function $f(E) = E - e \sin E - M$, as follows:

$$f(E + \Delta E) = \sum_{k=0}^{\infty} f^{(k)}(E) \frac{(\Delta E)^k}{k!} , \tag{4.146}$$

where $f^{(k)} \doteq \frac{\mathrm{d}^k f(E)}{\mathrm{d}E^k}$. When this infinite series is approximated by a finite number of terms, the accuracy of the approximation depends upon the step size, ΔE, as well as on the number of retained terms. When only the first two terms of the infinite series are retained, Newton's approximation results are given by

$$f(E+\Delta E) \simeq f(E) + f^{(1)}(E)(\Delta E). \tag{4.147}$$

With the help of Eq. (4.147), *Newton's method*[6] of solving Kepler's equation can be summarized as follows:

1. Given a mean anomaly, M, guess an initial value for the eccentric anomaly, E. (A good starting guess is $E = M + e\sin M$, although other, more refined estimates are possible.[7])
2. Calculate the change required in the value of E so that $f(E+\Delta E) = 0$, using

$$\Delta E = -\frac{f(E)}{f^{(1)}(E)} = \frac{-E + e\sin E + M}{1 - e\cos E}. \tag{4.148}$$

3. Update E, using $E = E + \Delta E$.
4. Calculate $f(E) = E - e\sin E - M$.
5. If the magnitude of $f(E)$ is less than or equal to a preselected small number, δ, called the *tolerance*, then the new value of E is acceptable. Otherwise, go back to step 2, and determine a new change in E.

This scheme normally converges in a few iterations to a very small tolerance. A larger number of iterations may be necessary when the eccentricity, e, is close to unity or when M is small. The number of iterations also depends upon the initial guess for E in the first step. It is rare for the scheme to require more than six iterations, even in extreme cases. One can easily write a program in MATLAB for solving Kepler's equation using Newton's method, as demonstrated in the following example.

Example 4.12. Determine the position and velocity of the space shuttle in Example 4.8 nine minutes after de-boost, neglecting the effects of the earth's atmosphere.

We begin by calculating the mean anomaly nine minutes after de-boost (apogee) as $M = n(t-\tau) = \pi + (0.001410782)(9)(60) = 3.903414864528$ rad. Next, we write a program called *kepler.m*, for the solution of Kepler's equation using Newton's method, which is tabulated in Table 4.2, and is executed as follows:

```
>> E=kepler(e,M)
     fE =   -2.089455640508220e-006
     i  =        1
     fE =   -1.421085471520200e-013
     i  =        2
     E  =   3.8249121296282
```

[6] Newton's method, when applied to the solution of a general transcendental equation, is called the *Newton–Raphson* method.

[7] See, for example, page 45 of Chobotov [9].

The specified tolerance, $\delta = 1 \times 10^{-10}$, is thus met in two iterations, with the eccentric anomaly calculated to be $E = 3.82491121296282$ rad (219.15127°). The true anomaly is then calculated according to Eq. (4.144) to be $\theta = 3.74977535060384$ rad (214.84630°). Then, by using the orbit equation, we may calculate the radius as $r = \frac{p}{1+e\cos\theta} = 6414.8109$ km, which implies an altitude of 36.6709 km. The position vector is uniquely determined by (r, θ). The speed at this point is calculated using the energy integral as $v = \sqrt{398600.14(\frac{2}{6414.8109} - \frac{1}{5850.6753})} = 7.493068197$ km/s. The flight-path angle is obtained from Eq. (4.119) to be $\phi = \tan^{-1}\frac{e\sin\theta}{1+e\cos\theta} = -0.07895315$ rad (−4.52368239°). The velocity vector is uniquely determined by (v, ϕ, θ).

Table 4.2. M-file *kepler.m* for Solving Kepler's Equation

```
function E=kepler(e,M)
%(c) 2006 Ashish Tewari
E=M+e*sin(M);
fE=E-e*sin(E)-M;
fpE=1-e*cos(E);
dE=-fE/fpE;
E=E+dE;
eps=1e-10; %tolerance
i=0;
while abs(fE)>eps
fE=E-e*sin(E)-M
fpE=1-e*cos(E);
dE=-fE/fpE;
E=E+dE;
i=i+1 %iteration number
end
```

The perifocal frame $(\mathbf{i_e}, \mathbf{i_p}, \mathbf{i_h})$ is a convenient axes system for representing the elliptical trajectory using Lagrange's coefficients of Eq. (4.127). We can directly solve the problem in Example 4.12 by determining $\mathbf{r}$ and $\mathbf{v}$ in the perifocal frame. This requires first determining the elliptical trajectory from $\mathbf{r_0}$ and $\mathbf{v_0}$, and then calculating the true anomaly from the trajectory parameters a, e, and τ, and Kepler's equation. However, the position and velocity vectors in an elliptical orbit can be expressed directly in terms of the eccentric anomaly, thereby eliminating the step requiring the calculation of the true anomaly. By substituting Eqs. (4.137) and (4.139) into Eqs. (4.120) and (4.121) for $\mathbf{r}$ and $\mathbf{v}$ in the perifocal frame, we have

$$\mathbf{r} = a(\cos E - e)\mathbf{i_e} + \sqrt{ap}\sin E\mathbf{i_p},$$
$$\mathbf{v} = -\frac{\sqrt{\mu a}}{r}\sin E\mathbf{i_e} + \frac{\sqrt{\mu p}}{r}\cos E\mathbf{i_p}. \tag{4.149}$$

Given the position and velocity at some time, t_0, we can employ the Lagrange coefficients (Eq. (4.127)) expressed as functions of the eccentric anomaly to determine the current position and velocity:

$$f \doteq 1 + \frac{a}{r_0}[\cos(E - E_0) - 1],$$

$$g \doteq \frac{a\alpha_0}{\mu}[1 - \cos(E - E_0)] + r_0\sqrt{\frac{a}{\mu}}\sin(E - E_0),$$

$$\dot{f} \doteq \frac{\mathrm{d}f}{\mathrm{d}t} = -\frac{\sqrt{\mu a}}{r r_0}\sin(E - E_0), \tag{4.150}$$

$$\dot{g} \doteq \frac{\mathrm{d}g}{\mathrm{d}t} = 1 + \frac{a}{r}[\cos(E - E_0) - 1], \tag{4.151}$$

where $\alpha_0 = \mathbf{r_0} \cdot \mathbf{v_0}$. Note that Eqs. (4.150) involve the *difference* in the eccentric anomaly, $E - E_0$, which is convenient.

Example 4.13. Determine the perifocal trajectory of the space shuttle in Example 4.8 from de-boost until earth impact, neglecting the effects of the earth's atmosphere, and assuming no further maneuvering.

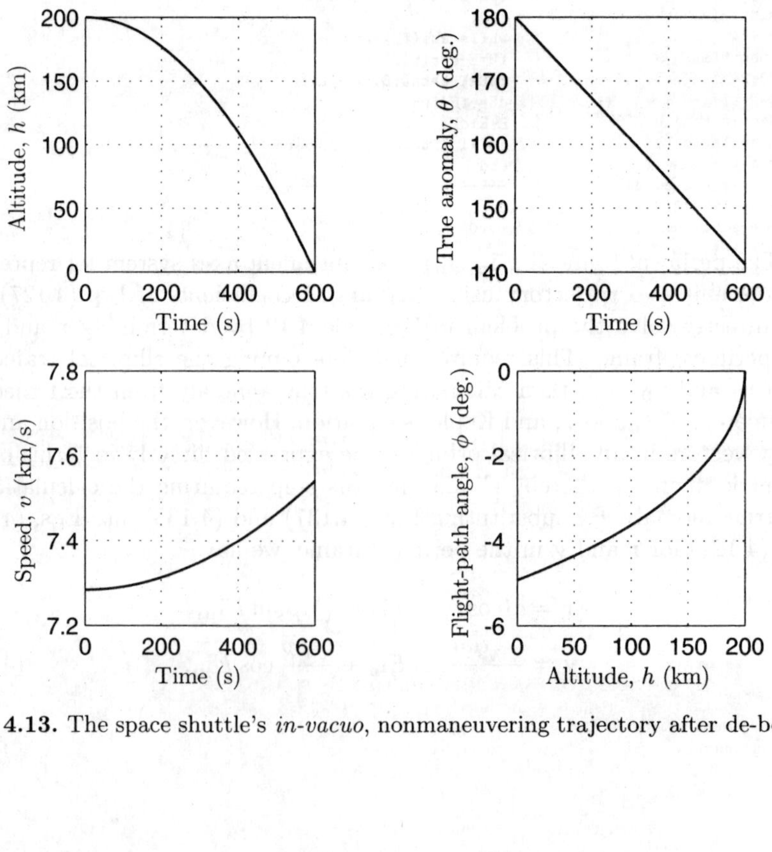

Fig. 4.13. The space shuttle's *in-vacuo*, nonmaneuvering trajectory after de-boost.

Table 4.3. M-file *trajE.m* for Determining Elliptical Trajectory

```
% This code requires 'kepler.m'
% (c) 2006 Ashish Tewari
function [R,V]=trajE(mu,t0,R0,V0,t)
eps=1e-10;
r0=norm(R0);
v0=norm(V0);
alpha=dot(R0,V0);
H=cross(R0,V0);
h=norm(H);
p=h*h/mu;
ecv=cross(V0,H)/mu-R0/r0;
e=norm(ecv);
ecth0=p/r0-1;
esth0=norm(cross(ecv,R0))/r0;
if abs(ecth0)>=eps;
th0=atan(esth0/ecth0);
if ecth0<0
    if esth0>=0;
    th0=th0+pi;
    end
elseif esth0<0
    th0=th0+2*pi;
end
elseif esth0>=0
    th0=pi/2;
else
    th0=3*pi/2;
end
ainv=-(v0*v0)/mu+2/r0;
a=1/ainv;
n=sqrt(mu/a^3);
E0=2*atan(sqrt((1-e)/(1+e))*tan(0.5*th0));
tau=t0+(-E0+e*sin(E0))/n;
M=n*(t-tau);
E=kepler(e,M);
r=a*(1-e*cos(E));
f=1+a*(cos(E-E0)-1)/r0;
g=a*alpha*(1-cos(E-E0))/mu+r0*sqrt(a/mu)*sin(E-E0);
fd=-sqrt(mu*a)*(sin(E-E0))/(r*r0);
gd=1+a*(cos(E-E0)-1)/r;
R=f*R0+g*V0;
V=fd*R0+gd*V0;
```

We begin by writing a MATLAB program for determining the elliptical trajectory from the initial position and velocity, employing Lagrange's coefficients according to Eqs. (4.150). Such a program is tabulated in Table 4.3. Next, the program is executed and the trajectory plotted using the following MATLAB statements:

```
>> mu=398600.4;
>> r=[-6578.14,0,0]';v=[0,-sqrt(398600.4/6578.14)+0.5,0]';%initial position velocity
>> t=0; i=0;
>> dt=1; %time step size (1 s)
>> z=norm(r)-6378.14; %altitude (km)
>>  while z>=0;[r,v]=trajE(mu,t,r,v,t+dt);i=i+1;t=t+dt;...
    z=norm(r)-6378.14;...
    R(:,i)=norm(r);V(:,i)=norm(v);T(:,i)=t;...
    costheta=dot(r,[1 0 0]')/norm(r);...
    theta=acos(costheta);...
    cosphi=norm(cross(r,v))/(norm(r)*norm(v));...
    sinphi=dot(r,v)/(norm(r)*norm(v));...
    phi=asin(sinphi);...
```

```
    th(:,i)=theta;ph(:,i)=phi;...
    end
>> subplot(221),plot(T,R-6378.14),grid,hold on,...%altitude vs. time
subplot(222),plot(T,th*180/pi),grid,...%true anomaly vs. time
hold on, subplot(223),plot(T,V),grid,...%speed vs. time
hold on,subplot(224),plot(R-6378.14,ph*180/pi),grid%flt.-path angle vs. altitude
```

The resulting plots are shown in Fig. 4.13.

4.5.4 Position and Velocity in a Hyperbolic Trajectory

An analog of Kepler's equation can be obtained for hyperbolic orbits by introducing the *hyperbolic anomaly*, H, and the hyperbolic functions. The parametric equation of a hyperbola centered at $x = 0,\ y = 0$, with semi-major axis a, can be written as $x = a\cosh H$, or $r\cos\theta = a(\cosh H - e)$. It follows that the magnitude of the radius vector is $r = a(1 - e\cosh H)$; consequently, the relationship between the hyperbolic anomaly and the true anomaly is given by

$$\cos\theta = \frac{\cosh H - e}{1 - e\cosh H}\,, \tag{4.152}$$

or,

$$\sin\theta = \pm\frac{\sqrt{e^2-1}\sinh H}{1 - e\cosh H}\,. \tag{4.153}$$

Both signs are possible for the term on the right-hand side of Eq. (4.153). If we take the negative sign, we have

$$\sin\theta = -\frac{\sqrt{e^2-1}\sinh H}{1 - e\cosh H}\,, \tag{4.154}$$

and we can write the second parametric equation for the hyperbola as $y = -b\sinh H$, where $b = a\sqrt{e^2-1}$ is the semi-minor axis. Another relationship between H and θ can be obtained by using the half-angle identities:

$$\tan\frac{\theta}{2} = \sqrt{\frac{1+e}{e-1}}\tanh\frac{H}{2}\,. \tag{4.155}$$

Employing Eqs. (4.120) and the foregoing relations, we can write the position and velocity in a hyperbolic orbit of eccentricity e as follows:

$$\begin{aligned}\mathbf{r} &= a(\cosh H - e)\mathbf{i_e} - \sqrt{-ap}\sinh H\mathbf{i_p}\,,\\ \mathbf{v} &= -\frac{\sqrt{-\mu a}}{r}\sinh H\mathbf{i_e} + \frac{\sqrt{\mu p}}{r}\cosh H\mathbf{i_p}\,.\end{aligned} \tag{4.156}$$

Upon differentiating Eq. (4.152), we have

$$\mathrm{d}\theta = \frac{\sin\theta}{\sinh H}\mathrm{d}H = -\frac{b}{r}\mathrm{d}H\,. \tag{4.157}$$

From the law of areas, we have

$$r^2 \mathrm{d}\theta = h\mathrm{d}t \,. \tag{4.158}$$

On substituting Eq. (4.157) into Eq. (4.158), we can write

$$(e \cosh H - 1)\mathrm{d}H = \sqrt{-\frac{\mu}{a^3}}\mathrm{d}t \,, \tag{4.159}$$

which, upon integration from $H = 0, t = \tau$, yields

$$e \sinh H - H = n(t - \tau) \,, \tag{4.160}$$

where the *hyperbolic mean motion, n*, is given by

$$n = \sqrt{-\frac{\mu}{a^3}} \,. \tag{4.161}$$

Equation (4.160) is the analog of Kepler's equation for a hyperbolic trajectory[8], and can be solved numerically using Newton's method in a manner similar to Kepler's equation.

Example 4.14. Determine the perifocal position and velocity of the spacecraft in Example 4.9, 10 hours after its altitude, flight-path angle, and speed relative to earth are observed to be 5000 km, 10°, and 10 km/s, respectively.

In Example 4.9, the trajectory parameters for the spacecraft were determined from the initial position and velocity to be $a = -13,315.1949$ km, $p = 31,499.7955$ km, and $e = 1.8345852$. Hence, the hyperbolic anomaly for the initial position is obtained as follows:

$$H_0 = \cosh^{-1}\left(\frac{1}{e} - \frac{r_0}{ae}\right) = 0.147296252 \,\text{rad},$$

$$n = \sqrt{-\frac{\mu}{a^3}} = 4.109108 \times 10^{-4} \,\text{rad/s},$$

which yields the time of periapsis as follows:

$$\tau = \frac{H_0 - e \sinh H_0}{n} = -301.54837 \,\text{s},$$

and the mean anomaly is calculated as $M = n(t - \tau) = 4.109108 \times 10^{-4}(10 * 3600 + 301.54837) = 14.91669882225$ rad.

Now we are ready to solve the hyperbolic analog of Kepler's equation using the MATLAB file *keplerhyp.m* listed in Table 4.4. Note that in this program, we have used the mean anomaly to be the initial guess for the hyperbolic anomaly. Other, more sophisticated techniques are available[9] to make the initial guess for H such that the iterations for a given tolerance are minimized. The program *keplerhyp.m* is executed, resulting in the following:

---

[8] If we take the positive sign in Eq. (4.153), we would get $H - e \sinh H = n(t - \tau)$.

[9] For example, see Vallado [8], pp. 78–83.

```
>>H=keplerhyp(e,M)
i=1 fH = 1.014952224618536e+006 H = 12.91673669562873
i=2 fH = 3.733730704786804e+005 H = 11.91680855810042
i=3 fH = 1.373495542042516e+005 H = 10.91699660866798
i=4 fH = 5.052161961527040e+004 H = 9.9174879137365O
i=5 fH = 1.857967289437311e+004 H = 8.91876908491407
i=6 fH = 6.829154325531749e+003 H = 7.92210179750439
i=7 fH = 2.506650999181199e+003 H = 6.93073915098228
i=8 fH = 9.167713602887050e+002 H = 5.95297547031013
i=9 fH = 3.321910110827640e+002 H = 5.00942645299948
i=10    fH = 1.174954146079913e+002 H = 4.14823643033086
i=11    fH = 39.00559087243553        H = 3.46512025582047
i=12    fH = 10.92479558757285        H = 3.07995585735341
i=13    fH = 1.91912604037694         H = 2.97894975852874
i=14    fH = 0.09824238098464         H = 2.97320027715255
i=15    fH = 2.968357362309604e-004 H = 2.97318279979355
i=16    fH = 2.732342352373962e-009 H = 2.97318279963267
i=17    fH = 0                    H = 2.97318279963267
```

Hence, the hyperbolic anomaly converges to double precision (tolerance less than 10^{-16}) as $H = 2.97318279963267$ rad in 17 iterations. Finally, we compute the perifocal position and velocity as follows:

```
>>R=a*[cosh(H)-e  -sqrt(e^2-1)*sinh(H) 0]' %position vector (km)
R =
   -106095.654013018
   199708.803081263
         0
>> r=norm(R)  %radius (km)
r = 2.261413138473391e+005
>> V=sqrt(-mu*a)*[-sinh(H)  sqrt(e^2-1)*cosh(H) 0]'/r %velocity vector (km/s)
V =
   -3.14146550904969
   4.85717833269184
         0
>> v=norm(V) %speed (km/s)
v =   5.78454725109234
>> theta=acos(a*(cosh(H)-e)/r) %true anomaly (rad.)
theta =   2.05913165393277
>> h=norm(cross(R,V)) %angular momentum magnitude (km^2/s)
h =   1.120528048685832e+005
>> phi=acos(h/(r*v))  %flight-path angle (rad)
phi = 1.48503213738220
```

Table 4.4. M-file *keplerhyp.m* for Determining Hyperbolic Anomaly

```
%(c) 2006 Ashish Tewari
function H=keplerhyp(e,M);
H=M;
fH=e*sinh(H)-H-M;
fpH=e*cosh(H)-1;
dH=-fH/fpH;
H=H+dH;
eps=1e-10;
i=0;
while abs(fH)>eps
i=i+1
fH=e*sinh(H)-H-M
fpH=e*cosh(H)-1;
dH=-fH/fpH;
H=H+dH
end
```

Lagrange's coefficients for a hyperbolic orbit can be easily derived from Eqs. (4.127) to be the following:

$$f = 1 + \frac{a}{r_0}[\cosh(H - H_0) - 1],$$

$$g = t - t_0 - \frac{\sinh(H - H_0) - (H - H_0)}{n},$$

$$\dot{f} = -\frac{\sqrt{-\mu a}}{r r_0}\sinh(H - H_0), \tag{4.162}$$

$$\dot{g} = 1 + \frac{a}{r}[\cosh(H - H_0) - 1]. \tag{4.163}$$

4.5.5 Parabolic Escape Trajectory

Whereas an elliptical orbit represents a closed trajectory with negative relative energy, a hyperbolic trajectory is that of escape from (or arrival at) a planet's gravity with positive relative energy. The boundary between these two trajectories is that of the parabolic trajectory, which has zero relative energy. A parabolic orbit is thus the minimum energy trajectory for escaping the gravitational influence of a planet. While not practical for interplanetary travel (due to a zero velocity at infinite radius), the parabolic trajectory is sometimes a valuable mathematical aid for quickly determining the minimum fuel requirements for a given mission. Equation (4.134) for a parabolic orbit is simply

$$\frac{1}{4}\sec^4\frac{\theta}{2}\mathrm{d}\theta = \sqrt{\frac{\mu}{p^3}}\mathrm{d}t\,, \tag{4.164}$$

which, on integration, results in

$$\tan^3\frac{\theta}{2} + 3\tan\frac{\theta}{2} = 6\sqrt{\frac{\mu}{p^3}}(t - \tau)\,. \tag{4.165}$$

Equation (4.165) represents the parabolic form of Kepler's equation and is called *Barker's equation*. Fortunately, a closed-form, real solution to Barker's equation exists and is unique. It can be obtained by substituting

$$\tan\frac{\theta}{2} = \alpha - \frac{1}{\alpha}\,, \tag{4.166}$$

and solving the resulting quadratic equation for α^3, yielding

$$\tan\frac{\theta}{2} = (C + \sqrt{1 + C^2})^{\frac{1}{3}} - (C + \sqrt{1 + C^2})^{-\frac{1}{3}}\,, \tag{4.167}$$

where

$$C \doteq 3\sqrt{\frac{\mu}{p^3}}(t - \tau) \tag{4.168}$$

is an equivalent mean anomaly.

Example 4.15. Obtain the position and velocity of a spacecraft in a parabolic orbit, 10 hours after its velocity relative to earth is measured as 10 km/s, with the flight-path angle of 10°.

We begin by obtaining the initial radius as $r_0 = \frac{(2)(398600.4)}{10^2} = 7972.008$ km). Next, the angular momentum is calculated to be $h = (10)(7972.008)\cos 10° = 78,508.95286$ km^2/s, which yields the parameter as $p = \frac{78508.95286^2}{398600.4} = 15,463.2451$ km. The initial true anomaly is thus $\theta_0 = \cos^{-1}(\frac{p}{r_0} - 1) = 0.34906585$ rad. From Barker's equation, we can find the time of periapsis as

$$\tau = -\frac{\tan^3\frac{\theta_0}{2} + 3\tan\frac{\theta_0}{2}}{6\sqrt{\frac{\mu}{p^3}}} = -271.29927 \text{ s},$$

and the equivalent mean anomaly is calculated as $C = 3\sqrt{\frac{\mu}{p^3}}(36000 - \tau) = 35.72747082559$ rad. The true anomaly is finally obtained from Eq. (4.167) as

$$\theta = 2\tan^{-1}[(C + \sqrt{1 + C^2})^{\frac{1}{3}} - (C + \sqrt{1 + C^2})^{-\frac{1}{3}}] = 2.64068857 \text{ rad}.$$

We can now calculate the final radius, speed, and flight-path angle as $r = \frac{p}{1+\cos\theta} = 125,869.675$ km, $v = \sqrt{\frac{2\mu}{r}} = 2.5166528$ km/s, and $\phi = \cos^{-1}\frac{h}{rv} = 1.3203442868$ rad (75.65°).

4.6 Summary

A trajectory model is based upon the particle assumption of a flight vehicle where the rotational motion is disregarded. The resulting translational model has only three degrees of freedom. While the laws of motion are valid in an inertial frame, it is often necessary to express the velocity and acceleration in a moving frame, because the force acting on the vehicle is generally resolved in such a frame. Acceleration resolved in a moving frame consists of linear, angular, centripetal, and Coriolis accelerations, as well as the acceleration of the frame's origin. The flight dynamic equations motion in a moving frame are inherently nonlinear and require an iterative numerical solution procedure. A vehicle with variable mass is capable of producing thrust due to reaction of the ejected mass by Newton's third law of motion. The general motion of a body can be described by the translation of the center of mass and the rotation of the body about its center of mass. Energy and angular momentum are very useful in describing the general motion of a body. The work done by a conservative force is indepedent of the path the body follows. A system of N bodies in mutual gravitational attraction is a conservative system, whose analytical solution is possible only if $N = 2$. The integrals of the two-body problem satisfy Kepler's laws of elliptical planetary orbits, but are more general in that they can also describe open trajectories. While the shape of a two-body trajectory can be obtained in a closed form, the relative position as a function of time generally requires an iterative solution. Lagrange's coefficients provide a compact representation of both relative position and velocity of the two-body motion.

Exercises

4.1. A space object is being tracked by a ground-based radar situated in the orbital plane. When the radial line joining the radar and the center of mass of the object makes an angle $\theta = 20°$ with the vertical, the radial distance, radial speed, and angular speed relative to the station are measured as $r = 250$ km, $\dot{r} = 3$ km/s, and $\dot{\theta} = 1°$/s, respectively. Assuming that the object experiences only the gravitational acceleration, $g = 9.081$ m/s^2, at the given point, calculate the speed, radial acceleration, and angular acceleration of the object relative to the groundstation.

4.2. A wheel of radius R rolls without slipping on a horizontal plane. The velocity and acceleration of the center of the wheel are v_0 and a_0, respectively. Derive expressions for the velocity and acceleration of a point located on the wheel at radius r and angle θ, measured along the direction of rotation from the horizontal. For given r, R, and a_0, what are the values of v_0 and θ for which the particle experiences no acceleration?

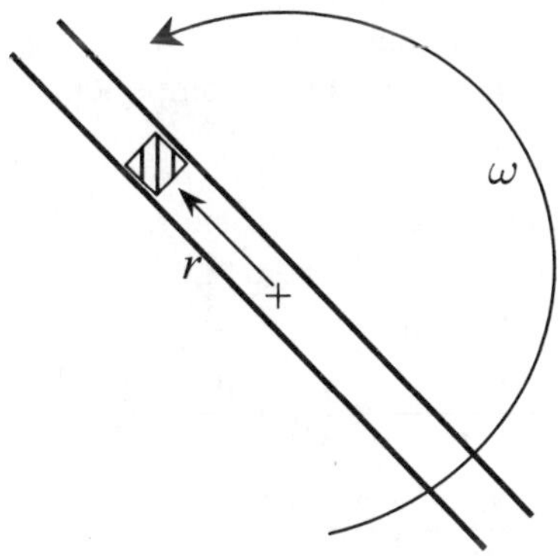

Fig. 4.14. A sliding body in a rotating slot.

4.3. A slot (Fig. 4.14) with a small sliding body inside is rotating about its midpoint with a constant angular speed, ω. The instantaneous postion, velocity, and acceleration of the body relative to the slot are given by r, $\dot{r}$, $\ddot{r}$, respectively. Derive expressions for the inertial velocity and acceleration of the sliding body.

4.4. The acceleration due to gravity, $\mathbf{g}$, of a spherical, nonrotating planet is given by Eq. (3.3). What would be the effective gravitational acceleration, $\mathbf{g_e}$, experienced by an observer standing on the planet at latitude δ if the planet were rotating at a rate ω from west to east? Express your answer in a north ($\mathbf{i_n}$), east ($\mathbf{i_e}$), down ($\mathbf{i_d}$) reference frame with the observer at its origin.

4.5. A particle of mass m is constrained to move along a parabolic path $y = x^2$. Apart from the force of constraint, the particle is acted upon by a force $-r^3\mathbf{r} + y^2\mathbf{i} - x^2\mathbf{j}$, where $\mathbf{r} = x\mathbf{i} + y\mathbf{j}$. If the particle passes through the point $(1, 1)$ with speed v, find its speed at the point $(0, 0)$.

4.6. Derive a potential function for each conservative force among the following (if any):
(a) $\mathbf{f} = \mathbf{i} - z\mathbf{j} + x\mathbf{k}$.
(b) $\mathbf{f} = yz\mathbf{i} + xz\mathbf{j} + xy\mathbf{k}$.
(c) $f = \sin\omega t$.
(d) $\mathbf{f} = x^2\mathbf{i} + y\mathbf{j}$.

4.7. Derive the equation of motion for a simple pendulum consisting of a massless rigid link of length L, suspended from a frictionless pivot, and a bob of mass m, constrained to move in a plane. If the constraint is removed and the pendulum is free to swing in any direction, find the force acting on the bob due to the rotation of the earth. [Hint: Assume a latitude, δ, and express the horizontal position of the bob by polar coordinates (r, θ).]

4.8. Assuming a spherical, nonrotating planet of radius R_0 and surface gravity g_0, derive the following equations of atmospheric flight in the vertical plane. Use the wind axes to resolve the force components, as discussed in Example 4.5.

$$\dot{h} = v\sin\phi,$$

$$\dot{\delta} = \frac{v\cos\phi}{R_e + h},$$

$$\dot{v} = \frac{f_T - D}{m} - \left(g - \frac{v^2\cos^2\phi}{R_e + h}\right)\sin\phi,$$

$$\dot{\phi} = \frac{L}{mv} - \left(\frac{g}{v} - \frac{v\cos^2\phi}{R_e + h}\right)\cos\phi,$$

$$g = g_0\left(\frac{R_0}{R_0 + h}\right)^2.$$

4.9. A spacecraft was observed with altitude, speed, and flight-path angle relative to earth of 1000 km, 10 km/s, and $-25°$, respectively. Determine the position and velocity 20 hours after the observation was taken.

4.10. Express the flight-path angle for an elliptical orbit in terms of the eccentric anomaly.

4.11. Taking the limiting case of a parabola, determine the smallest asymptote angle for a hyperbolic trajectory. What would be the asymptote angle in the limit that μ vanishes?

4.12. What is the maximum possible flight-path angle in a hyperbolic trajectory?

4.13. A hyperbolic earth departure trajectory has a perigee velocity of 15 km/s at an altitude of 300 km. Calculate the (a) hyperbolic excess velocity, (b) radius and velocity when the true anomaly is 100°, (c) time since perigee to a true anomaly of 100°.

4.14. Repeat Problem 4.9 with an observed speed of 12 km/s, and the other quantities unchanged.

4.15. A spacecraft's closest approach to Jupiter was at a periapsis radius of 300,000 km, and the hyperbolic excess velocity was 8.51 km/s. What was the angle through which the spacecraft's velocity vector was turned by Jupiter? (For Jupiter, $Gm_1 = 126,711,995.4\,\text{km}^3/\text{s}^2$.)

4.16. *Halley's* comet's last perihelion was on February 9, 1986. It has a semi-major axis of 17.9564 a.u. and eccentricity, $e = 0.967298$. Predict the next perihelion, and the current heliocentric position of the comet. (1 a.u. = $149,597,870.691$ km. For the sun, $Gm_1 = 132,712,440,018\,\text{km}^3/\text{s}^2$.)

5

Orbital Mechanics

5.1 Aims and Objectives

- To present orbital mechanics in a comprehensive manner, including orbital maneuvers, relative motion in orbit, and orbit determination for three-dimensional guidance.
- To introduce special coordinate frames (celestial, local horizon, planetary) that can be used to derive equations of translational motion in subsequent chapters.
- To numerically solve the three-dimensional Lambert problem, which is useful in designing trajectories for nonplanar rendezvous (either orbital or interplanetary).

5.2 Celestial Frame and Orbital Elements

In the previous chapter, we saw that a two-body trajectory is completely described by six scalar constants, which we shall call *orbital elements*. Several different sets of orbital elements can be chosen, depending upon the application. An obvious choice is the set formed by the elements of the initial position and velocity vectors, $\mathbf{r_0}, \mathbf{v_0}$, which, once specified, determine the orbit without ambiguity. Since the orbital motion is planar, one can fix the position and velocity by defining the plane of the motion and then specifying the two-dimensional position and velocity in the orbital plane. We have already employed various sets of scalar constants that give us the position in the orbital plane: (r, θ, τ), (v, ϕ, τ), (a, e, τ), or $(r_0, \theta - \theta_0, t - t_0)$, etc. What remains is to specify the orientation of the orbital plane by using at least three more scalar constants. In Chapter 2, we saw how the orientation of a coordinate frame can be described using various kinematical parameters (Euler angles, quaternion, modified Rodrigues parameters, etc.), each set having its advantages and disadvantages. Traditionally, Euler angles (Chapter 2) have

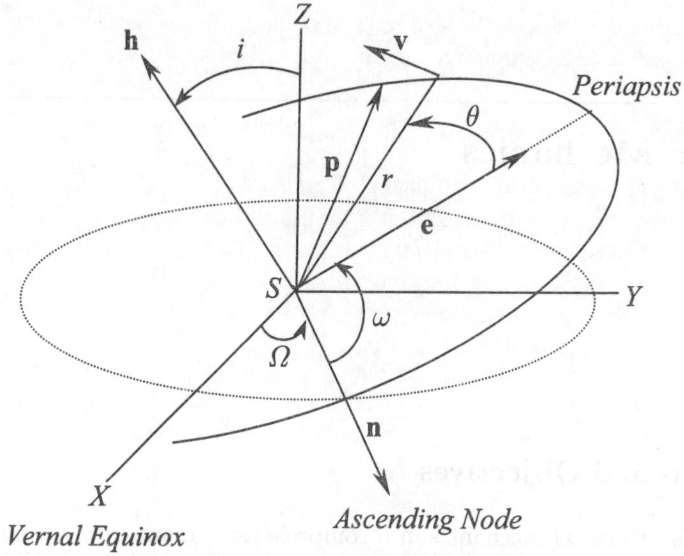

Fig. 5.1. Euler angle representation of the orbital plane.

been used to represent the orbital plane. Combined with Keplerian parameters for representing a conic section, a, e, and the time of periapsis, τ, the three Euler angles form the set of *classical orbital elements*. The Euler angles, Ω, ω, i, for describing the orientation of the perifocal orbital plane relative to a planet-centered, stationary coordinate frame, $(SXYZ)$, are shown in Fig. 5.1. The cartesian frame, $(SXYZ)$, is fixed relative to distant stars and is called a *celestial* frame. The axis SX is usually taken to be in the fixed direction of the *vernal equinox*, which indicates the location of the sun against a background of distant stars, as it crosses the equatorial plane from the south to the north. The vernal equinox thus indicates a direction along the intersection of the planet's equatorial plane and the plane of the orbit of the planet around the sun (called *ecliptic* plane). On earth, vernal equinox occurs at noon on the first day of spring, around March 21. The axis SZ is either the rotational axis of the planet or the axis normal to the plane of ecliptic, while SY completes the orthogonal triad. Normally, the equatorial plane is used as the reference plane, SXY, when an orbit close to a planet is concerned, while the ecliptic plane is used for interplanetary trajectories. We recall from Chapter 4 that the orbital plane is represented by the unit vectors $\mathbf{i_e}, \mathbf{i_p}, \mathbf{i_h}$, which indicate the respective directions of the eccentricity, parameter, and angular momentum vectors. The intersection of the orbital plane with the reference plane, SXY, yields the *line of nodes*. The *ascending node* is the name given to the point on the line of nodes where the orbit crosses the plane SXY from the south to the north. A unit vector, $\mathbf{n}$, pointing toward the ascending node makes an angle Ω with SX. The angle Ω is measured in the plane SXY in a counter-clockwise direction (Fig. 5.1), and represents a positive rotation of SX about SZ according to the right-hand rule (Chapter 2) to produce $\mathbf{n}$. It is called the *right ascension of the ascending node*. The *inclination*, i, is the angle between the orbital plane and SXY and is the positive rotation about $\mathbf{n}$ required to produce $\mathbf{i_h}$ from SZ. The angle ω represents a positive rotation of $\mathbf{n}$ about $\mathbf{i_h}$ to produce $\mathbf{i_e}$ in the orbital plane and is called the *argument of periapsis*. Using the Euler angle representation, $(\Omega)_3, (i)_1, (\omega)_3$ (Chapter 2), we can derive the rotation matrix representing the orientation of $\mathbf{i_e}, \mathbf{i_p}, \mathbf{i_h}$, in terms of $(SXYZ)$, as follows:

$$\mathsf{C} = \mathsf{C}_3(\omega)\mathsf{C}_1(i)\mathsf{C}_3(\Omega)\,. \tag{5.1}$$

However, it is more useful to transform the perifocal position and velocity (which are calculated using the methods of Chapter 4) to the celestial, cartesian frame, $(SXYZ)$. In order to do so, we require the inverse transformation

$$\begin{Bmatrix} \mathbf{I} \\ \mathbf{J} \\ \mathbf{K} \end{Bmatrix} = \mathsf{C}* \begin{Bmatrix} \mathbf{i_e} \\ \mathbf{i_p} \\ \mathbf{i_h} \end{Bmatrix}\,, \tag{5.2}$$

where the triad $\mathbf{I}, \mathbf{J}, \mathbf{K}$ denotes the frame, $(SXYZ)$, and $\mathsf{C}*$ is the following rotation matrix representing the orientation of the celestial frame relative to the perifocal frame:

$$\mathsf{C}* \doteq \mathsf{C}^T = \mathsf{C}_3{}^T(\Omega)\mathsf{C}_1{}^T(i)\mathsf{C}_3{}^T(\omega) = \begin{pmatrix} c*_{11} & c*_{12} & c*_{13} \\ c*_{21} & c*_{22} & c*_{23} \\ c*_{31} & c*_{32} & c*_{33} \end{pmatrix}\,, \tag{5.3}$$

where

$$c*_{11} = \cos\Omega\cos\omega - \sin\Omega\sin\omega\cos i, \tag{5.4}$$
$$c*_{12} = -\cos\Omega\sin\omega - \sin\Omega\cos\omega\cos i, \tag{5.5}$$
$$c*_{13} = \sin\Omega\sin i, \tag{5.6}$$
$$c*_{21} = \sin\Omega\cos\omega + \cos\Omega\sin\omega\cos i, \tag{5.7}$$
$$c*_{22} = -\sin\Omega\sin\omega + \cos\Omega\cos\omega\cos i, \tag{5.8}$$
$$c*_{23} = -\cos\Omega\sin i, \tag{5.9}$$
$$c*_{31} = \sin\omega\sin i, \tag{5.10}$$
$$c*_{32} = \cos\omega\sin i, \tag{5.11}$$
$$c*_{33} = \cos i. \tag{5.12}$$

It is easy to see that the Euler angle representation, $(\Omega)_3, (i)_1, (\omega)_3$, has singularities at $i = 0, \pm\pi$. Furthermore, the argument of periapsis, ω, is undefined for a circular orbit, $e = 0$. Thus, the utility of the classical orbital elements, $a, e, \tau, \Omega, \omega, i$, is limited to noncircular orbits, which do not lie in the plane SXY.

Example 5.1. For an earth orbit with $a = 8000$ km, $e = 0.5, \tau = -1000\ (s), \Omega = 60°, \omega = -35°$, and $i = 98°$, determine the celestial position and velocity at $t = 50$ min.

We begin by calculating the perifocal velocity and position as follows. First, the mean motion, $n = \sqrt{\frac{\mu}{a^3}} = 8.8234 \times 10^{-4}$ rad/s, and the mean anomaly, $M = n(t - \tau) = 3.52934307$ rad, are computed. Then, Kepler's equation is solved numerically (Chapter 4), to yield $E = 3.40106012608$ rad, leading to $\mathbf{r} = a(\cos E - e)\mathbf{i_e} + a\sqrt{1-e^2}\sin E\mathbf{i_p} = -11732.214\mathbf{i_e} - 1777.541\mathbf{i_p}$ km, and $\mathbf{v} = \frac{an(-\sin E\mathbf{i_e} + \sqrt{1-e^2}\cos E\mathbf{i_p})}{1 - e\cos E} = 1.22097\mathbf{i_e} - 3.983365\mathbf{i_p}$ km/s. The rotation matrix is computed using Eq. (5.3) as follows:

$$\mathsf{C}* = \begin{pmatrix} -0.0765 & 0.5086 & 0.8576 \\ 0.1448 & 0.8567 & -0.4951 \\ -0.9865 & 0.0863 & -0.1392 \end{pmatrix},$$

from which the celestial position and velocity follow:

$$\mathbf{r} = \mathsf{C}* \begin{Bmatrix} -11732.214 \\ -1777.541 \\ 0 \end{Bmatrix} = \begin{Bmatrix} -6.653 \\ -3221.591 \\ 11420.412 \end{Bmatrix} \text{ km}$$
$$= -6.653\mathbf{I} - 3221.591\mathbf{J} + 11420.412\mathbf{K} \text{ km},$$
$$\mathbf{v} = \mathsf{C}* \begin{Bmatrix} 1.22097 \\ -3.983365 \\ 0 \end{Bmatrix} = \begin{Bmatrix} -6.653 \\ -3221.591 \\ 11420.412 \end{Bmatrix} \text{ km/s}$$
$$= -2.1193\mathbf{I} - 3.2356\mathbf{J} - 1.5483\mathbf{K} \text{ km/s}.$$

5.2.1 Orbit Determination

Frequently, one is required to address the inverse problem of the one presented in Example 5.1, i.e., the determination of the orbit from measured position and velocity at a given instant. Finding the shape of the orbit, defined by a, e, τ, from the position and velocity vectors, has been addressed in Chapter 4. In order to determine the Euler angles, Ω, ω, i, from the measured position and velocity in the celestial frame, we can utilize the relationships obtained from the elementary rotations constituting the rotation matrix, C, of Eq. (5.1). The first step is to compute the unit angular momentum vector from the known position and velocity vectors, $\mathbf{r_o}, \mathbf{v_o}$, as follows:

$$\mathbf{i_h} = \frac{\mathbf{r_o} \times \mathbf{v_o}}{|\mathbf{r_o} \times \mathbf{v_o}|}. \tag{5.13}$$

Then the ascending node vector can be calculated by

$$\mathbf{n} = \frac{\mathbf{K} \times \mathbf{i_h}}{|\mathbf{K} \times \mathbf{i_h}|}, \tag{5.14}$$

and, using the elementary rotation C_3 about SZ, it follows that

$$\mathbf{n} = \mathbf{I}\cos\Omega + \mathbf{J}\sin\Omega \, , \tag{5.15}$$

from which the right ascension of ascending node, Ω, is obtained without quadrant ambiguity. The orbital inclination, i, is also obtained from the angular momentum unit vector simply as

$$\cos i = \mathbf{K} \cdot \mathbf{i_h} \, , \tag{5.16}$$

from which a unique i follows since $0 \leq i \leq \pi$. The argument of periapsis, ω, requires the computation of the eccentricity vector by

$$\mathbf{e} = \frac{\mathbf{v_0} \times (\mathbf{r_0} \times \mathbf{v_0})}{\mu} - \frac{\mathbf{r_0}}{r_0} \, , \tag{5.17}$$

which leads to the unit eccentricity vector, $\mathbf{i_e} = \frac{\mathbf{e}}{e}$. From the last elementary rotation, C_3, about $\mathbf{i_h}$, we have

$$\begin{aligned} \cos\omega &= \mathbf{n} \cdot \mathbf{i_e}, \\ \sin\omega &= \mathbf{i_h} \cdot (\mathbf{n} \times \mathbf{i_e}) \, , \end{aligned} \tag{5.18}$$

which yield a unique value of ω.

Example 5.2. A spacecraft was observed with the earth-centered, celestial position and velocity of $\mathbf{r} = -5000\mathbf{I} + 12,500\mathbf{K}$ km, and $\mathbf{v} = 5\mathbf{I} - 8\mathbf{J}$ km/s. Determine the classical orbital elements of the trajectory.

We solve this problem using the following MATLAB statements:

```
>>mu=398600.4; %gravitation constant for earth (km^3/s^2)
>> RO=[-5000;0;12500];VO=[5;-8;0]; %initial position (km), velocity (km/s)
>> H=cross(RO,VO),h=norm(H),iH=H/h %angular mom. vector, magnitude, unit vector

H = 100000     h = 1.245240940541227e+005     iH =  8.030574384788244e-001
    62500                                           5.019108990492652e-001
    40000                                           3.212229753915297e-001

>> N=cross([0;0;1],iH)/norm(cross([0;0;1],iH)) %ascending node vector

N =   -5.299989400031800e-001
       8.479983040050881e-001
       0

>> Omega=acos(N(1,1)) %right ascension of ascending node (rad.)

Omega =  2.129395642138459e+000

%Note: The correct quadrant of Omega is obtained as sin(Omega)>0 and
%       cos(Omega)<0.

>> i=acos(dot([0;0;1],iH))  % orbital inclination (rad.)

i =  1.243775706076561e+000

>> ec=cross(VO,H)/mu-RO/norm(RO),e=norm(ec),ie=ec/e %eccentricity vector

ec = -4.314183524376385e-001  e = 1.976596144782186e+000  ie = -2.182632772893419e-001
```

```
      -5.017556429948389e-001                                      -2.538483363530645e-001
       1.862539073273533e+000                                       9.422962187750184e-001

>> cosomega=dot(N,ie), sinomega=dot(iH,cross(N,ie)) %cosine, sine of omega

cosomega = -9.958365309694045e-002       sinomega =  9.950291935596002e-001

>> omega=acos(cosomega) %argument of perigee (rad)

omega =  1.670545312360925e+000

%Note: The correct quadrant of omega is obtained as sin(omega)>0 and
%      cos(omega)<0.

>> p=h^2/mu, a=p/(1-e^2) %parameter, semi-major axis (km)

p = 3.890174219594361e+004       a =  -1.338240382621893e+004

>> costheta0=dot(ie,R0)/norm(R0) %cosine of true anomaly

costheta0 =  9.559610212176732e-001

>> p=norm(R0)*(1+e*costheta0) %confirm parameter (km)

p =  3.890174219594361e+004

>> ip=cross(iH,ie) %unit parameter vector

ip =     5.544906602376137e-001
        -8.268291670873370e-001
        -9.430607702007031e-002

>> sintheta0=dot(ip,R0)/norm(R0) %sine of true anomaly

sintheta0 =  -2.934936556596461e-001

>> theta0=2*pi+asin(sintheta0) %true anomaly (rad)

theta0=  5.985305908740866e+000

>> H=2*atanh(tan(theta0/2)*sqrt((e-1)/(1+e))) %hyperbolic anomaly

H = -1.723212710379634e-001

>> a*(1-e*cosh(H))/norm(R0) %check hyperbolic anomaly

ans =   1

>> tau=(H-e*sinh(H))/sqrt(-mu/a^3) %time of perigee (s)

tau = 4.167937786907602e+002
```

Hence, the classical orbital elements are

$$\begin{aligned} a &= -13,382.404\,\text{km}, \\ e &= 1.9765961448, \\ \tau &= 416.7938\,\text{s}, \\ \Omega &= 122.0054^\circ, \\ \omega &= 95.7152^\circ, \\ i &= 71.2631^\circ. \end{aligned}$$

Note that a positive τ indicates that the spacecraft is approaching the earth (which is also evident from a true anomaly in the fourth quadrant).

Rather than using the position and velocity vectors at a given time to find the orbit, we can derive the orbital elements from the position vector at two different time instants. The determination of the orbit from two position vectors, $\mathbf{r}(t_1), \mathbf{r}(t_2)$, is known as *Lambert's problem* and is very useful in the navigation and rendezvous of spacecraft, and targeting of long-range, surface-to-surface missiles. Lambert's problem is a two-point, boundary-value problem, because its solution must satisfy conditions at two different points. Generally, a closed-form solution to such a problem is not possible, and we have to resort to an iterative numerical solution procedure. We shall have an occasion to discuss Lambert's problem later in this chapter.

5.3 Spherical Celestial Coordinates and Local Horizon

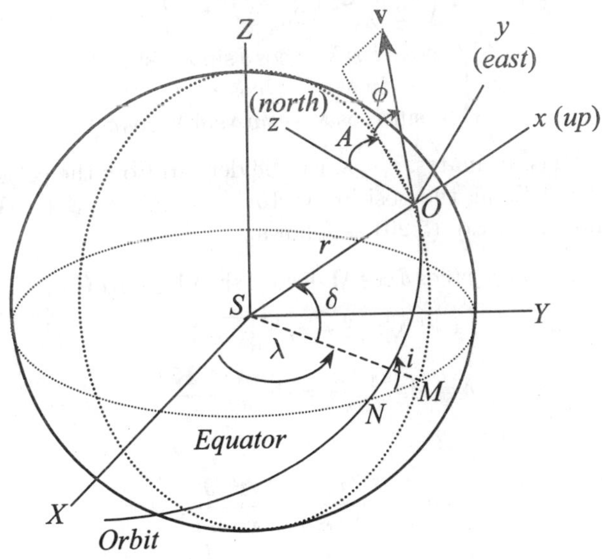

Fig. 5.2. Spherical celestial coordinates and the local horizon.

A spherical coordinate system based upon the celestial frame can be derived using the radial distance, r, and two angles, δ, λ, as shown in Fig. 5.2. The angle δ called the *declination*, or, *celestial latitude*, is the angle between the position vector, $\mathbf{r}$, with the equatorial plane, SXY (defined positive above the plane), while λ—called the *right ascension*, or *celestial longitude*—is the angle made by the projection of $\mathbf{r}$ on SXY with the vernal equinox direction,

SX, defined positive toward the east. In order to specify the velocity vector, $\mathbf{v}$, a moving *local horizon* frame ($oxyz$) (denoted by the triad $\mathbf{i}, \mathbf{j}, \mathbf{k}$), can be employed, as depicted in Fig. 5.2. The axis, ox, is along $\mathbf{r}$, while oy and oz point toward the east and north, respectively. The velocity vector, $\mathbf{v}$, makes an angle, ϕ—called the *flight-path angle*—with the local horizontal plane, oyz, defined positive above the plane. The horizontal projection of $\mathbf{v}$ makes an angle, A, with oz (north), called the *velocity azimuth*, and defined positive toward the east. The parameters $(r, \lambda, \delta, v, \phi, A)$ form an alternative set of orbital parameters for determining the position and velocity.

The orientation of the local horizon frame, $(oxyz)$, relative to the Cartesian, celestial frame, $(SXYZ)$, can be obtained using the Euler angle representation, $(\lambda)_3, (\frac{\pi}{2} - \delta)_2, (\frac{-\pi}{2})_2$. Therefore, the rotation matrix relating the two frames is obtained as

$$\begin{Bmatrix} \mathbf{i} \\ \mathbf{j} \\ \mathbf{k} \end{Bmatrix} = \mathsf{C_{LH}} \begin{Bmatrix} \mathbf{I} \\ \mathbf{J} \\ \mathbf{K} \end{Bmatrix}, \tag{5.19}$$

where

$$\begin{aligned} \mathsf{C_{LH}} &= \mathsf{C_2}\left(\frac{-\pi}{2}\right)\mathsf{C_2}\left(\frac{\pi}{2} - \delta\right)\mathsf{C_3}(\lambda) \\ &= \begin{pmatrix} \cos\delta\cos\lambda & \cos\delta\sin\lambda & \sin\delta \\ -\sin\lambda & \cos\lambda & 0 \\ -\sin\delta\cos\lambda & -\sin\delta\sin\lambda & \cos\delta \end{pmatrix}. \end{aligned} \tag{5.20}$$

The spherical coordinates, r, δ, λ, can be derived from the celestial, Cartesian coordinates defining the position vector, $\mathbf{r} = r_X\mathbf{I} + r_Y\mathbf{J} + r_Z\mathbf{K}$, by using the rotation matrix of Eq. (5.20) as follows:

$$\mathbf{r} = r\mathbf{i} = r(\cos\delta\cos\lambda\mathbf{I} + \cos\delta\sin\lambda\mathbf{J} + \sin\delta\mathbf{K}), \tag{5.21}$$

from which we have $r = \mid \mathbf{r} \mid = \sqrt{r_X{}^2 + r_Y{}^2 + r_Z{}^2}$,

$$\delta = \sin^{-1}\frac{r_Z}{r} = \sin^{-1}\frac{\mathbf{r}\cdot\mathbf{K}}{r}, \tag{5.22}$$

and

$$\begin{aligned} \sin\lambda &= \frac{r_Y}{r\cos\delta} = \frac{\mathbf{r}\cdot\mathbf{J}}{r\cos\delta}, \\ \cos\lambda &= \frac{r_X}{r\cos\delta} = \frac{\mathbf{r}\cdot\mathbf{I}}{r\cos\delta}. \end{aligned} \tag{5.23}$$

The spherical coordinates, v, ϕ, A, are derived from the celestial, Cartesian coordinates defining the velocity vector, $\mathbf{v} = v_X\mathbf{I} + v_Y\mathbf{J} + v_Z\mathbf{K}$, by expressing the same in the local horizon frame as

$$\begin{aligned} \mathbf{v} &= \mathsf{C_{LH}} \begin{Bmatrix} v_X \\ v_Y \\ v_Z \end{Bmatrix} \\ &= v(\sin\phi\mathbf{i} + \cos\phi\sin A\mathbf{j} + \cos\phi\cos A\mathbf{k}), \end{aligned} \tag{5.24}$$

resulting in $v = \mid \mathbf{v} \mid = \sqrt{v_X{}^2 + v_Y{}^2 + v_Z{}^2}$,

$$\phi = \sin^{-1}\frac{\mathbf{v}\cdot\mathbf{i}}{v}\,, \tag{5.25}$$

and

$$\sin A = \frac{\mathbf{v}\cdot\mathbf{j}}{v\cos\phi}, \tag{5.26}$$

$$\cos A = \frac{\mathbf{v}\cdot\mathbf{k}}{v\cos\phi}\,.$$

A useful relationship among the declination, δ, orbital inclination, i, and the velocity azimuth, A, is given by the following equation of *spherical trigonometry*[1] for the right spherical triangle formed by the arcs MO, ON, and NM (Fig. 5.2):

$$\cos i = \cos\delta \sin A\,, \tag{5.27}$$

which can also be derived by substituting Eqs. (5.21) and (5.24) into Eq. (5.16). Another relationship arising out of the same right spherical triangle is the following:

$$\tan i \sin(\lambda - \Omega) = \tan\delta\,. \tag{5.28}$$

It is clear from Eq. (5.27) that the minimum orbital inclination possible for a given declination is obtained for $A = 90°$, which denotes a flight path due east, and is equal to the local declination. Thus, it is not possible to launch a spacecraft into an orbit with an inclination smaller than the latitude of the launch site. Of course, as seen later in this chapter, the orbital inclination can be changed using propulsive maneuvers, but these are quite expensive and significantly increase the cost of launch. An zero inclination orbit can be obtained from an equatorial launch site without requiring a propulsive plane change. Such a launch site also gives the maximum velocity advantage due to the earth's rotation for an eastward launch. The market for geosynchronous satellite launches, which is currently the most lucrative, requires a zero orbital inclination, thus dictating that the launch sites be located as close to the equator as physically possible.

Example 5.3. For the spacecraft in Example 5.1, determine the declination, right ascension, flight-path angle, and velocity azimuth 50 min after the present time.

The celestial position and velocity vectors have already been calculated at $t = 50$ min to be $\mathbf{r} = -6.653\mathbf{I} - 3221.591\mathbf{J} + 11,420.412\mathbf{K}$ km and $\mathbf{v} = -2.1193\mathbf{I} - 3.2356\mathbf{J} - 1.5483\mathbf{K}$ km/s, respectively. We begin by computing

[1] Spherical trigonometry is a branch of mathematics devoted to the study of angular relationships among the arcs on the surface of a sphere (the common, plane trigonometry studies the angular relationships among the straight lines on a flat plane). A useful textbook on spherical trigonometry is by Smail [60].

the radius, $r = \mid \mathbf{r} \mid = 11,866.107$ km, and speed, $v = \mid \mathbf{v} \mid = 4.166$ km/s. From Eqs. (5.22) and (5.23), we have

$$\delta = \sin^{-1}\frac{11420.412}{11866.107} = 74.2467°,$$

$$\sin\lambda = \frac{-3221.591}{11866.107\cos(74.2467°)} = -0.9999979$$

$$\cos\lambda = \frac{-6.653}{11866.107\cos(74.2467°)} = -0.00206506,$$

which yield $\lambda = 269.8817°$. The rotation matrix of the local horizon frame is the following, according to Eqs. (5.19):

$$\mathsf{C_{LH}} = \begin{pmatrix} -0.0005607 & -0.2714952 & 0.9624396 \\ 0.9999979 & -0.0020651 & 0 \\ 0.0019875 & 0.9624376 & 0.2714958 \end{pmatrix}.$$

The velocity components in the local horizon frame are now obtained as

$$\mathbf{v} = \mathsf{C_{LH}}\begin{Bmatrix} -2.1193 \\ -3.2356 \\ -1.5483 \end{Bmatrix}$$
$$= -0.6104843\mathbf{i} - 2.1126542\mathbf{j} - 3.5386385\mathbf{k}.$$

Finally, by employing Eqs. (5.25) and (5.26), we get the flight-path angle and velocity azimuth as follows:

$$\phi = \sin^{-1}\frac{-0.6104843}{4.166} = -8.4259°,$$

$$\sin A = \frac{-2.1126542}{4.166\cos(-8.4259°)} = -0.51261604,$$

$$\cos A = \frac{-3.5386385}{4.166\cos(-8.4259°)} = -0.85861796,$$

which yield $A = 210.8382°$. Finally, we confirm that Eqs. (5.27) and (5.28) are satisfied for the calculated values of δ, λ, and A by checking $\cos i = -0.13917310096 = \cos\delta\sin A$ and $\tan\delta = 3.54495 = \tan i \sin(\lambda - \Omega)$.

5.4 Planet Fixed Frame

Instead of a celestial frame, it is often more useful to use a spherical coordinate system that is fixed to the planet and rotates with it. Such a planet-centered,

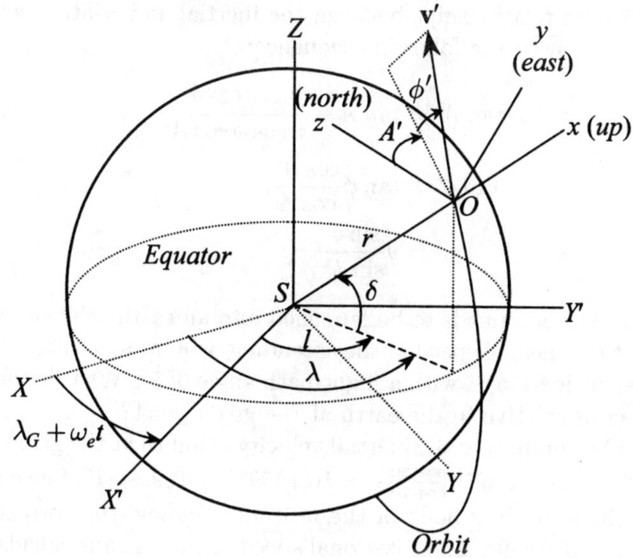

Fig. 5.3. Planet fixed, rotating coordinates and the local horizon.

rotating frame, $(SX'Y'Z)$, which shares the axis SZ with the celestial frame, $(SXYZ)$, is shown in Fig. 5.3. A set of spherical coordinates in the planet fixed frame is r, δ, l, denoting the radius, latitude (or declination), and longitude, respectively. For the velocity relative to the rotating frame, $\mathbf{v}'$, the spherical coordinates v', ϕ', A' can be employed, representing the relative magnitude, flight-path angle, and velocity azimuth measured in the local horizon frame $(oxyz)$. Of course, the relationship between the inertial velocity, $\mathbf{v}$, and the velocity relative to the rotating frame, $\mathbf{v}'$, is given by

$$\mathbf{v} = \mathbf{v}' + \boldsymbol{\omega} \times \mathbf{r} , \tag{5.29}$$

where $\boldsymbol{\omega} = \omega_e \mathbf{K}$ is the angular velocity of the planet, and $\mathbf{r} = r\mathbf{i}$. It is possible to obtain the relationship between the inertial and relative velocity coordinates. We begin by expressing the relative velocity as $\mathbf{v}' = v'(\sin\phi'\mathbf{i} + \cos\phi' \sin A'\mathbf{j} + \cos\phi' \cos A'\mathbf{k})$, and substitute it along with Eq. (5.24), into Eq. (5.29) to yield

$$\begin{aligned} & v(\sin\phi\mathbf{i} + \cos\phi \sin A\mathbf{j} + \cos\phi \cos A\mathbf{k}) \\ & \quad = v'(\sin\phi'\mathbf{i} + \cos\phi' \sin A'\mathbf{j} + \cos\phi' \cos A'\mathbf{k}) + \omega_e r \cos\delta\mathbf{j}. \end{aligned} \tag{5.30}$$

On comparing the respective vector components on both sides of Eq. (5.30), we have

$$\begin{aligned} v \sin\phi &= v' \sin\phi', \\ v \cos\phi \sin A &= v' \cos\phi' \sin A' + \omega_e r \cos\delta, \\ v \cos\phi \cos A &= v' \cos\phi' \cos A' . \end{aligned} \tag{5.31}$$

Thus, the explicit relationships between the inertial and relative velocity coordinates are given by the following sequence:

$$\tan A' = \tan A - \frac{\omega_e r \cos\delta}{v\cos\phi\cos A},$$
$$\tan\phi' = \tan\phi\frac{\cos A'}{\cos A}, \tag{5.32}$$
$$v' = v\frac{\sin\phi}{\sin\phi'}.$$

Example 5.4. A spacecraft is to be launched into an earth orbit of inclination 80° such that the inertial speed of the ascending orbit at an altitude of 200 km and latitude 10° is 8 km/s, with a flight-path angle of 5°. What is the velocity of the spacecraft relative to the earth at the given point?

We begin by computing the inertial velocity azimuth at the given point using Eq. (5.27) as $A = \sin^{-1}\frac{\cos(80°)}{\cos(10°)} = 10.1559°$, or 169.8441°. Since the given point is on the ascending node of the latitude crossing, the correct velocity azimuth is $A = 10.1559°$. The rotational speed of the earth is calculated from the *mean sidereal day* (23 hours, 56 minutes, and 4.09 seconds)[2] as follows:

$$\omega_e = \frac{2\pi}{23\times 3600 + 56\times 60 + 4.09} = 7.29211\times 10^{-5}\ \text{rad/s}.$$

Finally, we calculate the relative velocity coordinates from Eq. (5.32) by substituting $r = 6578.14$ km, $v = 8$ km/s, and $\phi = 5°$:

$$A' = \tan^{-1}\left(\tan A - \frac{\omega_e r\cos\delta}{v\cos\phi\cos A}\right) = 6.7815°,$$
$$\phi' = \tan^{-1}\left(\tan\phi\frac{\cos A'}{\cos A}\right) = 5.0438°,$$
$$v' = v\frac{\sin\phi}{\sin\phi'} = 7.9307\ \text{km/s}.$$

Note that the change in speed and velocity azimuth caused by the earth's rotation is significant, while that in the flight-path angle is relatively small.

The rotation matrix representing the orientation of $(SX'Y'Z)$ relative to $(SXYZ)$ is obtained simply as follows:

$$\begin{Bmatrix}\mathbf{I}'\\ \mathbf{J}'\\ \mathbf{K}\end{Bmatrix} = \mathsf{C}_{pf}\begin{Bmatrix}\mathbf{I}\\ \mathbf{J}\\ \mathbf{K}\end{Bmatrix}, \tag{5.33}$$

where

[2] The mean sidereal day is the average time the earth takes to complete one rotation against the background of distant stars, and is obtained by adding $\frac{360°}{365.25}$ to the 360° angle rotated in 24 hours relative to the sun (the *mean solar day*).

$$\mathsf{C_{pf}} = \mathsf{C_3}(\lambda - l) = \begin{pmatrix} \cos(\lambda - l) & \sin(\lambda - l) & 0 \\ -\sin(\lambda - l) & \cos(\lambda - l) & 0 \\ 0 & 0 & 1 \end{pmatrix} . \tag{5.34}$$

Thus, the coordinate transformation between planet fixed and local horizon frames is given by

$$\begin{Bmatrix} \mathbf{i} \\ \mathbf{j} \\ \mathbf{k} \end{Bmatrix} = \mathsf{C_{LH}}\mathsf{C_{pf}}^T \begin{Bmatrix} \mathbf{I}' \\ \mathbf{J}' \\ \mathbf{K} \end{Bmatrix} . \tag{5.35}$$

It is clear from above that the longitude, l, is an important planet fixed coordinate. Lines of constant longitude on the planet's surface are called *meridians*. The fixed stars cross each meridian at the same local time; thus, a meridian and its local time have been synonymous since the earliest days of navigation.

In order to utilize the planet fixed coordinates, the longitude, l, must be calculated at a given time, t, from the current right ascension, λ, and the right ascension of the zero longitude line, SX' (the *Greenwich meridian* on earth), λ_G, known at some previous time, $t = 0$, as follows (Fig. 5.3):

$$l = \lambda - \lambda_G - \omega_e t. \tag{5.36}$$

The right ascension of SX' is usually available from periodically published astronomical (or *ephemeris*) charts, such as the *American Ephemeris and Nautical Almanac* for the right ascension of the Greenwich meridian.

Example 5.5. Calculate the longitude of the spacecraft in Example 5.3 at $t =$ 50 min if the right ascension of the Greenwich meridian at $t = 0$ is 154°.

In Example 5.3, we had obtained the current right ascension, $\lambda = 269.8817°$. On substituting $\lambda_G = 154°$, $\omega_e = 0.004178074648°/\mathrm{s}$, and $t = (50)(60)$ s into Eq. (5.36), we have $l = 269.8817° - 154° - (0.004178074648)(3000) = 103.3475°$. The decrease in the longitude caused by the earth's rotation is $\omega_e t = 12.5342°$.

5.5 Single Impulse Orbital Maneuvers

It is often necessary to change a spacecraft's trajectory by the use of propulsive maneuvers. Since the duration of the rocket thrust applied in such a maneuver is negligible in comparison to the orbital time period (or the time scale $\frac{2\pi\sqrt{p^3}}{\sqrt{\mu}}$ for the open orbits), it is reasonable to assume that the velocity change occurs instantaneously at the point of thrust application. Hence, orbital maneuvers are usually regarded as being *impulsive*, and the velocity change is given by the magnitude and direction of a *velocity impulse*. The simplest impulsive maneuver is the case of intersecting initial and final trajectories, as shown in Fig. 5.4, where a single velocity impulse, $\Delta\mathbf{v}$, is sufficient to produce a velocity change from $\mathbf{v_i}$ to $\mathbf{v_f}$ at a given orbital position, $\mathbf{r}$. The vector $\mathbf{v_f}$

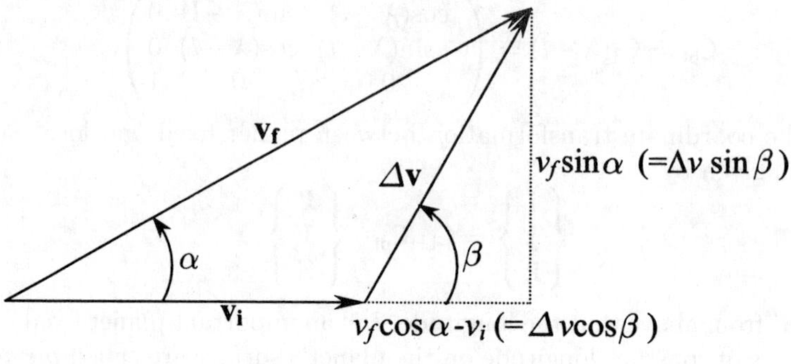

Fig. 5.4. Impulsive maneuver for intersecting initial and final trajectories.

makes an angle α with $\mathbf{v_i}$. From the vector triangle in Fig. 5.4, we see that the relationship among the magnitudes, $v_i, v_f, \Delta v$, is given by

$$\Delta v = \sqrt{v_i{}^2 + v_f{}^2 - 2 v_i v_f \cos\alpha} \; . \tag{5.37}$$

A general, single impulse maneuver may change both the shape and the plane of the orbit. Sometimes, it is necessary to change only the orbital shape without affecting the orbital plane. Such maneuvers are said to be *coplanar*. In a coplanar maneuver, there can be a simultaneous change in the speed, $(v_f - v_i)$, and the flight-path angle, α, leading to a new, coplanar trajectory. Often it is required to change the plane of the orbit without changing its shape. Such a maneuver is called a *plane change maneuver* and does not involve a modification of either the speed or the flight-path angle at the point of the impulse. From Eq. (5.37), it is clear that a plane change by angle α at a constant speed, v_i, requires the following impulse magnitude, applied at an angle $\beta = \frac{\alpha}{2} + 90°$ to $\mathbf{v_i}$:

$$\Delta v = 2 v_i \sin\frac{\alpha}{2} \; . \tag{5.38}$$

It is evident from Eq. (5.38) that the velocity impulse magnitude for a plane change by a given angle is directly proportional to the speed at which the change is performed. Hence, a plane change by even a small angle can require a large velocity impulse, which translates into a considerable increase in the required propellant mass (causing a reduction in the payload). For this reason, plane changes should be carried out at the smallest possible speed (e.g., apogee of an elliptical orbit). For a plane change by 60°, the impulse magnitude is equal to the flight speed.

Example 5.6. A spacecraft in a circular, earth orbit of altitude 500 km and inclination 10° has to be sent to an elliptical orbit with a perigee altitude of 200 km, an apogee altitude of 700 km, and an inclination of 5°. Calculate the magnitude of the velocity impulses required for the transfer.

It is clear that an impulse can be provided at the intersection of two coplanar orbits whose shapes are identical to those of the initial and final orbits. Since the final orbital plane is not uniquely specified, it is sufficient to send the spacecraft to any plane with the given inclination. Thus, we have a choice of where to apply the required inclination change. The most desirable point for making the plane change is at the apogee of the elliptical orbit, because it requires the least impulse magnitude. We calculate the shape of the final orbit, and the first velocity impulse at an intersection point, $r = r_i = 6878.14$ km, as follows:

$$
\begin{aligned}
e &= \frac{r_a - r_p}{r_a + r_p} = 0.03661319, \\
a &= \frac{r_p}{1-e} = 6828.14 \text{ km}, \\
v_p &= \sqrt{\frac{2\mu}{r_p} - \frac{\mu}{a}} = 7.9254818 \text{ km/s}, \\
v_i &= \sqrt{\frac{\mu}{r_i}} = 7.6126061 \text{ km/s}, \\
v_f &= \sqrt{\frac{2\mu}{r_i} - \frac{\mu}{a}} = 7.5846827 \text{ km/s}, \\
\cos\alpha = \cos\phi &= \frac{r_p v_p}{r_i v_f} = 0.999356306, \\
\Delta v_1 &= \sqrt{{v_i}^2 + {v_f}^2 - 2 v_i v_f \cos\alpha} = 0.27406656 \text{ km/s}\,.
\end{aligned}
$$

The transfer angle, α, is determined from $\alpha = \cos^{-1} 0.999356306 = \pm 2.0559°$, each sign denoting one of the two possible intersection points. The direction of the first impulse in the orbital plane is given by $\beta = \sin^{-1} \frac{v_f \sin\alpha}{\Delta v} = \pm 83.1252°$ (Fig. 5.4). Finally, we apply the second velocity impulse at the apogee, $r = r_a = a(1+e)$, of the elliptical orbit resulting from the first impulse, in order to change the orbital inclination by 5°:

$$
\begin{aligned}
v_a &= \sqrt{\frac{2\mu}{r_a} - \frac{\mu}{a}} = 7.3656256 \text{ km/s}, \\
\Delta v_2 &= 2 v_a \sin\frac{5°}{2} = 0.64256815 \text{ km/s}\,.
\end{aligned}
$$

The second impulse makes an angle $\beta = 92.5°$ with the initial orbital plane. The total velocity impulse magnitude for the orbital change is $\Delta v = \Delta v_1 + \Delta v_2 = 0.91663471$ km/s.

When the initial and final orbital planes are unambiguously defined by the given sets of Ω, ω, i, we cannot choose the point of plane change arbitrarily. For the general plane change, we must wait until the spacecraft reaches the line of nodes formed by the intersection of the two orbital planes, at which

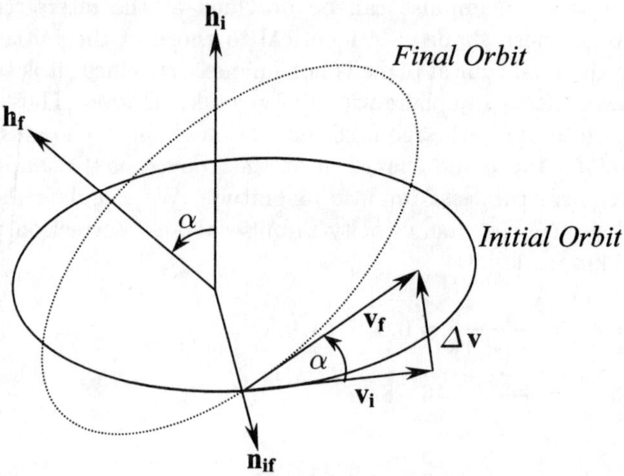

Fig. 5.5. Impulsive maneuver for a general plane change.

point a velocity impulse is applied to change the orbital plane. Of course, if the maneuver has to be performed by a single velocity impulse, the two orbits should also intersect at the line of nodes, which is true only for circular orbits of equal radii. Figure 5.5 describes the geometry of the general plane change in terms of the initial and final orbital angular momenta, $\mathbf{h_i}$ and $\mathbf{h_f}$, respectively. The nodal vector between the two planes denoting the point of application of the velocity impulse is given by

$$\mathbf{n_{if}} = \frac{\mathbf{h_i} \times \mathbf{h_f}}{| \mathbf{h_i} \times \mathbf{h_f} |} . \tag{5.39}$$

The impulsive maneuver is carried out at $\mathbf{r_i} = r_i \mathbf{n_{if}}$, from which the speed of plane change, v_i, is calculated. The cosine of the angle, α, between the two planes is obtained as follows:

$$\cos\alpha = \frac{\mathbf{h_i} \cdot \mathbf{h_f}}{h_i h_f} . \tag{5.40}$$

The direction of the plane change impulse relative to the initial velocity vector is given by $\beta = \frac{\alpha}{2} + 90°$, which results in the following expression for the velocity impulse:

$$\Delta\mathbf{v} = \Delta v(\cos\beta\frac{\mathbf{v_i}}{v_i} + \sin\beta\frac{\mathbf{h_i}}{h_i}) , \tag{5.41}$$

leading to

$$\mathbf{v_f} = \mathbf{v_i} + \Delta\mathbf{v} = \mathbf{v_i} + \Delta v(\cos\beta\frac{\mathbf{v_i}}{v_i} + \sin\beta\frac{\mathbf{h_i}}{h_i}) . \tag{5.42}$$

It is clear from Eqs. (5.38) and (5.42) that in order to achieve a plane change without affecting the speed (i.e., $v_f = v_i$), the angle β must be greater than 90°.

Example 5.7. A spacecraft is in an elliptical, earth orbit of $a = 6900$ km, $e = 0.6$ s, $\Omega = 120°$, $\omega = 25°$, and $i = 10°$. When the spacecraft is at its apogee, a velocity impulse is applied at an angle $\beta = 100°$ to the orbital plane, measured in a counter-clockwise direction from the initial velocity vector, such that there is no change in the speed. Determine the new orbit of the spacecraft.

Since the impulse is applied out of the orbital plane, at an angle greater than 90°, it does not affect the orbital speed, or the flight-path angle; thus a and e are unchanged, and the shape of the orbit is not modified. The magnitude of the impulse is obtained from Eq. (5.38)by substituting $\alpha = 2\beta - 180° = 20°$. A major component of the velocity impulse is in the direction of the initial angular momentum vector, $\mathbf{h}_i$; thus, it would tend to increase the orbital inclination, i, and decrease the right ascension of the ascending node, Ω. We begin the solution by deriving the angular momentum vector of the initial orbital planes from the perifocal apogee position and velocity, transformed into the celestial frame, by the rotation matrix, $\mathsf{C}*$. The nodal vector, $\mathbf{n}_{if}$, essentially lies toward the apogee. The final velocity at the orbit intersection is obtained by vectorially adding the velocity impulse to the initial velocity. Thus, the final angular momentum can be calculated, and the orbital elements defining the new orbital plane obtained. The necessary computations are performed using the following MATLAB statements and the program *rotation.m*, which is tabulated in Table 5.1:

```
>> mu=398600.4; %gravitation constant for earth (km^3/s^2)
>> a=6900; e=0.6; %semi-major axis (km) and eccentricity
>> n=sqrt(mu/a^3); %mean motion (rad/s)
>> ra=[-a*(1+e);0;0],va=[0;-a*n*sqrt(1-e^2)/(1+e);0] %position(km), velocity(km/s)

ra =  -11040          va =   0
          0                 -3.80026886920661
          0                  0

>> dtr=pi/180; Ci=rotation(10*dtr,120*dtr,25*dtr) %rotation matrix for initial orbit

   Ci =  -0.8136   -0.5617    0.1504
          0.5768   -0.8123    0.0868
          0.0734    0.1574    0.9848

>> rai=Ci*ra; vai=Ci*va; hi=cross(rai,vai) %angular momentum of initial orbit(km^/s)

   hi =  6309.34476083312
            3642.70189607715
            41317.5780750187

>> magdv=2*norm(va)*sin(10*dtr) %magnitude of velocity impulse (km/s)

   magdv =    1.31981952756419

>> dv=magdv*(cos(100*dtr)*vai/norm(vai)+sin(100*dtr)*hi/norm(hi)) %vel.change(km/s)

   dv =  0.0667421843481491
          -0.0733077336502738
           1.31609081842981

>> vf=vai+dv %final velocity (km/s)

   vf =  2.20117178698127
           3.01352761007753
           0.718009460772585

>> norm(vf)/norm(vai) %check that final and initial speeds are equal
```

```
ans = 1

>> hf=cross(rai,vf) %angular momentum of final orbit (km^/s)

hf = -2130.55310771777
     -8232.56883673061
      41084.120049037

>> norm(hf)/norm(hi) %check that angular momentum magnitude is constant

ans = 1

>> iH=hf/norm(hf);N=cross([0;0;1],iH)/norm(cross([0;0;1],iH)) %ascending node

N =  0.968105836625469
    -0.250541591540609
     0

>> Omega=asin(N(2,1)) %right ascension of ascending node (rad)

Omega = -0.253239648907155

>> i=acos(dot([0;0;1],iH)) %orbital inclination (rad)

i = 0.204102644403926

>> ec=cross(vf,hf)/mu-rai/norm(rai),e=norm(ec),ie=ec/e %eccentricity vec.

ec =  -0.488155025991245        e =  0.6  ie =  -0.813591709985409
       0.346072018114803                         0.576786696858004
       0.0440321346000229                        0.0733868910000383

>> cosomega=dot(N,ie), sinomega=dot(iH,cross(N,ie))%cosine, sine of omega

cosomega = -0.932151940077225       sinomega = 0.362067342645351

>> omega=acos(cosomega) %argument of perigee

omega = 2.77110789553964

%confirm Euler angles by checking whether the apogee position is correctly
%transformed from perifocal to celestial frame by comparing with the apogee
%position of the initial orbit (both must be the same).

>> Cf=rotation(i,Omega,omega);raf=Cf*ra, rai

raf = 8982.05247823891     rai = 8982.05247823891
     -6367.72513331237          -6367.72513331237
     -810.191276640425          -810.191276640422
```

The final orbital plane is thus described by $\Omega = -14.5096°$, $i = 11.6942°$, and $\omega = 158.7728°$, which implies an increase of about 1.7° in the inclination, a decrease of about 134.5° in right ascension of ascending node, and an increase of about 133.8° in the argument of perigee.

When both the shape and the plane of an orbit are to be changed, we can choose the sequence of the coplanar and plane change maneuvers, depending upon the speeds at which they are to be performed (plane change is more efficient at the smaller speed). However, when the coplanar and the plane change maneuvers are to be performed at the same radius, it is more efficient to combine the two maneuvers into a single impulse maneuver, rather than

Table 5.1. M-file *rotation.m* for Calculating the 3-1-3 Euler Angle Rotation Matrix

```
function C=rotation(i,Om,w)
%Rotation matrix of 3-1-3 Euler angles (radians)
%i=orbital inclination
%Om=right ascension of ascending node
%w=argument of periapsis
%(c) 2006 Ashish Tewari
L1=cos(Om)*cos(w)-sin(Om)*sin(w)*cos(i);
L2=-cos(Om)*sin(w)-sin(Om)*cos(w)*cos(i);
L3=sin(Om)*sin(i);
M1=sin(Om)*cos(w)+cos(Om)*sin(w)*cos(i);
M2=-sin(Om)*sin(w)+cos(Om)*cos(w)*cos(i);
M3=-cos(Om)*sin(i);
N1=sin(w)*sin(i);
N2=cos(w)*sin(i);
N3=cos(i);
C=[L1 L2 L3;M1 M2 M3;N1 N2 N3];
```

carrying out each separately. In such a case, the velocity impulses required for the separately performed individual maneuvers form two sides of a vector triangle, whose resultant is the impulse required for the combined maneuver (Fig. 5.4). An example is the combination of the *circularization* maneuver—required for converting an elliptical orbit into a coplanar, circular one—with the plane change maneuver at the apogee.

5.6 Multi-Impulse Orbital Transfer

When the initial and final orbits do not intersect, the orbital transfer cannot be made by a single impulse, and a multi-impulse maneuver becomes necessary. Even for intersecting, coplanar orbits, it is often more efficient to employ a multi-impulse transfer rather than a single impulse maneuver. It is evident from Eq. (5.37) that in a coplanar maneuver, the minimum velocity impulse magnitude for a given pair of initial and final speeds, v_i, v_f, corresponds to the case of $\alpha = 0$, which implies that the velocity impulse is applied in the direction of the initial (and final) velocity vector. Such an impulse is said to be *tangential*. Therefore, a coplanar maneuver involving the minimum possible total velocity change arising out of tangential velocity impulses requires the least propellant mass and is said to be *optimal* between a given pair of initial and final orbits. The trajectory connecting the initial and final orbits must essentially be elliptical, and the tangential velocity impulses can occur at the periapsis and the apoapsis of this emphtransfer ellipse.

Example 5.8. Calculate the smallest total velocity change required in a two-impulse orbital transfer from the circular earth orbit of 500 km altitude to the intersecting elliptical, earth orbit of Example 5.7.

The optimal maneuvers between coplanar circular and elliptical orbits involve the smallest total magnitude of tangential impulses applied at the initial and final orbits. Since the initial orbit is circular, any point can be chosen to apply a tangential impulse. However, in the final elliptical orbit, there are only two points where tangential velocity changes can take place, namely the perigee and the apogee. The smallest possible transfer ellipse ends at either the perigee or the apogee of the final elliptical orbit, and begins at a point in the circular orbit that is directly opposite to the terminating point. The required transfer ellipse in each case is shown in Fig. 5.6.

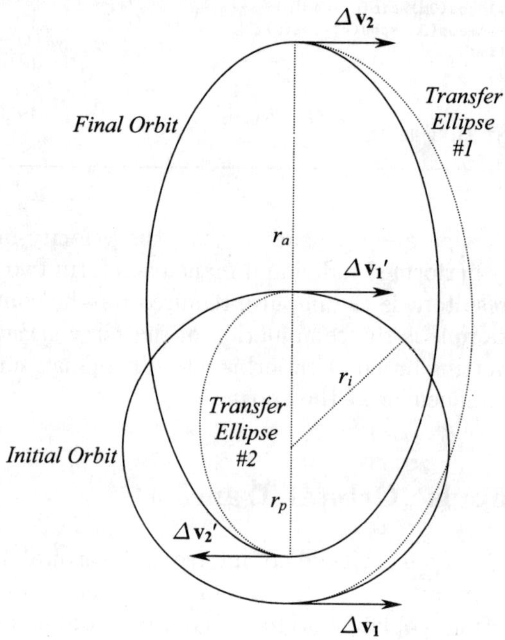

Fig. 5.6. Optimal transfers between coplanar, circular, and elliptical orbits.

When the initial and final orbits are circular, the optimal maneuver is called a *Hohmann transfer*. As shown in Fig. 5.7, a Hohmann transfer consists of two tangential velocity impulses, each applied when the spacecraft is at the initial and final radius, respectively.

The magnitudes of the two velocity impulses in a Hohmann transfer are easily obtained by merely subtracting the initial speed from the final speed at respective instants as follows:

$$\Delta v_1 = v_{f1} - v_{i1} = \sqrt{\frac{2\mu}{r_i} - \frac{\mu}{a}} - \sqrt{\frac{\mu}{r_i}}\,, \tag{5.43}$$

and

$$\Delta v_2 = v_{f2} - v_{i2} = \sqrt{\frac{\mu}{r_f}} - \sqrt{\frac{2\mu}{r_f} - \frac{\mu}{a}}\,, \tag{5.44}$$

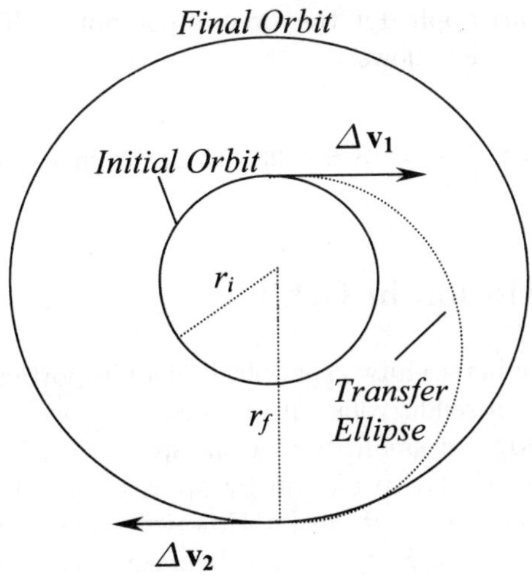

Fig. 5.7. Hohmann transfer between two circular, coplanar orbits.

where r_i, r_f are the initial and final circular orbit radii, respectively, and a refers to the semi-major axis of the transfer ellipse. From Fig. 5.7, it is clear that $a = \frac{r_i+r_f}{2}$. The time required for a Hohmann transfer, t_H, is half the period of the transfer ellipse and is given by

$$t_H = \pi\sqrt{\frac{a^3}{\mu}} = \frac{\pi(r_i + r_f)^{\frac{3}{2}}}{\sqrt{8\mu}} . \tag{5.45}$$

Example 5.9. Calculate the velocity impulses and the time required for a Hohmann transfer from a circular, earth orbit of altitude 250 km (parking orbit) to a *geosynchronous* orbit.

A geosynchronous orbit is a circular, equatorial earth orbit with the orbital period equal to the sidereal rotational rate of earth. Thus, a satellite placed in a geosynchronous orbit appears to be fixed above a given point on the earth's surface. The radius of the geosynchronous orbit is easily calculated from earth's sideral period, $T_s = 23$ hr, 56 min, 4.09 s as follows:

$$r_f = \left(\frac{T_s\sqrt{\mu}}{2\pi}\right)^{\frac{2}{3}} = 42,164.17 \text{ km}.$$

The semi-major axis of the transfer ellipse is thus $a = \frac{r_i+r_f}{2} = 24,396.155$ km. The two impulse magnitudes are then calculated by

$$\Delta v_1 = \sqrt{\frac{2\mu}{r_i} - \frac{\mu}{a}} - \sqrt{\frac{\mu}{r_i}} = 2.44 \text{ km/s},$$

$$\Delta v_2 = \sqrt{\frac{\mu}{r_f}} - \sqrt{\frac{\mu}{r_f} - \frac{\mu}{a}} = 1.472 \text{ km/s} .$$

Both the impulses are applied in the direction of motion. The Hohmann transfer time is computed as follows:

$$t_H = \pi\sqrt{\frac{a^3}{\mu}} = 18,961.08 \text{ s (5 hr, 16 min, 1.08 s)}.$$

5.7 Relative Motion in Orbit

The relative motion between two spacecraft is an important dynamic problem from the viewpoint of rendezvous with an orbiting space station, targeting of satellites, and ejecting of small items from spacecraft. Many texts on space dynamics model such relative motion by approximate, linearized equations, called *Clohessy–Wiltshire equations* [9]. However, such a model is invalid when considering moderately large separation between spacecraft. We shall adopt the more general approach of exact two-body equations of relative motion, expressed in a rotating frame attached to the spacecraft in the circular orbit (called the *target*). In such a case, the orbital equations of the two objects are simultaneously solved, and the relative separation and velocity obtained by vector subtraction.

Hohmann transfer can be utilized in a planar orbital rendezvous between two spacecraft, initially in concentric circular orbits of different radii. Let the target spacecraft be in the higher orbit. The initial angular separation (phase angle) of the two spacecraft is given by the lead angle $\Delta\theta$. Since the rendezvous in a Hohmann transfer must take place at the apogee (Fig. 5.7), the object spacecraft must wait for some time prior to launching, until the phase angle becomes just right ($\theta = \theta_H$) for a successful rendezvous. This ideal value of the phase angle is calculated from the initial and final orbital radii as follows:

$$\frac{(\pi - \theta_H)}{2\pi} T_f = \frac{\pi(r_i + r_f)^{\frac{3}{2}}}{\sqrt{8\mu}}, \tag{5.46}$$

or,

$$\theta_H = \pi\left[1 - \left(\frac{1 + \frac{r_i}{r_f}}{2}\right)^{\frac{3}{2}}\right]. \tag{5.47}$$

Clearly, $0 \leq \theta_H \leq 116.36°$. The waiting time, t_w, necessary for the object spacecraft depends upon the value of the actual phase angle $\theta_H + \Delta\theta$. The maximum value of t_w occurs when $\Delta\theta = 2\pi$ and is called the *synodic period*, T_s, given by

$$T_s = \frac{2\pi}{n_i - n_f}, \tag{5.48}$$

where n_i, n_f are the orbital frequencies of the initial and final orbits, respectively. If the difference between the two orbital radii is small, the synodic period can be very large. For example, a transfer between earth orbits of altitudes 185 km and 222 km involves a synodic period of approximately one week. The waiting period is related to the synodic period by

$$t_w = \frac{\Delta\theta}{2\pi} T_s \,. \tag{5.49}$$

Thus, we arrive at the concept of *launch window* for achieving the Hohmann transfer for rendezvous with a space station, or a planet.

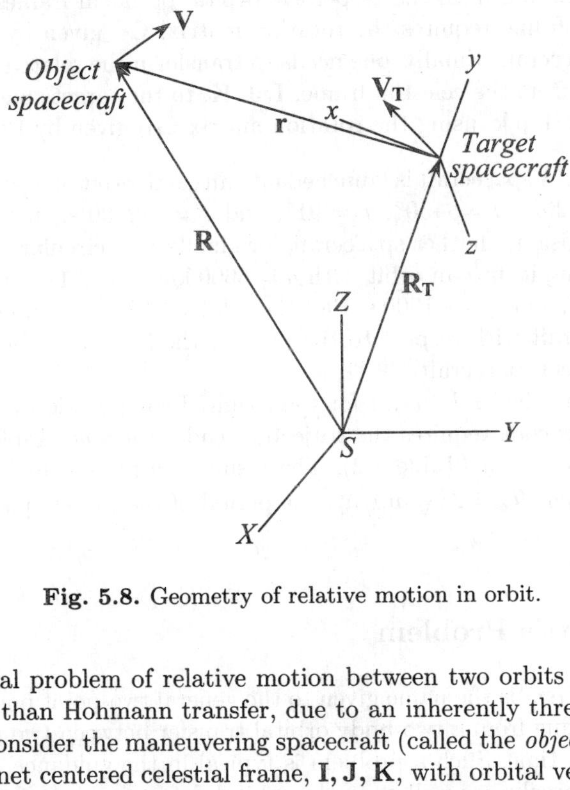

Fig. 5.8. Geometry of relative motion in orbit.

The actual problem of relative motion between two orbits is much more complicated than Hohmann transfer, due to an inherently three-dimensional character. Consider the maneuvering spacecraft (called the *object*), located by $\mathbf{R}$ in the planet centered celestial frame, $\mathbf{I}, \mathbf{J}, \mathbf{K}$, with orbital velocity $\mathbf{V}$. The target spacecraft is in an orbit defined by either the classical orbital elements, $a, e, i, \Omega, \omega, \tau$, or an initial condition, with instantaneous position and velocity, $\mathbf{R_T}, \mathbf{V_T}$. The relative position and velocity of the object are expressed in the local horizon frame fixed to the center of mass of the target, $\mathbf{i}, \mathbf{j}, \mathbf{k}$ (Fig. 5.8), as follows:

$$\begin{aligned} \mathbf{r} &\doteq \mathbf{R} - \mathbf{R_T} = x\mathbf{i} + y\mathbf{j} + z\mathbf{k}, \\ \mathbf{v} &\doteq \mathbf{V} - \mathbf{V_T} = \dot{x}\mathbf{i} + \dot{y}\mathbf{j} + \dot{z}\mathbf{k}. \end{aligned} \tag{5.50}$$

It is assumed that the initial conditions of the object at time $t = 0$ are known and given by the relative position and velocity, $\mathbf{r_0}, \mathbf{v_0}$. We can write a program for propagating the position of the two spacecraft in their respective orbits, using an appropriate method such as Lagrange's coefficients (Chapter 4):

$$\begin{Bmatrix} \mathbf{R} \\ \mathbf{V} \end{Bmatrix} = \begin{pmatrix} f & g \\ \dot{f} & \dot{g} \end{pmatrix} \begin{Bmatrix} \mathbf{R_o} \\ \mathbf{V_o} \end{Bmatrix} , \tag{5.51}$$

and

$$\begin{Bmatrix} \mathbf{R_T} \\ \mathbf{V_T} \end{Bmatrix} = \begin{pmatrix} f_T & g_T \\ \dot{f}_T & \dot{g}_T \end{pmatrix} \begin{Bmatrix} \mathbf{R_{TO}} \\ \mathbf{V_{TO}} \end{Bmatrix} . \tag{5.52}$$

The transformation from the respective orbital perifocal frames to the common celestial frame requires the rotation matrix, $\mathsf{C}*$, given by Eq. (5.3) for each each spacecraft. Finally, one needs to transform the relative position and velocity, $\mathbf{r}, \mathbf{v}$, from the celestial frame, $\mathbf{I}, \mathbf{J}, \mathbf{K}$, to the target spacecraft's local horizon frame, $\mathbf{i}, \mathbf{j}, \mathbf{k}$, using the rotation matrix $\mathsf{C_{LH}}$ given by Eq. (5.20).

Example 5.10. A spacecraft is launched into an earth orbit of $a = 6700$ km, $e = 0.01$ s, $\Omega = 120°$, $\omega = -50°$, $i = 91°$, and $\tau = -2000$ s. In order to avoid a possible collision, another spacecraft (originally in a circular orbit) fires its rockets, sending it into an orbit with $a = 6600$ km, $e = 0.1$ s, $\Omega = 110°$, $\omega = -25°$, $i = 10°$, and $\tau = -1000$ s. Plot the relative velocity and position of the second spacecraft with respect to the first for the first two orbital periods of the first (target) spacecraft.

A program called *relative.m* for the required computation is tabulated in Table 5.2. This code requires the trajectory code, *trajE.m* (Table 4.3), which in turn calls *kepler.m* (Table 4.2). The results are plotted in Fig. 5.9 up to $t = 2.25T_t$, where $T_t = 2\pi\sqrt{mu/a_t^3}$, the period of the target spacecraft.

5.8 Lambert's Problem

Lambert's problem is the name given to the general two-point boundary-value problem resulting from a two-body orbital transfer between two position vectors in a given time. Such a problem is typical in the guidance of spacecraft and ballistic missiles, as well as in the orbital determination of space objects from two observed positions separated by a specific time interval. Lambert's problem attracted the attention of the greatest mathematicians, such as Euler, Gauss, and Lagrange, and is responsible for advances in analytical and computational mechanics. We depict Lambert's problem in Fig. 5.10 by initial and final positions, P_i, P_f, given by $\mathbf{r_i}, \mathbf{r_f}$, and the transfer angle, θ, for the time of flight, $t_f - t_i$. Since the transfer orbit is coplanar with $\mathbf{r_i}, \mathbf{r_f}$, the two positions are uniquely specified through the radii, r_i, r_f, and the true anomalies, θ_i, θ_f. *Lambert's theorem* states that the transfer time is a function of the semi-major axis, a, of the transfer orbit, the sum, $r_i + r_f$, as well as of the chord, $c =| \mathbf{r_f} - \mathbf{r_i} |$, joining the two positions. For an elliptical transfer

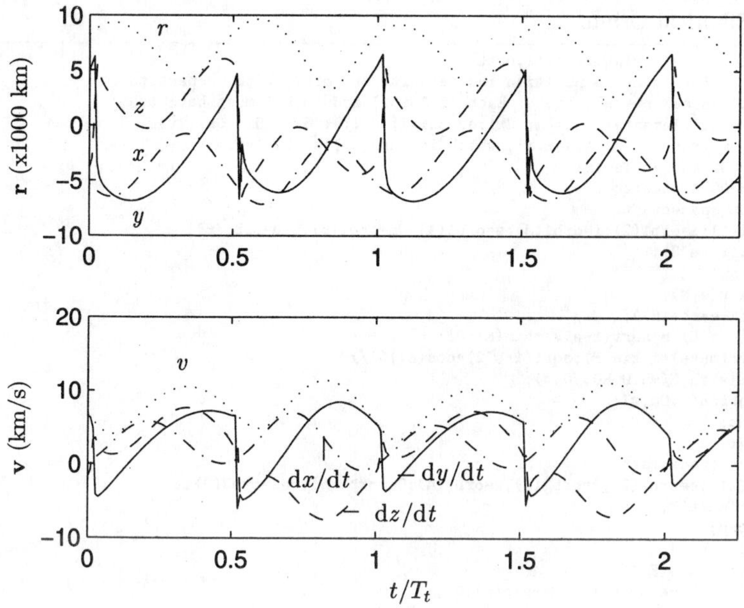

Fig. 5.9. Relative position and velocity of two spacecraft in elliptical orbits.

orbit, Lagrange derived the following equation [11] for the transfer time, using Kepler's equation:

$$\sqrt{\mu}(t_f - t_i) = 2a^{\frac{3}{2}}(\psi - \cos\phi \sin\psi) , \tag{5.53}$$

where

$$\begin{aligned} \cos\phi &= e \cos\frac{1}{2}(E_i + E_f) \\ \psi &= \frac{1}{2}(E_f - E_i) , \end{aligned} \tag{5.54}$$

and E_i, E_f denote the eccentric anomalies of the initial and final positions, repectively. Noting that

$$\begin{aligned} r_i + r_f &= 2a(1 - \cos\phi\cos\psi), \\ c &= 2a\sin\phi\sin\psi , \end{aligned} \tag{5.55}$$

the proof of Lambert's theorem can be obtained for the elliptical orbit. The following analogous expression was obtained by Euler for the parabolic transfer using Barker's equation (Chapter 4) and is relevant to orbital determination of comets:

$$6\sqrt{\mu}(t_f - t_i) = (r_i + r_f + c)^{\frac{3}{2}} \pm (r_i + r_f - c)^{\frac{3}{2}} , \tag{5.56}$$

Table 5.2. M-file *relative.m* for Computing Relative Motion between Two Spacecraft in Elliptical Orbits

```
function [r,v]=relative(orb1,orb2,t)
%program for relative position and velocity of 'orb2' with respect to
%'orb1' in elliptical orbits. Each of 'orb1' and 'orb2' must have the
%following elements: 1x1: a; 2x1:e; 3x1:i; 4x1:w; 5x1: Om; 6x1: tau.
%This program calls 'trajE.m', which in turn calls 'kepler.m'
%(c) 2005 Ashish Tewari
mu=398600.4; %earth
%target spacecraft:
a=orb1(1);e=orb1(2);i=orb1(3);w=orb1(4);Om=orb1(5);tau=orb1(6);
n=sqrt(mu/a^3);
M=-n*tau;
E=kepler(e,M);
r0=a*(1-e*cos(E));
R0=a*[cos(E)-e;sqrt(1-e^2)*sin(E);0];
V0=sqrt(mu*a)*[-sin(E);sqrt(1-e^2)*cos(E);0]/r0;
[Rt,Vt]=trajE(mu,0,R0,V0,t);
C=rotation(i,Om,w);
Rt=C*Rt;
Vt=C*Vt;
%Object spacecraft:
a=orb2(1);e=orb2(2);i=orb2(3);w=orb2(4);Om=orb2(5);tau=orb2(6);
n=sqrt(mu/a^3);
M=-n*tau;
E=kepler(e,M);
r0=a*(1-e*cos(E));
R0=a*[cos(E)-e;sqrt(1-e^2)*sin(E);0];
V0=sqrt(mu*a)*[-sin(E);sqrt(1-e^2)*cos(E);0]/r0;
[R,V]=trajE(mu,0,R0,V0,t);
C=rotation(i,Om,w);
R=C*R;
V=C*V;
%relative position and velocity:
r=R-Rt;
v=V-Vt;
rt=norm(Rt);
lat=asin(dot(Rt,[0;0;1])/rt);
slon=dot(Rt,[0;1;0])/(rt*cos(lat));
clon=dot(Rt,[1;0;0])/(rt*cos(lat));
long=atan(slon/clon);
if slon<0 && clon>0
    long=asin(slon);
elseif slon>0 && clon<0
    long=acos(clon);
end
CLH=INtoLH(lat,long);
r=CLH*r;
v=CLH*v;
```

where the positive sign is taken for the transfer angle, $\theta > \pi$, and the negative for $\theta < \pi$. Clearly, Lambert's theorem is also valid for the parabolic transfer. For a hyperbola, Lambert's theroem can be proved similarly by writing

$$\sqrt{\mu}(t_f - t_i) = 2(-a)^{\frac{3}{2}}(\cosh\phi \sinh\psi - \psi)\,, \tag{5.57}$$

where

$$\cosh\phi = e\cosh\frac{1}{2}(H_i + H_f),$$

$$\psi = \frac{1}{2}(H_f - H_i) \,, \tag{5.58}$$

and H_i, H_f denote the hyperbolic anomalies of the initial and final positions, repectively. The expressions for the sum of radii and chord for the hyperbolic case are

$$\begin{aligned} r_i + r_f &= 2a(1 - \cosh\phi \cosh\psi), \\ c &= -2a \sinh\phi \sinh\psi \,. \end{aligned} \tag{5.59}$$

For convenience in solving Lambert's problem, we define the semi-parameter, s, of the triangle SP_iP_f, as

$$s \doteq \frac{r_i + r_f + c}{2} \,. \tag{5.60}$$

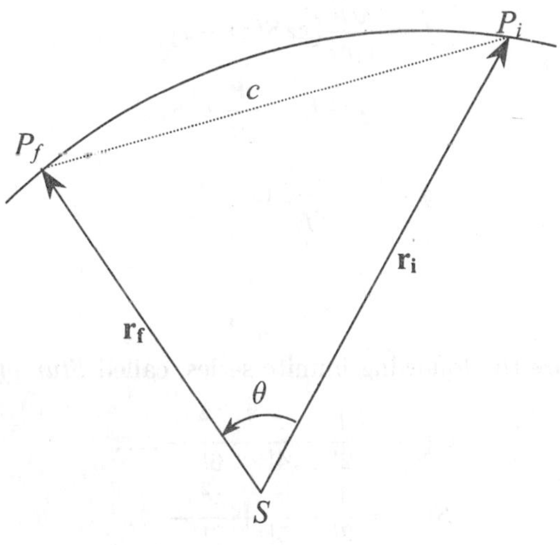

Fig. 5.10. Geometry of Lambert's problem.

Note that for $\theta = \pi$, we have $c = r_i + r_f$, or, $s = c$. We are usually interested in the minimum transfer energy, which corresponds to the smallest value of orbital energy $\epsilon = -\mu/2a$. Clearly, the minimum energy orbit involves the smallest positive value of the semi-major axis, a, joining the two points, P_i, P_f. Let this minimum energy elliptical orbit have the semi-major axis a_m. Then, it can be shown from geometry that $a_m = s/2$. Thus, the semi-parameter, s, has a physical significance in terms of the minimum energy of transfer.

A solution to Lambert's problem was first presented by Gauss in his *Theoria Motus* (1801), using three angular positions of the asteroid *Ceres* at three time instants. The numerical solution procedure adopted by Gauss for elliptical orbits was singular for $\theta = 180°$. A modern solution to Lambert's problem invariably involves a more sophisticated algorithm for an efficient implementation on a digital computer, without convergence and singularity issues. We shall utilize the universal variable approach by Battin [10] in deriving a general algorithm for Lambert's problem, that is applicable to elliptical, parabolic, and hyperbolic transfer trajectories. Let us define an auxiliary variable, x, by

$$x \doteq E\sqrt{a}\,, \tag{5.61}$$

where $E \doteq E_f - E_i$ is the difference between the eccentric anomalies of the final and initial positions, and a is the semi-major axis of the transfer orbit. In terms of x, the Lagrange's coefficients, f, g (Chapter 4) can be expressed as follows [10]:

$$\begin{aligned}
f &= 1 - \frac{x^2}{r_i}C(z), \\
\dot{f} &= \frac{\sqrt{\mu}}{r_i r_f}[xzS(z) - x], \\
g &= t_f - t_i - \frac{1}{\sqrt{\mu}}x^3 S(z), \\
\dot{g} &= 1 - \frac{x^2}{r_f}C(z),
\end{aligned} \tag{5.62}$$

where

$$z \doteq \frac{x^2}{a}\,, \tag{5.63}$$

and $C(z), S(z)$ are the following infinite series, called *Stumpff functions* [56]:

$$\begin{aligned}
C(z) &= \frac{1}{2!} - \frac{z}{4!} + \frac{z^2}{6!} - \dots, \\
S(z) &= \frac{1}{3!} - \frac{z}{5!} + \frac{z^2}{7!} - \dots .
\end{aligned} \tag{5.64}$$

It can be shown that Stumpff functions converge to the following for an elliptical transfer orbit ($z > 0$):

$$\begin{aligned}
C(z) &= \frac{1 - \cos\sqrt{z}}{2}, \\
S(z) &= \frac{\sqrt{z} - \sin\sqrt{z}}{z^{\frac{3}{2}}}\,,
\end{aligned} \tag{5.65}$$

whereas for a hyperbolic transfer ($z < 0$), we have

$$\begin{aligned}
C(z) &= \frac{\cosh\sqrt{-z} - 1}{-z}, \\
S(z) &= \frac{-\sqrt{-z} + \sinh\sqrt{-z}}{(-z)^{\frac{3}{2}}}\,.
\end{aligned} \tag{5.66}$$

Obviously, the parabolic case, $z = 0$, has $C(0) = \frac{1}{2}$, $S(0) = \frac{1}{6}$. In order to solve for the unknown variable x, another related auxiliary parameter, y is defined as follows:

$$y \doteq \frac{r_i r_f}{p}(1 - \cos\theta) \,, \tag{5.67}$$

where p is the parameter of the transfer orbit. Then, comparing Eq. (5.62) with the following definitions of Lagrange's coefficients (Chapter 4):

$$\begin{aligned} f &= 1 - \frac{r_f}{p}(1 - \cos\theta), \\ \dot{f} &= \sqrt{\frac{\mu}{p}}\left(\frac{1-\cos\theta}{\theta}\right)\left[\frac{1-\cos\theta}{p} - \frac{1}{r_i} - \frac{1}{r_f}\right], \\ g &= \frac{r_i r_f}{\sqrt{\mu p}}\sin\theta, \\ \dot{g} &= 1 - \frac{r_i}{p}(1 - \cos\theta) \,, \end{aligned} \tag{5.68}$$

we have

$$x = \sqrt{\frac{y}{C(z)}} \,, \tag{5.69}$$

where x is the solution to the following cubic equation:

$$\sqrt{\mu}(t_f - t_i) = A\sqrt{C}x + Sx^3 \,, \tag{5.70}$$

and A is the following:

$$A \doteq \sin\theta\sqrt{\frac{r_i r_f}{1 - \cos\theta}} \,. \tag{5.71}$$

In terms of the universal variables, it is now a simple matter to write

$$f = 1 - \frac{y}{r_i},$$

$$g = A\sqrt{\frac{y}{\mu}}, \tag{5.72}$$

$$\dot{g} = 1 - \frac{y}{r_f} \,. \tag{5.73}$$

The fourth Lagrange coefficient, $\dot{f}$, is obtained from the relationship $f\dot{g} - g\dot{f} = 1$ (Chapter 4). Finally, we can determine the initial and final velocities of the transfer as follows:

$$\begin{aligned} \mathbf{v_i} &= \frac{1}{g}(\mathbf{r_f} - f\mathbf{r_i}), \\ \mathbf{v_f} &= \dot{f}\mathbf{r_i} + \dot{g}\mathbf{v_i}) \,. \end{aligned} \tag{5.74}$$

Based upon the above given formulation, the following iterative algorithm can be devised for the solution to Lambert's problem:

(a) Based upon the initial and final position vectors, determine the transfer angle based upon whether the transfer orbit is *direct* ($i \leq 90°$) or *retrograde* ($i > 90°$). For direct orbits, the transfer angle is calculated as follows:

$$\theta = \begin{cases} cos^{-1}\left(\frac{\mathbf{r_i}\cdot\mathbf{r_f}}{r_i r_f}\right), & (\alpha > 0), \\ 2\pi - cos^{-1}\left(\frac{\mathbf{r_i}\cdot\mathbf{r_f}}{r_i r_f}\right), & (\alpha < 0), \end{cases} \tag{5.75}$$

where

$$\alpha \doteq \mathbf{K}\cdot(\mathbf{r_i}\times\mathbf{r_f}) \ . \tag{5.76}$$

(b) Calculate A using Eq. (5.71).
(c) Assume a value for z, usually a small, positive number.
(d) Compute the Stumpff functions, $C(z), S(z)$, from Eq. (5.64).
(e) Calculate y, x, and finally the transfer time using Eq. (5.70).
(f) If the transfer time is close to the desired value within a specified tolerance, then go to step (h).
(g) Estimate a new value of z by Newton's method, without taking into account the dependence of the Stumpff functions on z, and go back to step (d).
(h) The solution, z, has converged within the specified tolerance. Find Lagrange's coefficients through Eq. (5.72), and the initial and final velocities by Eq. (5.74).

The algorithm given above is programmed in the code *lambert.m* (Table 5.3). It is to be noted that the logic for transfer angle determination in step (a) must be reversed for a retrograde transfer orbit. Also note the fine tolerance of 10^{-9} s specified in the code for transfer time convergence. In most cases, such a small tolerance is achieved in about 20 iterations. Of course, increasing the tolerance results in a smaller number of iterations. Battin [11] presents an alternative procedure through the use of *hypergeometric* series evaluated by continued fractions, instead of the Stumpff functions. Such a formulation, although difficult to program, improves the convergence and renders the method insensitive to initial choice of z.

Example 5.11. It is decided to make a rendezvous between the two spacecraft of Example 5.10, by maneuvering the object spacecraft, such that it attains the same position and velocity as that of the target after 2000 sec. Find the maneuvering impulse magnitudes and directions in the celestial frame.

In order to solve the problem, we need the initial position and velocity of the object, as well as the final position and velocity of the target, which are easily computed from the given orbits to be the following:

$$\mathbf{R}(0) = \begin{pmatrix} -6069.39667 \\ 1845.10683 \\ 894.38397 \end{pmatrix} \text{km}, \quad \mathbf{V}(0) = \begin{pmatrix} -2.9185425 \\ -7.3934070 \\ 0.9294595 \end{pmatrix} \text{km/s},$$

Table 5.3. M-file *lambert.m* for Solving Lambert's Problem

```
function [p,Vi,Vf]=lambert(mu,Ri,Rf,t)
%Solution to Lambert's two-point boundary-value problem in space navigation.
%mu: gravitational constant of central mass
%Ri: initial position vector in celestial frame
%Rf: final position vector in celestial frame
%t: transfer time
%The transfer orbit is assumed to be direct (0<i<90 deg.)
%(For a retrograde orbit, the logic based upon sign of 'q' must be reversed.)
%(c) 2005 Ashish Tewari
ri=norm(Ri); rf=norm(Rf);
q=dot([0;0;1],cross(Ri,Rf));
theta=acos(dot(Ri,Rf)/(ri*rf)); %transfer angle
if q<0
    theta=2*pi-theta;
end
A=sqrt(ri*rf/(1-cos(theta)))*sin(theta);
z=0.01;
n=1;
[C,S]=stumpff(z,5);
y=ri+rf-A*(1-z*S)/sqrt(C);
x=sqrt(y/C);
tc=(x^3*S+A*sqrt(y))/sqrt(mu);
%Newton's iteration for 'z' follows:
while abs(t-tc)>1e-9
    fx=A*sqrt(C)*x+S*x^3-t*sqrt(mu);
    fxp=A*sqrt(C)+3*S*x^2;
    dx=-fx/fxp;
    x=x+dx;
    n=n+1;
    y=C*x*x;
    tc=(x^3*S+A*sqrt(y))/sqrt(mu);
    z=(1-sqrt(C)*(ri+rf-y)/A)/S;
    [C,S]=stumpff(z,20);
end
n
z
p=ri*rf*(1-cos(phi))/y;
f=1-y/ri;
g=A*sqrt(y/mu);
gd=1-y/rf;
Vi=(Rf-f*Ri)/g;
Vf=(gd*Rf-Ri)/g;
```

Table 5.4. M-file *stumpff.m* for Stumpff Functions Required in *lambert.m*

```
function [C,S]=stumpff(z,n)
%Stumpff functions evaluation:
C=0.5;
S=1/6;
for i=1:n
    C=C+(-1)^i*z^i/factorial(2*(i+1));
    S=S+(-1)^i*z^i/factorial(2*i+3);
end
```

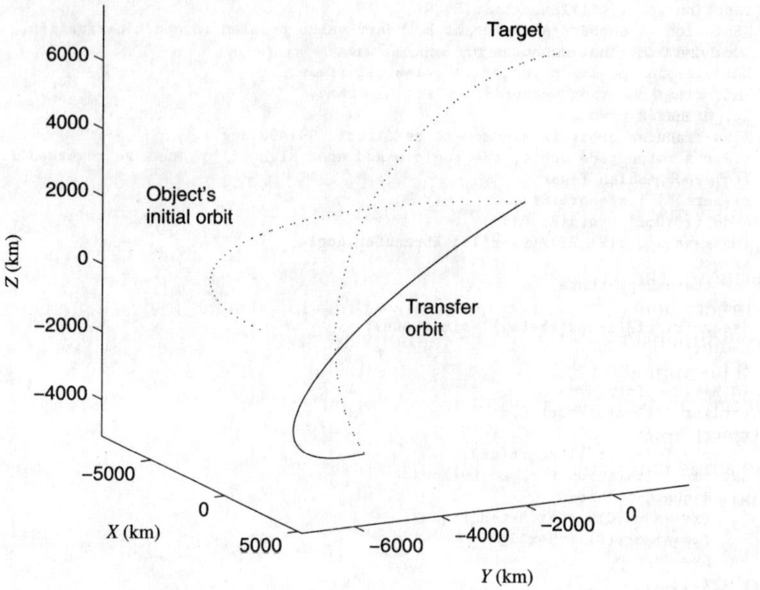

Fig. 5.11. Nonplanar orbital maneuver for rendezvous between two spacecraft using the solution to Lambert's problem.

$$\mathbf{R_T}(2000) = \begin{pmatrix} 2767.54394 \\ -4920.02168 \\ -3623.44589 \end{pmatrix} \text{km}, \qquad \mathbf{V_T}(2000) = \begin{pmatrix} -2.2105642 \\ 3.6039760 \\ -6.4403484 \end{pmatrix} \text{km/s} .$$

Thus, we have $\mathbf{R_i} = \mathbf{R}(0)$, $\mathbf{R_f} = \mathbf{R_T}(2000)$, and $t_f - t_i = 2000$ s. The Lambert's for this example is solved through the program *lambert.m* (Table 5.3) and the following MATLAB statements:

```
>> [p,Vi,Vf]=lambert(398600.4,Ri,Rf,2000)

theta = 2.31520925333295
n =     15
z = 5.59221293727655
p = 6.482521562308273e+003

Vi =     -2.73191506291483
     -5.72927382287410
     -4.76913693051223


Vf =     6.83337026980329
     2.23793401599097
     2.39519863911652

>> dv1=Vi-V  %first velocity impulse (t=0) (km/s)

dv1 =    0.18662745639649
     1.66413315097974
     -5.69859646882602
```

```
>> dv2=VT-Vf  %second velocity impulse (t=2000s) (km/s)

dv2 =   -9.04393442646690
         1.36604201944881
        -8.83554699120937
```

Therefore, for the transfer angle of $\theta = 2.3152$ rad (132.65°), the necessary impulse magnitudes are $\Delta v_1 = 5.939543$ km/s and $\Delta v_2 = 12.717142$ km/s, which are enormous due to the plane change involved in the maneuver. Furthermore, the target is in a retrograde orbit, while the object's initial orbit is direct; hence, a switching of direction takes place in the transfer orbit. A three-dimensional plot of the maneuver is depicted in Fig. 5.11.

The maneuver shown in this example is an extreme case, thereby requiring a large propellant expenditure. A practical orbital rendezvous takes place between spacecraft in nearly coplanar orbits, moving in the same direction. The same can be said for Lambert's problem arising out of interplanetary (or lunar) travel.

5.9 Summary

The classical orbital elements, $a, e, \tau, \Omega, \omega, i$, are very useful in describing the position and velocity of a spacecraft in the celestial coordinate frame. The problem of orbit determination involves determining the orbital elements from the measured position and velocity at a given time. When a planet fixed frame is employed, an appropriate coordinate transformation must be carried out taking into account the planetary rotation in terms of terrestrial and celestial longitudes. Impulsive orbital maneuvers are a reasonable approximation of the actual changes in the orbit caused by rocket thrusters. The simplest maneuver consists of a single-velocity impulse and is capable of changing both the shape and the plane of the orbit. Plane change maneuvers are the most expensive and must be performed at the smallest possible velocity. The most efficient two impulses, planar orbital maneuvers are those that involve only tangential impulses (such as Hohmann transfer between two circular orbits). An orbital rendezvous requires appropriate phasing between the object and the target spacecraft, as well as orbital transfer. A general three-dimensional rendezvous in a specified time results in a two-point boundary-value problem—called Lambert's problem—and forms the basis of orbital and interplanetary mission design. The solution of Lambert's problem invariably requires an iterative numerical procedure based upon a general formulation in terms of auxiliary variables and Lagrange's coefficients.

Exercises

5.1. The angle, ϵ, is the inclination of the celestial, equatorial frame, $(SXY'Z')$, relative to the celestial, ecliptic frame, $(SXYZ)$, about the common axis, SX, and is called *obliquity of ecliptic*. Derive the relationships between the ecliptic spherical coordinates, λ', δ', and the equatorial speherical coordinates λ, δ.

5.2. A spacecraft is returning from Mars with the following approach velocity relative to the earth in a geocentric (equatorial) frame:

$$\mathbf{v} = 3\mathbf{I} + 3\mathbf{J} + 2\mathbf{K}\ \text{km/s}\,,$$

where $\mathbf{K}$ is the north polar axis. Given that the spacecraft should enter the earth's atmosphere at altitude 100 km and latitude 8°, find the length, r_a, of the radial line from the earth's center that is perpendicular to an asymptote of the approach hyperbola. (This length is called the *aiming distance* and is the distance by which the spacecraft would miss the planet, were it not influenced by the planetary gravity.) Also, calculate the flight path angle of atmospheric entry.

5.3. A satellite is in an orbit of $e = 0.00132$, $i = 89.1°$ and $\omega = 261°$. Its perigee altitude is 917 km. What is the maximum height achieved by the spacecraft above the equatorial plane at any time in the orbit?

5.4. Using the coordinate transformation between the celestial and local horizon frames, derive the relationship between orbital inclination, inertial velocity azimuth, and declination given by Eq. (5.27).

5.5. A spacecraft is de-orbited from a circular, earth orbit of altitude 300 km and equatorial inclination 60°, by firing a retro-rocket that reduces the speed instantaneously by 200 m/s, without changing either the flight-path angle or the orbital inclination. The firing occurs at a declination of 45°, and a right ascension of 20°, when the spacecraft is moving toward higher latitudes. Determine the celestial position and velocity vectors, as well as declination, right ascension, flight-path angle, and inertial velocity azimuth 15 min after the firing of the retro-rocket.

5.6. A spacecraft is to be launched to a circular earth orbit of altitude 200 km from Cape Canaveral (latitude 28.5°). Assuming that the entire launch sequence can be approximated by a single-velocity impulse at the earth's surface, and that the atmospheric drag and thrust losses have no effect on the flight path, other than requiring an addition of 1 km/s to the magnitude of the initial velocity impulse, calculate the magnitude and direction of the launch velocity in the planet fixed frame.

5.7. A spacecraft is required to undergo a plane change by 20°, as well as a speed change from 7.3 km/s to 8.2 km/s, with a 5° increase in the flight-path angle. Determine the total velocity impulse required when (a) the speed and flight path are changed together, while the plane change is separately performed, and (b) all the three changes take place simultaneously.

5.8. A spacecraft is in an earth orbit of $e = 0.7$. When its altitude is 500 km and the speed is 7.4 km/s, a velocity impulse of magnitude 500 m/s is applied at an angle 60° from the initial flight direction, in order to increase the speed and flight-path angle simultaneously, but without changing the orbital plane. Determine the new orbit of the spacecraft.

5.9. An earth communications satellite has achieved an equatorial orbit, but the apogee altitude is 41,756 km while the eccentricity is 0.0661. What minimum total velocity change is required to place the satellite in the geosynchronous orbit?

5.10. Estimate the time of flight and total velocity change required for a Mars mission if the angular heliocentric positions of the earth and Mars at the time of launch, measured along the respective orbits from the intended rendezvous point, are 90° and 140°, respectively. Assume that Mars' orbital plane is inclined at 2° to the ecliptic, and its sidereal orbital period is 1.881 times that of the earth. Also, assume that the planetary orbits are circular, and the planetary gravitation on the spacecraft is negligible. (Take $\mu_{\text{sun}} = 1.32712440018 \times 10^{11}\ \text{km}^3/\text{s}^2$).

5.11. An *outer bi-elliptic transfer* is a three-impulse orbital transfer between two concentric circular orbits, such that all the impulses are applied tangentially, and the two transfer ellipses have semi-major axes greater than the radius of the larger circle. Show that an outer bi-elliptic transfer with an initial phase angle $\theta_H + \Delta\theta$ has the total transfer time given by

$$t = t_H + \frac{T_f}{2}\left(3 - \frac{\Delta\theta}{\pi}\right),$$

which is smaller than that required for Hohmann transfer if $\Delta\theta > 3\pi(1 - T_i/T_f)$.

5.12. Show that an outer bi-elliptic transfer between two concentric circular orbits is more economical than Hohmann transfer between the same orbits if the ratio of the two orbital radii is greater than 15.6.

5.13. It is possible to exactly solve the equations of relative motion between a target in a circular orbit, and an object in a general orbit, which is always coplanar with that of the target. Derive an algorithm for analytically determining the relative position and velocity after some time, given the initial *nondimensional* relative coordinates $x_0, y_0, \dot{x}_0, \dot{y}_0$, defined by

$$\mathbf{r} \doteq \frac{\mathbf{R} - \mathbf{R_T}}{R_T} = x\mathbf{i} + y\mathbf{j},$$

$$\mathbf{v} \doteq \frac{\mathbf{V} - \mathbf{V_T}}{V_T} = \dot{x}\mathbf{i} + \dot{y}\mathbf{j},$$

as well as the initial angular separation, θ_0.

5.14. A small experimental satellite is ejected radially outward with a speed $\frac{1}{10}$ the orbital speed, from a spacecraft in a circular earth orbit of period 120 min. Using the exact solution devised in Exercise 5.13, find the relative distance of the satellite after one complete orbital period of the spacecraft.

5.15. Repeat the calculation of Example 5.11 using a transfer time of 3000 s. What are the impulse magnitudes for this rendezvous maneuver?

5.16. The orbital elements of the planet Mars referred to the mean ecliptic plane are $a = 1.52372$ a.u., $e = 0.09331$ s, $\Omega = 49.55°$, $\omega = 336.011°$, and $i = 1.8498°$. Assuming Mars and earth last passed perihelion exactly 300 and 55 days ago, respectively, design a transfer orbit for a spacecraft launched from the earth ($a = 1$ a.u., $e = 0.01667$ s, $\omega = 103.059°$, and $i = 0°$) in order that it reaches Mars at the next Mars aphelion. Also, find the velocity impulses necessary to perform the flight. Neglect the effects of planetary gravitation on the spacecraft (1 a.u. = 149597870.691 km, $\mu_{\text{sun}} = 1.32712440018 \times 10^{11}$ km^3/s^2).
(Hint: Use *relative.m* with appropriate μ for relative positions of the earth and Mars at launch and arrival, respectively. Then solve the given Lambert's problem, and get the transfer orbital elements, as well as the initial and final velocity impulses.)

6

Perturbed Orbits

6.1 Aims and Objectives

- To model the effects of orbital perturbations caused by gravitational asymmetry due to oblateness and presence of a third body.
- To present a simple atmospheric model for predicting the life of a satellite in a low orbit.
- To introduce the concept of sphere of influence and the patched conic approach for the design and analysis of lunar and interplanetary missions.
- To solve the perturbed two-body problem using numerical integration (Cowell's and Encke's methods).

6.2 Perturbing Acceleration

The Keplerian motion presented in Chapter 5 is often perturbed by external disturbances, leading to deviation from the orbits obtained from the exact, closed-form solutions to the two-body problem. Examples of such disturbances include a nonuniform gravitational field due to an aspherical planetary mass (Chapter 3), atmospheric drag (Chapter 9), gravitational influence of a third body (Chapter 7), thrust of rocket engines (Chapter 8), and solar radiation pressure. Often, one can identify the largest disturbances acting on the spacecraft, and model the vehicle's motion accordingly. For example, a vehicle in a low orbit experiences the largest orbital perturbations due to atmospheric drag and nonspherical gravity field, while a satellite in a high earth orbit feels negligible atmospheric drag and planetary gravitational anomaly but an appreciable tug from lunar and solar gravitation, as well as some effects of solar radiation. A spacecraft on an interplanetary voyage is affected by gravitational attraction of a planet, near which it may be passing, apart from solar gravitation. Such a spacecraft, when flying toward the inner planets (Venus and Mercury), may experience a significant increase in solar radiation pressure. Our objective in this chapter is to present the modeling and simulation of the two-body orbits perturbed by various disturbances, as well as to introduce the strategy commonly adopted in designing interplanetary trajectories.

The equation of perturbed relative motion in the two-body problem (Chapter 4) can be expressed as follows:

$$\frac{d^2\mathbf{r}}{dt^2} + \frac{\mu}{r^3}\mathbf{r} = \mathbf{a_d} \; , \tag{6.1}$$

where $\mathbf{a_d}$ is the perturbing acceleration on mass m_2, whose position relative to m_1 is $\mathbf{r}$ (Fig. 4.8). The perturbing force, $m_2\mathbf{a_d}$, can be either conservative, [1] such as the gravitational anomalies due to aspherical shape and other bodies, or nonconservative, such as atmospheric drag, rocket thrust, and solar radiation pressure. We have seen in Chapters 3 and 4 how a conservative force can be expressed by the gradient of a scalar potential function, Φ, with respect to the position vector, $\mathbf{r}$. Thus, for a conservative perturbation, we have

$$\mathbf{a_d}^T = \frac{\partial \Phi}{\partial \mathbf{r}} \; . \tag{6.2}$$

This form of conservative acceleration allows us to express the solution to the perturbed problem in infinite series, such as those involving the *Legendre polynomials* (Chapter 3), and enables approximate computation. However, such a simple evaluation is not possible for a nonconservative disturbance.

6.3 Effects of Planetary Oblateness

Lagrange devised the following method of *variation of parameters* [11] for expressing the changes in the orbital parameters with time caused by a conservative perturbation. Upon substituting Eq. (6.2) into Eq. (6.1), we have

$$\frac{d\mathbf{v}}{dt} + \frac{\mu}{r^3}\mathbf{r} = \left(\frac{\partial \Phi}{\partial \mathbf{r}}\right)^T \; , \tag{6.3}$$

where

$$\mathbf{v} = \frac{d\mathbf{r}}{dt} \tag{6.4}$$

is the relative velocity. We can express the position and velocity as functions of the six scalar constants of the undisturbed two-body orbital motion comprising the vector $\mathbf{c}$ as follows:

$$\mathbf{r} = \mathbf{r}(t, \mathbf{c}), \qquad \mathbf{v} = \mathbf{v}(t, \mathbf{c}) \; . \tag{6.5}$$

Note that the orbital parameters, $\mathbf{c}$, are allowed to vary with time under the influence of the perturbation. Since the position and velocity are functions

[1] See Chapter 4 for the definition of a conservative force.

not only of time, but also of the varying parameters, $\mathbf{c}$, the equations of undisturbed motion can be expressed as follows:

$$\frac{\partial \mathbf{v}}{\partial t} + \frac{\mu}{r^3}\mathbf{r} = \mathbf{0}, \qquad \frac{\partial \mathbf{r}}{\partial t} = \mathbf{v}. \tag{6.6}$$

The velocity and acceleration of the disturbed motion are given by

$$\frac{d\mathbf{r}}{dt} = \frac{\partial \mathbf{r}}{\partial t} + \frac{\partial \mathbf{r}}{\partial \mathbf{c}}\frac{d\mathbf{c}}{dt}, \qquad \frac{d\mathbf{v}}{dt} = \frac{\partial \mathbf{v}}{\partial t} + \frac{\partial \mathbf{v}}{\partial \mathbf{c}}\frac{d\mathbf{c}}{dt}. \tag{6.7}$$

Considering the particular form of the linear differential equations, Eqs. (6.3) and (6.4), the following conditions must be satisfied by the time derivatives of the parameters:

$$\frac{\partial \mathbf{r}}{\partial \mathbf{c}}\frac{d\mathbf{c}}{dt} = \mathbf{0}, \qquad \frac{\partial \mathbf{v}}{\partial \mathbf{c}}\frac{d\mathbf{c}}{dt} = \left(\frac{\partial \Phi}{\partial \mathbf{r}}\right)^T. \tag{6.8}$$

These are then the six scalar ordinary differential equations to be solved for the orbital parameters, $\mathbf{c}(t)$. A slight manipulation of Eq. (6.8) yields

$$\mathsf{L}\frac{d\mathbf{c}}{dt} = \left(\frac{\partial \Phi}{\partial \mathbf{r}}\right)^T, \tag{6.9}$$

where L is the following six-dimensional square matrix, called the *Lagrange matrix*:

$$\mathsf{L} \doteq \left(\frac{\partial \mathbf{r}}{\partial \mathbf{c}}\right)^T \frac{\partial \mathbf{v}}{\partial \mathbf{c}} - \left(\frac{\partial \mathbf{v}}{\partial \mathbf{c}}\right)^T \frac{\partial \mathbf{r}}{\partial \mathbf{c}}. \tag{6.10}$$

It can be easily shown that L has the following properties:

$$\mathsf{L}^T = -\mathsf{L}, \qquad \frac{\partial \mathsf{L}}{\partial t} = \mathbf{0}. \tag{6.11}$$

Thus, L is a skew-symmetric matrix and does not depend explicitly on time. The latter feature of the Lagrange matrix is very useful in solving for the parameters, $\mathbf{c}(t)$. Selecting an appropriate set of orbital parameters, such as the classical orbital elements $\mathbf{c}^T = (a, e, i, \omega, \Omega, \tau)$, the equations of parametric variations can be expressed in the form of six scalar differential equations, called *Lagrange's planetary equations*.

Lagrange's planetary equations lend themselves easily to the determination of the effect of planetary oblateness, measured by the largest Jeffrey's constant, J_2 (Chapter 3). For a satellite in an elliptical orbit, it can be shown that a substitution of Eq. (3.32) into Eq. (6.9) results in

$$\begin{aligned}
\frac{d\bar{a}}{dt} &= 0, \\
\frac{d\bar{e}}{dt} &= 0, \\
\frac{d\bar{\omega}}{dt} &= \frac{3}{4} n J_2 \left(\frac{R_e}{p}\right)^2 (5\cos^2 i - 1), \\
\frac{d\bar{\Omega}}{dt} &= -\frac{3}{2} n J_2 \left(\frac{R_e}{p}\right)^2 \cos i,
\end{aligned} \tag{6.12}$$

where the bar represents the average of the particular parameter taken over a complete orbit. Thus, the planetary oblateness has no effect on the semi-major axis and eccentricity, while the line of apsides as well as the line of nodes rotate at fixed rates. For a direct orbit ($0 < i < 90°$), the line of nodes rotates *backwards*, such that the value of Ω continuously decreases. This behavior is called *regression of nodes*. The movement is in opposite direction for a retrograde orbit. For the *apsidal rotation*, there is a critical value of inclination, $i_c = \cos^{-1}\frac{1}{\sqrt{5}}$, which determines the direction of change of argument of periapsis, ω. If $i > i_c$, the line of apsides regresses. Clearly, both nodal and apsidal rotation rates diminish as the orbit size increases (larger p, smaller n) and are directly proportional to the oblateness, J_2.

6.3.1 Sun Synchronous Orbits

A beneficial effect of nodal regression is the *sun synchronous orbit* utilized effectively in all photographic mapping and observation satellites. The sun synchronous orbit yields a constant sun elevation angle at a given point, which is highly desirable for taking photographs of surface features. A satellite's orbital plane continuously departs from the line joining the sun and the planet (the direction of sun visible from the satellite). The sun synchronization is achieved by matching the rate of departure of the satellite's orbital plane from the sun's direction, $\frac{360°}{365.25} = 0.9856°/\text{day}$, with an equal and opposite rate of nodal regression, $\dot{\Omega}$, thereby rendering a constant sun direction relative to the orbital plane. Such a synchronous orbit would cross a given latitude at a fixed solar (local) time; thus, photographs of fixed points on the surface can be taken in approximately the same lighting conditions. Since J_2 is a positive number, a positive nodal regression rate can be achieved only by having $\cos i < 0$, implying $i > 90°$. Hence, a sun synchronous orbit is always retrograde. For planets such as the earth and Mars, the low sun synchronous orbits used in photography missions are nearly polar, which gives the additional advantage of covering the whole planet in a relatively small number of orbits.

Example 6.1. Calculate the orbital inclination of a sun synchronous earth satellite of $a = 6700$ km and $e = 0.01$.

The parameter and mean motion of the satellite are determined as follows:

$$p = a(1 - e^2) = 6699.33 \text{ km}, \qquad n = \sqrt{\frac{\mu}{a^3}} = 0.094231057 \text{ rad/s} .$$

For earth, $R_e = 6378.14$ km, $J_2 = 0.00108263$ (Chapter 3), which results in the following calculation of the orbital inclination by the last of Eq. (6.12):

$$\frac{2\pi}{(365.25)(24)(3600)} = -\frac{3}{2} n J_2 \left(\frac{R_e}{p}\right)^2 \cos i,$$

$$i = \cos^{-1}(-0.00143544) = 90.082245° .$$

As the size of the orbit increases, the inclination angle for sun synchronization also increases. For example, the *LANDSAT* series of earth observation satellites with a near circular orbit of 709 km altitude has a sun synchronous orbit of $i = 98.2°$, which crosses the equator at the 9:30 a.m. local time.

6.3.2 Molniya Orbits

Another class of special orbits designed with a consideration of oblateness effects are the *Molniya* orbits Russia uses for communication at high latitudes, where most of that country is confined. The requirement of remaining at high latitudes for long periods results in highly elliptical orbits ($e = 0.73$) of approximately a 12-hour period, with apogee near the north pole. For a Molniya orbit, it is crucial that the perigee latitude must not change with apsidal rotation due to oblateness. Hence, a Molniya orbit always has the critical inclination, $i = i_c = 63.435°$, for which $\dot{\omega} = 0$. This high inclination is conveniently achieved through a launch from Plesetsk located at a latitude of 62.8°. The nodal regression causes the orbital plane to change at a rate $\dot{\Omega} = -0.0024°$/day.

6.4 Effects of Atmospheric Drag

Spacecraft in a low-orbit experience a significant atmospheric drag, whose magnitude is proportional to the product of atmospheric density, ρ, and square of relative speed, $\dot{r}^2$. Since drag opposes the orbital motion, we can express the perturbed equation of motion by

$$\frac{d^2\mathbf{r}}{dt^2} + \frac{\mu}{r^3}\mathbf{r} = -q\dot{r}\dot{\mathbf{r}} , \tag{6.13}$$

where

$$q \doteq \frac{1}{2}\rho\frac{C_D A}{m_2}\,, \tag{6.14}$$

and C_D is the *drag coefficient* (Chapter 10) of the spacecraft based upon a reference area, A. Generally, the *free-molecular flow* assumption is valid at the orbital altitudes, which according to the methods of Chapter 10 produces a drag-coefficient value, $C_D \approx 2$, for most shapes, based upon the maximum cross-sectional area facing the flow. The dependence of drag on velocity, rather than on position, renders it a nonconservative force, which results in a decline of the orbital energy, $\epsilon = -\frac{\mu}{2a}$, and hence, the semi-major axis, a. Taking the scalar product of the deceleration due to drag with the relative velocity, we have the rate of change of orbital energy

$$\dot{\epsilon} = \frac{\mu}{2a^2}\dot{a} = -q\dot{r}\dot{\mathbf{r}}\cdot\dot{\mathbf{r}} = -qv^3\,, \tag{6.15}$$

or

$$\dot{a} = -\frac{2a^2}{\mu}qv^3\,, \tag{6.16}$$

where $v \doteq \dot{r}$. Clearly, the rate of decline of the orbit increases proportionally with q and diminishes as the orbit size increases. The atmospheric density at orbital altitudes can be approximated as an exponentially decaying function of the altitude, $z \doteq r - R_e$, whereby $\rho = \rho_0 e^{-\frac{z}{H}}$ (Chapter 9), and we can write

$$\dot{a} = -\frac{a^2 C_D A}{m_2 \mu}\rho_0 v^3 e^{-\frac{z}{H}}\,. \tag{6.17}$$

The rate of decline of the orbit due to atmospheric drag is an important factor in determining the life of the satellite in a low orbit. However, the prediction of a satellite's orbital life involves an accurate estimate of the atmospheric properties at high altitudes over long periods (years and decades), which is seldom possible due to random external disturbances of solar radiation and geomagnetic field (Chapter 9).

Taking the vector product of the deceleration due to drag with the relative velocity, we have the rate of change of angular momentum,

$$\dot{\mathbf{h}} = -\mathbf{r}\times q\dot{r}\dot{\mathbf{r}} = -q\dot{r}\mathbf{h}\,. \tag{6.18}$$

Now, it is to be noted that taking the time derivative of the equation $h^2 = \mathbf{h}\cdot\mathbf{h}$ results in the following:

$$h\dot{h} = \mathbf{h}\cdot\dot{\mathbf{h}}\,. \tag{6.19}$$

Substitution of Eq. (6.19) into Eq. (6.18) yields

$$h\dot{h} = -q\dot{r}\mathbf{h}\cdot\mathbf{h} = -q\dot{r}h^2\,, \tag{6.20}$$

or

$$\dot{h} = -q\dot{r}h\,. \tag{6.21}$$

Now, since $\mathbf{h} = h\mathbf{i_h}$, we have

$$\dot{\mathbf{h}} = \dot{h}\mathbf{i_h} + h\frac{\mathrm{d}\mathbf{i_h}}{\mathrm{d}t}\,. \tag{6.22}$$

Substituting Eqs. (6.18) and (6.21) into Eq. (6.22), we have

$$h\frac{\mathrm{d}\mathbf{i_h}}{\mathrm{d}t} = q\dot{r}h\mathbf{i_h} - q\dot{r}\mathbf{h} = \mathbf{0}, \tag{6.23}$$

$$\frac{\mathrm{d}\mathbf{i_h}}{\mathrm{d}t} = \mathbf{0}\,. \tag{6.24}$$

Therefore, there is no change in the orbital plane due to atmospheric drag; hence, the Euler angles, i, ω, Ω, defining the orbital plane, remain invariant with time.

The above-demonstrated invariance of the orbital plane due to drag can be advantageously utilized in effectively braking a spacecraft as it arrives at a planet from an interplanetary flight. Due to the exponential increase in the density with a decrease in the altitude, an elliptical orbit experiences the largest drag at its periapsis. The planetary atmospheres on the earth and Mars have a negligible density above about 150 km altitude. Thus, a highly elliptical orbit with a periapsis altitude less than 150 km around either the earth or Mars experiences a negative velocity impulse everytime it passes the periapsis, which remains fixed in space. Hence, if a spacecraft's arrival hyperbola is converted into a highly eccentric elliptical orbit, either by retro-rockets (called *orbit insertion burn*) or by an initial pass through the atmosphere (called *aerocapture*), the orbital eccentricity as well as the semi-major axis can be reduced by making successive passes through the atmosphere. Such an approach is referred to as *aeroassited orbital transfer*, or *aerobraking* [57], and has been employed in several spacecraft, such as the *Magellan* Venus mission, the *Mars Global Surveyor* (MGS), and the *Mars Odyssey* mission.

6.5 Third-Body Perturbation and Interplanetary Flight

The third-body gravitational perturbation on a two-body orbit is modeled in a manner similar to the nonspherical gravity of the primary body. Consider the mutual orbit of m_1, m_2 perturbed by the presence of m_3, as shown in Fig. 6.1. The equations of motion for the two orbiting bodies can be written as follows:

$$\begin{aligned}\ddot{\mathbf{R}}_1 - \frac{Gm_1}{r_{12}^3}\mathbf{r_{12}} - \frac{Gm_3}{r_{13}^3}\mathbf{r_{13}} &= \mathbf{0},\\ \ddot{\mathbf{R}}_2 + \frac{Gm_2}{r_{12}^3}\mathbf{r_{12}} - \frac{Gm_3}{r_{23}^3}\mathbf{r_{23}} &= \mathbf{0},\end{aligned} \tag{6.25}$$

where $\mathbf{R_1}, \mathbf{R_2}, \mathbf{R_3}$ denote the respective inertial position of the three bodies, and $\mathbf{r_{12}} \doteq \mathbf{R_2} - \mathbf{R_1}$, $\mathbf{r_{13}} \doteq \mathbf{R_3} - \mathbf{R_1}$, and $\mathbf{r_{23}} \doteq \mathbf{R_3} - \mathbf{R_2}$. Upon subtracting the two equations from each other, we obtain the equation of relative motion between m_1 and m_2 to be

$$\ddot{\mathbf{r}} + \frac{\mu}{r^3}\mathbf{r} = Gm_3 \left(\frac{1}{r_{23}^3}\mathbf{r_{23}} - \frac{1}{r_{13}^3}\mathbf{r_{13}} \right) , \tag{6.26}$$

where $\mathbf{r}, \mu$ have their usual nomenclature in the two-body problem. Since the third body is usually at a large distance compared to the separation between the orbiting bodies ($r_{13} \gg r$, $r_{23} \gg r$), we are often computing the difference between two nearly equal numbers in the term on the right-hand side. For convenience in such a computation of the disturbing acceleration, we define $\mathbf{r_{32}} \doteq -\mathbf{r_{23}}$ and a disturbance potential function, Φ, given by

$$\Phi \doteq Gm_3 \left(\frac{1}{r_{32}} - \frac{\mathbf{r} \cdot \mathbf{r_{13}}}{r_{13}^3} \right) . \tag{6.27}$$

Since it is true that

$$\frac{1}{r_{23}^3}\mathbf{r_{23}}^T - \frac{1}{r_{13}^3}\mathbf{r_{13}}^T = \frac{\partial}{\partial \mathbf{r}} \left(\frac{1}{r_{32}} - \frac{\mathbf{r} \cdot \mathbf{r_{13}}}{r_{13}^3} \right) , \tag{6.28}$$

we have the orbital equation in the same form as Eq. (6.3):

$$\ddot{\mathbf{r}} + \frac{\mu}{r^3}\mathbf{r} = \left(\frac{\partial \Phi}{\partial \mathbf{r}} \right)^T . \tag{6.29}$$

If we want to consider the disturbance caused by more bodies, we only have to add the gradients of disturbance potentials of the additional bodies on the right-hand side of Eq. (6.29). Thus, no qualitative insight has to be had by additional perturbation potentials, and we restrict ourselves to only the third-body disturbance. This is usually sufficient in most spacecraft missions, where at any given time the spacecraft experiences the gravitational attraction of no more than two primary bodies. As in the case of the nonspherical gravity perturbation [Eq. (3.23)], we expand the disturbance potential in an infinite series of Legendre polynomials as follows:

$$\Phi = \frac{Gm_3}{r_{13}} \left[1 + \sum_{n=2}^{\infty} \left(\frac{r}{r_{13}} \right)^n P_n(\cos\gamma) \right] , \tag{6.30}$$

where γ is the angle between $\mathbf{r}$ and $\mathbf{r_{13}}$, as depicted in Fig. 6.1. The primary advantage of Eq. (6.30) is the independence of the disturbance potential from $\mathbf{r_{32}}$ that is continuously changing in an orbit of m_2 relative to m_1. Therefore, the gradient of potential is easily calculated, and the perturbed equation of motion becomes

$$\ddot{\mathbf{r}} + \frac{\mu}{r^3}\mathbf{r} = G\frac{m_3}{r_{13}^2} \sum_{n=1}^{\infty} \left(\frac{r}{r_{13}} \right)^n \left[P'_{n+1}(\cos\gamma)\frac{\mathbf{r_{13}}}{r_{13}} - P'_n(\cos\gamma)\frac{\mathbf{r}}{r} \right] , \tag{6.31}$$

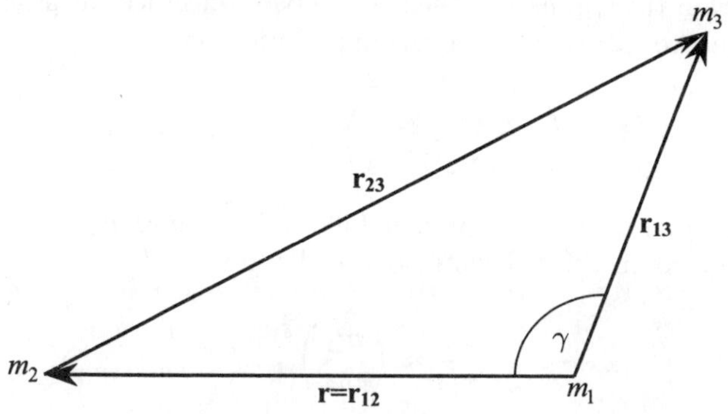

Fig. 6.1. The orbit of m_2 relative to m_1, perturbed by m_3.

where the prime indicates the derivative of a Legendre polynomial with respect to its argument. From the properties of Legendre polynomials (Chapter 3), we have the following useful recurrence formula:

$$P'_n(\nu) = \nu P'_{n-1}(\nu) + nP_{n-1}(\nu) \ . \tag{6.32}$$

Before addressing how the third-body perturbed orbital equation can be numerically tackled, let us discuss an important approximation made in designing interplanetary missions.

6.5.1 Sphere of Influence and Patched Conics

The unavailability of a closed-form solution to the perturbed two-body problem leads one to search for approximations involving the two-body trajectories. The concept of *sphere of influence* is such a method. Devised by Laplace when studying the motion of a comet passing close to Jupiter, the method assigns a spherical region to each planet, such that within the region the planet's gravity dominates, while outside it the sun's gravitation is predominant. We can express the motion of m_2 by either of the following equations of motion:

$$\ddot{\mathbf{r}} + \frac{G(m_1+m_2)}{r^3}\mathbf{r} = -Gm_3\left(\frac{1}{r_{32}^3}\mathbf{r}_{32} + \frac{1}{r_{13}^3}\mathbf{r}_{13}\right) ,$$
$$\ddot{\mathbf{r}}_{32} + \frac{G(m_2+m_3)}{r_{32}^3}\mathbf{r}_{32} = -Gm_1\left(\frac{1}{r^3}\mathbf{r} - \frac{1}{r_{13}^3}\mathbf{r}_{13}\right) . \tag{6.33}$$

We can derive the radius of the sphere of influence, r_s, for the planetary mass, m_1, by comparing the magnitudes of the perturbation on the right-hand sides of the two equations, Eq. (6.33). Which equation of motion offers the better model depends upon which of the two equations has a smaller magnitude of the ratio of the perturbation, $\mathbf{a_d}$, to the primary two-body acceleration. On the boundary of the region of influence, the two concerned ratios are identical. Equating the two ratios, we have the approximation

$$\frac{r_s}{r_{13}} = \left(\frac{m_1 + m_2}{m_3(m_2 + m_3)} \right)^{\frac{1}{3}} (1 + 3\cos^2\gamma)^{-\frac{1}{6}} , \tag{6.34}$$

which can be further approximated as follows, assuming $m_3 \gg m_1 \gg m_2$ and $r_{13} \gg r$, resulting in the following spherical region:

$$\frac{r_s}{r_{13}} \approx \left(\frac{m_1}{m_3} \right)^{\frac{2}{5}} . \tag{6.35}$$

For all planets, the sphere of influence is a good approximation and depends upon the ratio of the masses of the primary and disturbing bodies as well as the distance r_{13} between them. Thus, a small planet orbiting very close to the sun has a small sphere of influence, while that of a massive planet far away from the sun is a large region. In the solar system, the spheres of influence range from $r_s = 0.111 \times 10^6$ km for Mercury to $r_s = 80.196 \times 10^6$ km for Neptune. For a planetary subsystem, such as the earth–moon combination, the sphere of influence concept is invalid, and Eq. (6.34) provides a much better estimate of the nonspherical region of the moon's influence.

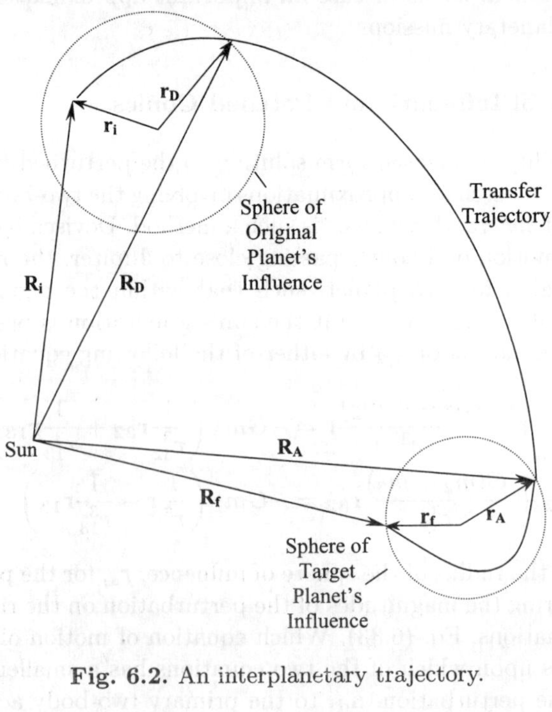

Fig. 6.2. An interplanetary trajectory.

The concept of sphere of influence enables a quick estimation of an interplanetary trajectory by smoothly connecting the two-body trajectories calculated by taking either m_1 (planet) or m_3 (the sun) as the primary body, depending upon whether the spacecraft (m_2) is inside or outside the region of influence of m_1. Such an approcah is called the *patched-conic approximation*. The main thrust in this method is the smooth patching of the two-body solutions at the boundary of the sphere of influence, because the actual trajectory does not encounter any discontinuity at that mathematical boundary. Generally, the design of an interplanetary trajectory requires aiming the spacecraft from a given heliocentric initial position, $\mathbf{R_i}$ (usually in the vicinity of a planet) to a final position, $\mathbf{R_f}$, close to another planet, in a given time, t_f, as shown in Fig. 6.2. Thus, the spacecraft transits from the spheres of influence of the two planets through the intervening solar gravitation region outside the two spheres. In order for us to design such a voyage using the patched-conic approach, the following steps are necessary:

(a) Given the initial relative velocity and position, $\mathbf{r_i}$, estimate the heliocentric position $\mathbf{R_D}$ and velocity of the spacecraft along the departure hyperbola, at the sphere of influence of the original planet. This step requires the solution of the two-body problem with the planet as the primary mass.
(b) Select a point of arrival, $\mathbf{R_A}$, at the sphere of influence of the target planet, and solve the Lambert's problem associated with $\mathbf{R_D}$, $\mathbf{R_A}$, and t_f, using the sun as the primary body.
(c) From $\mathbf{R_A}$ and the arrival velocity, estimate the hyperbolic trajectory relative to the target planet, and check whether it passes through the desired final relative position, $\mathbf{r_f}$.
(d) Iterate for $\mathbf{R_D}$ and $\mathbf{R_A}$ until the velocities from the relative hyperbolae match with those of the heliocentric transfer ellipse.

The patched-conic approach is a practical procedure for designing interplanetary missions. However, it is not sufficiently accurate to be used in the navigation of spacecraft, where even a small error in the heliocentric position results in the spacecraft completely missing the target planet. A spherical patched-conic approximation for a lunar mission is demonstrated by Battin [11]. In treating lunar missions by the patched-conic method, we take advantage of the fact that the moon's orbit lies within the earth's sphere of influence ($r_s = 0.924 \times 10^6$ km). Therefore, solar gravity would equally perturb the earth, the moon, and the spacecraft during such a mission. However, since the moon's region of influence cannot be regarded as being spherical, the patching of the conic solutions becomes complicated on the nonspherical boundary. For this reason, it is much simpler (and more accurate) to determine a lunar trajectory using the numerical solution of the restricted three-body problem (Chapter 7).

6.6 Numerical Solution to the Perturbed Problem

We are often interested in integrating the perturbed equation of motion, Eq. (6.1), in time, for obtaining the vehicle's position and velocity. Such a procedure becomes necessary when we want to predict long-term orbital position of a satellite and accurate position of an interplanetary spacecraft. Since the patched-conic method for third-body perturbation is an iterative procedure, and results in only a rough estimate of the actual trajectory, it is often preferable to integrate the equation of perturbed motion by a numerical procedure, such as the Runge–Kutta methods of Appendix A. Two distinct methods are available for obtaining the perturbed solution; they were initially devised for predicting cometary trajectories deviated by Jupiter:

(a) *Cowell's method* for a direct numerical integration of Eq. (6.1),
(b) *Encke's method* based upon deviation from the two-body orbital solution.

Of these, Cowell's direct integration method is reasonable when $\mathbf{a_d}$ is comparable (or larger than) in magnitude to the primary acceleration, $\frac{\mu}{r^2}$. However, in a perturbed problem, $\mathbf{a_d}$ is generally much smaller compared to $\frac{\mu}{r^2}$, which requires a larger number of integration intervals for a given accuracy. Apart from its inefficiency, Cowell's method may also experience a numerical instability due to the growth of truncation error over time, often leading to a divergence of the iterative procedure. Hence, Encke's method is considered the better choice for solving long-term trajectory problems, such as those associated with interplanetary navigation.

Encke's method is based upon the deviation from an instantaneous two-body solution, $\bar{\mathbf{r}}(t), \bar{\mathbf{v}}(t)$, projected froward in time from the current position and velocity, $\mathbf{r}(t_0), \mathbf{v}(t_0)$. The instantaneous conic solution, called *osculating orbit*, satisfies the following two-body orbital equation:

$$\frac{\mathrm{d}^2\bar{\mathbf{r}}}{\mathrm{d}t^2} + \frac{\mu}{\bar{r}^3}\bar{\mathbf{r}} = \mathbf{0}\,. \tag{6.36}$$

The elements of the osculating orbit are calculated at time t_0, from the true position and velocity at the given instant, such that

$$\begin{aligned} \bar{\mathbf{r}}(t_0) &= \mathbf{r}(t_0), \\ \bar{\mathbf{v}}(t_0) &= \mathbf{v}(t_0). \end{aligned} \tag{6.37}$$

At a slightly later time, $t = t_0 + \Delta t$, we can write

$$\begin{aligned} \mathbf{r}(t) &= \bar{\mathbf{r}}(t) + \boldsymbol{\alpha}(t), \\ \mathbf{v}(t) &= \bar{\mathbf{v}}(t) + \boldsymbol{\beta}(t). \end{aligned} \tag{6.38}$$

Subtracting Eq. (6.36) from the true equation of motion, Eq. (6.1), we have

$$\frac{\mathrm{d}^2\boldsymbol{\alpha}}{\mathrm{d}t^2} + \frac{\mu}{\bar{r}^3}\boldsymbol{\alpha} = \frac{\mu}{\bar{r}^3}\left(1 - \frac{\bar{r}^3}{r^3}\right)\mathbf{r} + \mathbf{a_d}\,, \tag{6.39}$$

which is, of course, subject to the initial conditions

$$\boldsymbol{\alpha}(t_0) = \mathbf{0},$$
$$\dot{\boldsymbol{\alpha}}(t_0) = \boldsymbol{\beta}(t_0) = \mathbf{0}. \tag{6.40}$$

Since r and $\bar{r}$ are nearly equal, the term in the brackets on the right-hand side of Eq. (6.39) may present numerical difficulties associated with machine round-off. In order to avoid such a problem, we may write [11]

$$1 - \frac{\bar{r}^3}{r^3} = -x\frac{3 + 3x + x^2}{1 + (1 + x)^{\frac{3}{2}}}\,, \tag{6.41}$$

where

$$x \doteq \frac{\boldsymbol{\alpha} \cdot (\boldsymbol{\alpha} - 2\mathbf{r})}{r^2}\,. \tag{6.42}$$

The numerical implementation of Encke's method can be expressed by the following steps:

(a) Compute $\bar{\mathbf{r}}(t), \bar{\mathbf{v}}(t)$ using the osculating orbit obtained at t_0 from Eq. (6.37). For this purpose, one can apply Lagrange's coefficients for the osculating orbit.
(b) Numerically integrate Eq. (6.39), and compute the true position and velocity by Eq. (6.38).
(c) If at any time, the magnitudes of $\boldsymbol{\alpha}, \boldsymbol{\beta}$ exceed specified tolerances, rectify the osculating orbit by making $t_0 = t$ and going back to step (a).
Due to the built-in rectification procedure, the method never faces the likelihood of numerical instability. We can use a sufficiently small time interval, Δt, to ensure an essentially forward marching solution, such as that with a finite-difference scheme.

Example 6.2. A spacecraft is on an outbound interplanetary voyage from the earth. The current heliocentric position and the inertial velocity of the spacecraft referred to the ecliptic are given by

$$\mathbf{R}(0) = \begin{pmatrix} -27 \\ 147.5 \\ 0.1 \end{pmatrix} \times 10^6 \text{ km}, \qquad \mathbf{V}(0) = \begin{pmatrix} -33 \\ -10 \\ 1 \end{pmatrix} ; \text{km/s}\,.$$

The earth's orbital elements computed from ephemeris charts (Chapter 5) for the present time are the following:

$$a = 149597870 \text{ km},$$
$$e = 0.01667,$$
$$\tau = -100 \text{ mean solar days}\,.$$

Determine the geocentric position and velocity of the spacecraft 100 mean solar days from now.

In this example, the spacecraft's initial position is well within the earth's sphere of influence. However, its velocity is quite large, implying that the sphere of influence would be crossed in a few days, resulting in a diminishing influence of the earth's gravity. For carrying out the given computation, we shall apply both Cowell's and Encke's methods and compare the results. While a fourth-order Runge–Kutta algorithm (Appendix A) is chosen to integrate the equations of perturbed motion with Cowell's method, a rather crude forward-difference approximation is applied for the time derivatives in a time-marching solution by Encke's method, with an update in the osculating orbit per day. The programs necessary in the two methods are encoded in *cowell.m* (Table 6.1), *encke.m* (Table 6.2), *perturb.m* (Table 6.3), and *disturb.m* (Table 6.4). In addition, a code named *orbit.m* (Table 6.5) is necessary for the orbital position of the earth in the ecliptic plane. The programs require the elliptical trajectory code, *trajE.m* (Table 4.3), and Kepler's equation solver, *kepler.m* (Table 4.2). The plots resulting from the two simulations are compared with the conic solution (unperturbed two-body solution) in Figs. 6.3–6.5. The heliocentric coordinates of the spacecraft (Fig. 6.3) and the inertial velocity components (Fig. 6.4) show a marked deviation from the conic solution. The two perturbed solutions are also significantly different from one another. The reason for the difference between the two solutions is the first-order, forward-difference approximation applied in Encke's method, whereas Cowell's method is applied with a fifth-order accuracy. When the radial distance and speed relative to the sun are compared in Fig. 6.5, we observe an accumulation of the various errors, resulting in a larger difference among the three methods. We can improve the accuracy of Encke's method by choosing a higher-order finite-difference approximation.

Table 6.1. M-file *cowell.m* for Integrating the Perturbed Equation of Motion

```
%Determination of spacecraft's heliocentric position and velocity
%after time 't' disturbed by a planet of gravity constant 'mu3' and
%'orb' orbit around the sun, by direct numerical integration (Cowell's method)
%(c) 2006 Ashish Tewari
global mu; mu=1.32712440018e11; %sun's gravity constant
global mu3; mu3=398600.4; %earth's gravity constant
global orb;orb=[149597870;0.01667;0;0;0;-100*24*3600];%earth's orbital elements
R0=1e8*[-0.27;1.475;0.001]; %spacecraft initial position
V0=[-33;-10;1]; %spacecraft's initial velocity
options=odeset('RelTol',1e-8);
[T,X]=ode45('perturb',[0 100*24*3600],[R0;V0]);
```

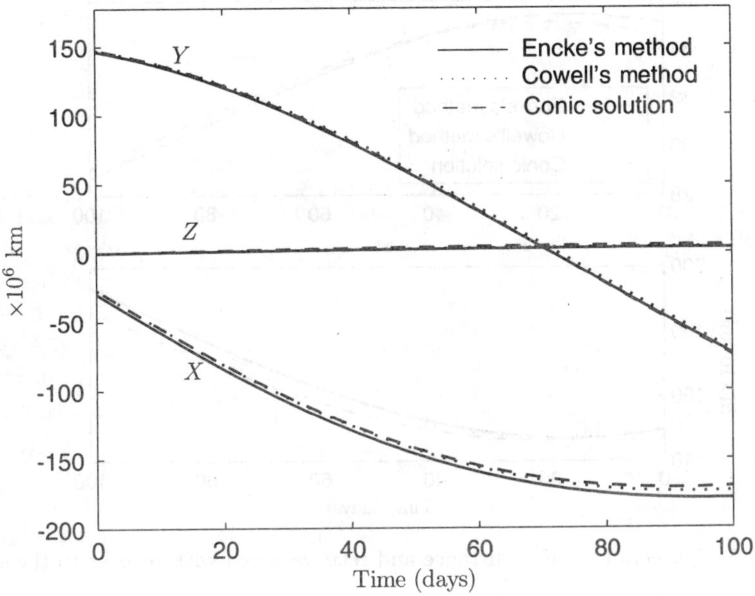

Fig. 6.3. Heliocentric position of a spacecraft on an interplanetary voyage.

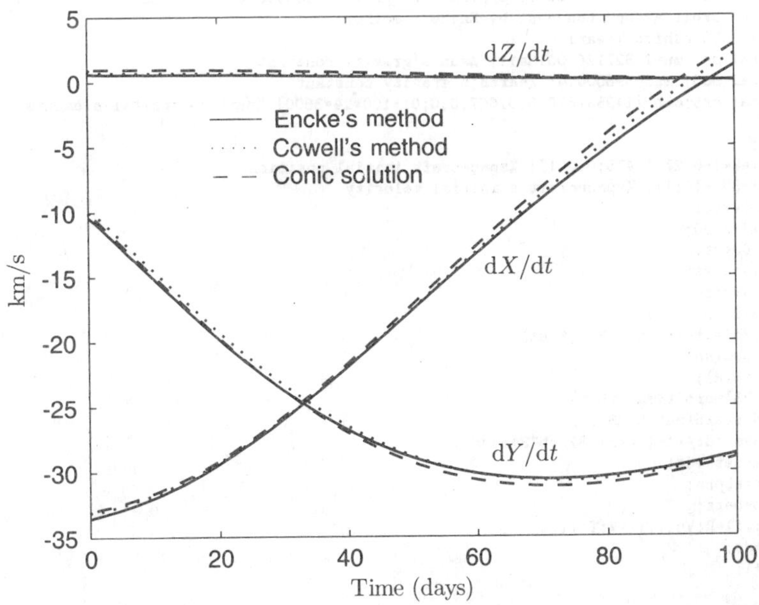

Fig. 6.4. Velocity relative to the sun of a spacecraft on an interplanetary voyage.

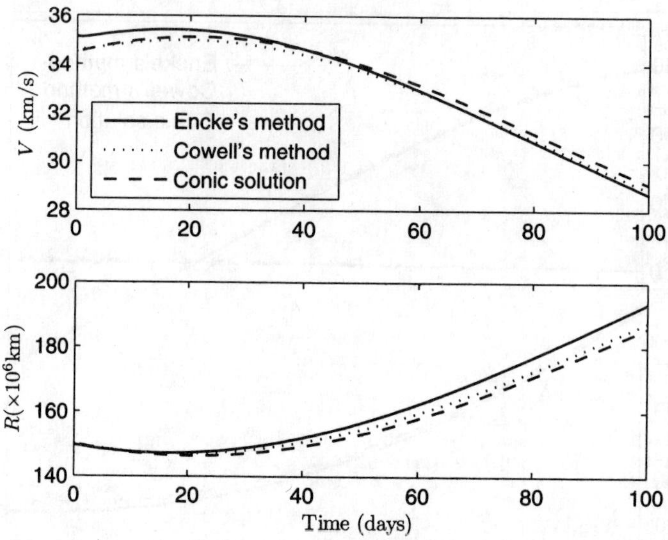

Fig. 6.5. Spacecraft's radial distance and relative speed with respect to the sun.

Table 6.2. M-file *encke.m* for Integrating the Perturbed Equation of Motion by Encke's Method

```
%Determination of spacecraft's heliocentric position and velocity
%after time 't' disturbed by a planet of gravity constant 'mu3' and
%'orb' orbit around the sun, by Encke's method
%(c) 2006 Ashish Tewari
global mu; mu=1.32712440018e11; %sun's gravity constant
global mu3; mu3=398600.4; %earth's gravity constant
global orb;orb=[149597870;0.01667;0;0;0;-100*24*3600];%earth's orbital elements
t=0;
i=1;
R0=1e8*[-0.27;1.475;0.001]; %spacecraft initial position
V0=[-33;-10;1]; %spacecraft's initial velocity
R=R0;V=V0;
dt=24*3600;
tf=100*dt;
while t<=tf
alpha=zeros(3,1);
beta=zeros(3,1);
[Rb,Vb]=trajE(mu,t,R,V,t+dt);
rb=norm(Rb);
r=norm(R);
[R3,V3]=orbit(mu,orb,t);
ad=disturb(mu3,R,R3);
beta=mu*dt*(1-(rb/r)^3)/rb^3+ad*dt;
alfa=beta*dt;
R=Rb+alpha;
V=Vb+beta;
Rs(:,i)=R;Vs(:,i)=V;Ts(i,1)=t;
t=t+dt;
i=i+1;
end
end
```

Table 6.3. M-file *perturb.m* for Specifying the Perturbed Equation of Motion

```
function xdot=perturb(t,x)
%(c) 2006 Ashish Tewari
global mu;
global mu3;
global orb;
%program for the perturbed equations of motion
R=x(1:3,1); %position of s/c relative to primary
r=norm(R);
xdot(1:3,1)=x(4:6,1);
[R3,V3]=orbit(mu,orb,t); %position and velocity of third body
ad=disturb(mu3,R,R3); %disturbing acceleration
xdot(4:6,1)=-mu*R/r^3+ad;
```

Table 6.4. M-file *disturb.m* for Calculating the Disturbance Acceleration

```
function a=disturb(mu3,R,R3)
%Program for calculating the disturbance acceleration 'a' caused by a third
%body on a two-body orbit
%mu3: gravitational constant of the disturbing body 'm3'
%R: position of mass 'm2' relative to primary mass 'm1'
%R3: position of 'm3' relative to 'm1'
%(c) 2006 Ashish Tewari
r=norm(R);
r3=norm(R3);
R23=R-R3;
r23=norm(R23);
fx=(r23/r3)^3-1;
a=-mu3*(R+fx*R3)/r23^3;
```

Table 6.5. M-file *orbit.m* for Calculating the Position and Velocity of the Disturbing Body

```
function [r,v]=orbit(mu,orb,t)
%program for position and velocity of a body in 'orb' elliptical orbit
%Elements of 'orb': 1x1: a; 2x1:e; 3x1:i; 4x1:w; 5x1: Om; 6x1: tau
%(c) 2006 Ashish Tewari
a=orb(1);e=orb(2);i=orb(3);w=orb(4);Om=orb(5);tau=orb(6);
n=sqrt(mu/a^3);
M=-n*tau;
E=kepler(e,M);
r0=a*(1-e*cos(E));
R0=a*[cos(E)-e;sqrt(1-e^2)*sin(E);0];
V0=sqrt(mu*a)*[-sin(E);sqrt(1-e^2)*cos(E);0]/r0;
[R,V]=trajE(mu,0,R0,V0,t);
if abs(i)>=1e-6
C=rotation(i,Om,w);
else
    C=eye(3);
end
r=C*R;
v=C*V;
```

6.7 Summary

The Keplerian motion is perturbed by external disturbances caused by a non-uniform gravitational field due to an aspherical planetary mass, gravitational influence of a third body, atmospheric drag, thrust of rocket engines, and solar radiation pressure. The conservative gravitational perturbations can be modeled using analytical methods and lead to precession of the orbital plane. Sun synchronous orbits take advantage of the regression of nodes due to orbital precession, while highly eccentric Molniya orbits must have a critical inclination such that there is no rotation of apsides caused by orbital precession. The nonconservative perturbations—such as atmospheric drag—change the shape of the orbit, resulting in a variation of semi-major axis and eccentricity. The sphere of influence is a useful concept for analyzing third-body perturbations and results in the patched-conic approximation for interplanetary trajectories. For a numerical solution of the exact perturbed equations of motion, two different integration procedures can be employed: direct numerical integration (Cowell's method), and deviation from the instantaneous two-body solution (Encke's method). Cowell's method is efficient when the perturbing acceleration is comparable (or larger) in order of magnitude to the primary acceleration. However, Encke's method is a better choice for long-range interplanetary missions as it is both efficient and robust with respect to numerical instabilities, on account of its built-in rectification procedure.

Exercises

6.1. The mean daily motion of Mars with respect to the sun is 0.524°/day. What is the inclination of a synchronous orbit of 300 km altitude around Mars?

6.2. Determine the elements of a Molniya orbit with a period of half-sidereal day and a perigee of 500 km.

6.3. The motion of a spacecraft under a constant thrust, $\mathbf{f}_T$, can be treated in a manner similar to the drag problem. Show that for a constant tangential thrust, f_T, the following closed-form expression can be derived for the orbital speed, v, of a vehicle originally in a circular orbit of radius r_0:

$$v \doteq \frac{\mathrm{d}s}{\mathrm{d}t} = \sqrt{2\frac{f_T}{m}s + \mu\left(\frac{2}{r} - \frac{1}{r_0}\right)}\,,$$

where s denotes the orbital arc length. Furthermore, show that if the acceleration due to thrust, $\frac{f_T}{m}$, is so small that the vehicle follows a nearly circular arc, the increasing radius of the spiral can be approximated by

$$r = \frac{r_0}{1 - \frac{2 f_T s}{m v_0^2}}\,,$$

where v_0 is the initial speed. Derive an expression for the total time required by such a spacecraft to escape the planet's gravity.

6.4. Repeat Exercise 6.3 using a constant *circumferential* (rather than tangential) thrust. Which approach is more efficient if a planetary escape is intended?

6.5. Write a program for numerically integrating the perturbed orbital equation for a spacecraft in a circular earth orbit of 400 km altitude, under a constant tangential *de-boost* thrust deceleration of 2 m/s^2 for a period of 500 s. Neglecting the mass variation caused by the firing of rocket motors, calculate the new orbital elements after the firing ceases.

6.6. Assuming that the atmospheric density varies exponentially with the altitude, h, such that

$$\rho = \rho_0 e^{-\frac{h}{H}} ,$$

where ρ_0 and H are atmospheric constants, derive an approximate expression for the life of a satellite in a low-altitude, circular orbit ($h_0 \ll R_e$) around a spherical planet of radius R_e. Take the drag coefficient, C_D, reference area, A, and mass, m, of the satellite.

6.7. Using the sixth-order R-K-N algorithm given in Appendix A, write a program for solving the problem of interplanetary flight given in Example 6.2 by Cowell's method. Compare the final position and velocity of the spacecraft with those computed in the example by the R-K-4(5) method.

7

The Three-Body Problem

7.1 Aims and Objectives

- To model the problem of three bodies in mutual gravitational attaction and present its general solution by Lagrange.
- To develop the restricted three-body problem for modeling the flight of a spacecraft under the gravitational influence of two massive bodies.
- To discuss the solvability, equilibrium points, and numerical solutions of the restricted three-body problem with examples of earth–moon–spacecraft trajectories.

7.2 Equations of Motion

The *three-body problem* refers to the dynamical system comprising the motion of three masses under mutual gravitational attraction. It usually arises when we are interested in either the two-body orbital perturbations caused by a distant third body, or the motion of a smaller body in the gravitational field formed by two larger bodies. The analysis of the former kind are undertaken when considering either the orbital motion of a moon around a planet perturbed by the sun or that of a satellite around the earth perturbed by the sun or moon. The latter model is generally applied to a comet under the combined gravity of the sun and Jupiter, a spacecraft under mutual gravity of the earth–moon system, or an interplanetary probe approaching (or departing) a planet. The three-body problem is thus a higher-order gravitational model compared to the two-body problem considered in Chapters 4 and 5 and is a much better approximation of the actual motion of a spacecraft in the solar system.

The three-body problem has attracted attention of mathematicians and physicists over the past 300 years, primarily due to its promise of modeling the erratic behavior of the moon. The first analyses of the three-body problem were undertaken by Newton and Euler, but the first systematic study of its equilibrium points was presented by *Lagrange*, both in the 18th century. One could write the equations of motion for the three-body problem using the N-body equations derived in Chapter 4 with $N = 3$ as follows:

$$\frac{\mathrm{d}^2\mathbf{R_i}}{\mathrm{d}t^2} = G\sum_{j\neq i}^{3} \frac{m_j}{R_{ij}^3}(\mathbf{R_j} - \mathbf{R_i}) \qquad i = 1, 2, 3, \tag{7.1}$$

where G is the universal gravitational constant, $\mathbf{R_i}$ denotes the position of the center of mass of body i, and $R_{ij} \doteq \mid \mathbf{R_j} - \mathbf{R_i} \mid$ denotes the relative separation of the centers of mass of bodies *i,j*. The potential energy of the three-body system is given by

$$\begin{aligned} V &\doteq \frac{1}{2}G\sum_{i=1}^{3} m_i \sum_{j\neq i}^{3} \frac{m_j}{R_{ij}} \\ &= -G\left(\frac{m_1 m_2}{r_{12}} + \frac{m_2 m_3}{r_{23}} + \frac{m_1 m_3}{r_{13}}\right), \end{aligned} \tag{7.2}$$

while its kinetic energy is the following:

$$T \doteq \frac{1}{2}\sum_{i=1}^{3}\sum_{j\neq i}^{3} m_i \left(\frac{\mathrm{d}R_{ij}}{\mathrm{d}t}\right)^2 . \tag{7.3}$$

Since no external force acts upon the system, the total energy, $E = T + V$, is conserved and represents a scalar constant. Another six scalar constants are obtained by considering the motion of the center of mass, $\mathbf{r_c}$, which follows a straight line at constant velocity, $\mathbf{v_{co}}$, beginning from a constant initial position, $\mathbf{r_{co}}$, due to Newton's first law of motion (Chapter 4):

$$\mathbf{r_c} \doteq \frac{\sum_{i=1}^{3} m_i \mathbf{R_i}}{\sum_{i=1}^{3} m_i} = \mathbf{v_{co}}t + \mathbf{r_{co}}\,, \tag{7.4}$$

while three more scalar constants arise out of the conserved angular momentum $\mathbf{H}$, of the system about the center of mass. There are thus a total of only 10 scalar constants of the three-body problem, whereas we need 18 for a complete analytical solution. The three-body problem has therefore resisted attempts at a closed-form, general solution. However, Lagrange showed that certain particular solutions of the problem exist when the motion of the three bodies is confined to a single plane. Such a coplanar motion of bodies is the most common occurrence in the universe, such as the solar system. Before attempting Lagrange's particular solutions, let us rewrite the equations of motion in the following form:

$$\mathbf{f_i} = Gm_i\sum_{j\neq i}^{3} \frac{m_j}{R_{ij}^3}(\mathbf{R_j} - \mathbf{R_i}) \qquad i = 1, 2, 3, \tag{7.5}$$

where $\mathbf{f_i}$ is the net force experienced by the mass m_i due to the other two masses.

7.3 Lagrange's Solution

The particular solution by Lagrange is confined to a coplanar motion of the three masses. Let us choose a rotating coordinate frame fixed to the common center of mass such that the three bodies appear to be moving radially in the rotating frame. At any particular instant of time, the angle made by the rotating frame, $(\mathbf{i}, \mathbf{j})$, with an inertial frame, $(\mathbf{I}, \mathbf{J})$, is $\theta(t)$. Then the coordinate transformation between the two frames at that instant is given by

$$\begin{Bmatrix} \mathbf{I} \\ \mathbf{J} \end{Bmatrix} = \mathsf{C}(t) \begin{Bmatrix} \mathbf{i} \\ \mathbf{j} \end{Bmatrix}, \tag{7.6}$$

where the rotation matrix, $\mathsf{C}(t)$, is the following (Chapter 2):

$$\mathsf{C}(t) \doteq \begin{pmatrix} \cos\theta(t) & -\sin\theta(t) \\ \sin\theta(t) & \cos\theta(t) \end{pmatrix} . \tag{7.7}$$

Now, in the rotating frame, the location of the mass m_i at time t is given by

$$\mathbf{R}_\mathrm{i}(t) = a(t)\mathbf{R}_\mathrm{i}(0) , \tag{7.8}$$

where $\mathbf{R}_\mathrm{i}(0)$ is the initial location of the mass at $t = 0$. At any given instant, all three bodies share the same values of $\theta(t)$ and $a(t)$. However, due to the radial movement of the bodies, the angular speed, $\dot{\theta}$, of the frame keeps changing with time due to conservation of angular momentum. By substituting Eq. (7.8) into Eq. (7.5), we can express the net force experienced by m_i as

$$\mathbf{f}_\mathrm{i}(t) = \frac{\mathbf{f}_\mathrm{i}(0)}{a^2} , \tag{7.9}$$

and the skew-symmetric matrix $\mathsf{S}(\boldsymbol{\omega})$ of the frame's angular velocity, $\boldsymbol{\omega} = \dot{\theta}\mathbf{k}$, is the following (Chapter 2):

$$\mathsf{S}(\boldsymbol{\omega}) \doteq \mathsf{C}^T\dot{\mathsf{C}} = \begin{pmatrix} 0 & -1 \\ 1 & 0 \end{pmatrix} \dot{\theta} . \tag{7.10}$$

The net acceleration of m_i expressed in the rotating frame is simply the following (Chapter 4):

$$\frac{\mathrm{d}^2\mathbf{R}_\mathrm{i}}{\mathrm{d}t^2} = \mathsf{C}\left(\frac{\partial^2\mathbf{R}_\mathrm{i}}{\partial t^2} + 2\boldsymbol{\omega} \times \frac{\partial\mathbf{R}_\mathrm{i}}{\partial t} + \frac{\mathrm{d}\boldsymbol{\omega}}{\mathrm{d}t}\mathbf{R}_\mathrm{i} + \boldsymbol{\omega} \times [\boldsymbol{\omega} \times \mathbf{R}_\mathrm{i}]\right) , \tag{7.11}$$

which can be written as follows (Chapter 2):

$$\frac{\mathrm{d}^2\mathbf{R}_\mathrm{i}}{\mathrm{d}t^2} = \mathsf{C}\left(\frac{\partial^2\mathbf{R}_\mathrm{i}}{\partial t^2} + 2\mathsf{S}(\boldsymbol{\omega})\frac{\partial\mathbf{R}_\mathrm{i}}{\partial t} + \dot{\mathsf{S}}(\boldsymbol{\omega})\mathbf{R}_\mathrm{i} + \mathsf{S}^2(\boldsymbol{\omega})\mathbf{R}_\mathrm{i}\right) , \tag{7.12}$$

or, substituting Eqs. (7.8) and (7.9), we have

$$m_i \left[\left(\frac{\mathrm{d}^2 a}{\mathrm{d}t^2} - a\dot{\theta}^2 \right) \mathsf{I} + \frac{1}{a} \frac{\mathrm{d}(a^2\dot{\theta})}{\mathrm{d}t} \mathsf{J} \right] \mathbf{R_i}(0) = \frac{\mathbf{f_i}(0)}{a^2}, \tag{7.13}$$

where I is the identity matrix and

$$\mathsf{J} = \begin{pmatrix} 0 & -1 \\ 1 & 0 \end{pmatrix}. \tag{7.14}$$

Since the net force experienced by m_i is toward the common center of mass (leading to its radial and rotary motion), we can express the force by

$$\mathbf{f_i}(t) = -m_i b^2 \mathbf{R_i}(t), \tag{7.15}$$

where b is a constant. In a planar motion with radial acceleration, we have

$$R_i^2\dot{\theta} = \text{constant}, \tag{7.16}$$

or,

$$\frac{\mathrm{d}a^2\dot{\theta}}{\mathrm{d}t} = 0. \tag{7.17}$$

Therefore, it follows from Eq. (7.13) that

$$\frac{\mathrm{d}^2 a}{\mathrm{d}t^2} - a\dot{\theta}^2 = -\frac{b^2}{a^2}. \tag{7.18}$$

Clearly, Eqs. (7.17) and (7.18) represent a conic section (Chapter 4) in polar coordinates, which is the equation of the relative motion of two bodies. Hence, each mass in the coplanar three-body problem traces a conic section about the common center of mass. This simple and elegant solution possesses several interesting cases and offers a valuable insight into an otherwise intractable problem.

From the general, coplanar motion given above, certain stationary solutions in terms of fixed locations of the masses in the rotating frame can be derived. These solutions represent equilibrium points of the three-body problem and describe the three bodies moving in concentric, coplanar circles. Obviously, for a stationary solution with respect to the rotating frame, we require a constant value of $a = 1$, which leads to a constant angular speed, $\omega = \dot{\theta}$, by angular momentum conservation. The equations of motion in such a case are written as follows:

$$\begin{gathered}
\left(\frac{\omega^2}{G} - \frac{m_2}{R_{12}^3} - \frac{m_3}{R_{13}^3} \right) \mathbf{R_1} + \frac{m_2}{R_{12}^3}\mathbf{R_2} + \frac{m_3}{R_{13}^3}\mathbf{R_3} = \mathbf{0}, \\
\frac{m_1}{R_{12}^3}\mathbf{R_1} + \left(\frac{\omega^2}{G} - \frac{m_1}{R_{12}^3} - \frac{m_3}{R_{23}^3} \right) \mathbf{R_2} + \frac{m_3}{R_{23}^3}\mathbf{R_3} = \mathbf{0}, \\
m_1\mathbf{R_1} + m_2\mathbf{R_2} + m_3\mathbf{R_3} = \mathbf{0}.
\end{gathered} \tag{7.19}$$

Particular equilibrium solutions of Eq. (7.19) include the *equilateral triangle* configurations of the three masses, wherein the masses are at the same constant distance from the common center of mass ($R_{12} = R_{13} = R_{23} = \rho$). Hence, the angular speed is given by

$$\dot{\theta}^2 = \frac{G(m_1 + m_2 + m_3)}{\rho^3} . \tag{7.20}$$

Another set of equilibrium points contains the *colinear* solutions, where the three masses share a straight line. Let the axial location of the three masses be given by x_1, x_2, x_3, where we assume $x_1 < x_2 < x_3$. Then the equations of motion, written in terms of the distance ratio $\alpha \doteq R_{23}/R_{12}$, yield the following [11]:

$$\begin{aligned} x_1 &= -R_{12}\frac{m_2 + (1+\alpha)m_3}{m_1 + m_2 + m_3}, \\ \omega^2 &= \frac{G(m_1 + m_2 + m_3)}{R_{12}^3(1+\alpha)^2}\frac{m_2(1+\alpha)^2 + m_3}{m_2 + (1+\alpha)m_3} \\ &(m_1+m_2)\alpha^5 + (3m_1 + 2m_2)\alpha^4 + (3m_1 + m_2)\alpha^3 \\ &\quad -(m_2 + 3m_3)\alpha^2 - (2m_2 + 3m_3)\alpha - (m_2 + m_3) = 0 . \end{aligned} \tag{7.21}$$

The last of Eq. (7.21), called the *quintic equation of Lagrange*, has only one positive root, α, which can be seen by examining the signs of the coefficients (they change sign only once). However, for each specific configuration of the three colinear masses, the positive value of α is different.

Example 7.1. Numerically determine the values of α for the colinear

(a) earth–moon–spacecraft system,
(b) sun–earth–moon system.

Assume the ratio of the moon's mass to the earth's mass is 1/81.3 and that of the earth's mass to the solar mass is 1/333,400. The spacecraft's mass is negligible in comparison with the masses of heavenly bodies.

The required numerical computation is easily performed using the intrinsic MATLAB function *root* and the last of Eq. (7.21) as follows:

```
>> m2=1/81.3; %moon between earth and spacecraft (m3=0,m1=1)
>> C=[1+m2  3+2*m2 3+m2 -m2 -2*m2 -m2];
>> roots(C)
   ans =      -1.4932 + 0.8637i
            -1.4932 - 0.8637i
             0.1678
            -0.0846 + 0.1310i
            -0.0846 - 0.1310i
>> m3=1/81.3; %spacecraft between earth and moon (m1=1,m2=0)
>> C=[1  3 3 -3*m3 -3*m3 -m3];
>> roots(C)
   ans =     -1.5014 + 0.8731i
            -1.5014 - 0.8731i
             0.1778
            -0.0875 + 0.1236i
            -0.0875 - 0.1236i
```

```
>> m3=1/81.3; %earth between spacecraft and moon (m1=0, m2=1)
>> C=[1  2 1 -1-3*m3 -2-3*m3 -1-m3];
>> roots(C)
   ans =        1.0071
          -0.5082 + 0.8660i
          -0.5082 - 0.8660i
          -0.9953 + 0.0785i
          -0.9953 - 0.0785i
```

Thus, for the earth–moon–spacecraft problem, $\alpha = 0.1678, 0.1778$, or 1.0071, depending upon whether the moon, the spacecraft, or the earth, respectively, lies between the other two bodies. For the sun–earth–moon system, the calculations of α are similarly carried out:

```
>> m3=333400; m2=1/81.3; %moon between earth and sun (m1=1)
>> C=[1+m2 3+2*m2 3+m2 -m2-3*m3 -2*m2-3*m3 -m2-m3];
>> roots(C)
   ans =       -50.4637 +86.2542i
          -50.4637 -86.2542i
           98.9396
          -0.5000 + 0.2887i
          -0.5000 - 0.2887i
>> m3=333400*81.3; m2=81.3; %earth between moon and sun (m1=1)
>> C=[1+m2 3+2*m2 3+m2 -m2-3*m3 -2*m2-3*m3 -m2-m3];
>> roots(C)
   ans =       -50.1385 +86.2542i
          -50.1385 -86.2542i
           99.2648
          -0.5000 + 0.2887i
          -0.5000 - 0.2887i
```

Thus, for the sun–earth–moon problem, $\alpha = 98.9396$ or 99.2648, depending upon whether the moon or the earth, respectively, lies between the other two bodies. The sun is so much more massive compared to the earth and the moon that the two values of α are very close together, the relative position of the earth and moon with respect to the sun makes little difference.

7.4 Restricted Three-Body Problem

When the mass of one of the three bodies, say m_3, is negligible in comparison with that of the other two bodies (called *primaries*), a simplification occurs in the three-body problem, where we neglect the gravitational pull of m_3 on both m_1 and m_2. In such a case, the motion of m_3 relative to the the primaries executing circular orbits about the common center of mass is referred to as the *restricted three-body problem*. The equations of motion of the restricted problem are usually nondimensionalized by dividing the masses by the total mass of the primaries, $m_1 + m_2$, and the distances by the constant separation between the primaries, R_{12}.

$$\begin{aligned}
\mu &\doteq \frac{m_2}{m_1 + m_2}, \\
r_1 &\doteq \frac{R_{13}}{R_{12}}, \\
r_2 &\doteq \frac{R_{23}}{R_{12}}.
\end{aligned} \tag{7.22}$$

Furthermore, the value of gravitational constant is nondimensionalized such that the angular velocity of the primaries, Eq. (7.20), becomes unity. Using the methods of Chapter 4, we can write the equations of motion of m_3 in terms of its position, $\mathbf{r} = x\mathbf{i} + y\mathbf{j} + z\mathbf{k}$, relative to a rotating frame $(\mathbf{i}, \mathbf{j}, \mathbf{k})$ with origin at the common center of mass (Fig. 7.1), and rotating with the nondimensional velocity, $\boldsymbol{\omega} = \mathbf{k}$ (Eq. (7.20)):

$$\begin{aligned}\ddot{x} - 2\dot{y} - x &= -\frac{(1-\mu)(x-\mu)}{r_1^3} - \frac{\mu(1-\mu+x)}{r_2^3},\\ \ddot{y} + 2\dot{x} - y &= -\frac{(1-\mu)y}{r_1^3} - \frac{\mu y}{r_2^3},\\ \ddot{z} &= -\frac{(1-\mu)z}{r_1^3} - \frac{\mu z}{r_2^3}.\end{aligned} \tag{7.23}$$

Note that we have not assumed that m_3 lies in the same plane as the primaries. The set of coupled, nonlinear, ordinary differential equations has proved unsolvable in a closed form over the past two centuries. However, we can obtain certain important analytical insights into the problem without actually solving it.

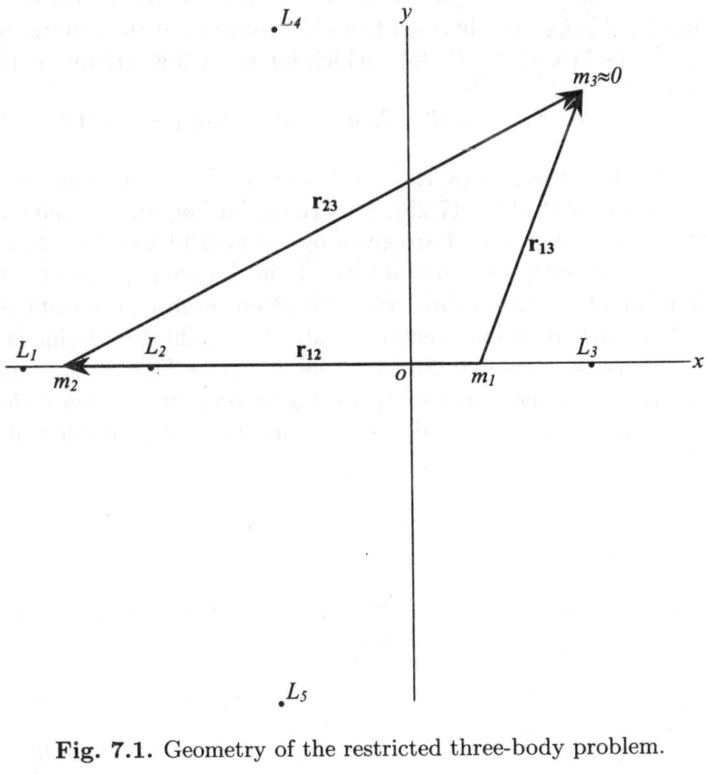

Fig. 7.1. Geometry of the restricted three-body problem.

7.4.1 Lagrangian Points and Their Stability

We have just seen that certain equilibrium solutions are possible to the general three-body problem. The equilibrium points for the restricted problem are called *Lagrangian points* (or *libration points*) and can be obtained by equating the time derivatives in Eq. (7.23) to zero, resulting in the following algebraic equations:

$$\begin{aligned} x &= \frac{(1-\mu)(x-\mu)}{r_1^3} + \frac{\mu(1-\mu+x)}{r_2^3}, \\ y &= \frac{(1-\mu)y}{r_1^3} + \frac{\mu y}{r_2^3}, \\ 0 &= \frac{(1-\mu)z}{r_1^3} + \frac{\mu z}{r_2^3}. \end{aligned} \tag{7.24}$$

From the last of Eq. (7.24), we have the result $z = 0$ for all Lagrangian points. Therefore, the equilibrium points always lie in the same plane as the primaries. As we have discovered above, the coplanar equilibrium solutions consist of the three possible colinear positions of m_3, called the *colinear Lagrangian points*, L_1, L_2, L_3, as well as the two equilateral triangle positions, called *triangular Lagrangian points*, L_4, L_5, with the two primaries (Fig. 7.1). The points L_1, L_2, L_3 are obtained from the solution of the quintic equation of Lagrange (the last of Eq. (7.21)), which for $m_3 = 0$ is written as follows:

$$r_2^5 + (3-\mu)r_2^4 + (3-2\mu)r_2^3 - \mu r_2^2 - 2\mu r_2 - \mu = 0\,. \tag{7.25}$$

As seen above, for the small values of μ that are typical in the solar system, we have only one real root of Eq. (7.25). The triangular Lagrangian points, L_4, L_5, correspond to $r_1 = r_2 = 1$ and are given by the coordinates $(\mu - \frac{1}{2}, \pm\sqrt{3}/2)$.

In order to investigate the stability of the Lagrangian points, we consider infinitesimal displacements, $\delta x, \delta y, \delta z$, from each equilibrium position, $x_0, y_0, 0$. If the displacements remain small, the equilibrium point is said to be stable, otherwise, unstable. Substituting $x = x_0 + \delta x, y = y_0 + \delta y, z = \delta z$ into the equations of motion, and neglecting second- and higher-order terms involving the small displacements, we have the following equation of motion for out-of-plane motion:

$$\delta\ddot{z} + \left[\frac{(1-\mu)}{r_1^3} + \frac{\mu}{r_2^3}\right]\delta z\,. \tag{7.26}$$

Now, for small displacements the denominator terms can be approximated through binomial expansion as follows:

$$\begin{aligned} r_1^{-3} &\approx [(x_0-\mu)^2 + y_0^2]^{-\frac{3}{2}} - 3r_{1_0}^{-5}[(x_0-\mu)\delta x + y_0\delta y], \\ r_2^{-3} &\approx [(x_0+1-\mu)^2 + y_0^2]^{-\frac{3}{2}} - 3r_{2_0}^{-5}[(x_0+1-\mu)\delta x + y_0\delta y], \end{aligned} \tag{7.27}$$

where r_{1_0}, r_{2_0} are the equilibrium values of r_1, r_2. Upon substitution of Eq. (7.27) into Eq. (7.26), we have

$$\delta\ddot{z} + c\delta z = 0 , \tag{7.28}$$

where

$$c \doteq (1-\mu)[(x_0-\mu)^2 + y_0^2]^{-\frac{3}{2}} + \mu[(x_0+1-\mu)^2 + y_0^2]^{-\frac{3}{2}} \tag{7.29}$$

is a constant. Thus, the out-of plane motion is decoupled from that occurring within the plane of the primaries. The general solution to Eq. (7.28) is written as

$$\delta z = \delta z_0 e^{\sqrt{-c}t} , \tag{7.30}$$

which represents a stable motion (constant amplitude oscillation) as long as $c \geq 0$, which is always satisfied. Hence, the out-of-plane small disturbance motion is unconditionally stable and can be safely ignored in a further analysis.[1]

For the small displacement, coplanar motion about a triangular Lagrangian point, say L_4, we can write

$$\begin{aligned} \delta\ddot{x} - 2\delta\dot{y} - \frac{3}{4}\delta x - \frac{3\sqrt{3}(\mu-\frac{1}{2})}{2}\delta y &= 0, \\ \delta\ddot{y} + 2\delta\dot{x} - \frac{3\sqrt{3}(\mu-\frac{1}{2})}{2}\delta x - \frac{9}{4}\delta y &= 0, \end{aligned} \tag{7.31}$$

whose general solution (Chapter 14) to initial displacement $\delta x_0, \delta y_0$ can be written as

$$\delta\mathbf{r} = \delta\mathbf{r_0} e^{\lambda t} , \tag{7.32}$$

where $\delta\mathbf{r} = (\delta x, \delta y)^T$, $\delta\mathbf{r_0} = (\delta x_0, \delta y_0)^T$, and λ is the *eigenvalue* [4] of the following matrix:

$$\mathsf{A} = \begin{pmatrix} 0 & 0 & 1 & 0 \\ 0 & 0 & 0 & 1 \\ \frac{3}{4} & \frac{3\sqrt{3}(\mu-\frac{1}{2})}{2} & 0 & 2 \\ \frac{3\sqrt{3}(\mu-\frac{1}{2})}{2} & \frac{9}{4} & -2 & 0 \end{pmatrix} . \tag{7.33}$$

The resulting characteristic equation for λ is thus the following:

$$\lambda^4 + \lambda^2 + \frac{27}{4}\mu(1-\mu) = 0 , \tag{7.34}$$

whose roots are

$$\lambda^2 = \frac{1}{2}[\pm\sqrt{1-27\mu(1-\mu)} - 1] . \tag{7.35}$$

[1] Battin [11] presents the general solution to the large displacement, out-of-plane, rectilinear motion about the center of mass ($x_0 = y_0 = 0$).

For stability, the values of λ must be purely imaginary, representing a harmonic oscillation about the equilibrium point. If the quantity in the square-root is negative, there is at least one value of λ with a positive real part (unstable system). Therefore, the critical values of μ representing the boundary between stable and unstable behavior of L_4 are those that correspond to

$$1 - 27\mu(1-\mu) = 0\,. \tag{7.36}$$

Hence, for the stability of the triangular Lagrangian points, we require either $\mu \leq 0.0385209$ or $\mu \geq 0.9614791$, which is always satisfied in the solar system. Therefore, we expect that the triangular Lagrangian points would provide stable locations for smaller bodies in the solar system. The existence of such bodies for the sun–Jupiter system has been verified in the form of *Trojan* asteroids. For the earth–moon system, $\mu = 0.01215$ is well within the stable region, hence one could expect the triangular points to be populated. However, the earth–moon system is not a good example of the restricted three-body problem because of an appreciable influence of the sun's gravity, which renders the triangular points of earth–moon system unstable. Hence, L_4, L_5 for the earth–moon system are empty regions in space, where one cannot park a spacecraft.

For a given system, one can determine the actual small displacement dynamics about the stable triangular Lagrangian points in terms of the eigenvalues, λ. For example, the earth–moon system has nondimensional natural frequencies 0.2982 and 0.9545, which indicate *long-period* and *short-period* modes of the small displacement motion. However, it is still left for us to investigate how much energy expenditure is necessary to reach the stable triangular points.

For the stability of the colinear Lagrangian points, we employ $y_0 = 0$, and the resulting equations of small displacement can be written as follows:

$$\begin{aligned}
\delta\ddot{x} - 2\delta\dot{y} - \left[\frac{2(1-\mu)}{(x_0-\mu)^3} + \frac{2\mu}{(x_0+1-\mu)^3} + 1\right]\delta x &= 0, \\
\delta\ddot{y} + 2\delta\dot{x} + \left[\frac{1-\mu}{(x_0-\mu)^3} + \frac{\mu}{(x_0+1-\mu)^3} - 1\right]\delta y &= 0\,.
\end{aligned} \tag{7.37}$$

The eigenvalues of this system are the eigenvalues of the following square matrix:

$$\mathsf{A} = \begin{pmatrix} 0 & 0 & 1 & 0 \\ 0 & 0 & 0 & 1 \\ \alpha & 0 & 0 & 2 \\ 0 & -\beta & -2 & 0 \end{pmatrix}, \tag{7.38}$$

where

$$\begin{aligned}
\alpha &= \frac{2(1-\mu)}{(x_0-\mu)^3} + \frac{2\mu}{(x_0+1-\mu)^3} + 1, \\
\beta &= \frac{1-\mu}{(x_0-\mu)^3} + \frac{\mu}{(x_0+1-\mu)^3} - 1\,.
\end{aligned} \tag{7.39}$$

The resulting characteristic equation for the eigenvalues, λ, is the following:

$$\lambda^4 + (4 + \beta - \alpha)\lambda^2 - \alpha\beta = 0 \,. \tag{7.40}$$

For the colinear Lagrangian points, $\alpha > 0$ and $\beta < 0$, which implies that the constant term in the characteristic equation is always positive. Therefore, there is always at least one eigenvalue with a positive real part, hence the system is unconditionally unstable. Thus, the colinear Lagrangian points of any restricted three-body system are always unstable. However, a spacecraft can successfully orbit a colinear point with small energy expenditure. Such an orbit around a Lagrangian point is termed a *halo orbit*, and several spacecraft have been designed to take advantage of halo orbits around the sun–earth L_1 and L_2 points for monitoring the interplanetary zone ahead of and behind the earth. For instance, a spacecraft near the sun–earth L_1 point can provide useful data about an approaching solar wind, and another about L_2 can explore the extent of the earth's magnetic field. Examples of spacecraft in halo orbits include the *International Sun-Earth Explorer (ISEE-3)*, the *Microwave Anisotropy Probe (MAP)*, the *Advanced Composition Explorer (ACE)*, and the *GEOTAIL* missions of NASA.

7.4.2 Jacobi's Integral

Jacobi derived the only additional scalar constant of the restricted three-body motion (out of the six necessary for a complete, closed-form solution), called *Jacobi's integral*. He defined a scalar function, J, by

$$J \doteq \frac{1}{2}(x^2 + y^2) + \frac{1-\mu}{r_1} + \frac{\mu}{r_2} \,. \tag{7.41}$$

The equation of motion of the infinitesimal mass can then be expressed as follows in the frame rotating with angular velocity $\boldsymbol{\omega} = \mathbf{k}$:

$$\ddot{\mathbf{r}} + \boldsymbol{\omega} \times \dot{\mathbf{r}} = \frac{\partial J}{\partial \mathbf{r}}^T \,, \tag{7.42}$$

where $\dot{\mathbf{r}} \doteq \dot{x}\mathbf{i} + \dot{y}\mathbf{j} + \dot{z}\mathbf{k}$ is the velocity of the mass relative to the rotating frame, and $(\partial J/\partial \mathbf{r})^T$ represents the gradient of J with respect to $\mathbf{r}$ (Chapter 3). By taking the scalar product of Eq. (7.42) with $\dot{\mathbf{r}}$, we have

$$\ddot{\mathbf{r}} \cdot \dot{\mathbf{r}} = \frac{1}{2}\frac{\mathrm{d}\dot{r}^2}{\mathrm{d}t} = \frac{\partial J}{\partial \mathbf{r}}^T \cdot \dot{\mathbf{r}} = \frac{\mathrm{d}J}{\mathrm{d}t} \,, \tag{7.43}$$

which is an exact differential and can be integrated to obtain

$$C = \frac{1}{2}v^2 - \frac{1}{2}(x^2 + y^2) - \frac{1-\mu}{r_1} - \frac{\mu}{r_2} \,. \tag{7.44}$$

Here, C is the constant of integration, called *Jacobi's constant*, and $v \doteq \dot{r}$ is the relative speed of the mass in the rotating frame. The Jacobi constant can thus be thought to represent a pseudo-energy of the mass m_3, which is a sum of the relative kinetic energy, the gravitational potential energy, and an additional pseudo-potential energy, $-1/2(x^2+y^2)$. The value of C at any point is a measure of its relative energy. The most useful interpretation of Jacobi's integral lies in the contours of zero relative speed, $v = 0$, which represent the boundary between a motion toward higher potential, or the return trajectory to the lower potential. On such a boundary, the mass must stop and turn back. Hence, the zero relative speed curves demarcate the regions accessible to a spacecraft and cannot be crossed.

Example 7.2. Plot the zero relative speed contours of constant C for $-1.5 \leq C \leq -2.5$ in the neighborhood of the Lagrangian points L_1, L_2, and L_4 for the earth–moon system with $\mu = 0.01215$. Analyze the energy requirements for reaching the given Lagrangian points.

The contours are plotted in Figs. 7.2 and 7.3, with the use of MATLAB statements such as

```
>> m=0.01215; %mu for earth-moon system
>> x=-1:0.02:1;y=0.5:0.005:1;
>> for i=1:size(x,2);for j=1:size(y,2);X(i,:)=x(i);
Y(:,j)=y(j);r1=sqrt((x(i)-m)^2+y(j)^2);
r2=sqrt((x(i)+1-m)^2+y(j)^2);
C(i,j)=-0.5*(x(i)^2+y(j)^2)-(1-m)/r1-m/r2;end;end
>> surfc(Y,X,C) %surface contours of constant C in nbd. of L4
```

The contours of C are seen in Fig. 7.2 to enclose a small spherical space in the vicinity of m_2, and a large cylindrical region around the origin with axis perpendicular to the orbital plane. We expect a similar spherical contour around the other primary, m_1. The closed zero-speed contours around a primary indicate that a flight from one primary to the other is impossible. As the value of C becomes less negative, the contours expand in size and may touch one another for some special values of C. For example, the zero-speed contours for $C = -1.594$ [which corresponds to $C(L_2)$] touch at the Lagrangian point L_2, thereby indicating the possibility of crossing over from one primary to the other through L_2 in a closed flight path, called *free-return trajectory*. At a larger value of C corresponding to $C(L_1)$, the region around L_2 opens up, indicating flight at nonzero relative velocity between the primaries, and the contours touch at L_1. For still larger C, corresponding to $C(L_3)$, the zero speed contours touch at L_3, while the region around L_1 is opened up, indicating an escape from the gravitational influence of the primaries. In such a case, contours enclose the triangular Lagrangian points, as depicted in Fig. 7.3, indicating that a flight to reach them is impossible from any of the primaries. As the value of C becomes very large, the forbidden regions around L_4, L_5 shrink, but never actually vanish, thereby denoting that the stable triangular Lagrangian points can be approached (but never reached) only with a very high initial energy.

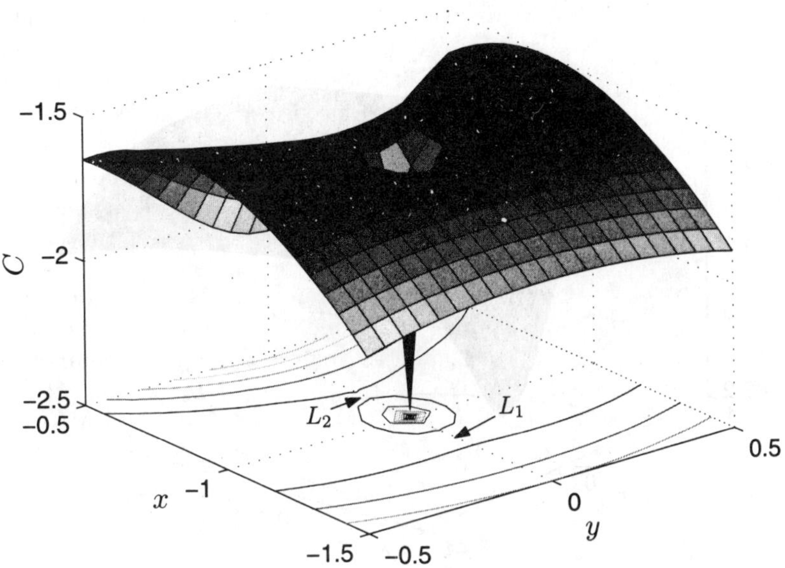

Fig. 7.2. Zero-speed contours of Jacobi's constant in the neighborhood of earth–moon L_1 and L_2.

7.4.3 Numerical Solution of the Restricted Problem

The restricted problem of three bodies appears more amenable to a solution than the general problem, due to the availability of an additional integral of motion, namely Jacobi's integral. However, the restricted three-body problem is unsolvable in a closed form because it does not possess the adequate number of scalar constants. With Jacobi's integral, the total number of scalar constants is only 13 whereas 18 are required for a closed-form solution. The unsolvability of the restricted three-body problem was also demonstrated in the 19th century by *Poincaré* using phase-space surfaces. The problem is also a member of a select group of mechanical systems, called *chaotic systems*, where a small change in the initial condition results in an arbitrarily large change in the response. Such systems are quite difficult to model, and their behavior is studied by a special branch of physics.

Numerical solutions to the problem have become possible since the availability of digital computers. When searching for a numerical solution, one desires a periodic behavior, which allows numerical integration of the equations of motion for only one time period, and a repetitive trajectory thereafter. Such solutions are encountered in the two-body orbits. However, in the restricted three-body problem, such periodic orbits are rarely encountered, and generally we have solutions changing forever with time. Numerical integration

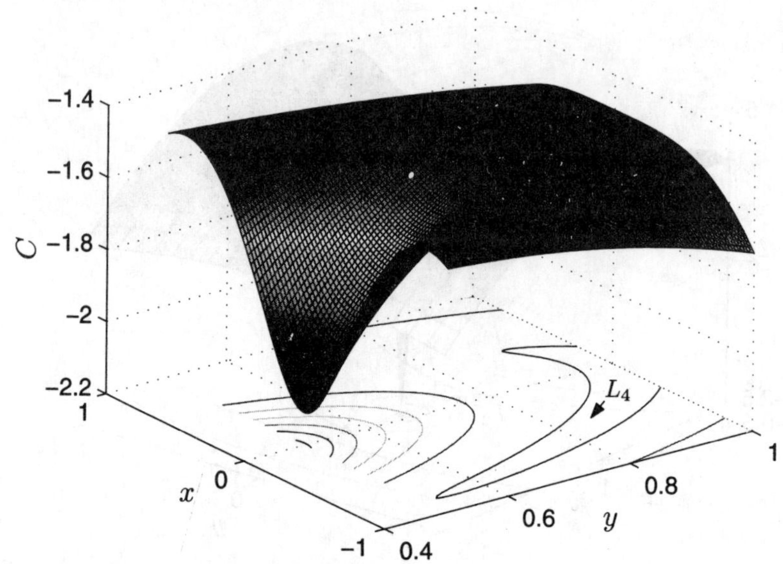

Fig. 7.3. Zero-speed contours of Jacobi's constant in the neighborhood of earth–moon L_4.

for such aperiodic solutions is frought with inaccuracies, due to truncation errors (Appendix A) that accumulate over time, and may grow to unacceptable magnitudes within a few orbital time periods of the primaries. Therefore, an accurate time-integration scheme, such as a higher-order Runge–Kutta technique (Appendix A), is employed with close tolerances to achieve reasonably accurate solutions.

Example 7.3. Simulate the trajectory of a spacecraft passing through the point $(0.1, 0)$ in the earth–moon system, with the following relative velocity components:

(a) $\dot{x} = 0,\ \dot{y} = 0.5,$
(b) $\dot{x} = -4,\ \dot{y} = 1,$
(c) $\dot{x} = -3.35,\ \dot{y} = 3,$
(d) $\dot{x} = -3.37,\ \dot{y} = 3,$
(e) $\dot{x} = -3.4,\ \dot{y} = 3,$
(f) $\dot{x} = -3.5,\ \dot{y} = 3,$
(g) $\dot{x} = -3.6,\ \dot{y} = 3.$

We begin by writing a MATLAB program for the restricted three-body equations of motion, called *res3body.m*, tabulated in Table 7.1. This code pro-

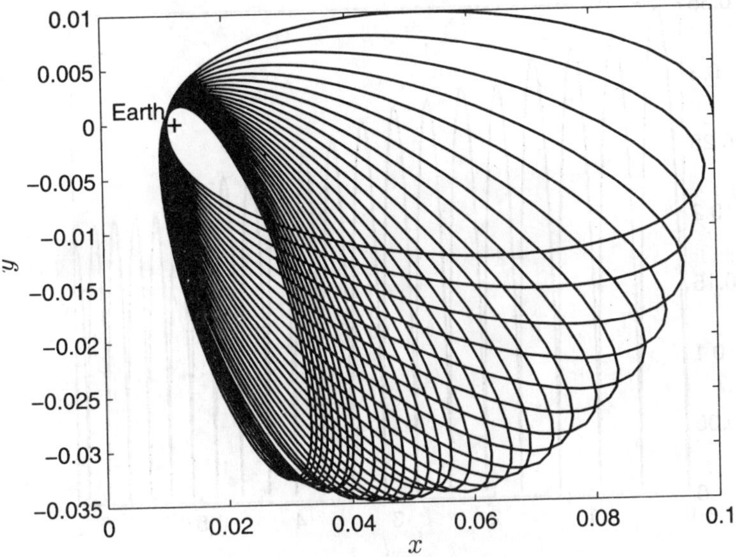

Fig. 7.4. Spacecraft trajectory with initial conditions $\dot{x}(0) = 0, \dot{y}(0) = 0.5$ [case (a)].

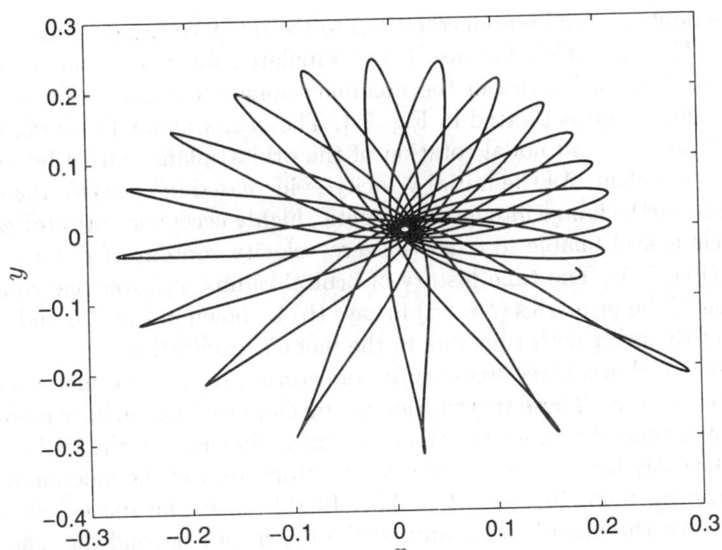

Fig. 7.5. Spacecraft trajectory with $\dot{x}(0) = -4, \dot{y}(0) = 1$ [case (b)].

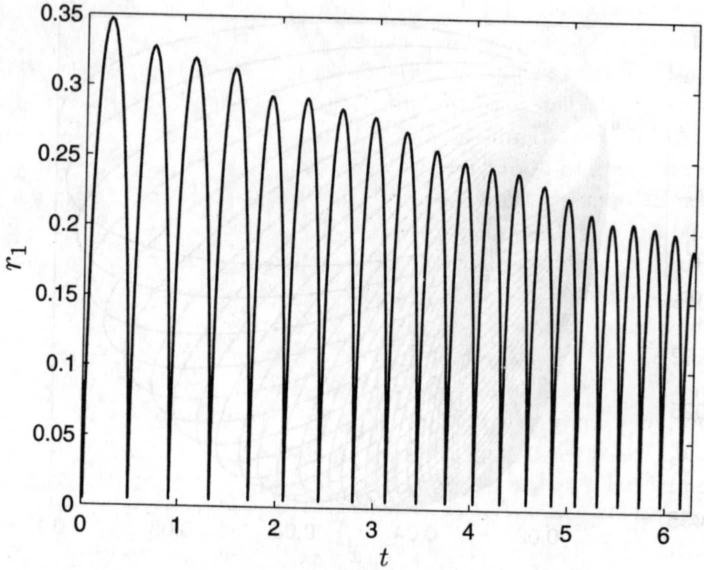

Fig. 7.6. Time variation of the orbital radius of spacecraft with $\dot{x}(0) = -4, \dot{y}(0) = 1$ [case (b)].

vides the nonlinear differential equations to the intrinsic Runge–Kutta solver, *ode45.m*. The trajectory for case (a) is simulated for $t = 1$ (one complete revolution of the primaries in the nondimensional time is $t = 2\pi$, which is about 30 days) and is plotted in Fig. 7.4. The orbits about the earth $(\mu, 0)$, with both apsidal and nodal rotation of the orbital plane caused by moon's gravity, are evident. As the initial velocity is increased in case (b), the orbits around the earth change into more energetic, highly eccentric trajectories, but the vehicle is still unable to cross the zero velocity contour of C for a lunar journey (Fig. 7.5). The time history of orbital radius, r_1, over one complete revolution of the primaries ($t = 2\pi$) in case (b) is shown in Fig. 7.6, indicating a decay of the orbit with time due to the moon's gravitation.

Figure 7.7 shows three free-return trajectories around the moon, arising out of cases (c)–(e). These trajectories can be employed for lunar exploration, without fuel expenditure for the return journey.[2] In case (c), the initial velocity is sufficiently large for a free-return trajectory around the moon and back to the starting point in about $t = 3.57$. In this case, the spacecraft passes slightly below the moon's orbit around the earth, but beyond L_2. The total flight time is reduced significantly in case (d) to about $t = 2.8$, when the space-

[2] The free-return concept formed the basis of Jules Verne's famous novel *From the Earth to the Moon* (1866), which was ultimately realized about a century later by the manned lunar exploration through the *Apollo* program.

craft passes around the moon, between L_1 and L_2. In the process, the return trajectory has a slightly larger kinetic energy due to the boost provided by the lunar *swing-by*. Lunar swing-by trajectories have been employed in cheaply boosting several spacecraft to the sun–earth Lagrangian points, such as the *ISEE-3, MAP*, and *ACE* missions of NASA [12]. In case (e), the round-trip flight time increases to about $t = 3.45$, as the spacecraft swings-by the moon at a greater distance, passing beyond L_1. For cases (f) and (g), plotted in Fig. 7.8, qualitatively different trajectories are observed. Case (f) is simulated for a long time ($t = 35$), demonstrating that the spacecraft makes a first pass of the moon at a large distance from L_1, but is unable escape the earth's gravity, which brings it closer to the moon in the next pass—crossing the earth–moon line near L_2—and ultimately brings the spacecraft into earth's orbit of ever-decreasing radius [somewhat similar to case (b)]. Case (f) illustrates a novel method of cheaply launching a satellite into the geosynchronous orbit ($r_1 \approx 0.1$) with the use of multiple lunar swing-bys. A similar approach of multiple planetary swing-bys has been found useful in reducing the mission cost of spacecraft bound for distant planets. As the initial energy is increased in case (g), the spacecraft does not return to earth but embarks on an escape trajectory from the earth–moon system, as depicted in Fig. 7.8. In such a trajectory, advantage is derived of gravity assist from the moon, thereby reducing the fuel expenditure, and hence, total cost of an interplanetary mission.

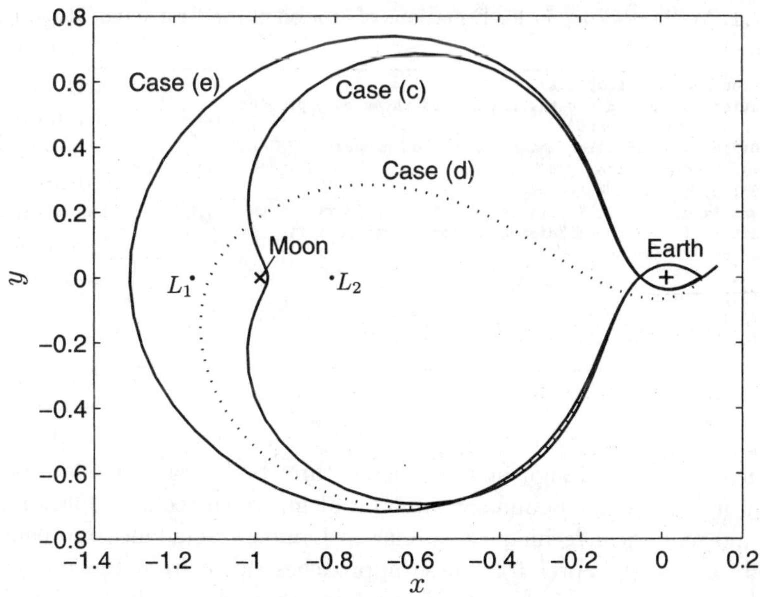

Fig. 7.7. Free lunar return trajectories [cases (c)–(e)].

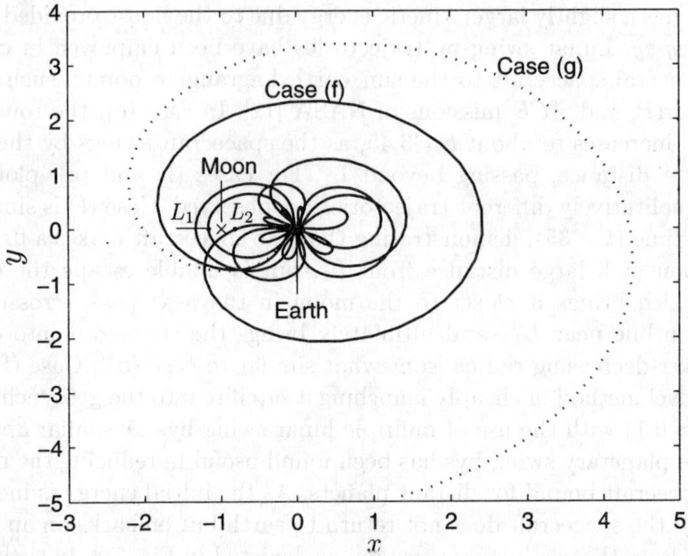

Fig. 7.8. Multiple lunar swing-by and escape trajectories [cases (f) and (g)].

Table 7.1. M-file *res3body.m* for Equations of Motion of the Restricted Three-Body Problem

```
function Xdot=res3body(t,X)
%Equations of motion for the restricted three-body problem.
%(c) 2006 Ashish Tewari
m=0.01215; %nondimensional mass of second primary
x=X(1); %x-coordinate
y=X(2); %y-coordinate
r1=sqrt((x-m)^2+y^2); %distance from first primary
r2=sqrt((x+1-m)^2+y^2); %distance from second primary
%Equations of motion:
Xdot(1,1)=X(3);
Xdot(2,1)=X(4);
Xdot(3,1)=x+2*Xdot(2,1)-(1-m)*(x-m)/r1^3-m*(x+1-m)/r2^3;
Xdot(4,1)=y-2*Xdot(1,1)-(1-m)*y/r1^3-m*y/r2^3;
```

The general mission design in a restricted three-body transfer involves the solution of a two-point boundary value problem, which requires special analytical and numerical techniques, similar to Lambert's guidance problem for two-body orbits (Chapter 5). These approaches range from the analytical search for particular periodic solutions passing through the given points [13], to the numerical patching of two-body Lambert solutions [14]. Such approximate solutions are indispensible and have served well in practical mission designs, keeping in mind the impossibility of solving the restricted three-body problem in a closed form (or even an approximate analytical form), thereby excluding the two-point boundary-value solutions of the actual problem.

7.5 Summary

The three-body problem arises when we are interested in either the two-body orbital perturbations caused by a distant third body or the motion of a smaller body in the gravitational field formed by two larger bodies. The physical conservation laws yield only 10 scalar constants of the three-body problem, whereas 18 are required for a complete analytical solution. Lagrange obtained particular solutions of the problem for a coplanar motion of the three bodies, including the colinear and equilateral triangle equilibrium point solutions with reference to a rotating frame. When the mass of one of the bodies is negligible compared to that of the other two bodies, the restricted three-body problem arises. Although the restricted problem is also unsolvable in a closed form, the stability of its equilibrium points—called Lagrangian points—can be analyzed using small disturbance approximation. While the colinear Lagrangian points of any restricted three-body system are always unstable, the equilateral points are conditionally stable (and the condition is always satisfied in the solar system). There exists an additional scalar constant of the restricted three-body motion, called *Jacobi's integral*, the zero relative speed contours of which provide an analysis of the accessible regions for the smallest body (spacecraft). Careful numerical integration of the restricted three-body equations of motion yields interesting trajectories and can be utilized to design and analyze lunar and interplantery missions.

Exercises

7.1. A spacecraft is located 6600 km away from the earth's center. What should be the minimum relative velocity of the spacecraft at this point in order to reach a triangular Lagrangian point of the earth–moon system?

7.2. Determine the locations of the Lagrangian points for the sun–Jupiter system with $\mu = 0.00095369$.

7.3. For the sun–Jupiter system, determine the nondimensional frequencies of small displacement motion of the *Trojan* asteroids about the triangular Lagrangian point.

7.4. Plot the zero relative speed contours of constant C for $-1.5 \leq C \leq -2.5$ in the neighborhood of the Lagrangian points L_1, L_2, and L_4 for the sun–Jupiter system.

7.5. Simulate the trajectory of a comet passing through the point $(0.1, 0)$ in the sun–Jupiter system, with a relative velocity of $\dot{x} = -3.1, \dot{y} = 3$, for a complete revolution of the primaries. What is the time period of the comet's solar orbit compared to the time period of the primaries?

7.6. Simulate the trajectory of a comet having undergone a close pass of Jupiter at the point $(-1, 0)$ in the sun–Jupiter system, with a relative velocity of $\dot{x} = 0, \dot{y} = 1.5$, for a complete revolution of the primaries. How much time elapses until the next pass of the comet by Jupiter, compared to the time period, T, of the primaries?

7.7. An approximate analysis of a comet's trajectory having passed close to Jupiter is possible using *Tisserand's criterion*. Using the inertial velocity and position of the comet, $\mathbf{V}, \mathbf{R}$, and the relationship between the inertial and relative velocities (Chapter 5), derive the following alternative expression for Jacobi's constant:

$$C = \frac{1}{2}V^2 - \boldsymbol{\omega} \cdot (\mathbf{R} \times \mathbf{V}) - \frac{1}{2}(x^2 + y^2) - \frac{1-\mu}{r_1} - \frac{\mu}{r_2} ,$$

where $\boldsymbol{\omega}$ denotes the angular velocity of the rotating frame attached to the primaries. Then, using the inertial angular momentum of the comet's trajectory, show that the following criterion is satisfied by the comet's orbital elements:

$$\frac{1}{a} + 2\sqrt{\frac{a(1-e^2)}{r_{12}^3}} \cos i = \text{constant}.$$

7.8. A comet is observed passing through the sun–Jupiter line at $x = -1$ with a relative velocity of 2 units. If the semi-major axis of the comet's subsequent orbit is 20 units, estimate its eccentricity. (Assume the comet's orbit is coplanar with that of Jupiter.)

8

Rocket Propulsion

8.1 Aims and Objectives

- To introduce the elements of rocket propulsion useful in flight dynamic modeling.
- To model the velocity impulse of single- and multi-stage rockets.
- To discuss the optimal selection of stage payload ratios of multi-stage rockets with examples.

8.2 The Rocket Engine

Rocket engines are indispensible for providing thrust to spacecraft. They are also essential for launch vehicles and missiles, due to their superior performance (thrust magnitude) as well as their relative simplicity of operation. Although efforts have been ongoing for quite sometime to replace rockets with air-breathing engines in the atmospheric phase of flight—due to the latter's higher efficiency—this alternative does not yet possess adequate reliability and cost-effectiveness for operation in long-range missiles and launch vehicles.

Rocket propulsion is based upon ejection of a propellant mass at a high speed, thereby achieving thrust generation by Newton's third law of motion (Chapter 4). The high exhaust speed of the propellant is obtained by first producing a gaseous (or plasma) mixture with a high *internal energy*, and then accelerating it in a *nozzle*. Theoretically, two categories of technologies are possible for generating the high internal energy required by the particles of the propellant: (a) *thermal*, which includes *chemical, nuclear*, and *thermonuclear reactions*, and (b) *electromagnetic*, including *magnetohydrodynamic* (MHD) devices for creating a hot plasma, and *ionic* for producing cold, ionized gas. Of these, the most common and practical method (albeit with the lowest efficiency) is that of a chemical rocket, wherein the propellants undergo a chemical reaction in order to liberate the stored chemical energy as the internal energy. Although ionic thrusters using cold, inert gas such as Argon have been built, their thrust is very low, and their application is limited to low-thrust, orbital transfers (e.g., the *Deep-Space* probes of NASA), which we will discuss in a later section. An MHD, nuclear, or thermo-nuclear rocket is far from being deployed (although some laboratory experiments have been carried out) due to technical and environmental reasons.

A nozzle is the primary component of any rocket and is responsible for converting the released internal energy into kinetic energy of the propellant mass. In a thermal rocket, the nozzle is a flow expansion device for expanding the propellant gas to a lower pressure. When the propellant is a charged gas, the nozzle consists of an *electromagnetic* or *electrostatic* field for accelerating the charged particles.

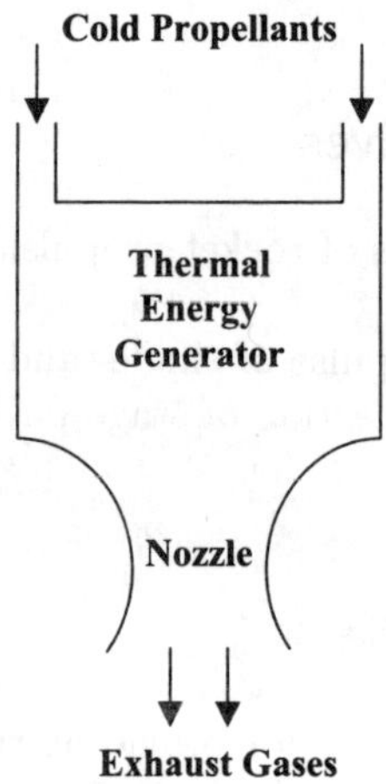

Fig. 8.1. Schematic diagram of a thermal rocket engine.

In this book, our discussion will be confined to thermal, chemical rockets. As depicted in Fig. 8.1, a thermal rocket has three primary components: the *propellants*, which can be in solid, liquid, or gaseous form; a thermal energy generator (*combustion chamber*); and a *nozzle.* The propellants of a chemical rocket consist of a *fuel* and an *oxidizer*. Since they must carry an oxidizer, the rockets are inherently less efficient than air-breathing engines, which derive the oxidizer required for combustion in the form of oxygen from the atmosphere. One can choose between solid and liquid propellants. While the solid propellants are easier to store, their chemical reaction cannot be *throttled*, or even shut-off, like the liquid propellants. Once ignited, solid propellants continue to burn until they are exhausted. The combustion chamber provides a mechanism for efficiently mixing and chemically combining the propellants. The chemical reaction carried out in the combustion chamber is called *combustion*, or *burning.*

A simple expression relating the internal energy per unit mass, e_i, to the maximum possible exhaust speed, v_e, of a propellant particle of mass, m, obtained in the nozzle of a thermal rocket, can be derived by assuming a perfect energy conversion

$$e_i = \frac{1}{2} m {v_e}^2 . \tag{8.1}$$

From the *kinetic theory*[1] of a perfect diatomic gas, we can write

$$e_i = \frac{5}{2} kT , \tag{8.2}$$

where $k = 1.38 \times 10^{-23}$ J/K is *Boltzmann's constant*, and T is the combustion temperature in Kelvin (K). From Eq. (8.1), we have

$$v_e = \sqrt{\frac{2e_i}{m}} , \tag{8.3}$$

which indicates that the largest possible exhaust velocity depends only upon the propellants employed, with distinctive internal energy and particle mass. A higher internal energy, and a lighter propellant, produce a larger exhaust speed. For example, the largest exhaust velocity for a given chemical propellant mass is that for the combination of liquid hydrogen (LH_2) and liquid fluorine LF. However, as the combustion product, hydrofluoric acid, is highly corrosive, the next best combination possible is that of liquid hydrogen and liquid oxygen (LO_2), whose e_i is roughly the same as that of LH_2/LF combination, but with a slightly heavier combustion product (water vapour).

From Eqs. (8.1) and (8.2), we can derive the following expression for the combustion temperature:

$$T = \frac{m {v_e}^2}{5k} , \tag{8.4}$$

which places a restriction on the maximum exhaust velocity due to material considerations. For example, in the combustion chamber of a rocket employing LH_2/LO_2, the temperature is approximately 6000 K. Due to the fact that the temperature increases with the square of the exhaust velocity, a significantly larger exhaust velocity with nonchemical, thermal technology (nuclear, or thermo-nuclear) is materially impossible, even though the lightest possible propellant (hydrogen gas) may be employed.

Being a variable mass vehicle, the reactive thrust, $\mathbf{f_T}$, of a rocket was derived in Chapter 4. Generally, the exhaust gas of a thermal rocket is not expanded to the ambient static pressure, p_a, causing an additional term in the thrust expression due to the aerostatic pressure of exhaust gas. The complete expression for a rocket's thrust is thus

---

[1] Kinetic theory of a perfect gas is a branch of statistical thermodynamics, which employs *quantum mechanics* for the microscopic description of gas particles. See, for instance, Hill [16].

$$\mathbf{f_T} = -\Delta m \frac{\mathrm{d}(\mathbf{v}+\mathbf{v_e})}{\mathrm{d}t} - \mathbf{v_e}\frac{\mathrm{d}\Delta m}{\mathrm{d}t} - A(p_e - p_a)\frac{\mathbf{v_e}}{v_e}\,, \tag{8.5}$$

where $\mathbf{v}$ is the velocity of the vehicle's center of mass, $\mathbf{v_e}$ is the velocity of the exhaust gas relative to the vehicle's center of mass, m and Δm are the instantaneous masses of the vehicle and the exhaust gas, respectively, A is the nozzle's exit area (normal to $\mathbf{v_e}$), and p_e is the static pressure of the exhaust gas. The direction of $\mathbf{v_e}$ is, by and large, opposite to that of $\mathbf{v}$ in normal operation, or along $\mathbf{v}$ for reverse thrust (*retro-rocket*). For controlling the trajectory and the attitude, all rockets employ swivelling of the nozzle to produce thrust deflection from the nominal direction. Since the exit velocity is generally constant in both magnitude and direction, we have

$$\mathbf{f_T} = -\Delta m \frac{\mathrm{d}\mathbf{v}}{\mathrm{d}t} - \mathbf{v_e}\frac{\mathrm{d}\Delta m}{\mathrm{d}t} - A(p_e - p_a)\frac{\mathbf{v_e}}{v_e}\,. \tag{8.6}$$

The nozzle design of a thermal rocket is crucial to its performance and efficiency. The magnitude of the thrust delivered in the rocket is directly proportional to the mass flow rate provided by a nozzle, which, in turn, is directly proportional to the nozzle's exit area. Thus, a larger nozzle delivers a higher thrust, for the same exhaust speed. The static pressure of the exhaust gas is rarely equal to the ambient static pressure at the nozzle exit. The thrust due to aerostatic pressure at nozzle exit is positive in magnitude whenever $p_e > p_a$ (an *underexpanded nozzle*). This is the situation at very high altitudes, where the ambient conditions are close to a vacuum ($p_a \approx 0$). However, in order to utilize this additional thrust, the flow expansion must occur inside rather than outside the nozzle. There is a loss of propulsive efficiency due to underexpansion of the flow inside the nozzle, as the flow kinetic energy is not fully converted to thrust at the nozzle wall. On the other hand, there could be a significant loss of thrust caused by an *overexpanded* nozzle at low altitudes, when $p_e < p_a$. This thrust loss is alleviated by a nozzle design that achieves complete expansion ($p_e = p_a$) at a suitably moderate altitude. Unlike airbreathing engines, the rocket nozzle is fixed in geometry; thus, a complete expansion is possible only at a specific altitude. A design for complete expansion at very high altitudes is infeasible due to weight and size considerations, while choosing a low altitude for complete expansion increases the range of altitudes where propulsive efficiency is lost by underexpansion. The novel concept of an *Aerospike* nozzle, which has exposed sides, was evolved by Rocketdyne in the 1970s and tested for the X-33 *Single-stage to orbit* (SSTO) program of NASA in the 1990s (now cancelled). This nozzle design has an inherent altitude compensation capability, not unlike a variable geometry nozzle, which, combined with a significant weight reduction, allows efficient operation in a large altitude range.

8.3 The Rocket Equation and Staging

The net equation of motion of a rocket-powered vehicle due to Newton's second law (Chapter 4) is rewritten as

$$\mathbf{f} + \mathbf{f_T} = (m - \Delta m)\frac{\mathrm{d}\mathbf{v}}{\mathrm{d}t}, \tag{8.7}$$

where $\mathbf{f}$ denotes the net external force acting on the vehicle's center of mass, arising due to gravity and aerodynamics. Upon substituting Eq. (8.6) into Eq. (8.7), and noting that $\frac{\mathrm{d}\Delta m}{\mathrm{d}t} = -\frac{\mathrm{d}m}{\mathrm{d}t}$, we can write

$$\mathbf{f} = m\frac{\mathrm{d}\mathbf{v}}{\mathrm{d}t} - \mathbf{v_e}\frac{\mathrm{d}m}{\mathrm{d}t} + A(p_e - p_a)\frac{\mathbf{v_e}}{v_e}. \tag{8.8}$$

In order to obtain an insight into rocket propulsion, it is usual to take the case of space flight (zero aerodynamic force) and a negligible acceleration due to gravity. These are valid assumptions for spacecraft maneuvers with small flight-path angles, for which the component of gravity along thrust direction is negligible. Furthermore, we assume that the exhaust gas is ejected in a direction opposite to the velocity vector ($\mathbf{v_e} = -v_e\frac{\mathbf{v}}{v}$), and a complete expansion of the exhaust gas to zero static pressure takes place. Therefore, the net external force, $\mathbf{f}$, vanishes, and the resultant motion of the vehicle is directly opposite to the direction of the ejected gas. Under these assumptions, Eq. (8.8) becomes the following scalar equation of motion:

$$-v_e\frac{\mathrm{d}m}{\mathrm{d}t} = m\frac{\mathrm{d}v}{\mathrm{d}t}. \tag{8.9}$$

We can integrate Eq. (8.9) between the limits of the initial and current values of the mass and vehicle speed, m_0, v_0, and m, v, respectively, to obtain

$$v - v_0 = v_e \ln\frac{m_0}{m}. \tag{8.10}$$

When the exhaust gas is ejected in the same direction as the velocity vector (a *retro-rocket*), $\mathbf{v_e} = v_e\frac{\mathbf{v}}{v}$, and Eq. (8.10) becomes $v - v_0 = -v_e \ln\frac{m_0}{m}$. Equation (8.10) is a simple, analytical expression relating the mass and change in the speed of a rocket and is called the *rocket equation*. The most important observation from Eq. (8.10) is the fact that the change in the rocket's speed is directly proportional to the relative exhaust speed, v_e, for a given decrease in the vehicle's mass. Since the rocket operation in space is confined to short bursts compared to the total flight time of the spacecraft (Chapter 5), it is often a good approximation to assume that the speed changes instantaneously. Thus, the rocket provides a a *velocity impulse* of magnitude $\Delta v = v - v_0$, which is given by the rocket equation. It is easy to see from Eq. (8.9) that since v_e is constant, $m\Delta v = -v_e\Delta m$; thus, the exhaust velocity is equal to the change in linear momentum per unit *mass* of the propellant consumed.

However, rather than using the exhaust speed, it is customary to define a *specific impulse*, I_{sp}, as the change in linear momentum per unit *weight* of the propellant consumed, leading to

$$I_{sp} = \frac{v_e}{g} , \tag{8.11}$$

where g is the acceleration due to gravity at standard sea level of the earth ($g = 9.81$ m/s^2). The various thermal rocket propellants are classified according to the maximum specific impulse they can produce in an ideal rocket engine [Eq. (8.3)]. Some practical values of I_{sp} are 180–270 s for solid propellants, 260–310 s for nitrogen tetra-oxide (NO_4)/mono-methyl hydrazine (MMH), or NO_4/unsymmetrical di-methyl hydrazine (UDMH) liquid, *hypergolic*[2] propellants, 300–350 s for kerosene/LO_2, 455 s for LH_2/LO_2, and 475 s for LH_2/LF . For exotic technologies, theoretical I_{sp} values using hydrogen gas as propellant range from about 5000 s with nuclear fission to 10,000 s with thermo-nuclear (fusion). However, the temperatures required in such high values of I_{sp} dictated by Eq. (8.4) cannot be withstood by present materials. If a cooling of the nuclear reactor by the propellant is employed in order to bring the temperatures to tolerable levels, the resulting specific impulse would be no larger than that of a chemical rocket. Similar considerations prohibit the current use of cold ion, or plasma thrusters in launch vehicles and impulsive maneuvers. Although the I_{sp} of an ion rocket can be as high as 50,000 s, its mass flow rate cannot be substantial due to temperature restrictions, and, consequently, the thrust is very small.

Example 8.1. Estimate the propellant mass and thrust required by an LH_2/LO_2 rocket of mass 2000 kg in a circular, earth orbit, in order to produce a velocity impulse of magnitude 700 m/s in 5 s in the direction of the velocity vector.

In a circular orbit, the gravity is normal to the velocity direction (zero flight-path angle). Assuming complete expansion, we can, therefore, apply the rocket equation. The exhaust speed is calculated from the specific impulse as $v_e = gI_{sp} = (9.81)(455) = 4463.55$ m/s, and the final mass is obtained from the rocket equation,

$$m = m_0.e^{-\frac{\Delta v}{v_e}} = 1709.706 \text{ kg} ,$$

which implies a propellant consumption of $\Delta m = 290.294$ kg. The mass flow rate is obtained as

$$\frac{\mathrm{d}m}{\mathrm{d}t} = -\frac{\Delta m}{\Delta t} = -58.0588 \text{ kg/s} .$$

Finally, the thrust is calculated by using Eq. (8.6) as follows:

[2] Hypergolic propellants are chemicals that do not require an ignition source for combustion, and react spontaneously with each other. Since it is easy to start and stop combustion using hypergolic propellants, they are often used in reaction control systems of spacecraft.

$$f_T = -v_e \frac{dm}{dt} = 259,148.495 \text{ N} .$$

The average acceleration due to thrust is $\frac{\Delta v}{\Delta t} = 140$ m/s^2, which is greater than 14 times the acceleration due to gravity at standard sea level.

Since the rocket equation does not consider either gravity, or the atmospheric forces, one must add a margin to the total velocity impulse obtained from Eq. (8.10) in order to address the losses due to gravity and atmospheric drag, and propulsive losses due to the maneuvering and static pressure difference at the nozzle exit. For example, a launch to the low earth orbit must add about 1.5 km/s to the approximately 8 km/s of required orbital velocity change, leading to a total required impulse of about 9.5 km/s from the rocket equation. The necessary margin increases to about 2.0 km/s for a launch to the geosynchronous orbit. Since rockets are launched almost vertically, the initial thrust of a rocket must be greater than its weight and drag combined. However, as the rocket ascends to orbital flight, this requirement of a large thrust is removed, and the assumptions of the rocket equation are nearly valid.

The total mass of a rocket consists of the propellant mass, the structural mass, and the payload mass. Since the propellant mass can be very large in a rocket, we must examine its efficiency in terms of the maximum payload delivered for a given velocity impulse. On an intuitive basis, we can see that a rocket's efficiency can improve if it can shed some of its structural mass (regarded as dead weight) during its operation. The rocket equation, Eq. (8.10), is valid only when all the propellant is consumed without a change in the structural mass. This is referred to as a *single-stage rocket*. An alternative approach is to build a rocket in several modules, called *stages*, each complete with its own propellant tanks, combustion chamber, and nozzles. Such a vehicle can discard a stage as soon as it has exhausted its propellants, thereby reducing the structural mass. This process of shedding burn-out modules is called *staging*. We shall first examine the limitations of a single-stage rocket, and then study the advantages offered by staging.

8.3.1 The Single-Stage Rocket

Consider a single-stage rocket of payload mass, m_L, structural mass, m_s, and propellant mass, m_p. Thus, the initial mass is $m_0 = m_L + m_s + m_p$, and the final mass after all the propellant has been consumed is $m_f = m_L + m_s$. The final to initial mass ratio is written as

$$\frac{m_f}{m_0} = \frac{m_s}{m_s + m_p} + \frac{m_p m_L}{m_0(m_s + m_p} = \sigma + (1 - \sigma)\lambda , \tag{8.12}$$

where

$$\sigma \doteq \frac{m_s}{m_s + m_p} \tag{8.13}$$

is called the *structural ratio*, and

$$\lambda \doteq \frac{m_L}{m_0} \tag{8.14}$$

is denoted the *payload ratio*. Note that the payload mass is not included in the definition of the structural ratio, because a rocket is generally capable of carrying different payloads with the same structure. From the rocket equation, Eq. (8.10), the total velocity impulse, using a single-stage rocket in space, and for a negligible gravity along the velocity direction, is given by

$$\Delta v \doteq v_f - v_0 = -v_e \ln[\sigma + (1-\sigma)\lambda] \,. \tag{8.15}$$

The higher the payload ratio, λ, for a given velocity impulse, the better is the rocket's efficiency (and utility). For a given set of propellants, a higher payload ratio is possible only by reducing the structural ratio, σ. Since a rocket must have a strong structure for storing the propellants and the payload, and for withstanding external loads, σ cannot be reduced below a certain limit. Usually, the net weight of the nozzles, guidance, control system, and other accessories is also included in the structural weight, which restricts the practical value of σ to be no smaller than about 0.05. The *performance* of a rocket is usually measured in terms of the fraction $\frac{\Delta v}{v_e}$. Of course, the highest performance is possible with a zero payload ratio. As the payload ratio is increased, the performance diminishes, and becomes zero for $\lambda = 1$. Hence, performance and efficiency are conflicting requirements for a single-stage rocket (as in any vehicle). Usually, the performance is specified in terms of the orbital radius for a given launch site. In order to meet such a requirement in the most efficient manner, the payload ratio should be maximized. The variation of the payload ratio of a single-stage rocket with the performance parameter, $\frac{\Delta v}{v_e}$, for specific structural ratios is shown in Fig. 8.2. It is evident from this figure that the maximum possible performance from a single-stage rocket occurs for the minimum possible structural ratio, $\sigma = 0.05$, and is limited to $\frac{\Delta v}{v_e} < 3.0$. With a payload ratio $\lambda = 0.01$ (the lowest practical efficiency), the performance drops down to $\frac{\Delta v}{v_e} = 2.82$, which is insufficient to launch a payload to a low earth orbit, unless *cryogenic* (very low temperature) propellants (such as LH_2/LO_2) are employed. Since cryogenic propellants are difficult to manufacture and store, their use in a single-stage rocket with a low payload ratio is very expensive. It is also currently infeasible to build large enough tanks to safely store a sufficient cryogenic propellant mass required for the single-stage operation, as demonstrated by the cancelled X-33 project of NASA.[3]

[3] The X-33 was an experimental scaled prototype of a larger re-usable, single-stage launch vehicle, called *VentureStar*, that was to be built under NASA contract by Lockheed Corporation. The wings and aerodynamic controls required for an airplane-like horizontal landing led to the X-33's structural ratio to be moderate at $\sigma = 0.108$, leading to a total LH_2/LO_2 propellant mass of about 95,000 kg. In order to keep the structural ratio low, the propellant tanks were to be constructed out of composite materials, which proved technologically challenging, leading to the project's cancellation in 2001.

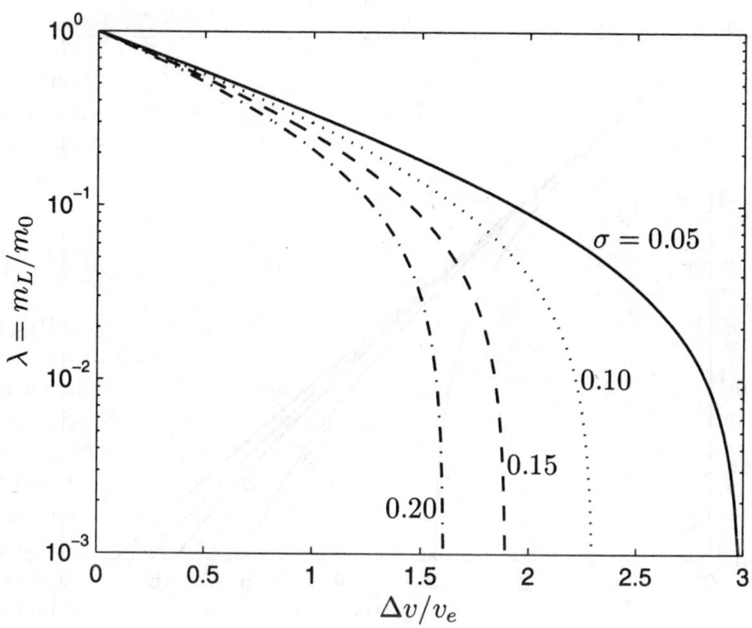

Fig. 8.2. Variation of payload ratio of a single-stage rocket with performance parameter $\frac{\Delta v}{v_e}$.

8.3.2 The Multi-Stage Rocket

Staging is necessary in order to overcome the performance drawback of a single-stage rocket. The rocket equation for a multi-stage rocket with a total of N stages can be written as follows:

$$\Delta v = -\sum_{k=1}^{N} v_{ek} \ln[\sigma_k + (1 - \sigma_k)\lambda_k] , \tag{8.16}$$

where the subscript k denotes quantities pertianing to the kth stage of the vehicle. It is assumed here that only one stage is burning at a given time, and each stage is discarded immediately after its propellant is consumed. Such a process is called *serial staging*. The initial mass of a given stage is the payload mass for the previous stage, i.e., the payload ratio of the kth stage is

$$\lambda_k = \frac{m_{0(k+1)}}{m_{0k}} , \tag{8.17}$$

resulting in the a rocket's payload ratio of

$$\lambda_T \doteq \frac{m_L}{m_{01}} = \prod_{k=1}^{N} \lambda_k . \tag{8.18}$$

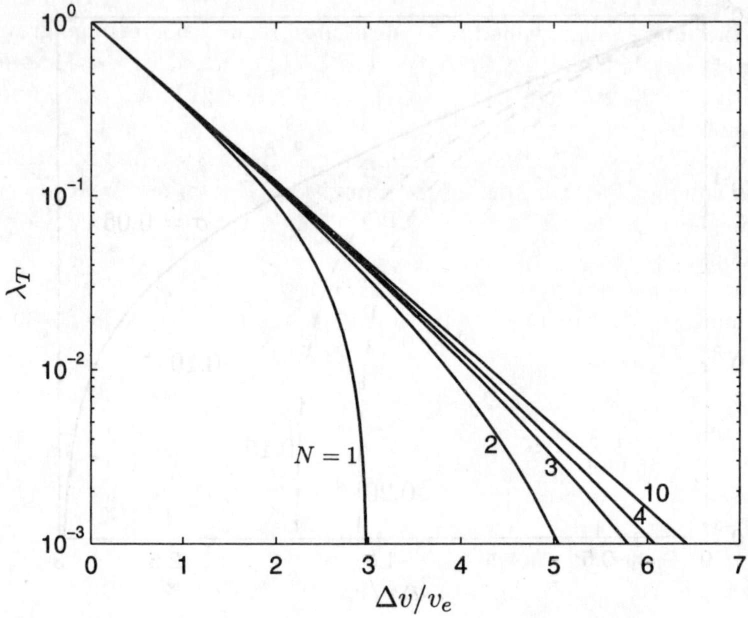

Fig. 8.3. Variation of total payload ratio of a multi-stage rocket with performance parameter $\frac{\Delta v}{v_e}$.

In order to see the performance benefits offered by staging, consider the simple case of N stages with the same exhaust velocity, structural ratio, and payload ratio, i.e., $v_{e1} = v_{e2} = \cdots = v_{eN} = v_e$, $\sigma_1 = \sigma_2 = \cdots = \sigma_N = \sigma$, and $\lambda_1 = \lambda_2 = \ldots = \lambda_N = \lambda$. The variation of the total payload ratio of such a rocket with the performance parameter, $\frac{\Delta v}{v_e}$, for a various number of stages, obtained using $\sigma = 0.05$, is shown in Fig. 8.3. It is clear that the total payload ratio, λ_T, increases with an increasing number of stages for a given performance. Conversely, with the same total payload ratio, it is possible to obtain a higher performance by increasing the number of stages. With $N = 2$ and $\lambda_T = 0.01$, a performance of $\frac{\Delta v}{v_e} = 3.86$ is possible. A two-stage rocket is capable of launching a payload to a low-earth orbit with $\lambda_T = 0.033$ using hydrocarbon liquid propellants. With $N = 3$ and $\lambda_T = 0.01$, a performance of $\frac{\Delta v}{v_e} = 4.1$ can be attained. Hence, staging offers a remarkable increase in performance over the single-stage rocket, for the same total payload ratio. However, the increase in performance is small with $N > 3$. Thus, even large rockets (such as the *Saturn-V* booster used in the *Apollo* manned, lunar missions during 1968–72) are limited to three stages, given the complexity involved in a large number of stages.

In a practical design, each stage employs a different propellant. The lower stages often use solid or liquid hydrocarbon propellants with a smaller specific impulse compared to the upper stages. This results in a large propellant mass for the initial stage, which requires a large thrust to overcome gravity at launch, and, in turn, large nozzles are required to produce the necessary thrust with an adequate mass flow rate. The mass flow required by the upper stages is much smaller (due to a reduced thrust requirement at near-orbital speeds), which enables the use of (expensive) high specific impulse propellants in smaller volumes for the upper-stage propulsion. The employment of cryogenic propellants in the smaller, upper stages improves the efficiency without an enormous cost, and thus most launch vehicles (especially for medium to high orbits) employ cryogenic upper stages. When the stages have different specific impulses, it is more efficient to employ a different payload ratio for each stage, as seen in the next section.

Example 8.2. A three-stage launch vehicle employs a liquid, hydrocarbon propellant in the first two stages with a specific impulse of 290 s, while the third stage uses a cryogenic LH_2/LO_2 propellant of specific impulse 455 s. All stages have the same structural ratio of $\sigma = 0.07$. The second and third stages' payload ratios are 1.2 and 0.65 times, respectively, that of the first stage. Calculate the propellant mass required to launch a payload of 1000 kg to the geosynchronous orbit, for which a total velocity increase of 13 km/s is required, considering gravity, drag, and thrust losses.

Since the payload ratios λ_2, λ_3 are referred to the unknown first-stage payload ratio, λ_1, we must express the rocket equation, Eq. (8.16), in terms of the nondimensional stage ratios $\beta_k \doteq \frac{v_{ek}}{v_{e1}}, \alpha_k \doteq \frac{\lambda_k}{\lambda_1}$ as follows:

$$\frac{\Delta v}{v_{e1}} = -\sum_{k=1}^{N} \beta_k \ln[\sigma_k + (1 - \sigma_k)\alpha_k]. \tag{8.19}$$

We begin by solving Eq. (8.19) with $N = 3$, for λ_1. Being a transcendental equation, Eq. (8.19) has to be solved by an iterative technique. As in the case of Kepler's equation (Chapter 4), we employ Newton's technique, which is incorporated in the MATLAB program *Nstage.m* (Table 8.1). Note that the initial guess for the first-stage payload ratio is $\lambda_1 = 0.1$, and a tolerance of 10^{-9} has been specified in the program for satisfying the rocket equation.

```
>> beta=[1;1;455/290]; sigma=0.07*[1;1;1]; alpha=[1;1.2;0.65]; %stage ratios
>> vf=13000/(9.81*290) %nondimensional speed change

   vf = 4.5696

>> p=Nstage(vf,beta,sigma,alpha);%Newton's iteration for first-stage payload ratio,p

   p =  0.204373458571974
   p =  0.247315220681014
   p =  0.25385430821876
   p =  0.254380998378698
   p =  0.254418303118114
   p =  0.254420916634985
   p =  0.254421099591889
   p =  0.254421112398925
   p =  0.254421113295418
   p =  0.254421113358172
```

Thus, a first-stage payload ratio of $\lambda_1 = 0.254421113$ satisfies the rocket equation up to the ninth decimal place. Therefore, the total payload ratio of the rocket is

$$\lambda_T = \lambda_1^N \prod_{k=1}^{N} \alpha_k = (0.254421113)^3(1.2)(0.65) = 0.01284559\,,$$

from which the initial rocket mass follows as $m_{01} = \frac{m_L}{\lambda_o} = 77,847.7296$ kg, with the second and third stage initial masses, $m_{02} = \lambda_1 m_{01} = 19,806.1060$ kg and $m_{03} = \lambda_2 m_{02} = 6046.9099$ kg, respectively. Since the structural ratio of each stage is identical, we have the same ratio of a stage's propellant mass to its structural mass of $\frac{m_{pk}}{m_{sk}} = \frac{1-0.07}{0.07} = 13.285714286$. The propellant masses of the stages are thus $m_{p3} = (m_{03} - m_L) * (1 - 0.07) = 4693.62617$ kg, $m_{p2} = (m_{02} - m_{03}) * (1 - 0.07) = 12,796.05244$ kg, and $m_{p1} = (m_{01} - m_{02}) * (1 - 0.07) = 53,978.7099$ kg, which add up to a total propellant mass of $m_{p1} + m_{p2} + m_{p3} = 71,468.3885$ kg. Hence, 91.8% of the initial mass of this rocket is the propellant mass.

Table 8.1. M-file *Nstage.m* for Calculating the Payload Ratio of a Multi-Stage Rocket

```
function p=Nstage(vf,beta,epsilon,alpha);
%Program for solving the multi-stage (N) rocket equation for the first-stage
%payload ratio, p. (Copyright @2006 by Ashish Tewari)
%beta=(Nx1) vector of ratios of specific impulses to that of the first stage
%(beta(k)=Isp_k/Isp_1); first element of beta should be 1.0
%alpha= (Nx1) vector of ratios of payload ratios to that of the first stage
%(alpha(k)=lambda_k/lambda_1); first element of alpha should be 1.0
%epsilon= (Nx1) vector of structural ratios of the stages
%vf=ratio of total velocity impulse to exhaust speed of first stage
%(vf=Delta_v/v_e1)
%(c) 2006 Ashish Tewari
N=size(beta,1);
p=0.1;
f=vf;
tol=1e-9;
for k=1:N
f=f+beta(k)*log(epsilon(k)+alpha(k)*(1-epsilon(k))*p);
end
while abs(f)>tol
f=vf;
fp=0;
for k=1:N
f=f+beta(k)*log(epsilon(k)+alpha(k)*(1-epsilon(k))*p);
fp=fp+alpha(k)*beta(k)/(epsilon(k)+alpha(k)*(1-epsilon(k))*p);
end
d=-f/fp;
p=p+d
end
```

8.3.3 Parallel Staging

A variation of the serial staging discussed above is *parallel staging*, where some stages are burned simultaneously. Parallel staging offers the advantage of getting rid of the propellant mass more quickly, thereby increasing the total efficiency when compared to serial staging. Therefore, all modern launch vehicles employ parallel staging of some kind, usually at the beginning of the launch phase when the propellant mass is the largest. A prime example of parallel staged vehicles is the *space shuttle*, which has two solid rocket boosters burning simultaneously with the main cryogenic engine as a combined stage. Parallel staging also offers the capability of modifying a rocket's performance by simply adding/removing a number of *strap-on* boosters outside the primary vehicle (called the *core vehicle*). The strap-on stages burn simultaneously with the first stage of the core vehicle. The structural simplicity and flexibility of strap-on parallel stages is a common design feature in launch vehicles and can be seen, for example, in the *Delta*, *Atlas*, and *Titan* rockets of the USA, the *Ariane* of Europe, the *Soyuz*, *Zenit*, and *Proton* of Russia, the *H-2* of Japan, and the *PSLV* and *GSLV* of India. The analysis of a parallel staged rocket is quite similar to the one presented above for a serial staged vehicle, the main difference being in the nomenclature of the stages and definitions of their structural and payload ratios. When the parallel boosters and the core first stage are burning simultaneously, they are taken together and called the *zeroth stage*, while the propellant remaining in the core's first stage after discarding the parallel boosters is called the *first stage* of the rocket. The thrust of the zeroth stage is written as follows:

$$f_T = -v_{eb}\frac{\mathrm{d}m_b}{\mathrm{d}t} - v_{e1}\frac{\mathrm{d}m_1}{\mathrm{d}t} = -v_{e0}\frac{\mathrm{d}m_0}{\mathrm{d}t}\,, \tag{8.20}$$

where the subscripts p and 1 refer to the quantities pertaining to the parallel boosters and the core first stage, respectively, and v_{e0} and m_0 are the average exhaust speed and total mass, respectively, of the rocket's zeroth stage. In order to continue the analogy with a serial staged rocket, it remains to define the equivalent structural and payload ratios. Let us assume that the core vehicle's first stage has a total initial mass, m_{01}, and a structural mass, m_{s1}. Of the total propellant mass, m_{p1}, of the first core stage, only m_{p10} is burned in parallel with the boosters in the zeroth-stage operation. The boosters burn a total propellant mass, m_{pb}, and have a total structural mass, m_{sb}. Thus, the structural and payload ratios of the zeroth-stage equivalent to a serial rocket are given by

$$\sigma_0 = \frac{m_{sb} + m_{s1}}{m_{sb} + m_{s1} + m_{pb} + m_{p10}}\,, \tag{8.21}$$

and

$$\lambda_0 = \frac{m_{01} - m_{p10}}{m_{00}}\,, \tag{8.22}$$

respectively, where $m_{00} = m_{01} + m_{sb} + m_{pb}$ is the initial mass of the zeroth stage (same as the initial mass of the entire rocket). Similarly, the equivalent ratios for the rocket's first stage are

$$\sigma_1 = \frac{m_{s1}}{m_{s1} + m_{p1} - m_{p10}}, \tag{8.23}$$

$$\lambda_1 = \frac{m_{02}}{m_{01} - m_{p10}}. \tag{8.24}$$

Of course, the initial mass of the first stage, $m_{01} - m_{p10}$, is related to the second stage's initial mass, m_{02}, by $m_{01} - m_{p10} = m_{02} + m_{s1} + m_{p1} - m_{p10}$, or $m_{01} = m_{02} + m_{s1} + m_{p1}$, which is the familiar relationship between the serial first and second stages of the core vehicle. We are now ready to write the rocket equation for the parallel staged vehicle as follows:

$$\Delta v = -\sum_{k=0}^{N} v_{ek} \ln[\sigma_k + (1 - \sigma_k)\lambda_k], \tag{8.25}$$

where the average exhaust velocity of the zeroth stage is given from Eq. (8.20) by

$$v_{e0} = \frac{v_{eb}\frac{dm_b}{dt} + v_{e1}\frac{dm_1}{dt}}{\frac{dm_0}{dt} + \frac{dm_1}{dt}}. \tag{8.26}$$

Upon comparing Eqs. (8.25) and Eq. (8.16), we find no difference other than the inclusion of the zeroth stage, with its average exhaust speed and equivalent ratios, and the modified first stage for the parallel staged vehicle.

Example 8.3. The three-stage rocket of Example 8.2 is modified by adding four solid rocket, strap-on boosters of specific impulse 200 s, total propellant mass 30,000 kg, and structural ratio 0.05. If 25,000 kg of the core vehicle's first stage is burned in parallel with the booster rockets, calculate the total velocity impulse possible from the modified vehicle.

In Example 8.2, we derived the core vehicle's stage masses as $m_{01} = 77,847.7296$ kg, $m_{02} = 19,806.1060$ kg, and $m_{03} = 6046.9099$ kg. The core stages have a common structural ratio of $\sigma_k = 0.07$, which yields $m_{s1} = 4062.9137$ kg. The total structural mass of the strap-on boosters is $m_{sb} = \frac{(30000)(0.05)}{1-0.05} = 1578.9474$ kg. With $m_{p10} = 25,000$ kg, and $m_{00} = 109,426.677$ kg, the equivalent ratios for the zeroth stage and the first stage are obtained as

$$\sigma_0 = \frac{m_{sb} + m_{s1}}{m_{sb} + m_{s1} + m_{pb} + m_{p10}} = 0.09303575,$$

$$\lambda_0 = \frac{m_{01} - m_{p10}}{m_{00}} = 0.48295106,$$

$$\sigma_1 = \frac{m_{s1}}{m_{s1} + m_{p1} - m_{p10}} = 0.1229635,$$

$$\lambda_1 = \frac{m_{02}}{m_{01} - m_{p10}} = 0.374776857.$$

The second and third stage payload ratios remain unchanged from their core values at $\lambda_2 = 0.305305336$ and $\lambda_3 = 0.165373724$, respectively, and their structural ratios are also unmodified. It remains to calculate the average exhaust velocity of the zeroth stage by Eq. (8.26). In order to do so, we assume that the parallel stages burn at constant mass flow rates, $\frac{dm_b}{dt} = \frac{30000}{t_0}$ and $\frac{dm_1}{dt} = \frac{25000}{t_0}$, where t_0 is the duration that the parallel stages are burning. These, when substituted into Eq. (8.26) along with the specific impulses of the parallel stages, yield

$$v_{e0} = (9.81)\frac{(25000)(290) + (30000)(200)}{25000 + 30000} = 2363.31818 \text{ m/s} ,$$

thereby leading to the following total speed change:

$$\Delta v = -\sum_{k=0}^{N} v_{ek} \ln[\sigma_k + (1 - \sigma_k)\lambda_k] = 13,393.7758 \text{ m/s} .$$

Thus, the additional booster rockets in parallel with the first stage increase the total velocity impulse by 393.7758 m/s. However, there is a decline in the efficiency measured by the total payload ratio, $\lambda_T = \frac{m_L}{m_{00}} = 0.0091$, when compared to $\lambda_T = 0.0128$ of the core rocket. An efficient harnessing of the performance increase can be achieved by sending a larger payload into the geosynchronous orbit ($\Delta v = 13$ km/s). We can calculate the largest payload the parallel-staged rocket can launch to a geosynchronous orbit by finding the new third-stage ratios, with unchanged structural and propellant masses of the lower stages, as follows:

$$\begin{aligned}\ln[\sigma_3 + (1 - \sigma_3)\lambda_3] &= -\frac{\sum_{k=0}^{2} v_{ek} \ln[\sigma_k + (1 - \sigma_k)\lambda_k] - 13000}{(9.81)(455)} \\ &= -1.4087930669 ,\end{aligned}$$

which implies $\sigma_3 + (1-\sigma_3)\lambda_3 = 0.2444381$. Now, since the structural and total masses of the third stage are unchanged, we have

$$(1 - \lambda_3)\sigma_3 = \frac{m_{s3}}{m_{03}} = 0.05842383896 .$$

Thus, we have two equations for the two unknowns, σ_3, λ_3, which are easily solved to yield $\lambda_3 = 0.1860142867$ and $\sigma_3 = 0.07177502$. Finally, we obtained the new payload mass to be

$$m_L = m_{03}\lambda_3 = 1124.8116 \text{ kg} .$$

Hence, there is an increase of 124.8116 kg, or 12.48%, in the payload to be launched to the geosynchronous orbit. The propellant mass of the third stage has to be decreased by 124.8116 kg to obtain this benefit, which increases the third stage's structural ratio slightly. The new launch efficiency given by $\lambda_T = \prod_{k=0}^{3} \lambda_k = 0.01028$ is better than that of the original mission ($\lambda_T = 0.0091$) with a payload of 1000 kg.

8.3.4 Mission Trade-Off

Modifications in the design of a rocket can significantly change its performance and efficiency. Such changes can occur naturally in the design process, where the payload tends to grow both in mass and in volume due to an updated mission. Also, there can be significant changes in the design of the stages over the lifetime of a rocket, due to improvements in the technology (generally, lower structural ratios and higher specific impulses). As seen in Example 8.3, an addition of strap-on booster stages is a practical way to address an increase in the payload, with an attendant loss of efficiency. If the efficiency loss is to be minimized, only small changes must be carried out in this manner. Thus, it becomes necessary to study the changes in the payload, or stage ratios, that can be tolerated within the opposing limits of efficiency and performance, without completely redesigning a vehicle. The process of absorbing small design changes by trading performance with efficiency, or vice versa, is called *mission trade-off*.

The most important parameters used in studying mission trade-off are the changes in the payload caused by modifications in the structural and payload masses of the stages, without any change in the total velocity impulse. This is reasonable, because (as seen in Examples 8.2 and 8.3) the total velocity impulse is the primary mision objective, defined by the final orbit; hence, no changes in it can be tolerated. The total velocity impulse can be expressed in terms of the initial and final masses of the stages as follows:

$$\Delta v = -\sum_{k=0}^{N} v_{ek} \ln \frac{m_{0k}}{m_{fk}} , \tag{8.27}$$

where

$$m_{0k} = m_L + m_{pk} + m_{sk} + \sum_{n=k+1}^{N} (m_{sn} + m_{pn}) , \tag{8.28}$$

and

$$m_{fk} = m_{0k} - m_{pk} = m_L + m_{sk} + \sum_{n=k+1}^{N} (m_{sn} + m_{pn}) . \tag{8.29}$$

The first trade-off parameter defining the change of payload due to a change in the structural mass of the kth stage is the partial derivative, $\frac{\partial m_L}{\partial m_{sk}}$. It is obtained by differentiating Eq. (8.27) with respect to m_{sk} and setting $\frac{\partial \Delta v}{\partial m_{sk}} = 0$:

$$\frac{\partial m_L}{\partial m_{sk}} = -\frac{\sum_{n=0}^{k} v_{en}\left(\frac{1}{m_{0n}} - \frac{1}{m_{fn}}\right)}{\sum_{j=0}^{N} v_{ej}\left(\frac{1}{m_{0j}} - \frac{1}{m_{fj}}\right)} . \tag{8.30}$$

For the final stage, $k = N$, we have $\frac{\partial m_L}{\partial m_{sN}} = -1$, which implies that an increase in the structural mass of the final stage is equal to the decrease in the payload mass, which is quite obvious. The decrease in the payload mass is less than the increase in the structural mass of a lower stage.

The second trade-off parameter is the partial derivative, $\frac{\partial m_L}{\partial m_{pk}}$, obtained by differentiating Eq. (8.27) with respect to m_{pk} and setting $\frac{\partial \Delta v}{\partial m_{pk}} = 0$ as follows:

$$\frac{\partial m_L}{\partial m_{pk}} = -\frac{\sum_{n=0}^{k} \frac{v_{en}}{m_{0n}}}{\sum_{j=0}^{N} v_{ej} \left(\frac{1}{m_{0j}} - \frac{1}{m_{fj}} \right)} . \tag{8.31}$$

Clearly, the parameter $\frac{\partial m_L}{\partial m_{pk}}$ is always positive, because an addition in the propellant mass of any stage leads to an increase in the payload. In Example 8.3 we saw that the addition of parallel boosters to the core rocket caused the mass of the zeroth stage to become nonzero, leading to an increase in the payload mass for the same total velocity change.

Example 8.4. Let us calculate the trade-off parameters for the three-stage core vehicle with strap-on boosters of Example 8.3, designed for the baseline mission of launching an 1134.2061 kg payload to the geosynchronous orbit, with a total velocity impulse of 13 km/s. Using the data provided in the previous two examples, we tabulate the two trade-off parameters for each stage in Table 8.2.

Table 8.2. Mission Trade-off Parameters for a Geosynchronous Launch Vehicle

k	v_{ek} (km/s)	m_{0k} (kg)	m_{fk}	$\frac{\partial m_L}{\partial m_{sk}}$	$\frac{\partial m_L}{\partial m_{pk}}$
0	2.3633	109426.677	54426.677	−0.008295	0.008209
1	2.8449	52847.730	23869.020	−0.033136	0.028669
2	2.8449	19806.106	7010.054	−0.132791	0.083263
3	4.4636	6046.910	1478.095	−1.0	0.363821

From Table 8.2 it is evident that the structural mass of the zeroth stage, or the first stage, trades for a very small payload, compared to that of the third stage. The negligible change in the payload caused by structural modifications in the zeroth stage enables good mission flexibility through the use of strap-on boosters of different designs. It is also apparent in Table 8.2 that the payload increase due to an increase in the propellant mass of the third stage is the largest, followed in descending order by that of the lower stages. For example, a 100 kg increase in the propellant mass of the third stage causes a payload increase of 36.38 kg. The same change in the zeroth stage would result in a payload increase of only 0.8 kg.

8.4 Optimal Rockets

From the foregoing discussion, it must be clear that the relative sizes of the stages have a profound impact on both performance and efficiency. For given mission performance and propellant for each stage, one can arrive at the arrangement of stages that provides the highest efficiency in terms of the total payload ratio, λ_T. Such a rocket, in which the total payload ratio is maximum, subject to a given total velocity impulse, is said to be *optimal*. The problem posed by the optimal, multi-stage rocket can have a unique solution, provided the structural ratio and specific impulse of each stage are specified. This is generally the case, because the mission specifications include the type of propellants as well as the structural strength and material requirements. Therefore, let us consider a rocket with a fixed total velocity impulse, for which the specific impulse and structural ratio of each stage are known, but the relative masses of the stages are open to adjustment. Let the payload ratios and exhaust velocities be nondimensionalized with respect to the initial stage (either the zeroth or the first), for which a subscript designation of 1 will be employed. The nondimensional rocket equation for an N-stage rocket can thus be written as follows:

$$\frac{\Delta v}{v_{e1}} = -\sum_{k=1}^{N} \beta_k \ln[\sigma_k + (1 - \sigma_k)\alpha_k] , \tag{8.32}$$

where $\beta_k \doteq \frac{v_{ek}}{v_{e1}}$, $\alpha_k \doteq \frac{\lambda_k}{\lambda_1}$. The optimization problem is then posed as follows:

Determine the relative payload ratios, α_k $(k = 2 \ldots N)$, such that the total payload ratio, $\lambda_T = \lambda_1 \prod_{k=1}^{N} \alpha_k$, is maximized, subject to the rocket equation, Eq. (8.31).

Generally, the solution of the constrained optimization problem posed above requires nonlinear programming methods that either compute the gradients of an objective function with respect to the vector of unknown $N-1$ variables, α_k $(k = 2 \ldots N)$, or perform a nongradient search for the maxima in an N-dimensional space. Without resorting to such methods, let us see whether we can determine the solution of the problem in a simpler manner. Since most rockets have either three or four stages, we are usually looking for only two or three optimization variables. For up to two variables, a graphical method can be employed to search for the maxima of the function, λ_T. However, we must ensure that the required variables, α_k $(k = 2 \ldots N)$, satisfy the rocket equation. In Example 8.2, we presented a computer program for iteratively solving the rocket equation for the first-stage payload ratio, given a set of α_k. Hence, when only two or three stages are concerned, we can adopt the following computational procedure for designing an optimal rocket:

1. Select a set of α_k $(k = 2 \ldots N)$.
2. Using an iterative scheme, solve for λ_1 that satisfies Eq. (8.31) with the given parameters.
3. Compute the total payload ratio, $\lambda_T = \lambda_1 \prod_{k=1}^{N} \alpha_k$, and plot it as a point

in the $(N-1)$ dimensional space defined by α_k $(k = 2 \ldots N)$ as Cartesian axes.
4. Repeat steps 2 and 3 with different sets of α_k $(k = 2 \ldots N)$ until a significant region in the $(N-1)$-dimensional space is covered.
5. Search for the maxima of λ_T in its graphical plot or through a comparison of its stored values at different points.
6. The values of α_k $(k = 2 \ldots N)$ corresponding to the maxima of λ_T are the optimal payload ratios of the rocket.

The search outlined above is usually rewarded by getting the maxima in a few iterations. The number of iterations required depend primarily upon the initial choice of α_k. With the use of a programming aid, such as MATLAB, the above algorithm is easily implemented.

8.4.1 Optimal Two-Stage Rocket

We will carry out the optimization presented above for designing an optimal two-stage rocket, with the use of MATLAB. The program, *Nstage.m*, given in Table 8.1, forms the basis of the procedure, wherein the payload ratio, λ_1, is computed for a given α_2, the only variable of optimization.

Example 8.5. Design an optimal two-stage rocket for launching a payload of 5000 kg into a low earth orbit, requiring a total velocity impulse of 9.5 km/s. The structural ratio of the first stage is 0.07, while that of the second stage is 0.05. The first stage employs a solid propellant with specific impulse 200 s. Examine two propellants as candidates for the second stage, namely UDMH/NO_4 and kerosene/LO_2, for which $\beta_2 = 1.5$ and 1.75, respectively.

Let us specify an initial value of $\alpha_2 = 0.05$ and vary it in steps of 0.0005 1000 times, computing λ_1 by solving the rocket equation for each value of α_2. This procedure is repeated for both values of β_2. The total payload ratio λ_T is plotted against α_2 and its maximum value obtained, for each value of β_2. The following MATLAB statements are employed for this purpose:

```
>> Isp1=200; vf=9500/(9.81*Isp1);epsilon=[0.07;0.05];beta=1.5; %parameters
>> A=[];P=[];mp=[];aopt=[];popt=[]; %storage matrices initialization
>> for j=1:2;                    %beta iteration
   a=0.05;                       %alpha initial value
           for k=1:1000;         %alpha iteration
           p=Nstage(vf,[1;beta],epsilon,[1;a]); %rocket equation iterative solution
           P(j,k)=p;
           A(j,k)=a;
           a=a+0.0005;
           end
plot(A(j,:),P(j,:).*P(j,:).*A(j,:)); hold on;
[mp(:,j),K]=max(P(j,:).*P(j,:).*A(j,:));
aopt(:,j)=A(j,K);
popt(:,j)=P(j,K);
beta=beta+0.25;
end
>> mp, aopt, popt  %optimal total payload, alpha, and first-stage payload ratios

   mp =          0.0101722144142938        0.0211756475762429
```

```
aopt =  0.245                       0.1165

popt =        0.203762711277902          0.426339319376855
```

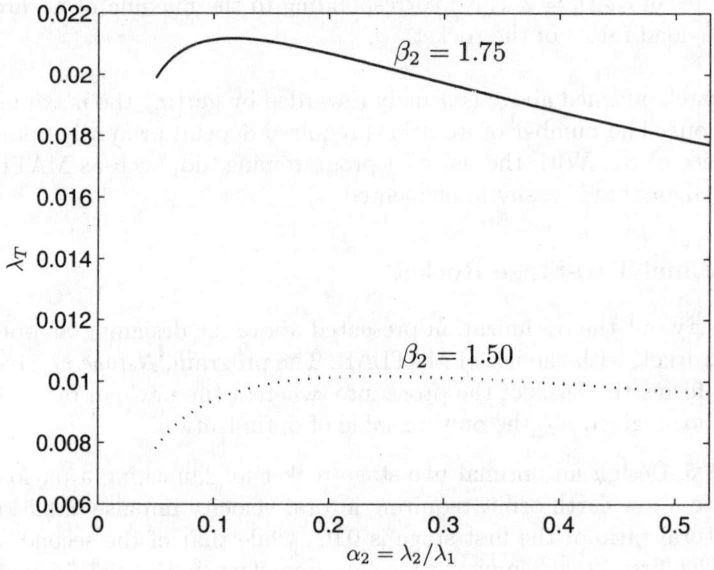

Fig. 8.4. Variation of total payload ratio of a two-stage rocket with $\alpha_2 = \frac{\lambda_2}{\lambda_1}$.

The plot of λ_T against α_2 is shown in Fig. 8.4 for the two given values of β_2. The kerosene/LO_2 ($\beta_2 = 1.75$) combination for the second stage results in a total payload ratio, $\lambda_T = 0.021176$, implying a total initial mass of $m_{01} = \frac{m_L}{\lambda_T} = 236,120.288$ kg. The lower specific impulse propellant, UDMH/NO_4, for the second stage yields, $\lambda_T = 0.010172$, which translates into a a total initial, first-stage mass of $m_{01} = \frac{m_L}{\lambda_T} = 491,535.058$ kg, which is more than twice that of the kerosene/LO_2 option. As shown in Fig. 8.5, the optimal second-stage payload ratios in the both cases are almost the same at $\lambda_2 = 0.05$, which implies the same second-stage propellant mass for both propellant options.

As seen above, the two-stage rocket prefers more propellant mass of the higher specific impulse. However, in general, the propellant with the higher specific impulse is more expensive per unit volume. Hence, in practice, the value of α_2 that produces the more cost-effective option is employed, rather than the one obtained here for the maximum total payload ratio.

A word of caution is necessary in the optimization algorithm presented above. If a very small value of α_2 is employed, the iterative solution of the

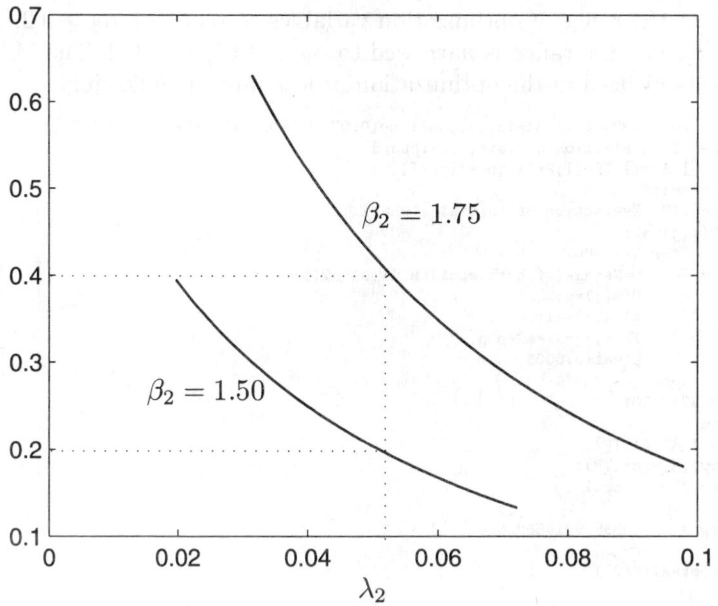

Fig. 8.5. First- and second-stage payload ratios, and their optimal values.

rocket equation may sometimes produce an optimal $\lambda_1 > 1.0$, which, of course, is absurd. This might happen when β_2 is large, which implies that there is a large difference in the specific impulses of the two stages. In such a case, one cannot employ the "optimal" solution, but a suitable value of α_2 must be chosen that leads to $\lambda_1 < 1.0$.

8.4.2 Optimal Three-Stage Rocket

With three stages, the algorithm for optimal staging involves two optimization variables, α_2, α_3. It is slightly more difficult to choose the initial values and range of variation for these variables, as compared to the two-stage rocket, but a good starting point is generally $\alpha_2 = \alpha_3 = 1$, i.e., equal payload ratios for all three stages. After a few trials, the possible range of optimal variables can be narrowed down to a small two-dimensional region.

Example 8.6. Design an optimal, three-stage rocket meeting the requirements presented in Example 8.5, such that the first stage uses a solid propellant with I_{sp1} = 200 s, the second stage employs UDMH/NO_4, while the third stage has the kerosene/LO_2 propellant combination. The structural ratios of the stages are $\sigma_1 = 0.07, \sigma_2 = 0.05$, and $\sigma_3 = 0.05$.

Using the mission data of Example 8.5, we have $\frac{\Delta v}{v_{e1}} = 4.842$, $m_L =$ 5000 kg, $\beta_2 = 1.5$, and $\beta_3 = 1.75$. With trial runs, it is noted that for

$\alpha_2 < 0.1, \alpha_3 < 0.2$, the absurd optimal value of $\lambda_1 \geq 1.0$ is obtained. Thus, we restrict the range of optimization variables to $\alpha_2 \geq 0.1, \alpha_3 \geq 0.2$. After a few more runs, the range is narrowed to $\alpha_2 \geq 0.17, \alpha_3 \geq 0.2$. The MATLAB statements utilized in the optimization process are the following:

```
>>Isp1=200;vf=9500/(9.81*Isp1);epsilon=[0.07;0.05;0.05];beta=[1;1.5;1.75];
>> a2=0.2; %selection of initial alpha_3
>> A1=[];A2=[];TP=[];P=[];mpc=[];K=[];
>> for j=1:5;
   a1=0.17; %selection of initial alpha_2
   A2(:,j)=a2;
        for k=1:300;
            p=Nstage(vf,beta,epsilon,[1;a1;a2]);
            P(k,j)=p;
            A1(k,:)=a1;
            TP(k,j)=a1*a2*p^3;
            a1=a1+0.0005;
        end
   a2=a2+0.001;
   end
>> surf(A2,A1,TP)
>> [mpc,K]=max(TP);
>> [mp,J]=max(mpc)

    mp =    0.0335890188803516     J =  1

>> a1opt=A1(K(J))

    a1opt =       0.1795

>> a2opt=A2(:,J)

    a2opt =       0.2

>> popt=P(K(J),J)

    popt =       0.97806478677034
```

A three-dimensional, perspective visualization of the total payload ratio against the optimization variables is shown in Fig. 8.6. The optimal solution is for the smallest possible α_3, which in this case is $\alpha_3 = 0.2$, and $\alpha_2 = 0.1795$, which satisfy the rocket equation with the first-stage payload ratio, $\lambda_1 = 0.978065$—a very large proportion. This implies an optimal first-stage propellant mass of only 3036.67 kg, which is almost negligible in comparison with the total propellant mass of 136, 599.96 kg. This is not surprising, as the optimal solution tends to reduce the mass of the smallest specific impulse propellant. In comparison, the second and third stages have propellant masses of 114, 030.646 kg and 19, 532.645 kg, respectively. The optimal total payload ratio $\lambda_T = 0.0336$ is quite large in comparison with that of the two-stage rocket (even with the higher specific impulse kerosene/LO_2 option for the second stage) in Example 8.5. The total initial mass of the first stage, $m_{01} = \frac{5000}{0.0336} = 148,858.1735$ kg, is less than one third of the lower initial mass (with the higher specific impulse second-stage propellant option) in Example 8.5. Thus, the three-stage design offers a significant improvement in efficiency over the optimal two-stage rocket for the same mission. Of course, it is tempting to do away with the first stage entirely, resulting in a two-stage rocket with

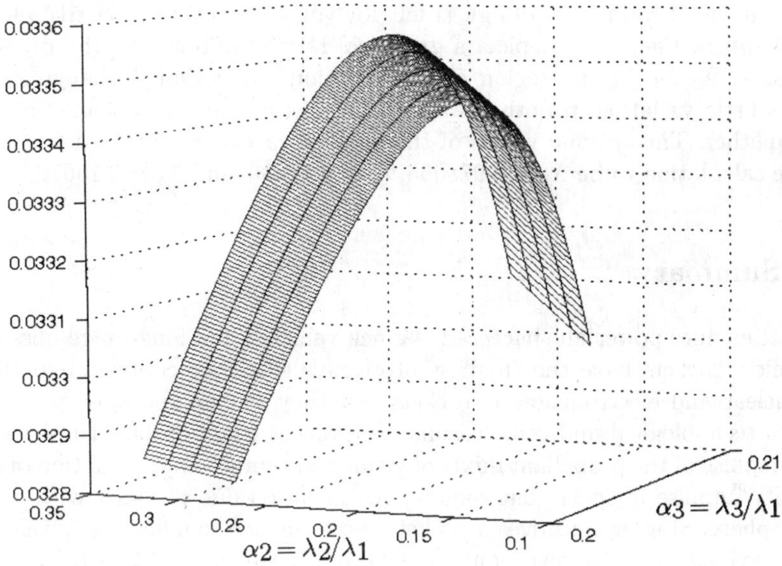

Fig. 8.6. Variation of total payload ratio of a three-stage rocket with α_2 and α_3.

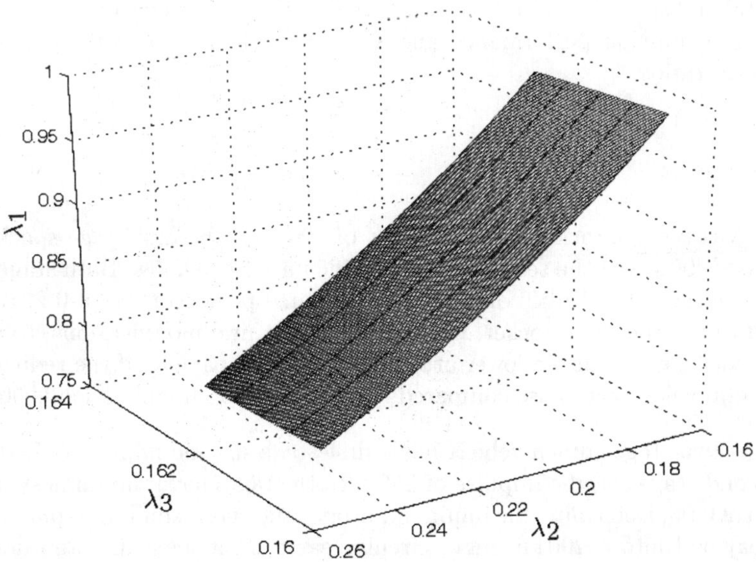

Fig. 8.7. Relationship among the stage payload ratios of a three-stage rocket.

UDMH/NO_4 in the first stage, and the kerosene/LO_2 propellant in the second stage, but it is to be seen whether the resulting two-stage vehicle will have a higher payload ratio (this design is left for you as an exercise at the end of the chapter). Figure 8.7 depicts a graphical relationship among the payload ratios, $\lambda_1, \lambda_2, \lambda_3$, in the region of optimization. Note that a change in λ_3 causes little variation in both λ_1, λ_2, which, in turn, are strong functions of one another. The optimal values of the payload ratios are obtained from the above calculation to be $\lambda_1 = 0.978065, \lambda_2 = 0.17036$, and $\lambda_3 = 0.156725$.

8.5 Summary

Rocket engines power all spacecraft, launch vehicles, and long-range missiles. Chemical rockets have the smallest propulsive efficiency compared to thermonuclear and electromagnetic rockets, yet they remain the most practical due to technological and cost reasons. The rocket equation based upon specific impulse of the propellant mixture yields an accurate approximation of the velocity impulse in space, but requires modification due to losses within the atmosphere. Staging improves a rocket's performance significantly, given the total payload ratio. However, only a small improvement is obtained by using more than three stages. It is important to study mission trade-off by calculating the changes in the payload caused by modifications in the structural and payload masses of the stages, without any change in the total velocity impulse. The design of optimal multi-stage rockets is a crucial problem in sub-orbital and space flight and requires a maximization of the total payload ratio for given mission performance and stage propellants, by varying the stage payload ratios.

Exercises

8.1. A two-stage rocket has a payload of 1000 kg, the first-stage specific impulse of 200 s, and the second-stage specific impulse of 455 s. Both stages have the same structural ratio of 0.07, and the same payload ratio of 0.2. It is desired to use the same rocket, without any structural modifications, to launch a heavier payload into a lower orbit. Find the new payload if the reduction in the total velocity impulse compared to that of the original mission is 500 m/s.

8.2. A two-stage launch vehicle has a first-stage specific impulse of 250 s, and a second-stage specific impulse of 350 s. Both stages have the same structural ratio of 0.05. Determine the minimum propellant mass required to place a 1000 kg payload into a 200 km high, circular earth orbit. (Assume an additional 1.5 km/s of velocity impulse required to overcome drag, gravity, and propulsive losses.)

8.3. Repeat Exercise 8.2 such that the total cost of launching a payload, $\sum_{k=1}^{3} c_{pk} m_{pk}$, is minimized, where c_{pk} is the cost per kilogram of the kth stage's propellant. Assume that $c_{p2} = 1.5c_{p1}$.

8.4. Repeat the the design of an optimal two-stage rocket with the mission requirements and structural ratios of Example 8.5 with UDMH/NO_4 in the first stage, and the kerosene/LO_2 propellant in the second stage. Compare the total payload ratio with that obtained for the optimal three-stage rocket of Example 8.6.

8.5. Repeat the the design of an optimal three-stage rocket with the mission requirements and structural ratios of Example 8.6, such that the solid propellant of the first stage is burned along with a fraction of the core second stage, as a combined parallel stage. Compare the total payload ratio with that obtained for the optimal three-stage rocket of Example 8.6.

9

Planetary Atmosphere

9.1 Aims and Objectives

- To develop a detailed planetary atmosphere model, including nondimensional aerodynamic parameters for modeling aerothermodynamic force, moment, and heat transfer during atmospheric flight.
- To present a 21-layer standard atmosphere for the earth—ranging from sea level to a geometric altitude of 700 km—for use in all simulations of atmospheric and transatmospheric trajectories in subsequent chapters.
- To present an exponential atmospheric model useful in obtaining analytical insight into planetary entry trajectories.

9.2 Introduction

The *atmosphere* is a thin layer of gases clinging to the planetary surface by gravitational attraction. All flight vehicles have to pass through the atmosphere at one stage or the other. The aerothermal loads on an atmospheric flight vehicle depend on the thermodynamic properties of the atmospheric gases, which, in turn, are variables of gravity, planetary rotation, chemical composition, solar radiation, and planetary magnetic field. Therefore, a model of thermodynamic properties of the atmosphere is crucial for the analysis and design of aerospace vehicles. At the low altitudes (less than 15 km on the earth) of operation of most aircraft, the atmosphere can be regarded to be in a thermal equilibrium, with negligible external influences (such as electromagnetic disturbances and chemical reactions). This layer of the atmosphere is, however, under constant perturbation by horizontal winds and two-phase nonequilibria due to the presence of vapours (water on the earth, carbon dioxide on Mars). These effects constitute all the local phenomena collectively called *weather*, such as evaporation, condensation, precipitation, lightning, and convective winds. Although weather is what fascinates and affects the earthbound creatures most, it has little impact on the mean thermodynamic properties affecting the flight vehicle. On the other hand, all thermodynamic properties vary greatly with the altitude, even when the atmosphere is at rest relative to the planet. Hence, all atmospheric models focus on the vertical variation of the thermodynamic variables and generally neglect the horizontal effects caused by weather and planetary rotation.

The basic variables representing the thermodynamic state of a gas are its *density*, *temperature*, and *pressure*. The density, ρ, is defined as the mass, m, per unit volume, v, and has units of kg/m^3. It can be defined at a given point in a gas by taking the limiting case of the volume tending to zero,

$$\rho \doteq \lim_{\Delta v \to 0} \frac{\Delta m}{\Delta v} = \frac{dm}{dv} . \tag{9.1}$$

The density affects the gas in two ways. First, the weight of a given volume of gas is directly proportional to its density. Thus, a stationary column of gas would exert a force directly proportional to the density. Second, the density determines the inertia of a flowing gas. A denser packet of gas is accelerated to a smaller speed when a force is applied on it. While the first effect is useful in deriving the loads on a vehicle due to a stationary gas (called *aerostatic loads*), the second comes into play whenever we are interested in understanding the loads created by the relative motion between the vehicle and the atmosphere (called *aerodynamic loads*).

A gas consists of a large number of infinitesimal particles (molecules), which are always in random relative motion due to mutual collisions, eventhough their net motion in any given direction (gas flow) is zero. The *temperature*, T, with units in *Kelvin* (*K*), is a measure of the average *kinetic energy* (Chapter 4) of a gas particle, expressed as $\frac{3}{2}kT$, where k is *Boltzmann's constant* and has the value $k = 1.38 \times 10^{-23}$ J/K. Thus, a gas at a higher temperature has particles whizzing past one another at a higher average speed. The absolute zero temperature, $T = 0$ K($-273°$c), denotes the case of all gas particles at rest.

The *pressure* exerted by a gas on a solid surface is defined as the net rate of change of normal momentum of the gas particles striking per unit area, A, of the surface. By Newton's second law, the exchange of normal momentum between the gas particles and the solid surface is responsible for a normal force, f_n, applied by the gas on the surface. Hence, the pressure at any given point on the surface is the following:

$$p \doteq \lim_{\Delta A \to 0} \frac{\Delta f_n}{\Delta A} = \frac{df_n}{dA} . \tag{9.2}$$

By this definition, it is not necessary that a solid surface be actually present at a point for the pressure to be defined at the same point. Hence, pressure can be regarded as the force per unit area that a hypothetical solid surface would encounter, were it present at the given point. By *Pascal's law* [22], a gas at rest has the same pressure at all the points. By contrast, a flowing gas has pressure varying from point to point. A gas flowing past a solid surface also imparts a tangential momentum that is responsible for the force of relative friction per unit area, called *shear stress*. We shall consider shear stress in Chapter 10 when defining the drag of atmospheric flight vehicles.

While it would be nice to have a mathematical model for the atmosphere that accounts for spatial (horizontal as well as vertical) and temporal (hourly to monthly) variation of the thermodynamic variables, it is neither practical nor necessary to do so. An aircraft is not designed to fly only at a particular place and time, hence horizontal and temporal atmospheric variations all over the world are averaged out annually into a *standard atmosphere* that contains a model of only the vertical variation of ρ, T, p. Such a standard atmosphere can then be universally applied to analyze and design vehicles, and serves as a common reference for the calibration of flight instruments as well as a benchmark for regulated operation of aircraft.

In the early days of aviation, there was little need to model the atmosphere above 30 km altitude due to the limited ceiling of operational aircraft of the day. However, after the second world war, there was a rapid increase in the altitudes of operation of aircraft, ballistic missiles, launch vehicles, and satellites. Thus, models of standard atmosphere after the 1950s extended well beyond the 100 km altitude limit considered sufficient for most vehicles.[1] However, there is a large uncertainty in the modeled properties at altitudes above 100 km due to the effects of solar radiation and sun-spot activity. These external influences imply that the atmosphere is no longer in a thermal equilibrium. Fortunately, most aircraft, missiles, and launch vehicles as well as entry vehicles are unaffected by such an uncertainty. However, the life of a satellite in a low orbit is drastically influenced by even small variations in the density at altitudes extending up to (and beyond) 200 km.

9.3 Hydrostatic Equilibrium

Most standard atmospheric models consider air to be a *perfect gas*, with the following *equation of state* relating the basic thermodynamic variables:

$$p = \rho R T, \tag{9.3}$$

where R is the *specific gas constant* for a particular atmospheric gas mixture. The gas constant is related to the *universal gas constant*, $\bar{R}(= 8314.32$ J/kg.mole.K), by

$$R = \frac{\bar{R}}{m}, \tag{9.4}$$

[1] The maximum atmospheric effect at high altitudes is felt by the atmospheric entry vehicles due to their extremely high speeds. For most entry vehicles, atmospheric deceleration is negligible in magnitude when compared to the gravitational acceleration for altitudes above 100 km.

where m is the molecular weight (in moles) of the gas. The molecular weight varies with the chemical composition of the gas. For a gas in a thermal equilibrium, the chemical composition is *frozen*, hence there is no variation in the molecular weight. An alternative form of the equation of state for a perfect gas is

$$\frac{p}{\rho^n} = \text{constant} , \tag{9.5}$$

where n, the *polytropic exponent*, is a constant as long as the gas is in a thermal equilibrium. When there is no heat transferred either to or from the atmosphere, the *adiabatic* condition holds, for which $n = \gamma$, the *specific heat ratio*. Apart from thermal equilibrium, which gives us constant values of R, γ, we can assume that the atmosphere is at rest in the vertical direction. This assumption is tantamount to neglecting vertical air currents that are responsible for the convective weather phenomena and is called *hydrostatic equilibrium*. A vertical column of gas in hydrostatic equilibrium has its pressure balanced by its weight per unit area. Thus, we can write the following equation of hydrostatic equilibrium for a thin slice of atmosphere of altitude $\mathrm{d}h$ and a unit cross-section area (Fig. 9.1):

$$\mathrm{d}p = -\rho g \mathrm{d}h , \tag{9.6}$$

where g denotes the acceleration due to gravity. Since we are interested in modeling the vertical variation of the atmospheric properties, it is often convenient to define a *geopotential altitude*, h_g, instead of the *geometric altitude*, h, as follows:

$$g_0 \mathrm{d}h_g \doteq g \mathrm{d}h , \tag{9.7}$$

where g_0 is the acceleration due to gravity at *standard sea level* ($h = 0$). Consequently, the hydrostatic equation becomes

$$\mathrm{d}p = -\rho g_0 \mathrm{d}h_g . \tag{9.8}$$

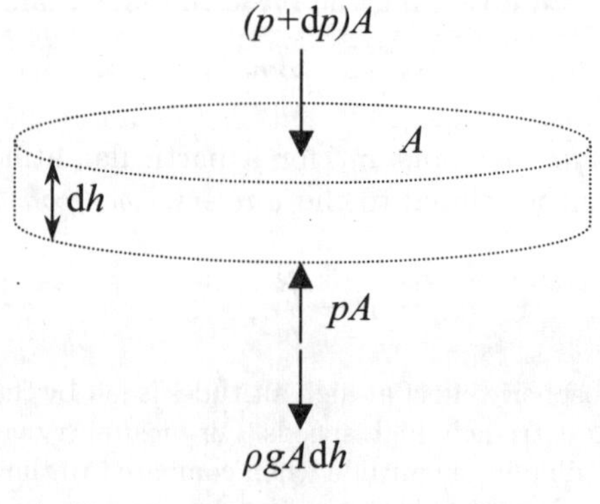

Fig. 9.1. A slice of atmosphere in hydrostatic equilibrium.

Note that the variation of g with altitude is absorbed into the geopotential altitude, such that

$$h_g = h\frac{r_0}{r_0 + h} . \tag{9.9}$$

For earth, h and h_g are virtually indistinguishable for $h \leq 65$ km. Hence, at the low altitudes of aircraft operation, we can substitute h for h_g.

9.4 Standard Atmosphere

A standard atmosphere is generally modeled as comprising consecutive layers of specified temperature variation with the altitude, $T(h)$. At low altitudes, the prevailing thermal and hydrostatic equilibrium can be represented by layers with a linear variation of the temperature with altitude, while the nonequilibrium pheonomena at high altitudes generally requires a nonlinear temperature model. The variation of molecular weight can be absorbed into the definition of a *molecular temperature*, T_m, as follows:

$$T_m \doteq T\frac{m_0}{m} , \tag{9.10}$$

where m_0 refers to the molecular weight at standard sea level. Clearly, while T_m and T are indistinguishable for the layers in which thermal equilibrium prevails, they differ at higher altitudes where the molecular weight is a variable. While being an artificial quantity, the molecular temperature is useful in that its relationship with the geopotential altitude is linear even for those layers where nonequilibrium pheonomena occur. Here we shall confine ourselves to the derivation of atmospheric properties in layers with a linear temperature variation. There are certain well-defined atmospheric strata in all standard atmospheric models for the earth, as depicted in Fig. 9.2. The *troposphere*, extending from the standard sea level up to $h = 11$ km (*tropopause*), sees a linearly decreasing temperature with the altitude, while the next higher layer, $11 < h \leq 47$ km, called the *stratosphere*, consists of three layers with constant (*isothermal*) and linearly increasing temperature at different rates, respectively. The operation of most aircraft is limited to the troposphere and the stratosphere. Immediately above the stratosphere lies the *mesosphere*, which extends up to $h = 86$ km and has an isothermal layer, along with two consecutive layers with linearly decreasing temperature. With the mesosphere, the assumption of thermal equilibrium ends, and the next stratum lying in the range $86 < h \leq 500$ km is called the *thermosphere*, which experiences a nonlinear decrease in the molecular weight with altitude due to thermal nonequilibrium and associated chemical reactions. The thermodynamic properties of thermosphere are strong functions of solar radiation, especially such periodic phenomena as the sun-spot activity. Beyond the thermosphere lies the *exosphere*, which is an indefinite region dominated by electromagnetic effects of an ionized gas due to the interaction between solar wind and the earth's

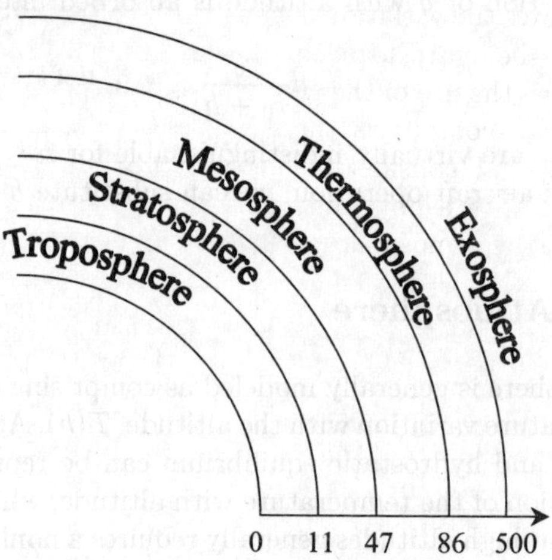

Fig. 9.2. The distinct atmospheric strata.

magnetic field. The exospheric temperature is usually considered constant, but its value varies wildly between different models.

Before considering a specific model, let us derive the basic atmospheric variables in the layers having thermodynamic equilibrium. The linear variation of the temperature with the altitude in such a layer is expressed as follows:

$$T = T_i + a(h - h_i) , \tag{9.11}$$

where the subscript i refers to the quantities at the base of the layer, and a is a constant called the *thermal lapse rate*. By substituting Eqs. (9.3) and (9.5) into Eq. (9.7), we have

$$a = \frac{\mathrm{d}T}{\mathrm{d}h} = -\frac{(n-1)}{n}\frac{g}{R} , \tag{9.12}$$

which directly relates the lapse rate to the ploytropic exponent. The lapse rate is crucial in determining the stability of the hydrostatic equilibrium of an atmospheric layer. A negative lapse rate implies a cooling-off of warm air as it rises due to a small thermal disturbance from the equilibrium condition, thereby causing the given air volume to become heavier and to descend back to its equilibrium level. Therefore, an atmospheric layer with $a < 0$ is thermally stable, and those with $a > 0$ are unstable. Whenever strong thermal gradients are established due to local convective activity in the troposphere, the prevailing negative lapse rate may become *inverted*, causing a temperature inversion and the formation of vertical air currents in the resulting locally unstable region.

In order to derive the variation of density and pressure with the altitude, we need to integrate the hydrostatic equation. In the past, analytical expressions were obtained with the use of the geopotential altitude, h_g. However, with the availability of digital computers, there is no particular advantage in doing so at present. Thus, we shall consider the geometric altitude, h, as the independent variable that results in a physically meaningful model. The acceleration due to gravity within the atmosphere ($h \ll r_0$) can be approximated as follows:

$$g = g_0 \left(\frac{r_0}{r_0 + h} \right)^2 \approx g_0 \left(1 - \frac{2h}{r_0} \right) . \tag{9.13}$$

Substitution of Eqs. (9.13) into Eq. (9.7) yields

$$\frac{dp}{p} = -\frac{g_0(1 - \beta h)}{R[T_i + a(h - h_i)]} , \tag{9.14}$$

where $r_0 \beta \doteq 2$. We can integrate Eq. (9.14) between (h_i, p_i) and (h, p) for a layer with $a \neq 0$ in order to obtain

$$p = p_i \left[1 + \frac{a(h - h_i)}{RT_i} \right]^{-\left\{ \frac{g_0}{aR} \left[1 + \beta \left(\frac{T_i}{a} - h_i \right) \right] \right\}} e^{\frac{\beta g_0}{aR}(h - h_i)} . \tag{9.15}$$

For an isothermal layer, $T = T_i$ and $a = 0$. Therefore, the integration of Eq. (9.14) leads to the following expression for pressure:

$$p = p_i e^{-\left[\frac{g_0 (h - h_i)}{RT_i} \right] \left[1 - \frac{\beta (h - h_i)}{2} \right]} . \tag{9.16}$$

The density can be easily derived from the temperature and pressure using the equation of state, Eq. (9.3), as follows:

$$\rho = \frac{p}{RT} . \tag{9.17}$$

The analytical expressions derived above can easily be programmed into a code for computing the temperature, pressure, and density at any point at $h <$ 86 km, where hydrostatic equilibrium is a valid assumption. The lapse rates of individual layers are specified in various standard atmosphere models. We shall follow the convention of the *1976 U.S. Standard Atmosphere* [17] in the range $0 \leq h \leq 86$ km. There are two layers in the *1976 U.S. Standard Atmosphere* above 86 km with nonlinear variation of temperature vs. altitude. Hence, above 86 km altitude, we shall employ the *1962 U.S. Standard Atmosphere* [18], which conveniently models all the layers with linearly varying temperature up to $h = 2000$ km. While the two models have a close agreement in the altitude range $0 \leq h \leq 86$ km, there is a large difference in the two models in the exospheric region, which is, however, of little concern to atmospheric flight vehicles. For a greater accuracy in the exospheric temperature, statistical models are usually employed that account for such phenomena

Table 9.1. Standard Atmosphere Derived from 1976 and 1962 U.S. Standard Atmospheres

i	h_i (km)	T_i (K)	R (J/kg.K)	a (K/km)
1	0	288.15	287.0	−6.5
2	11.0191	216.65	287.0	0.0
3	20.0631	216.65	287.0	1.0
4	32.1619	228.65	287.0	2.8
5	47.3501	270.65	287.0	0.0
6	51.4125	270.65	287.0	−2.8
7	71.8020	214.65	287.02	−2.0
8	86	186.946	287.02	1.693
9	100	210.02	287.84	5.0
10	110	257.0	291.06	10.0
11	120	349.49	308.79	20.0
12	150	892.79	311.80	15.0
13	160	1022.2	313.69	10.0
14	170	1103.4	321.57	7.0
15	190	1205.4	336.68	5.0
16	230	1322.3	366.84	4.0
17	300	1432.1	416.88	3.3
18	400	1487.4	463.36	2.6
19	500	1506.1	493.63	1.7
20	600	1506.1	514.08	1.1
21	700	1507.6	514.08	0.0

as solar flares and sun-spot activity. Such models are useful in predicting the drag (and life) of satellites in low orbits. The numerical values of the 21-layer, hybrid standard atmosphere are tabulated in Table 9.1 and encoded in the MATLAB M-file, *atmosphere.m* (Table 9.2). The highest layer in this model ranges between $700 \le h \le 2000$ km. The code *atmosphere.m* will be employed throughout the book for evaluation of atmospheric properties in the range $0 \le h \le 2000$ km.

Apart from the basic thermodynamic variables, the following additional parameters useful in determining aerothermal loads on a vehicle are calculated by the atmospheric model:

(a) the *speed of sound*, $a_\infty \doteq \sqrt{\gamma R T}$
(b) *Mach number*, $M \doteq \frac{v}{a_\infty}$, where v denotes the speed of the vehicle relative to the atmosphere
(c) the *dynamic viscosity coefficient*, μ, with the following *Sutherland's law*:

$$\mu = 1.458 \times 10^{-6} \frac{T^{\frac{3}{2}}}{T + 110.4} . \tag{9.18}$$

(d) *Prandtl number*,

$$Pr = \frac{\mu c_p}{k_T}\,, \tag{9.19}$$

where k_T is the *coefficient of thermal conductivity* of the perfect gas, and c_p is its *constant pressure specific heat*, which can be calculated by

$$c_p = \frac{R\gamma}{\gamma - 1}\,. \tag{9.20}$$

For perfect air, thermal conductivity is computed using the following empirical formula [19]:

$$k_T = \frac{2.64638 \times 10^{-3} T^{\frac{3}{2}}}{T + 245.4(10^{-\frac{12}{T}})} \text{ J/m.s.K}\,. \tag{9.21}$$

(e) *Knudsen number*, $Kn \doteq \frac{\lambda}{l_c}$, where λ is the *mean free path* of free-stream molecules and l_c is a characteristic length. The mean free path is based upon the collision diameter, σ, and is calculated by

$$\lambda = \frac{m}{\sqrt{2}\pi\sigma^2 \rho N_a}\,, \tag{9.22}$$

where m is the molecular weight in kg/mole, and $N_a = 6.0220978 \times 10^{23}$ is *Avogadro's number*.
(f) The flow regime parameter, d, is based upon Knudsen number. If $d = 1$, we have a *free-molecular* flow (Chapter 10), $d = 2$ represents a *continuum* flow, and $d = 3$ denotes the *transition* flow regime lying between the two limits.
(g) *Reynolds number*, $Re \doteq \frac{\rho v l_c}{\mu}$, representing the ratio of inertial and viscous forces.

Example 9.1. Using the standard atmosphere model given above, plot the variation of density and pressure in the altitude range $0 \le h \le 200$ km. The results are plotted in Fig. 9.3 in altitude increments of 1 km on a logarithmic scale using the program *atmosphere.m*. Clearly, both density and pressure become negligible at $h > 120$ km. An interesting feature is observed in both pressure and density plots: their variation in the altitude ranges $5 \le h \le 80$ km, $80 < h \le 120$ km, and $130 \le h \le 200$ km is piecewise linear on a log-scale with the altitude. This indicates that one can employ an exponential approximation for the density and pressure variation in the given regions, which amounts to the assumption of a constant temperature in each region [Eq. (9.16)]. Such a simple isothermal atmospheric model is often applied within selected altitude limits for analytical purposes.

Table 9.2. M-file *atmosphere.m* for the Earth's Standard Atmospheric Properties

```
function Y = atmosphere(h, vel, CL)
%(c) 2005 Ashish Tewari
R = 287; %sea-level gas constant for air (J/kg.K)
go = 9.806; %sea level acceleration due to gravity (m/s^2)
Na = 6.0220978e23; %Avogadro's number
sigma = 3.65e-10; %collision diameter (m) for air
S = 110.4;  %Sutherland's temperature (K)
Mo = 28.964; %sea level molecular weight (g/mole)
To = 288.15; %sea level temperature (K)
Po = 1.01325e5; %sea level pressure (N/m^2)
re = 6378.14e3; %earth's mean radius (m)
Beta = 1.458e-6; %Sutherland's constant (kg/m.s.K^0.5)
gamma = 1.405; %sea level specific-heat ratio
B = 2/re; layers = 21; Z = 1e3*[0.00; 11.0191; 20.0631; 32.1619;
47.3501; 51.4125;
    71.8020; 86.00; 100.00; 110.00; 120.00; 150.00; 160.00; 170.00; 190.00;
    230.00; 300.00; 400.00; 500.00; 600.00; 700.00; 2000.00];
T = [To; 216.65; 216.65; 228.65; 270.65; 270.65; 214.65; 186.946;
    210.65; 260.65; 360.65; 960.65; 1110.60; 1210.65; 1350.65; 1550.65;
    1830.65; 2160.65; 2420.65; 2590.65; 2700.00; 2700.0];
M = [Mo; 28.964; 28.964; 28.964; 28.964; 28.964; 28.962; 28.962;
28.880;
    28.560; 28.070; 26.920; 26.660; 26.500; 25.850; 24.690;
    22.660; 19.940; 17.940; 16.840; 16.170; 16.17];
LR = [-6.5e-3; 0; 1e-3; 2.8e-3; 0; -2.8e-3; -2e-3;
    1.693e-3; 5.00e-3; 1e-2; 2e-2; 1.5e-2; 1e-2; 7e-3; 5e-3; 4e-3;
    3.3e-3; 2.6e-3; 1.7e-3; 1.1e-3; 0];
rho0 = Po/(R*To); P(1) = Po; T(1) = To; rho(1) = rho0; for i =
1:layers
    if ~(LR(i) == 0)
        C1 = 1 + B*( T(i)/LR(i) - Z(i) );
        C2 = C1*go/(R*LR(i));
        C3 = T(i+1)/T(i);
        C4 = C3^(-C2);
        C5 = exp( go*B*(Z(i+1)-Z(i))/(R*LR(i)) );
        P(i + 1) = P(i)*C4*C5;
        C7 = C2 + 1;
        rho(i + 1) = rho(i)*C5*C3^(-C7);
    else
        C8 = -go*(Z(i+1)-Z(i))*(1 - B*(Z(i + 1) + Z(i))/2)/(R*T(i));
        P(i+1) = P(i)*exp(C8); rho(i+1) = rho(i)*exp(C8);
    end
end for i = 1:21
    if h < Z(i+1)
        if ~(LR(i)== 0)
            C1 = 1 + B*( T(i)/LR(i) - Z(i) );
            TM = T(i) + LR(i)*(h - Z(i));
            C2 = C1*go/(R*LR(i));
            C3 = TM/T(i);
            C4 = C3^(-C2);
            C5 = exp( B*go*(h - Z(i))/(R*LR(i)) );
            PR = P(i)*C4*C5; %Static Pressure (N/m^2)
            C7 = C2 + 1;
            rhoE = C5*rho(i)*C3^(-C7); %Density (kg/m^3)
        else
            TM = T(i);
            C8 = -go*(h - Z(i))*(1 - (h + Z(i))*B/2)/(R*T(i));
            PR = P(i)*exp(C8); %Static Pressure (N/m^2)
            rhoE = rho(i)*exp(C8); %Density (kg/m^3)
        end
        MOL = M(i) + ( M(i+1)-M(i) )*( h - Z(i) )/( Z(i+1) - Z(i) );
        TM = MOL*TM/Mo; %Kinetic Temperature
```

```
    asound = sqrt(gamma*R*TM); % Speed of Sound (m/s)
    MU = Beta*TM^1.5/(TM + S); % Dynamic Viscosity Coeff. (N.s/m^2)
    KT = 2.64638e-3*TM^1.5/(TM + 245.4*10^(-12/TM));
    Vm = sqrt(8*R*TM/pi); m = MOL*1e-3/Na; n = rhoE/m;
    F = sqrt(2)*pi*n*sigma^2*Vm;
    L = Vm/F; % Mean free-path (m)
    Mach = vel/asound; % Mach Number
    TO = TM*(1 + (gamma - 1)*Mach^2/2);
    MUO = Beta*TO^1.5/(TO + S);
    REO = rhoE*vel*CL/MUO;
    RE = rhoE*vel*CL/MU; % Reynold's Number
    Kn = L/CL; % Knudsen Number
    Kno = 1.25*sqrt(gamma)*Mach/REO;
%flow regime parameter
    if Kn >= 10
        d = 1; % free-molecule flow
    elseif Kn <= 0.01
        d = 2; % continuum flow
    else
        d = 3; % transition flow
    end
 Y = [TM; rhoE; Mach; Kn; asound; d; PR; MU; RE; KT];
 return;
 end
end
```

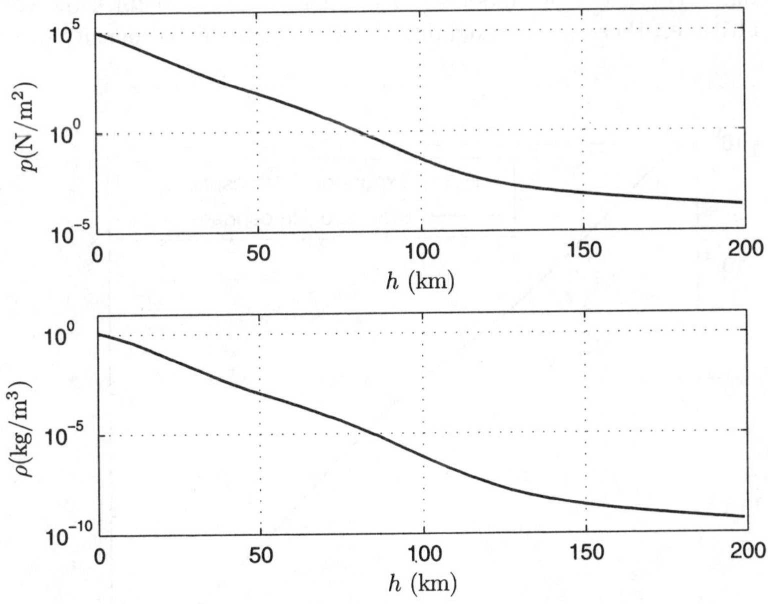

Fig. 9.3. The variation of density and pressure with altitude in the standard atmosphere.

9.5 Exponential Model for Planetary Atmospheres

As seen in the example given above, the complex variation of temperature, density, and pressure in a standard atmosphere can be simplified by employing an isothermal approximation in selected altitude regions with appropriately scaled average temperatures. Such an exponential model of density variation is very useful in determining the properties of planetary atmospheres and also enables a quick and accurate analysis. For an isothermal model, we can express the density variation with the altitude as follows:

$$\rho = \rho_0 e^{-\frac{h}{H}} , \tag{9.23}$$

where ρ_0 is a *base density* (not necessarily the sea level value), and H is a *scale height* that depends upon the selected average isothermal temperature. Both ρ_0 and H are chosen such that a good fit is obtained with the standard atmosphere in a given range of altitudes.

Example 9.2. Using an exponential atmosphere with $\rho_0 = 1.752$ kg/m^3 and $H = 6.7$ km, compare the density with that of the standard atmosphere in the range $5 \leq h \leq 120$ km.

The required comparison is shown in Fig. 9.4 and displays an excellent agreement between the two models in the range $5 \leq h \leq 50$ km. However, at higher altitudes, there is a marked deviation in the predicted density profiles,

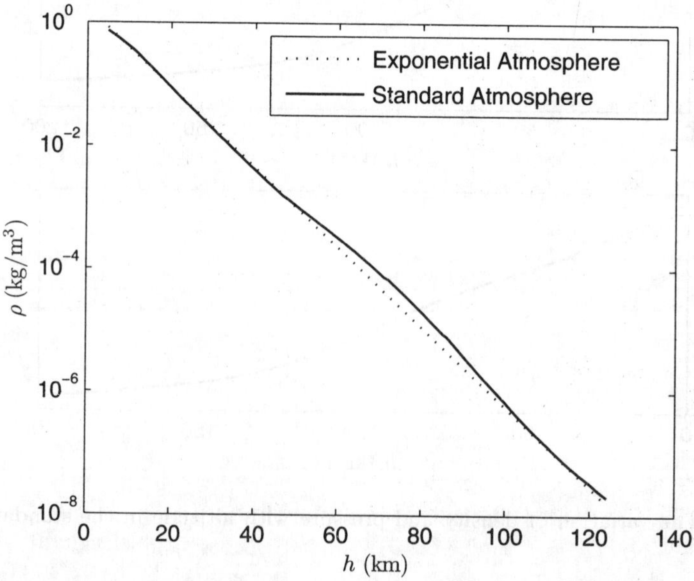

Fig. 9.4. Comparison of density profile with exponential and standard atmospheric models.

especially in the $50 < h \leq 100$ km range that is crucial for most entry vehicles. Thus, the exponential model is good enough for modeling flight below 50 km altitude, which is the realm of most airplanes. It is also adequate for analytically estimating the peak aerothermal loads encountered by entry and ascent vehicles, which occur at $h < 50$ km.

The exponential approximation is often employed by planetary probes for determining a rough atmospheric model of distant planets. Such probes are usually equipped with accelerometers for measuring deceleration caused by the atmospheric drag, which, ignoring the acceleration due to gravity, can be expressed as

$$-\dot{v} = \frac{1}{2}\rho_0 e^{-\frac{h}{H}} v_e^2 \frac{SC_D}{m} \,, \tag{9.24}$$

where v_e is the known entry speed and C_D is a predetermined drag coefficient based upon the reference area, S. By plotting deceleration against altitude (measured by radar) on a logarithmic scale, the constants ρ_0 and H can be determined. Examples of planetary atmospheric probes include the *Galileo*, *Viking*, *Pioneer*, and *Huygens* for the atmospheres of Jupiter, Mars, Venus, and Saturn's moon Titan, respectively.

9.6 Summary

The aerothermal loads on an atmospheric flight vehicle depend on the thermodynamic properties of the atmospheric gases, a careful model of which is thus crucial for the analysis and design of aerospace vehicles. Standard atmospheric models are based upon the hydrostatic and thermal equilibrium and consist of many layers, each having different variations of temperature with altitude. A linear variation of the temperature with geopotential altitude is a standard approximation employed in most atmospheric models. However, upper atmospheric layers (thermosphere and exosphere) do not have thermal and chemical equilibrium and hence have an inherently nonlinear variation of temperature with altitude. A suitable exponential model can be employed for fitting the variation of density with altitude in a specific (limited) range of altitudes. Such a model is valuable in imparting analytical insight into entry trajectories.

Exercises

9.1. For an earth's atmospheric entry vehicle of drag coefficient (Chapter 10), $C_D = 1.7$ at $h = 100$ km, reference area, $S = 10$ m^2, characteristic length, $l_c = 1.75$ m, and mass, $m = 300$ kg, compare the deceleration due to drag and acceleration due to gravity if the entry speed is $v = 8$ km/s. Use the standard atmospheric model of Table 9.1, as well as the exponential atmosphere approximation of Eq. (9.23).

9.2. Plot the dynamic pressure, Knudsen number, Reynold number, Mach number, and Prandtl number against altitude for the vehicle in Exercise 9.1, assuming that the speed varies linearly from $v = 7.8$ km/s at $h = 65$ km to $v = 3$ km/s at $h = 20$ km.

9.3. A satellite in a circular earth orbit of 200 km altitude has drag coefficient (Chapter 10) $C_D = 2.2$ based on reference area, $S = 200$ m^2, characteristic length, $l_c = 5$ m, and mass, $m = 1000$ kg. Using the method of Exercise 6.6, estimate the life of the satellite in orbit.

9.4. The average surface pressure, density, and temperature of Venus were estimated by four *Pioneer* probes to be $p_0 = 92 \times 10^5$ N/m^2, $\rho_0 = 65$ kg/m^3, and $T_0 = 737$ K, respectively. During the descent through the atmosphere, a constant thermal lapse rate of $a = -10.7$ K/km was measured. If the acceleration due to gravity and molecular weight at the surface are $g_0 = 8.88$ m/s^2 and $m_0 = 43.45$ g/mole, respectively, derive a standard atmospheric model for Venus. What would be the base density and scale height of the Venus atmosphere for a good fit with the standard atmosphere in the altitude range $0 \leq h \leq 50$ km?

10

Elements of Aerodynamics

10.1 Aims and Objectives

- To introduce aerodynamics, ranging from elementary concepts to models of viscous, hypersonic, and rarefied flows by computational fluid dynamics.
- To define the important aerothermodynamic variables and parameters necessary in atmospheric flight dynamic modeling.
- To build an appropriate model of aerodynamic force and moment vectors and coefficients for each flow regime.

10.2 Basic Concepts

Atmospheric flight involves creation of forces with the interaction of the atmosphere. Such an interaction can take the form of either *aerostatic* force of buoyancy that arises due to a lighter-than-air vehicle such as a hot-air balloon, or an *aerodynamic* force generated due to the motion of the vehicle relative to the atmosphere. As seen in the previous chapter, the aerostatic force can be easily modeled with the use of the hydrostatic equation and will not receive further discussion in the book. On the other hand, aerodynamic force requires a careful modeling involving the motion of a flowing gas relative to the vehicle, with all the necessary fluid dynamic and thermodynamic principles. It is our objective in the present chapter to high-light the important aerodynamic phenomena and their practical modeling for flight dynamic applications.

10.2.1 Aerodynamic Force and Moment

An object moving in a straight line relative to the atmosphere with a velocity vector, $\mathbf{v}$, generally experiences two force components: one opposite to the direction of motion ($\mathbf{v}$), called *drag, (D)*, and the other perpendicular to $\mathbf{v}$, called *lift, (L)*. It is to be noted that the relative flow the object experiences is directly opposite to the direction of relative motion of the object. Children are intimately acquainted with both of these force components through a hand extended outside a moving car's window. The relative magnitudes of the lift and drag are primarily dependent upon the shape of the object and its attitude relative to the direction of motion. For example, a flat plate kept either tangential or normal to the flight direction would experience only the drag, but also experiences lift when kept at an angle to the relative flow. In order to better model the aerodynamic force, it is necessary to introduce the concept of the *angle of attack* (α), which is defined as the angle made by $\mathbf{v}$ with a reference plane fixed to the vehicle. For a flat plate, the reference plane is the same as the plate, while for an airplane wing it is the *chord plane* obtained by joining the *leading edge* and the *trailing edge*,[1] as shown in Fig. 10.1. From this definition, it is clear that a flat plate with $\alpha \neq 0$ has $L \neq 0$, but with $\alpha = 0$ we have $L = 0$. Also, we can say that for generating large amounts of lift, flattened surfaces, called *lifting surfaces*, are essential. These surfaces are the wings and tails of an airplane and the fins of a missile. In contrast, the fuselage, nacelles, and missile bodies generate negligible lift at zero angle of attack. An airplane wing placed at an angle of attack to a uniform air flow experiences drag, D, lift, L, and a *pitching moment*, $\mathcal{M}$, at any given point on the wing (Fig. 10.1). The lift is useful in balancing the weight of the vehicle, whereby a sustained flight through the atmosphere becomes possible. On the other hand, drag is an undesirable force component, which causes a dissipation of total energy into heat. Drag is partly created by friction between adjacent layers of the fluid flowing past the vehicle. Since energy loss due to drag must be overcome by fuel expenditure, the drag force must be minimized for an efficient vehicle design. In order to understand drag a little better, let us consider the flow past a flat plate. When the plate is at $\alpha = 0$, drag is created solely by the relative friction between the plate and the

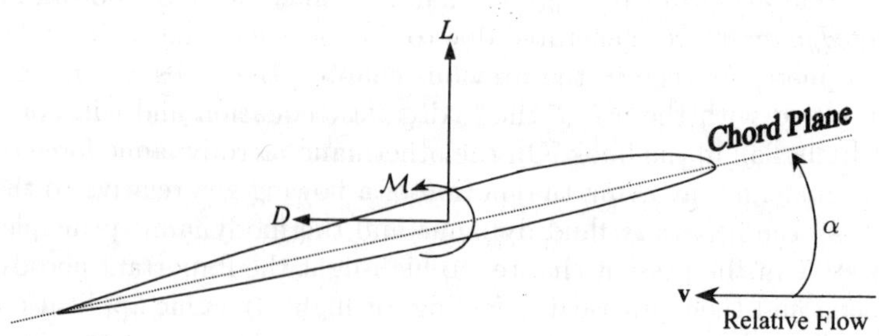

Fig. 10.1. Cross section of an airplane wing placed at an angle of attack to a uniform air flow.

[1] Leading and trailing edges refer to the lines on an object that first and last receive the relative air flow.

fluid particles[2] brushing past the plate [Fig. 10.2(a)]. Such a drag is called *skin friction* drag. The friction, being a nonconservative force (Chapter 4), leads to the conversion of a part of kinetic energy of the fluid particles into heat and is responsible for the deformation of the fluid elements in a continuum flow regime (Sec. 11.2) as depicted in Fig. 10.2(a). When the flat plate is placed normal to the relative flow ($\alpha = 90°$), as shown in Fig. 10.2(b), a fluid particle is incapable of making the right-angle turn for hugging the back of the plate. This leads to a phenomenon called *flow scparation*, where the fluid particles depart from the solid surface, thereby exerting neither frictional force nor normal pressure on the surface. The separated flow, however, undergoes a circulatory motion at the back of the plate, which causes no significant pressure change from the atmospheric pressure, p_a. Consequently, the plate experiences drag due to the difference in the pressure, p, exerted by the fluid particles on the front surface and the atmospheric pressure, p_a, on the back surface. The drag due to flow separation is therefore called *pressure drag* and can be regarded as the force that converts a part of the linear motion of the fluid particles (kinetic energy of translation) into useless rotary motion (rotational kinetic energy). For the flat plate at an intermediate angle, $0 < \alpha < 90°$, the drag consists of both skin friction and pressure drag. This is also the situation for an airplane wing, or any general object placed in a uniform flow. In a low-speed flight, the drag is caused by only the two mechanisms described above, while a high-speed flight involves a third mechanism that we will discuss a little later. Minimizing the drag is a principal objective in the design of atmospheric flight vehicles. However, an attempt to reduce the skin friction drag generally leads to an increase in the pressure drag, and vice versa. Therefore, a compromise is struck between the two conflicting requirements by selecting a shape that results in the total drag being minimized for a given flight condition. The most efficient design is a *streamlined* shape (such as that shown in Fig. 10.1) with a small external surface area, which allows a fluid particle to follow the surface to the largest possible extent, but without causing excessive skin friction. An important factor in airplane design is the creation of lift by wings. The streamwise cross section of an airplane wing is called an *airfoil*, as depicted in Fig. 10.1. The airfoil is shaped in a particular way for efficient lift creation. Notice the sharp trailing edge and a rounded leading edge in Fig. 10.1. Furthermore, the upper surface is much more curved than the lower surface. As a result of this difference in curvature, two fluid particles arriving simultaneously at the trailing edge after following the upper and lower surfaces would have traversed different distances in the same time. This leads to a larger average speed for the upper fluid particle and consequently (as we

[2] A fluid particle is an infinitesimal packet of fluid, whose motion in a steady flow can be described by a fixed line, called a *streamline*. In the continuum flow regime (discussed ahead), the fluid particle is the same as a fluid element, while for a rarefied flow, the term *particle* is understood to stand for an individual gas molecule.

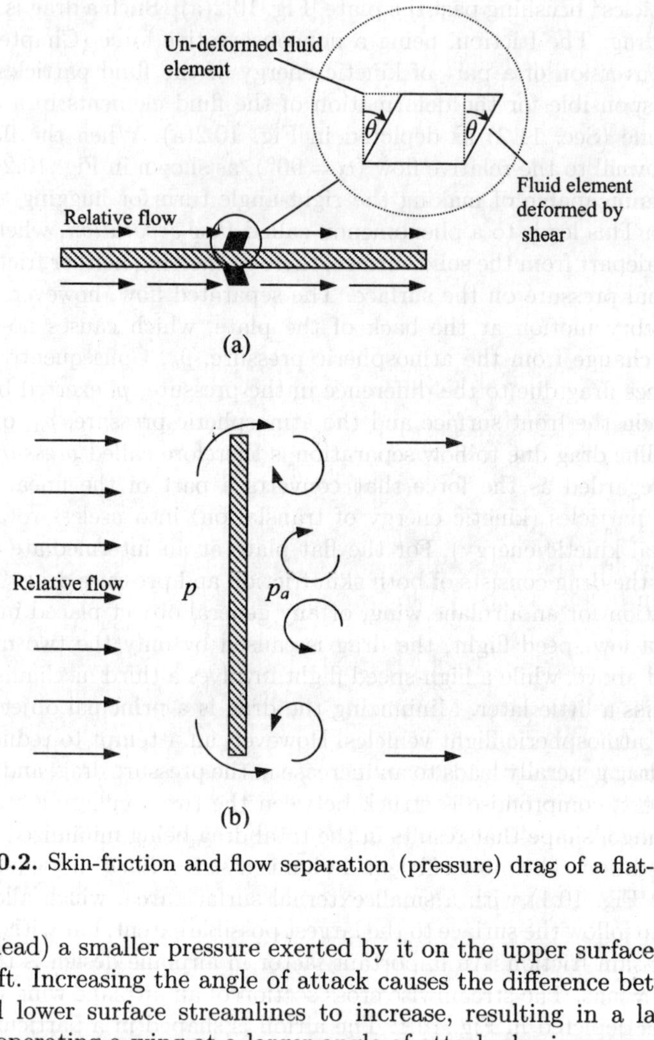

Fig. 10.2. Skin-friction and flow separation (pressure) drag of a flat-plate.

will see ahead) a smaller pressure exerted by it on the upper surface, thereby creating lift. Increasing the angle of attack causes the difference between the upper and lower surface streamlines to increase, resulting in a larger lift. However, operating a wing at a larger angle of attack also increases the drag due to flow separation, as the fluid particles find it increasingly difficult to follow a greatly curved path along the upper surface. There is a limiting angle of attack, $\alpha \leq \alpha_s$, called the *stalling angle of attack*, beyond which the flow on the upper surface becomes largely separated [Fig. 10.2(b)], leading to an almost complete loss of lift. This phenomenon, called *stall*, is encountered by all lifting surfaces.

The most efficient shape for a lifting surface can generate an adequate amount of lift at small values of α, in order to minimize the pressure drag. Such an optimization of the ratio $\frac{L}{D}$ forms the basis of airplane design. Up to the stalling angle of attack, most lifting surfaces have a linear variation of the lift with the angle of attack. Thus, we can write

$$L = L_0 + L_\alpha \alpha \ (\alpha \le \alpha_s) \, , \tag{10.1}$$

where L_0 is the lift at zero angle of attack, and L_α is the slope of the lift variation with α. For $\alpha > \alpha_s$, the variation of lift with angle of attack is nonlinear and depends primarily upon the shape of the leading edge, the flow speed, as well as the viscosity of the fluid. As a first-order approximation, one can ignore the nonlinear lift and regard the maximum lift created at the stalling angle of attack obtained simply as $L_{\max} = L_0 + L_\alpha \alpha_s$. For a post-stall flight, one can assume $L \approx 0$. Due to its dependence on the angle of attack, the drag of a vehicle can be split into two parts: (a) lift-independent (*parasite*) drag, which plays no part in lift generation, and (b) lift-dependent (*induced*) drag, which arises because of the lift creation. Clearly, parasite drag must be minimized in order to have the maximum possible lift-to-drag ratio. For most vehicles in the linear lift range, the drag can be expressed as a parabolic function of the lift:

$$D = D_0 + KL^2 \ (\alpha \le \alpha_s) \, , \tag{10.2}$$

where D_0 is the *parasite drag*, and the constant K is called the *lift-dependent drag factor*. Finally, the aerodynamic pitching moment at a given point in the vehicle can be expressed by the following linear dependence on the angle of attack:

$$\mathcal{M} = \mathcal{M}_0 + \mathcal{M}_\alpha \alpha \ (\alpha \le \alpha_s) \, , \tag{10.3}$$

where $\mathcal{M}_0$ is the moment at zero angle of attack, and $\mathcal{M}_\alpha$ is the rate of change of pitching moment with α. As discussed in Chapter 13, $\mathcal{M}_\alpha$ determines the *static stability* of an airplane's pitching motion ($\mathcal{M}_\alpha < 0$ for stability), and restricts the location of the airplane's center of mass. On the other hand, for an airplane to be in moment equilibrium (*trim*), we must have $\mathcal{M}_0 = 0$ for the whole airplane, which is usually achieved by using horizontal tail (or canard) surfaces. We will return to these concepts a little later.

The foregoing treatment of aerodynamic force and moment is restricted to the most common flight condition, wherein the motion occurs within the *plane of symmetry* of the vehicle. As discussed in Chapter 12, most atmospheric flight vehicles possess at least one plane of symmetry. When the motion of an atmospheric vehicle follows a general, three-dimensional curve, there is a component of the relative velocity normal to the plane of symmetry, called *sideslip*. The sideslip makes an angle, β, called the *sideslip angle* with the symmetry plane. Furthermore, the swirling motion of the flow caused by propellers and wing tips leads to an *aerodynamic bank angle*, σ. Due to the sideslip and aerodynamic bank, additional aerodynamic force components, called *side-force*, f_Y, and moment components, called *rolling moment*, $\mathcal{L}$, and *yawing moment*, $\mathcal{N}$, are generated. These components vary with the *sideslip angle*, β, in a manner quite similar to that of lift, drag, and pitching moment with the angle of attack, as we will see in Chapter 13. Hence, our discussion here of lift, drag, and pitching moment can be easily extended to the lateral aerodynamic force and moment.

It is important to emphasize that all of the relationships given above are valid only for a steady flow, where the time-dependent flow disturbances have died out. In the case of unsteady flow, there are additional terms describing the dependence of lift, drag, and pitching moment on the time rate of change of angle of attack, $\dot{\alpha}$, which we will discuss in Chapters 13 and 15.

10.3 Fluid Dynamics

Fluid is a term that encompasses all flowing matter, such as liquids, gases, and fine solid particles, and mixtures thereof. We can apply the basic principles of mechanics (Chapter 4) to a fluid element, after taking into account their special characteristics. Our primary concern is the flow of gaseous mixtures, such as air, hence our discussion in this section will be confined to gases. In Chapter 9, we saw how an element of air can be in static equilibrium under the influence of gravity and buoyancy. When the same element is moving past a vehicle, it exerts pressure and shear stress on the solid surfaces of the latter, thereby generating lift, drag, and pitching moment. Hence, our objective in the present section is to understand *fluid dynamics*, which is the science of moving fluids. For illustration, we will confine ourselves to unidirectional flows and understand that the same principles are applicable to flows in two and three dimensions (with a greater mathematical complexity). Our present analysis is largely restricted to steady flow, while unsteady fluid dynamics, important for understanding short-period rotational dynamics of a vehicle (Chapter 13), is discussed toward the end of the chapter.

10.3.1 Flow Regimes

Before continuing our discussion of fluid dynamics, it is necessary to classify the flow according to some characteristic *flow regimes*. Within each regime, the nature of the flow is markedly distinct, thereby requiring different modeling methods for the various regimes. The primary classification of flow is that of *continuum* and *rarefied* flow. The average distance separating any two gas molecules at any given time, called the *mean free path*, λ, is inversely proportional to atmospheric density, which itself varies with altitude (Chapter 9). At low altitudes, this average intermolecular distance is negligible, therefore a vehicle moving relative to the atmosphere is fairly large compared to λ, and thus encounters a continuous stream of molecules. In such a case, the fluid is modeled as a continuous medium, without any voids, and the principles of fluid dynamics based on this assumption are called *continuum fluid dynamics.* At high altitudes, the atmosphere becomes tenuous in nature, with the mean free path becoming of the order of meters, which is comparable to the vehicle's dimensions. Consequently, the vehicle experiences an intermittent collision with the molecules, and the fluid is modeled as a noncontinuum or rarefied medium. While the principles of continuum fluid dynamics are based on differential calculus, those of rarefied fluid dynamics involve a statistical approach based upon a stochastic model of fluid–solid interaction. The non-dimensional *Knudsen number*, defined as the ratio of the mean free path of undisturbed molecules and a characteristic length of the vehicle, $Kn \doteq \lambda/l_c$, is the governing parameter for rarefied flows. As pointed out in Chapter 9, flows with a small value of Kn are considered to be continuum flows, while those having a large Knudsen number fall into the rarefied category. Generally, $Kn \leq 0.01$ is considered the continuum regime, while $Kn > 0.01$ is called rarefied flow. As we shall find out, within each flow regime are additional subregimes, primarily based upon the particular mathematical modeling procedures.

10.3.2 Continuum Flow

The continuum flow assumption is based upon disregarding the intermolecular separation and treating the fluid as a continuous medium, whose volume can shrink to infinitesimal dimensions. Such a small packet of fluid is called a *fluid element* and is useful in deriving the governing equations of momentum and energy. *Continuity* of the flow is an important aspect of continuum fluid dynamics. Consider a fluid moving steadily in the direction indicated by the distance x. If we fix a volume in the space—called a *control volume*—indicated by dotted lines in Fig. 10.3(a), the net mass flow through the given control volume must be a constant by the law of mass conservation, which states that matter can be neither created nor destroyed in the process of merely flowing from one point to the other. Thus, we have

$$\dot{m} \doteq \rho A(x) v(x) = \text{constant} , \tag{10.4}$$

where $A(x)$ denotes the cross-section area of the control volume at a given distance x. Equation (10.4) is called the *continuity equation* for one-dimensional flow. The concept of control volume can be extended to a volume bounded by any set of streamlines, called a *streamtube*. In such a case, the area of cross section, $A(x)$, is determined solely by the shape of the streamtube, which in turn, depends upon the geometry of the flow. For example, the flow on the upper surface of an airfoil can be modeled by a converging-diverging streamtube, while that adjacent to the lower surface is treated as a converging streamtube. Gases are *compressible* fluids, which implies that their density can change from point to point under the influence of changing pressure and temperature. In contrast, liquids are largely *incompressible* in nature, thereby implying a roughly constant density despite small pressure and temeperature variations. A gas moving at sufficiently low speeds mimics the behavior of a liquid, because the pressure and temperature changes caused by its motion are not large enough to cause appreciable density variations. Hence, a low-speed gas flow is accurately regarded as an incompressible flow where $\rho \approx$ constant. For an incompressible flow, the continuity equation simply becomes

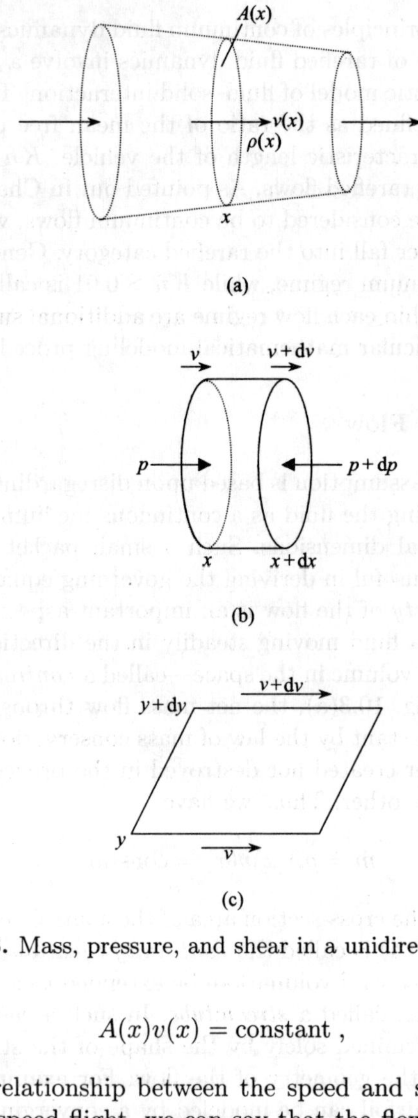

Fig. 10.3. Mass, pressure, and shear in a unidirectional flow.

$$A(x)v(x) = \text{constant}\ , \tag{10.5}$$

which is a direct relationship between the speed and streamtube area. For example, in a low-speed flight, we can say that the flow speed on the upper surface of an airfoil first increases and then decreases, while that on the lower surface continuously increases. However, in a compressible flow, we do not have such a simple area-velocity relationship.

The general form of the continuity equation for an unsteady, three-dimensional, continuum flow is the following:

$$\frac{\partial \rho}{\partial t} + \nabla \cdot (\rho \mathbf{v})\ , \tag{10.6}$$

where ∇ is the following vector gradient operator:

$$\nabla \doteq \mathbf{i}\frac{\partial}{\partial x} + \mathbf{j}\frac{\partial}{\partial y} + \mathbf{k}\frac{\partial}{\partial z}\,. \tag{10.7}$$

Clearly, the solution of the equation for a compressible flow is complicated by the fact that the changes in the density are also caused by changes in the pressure due to the fluid motion represented by $\mathbf{v}$. Hence, another set of equations is required for modeling the pressure variation due to change in velocity.

Consider a fluid element moving in the x-direction, as depicted in Fig. 10.3(b). The element has an infinitesimal thickness, Δx, and a unit cross-sectional area. We assume that the element is far enough from any solid interface, so that the rotation caused by shear stresses is negligible, and therefore, there is no energy lost as the rotational kinetic energy. Such an assumption is called the *inviscid flow* (or *irrotational flow*) approximation. We shall discuss below why the inviscid flow is a good assumption for most pressure calculations. Furthermore, the small size of the fluid element ensures that its weight is negligible in comparison with the force developed due to pressure differences across the element. This aerodynamic force is merely $\mathrm{d}p$ on the unit area and acts opposite to the direction of motion, as shown in Fig. 10.3(b). On applying Newton's second law of motion to the fluid element, we can write

$$\mathrm{d}p = -\rho \mathrm{d}x\frac{\mathrm{d}v}{\mathrm{d}t} = -\rho \mathrm{d}x\frac{\mathrm{d}v}{\mathrm{d}x}\frac{\mathrm{d}x}{\mathrm{d}t} = -\rho v \mathrm{d}x\frac{\mathrm{d}v}{\mathrm{d}x}\,, \tag{10.8}$$

or,

$$\mathrm{d}p = -\rho v \mathrm{d}v. \tag{10.9}$$

Equation (10.9) is called *Euler's equation* and represents the rate of change of momentum per unit volume in a steady, inviscid flow for a fluid element. This equation lets us calculate the pressure difference between any two points along a streamline and is thus useful for estimating lift and pitching moment. For example, ignoring the density changes, we can say that the flow on the upper surface of an airfoil results first in a reduction of pressure due to an accelerating flow and then in an increase in the pressure due to flow deceleration. However, Euler's equation does not give an accurate estimate of the drag due to the neglect of viscosity. In an unsteady, three-dimensional flow, Euler's equation is generalized as the following vector differential equation:

$$\nabla p = -\rho\frac{\partial \mathbf{v}}{\partial t} - \rho \mathbf{v}(\nabla \cdot v). \tag{10.10}$$

For the case of an incompressible, steady, inviscid flow, Euler's equation is directly integrated along a a streamline to yield the following result:

$$p + \frac{1}{2}\rho v^2 = p_0 = \text{constant}\,, \tag{10.11}$$

which is known as the *Bernoulli equation*. The Bernoulli equation is a quick and simple method for calculating the pressure distribution on a slowly flying airplane, for which the incompressible flow approximation is valid. It is also utilized for calibrating the *airspeed indicator* for indicating the speed through the atmosphere (called *airspeed*) of slow aircraft. In such a case, the airspeed indicator senses the barometric pressure difference between a *pitot probe* with an opening normal to the relative flow for measuring the *stagnation pressure*, p_0, and a *static vent* tangential to the relative flow for estimating the *static pressure*, p. The difference, $p_0 - p = 1/2\rho v^2$, is called the *dynamic pressure*, which is directly proportional to the square of the airspeed at a given altitude ($\rho =$ constant). The airspeed indicator is calibrated for a fixed density, usually the standard sea level value, and thus requires a density correction for converting the indicated airspeed to the true airspeed at any given altitude. A separate device, called the *altimeter*, for measuring p and converting it to a standard altitude, provides the pilot with the approximate standard altitude for making the necessary density correction for the airspeed. These two basic instruments have provided the fundamental navigational reference in all aircraft over the 100 years of manned atmospheric flight. However, it is important to remember that airspeed measurements based on the Bernoulli equation would be in gross error when flying at higher speeds for which incompressible flow assumption is invalid.

10.3.3 Continuum Viscous Flow and the Boundary Layer

A fluid element in close vicinity of a solid surface experiences a shear deformation due to fluid viscosity and a tangential velocity component relative to the surface. This situation is depicted in Fig. 10.3(c). The shearing action leads to a *shear stress* along the flow direction (x), τ_x, in the fluid elements, which is regarded as being directly proportional to the rate of deformation of the element, $\partial v/\partial y$, where y is the distance perpendicular to the flow. The linear relationship between the shear stress and rate of deformation is expressed as

$$\tau_x \doteq \mu \frac{\partial v}{\partial y}\,, \tag{10.12}$$

where μ is the coefficient of dynamic viscosity. Most liquids and gases obey the linear stress–strain relationship of Eq. (10.12) and are called *Newtonian fluids*.[3] The air flow past a solid surface experiences shear in a thin region close to the surface, called the **boundary layer** (Fig. 10.4). Within a boundary layer, the momentum equation must include the viscous term for shear stress, which makes the equation difficult to solve. However, because of the thin region of the boundary layer,[4] and due to the fact that the pressure normal to the layer remains essentially unchanged by shear, approximate closed

[3] However, fluids consisting of solid particles and slurries do not have such a simple behavior and fall into the category of non-Newtonian fluids.

[4] The boundary layer grows to a thickness less than 3 cm at $x = 3.5$ m on the wing of an airliner cruising near 11 km altitude.

form and semi-empirical solutions have been derived for the flow inside the boundary layer [21]. The character of the boundary layer flow may be either smooth and layer-like, called *laminar* flow, or rough and fluctuating, called *turbulent flow*. A boundary-layer flow usually begins as laminar, but eventually transitions into turbulent flow. The causes of transition are not very well understood but depend largely upon the surface roughness, flow speed, and viscosity. While the laminar flow has a smaller value of μ compared to the turbulent flow, thereby causing a smaller skin friction drag, the laminar boundary layer finds it more difficult (compared to turbulent flow) to adhere to the solid surface when the flow is decelerating. We recall that the flow separation is caused by a decelerating flow, which according to Euler's equation, implies an increasing pressure (or a positive pressure gadient), $\partial p/\partial x > 0$. Hence, a laminar layer separates more easily in a positive pressure gradient, compared to a turbulent layer, thereby creating a larger pressure drag. For this reason, some airplanes use small strategically placed rough patches, or protuberances, on top of the wings in order to facilitate transition of laminar flow into a turbulent flow, and thus reducing the chances of flow separation in the region of positive pressure gradient. A similar rationale is used in the dimpled golf balls, whose reduced pressure drag due to turbulence guarantees a better range. However, in regions of negative pressure gradient, it pays to have laminar flow to the largest possible extent by having an extra-smooth surface. Some low- to medium-speed airfoils are especially designed to maintain laminar flow over large portions, thereby reducing skin friction. This is primarily achieved by minimizing the region of positive pressure gradient.

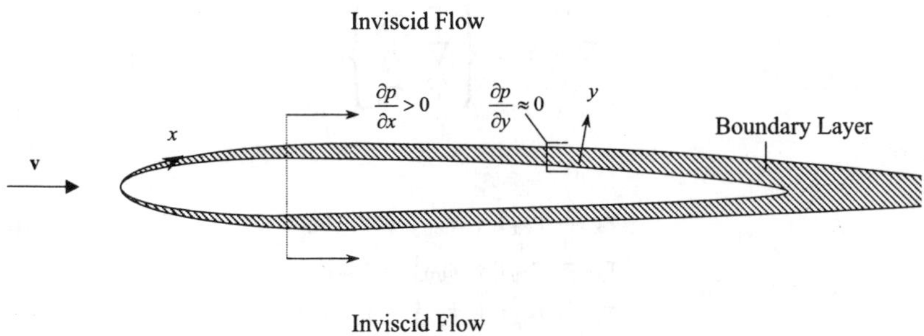

Fig. 10.4. The boundary layer and the outer, inviscid region.

Outside the boundary layer, the effects of viscosity are negligible, because the shearing action is virtually absent. Hence, the region external to the boundary layer (which is a large domain) is accurately regarded as being inviscid. From the mass conservation consideration, it can be seen that the boundary layer tends to grow in thickness, even over a flat plate at zero angle of attack. This growth is tantamount to a normal displacement of the inviscid flow region external to the layer. Furthermore, since the pressure normal to the boundary layer is transferred to the surface virtually unchanged, we can treat the outer edge of the boundary layer as the solid surface itself and apply Euler's equation for computing pressure by the inviscid approximation on the slightly thickened object. This two-layer approach is often utilized in computing pressure distribution on streamlined shapes. There is a feedback between the two solutions such that the thickness of the boundary layer influences the outer, inviscid region, while the pressure distribution computed by the inviscid assumption affects the growth and separation of the boundary layer, which is modeled by a different set of viscous equations [21]. Alternatively, a unified computational approach is applied in which the entire flow is modeled by the fully viscous governing equations, called *Navier–Stokes* equations [22], written as follows:

$$\rho\frac{\partial \mathbf{v}}{\partial t} + \rho\mathbf{v}(\nabla \cdot v) = -\nabla p + \nabla \odot \tau\,, \tag{10.13}$$

where τ is the following symmetric square matrix, called *stress tensor*, comprising the stresses acting on the fluid element:

$$\tau \doteq \begin{pmatrix} \tau_{xx} & \tau_{xy} & \tau_{xz} \\ \tau_{xy} & \tau_{yy} & \tau_{yz} \\ \tau_{xz} & \tau_{yz} & \tau_{zz} \end{pmatrix} . \tag{10.14}$$

The operator $\odot$ represents a dot product between the column vector, ∇, and the square matrix, τ, resulting in a column vector consisting of the scalar products between ∇ and the individual rows of τ:

$$\nabla \odot \tau \doteq \begin{Bmatrix} \nabla \cdot \boldsymbol{\tau}_{\mathbf{x}} \\ \nabla \cdot \boldsymbol{\tau}_{\mathbf{y}} \\ \nabla \cdot \boldsymbol{\tau}_{\mathbf{z}} \end{Bmatrix} , \tag{10.15}$$

where

$$\begin{aligned} \boldsymbol{\tau}_{\mathbf{x}} &\doteq \tau_{xx}\mathbf{i} + \tau_{xy}\mathbf{j} + \tau_{xz}\mathbf{k}, \\ \boldsymbol{\tau}_{\mathbf{y}} &\doteq \tau_{xy}\mathbf{i} + \tau_{yy}\mathbf{j} + \tau_{yz}\mathbf{k}, \\ \boldsymbol{\tau}_{\mathbf{z}} &\doteq \tau_{xz}\mathbf{i} + \tau_{yz}\mathbf{j} + \tau_{zz}\mathbf{k}. \end{aligned} \tag{10.16}$$

The stresses can be divided into *normal* stresses,

$$\begin{aligned} \tau_{xx} &\doteq \lambda(\nabla \cdot \mathbf{v}) + 2\mu\frac{\partial v_x}{\partial x}, \\ \tau_{yy} &\doteq \lambda(\nabla \cdot \mathbf{v}) + 2\mu\frac{\partial v_y}{\partial y}, \\ \tau_{zz} &\doteq \lambda(\nabla \cdot \mathbf{v}) + 2\mu\frac{\partial v_z}{\partial z}, \end{aligned} \tag{10.17}$$

and shear stresses,

$$\tau_{xy} \doteq \mu \left(\frac{\partial v_y}{\partial x} + \frac{\partial v_x}{\partial y} \right),$$
$$\tau_{yz} \doteq \mu \left(\frac{\partial v_y}{\partial z} + \frac{\partial v_z}{\partial y} \right), \tag{10.18}$$
$$\tau_{xz} \doteq \mu \left(\frac{\partial v_z}{\partial x} + \frac{\partial v_x}{\partial z} \right),$$

where $\mathbf{v} \doteq v_x\mathbf{i} + v_y\mathbf{j} + v_z\mathbf{k}$ is the velocity of the fluid element, and λ is the coefficient of *bulk viscosity*, approximated by $\lambda = -2/3\mu$. Usually, the viscosity coefficient, μ, due to turbulent flow is modeled by an appropriate statistical or empirical turbulence model. However, turbulence remains an ill-understood phenomenon.

The Navier–Stokes equations are the most difficult to solve among all fluid dynamic models and require enormous computational resources for an accurate solution. Their solution, subject to specific boundary conditions along the solid surface and on the far-field boundaries, requires iterative numerical procedures. We shall refer the reader to a textbook on computational fluid dynamics [23] for a deeper insight into the continuum, viscous flow modeling and computation aspects.

The governing parameter for viscous flow is the local *Reynolds number*, given by

$$Re_x \doteq \frac{\rho v x}{\mu}, \tag{10.19}$$

where x denotes distance along the streamwise direction on the solid surface, measured from the leading edge. The Reynolds number represents a non-dimensional ratio of linear momentum of a fluid element to its viscous force per unit mass in a steady flow. If the element is moving very rapidly, its linear momentum is quite large, and consequently it experiences a small acceleration due to viscosity. Thus, a larger Reynolds number indicates smaller viscous effects. However, if the Reynolds number is increased beyond a certain value for a given object, the nature of the viscous flow undergoes a change from laminar to turbulent. There exists a *critical Reynolds number*, Re_{cr} for each surface, such that $Re_x < Re_{\text{cr}}$ denotes the extent of laminar flow on the surface, while $Re_x \geq Re_{\text{cr}}$ represents the turbulent flow region. The point on the surface where $Re_x = Re_{\text{cr}}$ is called the *transition point.* As pointed out above, transition is a poorly understood phenomenon, hence Re_{cr} is difficult to predict for a given surface and can only be estimated experimentally. The skin friction in a boundary layer depends greatly upon its laminar or turbulent nature. Let us define a nondimensional *skin friction coefficient*, C_f, in a unidirectional flow as follows:

$$C_f \doteq \frac{\tau_{xy}}{\frac{1}{2}\rho v^2}. \tag{10.20}$$

By the boundary-layer approximation, the skin friction coefficient on a flat plate in an incompressible flow at zero angle of attack is given according to *Blasius* [21] as

$$C_f = \frac{0.664}{\sqrt{Re_x}} . \tag{10.21}$$

In a flat plate of chord length c fully immersed in incompressible turbulent flow at zero angle of attack, the skin friction coefficient is given by

$$C_f = \frac{0.074}{Re_c^{\frac{1}{5}}} , \tag{10.22}$$

where Re_c is the Reynolds number based on c. Clearly, the skin friction is larger for a turbulent flow, compared with the laminar flow of the same length. However, the turbulent boundary layer is much more resistant to the adverse pressure gradient that causes flow separation. The skin friction of most surfaces can be estimated by stretching out the external area exposed to the flow (called the *wetted area*) into an equivalent flat plate and then applying either Eq. (10.21) or Eq. (10.21) (or a combination thereof in case the flow is partly turbulent).

The Reynolds number is an important flow parameter that must be matched for a wind-tunnel test model and a full-scale vehicle, in order to faithfully reproduce the viscous flow on the latter. Most airplanes encounter turbulent flow in their normal cruising flight, and thus experimental investigation on a scaled model with the Reynolds number matching becomes crucial for accurate drag prediction. However, apart from the Reynolds number, there is another governing parameter in a viscous, compressible flow.

Invariably, friction is accompanied by heat transfer, which takes the form of *conduction* across the boundary layer in the y-direction, and *convection* along the flow direction, x. The heat transfer is usually modeled as conduction across the boundary layer, which changes the temperature at the boundary-layer edge. This temperature increase then affects the flow in the inviscid region through convection. At the very high speeds encountered during atmospheric entry, the third mode of heat transfer, namely *radiation*, also becomes important. The conductive rate of heat transfer per unit area in the boundary layer is given by

$$\dot{Q}_y = -k_T \frac{\partial T}{\partial y} , \tag{10.23}$$

where k_T is the *coefficient of thermal conductivity*. For a gas flow, k_T is obtained either from the kinetic theory of gases, which involves statistical thermodynamics, or from empirical formulas, such as Eq. (9.20) for perfect air [19]. The nondimensional parameter governing conductive heat transfer caused by viscous dissipation is the *Prandtl number*, Pr, defined by

$$Pr = \frac{\mu c_p}{k_T} , \tag{10.24}$$

where c_p is the constant-pressure specific-heat coefficient. The Prandtl number represents the ratio of heat generation by viscous effects and heat transfer by conduction. Pr is a function of temperature and pressure and has different values for different gases. For air up to a temperature of 600 K, $Pr \approx 0.71$. At higher temperatures, air is dissociated and consequently Pr becomes a function of temperature and pressure.

10.3.4 Continuum Compressible Flow

When modeling compressible gas flows, we must account for the variation of temperature caused by velocity gradients in the flow field. Such considerations are inconsequential in a constant-density (incompressible) flow and are given the name *gas dynamics*. Gas dynamics essentially involves energy conversion through thermodynamic principles. Our treatment of gas dynamics begins with the assumption of *perfect gas* for the atmospheric gas mixture, which follows the ideal *equation of state*,

$$p = \rho R T \,, \tag{10.25}$$

where R is the specific gas constant (Chapter 9). A perfect gas has negligible intermolecular cohesion, as well as a frozen chemical composition. The operation of most flight vehicles (with the exception of atmospheric entry vehicles) does not occur at speeds where perfect gas assumption does not hold. In addition to the equation of state, a perfect gas obeys the following relationships between R, the constant-pressure and constant-volume specific-heat coefficients, c_p, c_v, and their constant ratio, $\gamma \doteq c_p/c_v$:

$$\begin{aligned} c_v &\doteq \frac{R}{\gamma - 1}, \\ c_p &\doteq \frac{\gamma R}{\gamma - 1} \,. \end{aligned} \tag{10.26}$$

The specific-heat ratio, γ, depends upon the degrees of freedom in a gas molecule. For a monoatomic gas, $\gamma = 1.67$, while a diatomic gas has $\gamma = 1.41$. We can regard air to be a perfect diatomic gas with $R = 287$ J/kg.K.

The *speed of sound*, a, defined as the speed at which infinitesimal pressure disturbances travel in an otherwise undisturbed medium, is an important compressible flow variable. A governing parameter in an compressible flow is the *Mach number*, M, defined as the ratio of the flow speed and the speed of sound, $M \doteq v/a$, prevailing at a given point. A flow with $M < 1$ is said to be *subsonic*, while that with $M > 1$ is called *supersonic* flow. The boundary between subsonic and supersonic flow, $M \approx 1$, is termed *transonic* flow regime. The nature of the compressible flow is markedly different in each of these regimes.

In order to study compressible flow, we assume that a perfect gas element is free from all irreversible thermodynamic effects and does not experience appreciable heat transfer. Such a reversible, adiabatic flow conserves *entropy* and is called *isentropic flow*. The isentropic assumption is usually valid in the outer, inviscid region of a subsonic flow. In transonic and supersonic flow, the nonisentropic phenomenon of shock waves is present in the inviscid region. However, since shock waves are treated as discontinuities, we can still apply isentropic flow approximation in regions bounded by shock waves. For an isentropic flow of a perfect gas, we have

$$\frac{p}{\rho^\gamma} = \text{constant}. \tag{10.27}$$

Furthermore, in an isentropic flow of a perfect gas, the local speed of sound can be expressed as follows:

$$a = \sqrt{\gamma \frac{p}{\rho}} = \sqrt{\gamma R T}. \tag{10.28}$$

Since energy is conserved in an adiabatic flow, we can write the *energy equation* as follows:

$$c_p T + \frac{v^2}{2} = c_p T_0 = \text{constant}. \tag{10.29}$$

From the energy equation, it follows that the *stagnation temperature*, T_0, defined as the temperature of the flow brought to rest adiabatically, is constant. This is a powerful relation between the speed of an adiabatic flow and its temperature and can be applied even across a shock wave, because the latter is an adiabatic (though irreversible, and thus nonisentropic) process.

The energy equation in terms of the Mach number is the following:

$$\frac{T_0}{T} = 1 + \frac{\gamma - 1}{2} M^2 . \tag{10.30}$$

By substituting the equation of state and Eq. (10.27) into the energy equation, we can write the following additional relationships between the static and stagnation quantities in an isentropic flow:

$$\frac{\rho_0}{\rho} = \left(1 + \frac{\gamma - 1}{2} M^2\right)^{\frac{1}{\gamma - 1}} ,$$

$$\frac{p_0}{p} = \left(1 + \frac{\gamma - 1}{2} M^2\right)^{\frac{\gamma}{\gamma - 1}} . \tag{10.31}$$

We need the above-derived isentropic flow relations for the calculation of flow properties, given the local Mach number. However, the validity of these expressions is limited to an inviscid, adiabatic flow, without shock waves. An incompressible flow is practically regarded as that in which density changes caused by the flow do not exceed 5% in magnitude. Using the isentropic relation for density, we can easily show that for $\mid \rho_0 - \rho \mid \leq 0.05\rho_0$, we need $M \leq 0.3$, which is the flow regime of low-speed aircraft, and also the take-off and landing speed for all aircraft. When we apply the isentropic relations in a supersonic flow, we find that the usual intuitive variation of flow speed, v, with cross-section area, A, is no longer valid and is, in fact, reversed from that experienced at subsonic speeds. Thus, a supersonic flow with an *increasing* area is *accelerated*, while that with a *decreasing* A is *slowed down*. This flow behavior is caused by the extreme density variations with the speed in a supersonic flow. In order to understand the phenomenon of compressibility, let us consider an object moving at $M \leq 0.3$. The speed of motion of the object through the atmosphere is negligible compared to the speed of sound. Consequently, the small pressure disturbance created by the object's motion spreads out uniformly in all directions, almost at infinite speed compared with that of the object. This situation is depicted by uniformly expanding spherical pressure waves emanating from an almost stationary object, such as the waves caused by dropping a stone in a pool of water. Next, consider the same object moving at a larger speed, but still smaller than the speed of sound $(0.3 < M < 1)$. Now the waves propagating along the direction of motion are compressed into a smaller space in front of the object, while those at the back are stretched out due to the object's motion [Fig. 10.5(a)]. However, the fluid elements upstream of the object get an advance warning of the approaching object, which they try to accommodate by stepping aside, thereby causing a curvature in the streamlines—a hallmark of subsonic flow. As the Mach number of the object increases, the waves in the front are squeezed tighter together, until they all coalesce into a single wave at $M = 1$, as shown in Fig. 10.5(b). This limiting wave normal to the relative transonic flow, and comprising a finite pressure disturbance, which is obtained by adding a large number of small pressure disturbances, is called a *normal shock wave*. Since the shock wave and the object are moving at the same speed, no advance notice of the approaching object is available to an upstream fluid element, which experiences the entire pressure disturbance caused by the object only when the latter has reached the given point. Now consider the given object moving at a supersonic speed $(M > 1)$. In this case, the object has overtaken its own pressure disturbance. Hence, all pressure disturbances created by the object lag behind and are confined to a conical region behind the object, called the *shock cone*, or an *oblique shock wave*. Since the fluid elements upstream and abreast of the object are undisturbed, the supersonic flow is characterized by parallel streamlines [Fig. 10.5(c)]. The angle, β, made by the cone with the relative flow direction is a function of the Mach number as well as of the shape of the object. For the limiting case of an extremely slender object, which can create only an infinitesimal pressure disturbance, the shock wave is called a *Mach wave*, and $\beta \approx \sin^{-1}(1/M)$. The shock wave is a nonisentropic process. Behind the object moving at supersonic speed, we have the isentropic phenomenon of flow expansion, which is modeled by *expansion fan*, as discussed ahead.

Shock waves are extremely thin regions (spanning only a few mean free-path lengths) in a continuum flow. Hence, they are modeled as discontinuous

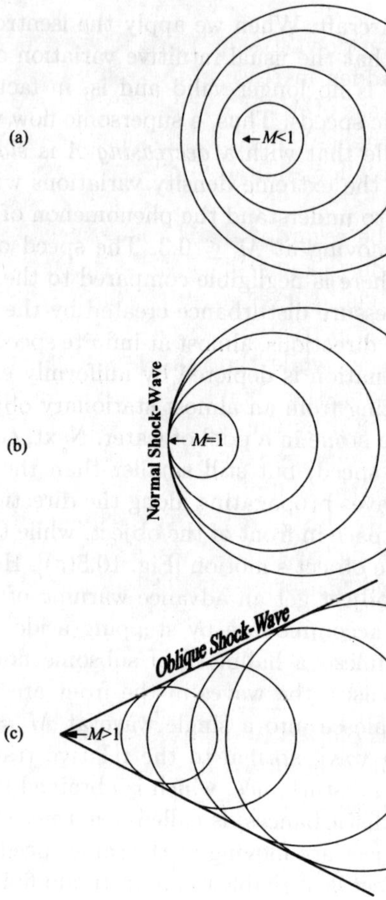

Fig. 10.5. Compressible flow and shock waves.

jumps in the pressure, density, and temperature. A normal shock wave is modeled as a one-dimensional, perfect gas flow, with a sudden, irreversible jump from the quantities upstream (indicated by subscript *1*) to those downstream (indicated by subscript *2*), according to the following expressions:

$$M_2^2 = \frac{2 + (\gamma - 1)M_1^2}{2\gamma M_1^2 - \gamma + 1},$$

$$\frac{p_2}{p_1} = 1 + \frac{2\gamma}{\gamma + 1}(M_1^2 - 1), \tag{10.32}$$

$$\frac{\rho_2}{\rho_1} = \frac{u_1}{u_2} = \frac{(\gamma + 1)M_1^2}{2 + (\gamma - 1)M_1^2}.$$

The temperature jump across the normal shock can be calculated using the above relations and the perfect gas equation of state. It is clear from above that for a normal shock, the downstream Mach number is always subsonic ($M_2 < 1$), and there is a sudden increase in the static pressure, density, and temperature. The limiting case of $M_1 = 1$ yields a normal Mach wave of zero strength (i.e., no pressure change). It is clear that the pressure jump across a normal shock is responsible for additional drag on the object, called *wave drag*.

An oblique shock is to be regarded merely as a normal shock slanted to the relative flow at an angle β and is caused by a sudden *inward* turn of the flow by a deflection angle, θ [Fig. 10.6(a)]. Hence, the flow component normal to the oblique shock is treated in exactly the same manner as the flow through a normal shock, while the tangential flow component passes through unchanged. Therefore, the oblique shock relations can be obtained by replacing M_1 in Eq. (10.32) by the normal Mach number, $M_1 \sin\beta$, and M_2 by the normal component, $M_2 \sin(\beta - \theta)$. The resulting equations display the fact that the strength of the oblique shock increases with the increase of either M_1, or β. However, M_1 and β are not independent quantities, but they are related through the flow deflection angle, θ. It can be shown [22] that the relationship among θ, β, and M_1 for an oblique shock is the following:

$$\tan\theta = 2\cot\beta \frac{M_1^2 \sin^2\beta - 1}{2 + M_1^2(\gamma + \cos 2\beta)} . \tag{10.33}$$

This implicit relationship is plotted in oblique shock charts in gas dynamics textbooks and gives the following important insight about the flow:

(a) For each value of M_1, there exists a maximum possible flow deflection, θ_m, for which the shock wave remains straight and attached to the solid surface. For a flow with $\theta > \theta_m$, we have a detached, curved shock wave. A detached shock is much stronger than an attached one, because a portion of it is normal to the flow and thus causes a much greater drag. Hence, the solid surface must be properly designed to avoid detached shock in its design range of Mach numbers.
(b) For $\theta < \theta_m$, two values for the shock angle, β, can satisfy Eq. (10.33). The larger of these is called the *strong shock solution*, while the other is the *weak shock solution*. Practically in all cases of straight, attached shock the weak solution prevails, which is beneficial for wave drag reduction. In a weak shock, the downstream Mach number, M_2, is always supersonic. For $\theta = 0$, such as a flat plate at zero angle of attack, the weak shock solution yields $\beta = \sin^{-1}(1/M_1)$, which is the case for a Mach wave, as noted above. However, in such a case the strong shock solution, with $\beta = 90°$, usually prevails, as depicted in Fig. 10.5(b) for $M_1 = 1$.
(c) In the limit of very high incident Mach number, M_1, the shock wave angle becomes small and approaches the limit, $\beta \to (\gamma + 1)\theta/2$. In such a case, the pressure, density, and temperature jump quantities become essentially invariant with Mach number. This is known as the *continuum hypersonic limit*, or the *Newtonian flow limit*. Usually, for $M_1 \geq 5$, we can assume hypersonic

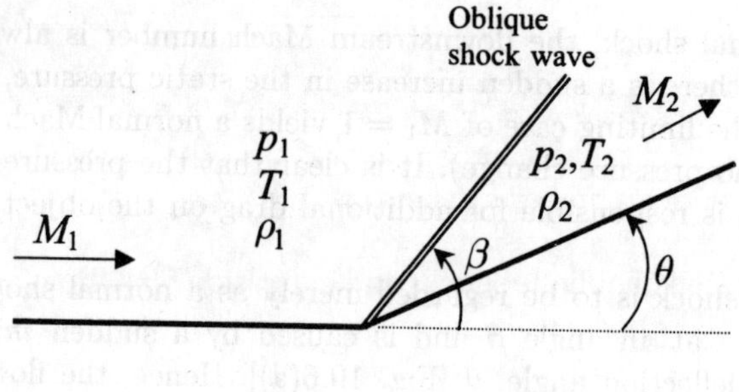

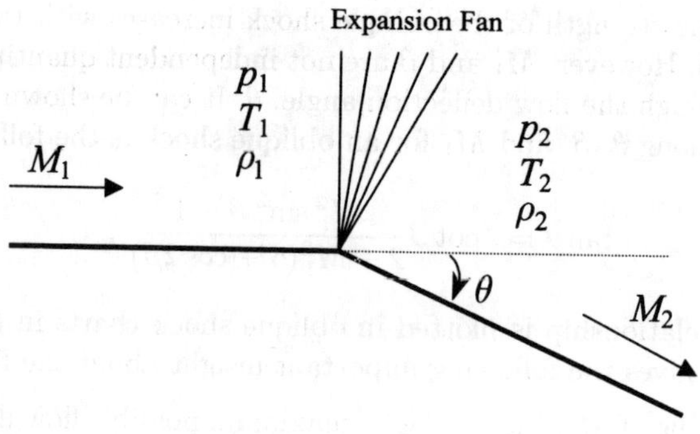

Fig. 10.6. An oblique shock-wave and an expansion fan.

flow and approximate the nondimensional jump quantities by appropriate constants. Such a behavior of flow quantities at high Mach numbers is referred to as the *Mach number independence principle*. When the supersonic flow deviates around an outward corner, it undergoes an expansion due to the increase in the flow area, as noted above. Such an expansion is modeled by an isentropic process through a set of Mach waves appearing at the expansion corner, called an *expansion fan*, and is shown in Fig. 10.6(b). Each wave in the expansion fan is a Mach wave, inclined at angle $\mu = \sin^{-1}(1/M)$ to the local flow, and causes an infinitesimal turning of the flow. The net effect of the expansion fan is a flow turned by the deflection angle, θ, with an attendant increase of Mach number from M_1 to M_2. The relationship among M_1, M_2, and θ in an expansion fan is given by

$$\theta = \nu(M_2) - \nu(M_1) , \tag{10.34}$$

where $\nu(M)$ is the following *Prandtl–Meyer function* [22]:

$$\nu(M) = \sqrt{\frac{\gamma+1}{\gamma-1}} \tan^{-1} \sqrt{\frac{\gamma-1}{\gamma+1}(M^2-1)} - \tan^{-1} \sqrt{M^2-1}. \tag{10.35}$$

Expansion charts and tables can be prepared based upon Eq. (10.34), in order to calculate the Mach number at the end of expansion.

Supersonic flow over most surfaces of aeronautical interest can be modeled using a combination of oblique shocks and expansion fans. However, difficulties arise in such a *shock expansion* model when there is an interaction between two shock waves, a shock wave and an expansion fan, or a reflection of a shock wave by another solid surface. In such a case, different computational techniques can be applied, such as the *method of characteristics* [22], or *finite-difference* solution of governing Euler's, continuity, and energy equations [23].

Finally, we note that viscous effects in a compressible flow are taken into account in a manner similar to that of incompressible flow. The boundary-layer approximation is still valid, although the viscous effects are now functions of the Mach number, as well as the Reynolds number. The same holds for a fully viscous model using the compressible Navier–Stokes equations, which, however, become even more intractable with regard to numerical solution. Certain well-known similarity rules exist between incompressible and compressible, subsonic flows on lifting surfaces, such as that by *Prandtl–Glauert* [22], which yield a quick and simple evaluation of pressure and shear stress. In supersonic and hypersonic flows, viscous effects are more complex and involve interaction between shock waves and the boundary layer. Such interactions are most amenable to either experimental studies or computational fluid dynamic modeling through the Navier–Stokes equations [22], [23].

10.3.5 Rarefied Flow

Rarefied flow can be described as the one in which the mean free path of molecules becomes comparable to the characteristic flow length. When describing the aerodynamics of complete vehicles, we encounter rarefied flow at high altitudes for which $Kn > 0.01$, based on a vehicle's characteristic length. However, sometimes rarefied flow is encountered when we study the flow on a smaller scale than that of the complete vehicle, such as the flow near the leading edge of an airfoil, even at low altitudes. The general rarefied flow is difficult to model due to the complex nature of the governing *Boltzmann equation* [25]. However, simpler empirical models are usually employed such as those based on a subdivision of rarefied flow into *free molecular flow* and *transition flow* regimes. For very large Knudsen numbers, such as $Kn > 10$, the gas particles are so widely separated that there is virtually no interaction among them. Hence, the incident gas stream is undisturbed by the object, and a simple statistical model for momentum and energy transfer can be constructed taking into account only the collision of a single gas particle with the solid surfaces. Such a flow is called the free molecular (or *collisionless* flow). There are two extremes of the particle–surface interaction models, first devised by *Maxwell* in 1879: (a) *specular reflection* in which the particles are assumed to be perfectly elastic spheres, with no change in their tangential momentum due to surface interaction, and (b) *diffuse reflection* that accounts for surface roughness, and thus transfer of both tangential and normal momentum. Interaction of most engineering surfaces with air approximately follows the diffuse model. Maxwell treated the reflected particles in the diffuse model according to a random statistical behavior, such that all directions were equally probable, and the speed of a reflected particle was independent of that of the incident particles, but obeys a *Maxwellian distribution* that depends only on the temperature of the reflected particle. A flat elemental surface inclined at angle θ to a free molecular flow with the Maxwellian velocity distribution has the following pressure, shear stress, and rate of heat transfer per unit area [24]:

$$\Delta p = q\left\{\left(\frac{\sin\theta}{s\sqrt{\pi}} + \frac{1}{2s^2}\sqrt{\frac{T_w}{T}}\right)e^{-s^2\sin^2\theta} + \left(\frac{1}{2s^2} + \sin^2\theta + \frac{\sqrt{\pi}\sin\theta}{2s}\sqrt{\frac{T_w}{T}}\right)[1 + erf(s\sin\theta)]\right\}, \quad (10.36)$$

$$\Delta\tau = q\frac{\cos\theta}{\sqrt{\pi}}\left(\frac{e^{-s^2\sin^2\theta}}{s}\sqrt{\pi}\sin\theta[1 + erf(s\sin\theta)]\right), \quad (10.37)$$

$$\dot{Q} = \frac{\rho}{4\beta^3\sqrt{\pi}}\left[\left(s^2 + \frac{\gamma}{\gamma-1} - \frac{\gamma+1}{2(\gamma-1)}\frac{T_w}{T}\right) \times \{e^{-s^2\sin^2\theta} + s\sqrt{\pi}\sin\theta[1 + erf(s\sin\theta)]\} - \frac{1}{2}e^{-s^2\sin^2\theta}\right], \quad (10.38)$$

where ρ, T are density and temperature of incident stream, T_w is the temperature of the solid surface, s is the *molecular speed ratio* of incident stream given by

$$s \doteq v\beta = \frac{v}{\sqrt{2RT}}, \quad (10.39)$$

and $erf(x)$ is the *error function* given by

$$erf(x) = \frac{2}{\sqrt{\pi}}\int_0^x e^{-y^2}dy. \quad (10.40)$$

With the use of Eqs. (10.36) and (10.37), we can estimate the aerodynamic force and moment on a general object by employing a *flat-panel approximation* wherein a number of flat panels are used, each at a specific inclination to the flow, such that the force and moment caused by pressure and shear stress on the elemental panels are summed vectorially.

In the rarefied transition regime, $0.01 < Kn < 10$, a *bridging relation* is used to interpolate between the quantities in the continuum and free-molecular limits. Such interpolations often take the form of exponential variation of the given quantity with the Knudsen number, which are written as linear functions of the logarithm of Kn [24]. We shall employ one such bridging relation in the simulation of a ballistic entry vehicle in Chapter 12.

10.4 Force and Moment Coefficients

We have seen that the aerodynamic force and moment are created due to the combined effects of pressure and shear stress distributions over the vehicle's surfaces. The magnitudes of the aerodynamic force and moment depend upon the size of the vehicle, its dynamic pressure, angle of attack, sideslip angle, aerodynamic bank angle, as well as the nondimensional governing flow parameters of a particular flow regime, such as the Knudsen, Reynolds, Prandtl, and Mach numbers. Usually, it is beneficial to remove the dependence on size while studying aerodynamic effects. This leads to the introduction of nondimensional force and moment coefficients. The force coefficients are rendered nondimensional by dividing them by the product of the *free-stream*[5] dynamic pressure, q_∞, and a reference area, S, which is often the wing-platform area of a lifting vehicle, and the base area of a nonlifting (ballistic) vehicle. For nondimensional moment coefficients, the product $q_\infty S l_c$ is employed, where l_c is a characteristic length. In case of pitching moment, $l_c = \bar{c}$, a *mean chord* of the wing, while $l_c = b$, the wing span for the rolling and yawing moments. From Eqs. (10.1)–(10.3), the lift, drag, and pitching moment coefficients of a generic flight vehicle (airplane) in steady flow are expressed as follows:

$$C_L \doteq \frac{L}{q_\infty S} = C_{L_0} + C_{L_\alpha}\alpha \ \ (\alpha \leq \alpha_s) \,, \tag{10.41}$$

$$C_D \doteq \frac{D}{q_\infty S} = C_{D_0} + K C_L^2 \ \ (\alpha \leq \alpha_s) \,, \tag{10.42}$$

and

$$C_m \doteq \frac{\mathcal{M}}{q_\infty S \bar{c}} = C_{m_0} + C_{m_\alpha}\alpha \ \ (\alpha \leq \alpha_s) \,, \tag{10.43}$$

where α_s is the stalling angle of attack at which the maximum lift coefficient is $C_{L_{\max}}$. The nondimensional *lift curve slope*, $C_{L_\alpha} \doteq L_\alpha / q_\infty S$, *lift coefficient at zero angle of attack*, C_{L_0}, the *longitudinal static margin* (Chapter 13), $C_{m_\alpha} \doteq \mathcal{M}_\alpha / q_\infty S \bar{c}$, pitching moment coefficient at zero angle of attack, C_{m_0}, the *parasite drag coefficient*, C_{D_0}, and the *lift-dependent drag factor*, K, are the parameters defining the aerodynamic characteristics of a given object. These parameters depend upon Kn, Re, and M. The dependence of these nondimensional aerodynamic coefficients on the governing nondimensional flow parameters is termed *similarity rule*, which is crucial in experimental investigation of vehicles through their appropriately scaled, wind-tunnel models. The aerodynamic behavior of an entire vehicle can be described by

[5] Free stream refers to the undisturbed air flow far upstream of the vehicle. The flow quantities in the free stream are denoted by the subscript ∞.

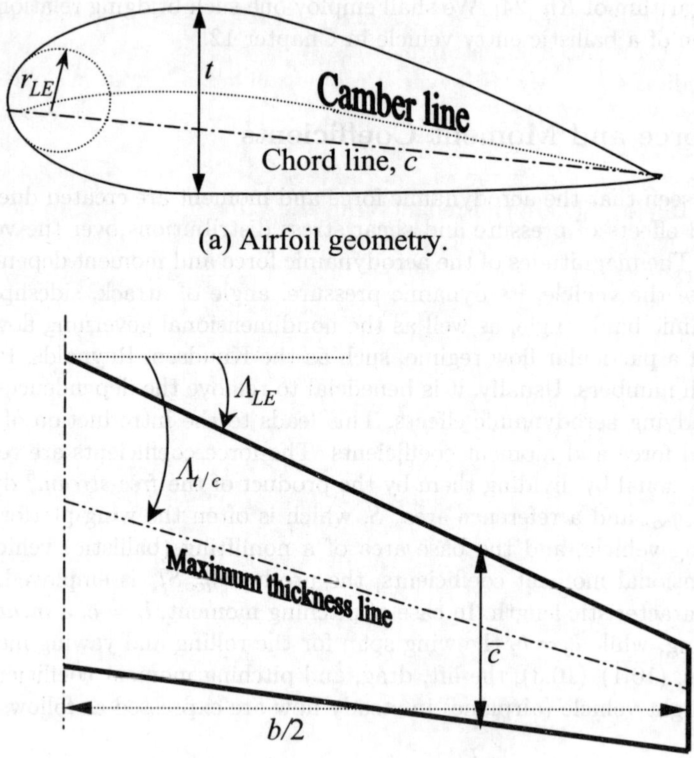

(a) Airfoil geometry.

(b) Wing planform geometry.

Fig. 10.7. A typical wing geometry.

Eqs. (10.41)–(10.43), with the appropriate values of the nondimensional constants. It must be noted that the fuselage and nacelles do not contribute much to the lift; hence their drag is primarily the parasite drag, C_{D_0}. Consequently, the lifting parameters, namely C_{L_α}, C_{L_0}, K, receive their main contribution from the lifting surfaces (wings, tails, and canards). However, the pitching moment coefficient is caused by both lifting surfaces and nonlifting bodies. The values of the aerodynamic coefficients are either established experimentally or estimated through numerical and analytical methods. We shall briefly consider the modeling of aerodynamic behavior in each flow regime.

At low speeds of operation of small aircraft, $M < 0.3$, we can regard the coefficients $C_{L_\alpha}, C_{L_0}, C_{m_\alpha}, C_{m_0}, C_{D_0}, K$ as constants. The incompressible regime has received attention since the earliest days of flying, beginning with the Wright brothers themselves who built a small wind tunnel to investigate various wing airfoil shapes. A vast compendium of analytical literature has become available over the last century, which has been reinforced by experimental and computational research, leading to simple empirical relations. For an infinite span, straight wing in incompressible flow, ideal flow theory predicts $C_{L_\alpha} \approx 2\pi$. However, a real airfoil has a lift curve slope about ten-percent smaller than the ideal value. The exact value of two-dimensional lift-curve slope depends greatly upon the airfoil geometry, which is governed by *thickness ratio*, t/c, shape of the *camber line*,[6] and *leading edge radius*, r_{LE} [Fig. 10.7(a)]. There is a reduction in the value of C_{L_α} due to the finite wing span, b, due to the presence of *wing-tip vortices*.[7] A nondimensional geometric parameter, $A \doteq b^2/S$, called the *aspect ratio*, governs the strength of the wing-tip vortices. For a finite-span wing, the lift-curve slope is a complex function of the wing planform geometry (apart from the airfoil shape), and depends upon the aspect ratio and the *sweep angles*, $\Lambda_{LE}, \Lambda_{t/c}$ [Fig. 10.7(b)]. The value of C_{L_α} increases by increasing A, reducing the wing sweep, and increasing the airfoil thickness, t/c. The value of C_{L_0} at low speeds is determined by a combination of the airfoil camber and the lift curve slope. A symmetrical airfoil has $C_{L_0} = C_{m_0} = 0$, whereas a positively cambered airfoil has $C_{L_0} > 0$, but $C_{m_0} < 0$. While symmetrical airfoils are used in tail (or canard) surfaces, the wings usually employ positively cambered airfoils for a better lift.

The lift-dependent drag (also called *induced drag*) is inversely proportional to the aspect ratio. Hence, the following expression is used for the incompressible value of K:

$$K = \frac{1}{\pi A e} , \tag{10.44}$$

where e is a measure of the lifting efficiency of the wing, and is called *Oswald's span efficiency factor*. For most wings, $0.6 \leq e \leq 0.85$, at low speeds. Often, a *wing twist* is employed to reduce the lift-induced drag as well as to improve the flow behavior near the stall, which results in an airfoil close to the wing-tip having a smaller geometric angle of attack, compared to that near the wing root (*washout*).

The value of C_{m_0} primarily depends upon the wing airfoil camber and the tail (or canard) lift at zero angle of attack. A positively cambered wing by itself has a negative C_{m_0}. However, in order to fly at positive angles of

[6] *Camber* refers to the difference in the positive and negative curvatures, respectively, of the upper and lower surfaces of an airfoil. A positively cambered airfoil has a greater positive curvature of the upper surface. Camber line is a line joining the leading and trailing edges such that it lies equidistant from the upper and lower surfaces. A *symmetrical* airfoil has identical chord and camber lines.

[7] Wing-tip vortices refer to the rotary flow due the leakage around wing tips of high-pressure air from the lower surface to the lower-pressure region of the upper surface. Due to this leakage of low, there is a loss of lift, because of reduction in the pressure difference between the lower and upper surfaces, when compared to a wing of infinite span. There is also an increase in the lift-dependent drag due to wing tip vortices, because the wing must operate at a higher angle of attack to overcome the loss in lift.

attack, a stable aircraft must have $C_{m0} > 0$, which is acheived through the presence of a horizontal tail (or canard). The pitch stability of an airplane is determined by the sign of C_{m_α}. For static stability, $C_{m_\alpha} < 0$, otherwise the angle of attack would diverge after even a small disturbance. The value of C_{m_α} is primarily influenced by the wing and the horizontal stabilizer (tail or canard).

All aerodynamic coefficients display a significant variation with the flight Mach number. In the range $0.3 < M < 0.8$, most lifting surfaces obey the Prandtl–Glauert similarity rule [22], which relates the incompressible and compressible *pressure coefficients*, $C_p \doteq (p - p_\infty)/q_\infty$, on an airfoil by

$$C_p = \frac{C_p|_{M=0}}{\sqrt{1-M^2}} \qquad (0.3 < M < 0.8). \tag{10.45}$$

Consequently, the lift-related coefficients $(C_{L_\alpha}, C_{L_0}, K)$ display an increase in magnitude with the subsonic Mach number. In the supersonic regime, a similar relationship holds between the pressure coefficient at $M = \sqrt{2}$ and that at a different supersonic Mach number according to *Ackeret's similarity rule*, given by

$$C_p = \frac{C_p|_{M=\sqrt{2}}}{\sqrt{M^2-1}} \qquad (M > 1.2). \tag{10.46}$$

Thus, the coefficients C_{L_α}, C_{L_0} undergo a decrease from their value at $M = 1.2$. In the transonic range, $0.8 < M < 1.2$, often a smooth curve is fitted between the subsonic and supersonic values, such that the maximum value of the lift-related coefficient occurs around $M \approx 1$. The Prandtl–Glauert and Ackeret rules are applicable to wings with moderate aspect ratio and sweep. For a highly swept wing of low aspect ratio (such as those found on some fighter aircraft), the lifting characteristics become nonlinear, wherein Eqs. (10.41)–(10.43) are invalid. Such a wing has flow that separates right at the leading edge, but re-attaches at a downstream location, thereby creating a *leading-edge vortex*. The leading-edge vortex appreciably increases the lift of a low aspect-ratio wing, and thus helps improve the flight performance. However, such vortex dominated flows are rather difficult to predict theoretically, and essentially require either an experimental investigation, or the solution of Euler's or Navier–Stokes equations. The *leading-edge suction analogy* analogy by *Polhamus* [26], however, gives a quick estimate of the nonlinear, vortex lift at a particular angle of attack. Leading-edge suction (LES) is a thrust force acting on the leading edge of a round-nosed airfoil due to the streamline curvature around the leading edge. While leading-edge suction occurs only for well-rounded leading edges at small to moderate sweep, its loss in a sharp, highly swept leading edge results in the vortex lift equal in magnitude to the loss of LES.

The lift-dependent drag factor is especially difficult to model at supersonic speeds, due to its dependence upon leading-edge suction. Generally, K increases significantly for $M > 1.2$ due to a drastic reduction in the Oswald's efficiency, e. The value of e, however, is difficult to predict at supersonic speeds. The alternative semi-empirical method based upon LES allows for the variation of K with Mach number and lift coefficient, C_L. A 100% LES would completely prevent flow separation and yield the maximum Oswald's efficiency ($e = 1$). However, a thin airfoil with a sharp leading edge would have almost no LES, for which the flow separates right at the leading edge, forming a leading-edge vortex, and thus the largest possible value of lift-induced drag, for which $K = 1/C_{L_\alpha}$. Since the actual amount of LES would lie somewhere between these two extremes, we can write

$$K = sK_{100} + (1 - s)K_0 , \tag{10.47}$$

where s is the actual fraction of LES, $K_{100} = 1/(\pi A)$ and $K_0 = 1/C_{L_\alpha}$. Leading-edge suction, s, primarily depends upon the Mach number normal to the leading edge. A sonic or supersonic leading edge cannot have LES, and thus has $s = 0$. For a subsonic leading edge, s depends upon the leading-edge radius, r_{LE}, as well as the lift coefficient of the wing. This dependence of s upon C_L is difficult to predict and often requires either a computational (Euler's or Navier–Stokes) model, or an experimental study.

The variation of the parasite drag coefficient, C_{D_0}, with Mach number is mainly caused by the presence of shock waves. As a result, there is little variation in C_{D_0} below a certain subsonic *critical Mach number*, M_{cr}, which is defined as the flight Mach number for which local sonic flow first appears on the given object. For an airfoil, this happens on the upper surface, close to the point of maximum thickness. As the Mach number is increased beyond M_{cr}, the locally sonic flow quickly expands into a supersonic bubble, thereby causing a standing normal shock, which causes an increase in flow separation as well as a wave drag. The flight Mach number for which the shock wave first appears on the object, leading to an almost exponential rise in C_{D_0} with Mach number, is called the *drag-divergence Mach number*, M_{DD}. The value of C_{D_0} continues rising with the Mach number in the region $M_{DD} < M \leq 1$, reaching a maximum at $M \approx 1$. Thereafter, there is a decline in C_{D_0} with the Mach number, due to the movement of the normal shock to the leading edge and its transformation into an attached oblique shock. Such variation of C_{D_0} with the Mach number is depicted in Chapter 12 for a fighter aircraft and an atmospheric entry vehicle. Due to the transonic drag rise, many subsonic vehicles are unable to cross into the supersonic regime due to thrust limitations. This was the reason why many people considered sonic flight impossible before 1947. However, the peak C_{D_0} can be significantly reduced by a combination of slender bodies and thin lifting surfaces, such that the total cross-section area changes smoothly with the downseam station. This concept of reducing wave drag by a smooth volume distribution is known as *area ruling*, and is employed in designing supersonic vehicles that must not only cross the transonic region but also cruise at $M > 1$. Examples of such vehicles include the Anglo-French *Concorde*, the American *SR-71*, and many supersonic cruise missiles (such as the Indo-Russian *Brahmos-3*).

Subsonic and supersonic aerodynamic characteristic of commonly employed shapes have been thoroughly investigated and carefully documented in texts and reports [27], [28]. Consequently, it has become possible to conceptually design vehicles based solely on semi-empirical relations and graphical data. However, an accurate prediction of aerodynamic force and moment still require experimental investigation and computational analysis based upon the governing Euler or Navier–Stokes equations.

When crossing into the hypersonic regime, we find a lack of empirical data, due to the complex aerothermochemistry prevailing at those speeds. The hypersonic region is often considered to begin at $M \approx 5$ and is dominated by phenomena absent at lower speeds. It is clear from the inviscid oblique shock relations that the shock-wave angle, β, becomes small for a given value of deflection angle, θ, and approaches $\beta \approx 1.2\theta$ for undissociated air. Due to the vicinity of the shock wave to the body surface, strong *entropy gradients* and *viscous interactions* with the appreciably thickened boundary layer take place, leading to complex flow patterns and a rise in temperature and pressure. These effects, combined with the temperature rise across the strong shock wave, lead to *high-temperature effects*, which are responsible for exciting the vibrational motion, and ultimately cause their dissociation and ionization. Due to these endothermic phenomena, a further rise in the temperature is prevented, but the flow field becomes enormously complicated, requiring multispecies continuity and energy equations for modeling the chemical reactions, as well as nonperfect, real gas effects in the momentum equations. Furthermore, the high speed causes the chemical reactions to occur with different rates in different directions, leading to a *chemical nonequlibrium*. At the large speeds of atmospheric entry vehicles ($M > 15$), the flow of ionized gases (*plasma*) leads to electric currents, and the associated electromagnetic considerations bring in additional nonlinear, governing partial differential equations (*Maxwell's equations*), which must be solved in tandem with either the Navier–Stokes or Boltzmann equation, depending upon the Knudsen number of the flow. Consequently, much of aerothermodynamic and plasmadynamic modeling required in very high-speed flows falls in the realm of physics rather than aerodynamics.

The hypersonic continuum regime in the range $5 \leq M \leq 10$ is often amenable to the *Newtonian approximation*, which predicts the pressure coefficient on an elemental *windward* surface inclined at an angle θ to the oncoming flow

$$C_p = 2\sin^2\theta \,. \tag{10.48}$$

The *leeward* side of the same surface does not experience any molecular collisions, and thus has $C_p = 0$. This flow model based upon particle impacts, wherein the entire normal momentum of the impinging gas is absorbed by the solid surface, was proposed by Newton to (incorrectly) model the low-speed flow, but is quite accurate in the continuum hypersonic regime. In the limiting case of $M \to \infty$, the oblique shock relations with $\gamma = 1$ yield the same expression as the Newtonian approximation. Also, it is interesting to note that the Newtonian approximation can also be obtained from the free-molecular pressure distribution in the limit $M \to \infty$, with zero tangential momentum transfer ($\Delta\tau = 0$). While the Newtonian model predicts a zero shear stress, we can employ a suitable boundary-layer approximation based upon semi-empirical relations [29], in order to estimate the shear stress on the surface, in combination with the Newtonian approximation. When integrated over a given surface, the pressure and shear-stress distribution yield aerodynamic force and moment coefficients with a reasonable accuracy. A modification of the Newtonian approximation is used for an object with a blunt nose, due to the essentially normal shock prevailing near the stagnation point, $\theta = 0$. The normal shock relations [Eq. (10.32] result in

$$C_p = C_{p_o} \sin^2\theta \ , \tag{10.49}$$

where[8]

$$C_{p_o} \doteq \frac{p_{02} - p_1}{\frac{1}{2}\gamma p_1 M_1^2} = \frac{2}{\gamma M_1^2}\left(\frac{p_{02}}{p_1} - 1\right) . \tag{10.50}$$

The modified Newtonian pressure distribution compares well with experimental measurements, especially near the stagnation region. The success of both Newtonian and modified Newtonian approximations in modeling hypersonic flow is entirely due to the extreme vicinity of the shock wave to the body, which results in a momentum exchange approximately normal to the surface.

10.5 Summary

Aerodynamic force and moment vectors arise out of pressure and shear-stress distribution on the external surface of a flight vehicle. While all shapes immersed in a flow experience drag, generating large amounts of lift and moment components, flattened lifting surfaces—such as the wings and tails of an airplane and fins of a rocket—are necessary. Flight efficiency through the atmosphere is improved by maximizing the lift-to-drag ratio. The most efficient design for an atmospheric vehicle is a streamlined shape with a small external surface area, which allows a fluid particle to follow the surface to the largest possible extent, but without causing excessive skin friction. The arrangement of lifting and control surfaces should be such that moment equilibrium (trim) can be achieved in all possible flight conditions for which the vehicle has been designed. A review of basic fluid mechanics and thermodynamics is necessary for understanding the aerodynamic force, moment, and thermal gradient. The conservation principles of mass, momentum, and energy conservation form the

[8] Note that in a compressible flow, the dynamic pressure can be expressed as

$$q = \frac{1}{2}\rho v^2 = \frac{1}{2}\gamma p M^2 \ .$$

basis of fluid mechanics and thermodynamics. According to the nature of the flow a vehicle experiences, certain simplifying assumptions are employed in the governing differential equations and their boundary conditions, resulting in specific flow regimes. Each flow regime is classified according to the range of important nondimensional parameters, such as the Knudsen, Mach, and Reynolds numbers. Useful modeling assumptions in continuum flow include the concepts of boundary layer and shock wave, whereas in the rarefied regime the free-molecular flow and bridging relations for transition flow are valuable approximations. Certain nondimensional aerodynamic force and moment coefficients (useful in flight dynamic calculations) are derived from the vehicle's geometry and the governing flow parameters in each regime, using either empirical and computational models or suitably scaled experimental results.

Exercises

10.1. Estimate the indicated and true airspeed of an airplane flying straight and level at a standard altitude of 3 km if the pitot-static system on the airplane registers the stagnation and static pressures of $p_0 = 7.2 \times 10^4$ N/m^2 and $p = 6.95 \times 10^4$ N/m^2, respectively. What is the Mach number of the airplane?

10.2. The velocity profile in the normal direction, y, in an incompressible boundary layer at a streamwise location on an airplane wing is given by

$$\frac{v}{v_e} = 1 - \left(1 - \frac{y}{\delta}\right)^2 ,$$

where v_e is the flow speed at the edge of the boundary layer, and δ is the boundary-layer thickness. Derive an expression for the skin friction coefficient on the wing surface at the given location.

10.3. Suppose the airplane in Exercise 10.2 is flying at a standard altitude of 5 km, and the specified streamwise location is at $x = 25$ cm from the leading edge, where $v_e = 85$ m/s and $\delta = 0.3$ cm. Calculate the surface shear stress and the local Reynolds number at the given point, assuming the coefficient of viscosity at the concerned altitude is $\mu = 1.628 \times 10^{-5}$ N.s. How does the shear stress compare with the equivalent flat-plate models for laminar and turbulent boundary layers?

10.4. Write the Navier–Stokes equations for an incompressible, two-dimensional flow past a circular cylinder of radius r_0. Express the equations in the polar coordinates, (r, θ), where r is the radial distance of the fluid element from the center, and θ is the angle measured from the relative flow direction.

10.5. Answer either *true* or *false* for each of the following:
(a) An adiabatic flow is also an isentropic flow.

(b) An isentropic flow is also an adiabatic flow.
(c) Energy in a nonisentropic flow can be conserved.
(d) Stagnation temperature in a nonisentropic process can be a constant.

10.6. Repeat Exercise 10.1 if the stagnation and static pressures measured by the pitot-static system are $p_0 = 7.6 \times 10^4$ N/m^2 and $p = 6.8 \times 10^4$ N/m^2, respectively, assuming
(a) subsonic flow
(b) supersonic flow.

10.7. Using the program *atmosphere.m* (Chapter 9), estimate the thickness of a normal shock wave at standard sea level and 100 km altitude, assuming that the thickness is equal to five molecular mean free-path lengths.

10.8. Calculate the static pressure, temperature, density, and Mach number behind an oblique shock wave on a wedge of semi-vertex angle 15° flying at a Mach number 2.3 at a standard altitude of 12 km. What is the minimum Mach number of the wedge at which the shock wave will remain attached to it?

10.9. Using the shock-expansion method, derive the expressions for the lift and wave drag coefficients on a flat plate flying at a supersonic Mach number and a small angle of attack.

10.10. For a sharp, right circular cone of semivertex angle θ, derive an expression for the total drag coefficient at zero angle of attack in the free-molecular flow regime. Compare the derived expression with that in the continuum hypersonic regime using the Newtonian approximation.

10.11. For a sphere, derive an expression for the total drag coefficient in the free-molecular flow regime. Compare the result with that obtained in the continuum hypersonic regime using the modified Newtonian approximation.

11

Airbreathing Propulsion

11.1 Aims and Objectives

- To introduce elements of airbreathing propulsion from the viewpoint of flight dynamic modeling of the thrust vector and the rate of fuel consumption.
- To discuss the characteristics and operational limitations of airbreathing engines—such as piston propeller, turbine, and ramjet—in a comprehensive and self-contained manner.
- To present a numerical model for the thrust and specific fuel consumption of a low-bypass turbofan engine to be utilized in subsequent simulation examples of airplane trajectories.

Airbreathing propulsion is the name given to the means employed for generating thrust, using the atmosphere as a medium as well as the source of oxygen for combustion. Most atmospheric vehicles employ airbreathing propulsion due to its higher efficiency compared with rocket propulsion (Chapter 8). Airbreathing engines can be subdivided into two broad categories: those that employ *propellers* for thrust generation, and those creating thrust through *jet* exhaust of gases. Both propeller and jet engines "push" the atmosphere "backwards" in order to generate thrust force in the forward (flight) direction by Newton's third law of motion (Chapter 4). We begin the analysis of airbreathing engines by modeling them as ideal momentum exchange devices.

11.2 Ideal Momentum Theory

We can understand the thrust generation by both propeller and jet engines by simply modeling them as mechanisms of pure momentum (and energy) transfer, without any losses caused by friction, flow rotation, and heat transfer. Such an ideal model assumes air to be a perfect gas flowing at a speed small

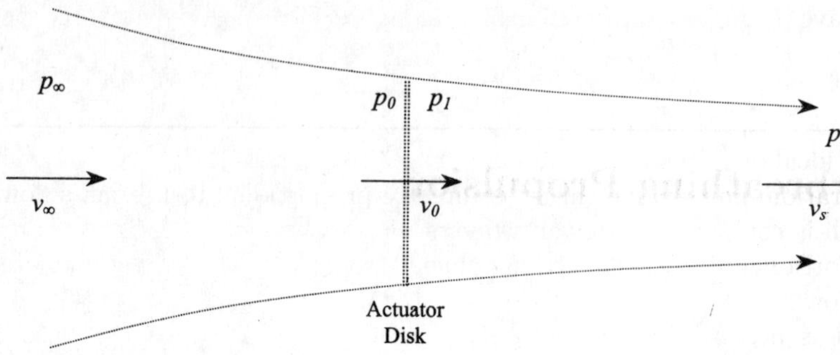

Fig. 11.1. Flow through an ideal airbreathing engine modeled as actuator disk.

enough for the flow to be treated as being incompressible (ρ = constant), and the engine is regarded as an *actuator disk* of small thickness that causes a sudden increase in the static pressure of the relative air flow. This situation is depicted in Fig. 11.1, where an actuator disk of cross-sectional area A is placed in a free stream of speed, v_∞, and static pressure, p_∞. As the flow approaches the disk, it accelerates to a speed v_0 near the disk. Immediately upstream of the disk, the static pressure, p_0, is obtained using the Bernoulli equation (Chapter 10) as follows:

$$p_0 = p_\infty + \frac{1}{2}\rho(v_\infty^2 - v_0^2) \,. \tag{11.1}$$

Since $v_0 > v_\infty$, it follows that $p_0 < p_\infty$. The disk imparts linear momentum to the flow such that its static pressure immediately downstream of the disk increases to $p_1 > p_0$. Far downstream of the disk, the flow must expand to the prevailing atmospheric pressure, p_∞, leading to a *slipstream speed*, v_s. Hence, the static pressure p_1 is related to the atmospheric pressure by another application of the Bernoulli equation downstream of the disk:

$$p_1 = p_\infty + \frac{1}{2}\rho(v_s^2 - v_0^2) \,. \tag{11.2}$$

The thrust the disk experiences is simply

$$f_T = A(p_1 - p_0) = \frac{\rho A}{2}(v_s^2 - v_\infty^2) = \rho A\frac{(v_s + v_\infty)}{2}(v_s - v_\infty) \,. \tag{11.3}$$

However, Newton's second law of motion also relates the thrust to the net rate of change of linear momentum of the fluid passing through the disk, or

$$f_T = \dot{m}(v_s - v_\infty) \,, \tag{11.4}$$

where $\dot{m}$ is the mass flow rate of the fluid, given by

$$\dot{m} = \rho A v_0 \,. \tag{11.5}$$

Upon substitution of Eq. (11.5) into Eq. (11.4), and comparison with Eq. (11.3), we have the important relationship among the flow speed at the three stations:

$$v_0 = \frac{(v_s + v_\infty)}{2} \,. \tag{11.6}$$

This ideal relationship gives an important insight into the mechanism of thrust generation, namely the thrust is directly proportional to the mass flow rate, which is constant for a given value of v_0. Furthermore, the thrust is directly proportional to the net speed increment imparted by the engine to the relative air flow.

It is necessary to define a *propulsive efficiency*, η_p, as the ratio of the *thrust-power* output,[1] $f_T v_\infty$, to the *total power* delivered by the engine. It is important to note that the total power of the engine in the ideal case is the sum of the thrust power and the *power lost* as the kinetic energy of the slipstream, $1/2\dot{m}(v_s - v_\infty)^2$. Therefore, the ideal propulsive efficiency is given by

$$\begin{aligned} \eta_p &\doteq \frac{f_T v_\infty}{f_T v_\infty + \frac{1}{2}\dot{m}(v_s - v_\infty)^2} \\ &= \frac{2v_\infty}{v_\infty + v_s} \,. \end{aligned} \tag{11.7}$$

The propulsive efficiency is a measure of the success in converting the total mechanical power delivered by the engine into thrust. Clearly, η_p decreases with an increasing value of the slipstream speed, v_s. Hence, from Eqs. (11.5) and (11.7), we see that the propulsive efficiency is maximized by increasing the mass flow rate for a given thrust, which minimizes v_0 (thus, v_s). A propeller engine with a large propeller (thus, a large flow area) delivers its thrust at the maximum possible propulsive efficiency, due to its smallest slipstream velocity. However, large propellers have serious limitations in terms of flight speed, v_∞, as discussed ahead. Furthermore, the propulsive efficiency is an incomplete criterion. All engines develop their mechanical power through a combustion of fuel. Hence, the *overall efficiency*, η_o, defined as the ratio of thrust power output and the energy spent in burning fuel per unit time, is a much better criterion for an engine's efficiency and is expressed as follows:

$$\eta_o \doteq \frac{f_T v_\infty}{\dot{m}_f h_f} \,, \tag{11.8}$$

where $\dot{m}_f$ is the mass flow rate of fuel through the engine, and h_f is the heat developed by a unit mass of fuel, called the *heating value* (or *calorific value*) of the fuel. The overall efficiency is expressed as the product of η_p and the *thermal efficiency*, η_t, defined as the ratio of mechanical power developed and the heating rate by fuel combustion,

[1] Thrust power is the rate of work done by the thrust as the vehicle moves forward through the atmosphere at a constant speed, v_∞.

$$\eta_t \doteq \frac{f_T v_\infty + \frac{1}{2}\dot{m}(v_s - v_\infty)^2}{\dot{m}_f h_f} = \frac{P_{\text{esh}}}{\dot{m}_f h_f}, \tag{11.9}$$

where $P_{\text{esh}} \doteq f_T v_\infty + \frac{1}{2}\dot{m}(v_s - v_\infty)^2$ is the total mechanical power developed, called the *equivalent shaft power*, and is the sum of thrust power and the power lost in accelerating the flow. In the actual case, there is an additional term in the expression for P_{esh} representing the power lost as the rotational kinetic energy of the slipstream. When the flight Mach number is greater than about 0.3, the flow compressibility, a smaller pressure difference is developed across the propeller due to the loss of energy in compressing the flow, leading to a decrease of thrust from the incompressible value with a given shaft power and, thus, a decline in the propulsive efficiency.

The net mass flow through the engine is a sum of fuel and air mass flows, $\dot{m} = \dot{m}_f + \dot{m}_a$. Therefore, the ideal thermal efficiency can be approximated as follows, in terms of the *fuel–air ratio*, $f \doteq \dot{m}_f/\dot{m}_a \ll 1$:

$$\eta_t \approx \frac{v_\infty(v_s - v_\infty)}{f h_f}. \tag{11.10}$$

The thermal efficiency can be increased for a given fuel by decreasing the fuel mass per unit equivalent shaft power, called the *power-specific fuel consumption*. However, with the available fuels, thermal efficiency is restricted to $\eta_t \leq 0.3$ in the earth's atmosphere. Hence, for improving the overall efficiency, we are left with maximizing the propulsive efficiency.

11.3 Propeller Engines

A propeller is a mechanical device for converting the mechanical power developed by the engine into thrust, with the use of a set of rotating blades that produce lift normal to the plane of rotation. By placing the propeller's plane normal to the flight path, the net lift of all the blades translates into a thrust for the vehicle. The propeller is closest to the actuator disk idealization due to its relatively thin work volume. However, the slipstream rotation a propeller causes is usually large and often represents an appreciable portion of the lost mechanical power.

In the early days of aviation (1903–1950), an *internal combustion* (or *piston*) engine was the only practical means available of powering a propeller. A piston engine works on a closed thermodynamic cycle, wherein the reciprocating motion of the piston is timed through a camshaft to open and close the intake and exhaust valves as well as to compress fuel–air mixture prior to a rapid combustion. The reciprocating motion of the piston powered by combustion is converted into a rotary motion of the propeller (crank) shaft by connecting rods. Such an engine often consists of multiple cylinders, and the piston of each cylinder is synchronized with those of the other cylinders through sequential ignition of the fuel–air mixture inside each cylinder, thereby producing a smooth power delivery to the propeller. The piston engines are classified into two distinct categories according to the type of thermodynamic cycle: (a) *petrol* (or *gasoline*) engines based upon the *Otto cycle*, and (b) *diesel* engines based upon the *Diesel cycle*. The petrol engines are further classified into *two-stroke* or *four-stroke* engines, depending upon the number of distinct strokes, or piston movements, occurring inside the cylinder during one complete revolution of the shaft. A four-stroke engine has distinctly defined intake, compression, power (combustion), and exhaust strokes, whereas in a two-stroke engine the intake and compression are carried out simultaneously in a stroke, and combustion and exhaust are combined into the other stroke. Other classifications arise out of the number of cylinders per engine and their arrangement (*radial, in-line, V*, or *rotary*).

The mechanical power delivered to the propeller shaft, P_{sh}, is directly proportional to the number of revolutions per second, n. For a given value of n, P_{sh} is directly proportional to the *mean-effective pressure*, p_m, acting on the piston's face during the power stroke. The mean effective pressure, in turn, is roughly proportional to the *intake manifold pressure*, which is the static pressure at the intakes of all the cylinders connected together by a common vent, called an *intake manifold*. An engine with the intake manifold directly open to the atmosphere is said to be *normally aspirated*. Clearly, the intake manifold pressure of a normally aspirated engine is approximately equal to the atmospheric pressure prevailing at a given altitude. Consequently, a normally aspirated engine sees its power drop off almost in direct proportion with the atmospheric pressure, p_∞. Airplanes powered by such engines have their maximum operating altitude (*ceiling*) limited to about 5 km. In order to improve the power at any given altitude, a *turbo-charger* is often employed, which is a small turbine-driven pump for increasing the air pressure in the intake manifold. With the use of a turbo-charger, the piston-powered airplane's performance is significantly boosted, leading to ceilings of about 11 km. The fighters, bombers, and airliners of the 1940—1950 period were usually equipped with large turbo-charged (or *supercharged*) piston engines. The variation of P_{sh} with altitude is usually linear for a normally aspirated engine, and piecewise linear for the same engine with a turbo-charger, as shown in Fig. 11.2. As pointed out earlier, P_{sh} is also proportional to the thermal efficiency, η_t, of the engine. The nonisentropic process by which P_{sh} is developed includes the heat lost during combustion, as well as frictional losses due to the engine's moving parts. The size of a piston engine is measured in terms of the total volume displaced by the movement of the pistons in all the cylinders. As the displacement volume in each cylinder increases, the net heat loss during combustion also increases. This indicates that the thermal efficiency can be improved by employing a larger number of smaller cylinders, for a given total power. However, as the number of cylinders in an engine increases,[2] the

---

[2] The giant *Wright R-3350 "Cyclone"* engine of the *Lockheed Super-Constellation* airliner of the 1950s had as many as 18 cylinders per engine.

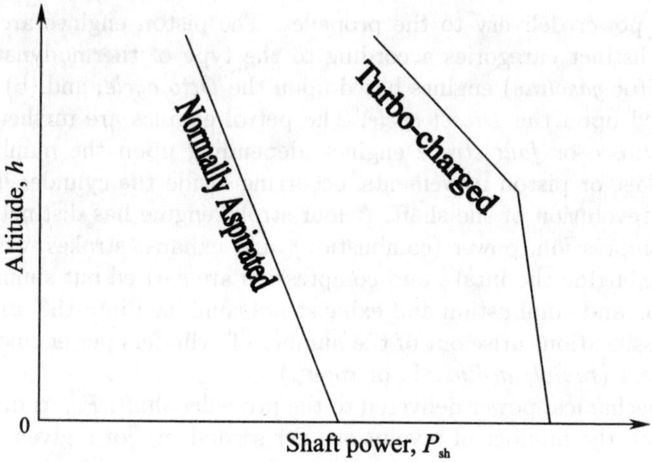

Fig. 11.2. Variation of shaft power of a piston engine with altitude.

frictional losses also increase, thereby providing an upper limit to the thermal efficiency.

The thrust developed by a propeller is usually expressed in terms of the propeller's propulsive efficiency, η_p, as follows [Eq. (11.7)]:

$$f_T = \eta_p \frac{P_{sh}}{v_\infty} . \tag{11.11}$$

Since there is no change in the shaft power of the internal combustion engine with flight speed, the thrust the propeller delivers is inversely proportional to v_∞, provided the propulsive efficiency remains constant. The angle made by the blade airfoil's chord line with the plane of rotation is called the *blade-pitch angle*, β. The propeller blades are twisted such that β is a function of the radial location, r. This is necessary in order that the airfoils at all radial locations receive the airflow, which has a varying relative speed (thus, angle of attack) from the root to the tip, at roughly the same angle of attack [Fig. 11.3(a)]. Therefore, a twisted propeller blade can be regarded as a rotating wing, whose lift is approximately proportional to the angle of attack up to the stall condition. The angle of attack at a given radial location can be approximated by $\alpha \approx \beta - v_\infty/(nr)$, where n is the angular speed of the propeller. A propeller with a fixed twist distribution, $\beta(r)$—called a *fixed-pitch propeller*—has η_p varying with the flight speed in a nonlinear fashion, such that the maximum thrust is produced for a specific value of the ratio, v_∞/n. Consequently, the propulsive efficiency is usually plotted as a function of the nondimensional *advance ratio*, $J \doteq v_\infty/nd$, as depicted in Fig. 11.3(b), where d denotes the propeller diameter. The figure depicts a *variable-pitch propeller*, in which $\beta(r)$ can be varied in-flight through mechanical adjustment, such that maximum efficiency occurs at different advance ratios. For airplanes having a

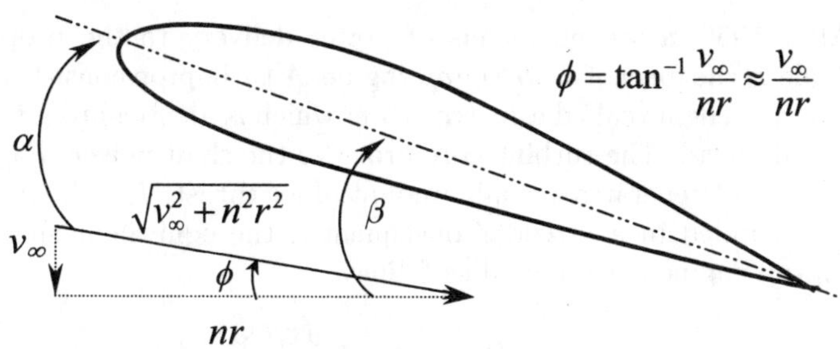

(a) A typical blade element.

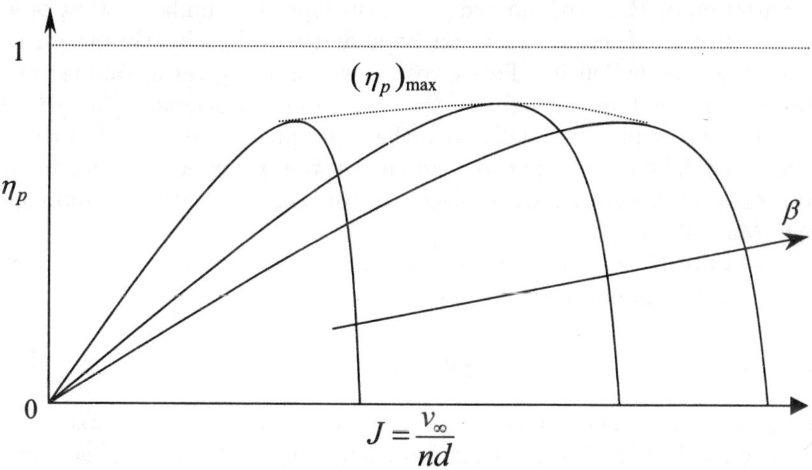

Fig. 11.3. Propeller pitch, advance ratio, and optimum propulsive efficiency.

significantly large speed range, a closed-loop control system is often employed, which maintains a constant angular speed, n, and automatically adjusts the blades to their optimum pitch distribution depending upon the prevailing flight speed (thus, advance ratio). Such an automatic mechanism is called a *constant-speed propeller*. Clearly, a constant-speed propeller always operates close to its maximum efficiency, and thus η_p can be regarded as being constant, irrespective of the flight speed. The design and analysis of a propeller blade for its propulsive efficiency require a modeling of the aerodynamics of a typical blade element and are called *blade-element theory*. Typically, the maximum propeller efficiency lies in the range 0.8–0.85. At very low speeds, such as during take-off, the thrust a propeller develops is termed the *static propeller thrust* and cannot be obtained from the efficiency vs. advance-ratio charts due to the singularity in Eq. (11.11) at zero speed. Separate charts are provided by the manufacturer for the variation of the static thrust with the shaft power, usually in the nondimensional coefficient form.

After 1950, a second means of power delivery to the propeller became available in the form of a *turboprop* engine. A turboprop consists of an external combustion engine called a *turbine core*, which is identical to a *turbojet* engine discussed ahead. The turbine core provides the shaft power, P_{sh}, to drive the propeller and generates a small amount of jet thrust, f_{T_j}, though the exhaust gases expanded in a nozzle. Consequently, the equivalent shaft power of a turboprop engine is expressed as follows:

$$P_{\text{esh}} = P_{\text{sh}} + \frac{f_{T_j} v_\infty}{\eta_p} . \tag{11.12}$$

In a typical case, the jet thrust is about 10–15% of the total turboprop thrust. The variation of P_{esh} with altitude for a turboprop is similar to that of a jet engine's thrust. Generally, P_{esh} can be assumed to be directly proportional to the atmospheric density. This gives a turboprop a great advantage over a normally aspirated piston engine of the same power rating, since the latter has its power lapsing proportionally with the atmospheric pressure. A turboprop is also much lighter compared to a similarly rated piston engine. Due to these advantages, turboprops have replaced piston engines in the medium-speed range (Mach 0.4–0.7).

A measure of the thermal efficiency of a propeller engine is its power-specific fuel consumption (PSFC), given by

$$c_P = \frac{\dot{m}_f}{P_{\text{esh}}} . \tag{11.13}$$

An engine with a smaller value of c_P consumes less fuel for each horsepower of P_{esh} developed and is thus more efficient. Generally, turboprop engines achieve much smaller PSFC than piston engines and are thus regarded as being more fuel efficient. When combined with their smaller size (and weight) per unit power, this makes turboprop engines an attractive choice for powering low- to medium-speed airplanes. A variation of the turboprop is the *turboshaft* engine, typically used in rotorcraft, wherein no direct jet thrust is developed; therefore, $P_{\text{esh}} = P_{\text{sh}}$.

A propeller suffers from the effects of compressibility at high flight speeds, which typically causes a resultant sonic flow at the tips, even when the flight Mach number is less than 0.7. The sonic flow at the tips plays havoc with the propeller efficiency and also creates serious structural vibration problems, often leading to outright structural failure of the blades. Consequently, the operation of propeller airplanes is generally restricted to $M < 0.7$.[3]

[3] The Russian *Tu-95* bomber and its *Tu-142* maritime reconnaissance version designed in the 1950s are the fastest operational propeller airplanes, with a flight Mach number of 0.87 at 40,000 ft. altitude. This speed was achieved through swept-back wings and a special propeller design, in which each engine drives two, four-blade, counter-rotating propellers. The aircraft is powered by four legendary *Kuznetsov NK-12* turboprops (the largest ever built), each generating a maximum sea level equivalent shaft power of about 15,000 HP.

11.4 Jet Engines

At flight speeds above Mach 0.7, jet engines are usually the only option for powering airplanes. A jet engine is an external combustion engine, where the inlet airflow is compressed, mixed with fuel and burnt, and the exhaust gases are expanded in a nozzle to produce thrust in a manner similar to a rocket engine. The jet engines are classified into two broad categories: (a) the *ramjets* and (b) *turbine* engines. A ramjet is operated at a flight speed sufficiently high for direct compression of air in a diffuser. Consequently, it does not require a mechanical compressor and the associated moving parts. A turbine engine, however, is capable of operation even at zero speed, because it contains a mechanical compressor to carry out the necessary compression before combustion. The compressor is rotated using a turbine, which absorbs a part of the kinetic energy of the exhaust and converts it into rotary motion like a pinwheel. The turbine engines are further classified into *turbojet*, *turbofan*, *turboprop*, and *turboshaft*, all of which share the common *turbine core* consisting of a compressor, combustion chamber, and turbine, but have additional features. We briefly introduced the turboprop and turboshaft, and now we will describe the remaining jet engine variants.

A typical jet engine departs from the ideal actuator disk assumption in that the flow undergoes an appreciable acceleration over the finite length of the engine. Consequently, the thin disk approximation, Eq. (11.4), is replaced by the following *jet equation*, which allows the exhaust static pressure, p_e, to be different from the atmospheric pressure:

$$f_T = \dot{m}(v_e - v_\infty) + A_e(p_e - p_\infty) , \tag{11.14}$$

where v_e and A_e refer to the relative flow speed and area in the exit plane of the nozzle. The jet engines usually have underexpanded nozzles (Chapter 8) ($p_e > p_\infty$), hence a small amount of thrust is added due to the exhaust pressure difference (provided a variable geometry nozzle is employed to allow the expansion to occur inside rather than outside the nozzle).

A jet engine's overall efficiency is indicated by its *thrust-specific fuel consumption* (TSFC), defined by

$$c_T = \frac{\dot{m}_f g_0}{f_T} , \tag{11.15}$$

where $g_0 = 9.8$ m/s^2 is the acceleration due to gravity at standard sea level. The standard unit of TSFC is 1/hr. Clearly, a smaller value of TSFC represents a more efficient engine.

11.4.1 Ramjet Engines

A ramjet is the simplest of all engines in construction. It consists of a long, hollow tube, with an intake and diffuser at one end and a nozzle at the other

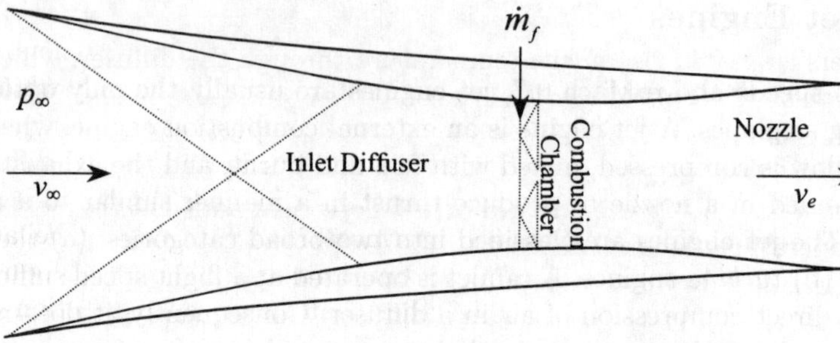

Fig. 11.4. A supersonic ramjet engine.

(Fig. 11.4). In between lies the combustion chamber with a suitable apparatus for efficiently mixing and burning the fuel with the rapidly flowing air at nearly constant pressure. Some have likened the design of a ramjet combustion chamber to an attempt to light a candle in the middle of a wind-storm. However, the problem of combustion is less severe at subsonic speeds when compared to that at supersonic speeds. Considering a subsonic ramjet first, we can express the isentropic diffusion process by the following pressure rise within the inlet (Chapter 10):

$$\frac{p_0}{p_\infty} \approx \left(1 + \frac{\gamma - 1}{2} M_\infty^2\right)^{\frac{\gamma}{\gamma-1}} . \tag{11.16}$$

Here we assume that the rise in the pressure in the diffuser is large enough for the flow to be slowed down to an incompressible speed $M \approx 0$ at the end of the compression process. The ram compression is quite poor ($\frac{p_0}{p_\infty} \leq 1.28$) below Mach 0.6, which indicates that a ramjet needs to be flown at $M > 0.6$ just to start its operation, which typically requires $\frac{p_0}{p_\infty} > 1.3$. Another disadvantage of the ramjet in subsonic operation is the almost inverse proportionality of the TSFC with Mach number at $M < 1$. The typical subsonic TSFC values of a ramjet lie between 3.5–5, but rapidly decline to less than half at supersonic flight speeds. Hence, a ramjet is ideally suited for supersonic operation, with the minimum TSFC occurring near $M = 3.5$.

In supersonic operation, we can no longer assume isentropic diffusion, but a staged compression via a series of oblique shocks in a carefully designed supersonic inlet. There are two options for the supersonic operation of a ramjet: (a) *subsonic combustion* and (b) *supersonic combustion*. The subsonic combustion requires a deceleration of the flow across a final normal shock wave to subsonic speed. Since a shock wave is considered as an adiabatic process (Chapter 10), we can express the stagnation temperature prior to combustion as follows using the energy equation:

$$T_0 = T_\infty \left(1 + \frac{\gamma - 1}{2} M_\infty^2\right) . \tag{11.17}$$

The energy equation indicates that for $M = 5$, the subsonic combustion ramjet encounters a six-fold rise in the temperature through the diffuser, which can be tolerated by the structure, since there are no moving parts. However, at larger Mach numbers, say at $M = 5.85$, the temperature rise in the diffuser is about eight-fold, which may cause structural failure in either the walls of the combustion chamber or the nozzle. Furthermore, heat added to the fuel at such high temperatures generally causes a decomposition of the hydrocarbon fuels, thereby leading to an endothermic (rather than exothermic) reaction and a negative thrust. For these reasons, subsonic combustion ramjets are limited in operation to below Mach 5. Therefore, in a hypersonic application, the only option available is a supersonic combustion ramjet (*Scramjet*) engine, which does not require a diffusion of the inlet flow to subsonic speeds, and thus avoids the high temperature prior to combustion. A Scramjet engine requires an efficient mixing and burning of fuel with a supersonic airflow passing through the combustion chamber. The design of a suitable combustion technology is, therefore, the crux of Scramjet development and has been an area of active research for the last two decades.[4]

The greatest utility of a ramjet is thought to be its use as a replacement for rocket engines in atmospheric flight, especially in hypersonic and single-stage to orbit (SSTO) launch operations. However, its practical use may be most efficient in a hybrid *turbo-ram-rocket* engine, which sequentially switches between turbine, ramjet, and rocket operations.

11.4.2 Turbojet and Turbofan Engines

A modern airplane requires efficient and versatile operation in the high subsonic to supersonic flight regime, which is best fulfilled by turbojet and turbofan engines. A revolution in the aeronautical technology, equal in magnitude to the invention of the airplane by Wright Brothers, occurred in the 1930s with the advent of the first turbojet-powered airplanes. As a result, the propeller engine was largely relegated to low-speed, general aviation aircraft and replaced by the turbojet engines in the military and civilian transports, fighters, and bombers at the end of the 1950s. The turbojet (shown in Fig. 11.5) is the primary jet engine consisting of a relatively lightweight and compact turbine core comprising an inlet diffuser, a turbine-driven compressor, a combustion chamber, and a nozzle (Fig. 11.5). Usually, the compressor and turbine consist of several stator and rotor stages in order to prevent the losses due to large flow angles. The operation of a turbojet is based upon the ideal *Brayton cycle* [31], which consists of an isentropic compression through the subsonic

[4] NASA's *X-43A* pilot-less, technology demonstrator experimental vehicle flew to $M = 9.6$ with a Scramjet engine in November 2004, which is currently the record for the fastest operation of an airbreathing vehicle. It was launched from a *Pegasus* rocket booster at 33 km altitude, and then sustained its speed by Scramjet operation for 10 seconds. The *Pegasus* rocket itself was launched from a *B-52* bomber at an approximate altitude of 12 km.

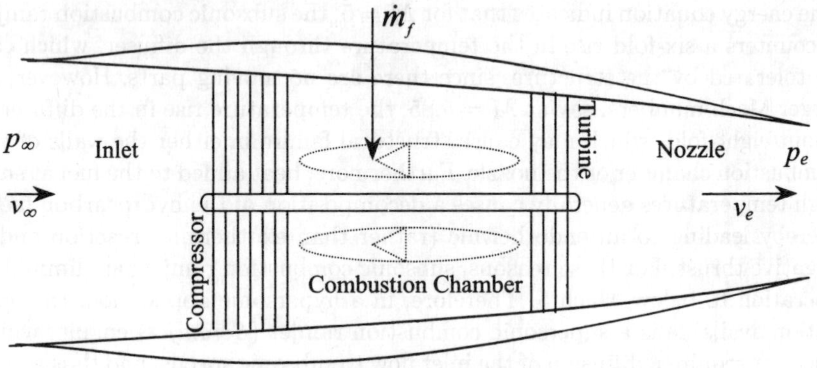

Fig. 11.5. A turbojet engine.

inlet and compressor, a constant pressure combustion, and an isentropic expansion through the turbine and nozzle. In supersonic operation, a supersonic inlet is added, which comprises either a two-dimensional ramp or a movable axisymmetric spike, in order to produce a series of oblique shock waves, culminating in a weak normal shock, which achieves the flow deceleration to subsonic speeds. Often, turbojets employ an additional burning of the fuel after the turbine, in an *afterburner* for boosting thrust during take-off, climb, or acceleration through the transonic regime. Turbojet engines can achieve a TSFC of 1–1.2/hr without afterburner, and 2–2.5/hr with afterburner. The almost doubling of TSFC in afterburner operation restricts its use for only brief intervals. In order to improve the overall efficiency of a turbojet, some airflow is bypassed outside the turbine core with the use of a *fan*, which helps in a simultaneous reduction of both speed and temperature of the exhaust gases. The resulting modification is termed a *turbofan* engine and is schematically depicted in Fig. 11.6. The ratio of the bypassed mass flow rate to that passing through the core is called the *bypass ratio*. Using a higher bypass ratio results in a smaller TSFC. Therefore, the modern airliners, such as the *Boeing-777* and *Airbus-A340*, utilize bypass ratios of 7–8, which reduces the TSFC to below 0.45/hr, and directly translates into an economical, long-range operation. However, a high-bypass ratio requires a large fan in front of the compressor, which creates problems associated with size, weight, and drag. Hence, a large bypass ratio is not feasible in fighter-type aircraft, which need a compact and lightweight powerplant. Consequently, the bypass ratios used in such airplanes do not exceed 1.5. In modern airplanes, turbojets have been replaced by turbofans, which, in turn, have evolved into lighter and more efficient powerplants through new materials and digital control for optimum fuel delivery. However, the turbojet and turbofan are generally limited to operation below Mach 3 due to the restriction arising out of *turbine-inlet temperature*. As the flight speed increases beyond $M = 3$, the temperature after the combustion chamber may become sufficiently high for causing structural failure in the

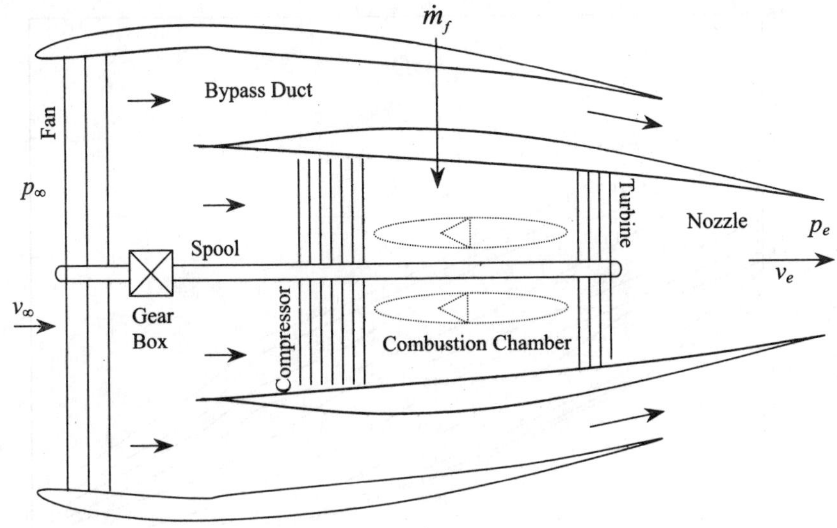

Fig. 11.6. A turbofan engine.

rapidly spinning turbine blades, which are under high thermal and centrifugal stress. Well before reaching this limit, a turbofan is already operating with an increased TSFC due to sonic flow at fan tips. Hence, a turbofan is typically restricted to flight Mach numbers less than about 2.5.

Both the turbojet and turbofan display a negligible variation of thrust with either flight speed or Mach number in normal subsonic cruise conditions. This is evident from the thrust equation, Eq. (11.14), which shows that an increase of speed leads to an increase of $\dot{m}$, but a decrease of $v_e - v_\infty$, which tends to render the thrust invariant with speed. However, there is an appreciable variation of the thrust with flight speed in the low-speed (static) limit, and with Mach number in supersonic operation (largely due to the supersonic inlet). Since the underexpanded nozzle produces only a relatively small thrust, an increase of altitude leads to the thrust declining in direct proportion with the atmospheric density due to the reduction in the mass flow rate, $\dot{m} = \rho A v$.

The turboprop can be derived from the turbofan by replacing the fan with a propeller and removing the outer wall of the bypass duct. Thus, the size of the propeller determines the bypass ratio. Clearly, a turboshaft, which has a negligible thrust due to jet exhaust, represents the limiting case of infinite bypass ratio.

While the ideal jet equation is useful in understanding how jet engines work, the actual operation may significantly depart from the ideal behavior due to losses in the diffuser, compressor, and nozzle. Consequently, detailed engine performance charts are necessary for accurately modeling a turbine engine. Such charts are usually provided by the manufacturer, or can be estimated by an actual cycle analysis [32].

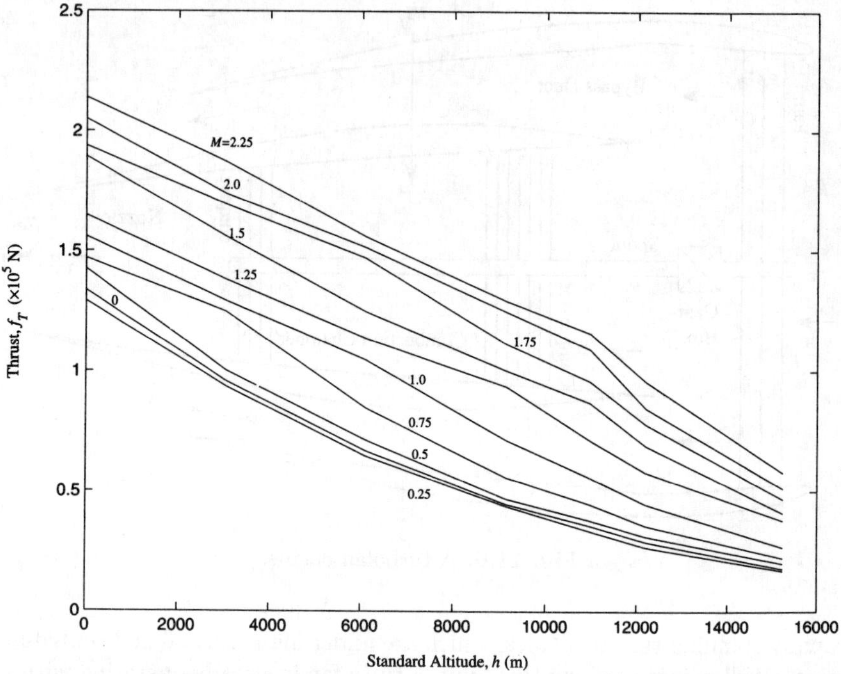

Fig. 11.7. Variation of maximum power thrust with altitude and Mach number for a low-bypass, afterburning turbofan.

An additional loss in thrust and an increase in specific fuel consumption occur when an engine is installed in an airframe. Such losses occur mainly due to inlet and nozzle design for matching the pressures with those expected in flight and are termed *installation losses*. They can be roughly estimated by semi-empirical expressions [33], but require detailed experimental tests for an accurate evaluation. Many an airplane has observed unsatisfactory performance due to unexpectedly high or poorly modeled installation losses.

Example 11.1. Consider a low-bypass turbofan engine with an afterburner [33]. The standard sea level static thrust of the engine with full power and afterburner (called *military power*) is $f_{T0} = 133,636.36$ N, while the TSFC at the same condition is $c_T = 1.64$/hr. The variation of the maximum thrust with altitude for specific Mach numbers is plotted in Fig. 11.7, while the equivalent plot of TSFC variation at maximum power is given in Fig. 11.8. Write a MATLAB program for calculating the thrust and TSFC at a specific pair of altitude and a Mach number by interpolation of the given engine performance data.

The necessary program is tabulated in Table 11.1 and uses data picked from the performance charts, Figs. 11.7 and 11.8, at selected values of altitude and Mach number.

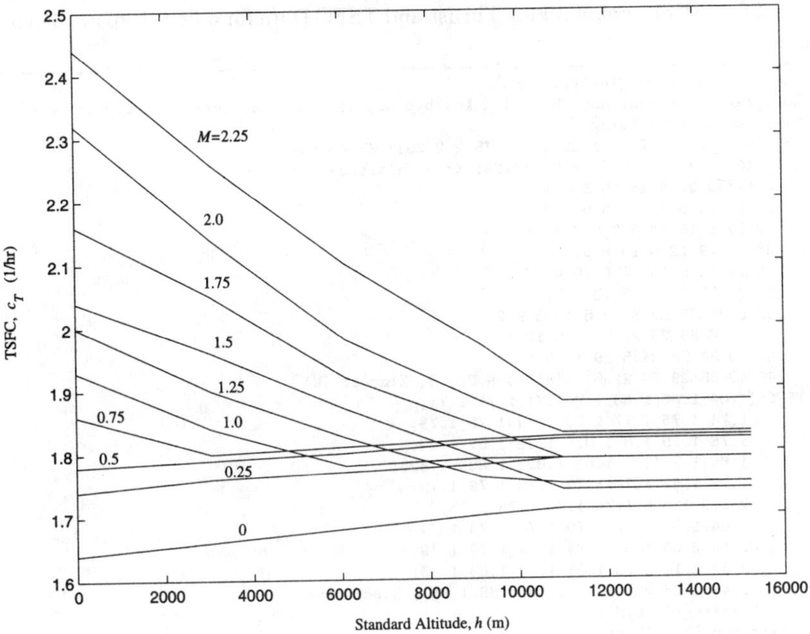

Fig. 11.8. Variation of maximum power TSFC with altitude and Mach number for a low-bypass, afterburning turbofan.

11.5 Summary

Airbreathing engines—propeller and jet—generate thrust by the reaction of exhaust gases against the atmosphere, and the ideal momentum theory gives an important insight into the thrust creation mechanism. Detailed models, however, are necessary to accurately represent the performance, efficiency, and operational characteristics (variation of thrust and efficiency with speed and altitude) of a specific engine. Propeller engines are most efficient at low subsonic speeds and up to moderate altitudes, while turbojet and ramjet engines achieve their maximum efficiency at high-subsonic and supersonic speeds, respectively, at altitudes near, or higher than, the tropopause. At hypersonic speeds, a supersonic combustion ramjet (or Scramjet) is the only practical airbreathing propulsion alternative. The efficiency of a turbine engine is increased by having a higher bypass ratio. A realistic engine model must include the installation losses and off-design performance, which are either obtained from experimental data (manufacturer's charts) or estimated by an empirical, actual cycle analysis.

Table 11.1. M-file *engine.m* for Thrust and TSFC Data for a Low-Bypass, Turbofan Engine.

```
function [T,cT]=engine(alt,mach)
%program for thrust and TSFC of a low-bypass, afterburning turbofan at
%maximum power setting
M=[0 0.25 0.5 0.75 1 1.25 1.5 1.75 2 2.25]; %Mach number
h=[0 10 20 30 36 40 50]*1000/3.28; %std. altitude (m)
Thrust=[30 21.5 15 10 8 6.5 4;
   29 21 14.5 9.8 7.5 6 3.8;
   32 22.5 16 10.5 8.5 7 4.5;
   33 28 19 12.5 10 8 5;
   35 29 23.5 16 12.5 10 6;
   37 31 25.5 21 16 13 8.5;
   42.5 35 28 22.5 19.5 15.5 9.2;
   43.5 38 33 25 21.5 17.5 10.5;
   46 39 34 28 24.5 19 11.5;
   48 42 35 29 26 21.5 13]*1000*9.8/2.2; %thrust (N)
TSFC=[1.64 1.66 1.68 1.7 1.71 1.71 1.71;
      1.74 1.76 1.77 1.78 1.79 1.79 1.79;
      1.78 1.79 1.8 1.815 1.82 1.82 1.82;
      1.86 1.8 1.81 1.82 1.825 1.825 1.825;
      1.93 1.84 1.78 1.79 1.79 1.79 1.79;
      2 1.9 1.825 1.76 1.75 1.75 1.75;
      2.04 1.96 1.87 1.79 1.74 1.74 1.74;
      2.16 2.05 1.92 1.84 1.79 1.79 1.79;
      2.32 2.14 1.98 1.88 1.83 1.83 1.83;
      2.44 2.26 2.1 1.97 1.88 1.88 1.88]; %(per hour)
  [X,Y]=meshgrid(h,M);
  T=interp2(X,Y,Thrust,alt,mach);
  cT=interp2(X,Y,TSFC,alt,mach);
```

Exercises

11.1. Compare the maximum equivalent shaft power of a piston engine and a turbine engine, both rated at the maximum sea level power of 2000 HP, when operating at a standard altitude of 11 km.

11.2. The ideal momentum theory can be applied to a helicopter's rotor blade in order to estimate the net thrust in a vertical flight. Assuming a steady, vertical ascent of the helicopter with a rotor diameter d, derive an expression for the ideal thrust. Also, derive the ideal thrust in a *hover* (stationary flight), assuming a flow speed, v_0, through the rotor, and a slipstream speed, v_s. What is the mass flow rate through the rotor of a *Bell 206B* helicopter with $d = 10.16$ m and mass 1450 kg, hovering at standard sea level?

11.3. Calculate the thrust of a turbojet engine with inlet and exit areas, 1.0 and 0.5 m^2, respectively, with pressure and relative flow speed at exit plane of 500 m/s and 22800 N/m^2, respectively, when flying at 11 km standard altitude at Mach 0.85.

11.4. Estimate the fuel mass flow rate of the engine in Exercise 11.3 if the TSFC at the given flight condition is 0.7 per hour.

11.5. The engine in Exercise 11.3 is to be modified by adding a fan, such that the TSFC at the given flight condition is reduced to 0.5 per hour, without changing the thrust, exit speed, and the fuel–air ratio of the turbine core. Estimate the bypass ratio of the modified engine, assuming that the exit speed of the bypass flow is 270 m/s and its density is roughly the same as the atmospheric density.

12

Atmospheric and Transatmospheric Trajectories

12.1 Aims and Objectives

- To derive the general equations of motion of translational flight in the planet-fixed frame from first principles in a systematic fashion. These equations govern the flight of all aerospace vehicles (airplanes, rockets, spacecraft, and entry vehicles).
- To present three-degree-of-freedom atmospheric flight models, including planetary form, rotation, aerodynamics, and propulsion.
- To simulate the important atmospheric and transatmospheric trajectories with detailed analytical insight into airplane flight, rocket ascent, and planetary entry.

12.2 Equations of Motion

Atmospheric flight is dominated by the presence of atmospheric forces, which can be divided into aerostatic and aerodynamic categories. While the aerostatic force does not require a special frame of reference, the aerodynamic force, with which we are primarily concerned, arises solely due to the motion of the vehicle relative to the atmosphere, and thus requires a frame fixed to the atmosphere. Since a planet's atmosphere rotates with it, we shall employ a planet-fixed reference frame for expressing the equations of atmospheric flight. Such a planet-centered, rotating frame ($SXYZ$), with axes represented by the unit vectors, $\mathbf{I}, \mathbf{J}, \mathbf{K}$, respectively, is shown in Fig. 12.1. We had employed a similar frame in Chapter 5 for describing orbital flight, using a set of spherical coordinates, r, δ, λ, denoting the radius, latitude, and longitude, respectively. Unlike Chapter 5, we will not use primed symbols here to denote the relative quantities. The velocity relative to the rotating frame, $\mathbf{v}$, can be expressed in terms of the the spherical coordinates v, ϕ, A, representing the relative magnitude, flight-path angle, and velocity azimuth, respectively, measured in the local horizon frame, $(oxyz)$, which has axes ox ($\mathbf{i}$), oy ($\mathbf{j}$), and oz ($\mathbf{k}$), along local vertical (up), local east, and local north, respectively (Fig. 12.1):

$$\mathbf{v} = v(\sin\phi\mathbf{i} + \cos\phi\sin A\mathbf{j} + \cos\phi\cos A\mathbf{k}). \tag{12.1}$$

The expression for the inertial velocity $\mathbf{v_I}$ is given by

$$\mathbf{v_I} = \mathbf{v} + \boldsymbol{\omega}\times\mathbf{r} = \mathbf{v} + \omega r\mathbf{K}\times\mathbf{i}\,, \tag{12.2}$$

where $\boldsymbol{\omega} = \omega\mathbf{K}$ is the angular velocity of the planet, and $\mathbf{r} = r\mathbf{i}$. As seen in Chapter 5, the coordinate transformation between the planet-centered and local horizon frames is given by

$$\begin{Bmatrix}\mathbf{i}\\ \mathbf{j}\\ \mathbf{k}\end{Bmatrix} = \mathsf{C_{LH}}\begin{Bmatrix}\mathbf{I}\\ \mathbf{J}\\ \mathbf{K}\end{Bmatrix}, \tag{12.3}$$

where

$$\mathsf{C_{LH}} = \mathsf{C_2}\left(\frac{-\pi}{2}\right)\mathsf{C_2}\left(\frac{\pi}{2}-\delta\right)\mathsf{C_3}(\lambda) \tag{12.4}$$

$$= \begin{pmatrix}\cos\delta\cos\lambda & \cos\delta\sin\lambda & \sin\delta\\ -\sin\lambda & \cos\lambda & 0\\ -\sin\delta\cos\lambda & -\sin\delta\sin\lambda & \cos\delta\end{pmatrix}.$$

Thus, $\mathbf{K}\times\mathbf{i} = \cos\delta\mathbf{j}$, and we have

$$\mathbf{v_I} = \mathbf{v} + \omega r\cos\delta\mathbf{j}\,. \tag{12.5}$$

The time derivative of the inertial velocity is the inertial (total) acceleration, $\mathbf{a_I}$, which is required to write the dynamic equations of translational motion according to Newton's second law. The inertial acceleration is thus calculated in the local horizon frame as follows:

$$\mathbf{a_I} \doteq \frac{\mathrm{d}\mathbf{v_I}}{dt} = \dot{\mathbf{v}} + \omega(\dot{r}\cos\delta - r\dot{\delta}\sin\delta)\mathbf{j} + \omega r\cos\delta(\omega\mathbf{K}\times\mathbf{j})\,, \tag{12.6}$$

where the dot represents the time derivative, $\frac{\mathrm{d}}{\mathrm{d}t}$. Before simplifying this equation any further, we write the relative velocity as follows:

$$\mathbf{v} = \dot{r}\mathbf{i} + \boldsymbol{\Omega}\times(r\mathbf{i})\,, \tag{12.7}$$

where

$$\boldsymbol{\Omega} \doteq \Omega_x\mathbf{i} + \Omega_y\mathbf{j} + \Omega_z\mathbf{k} \tag{12.8}$$

is the angular velocity of the local horizon frame $(oxyz)$ relative to the planet centered frame $(SXYZ)$. Upon substitution of Eq. (12.8) into Eq. (12.7), we have

$$\mathbf{v} = \dot{r}\mathbf{i} + r\Omega_z\mathbf{j} - r\Omega_y\mathbf{k}\,. \tag{12.9}$$

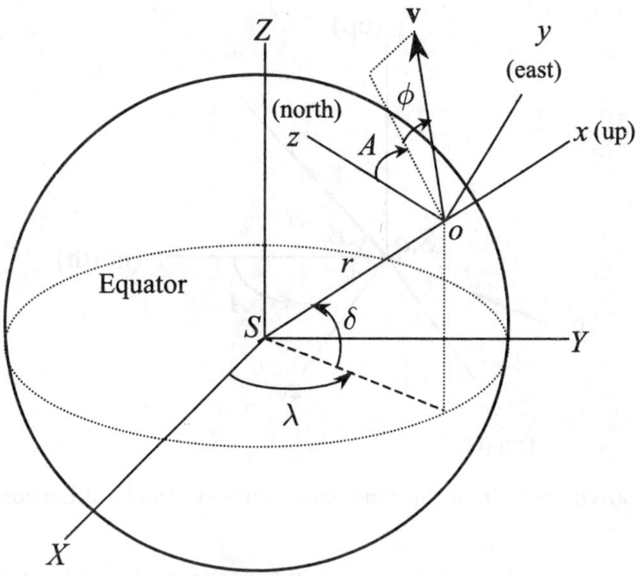

Fig. 12.1. Planet-fixed and local horizon frames for atmospheric flight.

A comparison of Eqs. (12.9) and (12.1) results in the following kinematic relationships:

$$\dot{r} = v \sin\phi, \tag{12.10}$$

$$\Omega_y = -\frac{v}{r}\cos\phi\cos A, \tag{12.11}$$

$$\Omega_z = \frac{v}{r}\cos\phi\sin A. \tag{12.12}$$

From the coordinate transformation of Eq. (12.4), it is clear that (Chapter 2)

$$\mathbf{\Omega} = \dot{\lambda}\mathbf{K} - \dot{\delta}\mathbf{j} = \dot{\lambda}\sin\delta\mathbf{i} - \dot{\delta}\mathbf{j} + \dot{\lambda}\cos\delta\mathbf{k}\,, \tag{12.13}$$

which, upon comparison with Eqs. (12.11) and (12.12), yields

$$\dot{\delta} = \frac{v}{r}\cos\phi\cos A, \tag{12.14}$$

$$\dot{\lambda} = \frac{v\cos\phi\sin A}{r\cos\delta}\,. \tag{12.15}$$

Equations (12.10), (12.14), and (12.15) are the kinematic equations of motion relative to a rotating planet. Once the relative velocity vector, (v, ϕ, A), is determined from the solution of the dynamic equations (derived ahead), the position vector, (r, δ, λ), is calculated from the kinematic equations, thereby completing the solution for the trajectory. It is to be noted that the same kinematic equations can be employed in orbital mechanics. Hence, the kinematic

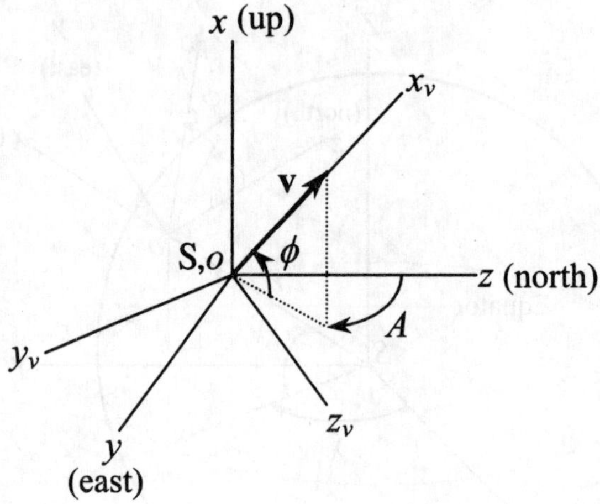

Fig. 12.2. Orientation of the wind axes relative to the local horizon frame.

equations of motion bridge the gap between the trajectory models for atmospheric and space flight. It now remains to write the dynamic equations of translational motion. As seen in Chapter 10, the aerodynamic forces are generally resolved in a right-handed, coordinate frame $(Sx_vy_vz_v)$, which has axes Sx_v $(\mathbf{i_v})$, Sy_v $(\mathbf{j_v})$, and Sz_v $(\mathbf{k_v})$, along the instantaneous, relative velocity vector, $\mathbf{v}$, and two mutually perpendicular directions normal to $\mathbf{v}$, respectively. This frame can be conveniently used to write the vehicle's equations of motion by fixing its origin at the planet's center, while orienting its axes along and perpendicular to $\mathbf{v}$. Since the frame $(Sx_vy_vz_v)$ has axes in the directions opposite and normal to that of the relative wind velocity, $-\mathbf{v}$, it is called the *wind axes* frame. Since atmospheric flight vehicles commonly have a plane of symmetry, it is convenient to choose the axis Sy_v $(\mathbf{j_v})$, normal to the plane of symmetry. The orientation of the wind axes,$(Sx_vy_vz_v)$, relative to the local horizon frame $(oxyz)$, is depicted in Fig. 12.2. Hence, the two frames are related by

$$\begin{Bmatrix} \mathbf{i_v} \\ \mathbf{j_v} \\ \mathbf{k_v} \end{Bmatrix} = \mathsf{C_W} \begin{Bmatrix} \mathbf{i} \\ \mathbf{j} \\ \mathbf{k} \end{Bmatrix}, \tag{12.16}$$

where

$$\begin{aligned} \mathsf{C_W} &= \mathsf{C_1}\left(\frac{-\pi}{2}\right)\mathsf{C_3}\left(\frac{\pi}{2}-\phi\right)\mathsf{C_1}\left(\frac{\pi}{2}-A\right) \\ &= \mathsf{C_2}\left(\phi-\frac{\pi}{2}\right)\mathsf{C_1}(-A) \\ &= \begin{pmatrix} \sin\phi & \cos\phi\sin A & \cos\phi\cos A \\ 0 & \cos A & -\sin A \\ -\cos\phi & \sin\phi\sin A & \sin\phi\cos A \end{pmatrix}. \end{aligned} \tag{12.17}$$

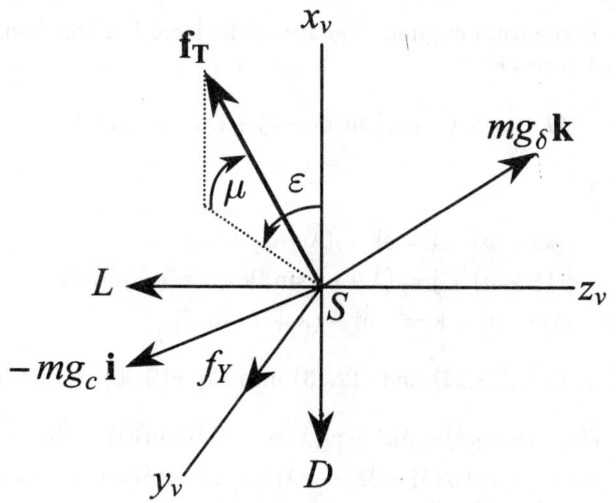

Fig. 12.3. External force resolved in the wind axes.

The dynamic equations of translational motion in the vector form are obtained from Newton's second law expressed as

$$\mathbf{f} = m\mathbf{a_I} = m\frac{d\mathbf{v_I}}{dt}, \tag{12.18}$$

where $\mathbf{f}$ denotes the net external force, and m stands for the total mass of the vehicle. Since $\mathbf{f}$ consists of the aerodynamic force vector, $(-D\mathbf{i_v}+f_Y\mathbf{j_v}-L\mathbf{k_v})$ (Chapter 10), thrust, $\mathbf{f_T}$ (Chapter 11), and gravity force, $-mg_c\mathbf{i} + mg_\delta\mathbf{k}$, (Chapter 3),[1] resolved in the wind axes (as shown in Fig. 12.3), we have

$$\begin{aligned}\mathbf{f} = &(f_T \cos\epsilon\cos\mu - D - mg_c\sin\phi + mg_\delta\cos\phi\cos A)\mathbf{i_v}\\ &+(f_T\sin\mu + f_Y - mg_\delta\sin A)\mathbf{j_v}\\ &+(-f_T\sin\epsilon\cos\mu - L + mg_c\cos\phi + mg_\delta\sin\phi\cos A)\mathbf{k_v}\,. \end{aligned}\tag{12.19}$$

Now, since the force vector is conveniently resolved along the wind axes, we must transform the inertial acceleration of the vehicle's center of mass, $\mathbf{a_I}$ [Eq. (12.6)] in the wind axes frame in order to write the dynamic equations of motion, Eq. (12.18), in a scalar form. We begin by resolving the individual terms on the right-hand side of Eq. (12.6) in the local horizon frame as follows:

$$\begin{aligned}\dot{\mathbf{v}} = &\ddot{r}\mathbf{i} + \dot{r}(\Omega_z\mathbf{j} - \Omega_y\mathbf{k}) + r(\dot{\Omega}_z\mathbf{j} - \dot{\Omega}_y\mathbf{k})\\ &+\dot{r}(\boldsymbol{\Omega} + \boldsymbol{\omega}) \times \mathbf{i} + r\Omega_z(\boldsymbol{\Omega} + \boldsymbol{\omega}) \times \mathbf{j}\\ &-r\Omega_y(\boldsymbol{\Omega} + \boldsymbol{\omega}) \times \mathbf{k}\,, \end{aligned}\tag{12.20}$$

[1] Recall that in Chapter 3 we denoted acceleration due to gravity by $\mathbf{g} = g_r\mathbf{i_r}+g_\phi\mathbf{i}_\phi$, expressed here as $\mathbf{g} = -g_c\mathbf{i}+g_\delta\mathbf{k}$, where $g_c = -g_r$, $g_\delta = -g_\phi$, $\mathbf{i} = \mathbf{i_r}$, and $\mathbf{k} = -\mathbf{i}_\phi$.

where $\boldsymbol{\Omega}+\boldsymbol{\omega}$ is the total angular velocity of the local horizon frame, obtained from Eq. (12.13) to be

$$\boldsymbol{\Omega}+\boldsymbol{\omega} = (\dot{\lambda}+\omega)\sin\delta\mathbf{i} - \dot{\delta}\mathbf{j} + (\dot{\lambda}+\omega)\cos\delta\mathbf{k} . \tag{12.21}$$

Thus, we have

$$\begin{aligned}
(\boldsymbol{\Omega}+\boldsymbol{\omega}) \times \mathbf{i} &= \dot{\delta}\mathbf{k} + (\dot{\lambda}+\omega)\cos\delta\mathbf{j}, \\
(\boldsymbol{\Omega}+\boldsymbol{\omega}) \times \mathbf{j} &= (\dot{\lambda}+\omega)\sin\delta\mathbf{k} - (\dot{\lambda}+\omega)\cos\delta\mathbf{i}, \\
(\boldsymbol{\Omega}+\boldsymbol{\omega}) \times \mathbf{k} &= -\dot{\delta}\mathbf{i} - (\dot{\lambda}+\omega)\sin\delta\mathbf{j}.
\end{aligned} \tag{12.22}$$

Substitution of Eqs. (12.22) and (12.13) into Eq. (12.20) results in

$$\begin{aligned}
\dot{\mathbf{v}} = \ddot{r}\mathbf{i} &+ \dot{r}(\dot{\lambda}\cos\delta\mathbf{j} + \dot{\delta}\mathbf{k}) + r[(\ddot{\lambda}\cos\delta - \dot{\lambda}\dot{\delta}\sin\delta)\mathbf{j} + \ddot{\delta}\mathbf{k}] \\
&+ \dot{r}[(\dot{\lambda}+\omega)\cos\delta\mathbf{j} + \dot{\delta}\mathbf{k}] + r\dot{\lambda}(\dot{\lambda}+\omega)\cos\delta(\sin\delta\mathbf{k} - \cos\delta\mathbf{i}) \\
&- r\dot{\delta}[\dot{\delta}\mathbf{i} + (\dot{\lambda}+\omega)\sin\delta\mathbf{j}] .
\end{aligned} \tag{12.23}$$

The last term on the right-hand side of Eq. (12.6) is expressed in the local horizon frame as follows:

$$\omega r\cos\delta(\omega\mathbf{K}\times\mathbf{j}) = \omega r(\dot{\lambda}+\omega)\cos\delta(\sin\delta\mathbf{k} - \cos\delta\mathbf{i}). \tag{12.24}$$

Collecting all the terms from Eqs. (12.23) and (12.24), and substituting them into Eq. (12.6), we have

$$\begin{aligned}
\mathbf{a_I} = &[\ddot{r} - r\dot{\delta}^2 - r(\dot{\lambda}+\omega)^2\cos^2\delta]\mathbf{i} \\
&+[r\ddot{\lambda}\cos\delta + 2\dot{r}(\dot{\lambda}+\omega)\cos\delta - 2r\dot{\delta}(\dot{\lambda}+\omega)\sin\delta]\mathbf{j} \\
&+[r\ddot{\delta} + 2\dot{r}\dot{\delta} + r(\dot{\lambda}+\omega)^2\sin\delta\cos\delta]\mathbf{k} .
\end{aligned} \tag{12.25}$$

The components of the inertial acceleration in the local horizon frame—given by Eq. (12.25)—are denoted in short hand by a_x, a_y, a_z. These are more useful when expressed in terms of the relative velocity components by substituting the kinematic equations, Eqs. (12.10), (12.14), and (12.15), into Eq. (12.25), yielding:

$$\begin{aligned}
a_x &= \dot{v}\sin\phi + v\dot{\phi}\cos\phi - \frac{v^2}{r}\cos^2\phi - 2\omega v\cos\phi\sin A\cos\delta - r\omega^2\cos^2\delta, \\
a_y &= \dot{v}\cos\phi\sin A - v(\dot{\phi}\sin\phi\sin A - \dot{A}\cos\phi\cos A) \\
&\quad +2\omega v(\sin\phi\cos\delta - \cos\phi\cos A\sin\delta) \\
&\quad +\frac{v^2}{r}\cos\phi\sin A(\sin\phi - \cos\phi\cos A\tan\delta), \\
a_z &= \dot{v}\cos\phi\cos A - v\dot{\phi}\sin\phi\cos A - v\dot{A}\cos\phi\sin A + \omega^2 r\sin\delta\cos\delta \\
&\quad +2\omega v\cos\phi\sin A\sin\delta + \frac{v^2}{r}\cos\phi(\sin\phi\cos A + \cos\phi\sin^2 A\tan\delta).
\end{aligned} \tag{12.26}$$

Finally, we make the coordinate transformation from the local horizon frame to the wind axes with the use of Eq. (12.16) as follows:

$$\begin{Bmatrix} a_{xv} \\ a_{yv} \\ a_{zv} \end{Bmatrix} = \mathsf{C_W} \begin{Bmatrix} a_x \\ a_y \\ a_z \end{Bmatrix} , \tag{12.27}$$

where $\mathbf{a_I} = a_{xv}\mathbf{i_v} + a_{yv}\mathbf{j_v} + a_{zv}\mathbf{k_v}$, which leads to

$$\begin{aligned} a_{xv} &= \dot{v} + \omega^2 r \cos\delta(\cos\phi\cos A\sin\delta - \sin\phi\cos\delta), \\ a_{yv} &= v\cos\phi\dot{A} - \frac{v^2}{r}\cos^2\phi\sin A\tan\delta - \omega^2 r\sin A\sin\delta\cos\delta \\ &\quad +2\omega v(\sin\phi\cos A\cos\delta - \cos\phi\sin\delta), \\ a_{zv} &= -v\dot{\phi} + \frac{v^2}{r}\cos\phi + 2\omega v\sin A\cos\delta \\ &\quad +\omega^2 r\cos\delta(sin\phi\cos A\sin\delta + \cos\phi\cos\delta) . \end{aligned} \tag{12.28}$$

Upon substituting the acceleration components from Eq. (12.28), and the force components from Eq. (12.19) (both resolved in the wind axes) into Eq. (12.18), we can write the dynamic equations of motion as follows:

$$\begin{aligned} m\dot{v} &= f_T\cos\epsilon\cos\mu - D - mg_c\sin\phi + mg_\delta\cos\phi\cos A \\ &\quad -m\omega^2 r\cos\delta(\cos\phi\cos A\sin\delta - \sin\phi\cos\delta), \\ mv\cos\phi\dot{A} &= m\frac{v^2}{r}\cos^2\phi\sin A\tan\delta + f_T\sin\mu + f_Y - mg_\delta\sin A \\ &\quad +m\omega^2 r\sin A\sin\delta\cos\delta \\ &\quad -2m\omega v(\sin\phi\cos A\cos\delta - \cos\phi\sin\delta), \\ mv\dot{\phi} &= m\frac{v^2}{r}\cos\phi + f_T\sin\epsilon\cos\mu + L - mg_c\cos\phi - mg_\delta\sin\phi\cos A \\ &\quad +m\omega^2 r\cos\delta(sin\phi\cos A\sin\delta + \cos\phi\cos\delta) \\ &\quad +2m\omega v\sin A\cos\delta . \end{aligned} \tag{12.29}$$

Here, the left-hand side of the first equation denotes the acceleration along the instantaneous flight path, whereas that of the second and third equations represents the centripetal acceleration caused by the curvature of the flight path in the local horizontal and local vertical directions, respectively. The centripetal and Coriolis acceleration terms (Chapter 4) due to planetary rotation appear on the right-hand side of these equations of motion, along with the respective components of the external force.

The complete set of governing equations for translation within the atmosphere consists of the kinematic equations [Eqs. (12.10), (12.14), (12.15)], and the dynamic equations [Eq. (12.29)]. All atmospheric trajectories must satisfy these equations. Their solution vector, $r(t), \lambda(t), \delta(t), v(t), \phi(t), A(t)$, yields the position and velocity as functions of time. However, a flight dynamic model is incomplete without gravity, atmospheric, aerodynamic, and propulsive models. While it is common to employ constant acceleration due to gravity (flat-planet approximation) or, at best, Newton's law of gravitation (spherical planet model), for atmospheric flight we shall take into account the radial and latitudinal gravity variations (nonspherical planet model) (Chapter 3). Our atmospheric model will also be as general as possible, such as the 21-layer, *U.S. Standard Atmosphere of 1976* for earth (Chapter 9). The aerodynamic model considers the variation of the aerodynamic force with position (altitude) and velocity. Such functional forms of nondimensional aerodynamic coefficients are employed as dictated by the flow regime (Chapter 10). Finally, the propulsion model must account for the variation of thrust with altitude and velocity, and the fuel consumption, which, in turn, determines the instantaneous vehicle mass. The additional set of nonlinear, ordinary differential equations resulting from these models is combined with the kinematic and dynamic equations derived previously.

Being nonlinear, coupled, ordinary differential equations, their integration in time, with given initial condition, is generally not possible in a closed form, but requires an iterative, numerical solution procedure, such as the Runge–Kutta methods presented in Appendix A. We shall now consider the specific atmospheric trajectories obtained by integrating the kinematic and dynamic equations of motion with appropriate models for aerodynamics and propulsion.

12.3 Airplane Flight Paths

Airplane flight is the most common example of atmospheric flight and consists, by and large, of horizontal (level), straight flight at a constant speed. Such a flight is known as the *cruise*. It is evident from above that a steady ($\dot{v} = 0$), horizontal flight ($\phi = 0$) requires that the terms on the right-hand side of the first two equations of Eq. (12.29) add up to zero. This is possible if the thrust is large enough to overcome the drag (which is the largest opposing term in the first equation) and acts along the velocity vector, while the lift balances the weight (the largest opposing term in the second equation). The problem of steady, horizontal flight through the air was first solved by the Wright brothers in 1903[2] by separately generating the aerodynamic lift by the wings, and the thrust from the propellers powered by an engine. This separation of lift and thrust was the crucial step missing from the failed attempts of earlier aeronauts, who adopted various complex mechanisms for the simultaneous creation of the two force components (such as by flapping wings). However, it is not sufficient to merely be able to fly horizontally. One also must be able

[2] The *Wright Flyer* of 1903 was the first successful airplane. It had a bi-plane wing and canard configuration of lifting surfaces, with two pusher, counter-rotating propellers driven by a 12.5 BHP piston engine (also designed and constructed by the Wright brothers). Due to its configuration and center of gravity, the airplane was longitudinally unstable and, therefore, very difficult to fly.

to control the direction of the horizontal flight (so as to reach the desired destination, while avoiding obstacles). If not, an aviator's plight would be no better than that of a balloonist, who is placed entirely at the mercy of the winds. The control of the horizontal flight direction, A, requires that the side force must create a horizontal centripetal acceleration, as dictated by the last equation of Eq. (12.29). In order to do so, the ingenious procedure of warping the wings to bank the aircraft,[3] so as to provide a component of the wing lift in the horizontal plane (side force), was another invention of the Wright brothers. Thus, the balancing of drag by thrust, weight by lift, and creation of horizontal centripetal acceleration by the side force obtained by banking the wings form the essence of airplane flight and are adopted even in this age of modern aviation.

From the foregoing discussion, it is clear that an airplane, by definition, possesses a plane of symmetry through its primary function of a straight and level flight. Although there are rare exceptions,[4] this symmetrical design incorporates streamlined bodies (fuselages and nacelles) for housing the payload, engines, and other essentials, along with a set of lifting surfaces, which are capable of acting as wings as well as stabilizing surfaces (tails, or canards). In the interest of low drag, it is important that in the normal operation of an airplane, the lifting surfaces and the streamlined bodies should be aligned with the relative flow direction, i.e., the relative velocity vector. Hence, the velocity vector should lie in the plane of symmetry, the axes of symmetry of the bodies should be parallel to this plane, while the essentially flat lifting surfaces must be approximately normal to the same plane. It is clear that such a symmetrical configuration would not, by its own, create a side force in the normal operation. However, a sideslip would cause a side force to be generated in addition to an increase in drag (Chapter 10), which is generally undesirable. The normal attitude of the airplane is thus without a sideslip, and is called *coordinated flight*. The airplane is carefully designed to achieve this equilibrium attitude through the shape of the bodies and arrangement of lifting surfaces, in the presence of a disturbance that would otherwise tend to create a sideslip. The important tendency of seeking an equilibrium with zero sideslip is called *weathercock*—or static, directional—stability (Chapter 13). In a coordinated flight, the airplane must be banked (i.e., rotated about the instantaneous velocity vector) in order to produce the necessary side force

---

[3] The wing-warping technique adopted in the Wright gliders and airplanes, was found to be cumbersome and later replaced by the more efficient *ailerons*. The ailerons, invented by Farman brothers in 1909 and used in all airplanes since then, are a pair of trailing-edge control surfaces, deployed in opposite directions on either side of the airplane, thereby creating a rolling moment.

[4] The *Boomerang* experimental, twin-propeller engine airplane designed by Burt Rutan has an unconventional, asymmetric configuration, with forward-swept wings of unequal sweeps and spans and two unsimilar fuselages in a "twin-boom" style. One engine is mounted ahead of the other, and both produce unequal thrusts for maintaining directional balance.

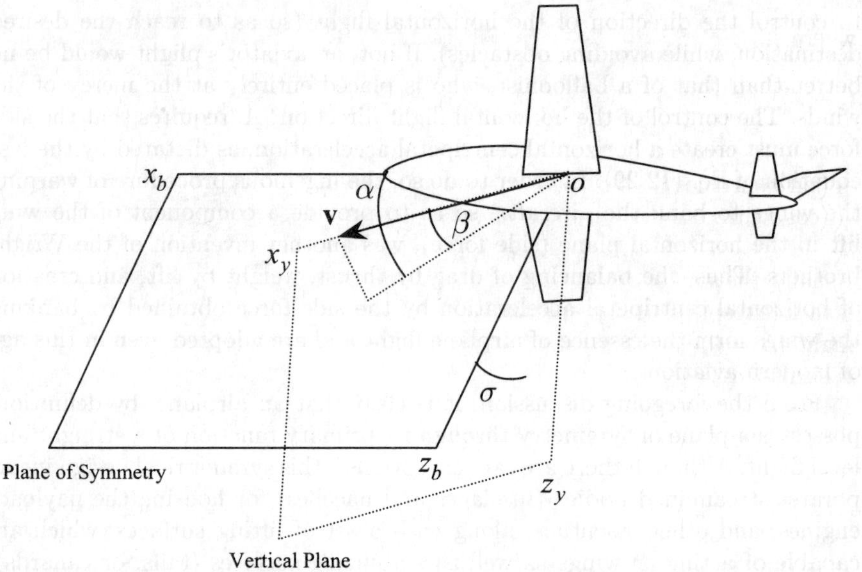

Fig. 12.4. The orientation of the airplane's plane of symmetry, ox_bz_b, relative to the vertical plane, ox_vz_v.

for making a horizontal turn. We must relate the "lift" produced by the lifting surfaces, to our definition of lift, drag, and side force as components of the net aerodynamic force resolved in the wind axes. For this purpose, we select a body-fixed frame, $ox_by_bz_b$, such that the ox_bz_b plane is the plane of symmetry. The airplane's sideslip angle, β, is the angle made by the relative velocity vector with the plane of symmetry. Furthermore, the airplane's net angle of attack, α, is defined as the angle between the projection of the relative velocity vector in the plane of symmetry and the axis ox_b. Finally, σ is the bank angle, defined as the angle between the plane of symmetry and the local vertical plane, ox_vz_v. The two planes, ox_bz_b and ox_vz_v, are depicted in Fig. 12.4. Then the net aerodynamic force can be expressed as

$$\mathbf{f_a} = -D\mathbf{i_v} + f_Y\mathbf{k_v} - L\mathbf{k_v} = -D_b\mathbf{i_b} + f_{Yb}\mathbf{j_b} - L_b\mathbf{k_b} \,, \tag{12.30}$$

where L_b, D_b, f_{Yb} are the lift, drag, and side force resolved in the body-fixed frame. The coordinate transformation between the body-fixed frame and the wind axes is given by

$$\begin{Bmatrix} \mathbf{i_v}' \\ \mathbf{j_v}' \\ \mathbf{k_v}' \end{Bmatrix} = \mathsf{C} \begin{Bmatrix} \mathbf{i_b} \\ \mathbf{j_b} \\ \mathbf{k_b} \end{Bmatrix} , \tag{12.31}$$

where

$$\mathsf{C} = \begin{pmatrix} \cos\alpha\cos\beta & \cos\alpha\sin\beta & -\sin\alpha \\ (\sin\sigma\sin\alpha\cos\beta - \cos\sigma\sin\beta) & (\sin\sigma\sin\alpha\sin\beta + \cos\sigma\cos\beta) & \sin\sigma\cos\alpha \\ (\cos\sigma\sin\alpha\cos\beta + \sin\sigma\sin\beta) & (\cos\sigma\sin\alpha\sin\beta - \sin\sigma\cos\beta) & \cos\sigma\cos\alpha \end{pmatrix}. \tag{12.32}$$

Usually, the airplane is capable of generating sufficient lift with a small angle of attack (except in large maneuvers). Hence, for the typical case of a coordinated flight ($\beta = 0, f_{Yb} = 0$), with small angle of attack, we can assume $\cos\alpha \approx 1, \sin\alpha \approx \alpha$. Thus, we have

$$\begin{aligned} D &\approx D_b + L_b\alpha\cos\sigma, \\ f_Y &\approx L_b\sin\sigma, \\ L &\approx L_b\cos\sigma - D_b\alpha. \end{aligned} \tag{12.33}$$

It is thus clear that a coordinated flight with zero bank angle is essentially confined to the vertical plane. If the bank angle and the angle of attack are selected to make $\dot{v} = \dot{\phi} = \dot{A} = 0$ in Eq. (12.29) while keeping $\beta = 0$, we have a steady, coordinated, horizontal turn.

For most airplanes, the thrust lies in the plane of symmetry. Exceptions to this rule are airplanes that employ thrust-vectoring for directional stability, such as the finless *B-2* stealth bomber,[5] or for creating a side force without banking.[6] Other exceptions are airplanes driven by a single propeller, which are designed with the axis of the propeller slightly offset from the longitudinal axis in order to reduce propulsive influence (propeller *p-effect*) on stability and control. Furthermore, if we exclude the aircraft using a part of the thrust for lift enhancement, such as the vertical/short take-off and landing (V/STOL) *Harrier* and *Sea-Harrier*[7] and the thrust-vectoring *F-22*[8] and *Sukhoi Su-30MKI*[9] fighter airplanes, the thrust can be considered to be aligned with

[5] The *B-2* bomber has a flying-wing design without vertical or horizontal tails, as dictated by low-radar observability (stealth) requirements.

[6] The generation of a side force without banking allows a greater horizontal maneuverability, called direct side-force turning, and has been flight-tested in the experimental *F-16D VISTA* fighter with a multi-axis, thrust-vectoring nozzle. NASA has also undertaken experimental investigations into multi-axis thrust-vectoring [35].

[7] The *Harrier* and *Sea Harrier* are British subsonic, V/STOL fighter aircraft, whose larger attack variants, *AV-8A* and *AV-8B*, respectively, are produced in the USA.

[8] The *F-22* is an advanced American fighter, capable of supersonic cruise and high maneuverability through thrust vectoring in the plane of symmetry.

[9] *Su-30MKI* is a highly maneuverable version of the Russian *Su-30* fighter, with thrust-vectoring nozzles and small canards, especially designed to meet Indian Air Force requirements.

the longitudinal axis, ox_b. Hence, in most cases we have $\epsilon = \alpha, \mu = \beta$. For a coordinated flight without thrust vectoring,

$$\mathbf{f_T} = f_T \cos\alpha \mathbf{i_v} + f_T \sin\alpha \mathbf{k_v} = f_T \mathbf{i_b} , \tag{12.34}$$

which implies that the thrust lies in the vertical plane.

Since airplane flight generally occurs close to the planetary surface at a small relative speed, the planetary rotational and curvature terms in Eqs. (12.10), (12.14), (12.15), and (12.29) are quite small. Hence, the traditional approach of studying airplane flight [34] by ignoring these terms—called the *flat-planet* approximation—is often valid. However, in considering long-duration airplane flight, the acceleration terms due to curvature and planetary rotation can cause a significant deviation of the flight path over time. We shall keep all the terms in the equations of motion, because of our objective of faithfully modeling the flight dynamics.

Before beginning simulations of airplane trajectories, let us briefly examine certain aspects of airplane flight derived analytically in an approximate analysis [34]. The approximate forms of Eqs. (12.10) and (12.29) indicate that a *steady climb* is possible by holding $\dot{v} = \dot{\phi} = 0$ with the use of $L \approx mg_c \cos\phi$ and $f_T - D \approx mg_c \sin\phi$. Such an equilibrium flight condition, however, cannot be held for long without adjusting the relative speed, due to the variation of the aerodynamic forces with altitude. Hence, in practice, a constant flight-path angle is achieved by having a *quasi-steady climb* in which the speed slowly changes with the altitude. This is also true for a descent. Thus, while being designed expressly for a level cruising flight, an airplane is capable of changing its altitude within its thrust capability. The maximum angle of climb at a given altitude clearly depends on the *specific excess thrust*, $\frac{f_T - D}{mg_c}$, while the maximum *rate of climb*, $\dot{r} = v \sin\phi$, is determined by the maximum *specific excess power*, $\frac{(f_T - D)v}{mg_c}$. Since the thrust of airbreathing engines used in airplanes diminishes with altitude more rapidly than the drag, an airplane cannot go on climbing indefinitely. An *absolute ceiling* is be defined as the altitude where the specific excess thrust and power become zero. However, a *service ceiling*, defined as the altitude where the quasi-steady rate of climb becomes 100 ft/min, is more practical in analyzing airplane performance.

A special kind of quasi-steady climb is the slow increase in altitude in a long-duration cruise due to the decrease in weight by fuel consumption. This is referred to as a *cruise climb*. While all airplanes are capable of quasi-steady climb in which the speed gradually decreases as the airplane climbs, modern fighter-type airplanes can increase their speed while climbing due to their high thrust-to-weight ratio. An accelerated climb, being an unsteady flight, is analyzed very differently from the quasi-steady climb. Apart from climbing, a measure of a fighter aircraft's performance is its capability of making horizontal and vertical turns, termed as *maneuverability*. Horizontal maneuverability is indicated by the maximum *sustained turn rate*, which is the steepest horizontal turn made without losing altitude, and the maximum *instantaneous turn rate*, where the airplane is allowed to lose altitude while turning. From the second equation of both Eq. (12.29) and Eq. (12.33), we can see that a conventional airplane without thrust vectoring can make a steady, coordinated turn of maximum instantaneous rate

$$\dot{A} \approx \frac{L_b \sin\sigma}{mv} , \tag{12.35}$$

where the wing lift in the body frame, L_b, is limited by aerodynamic and structural constraints. However, while turning at the maximum instantaneous rate, it is seldom possible to maintain level flight (because $f_T < D$), and the airplane loses altitude steadily [the first equation of Eq. (12.29) and Eq. (12.10)]. Since $f_T \approx D$ must be maintained to keep a constant altitude, it is clear that the sustained turn rate is also influenced by the thrust capability of the aircraft in addition to the aerodynamic and structural limitations. Thus, the sustained rate of turn cannot exceed the maximum instantaneous turn rate.

We shall now lay down the aerodynamic and structural limitations of airplane flight. From the stall limit of the maximum lift coefficient (Chapter 10), we have

$$C_L \doteq \frac{L_b}{\frac{1}{2}\rho v^2 S} \leq C_{L\,\max} , \tag{12.36}$$

where $C_{L\,\max}$ is the maximum lift coefficient of the airplane, based upon the total wing platform area, S. We know from Chapter 10 that $C_{L\,\max}$ depends upon the Mach number and the lifting configuration (deployment of high-lift devices and spoilers).

It is important to note that the airplane's maximum lift is generally less than the sum of the maximum lift forces of individual lifting surfaces (due to mutual interference) and the balance (*trim*) requirement (Chapter 13). Since the lifting surfaces are essentially flat, lightweight structures, their dynamic load arising out of the airplane's acceleration is limited by tensile strength limits. If the weakest part of the airplane (usually a lifting surface) can sustain, at most, an acceleration of $\bar{a}$, it is clear from Eqs. (12.18), (12.19), and (12.33)—as well as the fact that in a coordinated airplane flight $f_T - D \ll L - mg_c$—that

$$\mid \mathbf{f} \mid \approx \mid L_b - mg \mid \leq m\bar{a} , \tag{12.37}$$

where $g = \sqrt{g_c^2 + g_\delta^2}$. It is more common to express the structural constraints in terms of acceleration above due to gravity, with the use of a *load factor*, n, defined by $a_I = (n-1)g$. Hence, we have

$$\mid L_b \mid \leq mg\bar{n} , \tag{12.38}$$

where $\bar{n} = \frac{\bar{a}}{g} + 1$ is the *limit load factor* of the airplane. Hence, the *normal load* on the airplane, $L_b = nmg$, is restricted by the limit load factor and the *normal acceleration* by $\bar{a} - g$. Generally, the limit load factor incorporates a margin of safety and is less than the actual maximum load factor (called *design load factor*) by a *factor of safety* of 1.5–2. The aerodynamic and structural constraints are usually incorporated in a single diagram showing the allowable variation of load factor with speed, called a *v-n diagram*. Since an airplane requires a smaller magnitude of negative lift in its design life, compared to the maximum positive lift, a considerable structural weight can be saved by selecting a smaller-limit load factor for negative values of L_b. In terms of the load factor, the stall limit can be expressed as

$$C_L = \frac{nmg}{\frac{1}{2}\rho v^2 S} \leq C_{L\max}. \tag{12.39}$$

Another structural limitation is the maximum dynamic pressure, $q_{\max} = \frac{1}{2}\rho v_{\max}^2$, that can be safely tolerated by the structure, which gives the maximum speed, $v_{\max}$, at a given altitude. Clearly, the value of $v_{\max}$ is smallest at the sea level and increases with altitude. Since most airplane lifting surfaces are rather flexible, the dynamic pressure restriction comes from *aeroelastic* concerns of *divergence, control surface reversal*, and *flutter*. These phenomena provide a smaller dynamic pressure limit than that possible by assuming a rigid structure. In airplanes capable of high-subsonic and supersonic flight, there is also a Mach number limitation ($M < M_{\max}$) due to dynamic loads caused by shock waves that can destroy a propeller, and excite aeroelastic phenomena, such as *buzz* of control surfaces and air intakes. We will briefly study the modeling of aeroelastic effects in Chapter 15.

The modeling of thrust variation with speed and altitude is essential in a flight simulation. As seen in Chapter 11, the propeller thrust varies with both speed and altitude, whereas a jet engine's thrust can be considered constant with relative speed during cruise, and changes almost linearly with the atmospheric density. A turboprop engine displays a similar variation of the equivalent shaft power with density. However, the rate of decline in power of a normally aspirated piston engine with altitude is larger than that of a turboprop with the same maximum shaft power, which results in a smaller ceiling for a piston-powered airplane. As discussed in Chapter 11, the power-lapse rate of a piston engine with altitude can be reduced with the help of a turbo-charger. In all engines, the variation of thrust with Mach number must be properly taken into account. A rocket engine (Chapter 8) (rarely employed in airplanes) does not have a significant thrust variation with speed, altitude, and Mach number.

Finally, the drag description of an airplane is generally by a parabolic variation of the drag coefficient with the lift coefficient (Chapter 10),

$$C_D = C_{D0} + KC_L^2, \tag{12.40}$$

where C_{D0}, K are functions of the Mach number and aerodynamic configuration (deployment of high-lift devices, spoilers, brakes, and ground vicinity) (Chapter 10). The parabolic drag polar is applicable at both subsonic and supersonic speeds and breaks down in a stalled flight or through the transonic regime where normal shock waves are present.

12.3.1 Long-Range Cruising Flight

Consider the long-range cruising flight of an airplane, such as a modern airliner. In order to maximize the range with a given fuel mass, the airplane must fly at an altitude and speed where the drag and specific fuel consumption are minimized. All jet-powered airplanes normally cruise close to the tropopause (Chapter 9), $h = 11$ km, because the thrust-specific fuel consumption (TSFC) of jet engines remains small at the tropopause (Chapter 11) and increases rapidly at higher altitudes. On the other hand, propeller-engined airplanes have their power-specific fuel consumption (PSFC) minimized at approximatley half the jet airplane altitude. While the propeller airplanes must fly near a speed (or lift coefficient) that maximizes the lift-to-drag ratio, $\frac{L}{D}$, in order to maximize the range, the jet-powered airplanes have to maximize $v\frac{L}{D}$ (called the *Breguet range condition*) for the same objective [34]. Once the optimum speed and altitude are estimated, the cruise is carried out close to that condition. Modern airliners are equipped with a *flight management system* (FMS), which is a sophisticated autopilot that controls the airplane's attitude and speed to meet a desired trajectory, subject to aerodynamic, propulsive, and structural constraints. Usually, this translates into a coordinated flight without banking, at a constant angle of attack and a constant throttle setting. The FMS has access to real-time flow angles, airspeed, and altitude from a *central air-data acquisition system* (CADS), and airplane attitude, latitude, and longitude from *inertial measurement unit* (IMU) and *global positioning system* (GPS). Other navigation systems based on ground radio networks provide distance and angular positions (bearings) relative to the selected radio stations. An airplane unequipped with the FMS, CADS, or IMU has to be flown by less-sophisticated autopilots that can hold azimuth (heading), airspeed, and altitude, or follow a series of specific GPS *waypoints* (fixed enroute positions). In airplanes that do not have an autopilot, the pilot has to fly manually either by sight using landmarks and aeronautical charts by a procedure known as *dead reckoning*, or by available navigational aids (such as ground radio network and GPS). The former method can be employed only in clear weather situations according to *visual flight rules* (VFR), whereas the latter is indispensible in bad weather when landmarks are invisible, and is called flight by *instrument flight rules* (IFR). Even the most basic airplanes have a certain minimum set of instruments, such as the *airspeed indicator* (ASI), *altimeter*, *vertical speed indicator* (VSI), *turn, bank, and slip indicator*, *directional gyro*, and magnetic compass, in order to provide sufficient data for safely flying the airplane. For example, the pilot can maintain a level flight with the help of the altimeter, and ensure coordination, direction, and rate of turn with the help of turn, bank, and slip indicator. Furthermore, the optimum cruise condition can be ensured through the ASI. The ASI is an aneroid barometer calibrated to convert the dynamic pressure to an indicated airspeed, v_i, assuming a constant atmospheric density, ρ_0 (usually the standard sea level value). Since the lift coefficient in a straight and level flight can be expressed as

$$C_L \doteq \frac{mg}{\frac{1}{2}\rho v^2 S} = \frac{mg}{\frac{1}{2}\rho_0 v_i^2 S}, \tag{12.41}$$

the pilot can ensure an optimal lift coefficient by maintaining a constant indicated airspeed (after correcting for instrument errors), irrespective of the altitude. Thus, the ASI can serve as an indicator of the lift coefficient (hence, angle of attack). A magnetic compass, or a directional gyro, can be calibrated to accurately indicate the velocity azimuth with a known wind velocity. Since the early days of aviation, many long-range flights have been undertaken successfully with only the basic flight instruments.[10]

The variation of the airplane's mass during cruise is given by (Chapter 11)

$$\dot{m} = -\frac{c_T f_T}{g_0}, \tag{12.42}$$

where $g_0 = 9.8$ m/s^2. Since f_T, c_T are essentially constants, this implies a linear variation of mass with time. As the mass decreases, the airplane climbs and its speed changes in order to find equilibrium at the new altitude. These changes are, however, small in magnitude as the rate of change of mass is a small quantity. Due to a small variation in the Mach number, it is a good approximation to assume constant drag polar coefficients, C_{D0}, K. For the same reason, the Mach number variation of thrust and TSFC can be neglected during cruise.

Example 12.1. We will demonstrate in this example how a long-range, intercontinental flight is undertaken. Let us simulate the flight of a jet airliner with total wing planform area, $S = 223.0815$ m^2, and mean aerodynamic chord, $\bar{c} = 5.42$ m, which begins its level cruise at a standard altitude of 11 km over London's Heathrow airport ($\delta = 51.5°, \lambda = 0$), with an initial mass, $m = 84890.909$ kg, speed, $v = 270$ m/s, and velocity azimuth, $A = 287.2°$. Since the expected Mach number variation during cruise is small, it can be assumed that the drag polar is given by constant coefficients, $C_{D0} = 0.015, K = 0.08$. The maximum lift coefficient in the cruising configuration is $C_{L\max} = 1.2$, while the structural limitations are given by $\bar{n} = 3.5, q_{\max} = 17,200$ N/m$^2, M_{\max} = 0.985$. A constant power setting is maintained throughout the cruise, such that $f_T = 63,131.63$ N and $c_T = 0.54907$ 1/hr at $h = 11$ km, and varies in direct proportion to the atmospheric density thereafter. Let us assume that during cruise, the FMS maintains a zero sideslip, zero bank angle, and a constant angle of attack, such that the lift coefficient remains constant at $C_L = 0.2783$. Since the lift coefficient and bank angle are kept constant, there is no need to enforce the stall and load-factor limits in the simulation. However, the dynamic pressure

[10] The early aviation pioneers, such as Charles Lindbergh, who flew across the Atlantic in 1927, and Wiley Post, who flew around the world in 1931, both in small, single-propeller airplanes, relied upon the basic flight instruments and dead reckoning.

and Mach number limits are enforced by the FMS with the decrease of throttle setting, which reduces the thrust by 50% and the TSFC by 10% whenever either limit is violated. The airplane contains a fuel mass of 21, 222.66 kg, which is calculated to be sufficient for a six-hour cruise with initial thrust and TSFC. Of course, when the entire fuel is exhausted, the thrust falls to zero and the airplane begins a descent. We will carry out the simulation for $t = 5.73$ hr.

We begin by writing a MATLAB program called *airflight3dof.m* (Table 12.1), which calculates the time derivatives of all the motion variables, $r, \delta, \lambda, v, \phi, A, m$, according to the kinematic and dynamic equations of motion, mass variation, as well as aerodynamic and structural constraints. These time derivatives are provided as inputs to the MATLAB's intrinsic Runge–Kutta solver, *ode45.m*, at each time step. The atmospheric properties are calculated using the *1976 U.S. Standard Atmosphere*, programmed in *atmosphere.m* (Chapter 9), while the acceleration due to gravity of nonspherical earth is calculated by the program *gravity.m* (Chapter 3). Another program, called *runairflight3dof.m* (Table 12.2), specifies the initial condition and invokes *ode45.m*. The resulting plots of the motion variables and aerodynamic parameters are shown in Figs. 12.5–12.9.

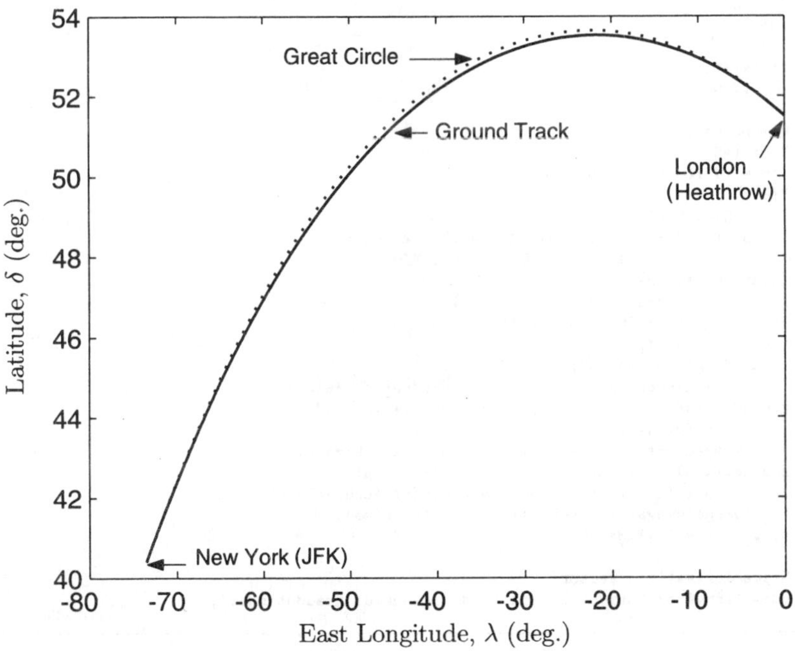

Fig. 12.5. The ground track of a jet airplane in a long-range cruise.

Table 12.1. M-file *airflight3dof.m* for Airplane's Governing State Equations

```
function xdot = airflight3dof(t,x)
global dtr; global S; global c; global rm; global omega; global CD0;
global K; global qmax; global Mmax; global CL; global fT0;
global tsfc0; global mfinal; global f8;
%acceleration due to gravity (nonspherical earth):
[g,gn]=gravity(x(3),x(2));
lo = x(1);la = x(2);
clo = cos(lo); slo = sin(lo);
cla = cos(la); sla = sin(la);
fpa = x(5); chi = x(6);
cfpa = cos(fpa); sfpa = sin(fpa);
cchi = cos(chi); schi = sin(chi);
%atmospheric properties:
if x(3)<rm
    x(3)=rm;
end
alt = x(3) - rm; %altitude
v  = x(4); %speed
atmosp = atmosphere(alt,v,c);
rho = atmosp(2); %atmospheric density
mach = atmosp(3); %Mach number
%aerodynamics and propulsion module:
Qinf = 0.5*rho*v^2;
if Qinf>=qmax || mach>=Mmax
    fT=fT0*0.85;
    tsfc=tsfc0*0.95;
else
    fT=fT0;
    tsfc=tsfc0;
end
m=x(7);
if m<=mfinal
    fT=0;
end
CD=CD0+K*CL^2;
D=Qinf*S*CD;
Xfo = fT*rho/0.3663-D;
Yfo = 0;
Zfo = Qinf*S*CL;
fprintf(f8,'\t%1.5e\t%1.5e\t%1.5e\t%1.5e\t%1.5e\n',...
                alt, mach, Qinf, Xfo, Yfo);
%state equations:
longidot= x(4)*cfpa*schi/(x(3)*cla);
latidot=  x(4)*cfpa*cchi/x(3);
raddot= x(4)*sfpa;
veldot= -g*sfpa+gn*cchi*cfpa+Xfo/m+...
     omega*omega*x(3)*cla*(sfpa*cla-cfpa*cchi*sla);
gammadot=(x(4)/x(3)-g/x(4))*cfpa-gn*cchi*sfpa/x(4)...
        +Zfo/(x(4)*m)+2*omega*schi*cla...
     +omega*omega*x(3)*cla*(cfpa*cla+sfpa*cchi*sla)/x(4);
headdot=x(4)*schi*tan(x(2))*cfpa/x(3)-gn*schi/x(4)...
     -Yfo/(x(4)*cfpa*m)-2*omega*(tan(x(5))*cchi*cla-sla)...
    +omega*omega*x(3)*schi*sla*cla/(x(4)*cfpa);
mdot=-tsfc*fT/(9.8*3600);

%time derivatives vector:
xdot=[longidot;latidot;raddot;veldot;gammadot;headdot;mdot];
```

Table 12.2. M-file *runairflight3dof.m* for Solving Airplane's State Equations

```
global dtr; dtr=pi/180; %degree to radian
global S; S=223.0815; %wing-planform area (m^2)
global c; c=5.42; %wing mean aerodynamic chord (m)
global rm; rm=6378140; %earth's radius (m)
global omega; omega=2*pi/(23*3600+56*60+4.0905);%earth's rot. speed (rad/s)
global CD0; CD0=0.015; %zero-lift drag coefficient
global K; K=0.08; %lift-dependent drag factor
global qmax; qmax=17200; %maximum dynamic pressure (N/m^2)
global Mmax; Mmax=0.985; %maximum Mach number
global CL; CL=0.2783; %constant lift coefficient
global fT0; fT0=63131.63; %initial thrust (N)
global tsfc0; tsfc0=0.54907; %initial TSFC (per hour)
global mfinal; mfinal=84890.909-21222.66; %zero fuel mass (kg)
global f8;f8=fopen('data8.mat','a');%file for aero-propulsive results
%Initial condition:
long = 0*dtr;           %initial longitude
lat = 51.5*dtr;         %latitude
rad=rm+11000;           %radius
vel=270;                %speed (m/s)
fpa=0;                  %flight-path angle
chi=287.2*dtr;          %velocity azimuth (from north)
m=84890.909;            %aircraft mass (kg)
%initial condition state vector:
init = [long; lat; rad; vel; fpa; chi; m];
%Runge-Kutta integration of state equations:
[t, o] = ode45('airflight3dof',[0, 5.73*3600], init);
fclose('all');
```

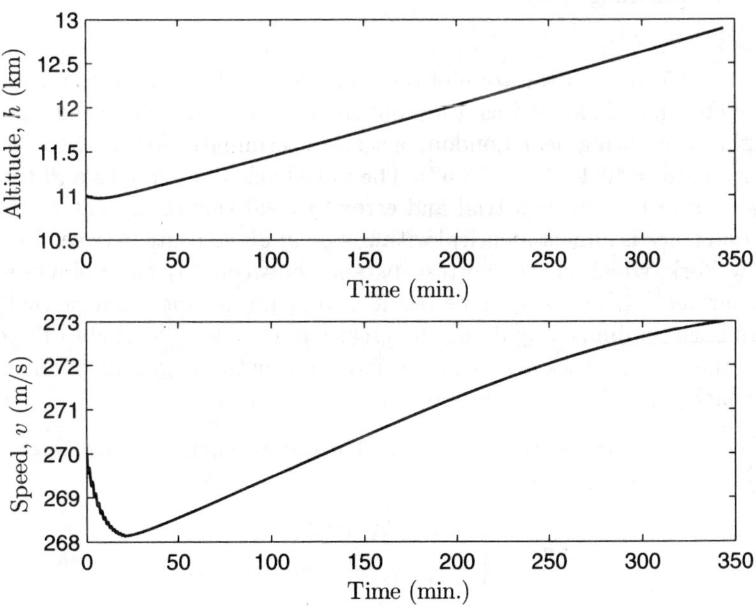

Fig. 12.6. The relative speed and altitude time history of a jet airplane in a long-range cruise.

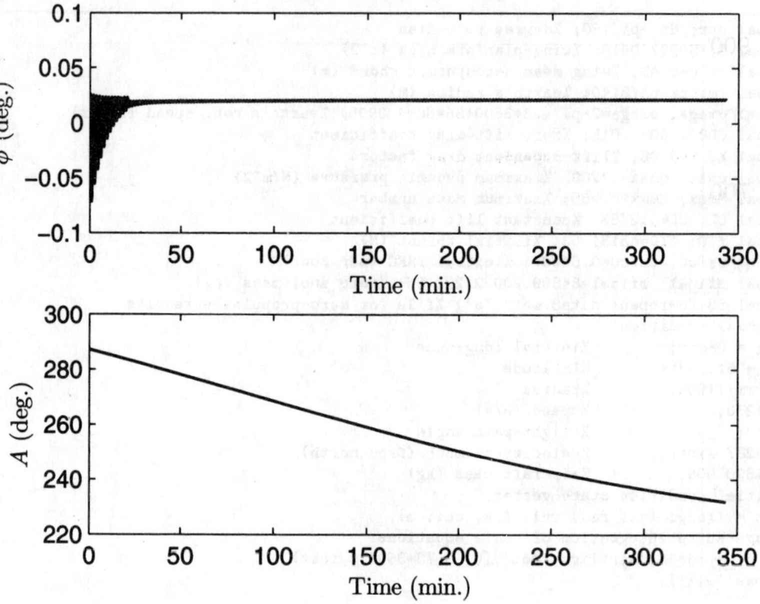

Fig. 12.7. The flight-path angle and velocity azimuth vs. the altitude of a jet airplane in a long-range cruise.

Figure 12.5 shows a variation of the longitude vs. latitude in the resulting flight. Such a plot is known as a *ground track* and is valuable in navigation. The flight, originating near London, is seen to terminate close to New York's JFK airport ($\delta = 40.4°, \lambda = -73.5°$). The initial velocity azimuth and time of flight were selected through trial and error to yield this result. Interestingly, the ground track is almost identical with the *great circle* route between London and New York, which is the shortest distance between any two points on the earth's surface[11] (Fig. 12.5). This result is very important, because fuel and time are both minimized by flying the great circle route. The slight difference between the ground track and the great circle is due to the gravity of an oblate earth, which pulls the track more toward the equator.

[11] The equation for the great circle passing between two points on a sphere, λ_1, δ_1 and λ_2, δ_2, is the following [2]:

$$\tan\delta = \begin{cases} \frac{\tan\delta_1 \sin(\lambda-\lambda_1+\beta)}{\sin\beta}, & \delta_1 \neq 0, \\ \frac{\tan\delta_2 \sin(\lambda-\lambda_1+\beta)}{\sin(\lambda_2-\lambda_1)}, & \delta_1 = 0, \end{cases}$$

where

$$\cot\beta = \frac{\tan\delta_2}{\tan\delta_1 \sin(\lambda_2-\lambda_1)} - \cot(\lambda_2-\lambda_1)\,.$$

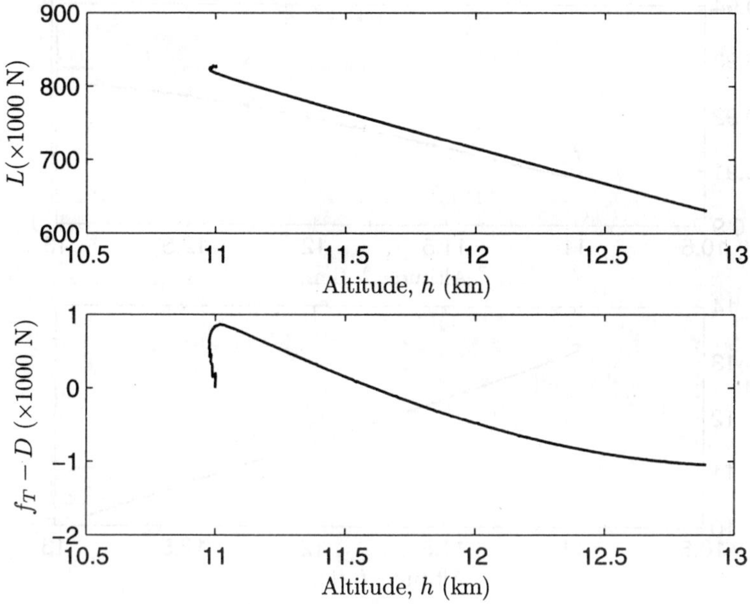

Fig. 12.8. The aerodynamic and thrust force components vs. the altitude of a jet airplane in a long-range cruise.

The altitude and speed time history are shown in Fig. 12.6, depicting a shallow climb to $h = 12.8$ km, a slight decrease, and then increase in the speed by $-2, +3$ m/s over a period of 5.73 hr. This is the typical cruise climb caused by fuel consumption. The change in the flight-path angle is seen to be negligible in Fig. 12.7, while the velocity azimuth almost steadily decreases by 55.2° over the flight. It is important to note that the airplane has made no turns; it has flown a straight and level trajectory. Therefore, the variation of azimuth and, thus, curvature in the ground track, are caused entirely by the centripetal and Coriolis acceleration due to the earth's curvature and rotation. The decrease in the lift is steady due to the linear weight reduction, and the difference between thrust and drag changes slightly due to the slight altitude increase, as seen in Fig. 12.8. The Mach number changes very slightly (by 0.01), while the dynamic pressure steadily decreases, and both remain well below their structural limits without the need of FMS throttle intervention (Fig. 12.9). This example shows the simplicity of cruising for long distances by merely maintaining a constant lift coefficient and throttle setting, where a straight and level flight is converted into a great circle route by the earth's rotation and curvature, without any maneuvers by the FMS. This navigation procedure is used in a long-range cruise. However, when a steady

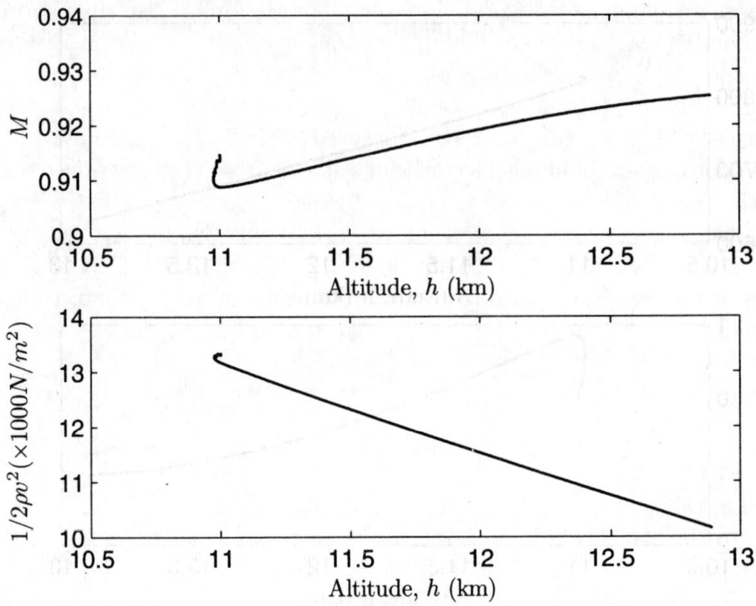

Fig. 12.9. The Mach number and dynamic pressure vs. the altitude of a jet airplane in a long-range cruise.

wind is present, the ground track deviates from the great circle route, thereby requiring FMS (or pilot) inputs to prevent such a deviation.

12.3.2 Effect of a Steady Wind on an Airplane Flight

The atmosphere is never completely at rest relative to the earth. At low altitudes of airplane flight, there can be strong horizontal air currents, called *winds*, capable of affecting the flight paths. Most airplanes are subjected to winds of various intensities during their flight. During cruise, the altitude is large enough for the atmospheric turbulence caused by the ground to die down, resulting in a wind of nearly constant speed and direction at a given altitude. Careful periodic observations of the wind velocities above selected weather stations yield hourly *winds-aloft* data at a series of altitudes, which the pilots use in flight planning. Often, the prevailing winds, such as the easterly *trade winds* in the subtropical belt near the equator, or the *westerlies* at the mid-latitudes, are strong enough to drastically change the ground track of airplanes, which may cause an increase in flight time, and even fuel starvation, if not properly compensated for. At the cruising altitudes of jet airplanes (near the tropopause), a very strong west-to-east wind is sometimes encountered, called a *jetstream*. This is a "river of air" confined to a narrow altitude range and width of a few kilometers, but covering almost the entire globe, with core winds reaching 500 km/hr. With the jetstream along the flight direction, an airplane is "helped along" to its destination, saving time and fuel. Consequently, many eastward flights are planned to encounter the jetstream, even at the cost of deviating from the great circle route. However, a jetstream crossing should be avoided due to the high *wind shear*[12] and the associated turbulence, which can cause severe structural damage to lifting surfaces.

In a coordinated flight with a constant angle of attack, an airplane does not "feel" the presence of a steady wind, as the aerodynamic force and moment, which depend only on the velocity relative to the atmosphere, are unaffected. However, a change in the wind strength, or direction, manifests itself into changes in the angle of attack, sideslip, and bank angle, causing a change in the aerodynamic force and moment. The ground track, on the other hand, is affected even by a steady wind. Let the aiplane's velocity relative to the earth be $\mathbf{v}$, while a wind of velocity,

$$\mathbf{v_w} = v_w(\sin A_w \mathbf{j} + \cos A_w \mathbf{k}) \; , \tag{12.43}$$

resolved in the local horizon frame, is blowing steadily. The airplane's velocity relative to the atmosphere, $\mathbf{v}'$, is given by

$$\begin{aligned} \mathbf{v}' &= \mathbf{v} - \mathbf{v_w} \\ &= v \sin\phi \mathbf{i} + (v\cos\phi \sin A - v_w \sin A_w)\mathbf{j} \\ &\quad + (v\cos\phi\cos A - v_w \cos A_w)\mathbf{k} \; . \end{aligned} \tag{12.44}$$

We can express the relative velocity in the wind axes as $\mathbf{v}' = v'\mathbf{i_v}'$, where

$$v' = \sqrt{v^2 + v_w^2 - 2vv_w \cos\phi \cos(A - A_w)} \; . \tag{12.45}$$

The governing kinematic equations, Eqs. (12.10), (12.14), (12.15), as well as the acceleration due to gravity, the earth's curvature, and the rotation in the dynamic equations, Eq. (12.29), are unmodified. However, the aerodynamic and propulsive forces, which are resolved in the axes $\mathbf{i_v}, \mathbf{j_v}, \mathbf{k_v}$ in Eq. (12.29), are changed due to the wind, because of the change in airspeed [Eq. (12.45)], and the modified wind axes, $\mathbf{i_v}', \mathbf{j_v}', \mathbf{k_v}$. Let the aerodynamic and propulsive force in a coordinated flight be expressed as

$$\begin{aligned} \mathbf{f_a} + \mathbf{f_T} &= (f_T' - D')\mathbf{i_v}' - L'\mathbf{k_v} \\ &= (f_T - D)\mathbf{i_v} + f_Y \mathbf{j_v} - L\mathbf{k_v} \; , \end{aligned} \tag{12.46}$$

where prime denotes the wind-modified force magnitudes. Employing the coordinate transformation between $\mathbf{i_v}, \mathbf{j_v}, \mathbf{k_v}$ and $\mathbf{i_v}', \mathbf{j_v}', \mathbf{k_v}$, we have

[12] Wind shear is the name given to abrupt changes in the wind speed and direction, which can suddenly increase the angle of attack, sideslip, and bank angle, thereby putting enormous impulsive loads on the airplane. The high-altitude wind shear is also referred to as *clear-air turbulence* (CAT), as it is unaccompanied by clouds and, hence, gives no advance warning to the pilot.

$$\mathbf{i_v}' \cdot \mathbf{i_v} = \mathbf{j_v}' \cdot \mathbf{j_v} = \cos\beta \,, \tag{12.47}$$

where β is the effective sideslip angle caused due to the wind, given by

$$v' \cos\beta = v\sin^2\phi + (v\cos\phi\sin A - v_w \sin A_w)\cos\phi\sin A$$
$$+(v\cos\phi\cos A - v_w\cos A_w)\cos\phi\cos A \,. \tag{12.48}$$

Substituting Eq. (12.47) into Eq. (12.46), we have

$$(f_T - D) = (f'_T - D')\cos\beta,$$
$$f_Y = (f'_T - D')\sin\beta, \tag{12.49}$$
$$L = L'.$$

Therefore, the wind generates a side force in the original (no-wind) wind axes, even though the flight is coordinated in reference to the modified wind axes. If the angle of attack is kept unchanged in the presence of the wind, the lift coefficient would remain unmodified from its no-wind value.

Example 12.2. Let us repeat the simulation presented in Example 12.1 with a steady wind. In order to do so, we will modify the aerodynamic and propulsive force calculations in the program *airflight3dof.m* as follows:

```
v=x(4);
vp=sqrt(v^2+vw^2-2*v*vw*cfpa*cos(chi-Aw));
cosbeta=(v*sfpa^2+(v*cfpa*schi-vw*sw)*cfpa*schi...
        +(v*cfpa*cchi-vw*cw)*cfpa*cchi)/vp;
beta=acos(cosbeta);
Qinf = 0.5*rho*vp^2;
if Qinf>=qmax || machb>=Mmax
    fT=fT0*0.85;
    tsfc=tsfc0*0.95;
else
    fT=fT0;
    tsfc=tsfc0;
end
m=x(7);
if m<=mfinal
    fT=0;
end
CD=CD0+K*CL^2;
D=Qinf*S*CD;
Xfo = (fT*rho/0.3663-D)*cosbeta;
Yfo = (fT*rho/0.3663-D)*sin(beta);
Zfo = Qinf*S*CL;
```

We study two normally encountered wind conditions: (a) a wind speed of 50 km/hr from the southeast ($A_w = 135°$), and (b) a wind speed of 50 km/hr from the northwest ($A_w = 315°$). The resulting ground tracks are compared with that of zero wind in Fig. 12.10. It is clear that in both wind conditions, the airplane departs significantly from the zero-wind, great circle route. At the end of the 5.73 hr flight, the airplane has deflected by approximately (a) 489 km and (b) 508.2 km from the intended destination (New York's JFK airport). One can ensure that the original destination is reached in the presence of wind by either selecting a different initial azimuth and following a straight flight, or constantly adjusting the lift coefficient, thrust, and bank angle in order to follow the great circle route (such as the lift and bank modulation presented in Example 4.5). However, while the former option will increase both the time and fuel of flight due to its longer path (and may cause fuel starvation before reaching the destination), the latter requires a precise control of the flight path (called *closed-loop guidance*), which is generally feasible only with a computerized FMS. Therefore, a steady wind of even a relatively small strength can have an insidious effect on the flight if not properly planned for by making the necessary course corrections and carrying the extra fuel an adverse wind requires.

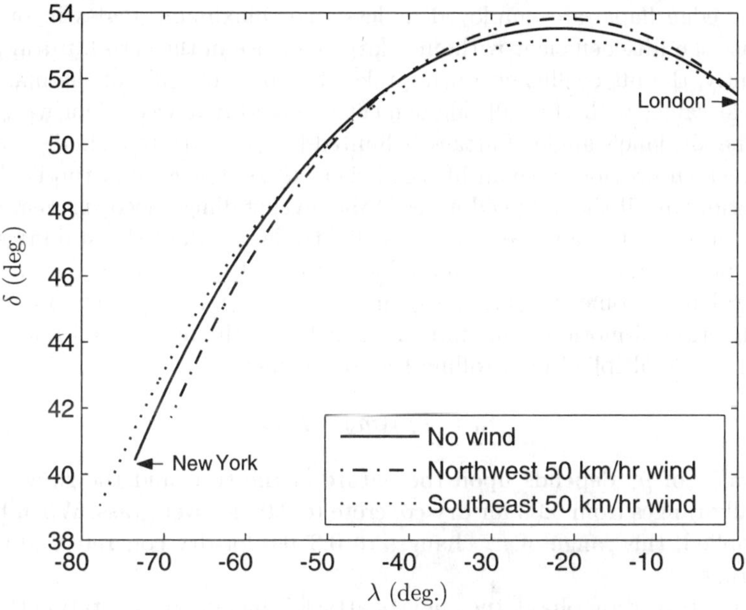

Fig. 12.10. The ground track of a jet airplane in a long-range cruise with a steady wind.

Another simulation of the wind's effect on an airplane flight was presented in Chapter 4 (Example 4.5), where we considered a ground-referenced, horizontal turn in the presence of a steady wind. Example 4.5 considers a turn close to a ground station, where it was unimportant to include the effects of the earth's curvature and rotation. In a manner similar to Example 4.5, we can derive the modulation in the lift coefficient and bank angle for maintaining either a desired straight ground track (great circle route) or a large-radius turn (such as a surveillance aircraft observing a facility from a distance), where the additional terms due to planetary curvature and rotation may be necessary.

By carrying out a simulation with the modulated lift and bank angle, we can test whether the airplane follows the desired, closed-loop guided path.

12.3.3 Take-Off Maneuver

The take-off is a complex maneuver, with a changing speed, flight-path angle, and variations in the aerodynamic and propulsive forces due to changes in airspeed, angle of attack, and vicinity of the ground.[13] A conventional take-off can be divided into three distinct phases: (a) *ground run*, where the airplane accelerates with all wheels in contact with the ground; (b) *transition* from level to a climbing attitude, at the end of which the airplane becomes airborne; and (c) a quasi-steady *climb*. At take-off, high-lift devices, such as leading- and trailing-edge flaps, are employed at less than maximum angles in order to increase the lift coefficient without a large increase in the zero-lift drag. Consequently, the lift coefficient during take-off is only 60–80% of the maximum possible value, with the full deployment of high-lift devices. During ground run, the airplane's angle of attack is limited by geometry to a small, constant value, which provides a small lift coefficient, C_{LG}. The ground run is allowed to continue until the airspeed exceeds the level stalling speed, v_s, usually by 10%. Although the airspeed becomes slightly larger than the stalling speed, the airplane remains on the ground due to the fact that $C_{LG} \ll C_{L\,\max}$. Until the airplane becomes airborne, a ground frictional force, $\mathbf{f}_\mu$, acts opposite to the direction of motion. This force is modeled as the net normal reaction on the wheels, multiplied by a rolling friction coefficient, μ_r,

$$\mathbf{f}_\mu = -\mu_r(mg - L)\mathbf{i_v} \,. \tag{12.50}$$

The value of μ_r depends upon the nature of the tires and the runway, and typically ranges from 0.03 for dry concrete to 0.08 for wet grass. When brakes are applied, this range of μ_r changes to 0.3–0.5 for dry concrete and 0.2 for wet grass.

In the transition phase, the angle of attack is increased by rotating the nose upward, such that the lift coefficient becomes close to (approximately 80%) the maximum lift coefficient. During this time, the airspeed has increased to approximately $1.2v_s$. This combination of airspeed and lift coefficient produces a lift greater than the weight, and the flight-path angle increases due to the larger-than-unity load factor, $n \approx (0.8)(1.2)^2 = 1.152$. Thus, the airplane assumes a positive flight-path angle at the end of transition. At this time, the pilot adjusts the lift coefficient such that a quasi-steady climb is established at $v \approx 1.2v_s$. The take-off maneuver is considered complete when all ground obstacles have been avoided, and a climb has been established. The airplane certifying agencies [such as the Federal Aviation Administration (FAA) of the

[13] The discussion presented here can be applied—with appropriate modifications—to a ship-board take-off and take-off from a liquid surface. Hence, the generic term "ground" is taken to refer to any flat surface.

U.S.] lay down specific definitions of the take-off distance and the requirements of a quasi-steady climb angle (or rate) for a safe take-off. These specifications dependent on the airplane category (such as *normal*, *utility*, or *transport*) and the number and type of engines. For example, the take-off distance of a single piston-propeller engined, normal airplane is defined under FAA regulation FAR Part 23 [33] as the distance from brakes release to the top of a 50-ft-high obstacle, and the safe rate of climb with landing gear retracted is considered to be 300 ft/min at standard sea level. Multi-engined airplanes have much more complex take-off specifications and include the distance and climb requirements with one engine inoperative. Here we present a strategy for simulating an airplane take-off, which can be applied with due modification to any realistic situation (presence of a steady wind, engine failure during take-off, etc.).

Generally, the effects of planetary curvature, rotation, and oblateness are neglected during take-off, which involves small relative speeds and duration. However, we shall keep all the terms in the equations of motion and study their possible effects in the ensuing trajectory. During the ground roll, the pilot ensures the changes in the heading (velocity azimuth, A) are negligible, by keeping the nose pointed along the runway center line, even in the presence of a cross-wind. Hence, we will neglect the variation in the heading and flight-path angle until the time the airplane becomes completely airborne. Thereafter, the trajectory is free to be modified by the effects of planetary rotation, curvature, oblateness, and wind. Generally, the pilot turns to a desired heading at the completion of take-off, in order to establish a cruise toward a destination, as discussed above. Sometimes, the time taken to rotate a heavy airplane during the transition period could be sufficiently large (1–2 seconds) to affect the total take-off distance. For this reason, a distinct *rotation* phase is often added immediately preceding the transition. An accurate estimation of the rotational maneuver is possible by taking into account the equation for the pitching motion (Chapter 13), with the required moment of inertia and pitching moment contributions from the wing, fuselage, tail, engines, and landing gear. We will postpone the discussion of rotational dynamics until Chapter 13, and assume an instantaneous change in the lift coefficient from C_{LG} to $0.8C_{L\,\max}$, which is reasonably accurate for most airplanes.

A *ground-effect* phenomenon is encountered during take-off and landing when the airplane's wings are in the vicinity of the ground (less than a wing span away). We saw in Chapter 10 that the lift-induced drag of a wing depends, in part, on the strength of the rotational flow near the wing tips (tip vortices). Due to the ground effect, the strength of the tip vortices diminish, thereby directly reducing the lift-induced drag. Thus, the lift-dependent drag factor, K, is modified by the ground effect to an effective value, K', whose semi-empirical expression can be derived to be the following [36]:

$$K' = K\frac{33(\frac{h}{b})^{\frac{3}{2}}}{1+33(\frac{h}{b})^{\frac{3}{2}}}\,, \qquad (12.51)$$

where b is the wing span, and h denotes the height of the wing plane above the ground.[14] During ground run and transition, $h \ll b$, thereby making K' much smaller than K. The reduced drag in the ground effect causes a tendency of accelerating and climbing more rapidly at take-off, or "floating" without touching down during landing, as compared to flights without ground effect. When $h > b$, the ground effect vanishes, and there is a tendency to sink at that point, which could be dangerous in some cases (some accidents are caused when airplanes are too heavy to be flown "outside" the ground effect, but they nevertheless take-off momentarily due to ground effect). This is the main reason why minimum climb requirements during take-off are specified for airplane certification.

The variation of thrust during take-off is an important modeling parameter. An airbreathing engines display a change of thrust with airspeed at small take-off velocities when compared to a static situation immediately prior to the beginning of take-off. For airplanes with propeller engines, this variation is modeled by taking into account the static thrust coefficient (Chapter 11), which is only a function of the power coefficient and is available from the propeller charts provided by the manufacturer. Since take-off normally occurs at the maximum throttle setting, the static thrust coefficient can be determined from the maximum power developed at the atmospheric conditions prevailing at the airport. As the airspeed becomes sufficiently large and constant (approximately $1.2v_s$), a model for the propeller efficiency in terms of the advance ratio (Chapter 11) is employed for predicting the thrust. Hence, a smooth interpolation from the constant static thrust to the thrust value at $v = 1.2v_s$ can be used for a propeller airplane. A similar approach can be applied for jet-engined airplanes, where the rated static sea level thrust can be corrected for atmospheric density and installation losses, the level flight thrust at the airport altitude and Mach number corresponding to $v = 1.2v_s$ is obtained from engine data, and a smooth interpolation is employed between the two points.

Example 12.3. Consider the take-off of a jet airliner with total wing planform area $S = 223.0815$ m^2, mean aerodynamic chord $\bar{c} = 5.42$ m, and span, $b = 41.2$ m, beginning at $v = 0$ at a standard altitude of 3 m (the height of the airplane's center of mass during ground-run) at $\delta = 51.45°, \lambda = 0$, with initial mass $m = 84,890.909$ kg and (heading) velocity azimuth $A = 270°$. The aerodynamic coefficients during take-off are $C_{L\max} = 1.6, C_{LG} = 0.1, C_{D0} = 0.015$, and $K = 0.055$, while the rolling friction coefficient between the tires and the runway is $\mu_r = 0.03$. The maximum throttle setting is maintained throughout the take-off such that the static thrust is $f_T = 211,128.17$ N and static TSFC, $c_T = 0.4$ 1/hr. Both the thrust and the TSFC vary linearly with airspeed until achieving a constant value of 90% of static thrust and 110% of static TSFC, respectively, at the level stalling speed, v_s. Thereafter, both

[14] Although wings are nonplanar surfaces, we can associate an average (or median) plane, such as the one containing the mean aerodynamic chords of the wings.

thrust and TSFC are assumed invariant with airspeed, but the thrust varies in direct proportion to the atmospheric density. The lift coefficient is maintained constant at $C_L = 0.8C_{L\max}$ after lift-off. The pilot maintains a straight-line path until the airplane becomes airborne, and thereafter no corrections in the heading are made. The landing gear and high-lift devices are not retracted in the take-off simulation. No wind is assumed (although a steady wind can be easily modeled using the approach given in the previous subsection). It is best to simulate the take-off in its first minute.

A MATLAB program called *takeoff.m* (Table 12.3) is written to compute the time derivatives of all motion variables, $r, \delta, \lambda, v, \phi, A, m$, according to the kinematic and dynamic equations of motion, mass variation, as well as the variation of aerodynamic and propulsive parameters during take-off. These time derivatives are provided as inputs to the MATLAB's intrinsic Runge–Kutta solver, *ode45.m*, at each time step. Another program, called *runtakeoff.m* (Table 12.4), specifies the initial condition and invokes *ode45.m*. As before, the standard atmospheric and nonspherical earth gravity models are provided by *atmosphere.m* (Chapter 9) and *gravity.m* (Chapter 3), respectively.

The resulting plots of the important motion variables and related quantities are shown in Figs. 12.11–12.15. It is evident from the airspeed vs. altitude plot, Fig. 12.11, that the stall boundary is not violated after the airplane becomes airborne. It reaches an altitude of approximately 245.45 m at the end of 60 s after brake release, and covers a total distance of about $x = 5000$ m down-range of the starting point in this duration (Fig. 12.12). Using the standard FAA definition of take-off distance for a jet transport with all engines operating at 115% of the distance covered along the runway until a 35-ft obstacle is cleared, we arrive at a figure of $(1.15)(2690) = 3093.5$ m as the total take-off distance. After the airplane has become airborne, its heading changes slightly due to the combined effects of nonspherical earth gravity and the earth's curvature and rotation. This change in heading causes a drift normal to the runway, called the *cross-range, y*. As shown in Fig. 12.13, the total cross-range drift at the end of 60 sec is less than 6.3 m. Such a drift would be much larger in the presence of a cross-wind if uncorrected. The flight-path angle (or the angle of climb), ϕ, during take-off is plotted in Fig. 12.14, which displays an overshoot to $\phi = 16°$, before stabilizing at a nearly constant value, $\phi \approx 8°$ at the end of about 50 sec. The transition from a level, accelerating flight to a quasi-steady climb with a nearly constant airspeed is also evident in Fig. 12.15, which plots the time history of airspeed, v, and rate of climb, $\dot{h} = v \sin\phi$. After 50 sec, a quasi-steady rate of climb of about 10 m/s is reached with a constant airspeed of approximately 74 m/s, which is quite close to the desired speed $[v = 1.2v_s = (1.2)(61.66) = 74 \text{ m/s}]$. This is indeed remarkable, as the pilot maintains a constant lift coefficient, $C_L = 0.8C_{L\max}$, after lift-off. Thus, the airplane automatically seeks to establish a quasi-steady climb after the lift coefficient and thrust are set, without any further input from the pilot. The initial oscillation in the flight-path angle is caused by

Table 12.3. M-file *takeoff.m* for an Airplane's Governing State Equations during Take-Off

```
function deriv = takeoff(t,o)
global dtr; global mu; global omega; global S; global c; global rm;
global CD0; global K; global b; global CLmax; global CLG;
global fT0; global tsfc0; global mur;
%acceleration due to gravity (nonspherical earth):
[g,gn]=gravity(o(3),o(2));
lo = o(1);la = o(2);
clo = cos(lo); slo = sin(lo);
cla = cos(la); sla = sin(la);
fpa = o(5); chi = o(6);
cfpa = cos(fpa); sfpa = sin(fpa);
cchi = cos(chi); schi = sin(chi);
%%%atmosphere determination
if o(3)<rm+3
    o(3)=rm+3;
end
alt = o(3) - rm;
v  = o(4);
atmosp = atmosphere(alt,v,c);
rho = atmosp(2);
machb = atmosp(3);
%%%end:atmosphere determination
[t alt v]
Qinf = 0.5*rho*v^2;
if alt<=b
    Keff=K*33*(alt/b)^1.5/(1+33*(alt/b)^1.5);
else
    Keff=K;
end
m=o(7);
vstall=sqrt(2*m*g/(1.2249*S*CLmax));
if v<=vstall
fT=fT0-0.1*fT0*v/vstall;
tsfc=tsfc0+0.1*tsfc0*v/vstall;
else
    fT=0.9*fT0;
    tsfc=1.1*tsfc0;
end
if v<1.2*vstall
    CL=CLG;
else
    CL=0.8*CLmax;
end
CD=CD0+Keff*CL^2;
D=Qinf*S*CD;
L=Qinf*S*CL;
if alt==3
    Xfo = fT-D-mur*(m*g-L);
else
    Xfo = fT*rho/1.2249-D;
end
Zfo = L;
Yfo = 0;
%trajectory equations follow:
longidot = o(4)*cfpa*schi/(o(3)*cla); %longitude
latidot =  o(4)*cfpa*cchi/o(3); %latitude
raddot = o(4)*sfpa;  %radius
veldot=-g*sfpa+gn*cchi*cfpa...
   +Xfo/m+omega*omega*o(3)*cla*(sfpa*cla-cfpa*cchi*sla);
if v<1.2*vstall && alt==3
    gammadot=0;
    headdot=0;
else
```

```
gammadot=(o(4)/o(3)-g/o(4))*cfpa-gn*cchi*sfpa/o(4)...
        +Zfo/(o(4)*m)+2*omega*schi*cla...
        +omega*omega*o(3)*cla*(cfpa*cla...
        +sfpa*cchi*sla)/o(4);
headdot =o(4)*schi*tan(o(2))*cfpa/o(3)...
        -gn*schi/o(4)-Yfo/(o(4)*cfpa*m)...
        - 2*omega*(tan(o(5))*cchi*cla-sla)...
        +omega*omega*o(3)*schi*sla*cla/(o(4)*cfpa);
end
mdot=-tsfc*fT/(9.8*3600);
%time derivatives:
deriv=[longidot;latidot;raddot;veldot;gammadot;headdot;mdot];
```

Table 12.4. M-file *runtakeoff.m* for Solving an Airplane's State Equations during Take-Off

```
global dtr; dtr = pi/180;
global mu; mu = 3.986004e14; %earth's grav. const (m^3/s^2)
global omega; omega=2*pi/(23*3600+56*60+4.0905); %(rad/sec)
global S; S =  223.0815; %wing planform area (m^2)
global c; c=5.42; %wing mean aerodynamic chord (m)
global rm; rm = 6378140 ; %earth's radius (m)
global CD0; CD0=0.02;
global K; K=0.055;
global b; b=41.2; %wing span (m)
global CLmax; CLmax=1.6;
global CLG; CLG=0.1; %lift coeff. during ground run
global fT0; fT0=211128.17; %sea level static thrust (N)
global tsfc0; tsfc0=0.4; %sea level static TSFC (per hour)
global mur; mur=0.03; %rolling friction coefficient
%Initial condition:
long = 0*dtr;          %initial longitude
lat = 51.45*dtr;       %latitude
rad=rm+3;              %radius (m)
vel=0.001;             %speed (m/s)
fpa=0;                 %flight-path angle
chi=270*dtr;           %velocity azimuth (from north)
m=84890.909;           %aircraft mass (kg)
init = [long; lat; rad; vel; fpa; chi; m];
[t, o] = ode45('takeoff',[0, 60], init);
fclose('all');
```

the fact that the airspeed at lift-off is much larger than that required for achieving a quasi-steady climb with the given lift coefficient. In the ensuing flight-path oscillation, the airspeed is reduced to the required value, and a constant climb angle is achieved, whose value is dictated by the specific excess thrust, $\phi \approx \frac{f_T - D}{mg}$. The mass variation is nearly linear, with the net change in mass during the first 60 sec (not plotted) of about 143 kg (0.17% of the initial mass).

The landing maneuver can be regarded as the reverse of the take-off, with the objective of bringing the airplane to rest on the ground from a quasi-steady descent. The landing approach is normally begun at a speed approximately 30% higher than the level stalling speed, v_s, with the maximum deployment of high-lift devices, and a small power setting. The power setting is adjusted in order to achieve the desired rate of descent in a high-drag configuration.

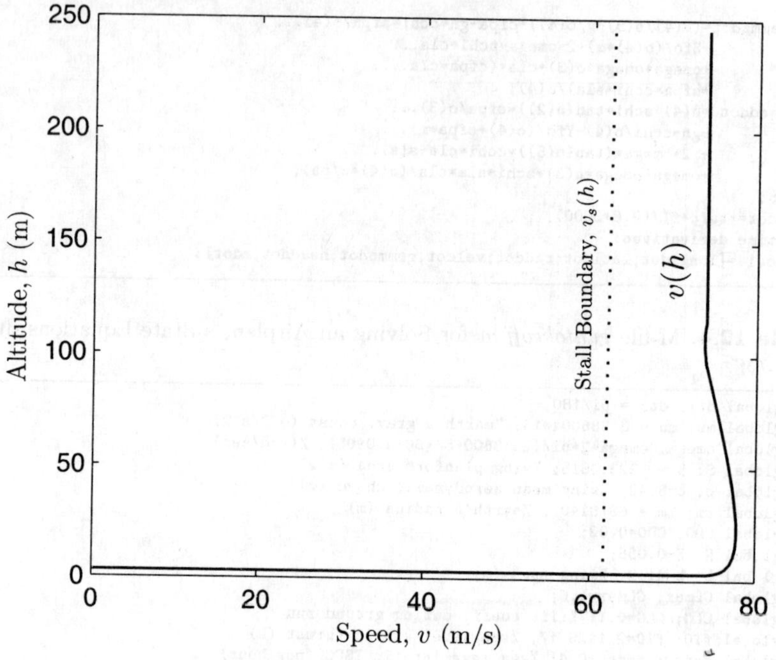

Fig. 12.11. The airspeed vs. altitude plot of a jet airplane during take-off.

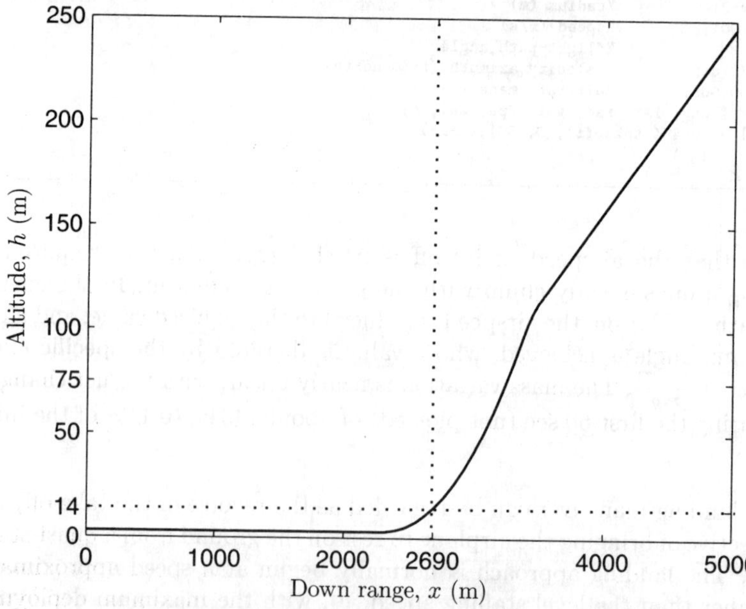

Fig. 12.12. The down range vs. altitude plot of a jet airplane during take-off.

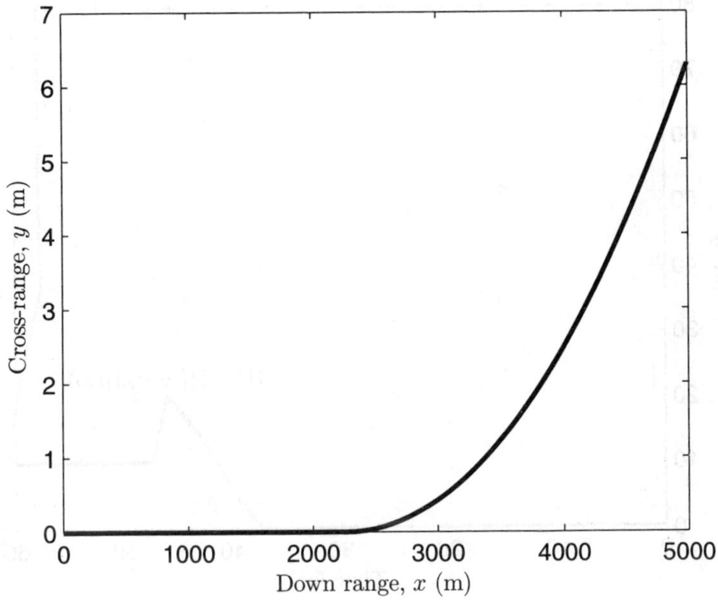

Fig. 12.13. The down range vs. cross-range plot of a jet airplane during take-off.

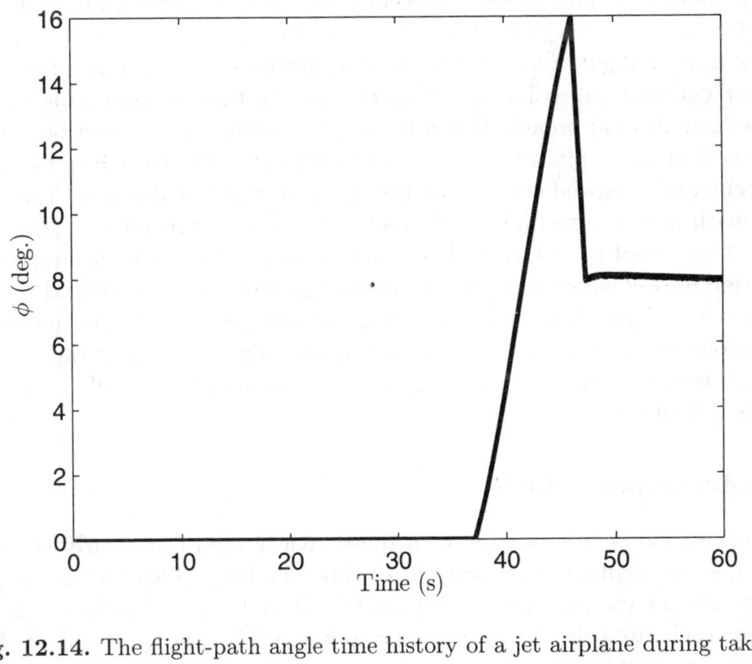

Fig. 12.14. The flight-path angle time history of a jet airplane during take-off.

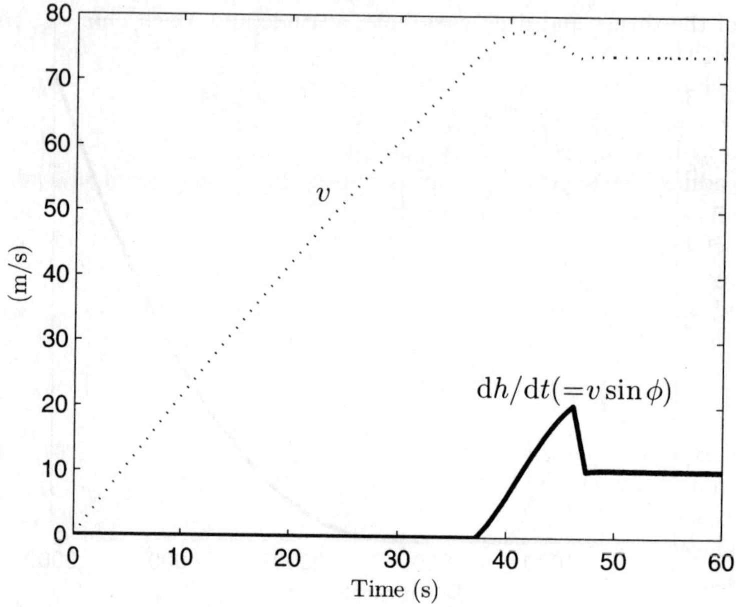

Fig. 12.15. The airspeed and rate of climb of a jet airplane during take-off.

Usually, there is also an angle of descent required by either a *glide-slope indicator* (GSI), which is a ground-based radio beacon, transmitting in a narrow beam at a fixed angle (about 3°) above the runway, or a *visual approach-slope indicator* consisting of color-coded lights near the runway indicating the approximate angle of approach. When the airplane arrives at a prescribed height above the runway, a pitch maneuver is executed in order to reduce the angle of descent and airspeed, such that the airplane touches down on the main wheels with a near-zero flight-path angle, ϕ, and an airspeed just above the level stalling speed ($v \approx 1.1v_s$). The touch down is followed by a brief period of rotation during which the airplane is brought to its normal attitude on the ground, with all wheels in contact with the runway. Finally, brakes and other retardation devices (such as reverse thrust) are applied to bring the airplane to rest. It is easy to modify the program given earlier for take-off, in order to simulate a landing maneuver.

12.3.4 Accelerated Climb

While all airplanes are capable of a quasi-steady climb at nearly constant airspeed, some airplanes can accelerate while climbing. This requires a high thrust-to-weight (or power-to-weight) ratio. Most fighter airplanes can perform the accelerated climb to supersonic speeds, with some even able to climb vertically for limited periods. The simulation of an accelerated climb of a fighter airplane requires that one should account for the simultaneous variation of the thrust and drag with both altitude and Mach number. The variation of the parasite drag coefficient, C_{D0}, with Mach number is possible by taking into account subsonic compressibility effects, as well as transonic and supersonic wave drag (Chapter 10). The lift-dependent drag factor, K, also varies with M due to compressibility and wave drag and is best modeled by the leading-edge suction analogy (Chapter 10). The amount of leading-edge suction, s, depends on whether the leading edge is subsonic as well as the lift coefficient, C_L. Modern fighter aircraft are powered by low-bypass, afterburning turbofan engines, the variation of whose thrust and TSFC with the altitude and Mach number was presented by a typical example in Chapter 11. Thus, we are well prepared to appropriately model an accelerated climb.

Example 12.4. Consider a modern fighter airplane of wing-planform area $S = 56.5$ m^2, span $b = 14.1512$ m, mean aerodynamic chord $\bar{c} = 3.9926$ m, initial mass $m = 17350$ kg, leading-edge sweep angle $\Lambda_{LE} = 50°$, and the maximum lift coefficient in clean configuration, $C_{L\max} = 1.5$. Structurally, the airplane is limited by $\bar{n} = 7.5, q_{\max} = 49,000$ N/m^2, $M_{\max} = 2.5$. The airplane is equipped with two low-bypass-ratio, afterburning turbofan engines whose characteristics of installed thrust and TSFC were given in Example 12.1 and coded in the M-file *engine.m*. The airplane begins a climb with maximum afterburning thrust setting after retracting the landing gear and flaps following a take-off, with initial altitude $h = 200$ m, position $\delta = 20°$, $\lambda = 70°$, relative speed $v = 69$ m/s, azimuth $A = 270°$, and flight-path angle $\phi = 40°$. It is best to simulate the climb for 80 seconds, during which the lift coefficient is modulated to follow the instantaneous quasi-steady climb value

$$C_L = \frac{2mg\cos\phi}{\rho S v^2}\,,$$

in order that the flight-path curvature remain small. The effects of steady wind are neglected, but can be easily incorporated.

A MATLAB program named *accelclimb.m* (Table 12.5) is written to compute the time derivatives of the motion variables, $r, \delta, \lambda, v, \phi, A, m$, which uses the programs *parasite.m* (Table 12.6) for C_{D0}, *liftddf.m* for K by the leading-edge suction analogy (Table 12.7), and *engine.m* for afterburning turbofan engine characteristics with maximum power, along with *atmosphere.m* (Chapter 9) and *gravity.m* (Chapter 3). The time derivatives are provided as inputs to *ode45.m*, which is invoked by another program, called *runaccelclimb.m* (Table 12.8), along with the relevant parameters and initial condition. The plot of parasite drag coefficient with Mach number is given in Fig. 12.16, while the variation of the zero-suction and full-suction values of the lift-dependent drag factor, K_0 and K_{100}, respectively, are plotted in Fig. 12.17. We recall from Chapter 10 that the net lift-dependent drag factor, K, depends upon the actual amount of the leading-edge suction, s, according to

$$K = sK_{100} + (1 - s)K_0\,.$$

Table 12.5. M-file *accelclimb.m* for an Airplane's State Equations during Accelerated Climb

```
function deriv = accelclimb(t,o)
global dtr; global mu; global omega; global S; global c;
global rm; global b; global CLmax; global sweep;
global f8;
%acceleration due to gravity (nonspherical earth):
[g,gn]=gravity(o(3),o(2));
lo = o(1);la = o(2);
clo = cos(lo); slo = sin(lo);
cla = cos(la); sla = sin(la);
fpa = o(5); chi = o(6);
cfpa = cos(fpa); sfpa = sin(fpa);
cchi = cos(chi); schi = sin(chi);
%%%atmospheric properties:
if o(3)<rm
    o(3)=rm;
end
alt = o(3) - rm;
v = o(4);
atmosp = atmosphere(alt,v,c);
rho = atmosp(2);
Qinf = 0.5*rho*v^2;
mach = atmosp(3);
%%%engine data:
[fT,tsfc]=engine(alt,mach);
fT=2*fT;
m=o(7);
%%%drag data:
CD0=parasite(mach);
CL=m*g*cfpa/(Qinf*S);
if CL>CLmax
   CL=0
end
K=liftddf(mach,CL,sweep,b^2/S);
[t alt mach CL]
if alt<=b
    Keff=K*33*(alt/b)^1.5/(1+33*(alt/b)^1.5);
else
    Keff=K;
end
CD=CD0+Keff*CL^2;
D=Qinf*S*CD;
%%%aero-propulsive force components:
L=Qinf*S*CL;
Xfo = fT-D;
Zfo = L;
Yfo = 0;
fprintf(f8,'\t%1.5e\t%1.5e\t%1.5e\t%1.5e\t%1.5e\n',...
        t,alt,CL,mach,Qinf);
%%%trajectory equations:
longidot = o(4)*cfpa*schi/(o(3)*cla);
latidot =  o(4)*cfpa*cchi/o(3);
raddot = o(4)*sfpa;  %Radius
veldot = -g*sfpa +gn*cchi*cfpa + Xfo/m...
         +omega*omega*o(3)*cla*(sfpa*cla-cfpa*cchi*sla);
gammadot=o(4)/o(3)-g/o(4))*cfpa-gn*cchi*sfpa/o(4)...
         +Zfo/(o(4)*m)+ 2*omega*schi*cla...
         + omega*omega*o(3)*cla*(cfpa*cla...
         + sfpa*cchi*sla)/o(4);
headdot = o(4)*schi*tan(o(2))*cfpa/o(3)...
         -gn*schi/o(4)-Yfo/(o(4)*cfpa*m)...
         - 2*omega*(tan(o(5))*cchi*cla - sla)...
         + omega*omega*o(3)*schi*sla*cla/(o(4)*cfpa);
mdot=-tsfc*fT/(9.8*3600);
%%%Time derivatives:
deriv = [longidot; latidot; raddot;...
        veldot; gammadot; headdot; mdot];
```

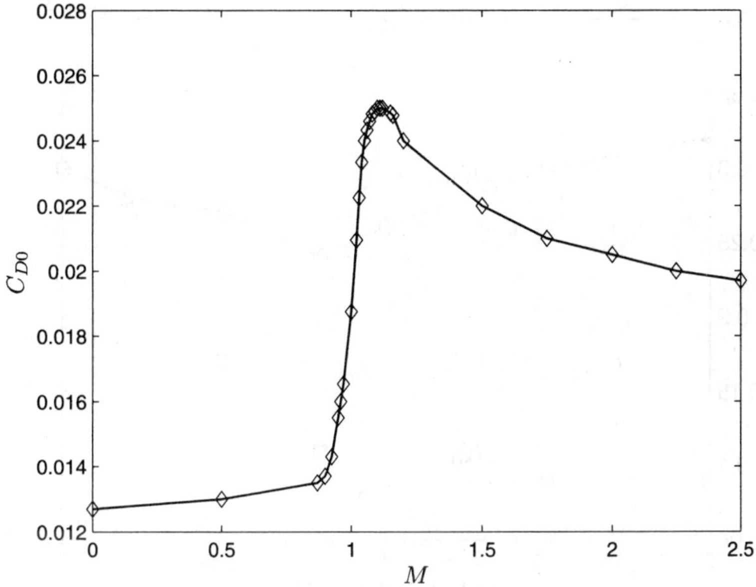

Fig. 12.16. The parasite drag coefficient of a jet fighter in clean configuration.

Table 12.6. M-file *parasite.m* for the Parasite Drag Coefficient of a Fighter Airplane

```
function cd0=parasite(mach)
%parasite drag coefficient of a fighter airplane (tabular data)
M=[0 0.5 0.87 0.9 0.925 0.95 0.96 0.97 1 1.02...
   1.03 1.04 1.05 1.06 1.07 1.08 1.09 1.1 1.11...
   1.12 1.15 1.16 1.2 1.5 1.75 2.0 2.25 2.5];
cd=[0.0127 0.013 0.0135 0.0137 0.0143 0.0155 0.016...
    0.01654 0.01875 0.02095 0.02225 0.02335 0.024...
    0.024325 0.0246 0.02482 0.0249 0.025 0.025 0.025...
    0.024855 0.02478 0.024 0.022 0.021 0.0205 0.02 0.0197];
cd0=interp1(M,cd,mach);
```

We have assumed the following variation of s with C_L:

$$s = \begin{cases} 0.9, & C_L \leq 0.1, \\ \frac{0.1}{C_L} - 0.1, & 0.1 < C_L \leq 1, \\ 0, & C_L > 1. \end{cases}$$

Of course, $s = 0$ when the leading edge is supersonic, i.e., $M > \sec \Lambda_E$.

The simulation results are shown in Figs. 12.18–12.22. The accelerated climb is evident from Fig. 12.18, which shows a continuous increase in both airspeed and altitude until $h = 11,390$ m, where the speed attains a maximum value of $v = 387$ m/s, followed by a decline in the speed to $v = 361.8$ m/s, at the maximum altitude of $h = 15,215$ m. The stall boundary dictated by

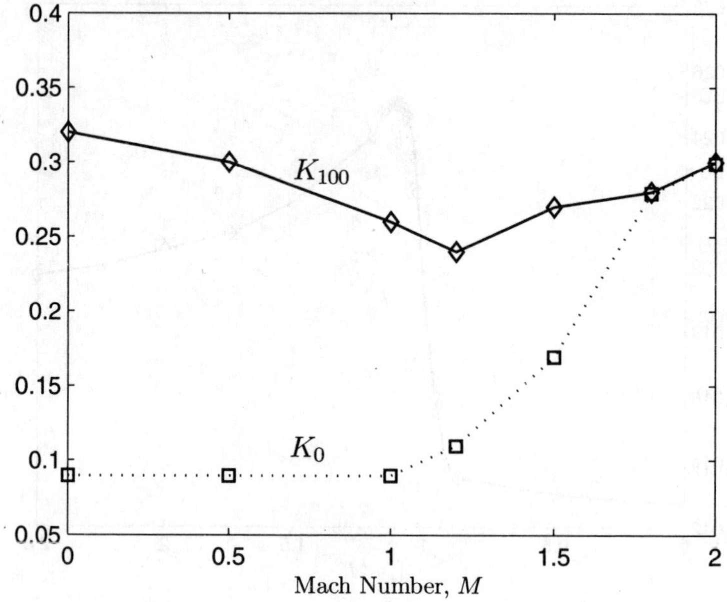

Fig. 12.17. The lift-dependent drag factors of a jet fighter with zero and full leading-edge suction.

Table 12.7. M-file *liftddf.m* for the Lift-Dependent Drag Factor of a Fighter Airplane

```
function K=liftddf(mach,CL,sweep,AR)
%lift-dependent drag factor with leading-edge suction
%method
M=[0 0.5 1 1.2 1.5 1.8 2];
k0=[0.32 0.3 0.26 0.24 0.27 0.28 0.3];
k100=[1 1 1 1.2222 1.8889 3.1111 3.3333]/(pi*AR);
K0=interp1(M,k0,mach);
K100=interp1(M,k100,mach);
if mach>1/cos(sweep)
    s=0;
elseif CL<=0.1
    s=0.9;
else
    s=0.1/CL-0.1;
end
if s<0
    s=0;
end
K=s*K100+(1-s)*K0;
```

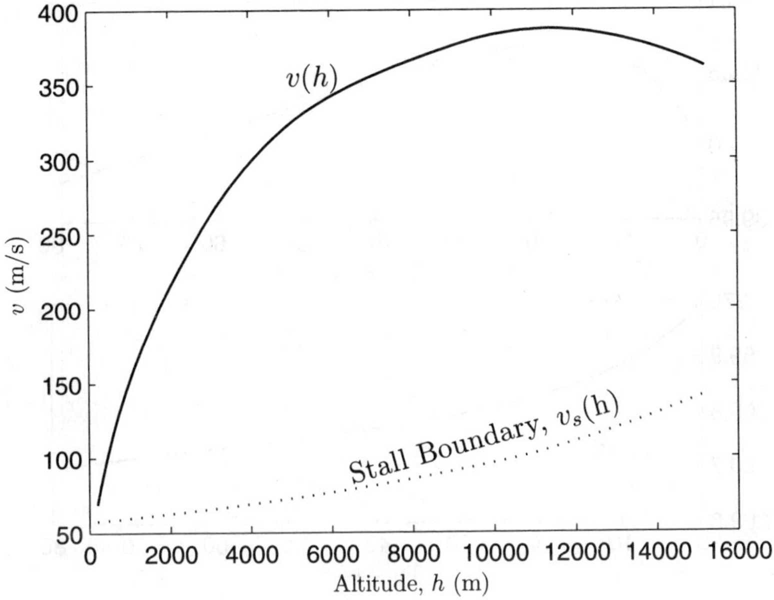

Fig. 12.18. The airspeed vs. altitude plot of a jet fighter during accelerated climb.

Table 12.8. M-file *runaccelclimb.m* for Solving an Airplane's State Equations during Climb

```
global dtr; dtr = pi/180;
global mu; mu = 3.986004e14; %(m^3/s^2)
global omega; omega = 2*pi/(23*3600+56*60+4.0905); %(rad/s)
global S; S =  56.5;  %ref area (m^2)
global c; c=3.9926; %mean aerodynamic chord (m)
global rm; rm = 6378140 ;  %Earth's radius (m)
global b; b=14.1512; % wing-span (m)
global CLmax; CLmax=1.5;
global sweep; sweep=50*pi/180; %LE sweep angle (rad.)
global f8; f8 = fopen('data8.mat', 'a');
%initial condition:
long =70*dtr;        %initial longitude (rad)
lat = 20*dtr;        %latitude (rad)
radint=rm+200;       %radius (m)
velint=69;           %speed (m/s)
fpaint=40*dtr;       %flight-path angle (rad)
chiint=270*dtr;      %velocity azimuth (from north) (rad)
m=17350;             %aircraft mass (kg)

%solution of state equations:
options=odeset('RelTol',1e-4);
init = [long; lat; radint; velint; fpaint; chiint; m];
[t, o] = ode45('accelclimb',[0, 80], init,options);
fclose('all');
```

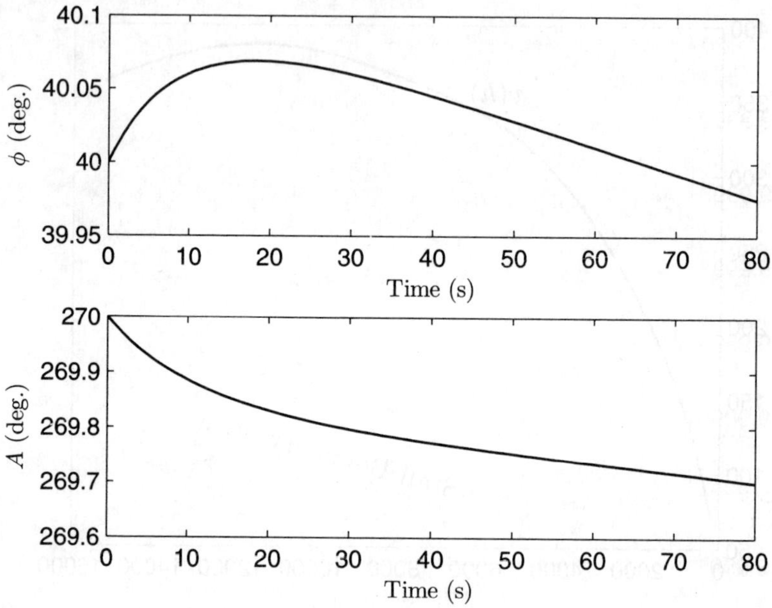

Fig. 12.19. The flight-path and heading angles of a jet fighter during accelerated climb.

the level stalling speed, v_s, also plotted in Fig. 12.18, is never violated during the climb.[15] The flight-path angle, ϕ, and the velocity azimuth (heading), A, show only a small variation in Fig. 12.19, thereby confirming an almost straight-line trajectory, whose ground track is plotted in Fig. 12.20. The speed and rate of climb time history are shown in Fig. 12.21, which depicts an almost uniformly increasing rate of climb, attaining a maximum value of $\dot{h} =$ 248.77 m/s at $t = 64.26$ s, and thereafter declining to $\dot{h} = 232.44$ m/s at $t = 80$ s. Figure 12.22 shows time-history plots of the Mach number, lift coefficient, dynamic pressure, q, and mass of consumed fuel. The Mach number attains a maximum of $M = 1.3119$ at $t = 65.06$ s, and then declines to $M = 1.2256$ at $t = 80$ s. The lift coefficient decreases for the first 39.25 s, when it reaches a minimum of $C_L = 0.05674$, and then increases to $C_L = 0.1768$ at the end of simulation. The dynamic pressure reaches a maximum value of $q = 39,430$ N/m^2 nearly at the same time as the lift coefficient becomes a minimum, and then declines almost steadily due to increasing altitude and a decreasing speed. This behavior of the dynamic pressure can be understood from the fact that the lift, $L = mg\cos\phi$, remains nearly constant due to a small variation in the mass and flight-path angle and, thus, q varies almost

[15] In a climb, the stall boundary is exactly given by $v_s \cos\phi$ rather than v_s. Thus, it is more conservative to restrict $v \geq v_s$.

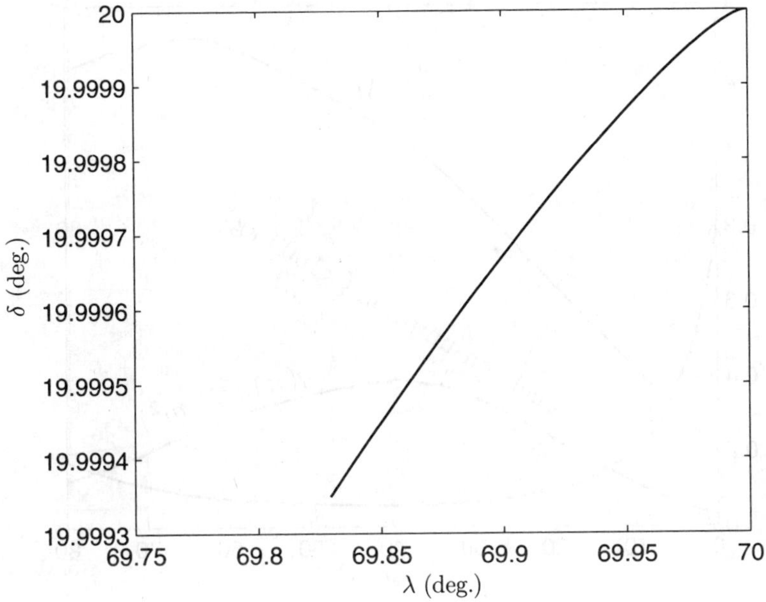

Fig. 12.20. The ground track of a jet fighter during accelerated climb.

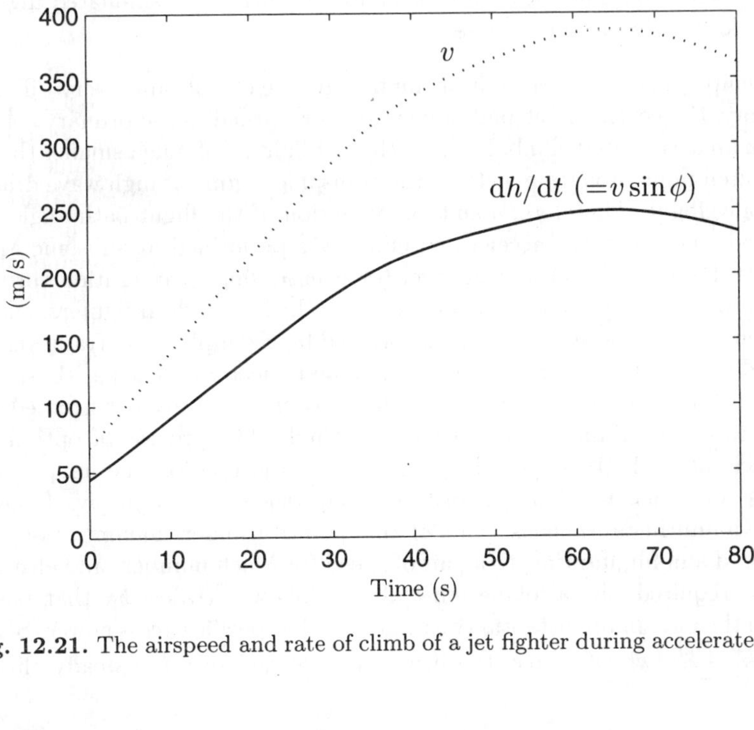

Fig. 12.21. The airspeed and rate of climb of a jet fighter during accelerated climb.

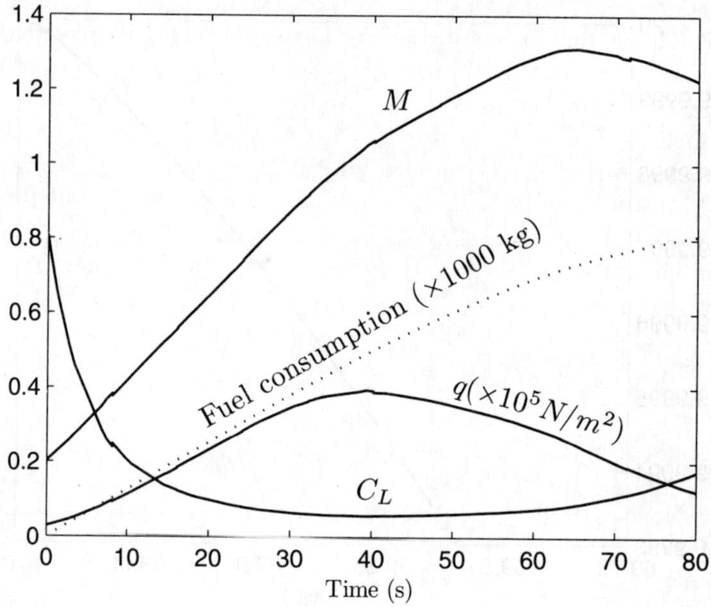

Fig. 12.22. Additional flight parameters of a jet fighter during accelerated climb.

inversely with C_L. The total fuel consumption during the simulated flight is 809.665 kg (4.67% of initial mass).

By employing a different initial climb angle, we can obtain a very different trajectory. Hence, the flight-path angle can be regarded as the primary control input in an accelerated climb. In the earlier jet fighters of much smaller thrust, it was often impossible to penetrate the transonic regime of high wave drag in a climbing flight. Therefore, a suitable variation of the flight-path angle was employed, such that the accelerated climb was performed at subsonic Mach numbers, followed by an almost level (or even diving) acceleration through the transonic regime, and a supersonic climb. In the modern fighters, such a variation of ϕ is unnecessary, as demonstrated by Example 12.4. Traditionally, the modulation of ϕ to achieve optimum climb trajectories, such as those that minimize the time or the fuel consumption, were carefully constructed out of quantities pertaining to a quasi-steady climb. The process of optimizing the accelerated climb consisted of plotting a trajectory on the airspeed vs. altitude (or Mach number vs. altitude) map, such that it always followed the maximum instantaneous value of the quantity under maximization. For example, if a minimum time to a pair of speed (or Mach number) and altitude, (h, v), is required, the airplane must always follow a trajectory that passes through the maximum instantaneous values of the specific excess power (SEP), $P_s = \frac{(f_T - D)v}{mg_c}$ (which is also the instantaneous rate of quasi-steady climb).

The pair (h, v) thus represents the total *specific energy level* (sum of potential and kinetic energies per unit mass) of the airplane, whose rate of change is the SEP. The minimum-time trajectory is thus obtained by plotting a path on the h vs. v (or M) diagram, where the contours of the maximum SEP are tangential to the contours of the specific energy level, $h + \frac{v^2}{2g_c}$. Similarly, the minimum-fuel trajectory can be constructed out of line joining the points of tangency of energy-level contours and the contours of *fuel-specific energy*, $\frac{P_s}{c_T f_T}$ [33]. However, with the direct simulation of accelerated climb (presented above), the optimization process is greatly simplified and consists of variation in the initial condition. For instance, the trajectory obtained in Example 12.4 is very close to the minimum-time trajectory to $h = 15,200$ m, $M = 1.2$.

12.3.5 Maneuvers and Supermaneuvers

Maneuver is the name given to a deliberately curved flight path through an aero-propulsive force vector normal to the flight path. As discussed earlier in the chapter, a desired general flight-path curvature can be achieved either by a combination of lift modulation and banking in a coordinated flight, or through direct side force and normal force generated by thrust vectoring in an airplane equipped accordingly. Such maneuvers can be performed by either airplanes or missiles; hence we make no further distinction between the two kinds of atmospheric vehicles.[16] Normally, an airplane maneuver is confined to either the vertical or the horizontal plane. In both the cases, the necessary centripetal acceleration is produced by a lift force greater than the weight. This is accomplished by flying at a large angle of attack, thereby increasing the lift coefficient. The maximum lift coefficient is limited by aerodynamic (stall) and structural constraints, as discussed above. Certain well-defined analytical results are possible by studying approximate equations of motion in a maneuvering flight [34]. It is a common practice to neglect airspeed variations during a horizontal turn, which is called a steady turn approximation. In a vertical turn (also known as a pitch maneuver), either the lift coefficient or the load factor can be assumed constant. Furthermore, there is a special class of analytical trajectories called *conservative maneuvers*, in which the non-conservative force (drag) is balanced by the thrust force ($f_T = D$). If the drag is neutralized, there is no other force opposing the motion, hence the total energy (specific energy level) is conserved. In a conservative horizontal turn, the airplane maintains its speed and altitude, while a conservative vertical turn is an exchange between potential energy (altitude) and kinetic energy (speed). We refer the reader to Miele [34] for an excellent treatment of conservative maneuvers.

Since the earliest days of flying, both vertical and horizontal maneuvers have been extensively studied and adopted in aerobatic and fighter airplanes.

[16] Certain air-to-air missiles are spin-stabilized and create a normal aerodynamic force out of the *Magnus effect* due to the spinning motion in a sideslip. Their modeling must include the rolling motion, and is thus postponed until Chapter 13.

An air-combat fighter airplane is primarily designed to turn horizontally at a high rate in order to achieve the firing-position advantage over an enemy. Since a loss of energy during turning is disadvantageous, a conservative turn is desirable whenever possible. However, due to thrust limitation, we often have $f_T < D$ in the high-drag flight of the maximum lift coefficient and load factor, which is required for the maximum instantaneous turn rate given by Eq. (12.35). The airspeed at its minimum possible value (the stalling speed) for the maximum possible load factor, $\bar{n}$, that maximizes the instantaneous turn rate, is called the *corner speed*, because it occurs at the intersection of the curves for $C_L = C_{L\,\max}$ and $n = \bar{n}$ on a *v-n* diagram, and is given by

$$v_c = \sqrt{\frac{2\bar{n}mg}{\rho S C_{L\,\max}}}. \tag{12.52}$$

It is clear the corner speed is fixed for a given altitude. Since a turn at the corner speed is often nonconservative, neither the speed nor the altitude will remain constant during the maximum instantaneous turn. Turning at the maximum possible rate is often a matter of survival and is thus indispensible. For this reason, a dog-fight that begins at a high altitude normally ends near the sea level. This limitation of air-combat maneuverability by coordinated turning has plagued all fighter airplanes from the *Sopwith Camel* and *Albatross D-V* of the first world war, to the *F-16* and *F-22* of the present day.[17]

In order to increase the maneuverability of airplanes without an excessive loss of energy, the concept of *supermaneuvers* is under investigation. A supermaneuver refers to a maneuver assisted by thrust vectoring, where thrust components enhance the lift and the side force. Since it is no longer necessary to balance both weight and centripetal acceleration by aerodynamic lift in a supermaneuver, the stalling restriction is effectively removed as a constraint. Furthermore, since the wings have to produce less lift in a supermanuever, the load factor on the wing is reduced, thereby also alleviating the structural constraint. However, a supermaneuverable airplane must have a much greater thrust (due to a high post-stall drag) than a conventional airplane of similar size, as well as a capability to provide balancing moments either by additional lifting surfaces (such as canards) or by small jet thrusters, and should be equipped with an automatic flight control system (for the rapid modulation and precision required in thrust vectoring while maneuvering). NASA's *X-31* technology-demonstrator airplane explored the concept of supermaneuverability by having a paddle-type thrust-vectoring nozzle, canard surfaces, and a computerized, fly-by-wire, flight control system. It demonstrated successful thrust-induced maneuvering at angles of attack of about 70° [37], while the stalling angle of attack of the airplane was approximately 30°. The X-31

[17] Although the close combat of the former era has been largely replaced by the *beyond visual range* (BVR) combat with radar-guided missiles, the guns have not been entirely removed from the modern fighters. This implies that the classic combat is still a part of a modern fighter's mission.

also displayed the advantage of *fuselage pointing*, which is the name given to willfully pointing the airplane's nose in a given direction without turning the velocity vector by a maneuver. Fuselage pointing is very valuable in air combat, as it enables aiming and firing of weapons in a direction different from the velocity vector.

We shall now simulate the conventional maneuvers and a supermaneuver of a fighter airplane.

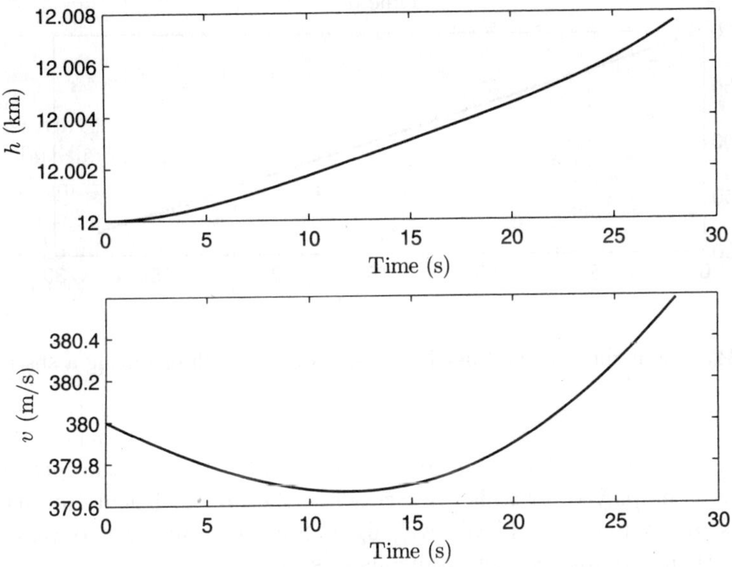

Fig. 12.23. The airspeed and altitude time history of a jet fighter during a sustained turn.

Example 12.5. Consider a horizontal maneuver of the fighter airplane of Example 12.4, beginning from a straight and level flight toward the west at initial altitude $h = 12$ km, position $\delta = 20°$, $\lambda = 70°$, relative speed $v = 450$ m/s, and mass $m = 16,350$ kg. We consider the following options:

(a) A turn with a constant bank angle $\sigma = 77.07°$, during which the lift coefficient is modulated in order to maintain a nearly level flight with a constant load factor, $n = \sec\sigma$, such that

$$C_L = \frac{2nmg}{\rho S v^2} \, .$$

(b) A nearly level turn with the maximum lift coefficient, $C_{L\,\max} = 1.5$, at $v < v_c$, and the bank angle modulated to achieve the maximum possible load

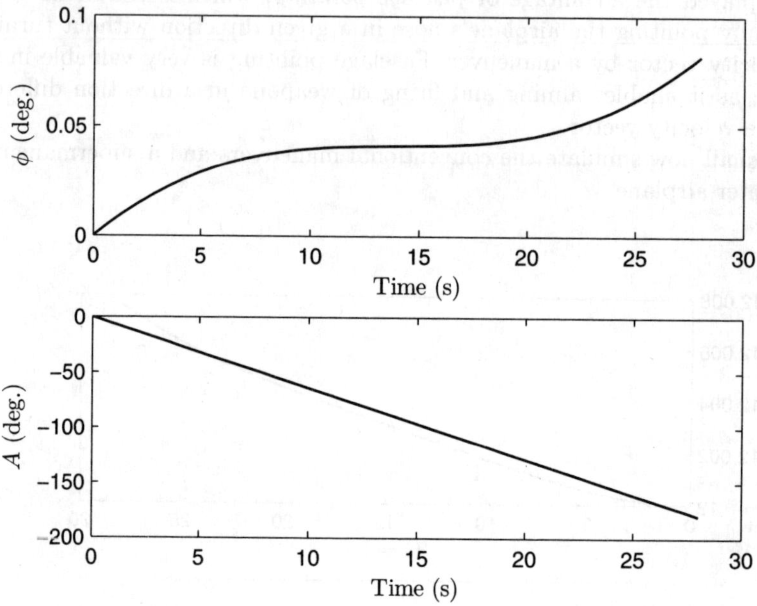

Fig. 12.24. The flight-path and heading angles of a jet fighter during a sustained turn.

factor (limited by $\bar{n} = 7.5$) at a given instant. For $v > v_c$, the maximum bank angle is employed for which $n = \bar{n}$, while the lift coefficient (limited by $C_{L\,\max}$) is modulated to achieve level flight. Such a profile would yield the maximum turn rate for the instantaneous airspeed.
(c) A supermaneuver with $\mu = 0$ and the angle ϵ modulated in such a way that the level turn of case (b) occurs at an enhanced rate. The angle ϵ is held constant at different values for airpseeds above and below the corner speed v_c, such that the lift coefficient is enhanced for $v > v_c$, while the load factor is enhanced for $v < v_c$.

The structural limitations of Example 12.4 are valid in all the cases.

We present a MATLAB program *maneuver.m* for case (a) in Table 12.9 and assume that the global variables and the initial condition are duly specified in a calling program. The resulting plots for case (a) simulated for 28 seconds are shown in Figs. 12.23–12.26. The turn at a sustained rate is evident from Fig. 12.23, showing that the altitude and airspeed are nearly constant during the almost 180° turn (Fig. 12.24), which gives a nearly circular ground track (Fig. 12.25) and almost constant values of M, C_L, and q (Fig. 12.26). The average turn rate in case (a) is $\frac{180°}{28} = 6.4286°/\text{s}$. For the simulation of cases (b) and (c), the program *maneuver.m* is modified to *supmaneuver.m* (Table 12.10), where the bank angle is no longer constant, but depends upon

Table 12.9. M-file *maneuver.m* for an Airplane's State Equations during a Horizontal Turn

```
function deriv = maneuver(t,o)
global dtr; global mu; global omega; global S; global c;
global rm; global b; global CLmax; global sweep;
global n; global bank;
global f8; %data file for additional flight parameters
%acceleration due to gravity (nonspherical earth):
[g,gn]=gravity(o(3),o(2));
lo = o(1);la = o(2);
clo = cos(lo); slo = sin(lo); cla = cos(la); sla = sin(la);
fpa = o(5); chi = o(6);
cfpa = cos(fpa); sfpa = sin(fpa); cchi = cos(chi); schi = sin(chi);
%atmospheric properties:
if o(3)<rm
    o(3)=rm;
end
alt = o(3) - rm;
v  = o(4);
atmosp = atmosphere(alt,v,c);
rho = atmosp(2);
Qinf = 0.5*rho*v^2;
mach = atmosp(3);
if mach<0
mach=0;
elseif mach>2.25
mach=2.25;
end
%aero-propulsive parameters:
[fT,tsfc]=engine(alt,mach);
fT=2*fT;
m=o(7);
CD0=parasite(mach); CL=n*m*g/(Qinf*S);
if CL>CLmax        %stall condition
    CL=0
end
K=liftddf(mach,CL,sweep,b^2/S);
%ground effect:
if alt<=b
    Keff=K*33*(alt/b)^1.5/(1+33*(alt/b)^1.5);
else
    Keff=K;
end
CD=CD0+Keff*CL^2;
D=Qinf*S*CD; L=Qinf*S*CL;
Xfo = fT-D; Zfo = L*cos(bank); Yfo = L*sin(bank);
fprintf(f8,'\t%1.5e\t%1.5e\t%1.5e\t%1.5e\t%1.5e\n',t,alt,CL,mach,Qinf);
%state equations:
longidot = o(4)*cfpa*schi/(o(3)*cla);
latidot =  o(4)*cfpa*cchi/o(3);
raddot = o(4)*sfpa;
veldot = -g*sfpa +gn*cchi*cfpa + Xfo/m...
       + omega*omega*o(3)*cla*(sfpa*cla-cfpa*cchi*sla);
gammadot=(o(4)/o(3)-g/o(4))*cfpa-gn*cchi*sfpa/o(4)...
         +Zfo/(o(4)*m)+ 2*omega*schi*cla...
         +omega*omega*o(3)*cla*(cfpa*cla + sfpa*cchi*sla)/o(4);
headdot=o(4)*schi*tan(o(2))*cfpa/o(3)-gn*schi/o(4)...
     -Yfo/(o(4)*cfpa*m)-2*omega*(tan(o(5))*cchi*cla-sla)...
     +omega*omega*o(3)*schi*sla*cla/(o(4)*cfpa);
mdot=-tsfc*fT/(9.8*3600);
%time derivatives:
deriv=[longidot; latidot; raddot; veldot; gammadot; headdot; mdot];
```

Table 12.10. M-file *supmaneuver.m* for an Airplane's State Equations during a Horizontal Supermaneuver

```
function deriv = maneuver(t,o)
global dtr;global mu;global omega;global S;global c;global rm;
global b;global CLmax;global sweep;global n;
global f8;    %data file for additional flight parameters
%acceleration due to gravity (nonspherical earth):
[g,gn]=gravity(o(3),o(2));
lo = o(1);la = o(2);
clo = cos(lo); slo = sin(lo); cla = cos(la); sla = sin(la);
fpa = o(5); chi = o(6);
cfpa = cos(fpa); sfpa = sin(fpa); cchi = cos(chi); schi = sin(chi);
%atmosphere determination:
if o(3)<rm
   o(3)=rm;
end
alt = o(3) - rm;
v=o(4);
m=o(7);
atmosp = atmo_sre(alt,v,c);
rho = atmosp(2);
mach = atmosp(3);
if mach<0
   mach=0;
elseif mach>2
   mach=2;
end
Qinf = 0.5*rho*v^2;
[fT,tsfc]=engine(alt,mach);
fT=2*fT;
CD0=parasite(mach);
vc=sqrt(2*n*m*g/(rho*S*CLmax)); %corner speed (m/s)
if v>=vc
   nb=n;
   epsilon=13.5*dtr;
   CL=n*(m*g-fT*sin(epsilon))/(Qinf*S);
   if CL>CLmax
   CL=CLmax;
   epsilon=0;
   end
else
   CL=CLmax;
   epsilon=32*dtr;
   nb=Qinf*S*CLmax/(m*g-fT*sin(epsilon));
   if nb>n
   nb=n;
   epsilon=0;
   end
end
if nb>1
bankb=acos(1/nb);
else
   bankb=0;
end
K=liftddf(mach,CL,sweep,b^2/S);
if alt<=b
   Keff=K*33*(alt/b)^1.5/(1+33*(alt/b)^1.5);
else
   Keff=K;
end
CD=CD0+Keff*CL^2;
D=Qinf*S*CD; L=Qinf*S*CL;
Xfo = fT*cos(epsilon)-D; Zfo = fT*sin(epsilon)+L*cos(bankb);
Yfo = L*sin(bankb);
```

```
%trajectory equations:
longidot = v*cfpa*schi/(o(3)*cla);
latidot =  v*cfpa*cchi/o(3);
raddot = v*sfpa;  %Radius
veldot = -g*sfpa +gn*cchi*cfpa + Xfo/m...
       + omega*omega*o(3)*cla*(sfpa*cla-cfpa*cchi*sla);
gammadot = (v/o(3)-g/v)*cfpa-gn*cchi*sfpa/v...
        + Zfo/(v*m) + 2*omega*schi*cla...
        + omega*omega*o(3)*cla*(cfpa*cla + sfpa*cchi*sla)/v;
headdot = v*schi*tan(o(2))*cfpa/o(3)-gn*schi/v...
        - Yfo/(v*cfpa*m)- 2*omega*(tan(o(5))*cchi*cla - sla)...
        + omega*omega*o(3)*schi*sla*cla/(v*cfpa);
mdot=-tsfc*fT/(9.8*3600);
fprintf(f8,'\t%1.5e\t%1.5e\t%1.5e\t%1.5e\t%1.5e\n',t,CL,mach,nb,headdot);
%time derivatives:
deriv = [longidot; latidot; raddot; veldot; gammadot; headdot; mdot];
```

the airspeed. Furthermore, the lift coefficient is also modulated with airspeed, as discussed above. The simulation for case (b) is obtained by setting $\epsilon = 0$ at all times in *supmaneuver.m*, while in case (c) $\epsilon = 13.5°$ for $v \geq v_c$, and $\epsilon = 32°$ for $v < v_c$. These values of ϵ are selected by trial and error to yield the maximum turn angle in a nearly level flight. Cases (b) and (c) are simulated for 45 seconds, and their results are compared in Figs. 12.27–12.31. It is evident in Fig. 12.27 that the total change in azimuth by the

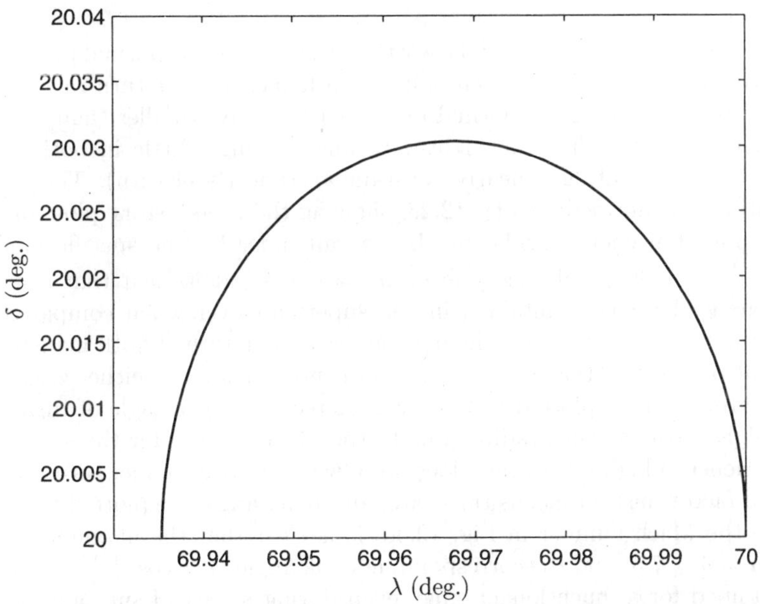

Fig. 12.25. The ground track of a jet fighter during a sustained turn.

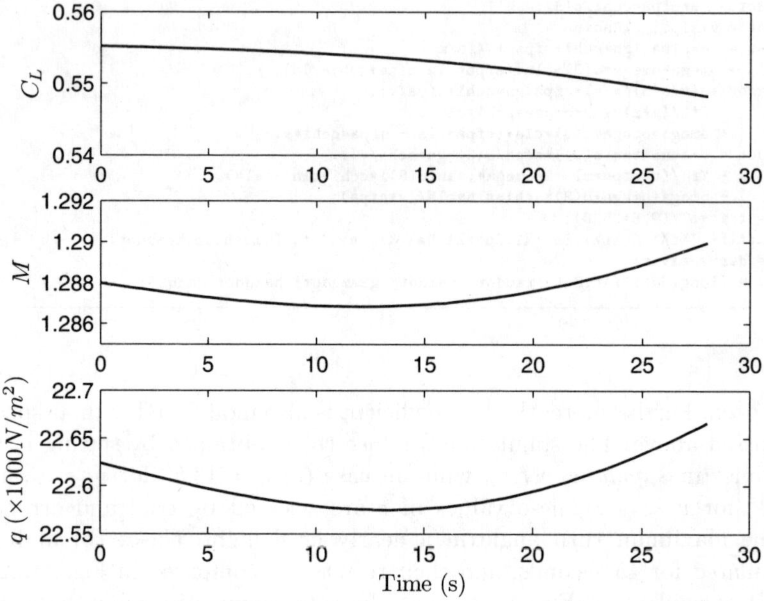

Fig. 12.26. The Mach number, lift-coefficient, and dynamic pressure of a jet fighter during a sustained turn.

supermaneuver [case (c)] is much larger than that due to the normal maneuver [case (b)] in the same time, which indicates a higher average turn rate of the former. The airspeed in the normal maneuver is always smaller than that in the supermaneuver, while the flight-path angle changes little in both cases ($| \phi | \leq 0.6°$), indicating a nearly constant altitude (level turn). The latter observation is confirmed by Fig. 12.28, showing the time-history plots of the specific potential energy (indicated by the altitude, h), the specific kinetic energy, $\frac{v^2}{2g}$, and the total energy level, $h_e = h + \frac{v^2}{2g}$. It is clear that a much larger energy level is maintained in the supermaneuver when compared to that in the normal maneuver. The nonconservative nature of both maneuvers is evident by the fact that the energy level is not constant in either case.

The ground track plotted in Fig. 12.29 depicts a larger angle of turn and also a higher average turn radius (due to the higher speed) for the supermaneuver. Hence, a higher turn rate does not always imply a smaller turn radius. The load factor, instantaneous turn rate, and bank angle are plotted with respect to the Mach number in Fig. 12.30. It is clear that the maximum load factor, $n = \bar{n} = 7.5$ (and the corresponding bank angle, $\sigma = \cos^{-1}\frac{1}{\bar{n}} = 82.34°$) is maintained for a much longer time, even during a part of subsonic flight, in the supermaneuver, whereas the normal maneuver has $n = \bar{n}$ only at supersonic speeds. Also, the load factor of the supermaneuver is always higher

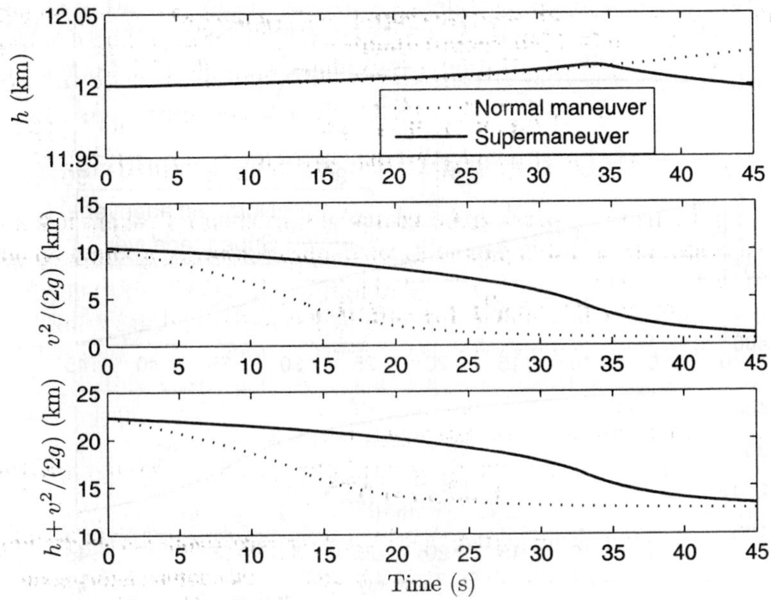

Fig. 12.27. The energy level vs. altitude of a jet fighter during instantaneous turns with and without thrust vectoring.

than that of the normal maneuver. Hence, the supermaneuver maintains a high instantaneous turn rate over a larger portion of the flight compared to the normal maneuver, leading to a higher average turn rate. The Mach number and lift coefficient time-history plots are shown in Fig. 12.31. The slower trajectory of the normal maneuver (smaller M, larger C_L) compared to the supermaneuver is clear in this figure. While the normal maneuver is performed at the stalling speed for almost 30 seconds (out of 45), the supermaneuver requires $C_L = C_{L\max}$ for only the last 11 seconds. This is due to the enhanced lift provided by thrust vectoring, precluding the necessity of flying at the stalling speed most of the time.

The supermaneuver can be further enhanced by employing a sideways thrust deflection ($\mu \neq 0$), thereby increasing the instantaneous rate of turn. However, since the deflected thrust decreases the forward thrust component, $f_T \cos\epsilon \cos\mu$, the net deceleration would increase in this case, leading to a smaller speed and loss of energy level. Hence, the sideways deflection of thrust is employed only when the available thrust, f_T, is very much larger than the drag.

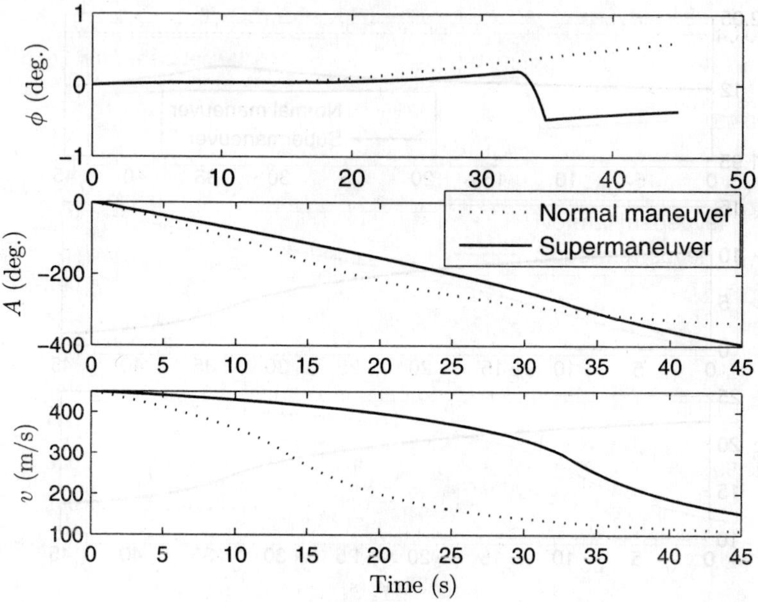

Fig. 12.28. The flight-path angle, heading angle, and airspeed of a jet fighter during instantaneous turns with and without thrust vectoring.

12.4 Entry Trajectories

The problem of atmospheric entry from an orbital or suborbital trajectory is of primary interest in manned space missions, aeroassisted orbital transfers, robotic lander and sample return missions, as well as in the flight of surface-to-surface missiles. The ubiquitous meteors are the earliest known and most common entry objects [38], which awed the first humans as *shooting stars.* They were recognized as *rocks from space* only in the late 19th century. The study of meteoric trajectories first employed an approximate form of equations of motion derived in the previous section. Apart from the equations of motion, it is also necessary to include the aerothermal loads (Chapter 10) caused by the high speeds of atmospheric entry, which, in turn, are capable of modifying the shape and size of the object or even of disintegrating it into several smaller objects. Therefore, when modeling a re-entry trajectory, one must take into account the loads caused by high heating rates and dynamic pressures, as well as the structural and material changes brought upon by these loads, in the form of additional differential equations. While it is beyond the scope of this book to derive such equations in a general case, we mention here that the aerothermal loads can be estimated by statistical (or semi-empirical) aerothermal models or, more accurately, by sophisticated computational aerothermodynamic models, as pointed out in Chapter 10.

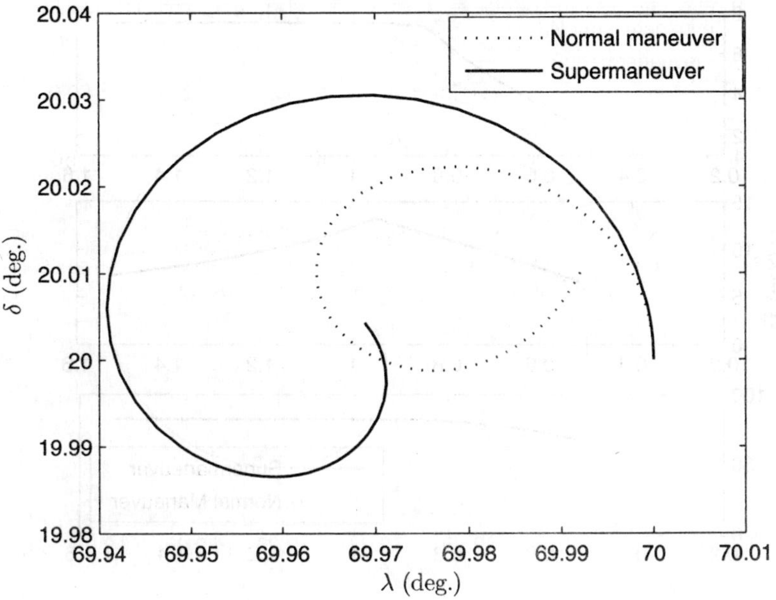

Fig. 12.29. The ground track of a jet fighter during instantaneous turns.

Since the objective of a re-entry vehicle is to slow down, while safely dissipating the kinetic energy of entry into the atmosphere, we shall begin our analysis of re-entry trajectories by dropping the propulsive terms from Eq. (12.29), resulting in

$$\begin{aligned} m\dot{v} &= -D - mg_c \sin\phi + mg_\delta \cos\phi \cos A - m\omega^2 r \cos\delta(\cos\phi \cos A \sin\delta \\ &\quad - \sin\phi\cos\delta), \\ mv\cos\phi\dot{A} &= m\frac{v^2}{r}\cos^2\phi \sin A \tan\delta + f_Y - mg_\delta \sin A \\ &\quad + m\omega^2 r \sin A \sin\delta\cos\delta \\ &\quad - 2m\omega v(\sin\phi\cos A\cos\delta - \cos\phi\sin\delta), \\ mv\dot{\phi} &= m\frac{v^2}{r}\cos\phi + L - mg_c\cos\phi - mg_\delta \sin\phi\cos A \\ &\quad + m\omega^2 r\cos\delta(sin\phi \cos A \sin\delta \\ &\quad + \cos\phi\cos\delta) + 2m\omega v \sin A\cos\delta\ . \end{aligned} \tag{12.53}$$

The kinematic equations, Eqs. (12.10), (12.14), and (12.15), remain unchanged. The aerodynamic forces of lift, L, drag, D, and side force, f_Y, chiefly determine the atmospheric entry trajectory. These forces, in turn, are dependent upon the vehicle's shape and size, as well as on its relative speed, altitude, and attitude. Since the relative speed, altitude, and attitude vary with time,

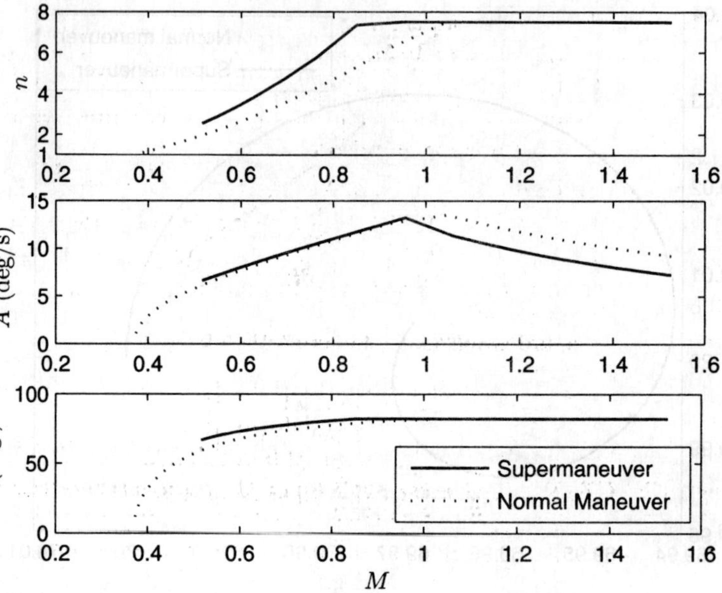

Fig. 12.30. The load factor, turn rate, and bank angle of a jet fighter during instantaneous turns.

we must specify aerodynamic relationships as additional equations for determining the instantaneous aerodynamic forces as functions of time. As pointed out in Chapter 4, it is generally a good approximation to specify the attitude parameters as external inputs to the translational motion, due to the smaller time scales of rotational maneuvers. However, for greater accuracy, both translational and rotational equations of motion must be solved simultaneously to yield a six-degree-of-freedom simulation (Chapter 15).

There are two broad categories of atmospheric entry trajectories: (a) the *ballistic entry*, where the lift and side force are negligible in comparison with the drag, and (b) the *maneuvering entry*, wherein the lift and side force are significant and are used to change the flight-path direction. If the vehicle's shape is nearly axisymmetric with respect to the velocity vector, such as in the ballistic missile warheads, and the *Soyuz, Mercury, Gemini*, and *Apollo* manned capsules, the assumption of ballistic re-entry is valid. In a ballistic re-entry, either the vehicle's shape and the center of gravity—through deliberate design—ensure static aerodynamic stability, essentially aligning the vehicle's axis of symmetry along the flight direction (somewhat like a *badminton shuttle*), or an active attitude control system can be employed for the same purpose. This is important for meeting the ballistic constraint, because even a re-entry capsule, shaped like an inverted bottle, can produce an appreciable lift (or side force) when operated at an angle of attack (or sideslip). The

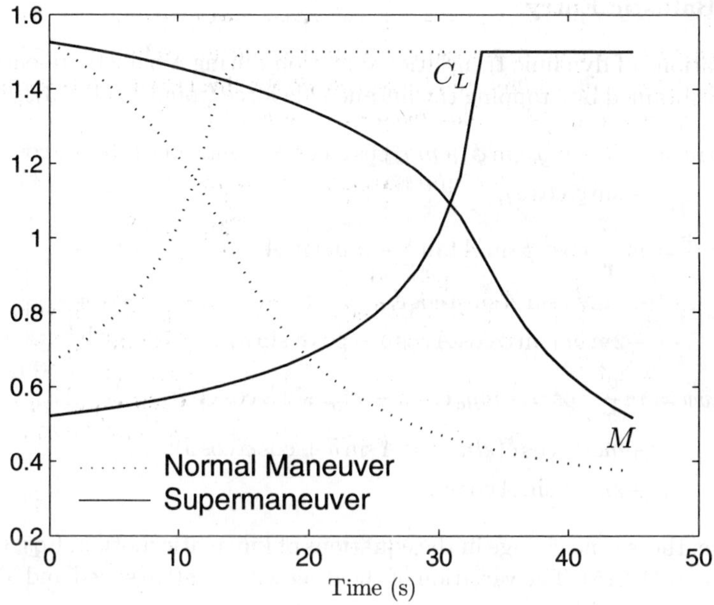

Fig. 12.31. The Mach number and lift coefficient of a jet fighter during instantaneous turns.

Apollo re-entry capsules employed such a lifting maneuver for controlling the initial flight path at re-entry. NASA's *space shuttle* is a maneuvering re-entry vehicle, which employs large bank angles to produce both lift and side force from its wings during re-entry. By changing the bank angle in a specific manner, the re-entry flight path can be controlled. Up to a certain extent, the lift coefficient can also be varied by changing the angle of attack. The modulation of the lift and side force through the bank angle and angle of attack forms the basis for the re-entry *guidance and control.*

In any entry trajectory, the rate of heat transfer is of utmost important. The total rate of convective heat transfer over an a ballistic entry vehicle, $\dot{Q}$, can be related semi-empirically [19] to its drag coefficient as follows:

$$\dot{Q} = \frac{1}{2}\rho v^3 \frac{SC_D}{20} , \tag{12.54}$$

which implies that the heating rate increases with the cube of the relative speed. Equation (12.54) entails a boundary-layer approximation (Chapter 10), with the assumption that about one tenth of the vehicle's drag is caused by the skin friction (which is responsible for heat transfer from the atmosphere to the vehicle). As we know from Chapter 10, the remaining drag is pressure-induced and includes the wave drag and flow separation drag. A lifting entry vehicle would have a much larger portion of drag caused by skin friction.

12.4.1 Ballistic Entry

The equations of dynamic translational motion during a ballistic re-entry can be easily obtained by dropping the lift and side force from Eq. (12.53), yielding

$$\begin{aligned} m\dot{v} &= -D - mg_c \sin\phi + mg_\delta \cos\phi \cos A - m\omega^2 r \cos\delta(\cos\phi \cos A \sin\delta \\ &\quad - \sin\phi \cos\delta), \\ mv\cos\phi \dot{A} &= m\frac{v^2}{r}\cos^2\phi \sin A \tan\delta - mg_\delta \sin A \\ &\quad + m\omega^2 r \sin A \sin\delta \cos\delta \\ &\quad - 2m\omega v(\sin\phi \cos A \cos\delta - \cos\phi \sin\delta), \\ mv\dot{\phi} &= m\frac{v^2}{r}\cos\phi - mg_c\cos\phi - mg_\delta \sin\phi \cos A \\ &\quad + m\omega^2 r \cos\delta(sin\phi \cos A \sin\delta + \cos\phi\cos\delta) \\ &\quad + 2m\omega v \sin A \cos\delta\,. \end{aligned} \tag{12.55}$$

As before, there is no change in the equations of kinematic motion, Eqs. (12.10), (12.14), and (12.15). The variation of the drag with relative speed and altitude can be represented by

$$D = \frac{1}{2}\rho v^2 S C_D(Kn, \mathrm{M}, \mathrm{Re}), \tag{12.56}$$

where ρ is the atmospheric density, S is a reference area for aerodynamics, C_D is the drag coefficient based on S, being a function of the Knudsen number, Kn, the Mach number, M, and the Reynolds number, Re, which are the governing nondimensional flow parameters (Chapter 10). Different expressions are employed for C_D in each flow regime, such as the free molecular ($Kn > 10$), rarefied transitional ($0.1 < Kn < 10$), or continuum ($Kn < 0.1$) based upon the Knudsen number, hypersonic continuum ($M \gg 1$), supersonic continuum ($M > 1$), transonic continuum ($M \approx 1$), or subsonic continuum ($M < 1$) in terms of the Mach number, and laminar continuum, or turbulent continuum in terms of the Reynolds number. In each regime, the drag coefficient is expressed as a function of the defining flow parameter. In a typical atmospheric entry trajectory, the critical Reynolds number is easily crossed, and we are primarily dealing with turbulent flow in the continuum regime. Furthermore, the drag coefficient becomes independent of the Mach number for speeds larger than a critical hypersonic Mach number in the continuum regime (Chapter 10). In the transitional, rarefied flow regime, a suitable interpolation of the drag coefficient with the Knudsen number is employed. It would be sufficient for our purpose of modeling and simulation to assume that the functions defining the drag coefficient are known, and thus there is no difficulty in calculating C_D at a given combination of r and v. For an analytical insight into ballistic entry, further simplifying assumptions can be made, such as a flat, nonrotating planet, planar motion, etc., which we shall not employ here, as they depart from our objective of accurate modeling. A reader can refer to the excellent textbooks devoted to analytical atmospheric entry mechanics, [39], [40], and [19].

Example 12.6. Simulate the ballistic entry into the earth's atmosphere of a space capsule of mass 350 kg from an orbit of perigee altitude 200 km, eccentricity 0.2, inclination 80°, argument of perigee, 265°, and right ascension of ascending node, 100°. The de-boost impulse is applied tangentially when the right ascension and east longitude are 0° and −10°, respectively. Carry out the simulation with different de-boost impulse magnitudes in the range 0.1 to 0.2 times the perigee speed. The capsule possesses longitudinal static stability that tends to achieve an equilibrium attitude with zero angle of attack and sideslip, for which condition its drag coefficient (referred to the base area, $S = 4\ \text{m}^2$) is the following:

$$\begin{aligned}
&C_D = C_{Dc} \quad (Kn < 0.0146)\\
&C_D = C_{Dfm} \quad (Kn > 14.5),\\
&C_D = C_{Dc} + (C_{Dfm} - C_{Dc})\left(\frac{1}{3}\log_{10}\frac{Kn}{\sin 30^\circ} + 0.5113\right)\\
&\qquad (0.0146 < Kn < 14.5),
\end{aligned}$$

where C_{Dc} is the drag coefficient in the continuum limit, plotted in Fig. 12.32 as a function of Mach number, C_{Dfm} is the drag coefficient in the free-molecular flow limit with cold-wall approximation (Chapter 10), given by

$$C_{Dfm} = 1.75 + \frac{\sqrt{\pi}}{2s},$$

with $s = \frac{v}{\sqrt{2RT}}$ denoting the molecular speed ratio (Chapter 10), and the Knudsen number, Kn, is based on a nose radius of 0.5 m. Note that in Fig. 12.32, the hypersonic Mach number independence principle (Chapter 10) is valid; thus C_{Dc} becomes essentially invariant with M for $M > 8$. Before we begin the simulation, we must obtain the initial condition for ballistic entry. This involves the assumption of Keplerian motion (no drag and a spherical earth gravity) for the orbit. The semi-major axis of the initial orbit is easily calculated as

$$a = \frac{r_p}{1-e} = \frac{6578.14}{0.8} = 8222.675\ \text{km},$$

from which the perigee speed is obtained as

$$v_p = \sqrt{\mu\left(\frac{2}{r_p} - \frac{1}{a}\right)} = 8.527229115\ \text{km/s}.$$

It still remains to determine the radius, r_0, latitude, δ_0, and inertial velocity when the spacecraft reaches the de-boost point. The latitude of the de-boost

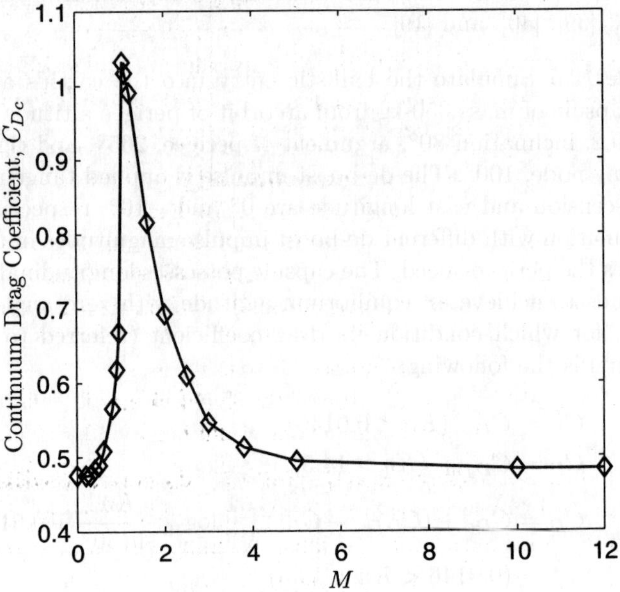

Fig. 12.32. Variation the capsule's continuum drag-coefficient with Mach number.

point, δ_0, can be obtained by considering its zero right ascension and spherical trigonometry (Chapter 5) as follows:

$$\delta_0 = \tan^{-1}[-\tan(80°)\sin(100°)] = -79.8489182889° .$$

The rotation matrix representing the orientation of the celestial frame relative to the perifocal frame (Chapter 5) is calculated to be the following:

$$\mathsf{C}* \doteq \mathsf{C}_3{}^T(100°)\mathsf{C}_1{}^T(80°)\mathsf{C}_3{}^T(265°) = \begin{pmatrix} 0.185493763 & -0.158082884 & 0.969846310 \\ -0.055792706 & 0.983688329 & 0.171010072 \\ -0.981060262 & -0.085831651 & 0.173648178 \end{pmatrix} ,$$

from which a relationship between the latitude, δ_0, and the true anomaly, θ_0, of the de-boost point is obtained to be

$$r_0 \begin{Bmatrix} \cos\delta_0 \\ 0 \\ \sin\delta_0 \end{Bmatrix} = r_0 \mathsf{C}* \begin{Bmatrix} \cos\theta_0 \\ \sin\theta_0 \\ 0 \end{Bmatrix} ,$$

which yields

$$\cos\theta_0 = 0.998395412021, \qquad \sin\theta_0 = 0.056626859843 ,$$

or, $\theta_0 = 3.2462165419°$. Thus, we have the radius and inertial speed immediately before de-boost as

$$r_0 = \frac{a(1-e^2)}{1+e\cos\theta_0} = 6579.89967 \text{ km},$$

$$v{*}_0 = \sqrt{\mu\left(\frac{2}{r_0} - \frac{1}{a}\right)} = 8.5253285295 \text{ km/s}.$$

The perifocal velocity at the de-boost point is given by

$$\begin{aligned}\mathbf{v{*}_0} &= \sqrt{\frac{\mu}{p}}[-\sin\theta_0 \mathbf{i_e} + (e+\cos\theta_0)\mathbf{i_p}] \\ &= -0.402391833995\mathbf{i_e} + 8.515826873757\mathbf{i_p},\end{aligned}$$

which yields the following celestial velocity at the de-boost point:

$$\mathbf{v{*}_0} = \mathsf{C}{*}\begin{Bmatrix} -0.40239183995 \\ 8.515826873757 \\ 0 \end{Bmatrix} = \begin{Bmatrix} -1.42084764924 \\ 8.3993700404 \\ -0.33615683771 \end{Bmatrix}.$$

In order to compute the celestial velocity azimuth and elevation at the de-boost point, we require the rotation matrix relating the local horizon frame to the celestial frame (Chapter 5),

$$\mathsf{C_{LH}} = \begin{pmatrix} \cos\delta_0 & 0 & \sin\delta_0 \\ 0 & 1 & 0 \\ -\sin\delta_0 & 0 & \cos\delta_0 \end{pmatrix},$$

or,

$$\mathsf{C_{LH}} = \begin{pmatrix} 0.176244384228 & 0 & -0.984346441568 \\ 0 & 1 & 0 \\ 0.984346441568 & 0 & 0.176244384228 \end{pmatrix},$$

which leads to the following inertial velocity in the local horizon frame:

$$\mathbf{v{*}_0} = \mathsf{C_{LH}}\begin{Bmatrix} -1.42084764924 \\ 8.3993700404 \\ -0.33615683771 \end{Bmatrix} = \begin{Bmatrix} 0.08047836799 \\ 8.39937004035 \\ -1.45785208241 \end{Bmatrix}.$$

The inertial velocity azimuth, $A*$, and flight-path angle, $\phi*$, are thus calculated as follows:

$$\phi* = \sin^{-1}\frac{0.08047836799}{8.5253285295} = 0.540875°,$$

$$\sin A* = \frac{8.39937004035}{8.5253285295\cos(0.540875°)} = 0.98526928065,$$

$$\cos A* = \frac{-1.45785208241}{8.5253285295\cos(0.540875°)} = -0.17101007166,$$

or, $A* = 99.846552°$.

Let us take the tangential de-boost impulse to be alternately 8.95, 10, and 15%, of the perigee speed. For instance, in the first case of 10% tangential de-boost, the spacecraft's inertial speed is instantaneously decreased by $\Delta v = 0.8527229115$ km/s at the point of de-boost. Hence, the inertial speed immediately after de-boost is $v{*}_0 = 7.67260561806$ km/s. Then we transform the inertial velocity immediately following de-boost into the velocity relative to the atmosphere as follows:

$$
\begin{aligned}
A &= \tan^{-1}\left(\tan A* - \frac{\omega r_0 \cos\delta_0}{v *_0 \cos\phi * \cos A*}\right) = 99.955734^\circ,\\
\phi &= \tan^{-1}\left(\tan\phi * \frac{\cos A}{\cos A*}\right) = 0.54681217^\circ,\\
v_0 &= v *_0 \frac{\sin\phi*}{\sin\phi} = 7.58930433867 \text{ km/s}.
\end{aligned}
$$

Similarly, for the other two cases, we have

$$
\begin{aligned}
\Delta v &= 0.15 v_p = 1.27908436725 \text{ km/s},\\
A &= 99.962233^\circ,\\
\phi &= 0.54716522^\circ,\\
v_0 &= 7.16294370544 \text{ km/s},
\end{aligned}
$$

and

$$
\begin{aligned}
\Delta v &= 0.0895 v_p = 0.7631870057925 \text{ km/s},\\
A &= 99.954461^\circ,\\
\phi &= 0.5467426866^\circ,\\
v_0 &= 7.67884008322 \text{ km/s}.
\end{aligned}
$$

Clearly, the flight-path angle and the azimuth change little by a changing impulse magnitude. We are now ready to begin the simulation of the ballistic entry trajectory.

A MATLAB program called *reentry.m* (Table 12.11) is written to compute the time derivatives of all the motion variables, $r, \delta, \lambda, v, \phi, A$, according to the kinematic and dynamic equations of motion, as well as the variation of aerodynamic parameters during atmospheric entry. These time derivatives are provided as inputs to the MATLAB's intrinsic Runge–Kutta solver, *ode45.m*, at each time step. Another program, called *runreentry.m* (Table 12.12), specifies the initial condition and invokes *ode45.m*. As before, the standard atmospheric and nonspherical earth gravity models are provided by *atmosphere.m* (Chapter 9) and *gravity.m* (Chapter 3), respectively. The simulations are carried out in each case until the capsule reaches an altitude of approximately 5 km, after which the capsule lands gently under parachutes (not modeled here). The results of the simulation are plotted in Figs. 12.33–12.37. Figures 12.33 and 12.34 show the position and velocity of the capsule relative to the earth for

Table 12.11. M-file *reentry.m* for State Equations during a Ballistic Entry

```
function Y = reentry(t,o)
global dtr, global mu; global S; global c; global m;
global rm; global omega; global Gamma;
global f8; %Data tape for storing aerothermal results
%acceleration due to gravity (nonspherical earth):
[g,gn]=gravity(o(3),o(2));
lo = o(1);la = o(2);
clo = cos(lo); slo = sin(lo);
cla = cos(la); sla = sin(la);
fpa = o(5); chi = o(6);
cfpa = cos(fpa); sfpa = sin(fpa);
cchi = cos(chi); schi = sin(chi);
%atmospheric properties and flow parameters:
if o(3)<rm
   o(3)=rm;
end
alt = o(3) - rm;
v  = o(4);
atmosp = atmosphere(alt, v, c);
rho = atmosp(2);
Qinf = 0.5*rho*v^2;
mach = atmosp(3);
[t alt v mach]
Kn=atmosp(4);
CDC=conticap(mach);
s = mach*sqrt(Gamma/2);
CDFM=1.75+sqrt(pi)/(2*s);
iflow=atmosp(6);
if iflow==2      %continuum regime
    CD=CDC;
elseif iflow==1 %free-molecular flow regime
    CD=CDFM;
else              %transition regime
    CD = CDC + (CDFM - CDC)*(0.333*log10(Kn/sin(pi/6))+0.5113);
end
Xfo=-Qinf*S*CD;
Qdot=Qinf*v*S*CD/20; %rate of heat transfer (W)
%trajectory equations follow:
longidot = o(4)*cfpa*schi/(o(3)*cla);
latidot =  o(4)*cfpa*cchi/o(3);
raddot = o(4)*sfpa;
veldot = -g*sfpa +gn*cchi*cfpa+Xfo/m...
        +omega*omega*o(3)*cla*(sfpa*cla-cfpa*cchi*sla);
gammadot = (o(4)/o(3) - g/o(4))*cfpa-gn*cchi*sfpa/o(4)...
        + 2*omega*schi*cla+ omega*omega*o(3)*cla*(cfpa*cla...
        + sfpa*cchi*sla)/o(4);
headdot = o(4)*schi*tan(o(2))*cfpa/o(3)-gn*schi/o(4)...
         - 2*omega*(tan(o(5))*cchi*cla - sla)...
         + omega*omega*o(3)*schi*sla*cla/(o(4)*cfpa);
Y = [longidot; latidot; raddot;...
 veldot; gammadot; headdot]; %time derivatives
if alt<=120e3
fprintf(f8,'\t%1.5e\t%1.5e\t%1.5e\t%1.5e...
        \t%1.5e\t%1.5e\n',alt,CD,mach,veldot,Qinf,Qdot);
end
```

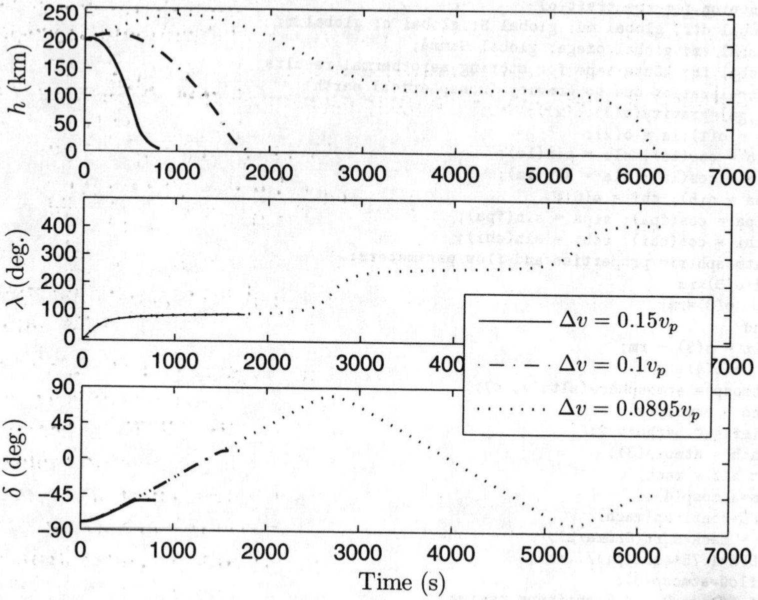

Fig. 12.33. The altitude, longitude, and latitude of a ballistic entry capsule.

Table 12.12. M-file *runreentry.m* for Integrating the State Equations for a Ballistic Entry

```
global dtr; dtr = pi/180;
global mu; mu = 3.986004e14;
global S; S = 4;
global c; c=0.5;
global m; m =  350;
global rm; rm = 6378140;
global omega; omega = 2*pi/(23*3600+56*60+4.0905);
global Gamma; Gamma=1.41;
global f8; f8 = fopen('data8.mat', 'a');
long = -10*dtr;               %initial longitude
lat = -79.8489182889*dtr; %initial latitude
rad= 6579.89967e3;  %initial radius (m)
vel= 7589.30433867; %initial velocity (m/s)
fpa= 0.54681217*dtr;%initial flight-path angle
chi= 99.955734*dtr; %initial heading angle (measured from north)
options = odeset('RelTol', 1e-8);
orbinit = [long; lat; rad; vel; fpa; chi];
[t, o] = ode45('reentry',[0, 1750],orbinit,options);
fclose('all');
```

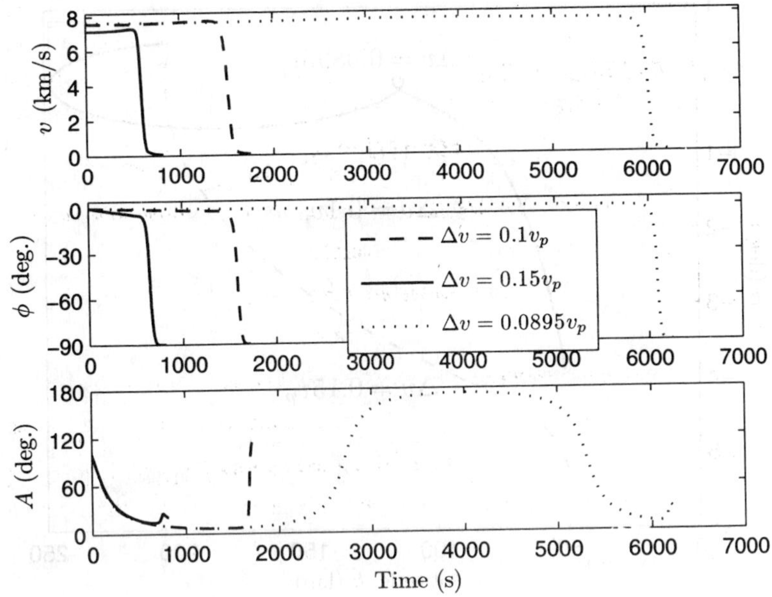

Fig. 12.34. The relative speed, flight-path angle, and heading angle of a ballistic entry capsule.

the three trajectories. The trajectory with the smallest de-boost impulse magnitude, $\Delta v = 0.0895v_p$, is seen to have the longest flight time (6250 s), which is more than triple the flight time for $\Delta v = 0.1v_p$ (1750 s). As the de-boost impulse is increased to $\Delta v = 0.15v_p$, we do not observe the same decrease in the flight time (Fig. 12.33). The ground tracks traveled in the three cases differ very slightly, and the landing point is a point on the each track where the altitude becomes approximately 5 km, after which the flight path is nearly vertical ($\phi \approx -90°$) (Fig. 12.34). Therefore, the landing point is dictated by the time of flight, which in turn strongly depends on the de-boost impulse magnitude. There is a large difference between the landing points of $\Delta v = 0.0895v_p$ ($\delta = -42.62°, \lambda = 55.5°$) and $\Delta v = 0.1v_p$ ($\delta = 8.52°, \lambda = 85.3°$) even though the difference in the velocity is quite small.

A plot of flight-path angle vs. altitude, Fig. 12.35, indicates the fundamentally different ballistic entry in each case. For $\Delta v = 0.0895v_p$, the flight-path angle undergoes a loop at $h \approx 110$ km, after which the altitude starts increasing until ϕ again becomes negative. This is known as a *skip entry*, wherein the vehicle does not have a sufficiently large, negative flight-path angle for a normal ballistic entry and is similar to a stone skipping a pond's surface when launched at a particularly shallow angle. Due to the skip, the capsule travels further along the ground track before dipping down again for a normal

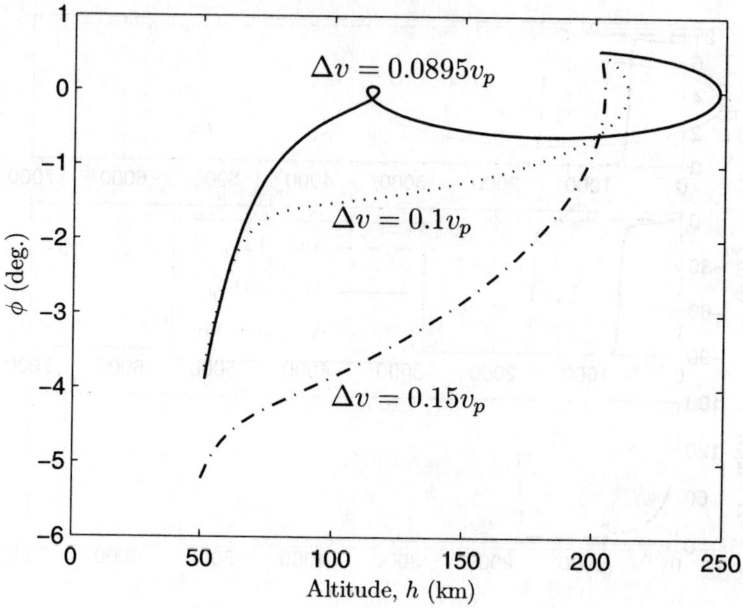

Fig. 12.35. The flight-path angle as a function of altitude during ballistic entry.

entry. The skip is evident in Fig. 12.33 from about 3800 s to 5300 s into the flight, during which the altitude stays above 110 km. The case $\Delta v = 0.0895v_p$ leads to only a "modest" skip, while a smaller de-boost impulse magnitude would result in a more extreme skip, where the ensuing altitude and time of flight are increased drastically, even leading to several successive skips. Such an extreme trajectory would be undesirable if unplanned because of the large dispersion of the landing point as well as an increase in the flight time (which is unsuitable in a manned mission due to limited oxygen supply). The other two cases show a normal entry, albeit at different initial flight-path angles. The values of ϕ for entry with $\Delta v = 0.15v_p$ are always much smaller than those of the $\Delta v = 0.1v_p$ entry (Fig. 12.35).

From the above results, it is clear that the conditions at an altitude of approximately 120 km (where the earth's atmosphere begins to cause an appreciable deceleration) are extremely important for an entry trajectory. Hence, for most entry vehicles, $h = 120$ km can be regarded as the *entry altitude* at which the speed and flight-path angle must be carefully managed (by a suitable de-boost impulse) to achieve a desired landing point and time of flight.

Let us now consider the approximate aerothermal loads during entry. The Mach number and drag coefficient, plotted in Fig. 12.36, have almost the same variation with the altitude in all three cases. The value of C_D decreases almost linearly with the altitude from the near free-molecular flow condition at

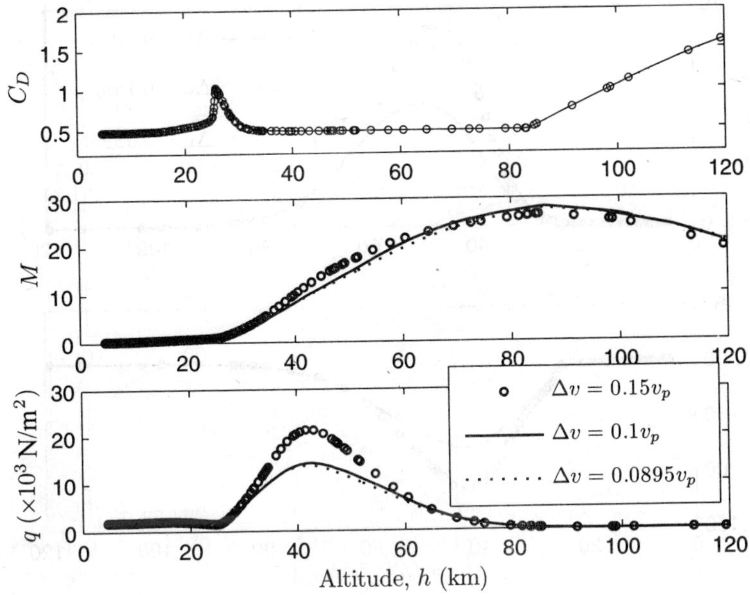

Fig. 12.36. The drag coefficient, Mach number, and dynamic pressure vs. altitude during ballistic entry.

$h = 120$ km, until $h = 83$ km, and then remains constant in the continuum hypersonic regime, $32 \leq h < 83$ km, during which the Mach number remains above 5. At lower altitudes, the drag coefficient shows an increase through the supersonic regime, reaching a peak at $M \approx 1$, and then declining through the subsonic regime. The dynamic pressure, $q = \frac{1}{2}\rho v^2 = \frac{1}{2}\gamma p M^2$, also plotted in Fig. 12.36, is much higher for the trajectory with $\Delta v = 0.15 v_p$ than that for the other two cases. The smallest dynamic pressure is observed for the skip entry trajectory ($\Delta v = 0.0895 v_p$), indicating a decay in the speed during skip, leading to entry with a smaller kinetic energy. Since the aerothermal loads are directly proportional to the dynamic pressure, it is expected that they would be smallest in magnitude for the skip entry, and largest for the entry with the maximum de-boost impulse ($\Delta v = 0.15 v_p$). These deductions are confirmed in Fig. 12.37, which contains plots of the rate of heat transfer, $\dot{Q}$, and axial acceleration, $\dot{v}$, vs. altitude. These two parameters are the primary indicators of the structural loads on the vehicle, and their maximum magnitudes are very important in design. The entry with the largest de-boost impulse ($\Delta v = 0.15 v_p$) achieves a maximum heating rate of $\dot{Q} = 9.2 \times 10^6$ W at an altitude of approximately 47 km, and the maximum axial deceleration of -117 m/s^2 at the altitude of 42.5 km (which is also the altitude for the maxi-

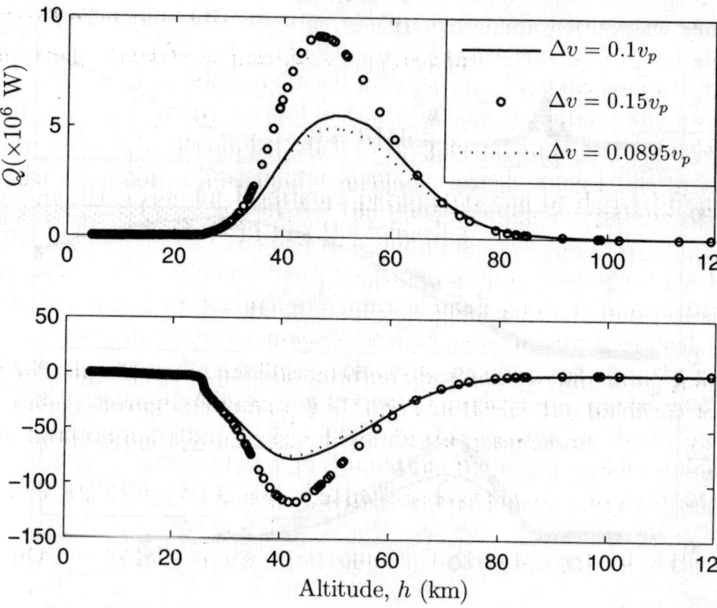

Fig. 12.37. The rate of heat transfer and deceleration vs. altitude during ballistic entry.

mum dynamic pressure).[18] In comparison, the skip entry yields a maximum heating rate of only $\dot{Q} = 4.95 \times 10^6$ W at 49 km altitude, and the maximum axial deceleration of -75 m/s^2 at the altitude of 42 km. Although the magnitudes of the aerothermal loads are approximate, their trend with altitude is quite accurate. The smaller aerothermal loads of a skip entry, combined with its longer range and endurance after de-boost, make it an attractive flight path for various purposes. For example, during World War II the Swiss engineer Eugen Sänger proposed a bomber of global range using successive skips, whereas Walter Hohmann originated *aeroassisted orbital transfer* and *aerobraking* of spacecraft from a high-energy trajectory to one with a lower energy, using successive passes through the atmosphere. While Sänger's *antipodal bomber* did not materialize, Hohmann's concept of aeroassisted orbital transfer has been suceessfully implemented in various interplanetary missions, such as NASA's *Magellan* mission to Venus in 1992, and *Mars Global Surveyor* (MGS) in 1997. The *Apollo* earth-return capsule employed a partial skip entry in order to reduce the excessive aerothermal loads. A precise attitude control was employed to make the small lift force point downward, such that the capsule did not leave the atmosphere on an extreme skip trajectory (which

[18] A good estimate of the altitudes for maximum deceleration and heat transfer can also be obtained from an approximate analysis, such as by Loh [39].

would have been disastrous for the astronauts). The resulting "double-dip" trajectory was quite similar in trend to the $\Delta v = 0.0895v_p$ case below 120 km altitude. The chief drawback of a skip entry is the extreme sensitivity of the shallow trajectory to the speed and flight-path angle at entry. For example, a change by $\pm 0.001°$ in the flight-path angle at 120 km altitude for the $\Delta v = 0.0895v_p$ case presented above would cause a dispersion in the landing point by ± 520 km. It is seldom possible to maintain such accurate control of the flight path at the point of entry, thereby implying an unacceptably high uncertainty in the landing point in the case of a skip entry. For this reason, the classical nonskipping entry (such as the $\Delta v = 0.15v_p$ case) is usually used, which is less sensitive to variations in the flight path at entry, and where the higher aerothermal loads are absorbed with a thicker heat shield.[19]

12.4.2 Maneuvering Entry

The main disadvantage of a ballistic entry vehicle is its lack of maneuverability within the atmosphere. Furthermore, the high heating rates in a normal ballistic entry restrict the vehicle to a single mission. In order to have some control over the entry trajectory, as well as to enter at a smaller negative flight-path angle (which reduces the aerothermal loads), a re-usable vehicle must necessarily produce lift. This requirement can be met by wings, such as in NASA's space shuttle, or by a more structurally robust *lifting-body* configuration, such as the *Lockheed X-33.* Although the hypersonic lift-to-drag ratios of such lifting entry vehicles are modest (less than or equal to unity), they are capable of greatly modifying the trajectory, thereby enabling control of the landing point. Another advantage of the lifting configuration is the added ability to lose speed through maneuvering. The lift generation in a maneuvering entry vehicle entails a larger portion of the drag due to skin friction, because of its flatter shape, compared to a similar-sized ballistic entry vehicle at the same angle of attack. Hence, the rate of heat transfer at a given speed is higher for the lifting vehicle. Therefore, a shallow trajectory, and a rapid bleeding of the entry speed through banking maneuvers, becomes an indispensible requirement of a lifting entry vehicle. The space shuttle uses such a maneuvering entry, through a series of sharply banked *S-turns* at entry for increased deceleration. Of course, a banking trajectory would increase the normal acceleration (load factor), as seen above. However, the increase in aerothermal loading can be spread out over a larger portion of the trajectory by a suitable adjustment of the bank angle, which would maximize deceleration without exceeding the limits on load factor and heating rate. Therefore, the maneuvering entry quickly becomes an optimal guidance problem, whose

[19] The first entry capsule, the *Vostok*, was a simple spherical design that used a heat shield all around the vehicle. The later bottle-shaped designs, *Mercury, Gemini, Soyuz*, and *Apollo*, as well as the sphero-conical ballistic missile warheads, have used heat shielding of selected portions of the body, which necessitate either some form of attitude control or an inherent attitude stability during entry.

solution involves a series of carefully executed turns. Such an optimal trajectory is flown by an automatic flight control system every time the space shuttle returns from orbit.

The normal acceleration magnitude of a manuevering vehicle can be expressed as follows:

$$a_n = v\sqrt{\dot{\phi}^2 + \dot{A}^2}\,, \tag{12.57}$$

which is related to the *normal load factor*, $n_n \doteq \frac{a_n}{g} + 1$, where g is the magnitude of acceleration due to gravity. One can also define an *axial load factor*, $n_a \doteq \frac{\dot{v}}{g} + 1$. Generally, we have $n_n \gg 1$ and $\mid n_a \mid \gg 1$ during entry, which allows the gravitational acceleration to be ignored in an approximate analysis [19].

Example 12.7. Consider a maneuvering entry vehicle derived from the capsule of Example 12.6 such that the drag coefficient referred to the same area is unmodified, but some lift is generated through flattened portions of the body in the design entry attitude. Simulate the entry of the lifting vehicle from the same orbit and initial position as in Example 12.6, with a de-boost impulse magnitude of $\Delta v = 0.15v_p$, such that a constant lift-to-drag ratio of 0.5 is maintained throughout the flight. It can be assumed that the skin friction drag is one fifth the total drag at any given point in the lifting entry trajectory. This implies that the rate of heat transfer can be approximated by $\dot{Q} = 0.1qvSC_D$, where q is the dynamic pressure.

Simulate the following maneuver profiles:

(a) No banking is carried out during entry ($\sigma = 0$).
(b) The bank angle is held at $\sigma = 90°$ during $450 \leq t < 500$ s, and reversed to $\sigma = -90°$ during $500 \leq t < 550$ s. At all other times, $\sigma = 0$. In order to perform the simulation for lifting entry, we modify the program *reentry.m* to *liftentry.m* (Table 12.13), which has provision for lift and side force generated by banking and a nonzero lift coefficient. The atmospheric, drag, and gravity models are unchanged. The program evaluates the normal acceleration according to Eq. (12.57), in addition to the axial acceleration and heating rate, at altitudes below 120 km.

The resulting plots of simulation data for the two maneuver profiles are contained in Figs. 12.38–12.42. In these plots, the time is measured from the de-boost point, and the trajectory for $t < 400$ s is the same as the ballistic case (Example 12.6). Figures 12.38 and 12.39 show the position and velocity of the vehicle. It is clear that the lifting trajectories have larger flight time and (positive) flight-path angle compared to the ballistic trajectory discussed in Example 12.6. Consequently, the vehicle covers a longer distance along the ground track. This increase in the range due to lift is often considered useful in a military mission, such as the cancelled *Dyna-Soar* design project of the U.S. Air Force in the 1960s. The stretched flight of a lifting entry is sometimes referred to as a *gliding entry*. However, a more desirable trajectory is the one in which the vehicle loses both speed and altitude quickly, in preparation for

Table 12.13. M-file *liftentry.m* for State Equations during a Maneuvering Entry

```
function Y = liftentry(t,o)
global dtr; global mu; global S; global c; global m; global rm;
global omega; global Gamma;
global f8; %data tape for storing aerothermal results
[g,gn]=gravity(o(3),o(2)); %acceleration due to gravity
lo = o(1);la = o(2);
clo = cos(lo); slo = sin(lo);
cla = cos(la); sla = sin(la);
fpa = o(5); chi = o(6);
cfpa = cos(fpa); sfpa = sin(fpa);
cchi = cos(chi); schi = sin(chi);
%atmospheric properties and flow parameters:
if o(3)<rm
    o(3)=rm;
end
alt = o(3) - rm;
v  = o(4);
atmosp = atmosphere(alt, v, c);
rho = atmosp(2);
Qinf = 0.5*rho*v^2;
mach = atmosp(3);
[t alt v mach]
Kn=atmosp(4);
CDC=conticap(mach);
s = mach*sqrt(Gamma/2);
CDFM=1.75+sqrt(pi)/(2*s);
iflow=atmosp(6);
if iflow==2
    CD=CDC;
elseif iflow==1
    CD=CDFM;
else
    CD=CDC+(CDFM-CDC)*(0.333*log10(Kn/sin(pi/6))+0.5113);
end
CL=0.5*CD;
if t>=450 && t<=500
   %bank=pi/2;
   bank=0;
elseif t>500 && t<=550
    %bank=-pi/2;
    bank=0;
else
    bank=0;
end
Xfo=-Qinf*S*CD;
Zfo=Qinf*S*CL*cos(bank);
Yfo=Qinf*S*CL*sin(bank);;
%trajectory equations follow:
longidot = o(4)*cfpa*schi/(o(3)*cla); %longitude
latidot =  o(4)*cfpa*cchi/o(3); %latitude
raddot = o(4)*sfpa; %radius
veldot =-g*sfpa +gn*cchi*cfpa + Xfo/m+...
       omega*omega*o(3)*cla*(sfpa*cla-cfpa*cchi*sla);
gammadot=(o(4)/o(3)-g/o(4))*cfpa-gn*cchi*sfpa/o(4)+...
          Zfo/(o(4)*m) + 2*omega*schi*cla...
          + omega*omega*o(3)*cla*(cfpa*cla...
          + sfpa*cchi*sla)/o(4);
headdot=o(4)*schi*tan(o(2))*cfpa/o(3)-gn*schi/o(4)...
          - Yfo/(o(4)*cfpa*m)...
          - 2*omega*(tan(o(5))*cchi*cla - sla)...
          +omega*omega*o(3)*schi*sla*cla/(o(4)*cfpa);
Y = [longidot ; latidot ; raddot ;...
      veldot ; gammadot ; headdot ];
if alt<=120e3
Qdot=Qinf*v*S*CD/10;
transacc=v*sqrt(gammadot^2+headdot^2);
fprintf(f8,'\t%1.5e\t%1.5e\t%1.5e\t%1.5e...
  \t%1.5e\t%1.5e\n',alt,mach,veldot,transacc,Qinf,Qdot);
end
```

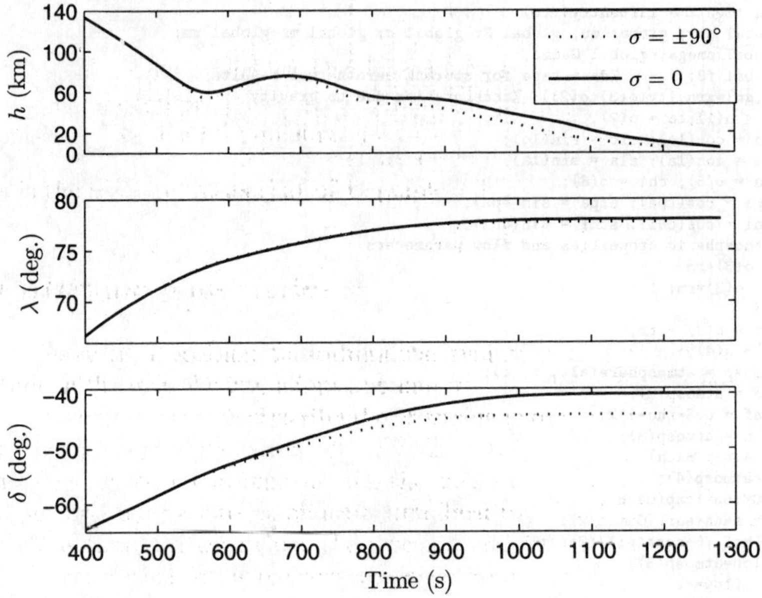

Fig. 12.38. The altitude, longitude, and latitude in maneuvering entry.

a landing (such as NASA's space shuttle). This is seen to be accomplished by the extremely sharp banking profile of $\sigma = \pm 90°$ in the period $450 \leq t \leq 550$ s, which causes a reduction in the total flight time by about 70 s, with an attendant decrease in the speed by more that 1 km/s in the crucial altitude range of $40 \leq h \leq 80$ km, compared to the trajectory with $\sigma = 0$. However, the flight-path angle in the same altitude range becomes more negative due to banking, which brings the vehicle deeper into the atmosphere, causing an increase in the peak dynamic pressure near 60 km altitude (Fig. 12.40), leading to a sharp increase in the axial and normal acceleration (Fig. 12.41), as well as the rate of heat transfer (Fig. 12.42), when compared to the trajectory without banking.

The advantage of lifting entry is evident from a comparison of the dynamic pressure, axial deceleration, and heating rate for the ballistic case (Example 12.6). When no banking is employed, the dynamic pressure and maximum deceleration are cut to one third, and the maximum heating rate is about one mega-Watt lower when compared to the ballistic entry with the same mass, reference area, and drag coefficient (despite the doubling of the heating rate at a given speed due to the higher skin friction of the lifting vehicle). This is due to the higher altitude of the lifting trajectory at any given speed in the hypersonic portion of flight. The reduction of aerothermal loads is thus the primary advantage of a maneuvering entry trajectory, which enables a re-

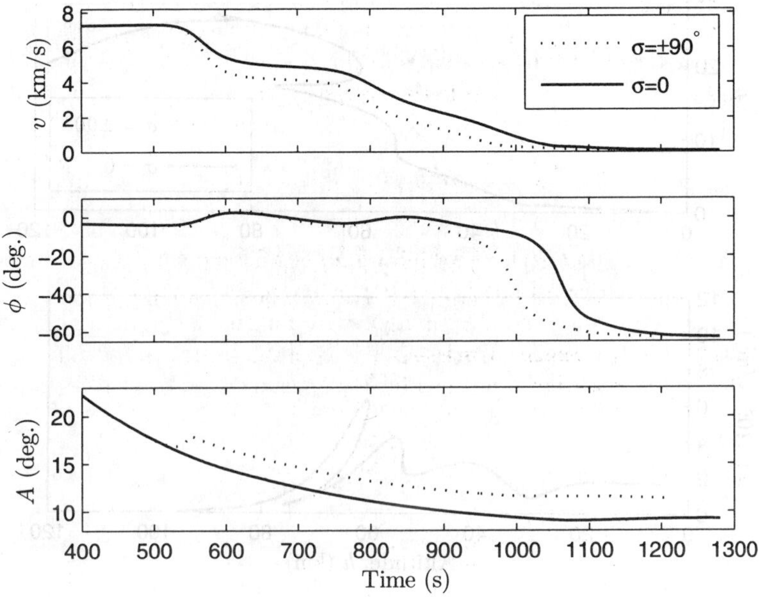

Fig. 12.39. The relative speed, flight-path angle, and heading angle in maneuvering entry.

usable entry vehicle through a permanent *thermal protection system* (such as the heat soaking tiles of the space shuttle). The amount of banking allowable in the actual entry trajectory is determined by the design axial and normal load factors, as well as the rate of heat transfer safely tolerated by the thermal protection system.

12.5 Rocket Ascent Trajectories

Finally, we draw our attention to the ascent trajectories of rocket-powered vehicles. A rocket ascent is fundamentally different from the accelerated climb of the high-performance fighter airplane, mainly due to the former's more rapid decrease in mass, and a nearly constant thrust, with altitude. For these reasons, a rocket can deliver a payload to an orbital trajectory in a fraction of the time an airplane needs to reach its cruising altitude. The simplest rocket ascent is the vertical trajectory of a *sounding rocket*, which is commonly employed in weather observations of upper-atmospheric strata, and for which a closed-form solution is easily carried out. However, the curving ascent of a spacecraft launch vehicle, from $\phi = 90°$ to $\phi \approx 0$, is much more interesting from a modeling viewpoint. It is important to note that such an ascent must be performed at zero lift, because even a slight build-up of normal load factor

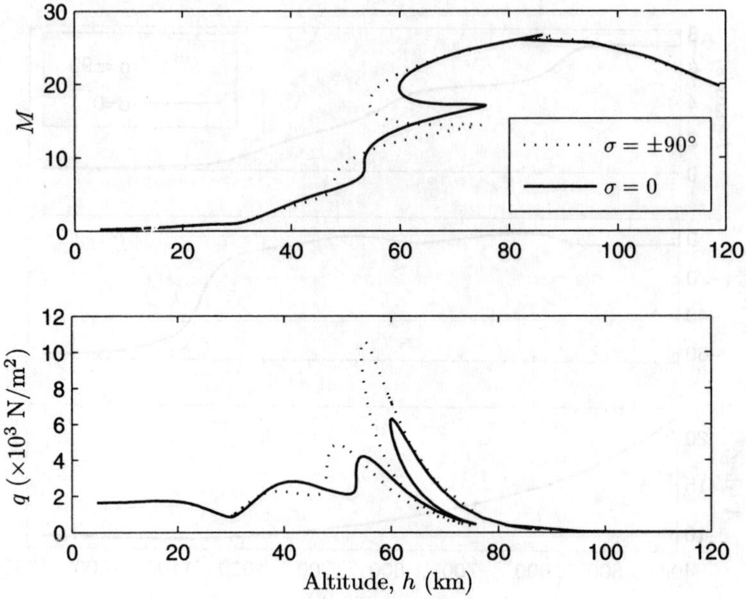

Fig. 12.40. The Mach number and dynamic pressure vs. altitude during maneuvering entry.

would cause the thin, shell-like structure of the launch vehicle (often almost completely full of liquid propellants) to disintegrate. Therefore, since maneuvering is completely ruled out during launch, it is natural to inquire what causes the launch vehicle to change its flight path from vertical to nearly horizontal. The answer lies in the combined effects of gravity and planetary rotation. The naturally curving trajectory from the planetary surface to orbit is called a *gravity turn*.

An attitude control system is an indispensible feature of the launch vehicle, which always maintains a zero angle of attack despite small atmospheric disturbances. The attitude control of a rocket is carried out through thrust-vectoring by a gyro-based feedback mechanism and a control law that maintains the vehicle at its open-loop, inherently unstable equilibrium (Chapter 14). In addition to the attitude control, a guidance system is required for maintaining a near-vertical flight in the first few seconds after launch, during which the launch tower is cleared, and the trajectory is unstable due to a near-zero flight speed. The gravity turn is initiated by discontinuing the automatic guidance system as soon as the vehicle has accelerated sufficiently for $|\phi|$ to become small [Eq. (12.29)]. However, the attitude control is always performing its task of maintaining zero angles of attack and sideslip throughout the launch. Since we have adopted a complete set of translational equations in

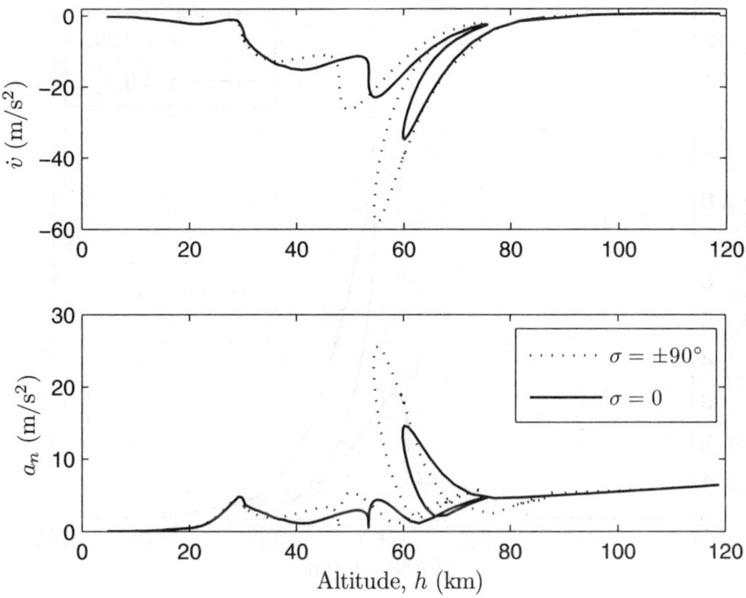

Fig. 12.41. The axial and normal acceleration vs. altitude during maneuvering entry.

this chapter, we are well equipped to model the *gravity turn* trajectories of launch vehicles without any further assumption or modification.

Before considering a simulation example, it is necessary to point out some important aspects of a rocket. In Chapter 8, we considered the optimal selection of stage payload ratios that maximized the total payload ratio for a given total velocity impulse. However, the rocket equation of Chapter 8 assumed zero drag and gravity, as well as an impulsive thrust. Neither of these assumptions is valid in the actual rocket performance. Therefore, additional design parameters become important, namely the duration of operation of each stage, which depends in turn on the mass flow rate (thus, the thrust) developed in the nozzle. The best performance is usually obtained if the burn times are in the same proportion as the specific impulses. The selection of stage burn time requires simulating the actual trajectory in the presence of drag, gravity, and propulsive losses due to nonimpulsive burning. Furthermore, the launch direction is another important parameter for optimizing a rocket's performance. Due to the planet's rotation, there is a particular velocity azimuth at launch (depending upon the latitude of launch site) that maximizes the orbit achieved at burn-out. This optimal direction is again selected through simulated trajectories. We illustrate a rocket launch by the following example.

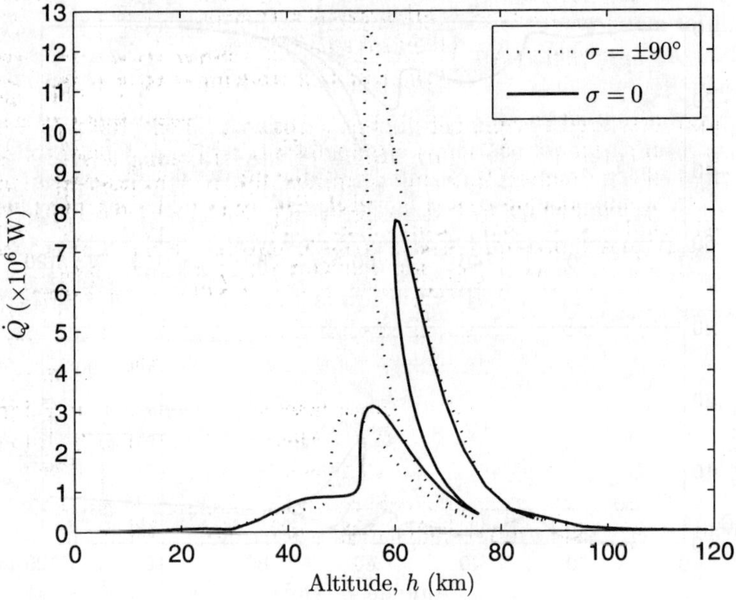

Fig. 12.42. The rate of heat transfer vs. altitude during maneuvering entry.

Example 12.8. Consider an optimal, two-stage, serial rocket for launching a 350 kg payload to a circular earth orbit. The first stage employs a solid propellant with specific impulse 200 s, while the second stage has kerosene/LO_2 with a specific impulse 350 s. The structural ratio of the first stage is 0.07, while that of the second stage is 0.05. Design the optimal payload ratios of the stages using the methods of Chapter 8, for the total velocity impulses of (a) 9.5 and (b) 10.5 km/s, respectively. In the resulting rockets, design (a) has first- and second-stage burn times of 50 and 87.5 s, respectively, whereas the burn times are doubled in design (b). Simulate the actual launch trajectories for each design, assuming that the thrust loss due to incorrect expansion of exhaust gases (Chapter 8) is negligible, and the parasite drag coefficient of the payload, C_{Dp}, is the same as that of the ballistic capsule of Example 12.6, while that of the first and second stage is $5C_{Dp}$ and $2.2C_{Dp}$, respectively, in design (a) and $8C_{Dp}$ and $3C_{Dp}$, respectively, in design (b). The launch is to be carried out from Cape Canaveral ($\delta = 28.5°, \lambda = -80.55°$) with an initial relative flight path of $\phi = 90°$, $A = 170°$, which translates into the best inertial launch direction (due east) for taking advantage of the earth's rotation.

An optimal design of this two-stage rocket for $\Delta v = 9.5$ km/s was carried out in Example 9.5 using the program *Nstage.m*. The relevant case is that of $\beta_2 = 1.75$, for which the optimal total payload ratio, $\lambda_T = 0.021176$, and the optimal first-stage payload ratio, $\lambda_1 = 0.4263393$, were obtained, implying a first-stage initial mass, $m_{01} = \frac{350}{\lambda_T} = 16,528.420$ kg, and a second-stage initial mass, $m_{02} = \lambda_1 m_{01} = 7046.715$ kg. This leads to the first-stage propellant mass, $m_{p1} = (m_{01} - m_{02}) * (1 - 0.07) = 8817.985$ kg, and the second-stage propellant mass, $m_{p2} = (m_{02} - m_L) * (1 - 0.05) = 6361.880$ kg. Assuming constant mass flow rates during each stage, we get the mass exhaust rates of the two stages to be $\dot{m}_1 = \frac{m_{p1}}{50} = -176.3597$ kg/s and $\dot{m}_2 = \frac{m_{p2}}{87.5} = -72.7072$ kg/s. Assuming no thrust losses due to overexpansion, we have the approximate thrust developed in the respective stages to be $f_{T1} = -\dot{m}_1 g_0 I_{sp1} = 346,017.7314$ N and $f_{T2} = -\dot{m}_2 g_0 I_{sp2} = 249,640.1712$ N for design (a).

In design (b) with $\Delta v = 10.5$ km/s, the optimal case is for $\alpha_2 = 0.1582$ ($\lambda_1 = 0.267767$) (Fig. 12.43), yielding $\lambda_T = 0.011342824$. This results in $m_{01} = 30,856.513$ kg, $m_{02} = 8262.362$ kg, $m_{p1} = 21,012.561$ kg, $m_{p2} = 7516.744$ kg, $f_{T1} = 412,266.44$ N and $f_{T2} = -\dot{m}_2 g_0 I_{sp2} = 147,478.51$ N.

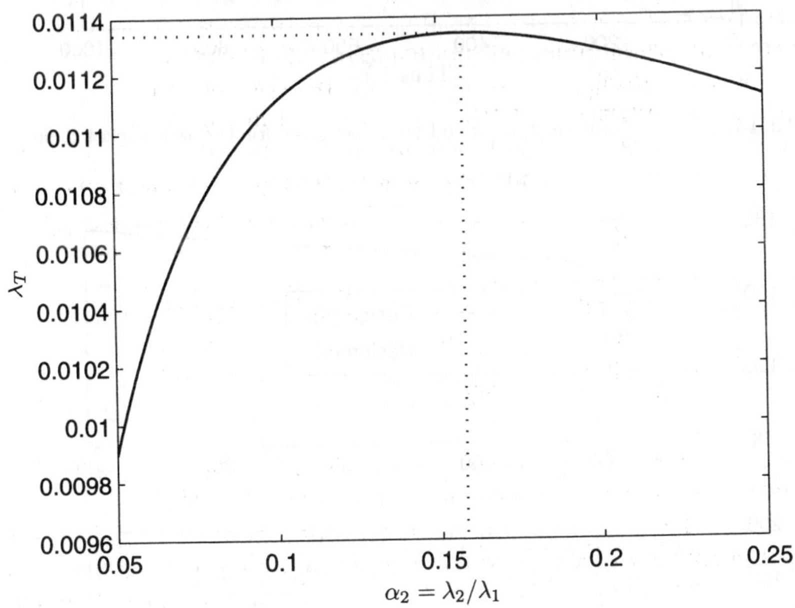

Fig. 12.43. Payload ratio optimization for the two-stage rocket with Δv = 10.5 km/s.

The ascent trajectories of the two designs are simulated through the program *rocket.m*, which is tabulated in Table 12.14. This program supplies the time derivatives of the state vector, $r, \delta, \lambda, v, \phi, A$, for integration by an appropriate procedure, such as the Runge–Kutta method. The initial condition and the global parameter values must be declared in the calling program.

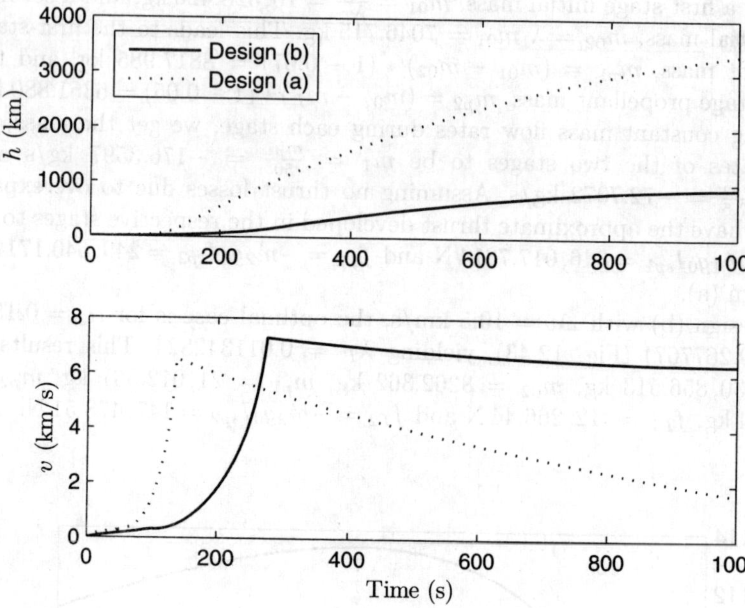

Fig. 12.44. Altitude and relative speed time histories for the two rocket designs.

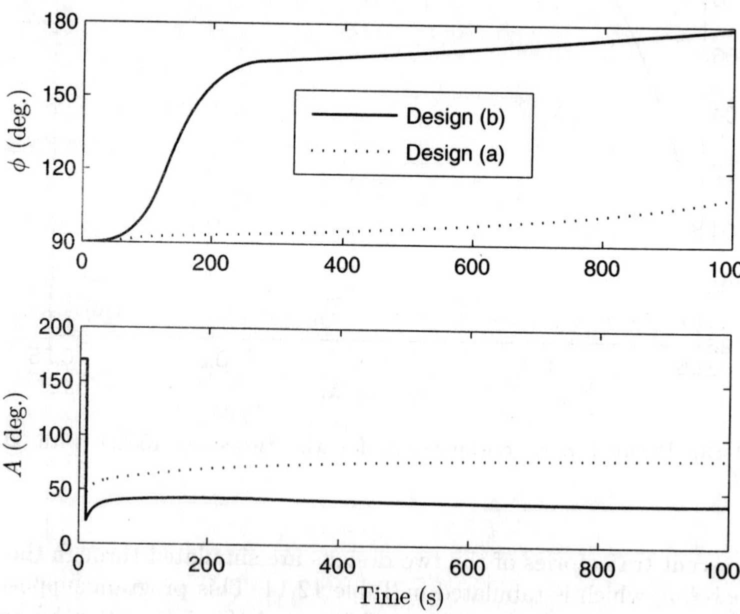

Fig. 12.45. Relative flight-path angle and azimuth for the two rocket designs.

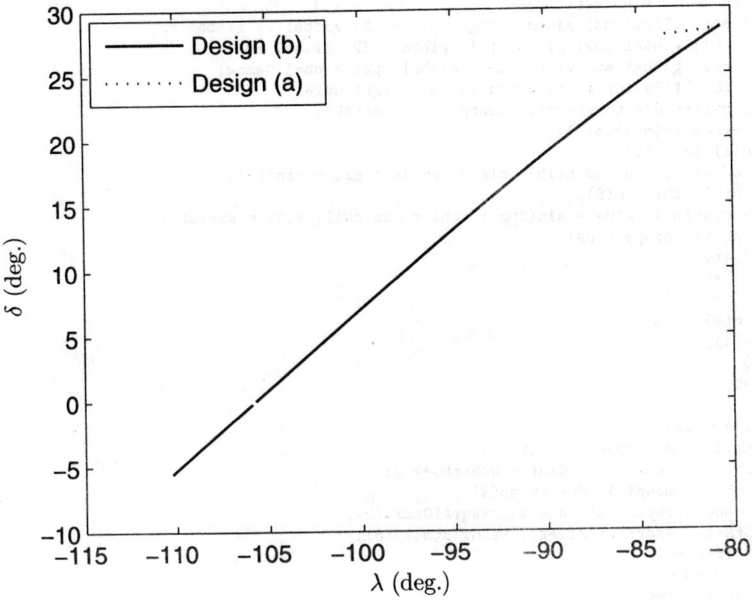

Fig. 12.46. Ground tracks for the two rocket designs.

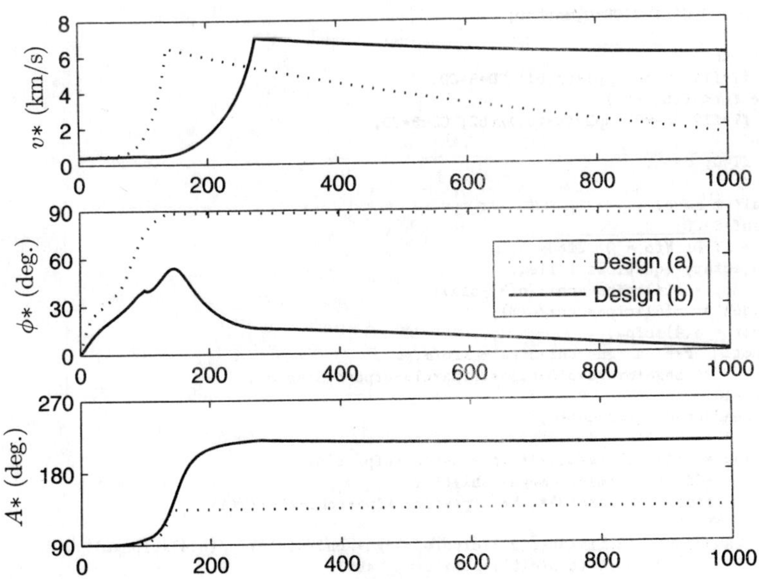

Fig. 12.47. Inertial velocity components for the two rocket designs.

Table 12.14. M-file *rocket.m* for State Equations during Ascent of a Launch Vehicle

```
function deriv = rocket(t,o)
global dtr; global mu; global omega; global S; global c; global rm;
global tb1; global tb2; global fT1; global fT2; global m01;
global m02; global mL; global mp1; global mp2; global Gamma;
global f8; %file for storing additional flight data
%acceleration due to gravity (nonspherical earth):
[g,gn]=gravity(o(3),o(2));
lo = o(1);la = o(2);
clo = cos(lo); slo = sin(lo); cla = cos(la); sla = sin(la);
fpa = o(5); chi = o(6);
cfpa = cos(fpa); sfpa = sin(fpa); cchi = cos(chi); schi = sin(chi);
%atmospheric properties:
if o(3)<rm
    o(3)=rm;
end
alt = o(3) - rm;
v  = o(4);
if v<0
    v=0;
end
if alt<=2000e3
    atmosp = atmosphere(alt,v,c);
    rho = atmosp(2);      Qinf = 0.5*rho*v^2;
    mach = atmosp(3); Kn=atmosp(4);
    CDC=conticap(mach); s = mach*sqrt(Gamma/2);
    CDFM=1.75+sqrt(pi)/(2*s); iflow=atmosp(6);
    if iflow==2
        CD=CDC;
    elseif iflow==1
        CD=CDFM;
    else
        CD =CDC+(CDFM-CDC)*(0.333*log10(Kn/sin(pi/6))+0.5113);
    end
else
    rho=0;Qinf=0;CD=0;mach=0;
end
if t<=tb1
    fT=fT1; m=m01-mp1*t/tb1; CD=8*CD;
elseif t<=(tb1+tb2)
    fT=fT2; m=m02-mp2*(t-tb1)/tb2; CD=3*CD;
else
    fT=0; m=mL;
end
[t alt m mach]
D=Qinf*S*CD;
Xfo = fT-D; Yfo = 0; Zfo = 0;
%trajectory equations follow:
longidot = o(4)*cfpa*schi/(o(3)*cla);
latidot =  o(4)*cfpa*cchi/o(3);
raddot = o(4)*sfpa;
veldot = -g*sfpa +gn*cchi*cfpa + Xfo/m...
         +omega*omega*o(3)*cla*(sfpa*cla-cfpa*cchi*sla);
if t<=10;
    headdot=0; gammadot=0;
else
gammadot=(o(4)/o(3)-g/o(4))*cfpa-gn*cchi*sfpa/o(4)...
        +Zfo/(o(4)*m)+2*omega*schi*cla...
        +omega*omega*o(3)*cla*(cfpa*cla+ sfpa*cchi*sla)/o(4);
    if abs(cfpa)>1e-6
    headdot=o(4)*schi*tan(o(2))*cfpa/o(3)-gn*schi/o(4)-Yfo/(o(4)*cfpa*m)...
            -2*omega*(tan(o(5))*cchi*cla - sla)...
            +omega*omega*o(3)*schi*sla*cla/(o(4)*cfpa);
```

```
        else
        headdot=0;
        end
    end
    %Time derivatives:
    deriv = [longidot; latidot; raddot; veldot; gammadot; headdot];
    if alt<=120e3
    Qdot=Qinf*v*S*CD/20;
    fprintf(f8,'\t%1.5e\t%1.5e\t%1.5e\t%1.5e\t%1.5e\t%1.5e\t%1.5e\n',...
            t,alt,m,mach,veldot,Qinf,Qdot);
    end
```

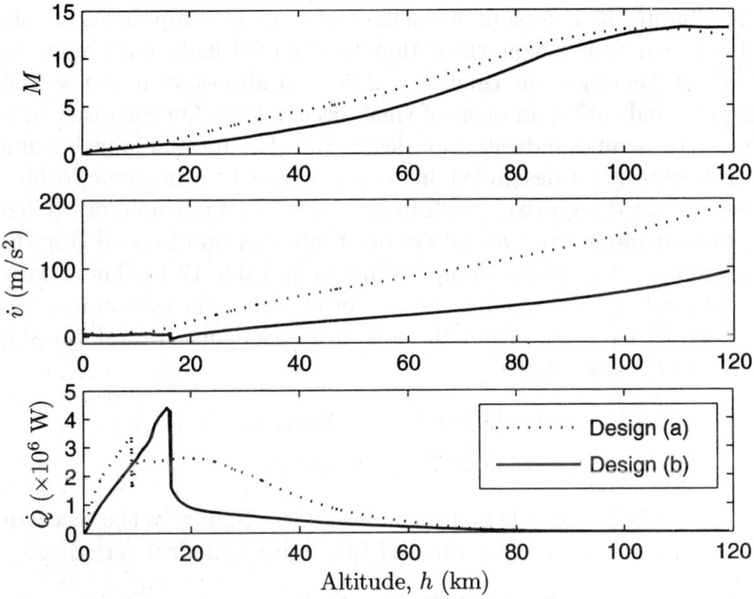

Fig. 12.48. Mach number, dynamic pressure, and heating rate until burn-out for the two rocket designs.

The altitude and relative speed time-history plots are shown in Fig. 12.44. Design (a) reaches a higher altitude but a lower relative speed at second-stage burn-out compared to design (b). Also, the rate of decrease of speed after burn-out is higher for design (a), indicating an orbit of higher eccentricity when compared to that of design (b). Consequently, the relative speed at the end of 1000 s for design (a) is about one fourth that of design (b). Our deductions about orbital eccentricity are confirmed in the flight-path angle plot in Fig. 12.45, which also depicts the natural variation in ϕ and A due to gravity and the earth's rotation (gravity turn). While design (b) achieves a nearly horizontal attitude at $t = 1000$ s, design (a) is still traveling almost vertically at that time. The variation of azimuth, A, in Fig. 12.45 is quite rapid initially but changes much more gradually due to gravity turn and settles down at $A \approx 81°$ for design (a) and $A \approx 39°$ for design (b) at $t = 1000$ s. The easterly launch direction is optimum for design (a), making the maximum use of the earth's rotation. The ground track covered in 1000 s after launch is plotted in Fig. 12.46. The much smaller ground track for design (a) in the same flight time confirms the near-vertical flight path of that case. It must be remembered that the ground track is relative to a rotating earth and is thus different from the orbital projection in the celestial frame (Chapter 5). The characteristic of the resulting orbit in each case can be obtained by converting the relative velocity into the inertial velocity components, $v*, \phi*, A*$, as explained in Chapter 5. The plots of the inertial velocity components are shown in Fig. 12.47, where it is confirmed that the inertial flight-path angle varies from about 16° at burn-out time, $t = 275$ s, to almost zero at $t = 1000$ s, for design (b), indicating an orbit of small eccentricity. On the other hand, a highly eccentric orbit is indicated for design (a). The inertial launch azimuth, A^*, is south-easterly for design (a), indicating a direct (or *prograde*) orbit, and south-westerly for design (b), which implies a *retrograde* trajectory (Chapter 5). The position and inertial velocity at orbit initialization (second-stage burn-out) are tabulated for the two launch vehicles in Table 12.15. These data are useful in determining the ensuing orbits. For example, the orbital inclination, i, is obtained in each case from the spherical trigonometric relationship of Chapter 5, $\cos i = \cos\delta \sin A*$:

$$i = 51.0941395°, \qquad \text{Design(a)},$$
$$i = 125.013428°, \qquad \text{Design(b)},$$

which confirms the retrograde motion in design (b). Similarly, the eccentricity, e, and semi-major axis, a, are estimated (assuming spherical earth) to be

$$a = 5008.815276\,\text{km}, \qquad \text{Design(a)},$$
$$a = 5471.659128\,\text{km}, \qquad \text{Design(b)},$$

and

$$e = 0.9996533, \qquad \text{Design(a)},$$
$$e = 0.3331435, \qquad \text{Design(b)},$$

which confirm our earlier remarks about the nature of each orbit. Design (a) is a near-parabolic orbit, which is an escape trajectory to infinite radius (and zero speed), while design (b) is an elliptical, earth-crossing (*suborbital*) trajectory. If we extend the simulation for design (b) to a larger time, we would observe a re-entry of the capsule (payload) into the earth's atmosphere. The time of re-entry at $h = 120$ km can be estimated from Kepler's equation (Chapter 4).

It is interesting to compare the aerothermal parameters during the ascents of the two designs. Figure 12.48 shows the altitude history plots of the

Table 12.15. Orbit Initialization Conditions for the Two Launch Vehicle Designs

	Design (a)	Design (b)
r (km)	6538.839304	6528.693364
λ (deg.)	−80.6096093	−82.7620887
δ (deg.)	28.4758704	26.3762164
v* (m/s)	6506.766314	7018.480802
ϕ* (deg.)	88.4154386	16.05878
A* (deg.)	134.3992248	219.824687

Mach number, acceleration, and rate of heat transfer in the two cases for $0 \leq h \leq 120$ km. It is evident that design (a) encounters a much higher acceleration due to its larger excess thrust, but a smaller heating rate because of its higher trajectory (smaller dynamic pressure). In essence, design (a) leaves the atmosphere much more rapidly, thereby experiencing smaller drag and heat transfer. Both trajectories display an almost-steady increase of Mach number with altitude. Due to its higher velocity at a particular altitude, the Mach number of design (a) is larger than that of design (b) in the range $15 \leq h \leq 110$ km.

12.6 Summary

Atmospheric flight is influenced by atmospheric forces—created by the motion of the vehicle relative to the atmosphere—and requires a frame fixed to the rotating planet. The kinematic equations of motion relative to a rotating planet are the same as those employed in orbital mechanics. The dynamic equations of motion are derived by expressing the inertial acceleration and force vectors in a *wind axes* frame, having axes in the directions opposite and normal to that of the relative wind velocity, and include the centripetal and Coriolis accelerations due to planetary rotation and flight-path curvature. The equations of atmospheric translation require accurate atmospheric, gravity, aerodynamic, and propulsion models and are inherently nonlinear in nature, requiring an iterative, numerical solution procedure, such as the Runge–Kutta methods. An airplane is a generic atmospheric vehicle with a plane of symmetry and separate mechanisms for generating aerodynamic lift and thrust necessary for a steady, level flight. Coordinated airplane flight requires that the side force for horizontal turns must be generated by banking the wings rather than by having a nonzero sideslip angle. The angle of attack is the primary variable employed in controlling the magnitude of the lift (and drag). Hence, airplane trajectories are governed by the angle of attack and the bank angle as well as the thrust magnitude and direction. Interesting airplane flight examples are long-range cruise, take-off, accelerated climb, maneuvers, and supermaneuvers. Transatmospheric flight-of-entry vehicles require the consideration of aerothermal loads due to excessive flight speeds. A ballistic entry vehicle is incapable of maneuvering within the atmosphere, whereas a lifting entry vehicle possesses moderate down-range and cross-range maneuvering capability, as well as reduced aerothermal loads by having a shallow trajectory. Rocket ascent trajectories are characterized by a rapidly decreasing mass and a nearly constant thrust, resulting in high rates of climb and transatmospheric acceleration. An interesting ballistic rocket ascent trajectory is the gravity turn most launch vehicles use for achieving a continuously curving flight in absence of lift.

Exercises

12.1. Derive the equations of motion for an atmospheric flight vehicle assuming a nonrotating, spherical planet. Such an approximate model is often employed for a preliminary analysis into the vehicle's trajectory.

12.2. Carrying forward the approximations of Exercise 12.1, derive the equations of translational motion assuming further that the vehicle maintains zero side force and bank angle and is flying in an absence of wind.

12.3. If the planetary curvature is ignored in Exercise 12.2 (*flat-planet approximation*), what is the nature of possible trajectories, and what are the relevant equations of motion? Such an approximation is commonly employed in analyzing short-range flights.

12.4. Using the assumptions of Exercise 12.3, with $\epsilon = 0$, the parabolic drag polar of Eq. (12.40) with constant parameters C_{D0}, K, and the ideal powerplant characteristics (Chapter 11), determine

(a) the lift coefficient required for the maximum gliding range ($f_T = 0$), assuming a small (constant) glide angle, $-\phi$.
(b) the lift coefficient required for the maximum cruising range of a jet airplane, assuming level flight and a constant thrust setting (f_T, c_T constants).
(c) the lift coefficient required for the maximum cruising range of a propeller airplane, assuming level flight and a constant power setting (P_{esh}, c_P constants).

12.5. Using the assumptions of Exercise 12.3, estimate how long a *Diamant-18* glider (weight 441 kg, aspect ratio 22.7, wing span 18 m, $K = 0.0171$, and the maximum L/D of 45) can stay airborne in a flight beginning at 1 km standard altitude.

12.6. Employing the assumptions given in the preamble to Exercise 12.4, estimate the following for *Boeing-737* jetliner at the cruising altitude of 11 km (data for the airplane during cruise are $C_{D0} = 0.017$, $K = 0.047$, $c_T = 0.75$ 1/hr, wing area 91.05 m^2, empty mass 27,310 kg, payload 16,000 kg, and take-off mass 52,400 kg):

(a) maximum lift-to-drag ratio during cruise.
(b) maximum range and corresponding lift-to-drag ratio.
(c) maximum endurance and corresponding lift-to-drag ratio.

12.7. With the assumptions of Exercise 12.4, derive the following expressions for the speeds at which level flight is possible for a jet airplane at a given altitude and thrust setting:

$$v = v_x\sqrt{x \pm \sqrt{x^2 - 1}}\,,$$

where

$$v_x = \sqrt{\frac{2mg\sqrt{K}}{\rho S\sqrt{C_{D0}}}}$$

is the level flight speed corresponding to the maximum lift-to-drag ratio, $(L/D)_{\max}$, and $x \doteq \frac{f_T(L/D)_{\max}}{mg}$.

12.8. How important are the acceleration terms due to the earth's rotation and curvature in the long-distance flight of a jet airliner? Try to answer this question by repeating the simulation of the trajectory of the airplane in Example 12.1 without the terrestrial rotation and curvature terms, and comparing the final position of the aircraft with that computed in the example.

12.9. Carry out the simulation of the jet airliner's trajectory given in Example 12.1 until 2 min after all the fuel is exhausted, i.e., $t = 6$ hr, 2 min. Assume that the FMS continues maintaining the constant lift coefficient.

12.10. Repeat Example 12.2 in the presence of a steady, 50 km/hr wind from the southeast. What is the initial heading such that the intended destination can be reached in a straight flight? What is the total time of flight?

12.11. Modify the programs of Example 12.1 for simulating the trajectory of a turboprop airplane with wing platform area, $S = 45$ m^2, mean aerodynamic chord, $\bar{c} = 2.49$ m, which begins its level cruise at a standard altitude of 18,000 ft over London's Heathrow airport ($\delta = 51.5°, \lambda = 0$, with an initial mass, $m = 22,000$ kg, speed, $v = 140$ m/s, and velocity azimuth, $A = 100°$). The drag polar is given by constant coefficients, $C_{D0} = 0.02, K = 0.055$. The structural limitations are given by $q_{\max} = 9000$ N/m^2, $M_{\max} = 0.51$. A constant power setting is maintained throughout the cruise such that $P_{\text{esh}} =$ 1800.93 SHP and $c_P = 0.61$ kg/hr/SHP at $h = 18,000$ km, and P_{esh} varies in direct proportion to the atmospheric density thereafter. Assume that during cruise, the autopilot maintains a zero sideslip, zero bank angle, and a constant angle of attack, such that the lift coefficient remains constant at $C_L = 0.45$. The total fuel mass is 3300 kg, sufficient for a 3−hr cruise. Run the simulation for a 2.75-hr cruise, and plot the state variables and relevant aerodynamic and propulsive parameters.

12.12. Repeat Exercise 12.8 for the airplane flight of Exercise 12.11.

12.13. Carry out the simulation of a New York–London cruise of a *Concorde* supersonic airliner ($S = 358.25$ m^2, mean aerodynamic chord, $\bar{c} = 14.02$ m) initial mass, $m = 175,000$ kg, initial fuel mass 90,000 kg, at a constant Mach number, $M = 2$, and initial altitude 55,000 ft. Assume $C_{D0} = 0.02, K = 0.25$ at $M = 2$, and a constant TSFC, $c_T = 1.8$ 1/hr. The service ceiling of a *Concorde* is 60,000 ft, which limits the maximum altitude during the cruise-climb. Select the initial heading, A, such that a great-circle route in the zero-wind condition is obtained. The airplane is to be flown at the constant Mach number while maintaining a coordinated, straight flight. There is little possibility of structural constraints being violated in such a flight profile (in the absence of wind gusts). Plot the time history of all relevant flight parameters, including lift coefficient and fuel mass.

12.14. Repeat Exercise 12.13 in the presence of a steady, 100 km/hr wind from the northeast. Would the airplane reach London with the available fuel?

12.15. Assuming a quasi-steady climb with a small, constant climb angle and the preamble of Exercise 12.4, constant thrust (or power) setting, and a given altitude, derive the expressions for the maximum rate of climb (and the corresponding lift coefficient and speed) for

(a) a jet airplane.
(b) a propeller airplane.

12.16. For the airplane of Exercise 12.16, estimate the maximum quasi-steady rate of climb and the corresponding speed at standard sea level if the airplane is powered by two engines, each rated at sea level thrust of 66,708 N. How good is the assumption of small climb angle in this case?

12.17. Simulate the accelerated climb of a *Lockheed SR-71* reconnaisance aircraft[20] with wing planform area $S = 167.22$ m^2, wing span $b = 16.9461$ m, mean aerodynamic chord $\bar{c} = 9.868$ m, initial mass $m = 77,000$ kg, initial fuel mass 36,290 kg, beginning at $\delta = 40°, \lambda = -120°$, speed $v = 250$ m/s, flight-path angle $\phi = 30°$, and velocity azimuth $A = 40°$. Since the actual aero-propulsion data of this airplane are unavailable in literature, assume the engine thrust and TSFC can be obtained from *engine.m* by multiplying the values resulting from the code by 2.17 and 1.5, respectively. Furthermore, assume that the parasite drag coefficient, C_{D0}, and lift-dependent drag factor,

[20] The SR-71 *Blackbird* is the world's fastest operational airplane. Its first few years of existence in the 1960s were undisclosed, due to the secret nature of its mission of flying above Mach 3.2 and 80,000 ft altitude over hostile territories. Due to its low-observable technology, the SR-71 was routinely undetected, and its flight path was commonly beyond the reach of most surface-to-air missiles and interceptor fighters. Although the SR-71 is now in museum, most of its flight characterstics are held secret.

K, can be obtained from *parasite.m* and *liftddf.m* by multiplying the results by 0.5 and 2.04, respectively. The structural limitations can be approximated by $q_{\max} = 26,000$ N/m^2, $M_{\max} = 3.55$.

12.18. Develop the approximate equations of motion for the ground-run phase of the take-off maneuver such that the assumptions of Exercise 12.4 are valid. Furthermore, derive an approximate expression for the ground-run distance, s_G, assuming that the reciprocal of acceleration is a linear function of the square of speed, i.e., $1/\dot{v} \approx A + Bv^2$, where A, B are constants.

12.19. Airplane vertical maneuvers are often analyzed by assuming either (a) constant speed or (b) constant energy (conservative) flight. The load factor, n, is taken to be a constant in both cases. Solve the approximate equations of flight in a vertical plane with the assumptions of Exercise 12.4 for each case, given the initial condition v_0, h_0, ϕ_0.

12.20. A fighter airplane with $mg/S = 4000$ N/m^2, subsonic aerodynamic parameters ($M < 0.9$), $C_{L\max} = 1.6$, $C_{D0} = 0.012$, $(L/D)_{\max} = 12$, $n_{\max} = 9$, and a standard sea level thrust-to-weight ratio $(f_T/mg)_{SL} = 1.2$ executes a conservative pull-up of $n = 6$ from level flight at standard sea level. Using the analytical results of Exercise 12.19 and neglecting density variations during the maneuver, calculate the minimum initial speed v_0 required for completing a vertical loop without stalling. What is the altitude at the top of the loop?

12.21. The fighter airplane of Exercise 12.20 executes a constant speed pull-up of $v = 300$ m/s from level flight at standard sea level. What is the highest load factor possible for a complete vertical loop? (Ignore atmospheric density variations during the maneuver.)

12.22. Horizontal turns without thrust vectoring are usually analyzed by assuming steady and coordinated flight with a constant load factor, $n = \sec\sigma$. Using these assumptions, estimate (a) the maximum instantaneous and (b) the maximum sustained turn rates of the fighter airplane given in Exercise 12.20 at standard sea level. Also, calculate the corresponding speeds and load factors.

12.23. Repeat the calculations of Exercise 12.22 at a standard altitude of 11 km, assuming the thrust is directly proportional to atmospheric density, and the supersonic aerodynamic parameters ($1.5 < M < 1.8$) of the airplane are $C_{L\max} = 0.9$, $C_{D0} = 0.018$, $(L/D)_{\max} = 8$.

12.24. Repeat case (c) of Example 12.5 using a sideways thrust deflection by employing $\mu = 45°$ throughout the turn. What is the net change (if any) in the final azimuth, speed, and altitude from that presented in the example?

12.25. Use the nonrotating, flat-planet approximation (Exercise 12.3) to derive the approximate equations for a ballistic entry. Choose the altitude, h, as the independent variable, and range, $s = \sqrt{x^2 + y^2}$, square of speed, and flight-path angle as dependent variables.

12.26. Of the several possible analytical solutions to the approximate ballistic entry equations derived in Exercise 12.25, a popular one involves the assumption that the component of acceleration due to gravity along the flight path is negligible compared to the deceleration caused by drag (i.e., $g \sin \phi \ll D/m$). Furthermore, since entry speeds are quite large, the variation of the flight-path angle during ballistic entry is usually taken to be negligible. With these assumptions, along with an exponential atmospheric model, $\rho = \rho_0 e^{-h/H}$, and a constant drag coefficient, C_D, obtain a closed-form expression for the variation of speed with altitude, and estimate the maximum deceleration and rate of heat transfer [Eq. (12.54)] with the corresponding altitudes. Take the initial condition at entry to be (h_i, v_i, ϕ_i).

12.27. Repeat the ballistic entry simulation of Example 12.6 assuming that the de-boost is performed at the apogee of the given orbit, with a 10% impulsive reduction in the inertial speed.

12.28. Repeat the maneuvering entry simulation of Example 12.7 with a smooth variation of the bank angle during $450 \leq t < 550$ s, given by

$$\sigma(t) = 90^\circ \sin \frac{\pi(t-450)}{50} .$$

What is the difference, if any, in the resulting trajectory and the landing point location compared with cases (a) and (b) of Example 12.7?

12.29. Use the nonrotating, flat-planet approximation (Exercise 12.3) with $\epsilon = 0$ to develop the approximate equations for the vertical ascent of a rocket. Assuming an ideal, single-stage rocket, negligible drag, and a linear variation of mass with time, solve for the altitude and speed at burn-out, given the initial and final mass to be m_0 and m_f, respectively.

12.30. Calculate the remaining classical orbital elements (Chapter 5) of the resulting orbits for the two launch vehicle designs in Example 12.8, from the data of Table 12.15. Which of the two orbits has the higher energy, and why?

12.31. Repeat the launch trajectory simulation of design (b) in Example 12.8, and tabulate the resulting orbital elements using the following initial relative velocity azimuth:

(a) $A = 90^\circ$.
(b) $A = 0^\circ$.
(c) $A = 270^\circ$.

Is there any difference in the trajectories of cases (a) and (c)? Explain.

12.32. Extend the launch trajectory simulation of design (b) in order to predict the latitude and longitude of landing point (5 km altitude) after the capsule's re-entry.

13

Attitude Dynamics

13.1 Aims and Objectives

- To present the universal rotational dynamics model applicable to all aerospace vehicles, emphasizing the commonality between the stability and control characteristics of aircraft and spacecraft.
- To derive several attitude dynamics models based on the useful kinematic parameters introduced in Chapter 2.
- To introduce single-axis, open-loop, time-optimal impulsive maneuvers.
- To present a rigorous derivation of the attitude motion model for atmospheric flight.
- To model and simulate important aerospace attitude motion examples, ranging from spin-stabilized, rotor- and thruster-controlled spacecraft, to gravity gradient satellites, thrust-vectored rockets, and six-degree-of-freedom, inertia-coupled, fighter airplanes.

13.2 Euler Equations of Rotational Motion

Up to this point, we have largely confined our attention to the translational motion of flight vehicles, which is represented by the motion of the center of mass. The rotational motion of a vehicle is important for various reasons (aerodynamics, pointing of weapons, payload, or antennas, etc.) and governs the instantaneous attitude (orientation). In Chapter 2, we saw how the attitude of a coordinate frame can be described relative to a reference frame. It was evident that the instantaneous attitude depends not only upon the rotational kinematics, but also on rotational dynamics which determine how the attitude parameters change with time for a specified angular velocity. If we consider a flight vehicle to be rigid, a reference frame attached to the vehicle could be used to represent the vehicle's attitude. However, in such a case, the angular velocity cannot be an arbitrary parameter but must satisfy the laws of rotational dynamics that take into account the mass distribution of the vehicle. In this chapter we shall derive the governing equations of rotational dynamics, which are equivalent to Newton's laws for translational dynamics (Chapter 4).

In Chapter 4, the rotational dynamics of a body—taken to be a collection of particles of elemental mass, ∂m—was seen to be described by the following equation of motion derived from Newton's second law by taking moments about a point o, which is either stationary or the body's center of mass:

$$\mathbf{M} = \sum \left(\mathbf{r} \times \partial m \frac{d\mathbf{v}}{dt} \right) . \tag{13.1}$$

Here $\mathbf{v}$ is the total (inertial) velocity of the particle, $\mathbf{r}$ is the relative position of the particle with respect to the point o (which serves as the origin of a reference coordinate frame), and $\mathbf{M} \doteq \sum(\mathbf{r} \times \partial \mathbf{f})$ is the net *external torque* about o. In the derivation of Eq. (13.1), it has been assumed that all internal torques cancel each other by virtue of Newton's third law. This is due to the fact that the internal forces between any two particles constituting the body act along the line joining the particles.[1] By taking the limit $\partial m \to 0$, we can replace the summation over particles by an integral over mass, and write

$$\mathbf{M} = \int \left(\mathbf{r} \times \frac{d\mathbf{v}}{dt} \right) dm . \tag{13.2}$$

In Chapter 4, we also defined a particle's angular momentum by $\partial \mathbf{H} \doteq \mathbf{r} \times \partial m \mathbf{v}$. By integration, the total angular momentum of the body can be written as follows:

$$\mathbf{H} = \int \mathbf{r} \times \mathbf{v} dm . \tag{13.3}$$

Assuming that the body has a constant mass, let us differentiate Eq. (13.3) with time, leading to

$$\frac{d\mathbf{H}}{dt} = \int \mathbf{v} \times \mathbf{v} dm + \int \mathbf{r} \times \frac{d\mathbf{v}}{dt} dm . \tag{13.4}$$

The first term on the right-hand side of Eq. (13.6) is identically zero, while the second term is easily identified from Eq. (13.2) to be the net external torque, $\mathbf{M}$; thus, we have

$$\frac{d\mathbf{H}}{dt} = \mathbf{M} . \tag{13.5}$$

Note that in the above derivation, o is either a stationary point or the body's center of mass. When applied to the general motion of a flight vehicle, it is useful to select o to be the center of mass. In such a case, the moving reference frame, $(oxyz)$, is called a *body frame*.

Now, let us assume that the body is *rigid*, i.e., the distance between any two points on the body does not change with time. The rigid-body

[1] Most forces of interaction among particles obey this principle, with the exception of the magnetic force.

assumption—valuable in simplifying the equations of motion—is a reasonable approximation for the rotational dynamics of most flight vehicles. Then it follows that since the center of mass is a point fixed relative to the body (although it may not always lie on the body), the magnitude of the vector $\mathbf{r}$ is invariant with time. Hence, we can write the total (inertial) velocity of an arbitrary point on the rigid body located at $\mathbf{r}$ relative to o as follows:

$$\mathbf{v} = \mathbf{v_o} + \boldsymbol{\omega} \times \mathbf{r}\,, \tag{13.6}$$

where $\mathbf{v_o}$ denotes the velocity of the center of mass, o, and $\boldsymbol{\omega}$ is the angular velocity of the reference coordinate frame with the origin at o. Therefore, for a rigid body, Eqs. (13.3) and (13.6) lead to the following expression for the angular momentum:

$$\mathbf{H} = \int \mathbf{r} \times \mathbf{v_o} \mathrm{d}m + \int \mathbf{r} \times (\boldsymbol{\omega} \times \mathbf{r}) \mathrm{d}m\,. \tag{13.7}$$

The first term on the right-hand side of Eq. (13.7) can be expressed as

$$\int \mathbf{r} \times \mathbf{v_o} \mathrm{d}m = \left(\int \mathbf{r} \mathrm{d}m \right) \times \mathbf{v_o}\,, \tag{13.8}$$

which vanishes by the virtue of o being the center of mass ($\int \mathbf{r}\mathrm{d}m = 0$). Thus, we have

$$\mathbf{H} = \int \mathbf{r} \times (\boldsymbol{\omega} \times \mathbf{r}) \mathrm{d}m\,. \tag{13.9}$$

We choose to resolve all the vectors in the body frame with axes ox, oy, oz along unit vectors $\mathbf{i}, \mathbf{j}, \mathbf{k}$, respectively, such that

$$\mathbf{r} = x\mathbf{i} + y\mathbf{j} + z\mathbf{k}, \tag{13.10}$$

$$\boldsymbol{\omega} = \omega_x\mathbf{i} + \omega_y\mathbf{j} + \omega_z\mathbf{k}, \tag{13.11}$$

$$\mathbf{H} = H_x\mathbf{i} + H_y\mathbf{j} + H_z\mathbf{k}, \tag{13.12}$$

$$\mathbf{M} = M_x\mathbf{i} + M_y\mathbf{j} + M_z\mathbf{k}. \tag{13.13}$$

By substituting the vector components into Eq. (13.9) and simplifying, we have the following matrix-vector product for the angular momentum:

$$\mathbf{H} = \mathsf{J}\boldsymbol{\omega}\,, \tag{13.14}$$

where J is the *inertia tensor*, given by

$$\mathsf{J} \doteq \begin{pmatrix} \int (y^2 + z^2)\mathrm{d}m & -\int xy\mathrm{d}m & -\int xz\mathrm{d}m \\ -\int xy\mathrm{d}m & \int (x^2 + z^2)\mathrm{d}m & -\int yz\mathrm{d}m \\ -\int xz\mathrm{d}m & -\int yz\mathrm{d}m & \int (x^2 + y^2)\mathrm{d}m \end{pmatrix}. \tag{13.15}$$

Clearly, J is a symmetric matrix. In terms of its components, J is written as follows:

$$\mathsf{J} \doteq \begin{pmatrix} J_{xx} & J_{xy} & J_{xz} \\ J_{xy} & J_{yy} & J_{yz} \\ J_{xz} & J_{yz} & J_{zz} \end{pmatrix} . \tag{13.16}$$

The components of the inertia tensor are divided into the *moments of inertia*, J_{xx}, J_{yy}, J_{zz}, and the *products of inertia*, J_{xy}, J_{yz}, J_{xz}. Recall that $\boldsymbol{\omega}$ is the angular velocity of the reference coordinate frame, $(oxyz)$. This frame has its origin, o, fixed at the center of mass of the rigid body. However, if the axes of the frame are *not* fixed to the rigid body, the angular velocity of the body would be different from $\boldsymbol{\omega}$. In such a case, the moments and products of inertia would be time-varying. Since the main advantage of writing the angular momentum in the form of Eq. (13.14) lies in the introduction of an inertia tensor, whose elements describe the constant mass distribution of the rigid body, we want to have a constant inertia tensor. If we deliberately choose to have the axes of the body frame, $(oxyz)$, fixed to the body, and thus rotating with the same angular velocity, $\boldsymbol{\omega}$, as that of the body, the moments and products of inertia will be invariant with time. Such a reference frame, with axes tied rigidly to the body, is called a *body-fixed frame*. From this point forward, the body frame $(oxyz)$ will be taken to be the body-fixed frame. Hence, $\boldsymbol{\omega}$ in Eq. (13.14) is the angular velocity of the rigid body, and J is a constant matrix.

The equations of rotational motion of the rigid body can be obtained in the body-fixed frame by substituting Eq. (13.14) into Eq. (13.5) and applying the rule of taking the time derivative of a vector (Chapter 2):

$$\mathbf{M} = \mathsf{J}\frac{\partial \boldsymbol{\omega}}{\partial t} + \boldsymbol{\omega} \times (\mathsf{J}\boldsymbol{\omega}) , \tag{13.17}$$

where the partial time derivative represents the time derivative taken with reference to the body-fixed frame,

$$\frac{\partial \boldsymbol{\omega}}{\partial t} \doteq \begin{Bmatrix} \frac{d\omega_x}{dt} \\ \frac{d\omega_y}{dt} \\ \frac{d\omega_z}{dt} \end{Bmatrix} = \begin{Bmatrix} \dot{\omega}_x \\ \dot{\omega}_y \\ \dot{\omega}_z \end{Bmatrix} . \tag{13.18}$$

By replacing the vector product in Eq. (13.17) by a matrix product (Chapter 2), we can write

$$\mathbf{M} = \mathsf{J}\frac{\partial \boldsymbol{\omega}}{\partial t} + \mathsf{S}(\boldsymbol{\omega})\mathsf{J}\boldsymbol{\omega} , \tag{13.19}$$

where

$$\mathsf{S}(\boldsymbol{\omega}) = \begin{pmatrix} 0 & -\omega_z & \omega_y \\ \omega_z & 0 & -\omega_x \\ -\omega_y & \omega_x & 0 \end{pmatrix} . \tag{13.20}$$

Equation (13.19) represents three scalar, coupled, nonlinear, ordinary differential equations, called *Euler's equations of rotational dynamics*. These are the governing equations for rotational dynamics of rigid bodies, and their solution gives the angular velocity, $\boldsymbol{\omega}$, at a given instant. In Chapter 2, we derived the kinematic equations for the rotation of a coordinate frame in terms of various alternative attitude representations. These kinematic equations, along with Euler's equations of rotational dynamics, complete the set of differential equations needed to describe the changing attitude of a rigid body under the influence of a time-varying torque vector, $\mathbf{M}$. The variables of the rotational motion are thus the kinematical parameters representing the instantaneous attitude of a body-fixed frame, and the angular velocity of the rigid body resolved in the same frame.

13.3 Rotational Kinetic Energy

In Chapter 4, we derived the kinetic energy for a system of N particles

$$T = \frac{1}{2}mv_0^2 + \frac{1}{2}\sum_{i=1}^{N} m_i u_i^2 , \tag{13.21}$$

where u_i is the speed of the ith particle (of mass m_i) relative to the center of mass o, which is moving with a speed v_0. When applied to a body, the summation over particles is replaced by an integral over mass, and we have

$$T = \frac{1}{2}mv_0^2 + \frac{1}{2}\int u^2 \mathrm{d}m . \tag{13.22}$$

It is clear from Eq. (13.6) that for a rigid body, $u^2 = (\boldsymbol{\omega} \times \mathbf{r}) \cdot (\boldsymbol{\omega} \times \mathbf{r})$, and we can write

$$T = \frac{1}{2}mv_0^2 + \frac{1}{2}\int (\boldsymbol{\omega} \times \mathbf{r}) \cdot (\boldsymbol{\omega} \times \mathbf{r})\mathrm{d}m . \tag{13.23}$$

The same result could be obtained by using the following defining expression of the kinetic energy, and substituting Eq. (13.6) for a rigid body:

$$T = \frac{1}{2}\int \mathbf{v} \cdot \mathbf{v}\mathrm{d}m . \tag{13.24}$$

The first term on the right-hand side of Eq. (13.23) represents the kinetic energy due to the translation of the center of mass, whereas the second term denotes the kinetic energy of rotation about the center of mass. The expression for the rotational kinetic energy of the rigid body, T_{rot} can be simplified by utilizing the angular momentum [Eq. (13.9)], leading to

$$T_{\mathrm{rot}} \doteq \frac{1}{2}\int (\boldsymbol{\omega} \times \mathbf{r}) \cdot (\boldsymbol{\omega} \times \mathbf{r})\mathrm{d}m = \frac{1}{2}\boldsymbol{\omega} \cdot \mathbf{H} = \frac{1}{2}\boldsymbol{\omega}^T \mathsf{J}\boldsymbol{\omega} . \tag{13.25}$$

This expression for the rotational kinetic energy of a rigid body is very useful in simplifying Euler's equations.

The rotational kinetic energy is conserved if there is no external torque applied to the rigid body. This fact is evident by taking the time derivative of Eq. (13.25), and substituting Euler's equations, Eq. (13.18), with $\mathbf{M} = \mathbf{0}$:

$$\frac{dT_{\text{rot}}}{dt} = \frac{1}{2}\frac{d\boldsymbol{\omega}}{dt} \cdot \mathbf{H} + \frac{1}{2}\boldsymbol{\omega} \cdot \frac{d\mathbf{H}}{dt} = 0 \,. \tag{13.26}$$

Since $\mathbf{M} = \mathbf{0}$, the second term on the right-hand side of Eq. (13.26) vanishes due to Eq. (13.5), while the first term vanishes by the virtue of Eq. (13.18), which produces $\frac{d\boldsymbol{\omega}}{dt} \cdot \mathbf{H} = -\boldsymbol{\omega} \cdot (\boldsymbol{\omega} \times \mathbf{H}) = \mathbf{0}$. In the absence of an external torque (such as in spacecraft applications), the conservation of both rotational kinetic energy and angular momentum can be effectively utilized in obtaining analytical relationships between the angular velocity and the inertia tensor.

13.4 Principal Body Frame

As seen in Chapter 2, the translation of a frame is trivially handled by merely shifting the origin of the coordinate frame. In terms of the body-fixed frame, such a translation of the origin (center of mass of rigid body) would produce a modification of the inertia tensor easily obtained by the *parallel axes theorem* (discussed later in this chapter). However, a rotation of the body-fixed frame about the same origin is nontrivial. The body-fixed frame used above in deriving Euler's equations has an arbitrary orientation relative to the rigid body. There are infinitely many ways in which these axes can be fixed to a given rigid body at the center of mass. A great simplification in Euler's equations is possible by choosing a particular orientation of the body-fixed frame relative to the rigid body such that the products of inertia, J_{xy}, J_{yz}, J_{xz}, vanish. Such a frame is called the *principal body-fixed frame*. The inertia tensor resolved in the principal body frame is a diagonal matrix, $\mathsf{J_p}$. In order to derive the coordinate transformation that produces the principal frame, $\mathbf{i_p}, \mathbf{j_p}, \mathbf{k_p}$, from an arbitrary body-fixed frame, $\mathbf{i}, \mathbf{j}, \mathbf{k}$, consider the rotation matrix, $\mathsf{C_p}$, defined by

$$\begin{Bmatrix} \mathbf{i} \\ \mathbf{j} \\ \mathbf{k} \end{Bmatrix} = \mathsf{C_p} \begin{Bmatrix} \mathbf{i_p} \\ \mathbf{j_p} \\ \mathbf{k_p} \end{Bmatrix} . \tag{13.27}$$

The relationship between a vector resolved in the principal frame and the same vector in an arbitrary body-fixed frame is thus through the rotation matrix, $\mathsf{C_p}$. If we continue to denote the vectors resolved in the principal frame by the subscript p, we have

$$\boldsymbol{\omega} = \mathsf{C_p}\boldsymbol{\omega}_\mathbf{p} \,. \tag{13.28}$$

Now, since there is no change in the rotational kinetic energy caused by the coordinate transformation, we can utilize Eq. (13.25) and write

$$T_{\text{rot}} = \frac{1}{2}\boldsymbol{\omega}^T \mathsf{J}\boldsymbol{\omega} = \frac{1}{2}\boldsymbol{\omega}_\mathbf{p}{}^T \mathsf{J_p}\boldsymbol{\omega}_\mathbf{p} \,. \tag{13.29}$$

Upon substituting Eq. (13.28) into Eq. (13.29), and comparing the terms on both the sides of the resulting equation, we have

$$\omega_p{}^T \mathsf{J}_p \omega_p = \omega^T \mathsf{J} \omega = \omega_p{}^T \mathsf{C}_p{}^T \mathsf{J} \mathsf{C}_p \omega_p \,, \tag{13.30}$$

which, on applying the orthogonality property of the rotation matrix, yields

$$\mathsf{J}_p = \mathsf{C}_p{}^T \mathsf{J} \mathsf{C}_p \,. \tag{13.31}$$

Since J_p is a diagonal matrix, it easily follows [4] that the diagonal elements of J_p are the distinct eigenvalues of J, while C_p has the eigenvectors of J as its columns. Thus, Eq. (13.31) is the formula for deriving the inertia tensor in the principal frame and the coordinate transformation matrix, C_p, from the eigenvalue analysis of J.

Example 13.1. A rigid body has the following inertia tensor:

$$\mathsf{J} = \begin{pmatrix} 100 & 10 & 35 \\ 10 & 250 & 50 \\ 35 & 50 & 300 \end{pmatrix} \text{ kg.m}^2 \,.$$

Find the inertia tensor in the principal frame and the coordinate transformation matrix, C_p.

This problem is easily solved with the following MATLAB statements employing the intrinsic eigenvalue analysis function *eig.m*:

```
>> J=[100 10 35;10 250 50;35 50 300]; %inertia tensor
>> [Cp,Jp]=eig(J) %rotation matrix & principal inertia tensor

Cp =  0.9862      -0.0754      0.1473
     -0.0103       0.8605      0.5094
     -0.1651      -0.5039      0.8479


Jp =      94.0366        0          0
           0          219.8462      0
           0             0        336.1172

>> Cp'*J*Cp %check the rotation matrix

ans =     94.0366 0.0000         -0.0000
           0.0000        219.8462    -0.0000
        -0.0000       -0.0000      336.1172
```

Thus, the principal inertia tensor is

$$\mathsf{J}_p = \begin{pmatrix} 94.0366 & 0 & 0 \\ 0 & 219.8462 & 0 \\ 0 & 0 & 336.1172 \end{pmatrix} \text{ kg.m}^2 \,.$$

The computed rotation matrix represents the orientation of the currently employed body-fixed frame with respect to the principal frame.

The inertia tensor can be diagonalized by the foregoing procedure to produce the principal inertia tensor if and only if the principal moments of inertia are distinct, which is the case for an asymmetric object. For an axisymmetric body, two principal moments of inertia are equal, but it is neither necessary, nor feasible, to follow the above approach for obtaining the principal moments of inertia (since the principal axes are easily identified from symmetry). Hence, for all practical purposes we shall work only in the principal body frame, and the following discussion pertains to the principal body axes, without explicitly carrying the subscript p.

13.5 Torque-Free Rotation of Spacecraft

A spacecraft's rotational motion is generally in the absence of external torques. In order to analyze the rotational stability and control characteristics of spacecraft, it thus becomes necessary to study the torque-free motion of rigid bodies. Since the external torque is zero, the angular momentum of the rigid body about its center of mass (or a fixed point) is conserved by the virtue of Eq. (13.5). Thus, we can express Euler's equations for the torque-free motion ($\mathbf{M} = \mathbf{0}$) of a rigid body in the principal frame as follows:

$$\begin{aligned} J_{xx}\dot{\omega}_x + \omega_y\omega_z(J_{zz} - J_{yy}) &= 0, \\ J_{yy}\dot{\omega}_y + \omega_x\omega_z(J_{xx} - J_{zz}) &= 0, \\ J_{zz}\dot{\omega}_z + \omega_x\omega_y(J_{yy} - J_{xx}) &= 0, \end{aligned} \tag{13.32}$$

where the dot represents the time derivative, $\frac{\mathrm{d}}{\mathrm{d}t}$. For a general, asymmetric body possessing nonzero angular velocity components about all three axes, Eq. (13.32) is difficult to solve in a closed form, but is amenable to numerical integration in time.

Since a torque-free rigid body does not have a mechanism for energy dissipation, its rotational kinetic energy is conserved according to Eq. (13.26). However, a spacecraft is an imperfect rigid body, generally consisting of several rigid bodies rotating relative to each other (e.g., reaction wheels and control gyroscopes), as well as containing liquid propellants. The rotors and liquid propellants provide mechanisms for internal dissipation of the rotational kinetic energy through friction and sloshing motion, respectively. When analyzing the rotational stability of spacecraft, it is therefore vital to regard them as semirigid objects that continually dissipate kinetic energy until a stable equilibrium is achieved. For a semirigid body, Euler's equations remain valid (as the external torque remains zero), but the rotational kinetic energy is not conserved.

Before solving torque-free Euler equations for a general case, let us use them to analyze rotational stability characteristics of rigid spacecraft. Such an analysis would reveal the axes about which a stable rotational equilibrium can be achieved. The process of obtaining a stable equilibrium through constant speed rotation about a principal axis is called *spin stabilization*. Although spin stabilization is strictly valid only for a spacecraft, it can be applied approximately to some atmospheric flight vehicles that have a small aerodynamic moment about the spin axis, such as certain missiles and projectiles. A rifle bullet is a good example of a spin-stabilized object. Furthermore, spin stabilization is also the principle of operation of gyroscopic instruments, which are commonly used in aerospace vehicles.

Stability is a property of an equilibrium and can be defined in many ways. For our purposes, we shall define a stable equilibrium as the one about which a bounded disturbance does not produce an unbounded response. The disturbance can be regarded as the initial condition, expressed in terms of an initial deviation of the motion variables from the equilibrium. In a stability analysis, it is sufficient to study the response to a small initial deviation, because stability is not influenced by the magnitude of the disturbance.

13.5.1 Axisymmetric Spacecraft

When the spacecraft possesses an axis of symmetry, Euler's equations are further simplified. Consider a spacecraft rotating about its axis of symmetry, oz, called the *longitudinal axis*. Due to axial symmetry, $J_{xx} = J_{yy}$, and we have

$$\begin{aligned} J_{xx}\dot{\omega}_x + \omega_y\omega_z(J_{zz} - J_{xx}) &= 0, \\ J_{xx}\dot{\omega}_y + \omega_x\omega_z(J_{xx} - J_{zz}) &= 0, \\ J_{zz}\dot{\omega}_z &= 0. \end{aligned} \tag{13.33}$$

It is clear from Eq. (13.33) that the spacecraft is in a state of equilibrium whenever $\omega_x = \omega_y = 0$, called *pure spin* about the axis of symmetry. It is also evident from the last of Eq. (13.33) that $\dot{\omega}_z = 0$, or $\omega_z = n =$ constant, irrespective of the magnitudes of $\omega_x, \omega_y = 0$. Let us assume that the spacecraft was in a state of pure spin when a disturbance, $\omega_x(0), \omega_y(0)$, is applied at time $t = 0$. Let us examine the resulting motion of the spacecraft by solving the first two equations of Eq. (13.33), which are written in the following vector matrix form:

$$\begin{Bmatrix} \dot{\omega}_x \\ \dot{\omega}_y \end{Bmatrix} = \begin{pmatrix} 0 & -k \\ k & 0 \end{pmatrix} \begin{Bmatrix} \omega_x \\ \omega_y \end{Bmatrix}, \tag{13.34}$$

where $k = n\frac{(J_{zz} - J_{xx})}{J_{xx}}$. Equation (13.34) represents linear, time-invariant state equations (Chapter 14) whose solution with the initial condition, $\omega_x(0), \omega_y(0)$ at $t = 0$, is easily written in a closed form as follows:

$$\begin{Bmatrix} \omega_x(t) \\ \omega_y(t) \end{Bmatrix} = e^{\mathsf{K}t} \begin{Bmatrix} \omega_x(0) \\ \omega_y(0) \end{Bmatrix}, \tag{13.35}$$

where $e^{\mathsf{K}t}$ is the matrix exponential (Chapter 14), and

$$\mathsf{K} = \begin{pmatrix} 0 & -k \\ k & 0 \end{pmatrix} . \tag{13.36}$$

Using one of the methods of Chapter 14, we can write the matrix exponential by taking the inverse Laplace transform of the resolvent as follows:

$$e^{\mathsf{K}t} = \mathcal{L}^{-1}(s\mathsf{I} - \mathsf{K})^{-1} = \begin{pmatrix} \cos(kt) & -\sin(kt) \\ \sin(kt) & \cos(kt) \end{pmatrix} . \tag{13.37}$$

Therefore, the solution is given by

$$\omega_x(t) = \omega_x(0)\cos(kt) - \omega_y(0)\sin(kt),$$
$$\omega_y(t) = \omega_x(0)\sin(kt) + \omega_y(0)\cos(kt). \tag{13.38}$$

Equation (13.33) implies that the rotational motion of an axisymmetric, rigid spacecraft, disturbed from the equilibrium state of pure spin about the longitudinal axis by a disturbance $\omega_x(0), \omega_y(0)$, is oscillatory in the *oxy* plane (called the *lateral plane*), while the spin rate, $\omega_z = n$, remains unaffected. This causes a coning motion of the disturbed body about the axis of symmetry. An important characteristic of the solution given by Eq. (13.38) is easily seen to be the following:

$$\omega_{xy}^2 \doteq \omega_x^2 + \omega_y^2 = \omega_x^2(0) + \omega_y^2(0) = \text{ constant}, \tag{13.39}$$

which implies that the magnitude of the angular velocity component in the lateral plane is constant. This lateral angular velocity component, ω_{xy}, is responsible for the coning motion called *precession*. Since precession is a constant amplitude oscillation, whose magnitude is bounded by that of the applied disturbance, we say that the motion of a rigid spacecraft about its axis of symmetry is unconditionally stable. Figure 13.1 shows the geometry of precessional motion, where the angular velocity, $\boldsymbol{\omega}$, makes a constant angle, $\alpha = \tan^{-1}\frac{\omega_{xy}}{n}$, with the axis of symmetry, *oz*. Furthermore, the angular momentum, $\mathbf{H} = J_{xx}(\omega_x\mathbf{i} + \omega_y\mathbf{j}) + J_{zz}n\mathbf{k}$, makes a constant angle, $\beta = \tan^{-1}\frac{J_{xx}\omega_{xy}}{J_{zz}n}$, with the axis of symmetry, called the *nutation angle*. The axis of symmetry thus describes a cone of semivertex angle α, called the *body cone*, about the angular velocity vector, and a cone of semivertex angle β, called the *space cone*, about the angular momentum vector. Note that while $\boldsymbol{\omega}$ is a rotating vector, $\mathbf{H}$ is fixed in inertial space due to the conservation of angular momentum. In Fig. 13.1, $J_{xx} > J_{zz}$ is assumed, for which $\beta > \alpha$. While a rigid, axisymmteric spacecraft's precessional motion is unconditionally stable (as seen above), the same cannot be said for a semirigid spacecraft. Since most spacecraft carry some liquid propellants, they must be regarded as semirigid, wherein the angular momentum is conserved, but the rotational kinetic energy dissipates due to the sloshing of liquids caused by precession.

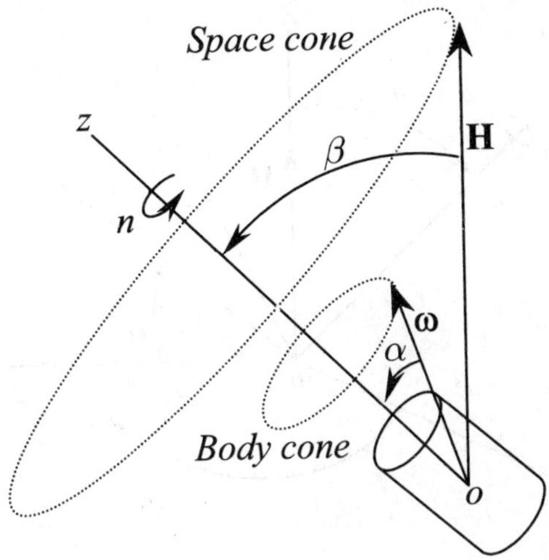

Fig. 13.1. Precession of an axisymmetric spacecraft.

Whenever energy dissipation is present in a dynamical system, there is a tendency to move toward the state of equilibrium with the lowest kinetic energy. In a state of pure spin about a principal axis, there is no energy dissipation because the liquids rotate with the same speed as the spacecraft. Thus, pure spin about a principal axis spin is a state of equilibrium for a semirigid spacecraft. For the torque-free rotation of spacecraft, the lowest kinetic energy is achieved for pure spin about the major principal axis. This can be seen from Eq. (13.25), while applying the law of conservation of angular momentum. Hence, the internal energy dissipation eventually converts the precessional motion into a spin about the major axis. Therefore, a semirigid spacecraft can be spin-stabilized only about its major axis. In applying the foregoing results to such a spacecraft, it is necessary that $J_{zz} > J_{xx}$. If the axis of symmetry is the minor axis, pure spin about it would eventually be converted into a tumbling motion about the major principal axis in the presence of inevitable disturbances and liquid propellants. This phenomenon was encountered in the first satellite launched by NASA, named *Explorer*, rendering the long cylindrical spacecraft useless after a few days in orbit. For this reason, all spinning satellites are designed to have the axis of symmetry as the major axis. One may study the attitudinal kinematics of axisymmetric spacecraft, spin stabilized about the longitudinal axis, by simultaneously solving the kinematic equations of motion (Chapter 2) with the Euler equations. Since the angular momentum vector, **H**, is fixed in space, an obvious choice of the reference inertial frame is with the axis **K** along **H**. The most commonly used kinematic parameters for spin-stabilized spacecraft are the $(\psi)_3, (\theta)_1, (\phi)_3$ Euler angles (Chapter 2). Since the spin axis of the precessing spacecraft is never exactly aligned with the angular momentum ($\theta \neq 0$), the singularity of this attitude

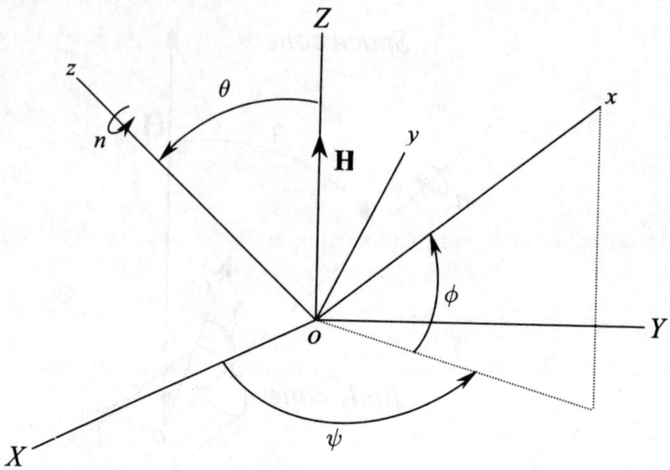

Fig. 13.2. Attitude of a precessing, axisymmetric spacecraft via 3-1-3 Euler angles.

representation at $\theta = 0, 180°$ is not encountered, thereby removing the main disadvantage of Euler's angle representation. Therefore, the constant nutation angle is given by $\beta = \theta$, and from Fig. 13.2 depicting Euler's angles, we have

$$\sin\theta = \frac{J_{xx}\omega_{xy}}{H} = \frac{J_{xx}\omega_{xy}}{\sqrt{J_{xx}\omega_{xy}^2 + J_{zz}n^2}},$$
$$\cos\theta = \frac{J_{zz}n}{H} = \frac{J_{zz}n}{\sqrt{J_{xx}\omega_{xy}^2 + J_{zz}n^2}}. \quad (13.40)$$

The general kinematic equations for the $(\psi)_3, (\theta)_1, (\phi)_3$ Euler angles were derived in Chapter 2 and are repeated here as follows:

$$\begin{Bmatrix} \dot{\psi} \\ \dot{\theta} \\ \dot{\phi} \end{Bmatrix} = \frac{1}{\sin\theta}\begin{pmatrix} \sin\phi & \cos\phi & 0 \\ \cos\phi\sin\theta & -\sin\phi\sin\theta & 0 \\ -\sin\phi\cos\theta & -\cos\phi\cos\theta & \sin\theta \end{pmatrix}\begin{Bmatrix} \omega_x \\ \omega_y \\ \omega_z \end{Bmatrix}. \quad (13.41)$$

Upon substitution of $\omega_z = n$, and Eq. (13.39) into Eq. (13.41) we have

$$\dot{\psi} = \frac{\omega_{xy}}{\sin\theta},$$
$$\dot{\theta} = 0, \quad (13.42)$$
$$\dot{\phi} = n - \frac{\omega_{xy}}{\tan\theta}.$$

Since both θ and ω_{xy} are constants [Eqs. (13.39) and (13.40)], the angular rates $\dot{\psi}$ and $\dot{\phi}$ are also constants, whose alternative expressions are obtained by substituting Eq. (13.40) into Eq. (13.42) as

$$
\begin{aligned}
\dot{\psi} &= \frac{\sqrt{J_{xx}\omega_{xy}^2 + J_{zz}n^2}}{J_{xx}}, \\
\dot{\theta} &= 0, \\
\dot{\phi} &= n(1 - \frac{J_{zz}}{J_{xx}}) = -k\,.
\end{aligned}
\tag{13.43}
$$

The angular rate $\dot{\psi}$ represents the frequency of precession and is called the *precession rate*, while $\dot{\phi}$ represents the total spin rate of the body in the inertial frame and is known as the *inertial spin rate*. If $J_{xx} > J_{zz}$, the axisymmetric body is said to be *prolate*, and $\dot{\psi}$ has the same sign as that of $\dot{\phi}$. For the case of an *oblate* body ($J_{xx} < J_{zz}$), the angular rates $\dot{\psi}$ and $\dot{\phi}$ have opposite signs. The solution for the Euler angles is easily obtained by integration of Eq. (13.43)—with the initial orientation at $t = 0$ specified as $\psi(0), \theta(0), \phi(0)$—to be the following:

$$
\begin{aligned}
\psi &= \psi_0 + \frac{\sqrt{J_{xx}\omega_{xy}^2 + J_{zz}n^2}}{J_{xx}}t, \\
\theta &= \theta(0), \\
\phi &= \phi(0) - kt = \phi(0) - n(1 - \frac{J_{zz}}{J_{xx}})t.
\end{aligned}
\tag{13.44}
$$

The angles ψ and ϕ thus vary linearly with time due to a constant precession rate, ω_{xy}.

13.5.2 Asymmetric Spacecraft

Let us assume that an asymmetric spacecraft is in a state of pure spin of rate n about the principal axis oz, prior to the time $t = 0$ when a small disturbance, $\omega_x(0), \omega_y(0)$, is applied. At a subsequent time, the angular velocity components can be expressed as $\omega_z = n + \epsilon$, and ω_x, ω_y. Since a small disturbance has been applied, we can treat $\epsilon, \omega_x, \omega_y$ as small quantities and solve Euler's equations. If the solution indicates that $\epsilon, \omega_x, \omega_y$ grow with time in an unbounded fashion, it will be evident that our assumption of small deviations remaining small is false, and we are dealing with an unstable equilibrium. Otherwise, we have a stable equilibrium. Hence, with the assumption of small deviation from equilibrium, we can write the approximate, linearized Euler equations as follows:

$$
\begin{aligned}
J_{xx}\dot{\omega}_x + n\omega_y(J_{zz} - J_{yy}) &\approx 0, \\
J_{yy}\dot{\omega}_y + n\omega_x(J_{xx} - J_{zz}) &\approx 0, \\
J_{zz}\dot{\epsilon} &\approx 0,
\end{aligned}
\tag{13.45}
$$

in which we have neglected second- (and higher-) order terms involving $\epsilon, \omega_x, \omega_y$. The first two equations of Eq. (13.45) can be written in the following vector matrix form:

$$\begin{Bmatrix} \dot{\omega}_x \\ \dot{\omega}_y \end{Bmatrix} = \begin{pmatrix} 0 & -k_1 \\ k_2 & 0 \end{pmatrix} \begin{Bmatrix} \omega_x \\ \omega_y \end{Bmatrix} , \tag{13.46}$$

where $k_1 = n\frac{(J_{zz}-J_{yy})}{J_{xx}}$ and $k_2 = n\frac{(J_{zz}-J_{xx})}{J_{yy}}$. Being in a linear, time-invariant state-space form, these approximate equations are solved using the matrix exponential as follows:

$$\begin{Bmatrix} \omega_x(t) \\ \omega_y(t) \end{Bmatrix} = e^{\mathsf{A}t} \begin{Bmatrix} \omega_x(0) \\ \omega_y(0) \end{Bmatrix} , \tag{13.47}$$

where $e^{\mathsf{A}t}$ is the matrix exponential denoting the state transition matrix (Chapter 14) and

$$\mathsf{A} = \begin{pmatrix} 0 & -k_1 \\ k_2 & 0 \end{pmatrix} . \tag{13.48}$$

The eigenvalues of A determine whether the ensuing motion will be bounded, and thus denote stability or instability. They are obtained as follows:

$$\mid s\mathsf{I} - \mathsf{K} \mid = s^2 + k_1 k_2 = 0 \, , \tag{13.49}$$

or,

$$s_{1,2} = \pm\sqrt{-k_1 k_2} \, . \tag{13.50}$$

From the eigenvalues of A, it is clear that two possibilities exist for the response: (a) $k_1k_2 < 0$, for which one eigenvalue has a positive real part, indicating exponentially growing (unbounded) motion, or (b) $k_1k_2 > 0$, for which both eigenvalues are imaginary, and the motion is a constant amplitude (bounded) oscillation about the equilibrium. Therefore, for stability we must have $k_1k_2 > 0$, which implies that either $(J_{zz} > J_{xx}, J_{zz} > J_{yy})$ or $(J_{zz} < J_{xx}, J_{zz} < J_{yy})$. Hence, spin stabilization of a rigid, asymmetric spacecraft is possible about either the major principal axis or the minor principal axis. This confirms our conclusion of the previous section, where the axisymmetric spacecraft (which, by definition, has only major and minor axes) was seen to be unconditionally stable. However, if we take into account the internal energy dissipation, the analysis of the previous section dictates that an asymmetric, semirigid spacecraft can be spin-stabilized only about the major axis.

There is a major difference in the stable oscillation of the asymmetric spacecraft from that of the axisymmetric spacecraft studied in the previous section. Due to the presence of a nonzero, bounded disturbance, ϵ, about the spin axis, the angular velocity component, $\omega_z = n + \epsilon$, does not remain constant in the case of the asymmetric body. This translates into a *nodding motion* of the spin axis, wherein the nutation angle, β, changes with time. Such a motion is called *nutation* of the spin axis and is superimposed on the precessional motion.[2]

[2] Certain textbooks and research articles on space dynamics use nutation interchangeably with precession, which is incorrect and causes untold confusion. The

Assuming $k_1k_2 > 0$, we have from Eq. (13.47),

$$e^{\mathsf{K}t} = \mathcal{L}^{-1}(s\mathsf{I} - \mathsf{K})^{-1} = \begin{pmatrix} \cos(\sqrt{k_1k_2}t) & -\sqrt{\frac{k_1}{k_2}}\sin(\sqrt{k_1k_2}t) \\ \sqrt{\frac{k_2}{k_1}}\sin(\sqrt{k_1k_2}t) & \cos(\sqrt{k_1k_2}t) \end{pmatrix} . \quad (13.51)$$

Therefore, the approximate, linearized solution for precessional motion for small disturbance is given by

$$\omega_x(t) = \omega_x(0)\cos(\sqrt{k_1k_2}t) - \omega_y(0)\sqrt{\frac{k_1}{k_2}}\sin(\sqrt{k_1k_2}t),$$

$$\omega_y(t) = \omega_x(0)\sqrt{\frac{k_2}{k_1}}\sin(\sqrt{k_1k_2}t) + \omega_y(0)\cos(\sqrt{k_1k_2}t). \quad (13.52)$$

In order to solve for the nutation angle, we must integrate the last equation of Eq. (13.32). However, by consistently neglecting the second-order term, $\omega_x\omega_y$, in this equation due to the assumption of small disturbance, we have obtained an erroneous result of $\dot{\epsilon} = 0$, or $\epsilon =$ constant in Eq. (13.45). Hence, the linearized analysis is insufficient to model the nutation of an asymmetric body. We must drop the assumption of small disturbance and numerically integrate the complete, torque-free, nonlinear Euler equations, Eq. (13.32), for an accurate simulation of the combined precession and nutation.[3]

The kinematic equations for the instantaneous attitude of the asymmetric spacecraft in terms of the Euler angles are given by Eq. (13.41), with the choice (as before) of the constant angular momentum vector as the **K**-axis of the inertial frame. By a simultaneous, numerical integration of the kinematic equations along with the nonlinear Euler's equations, we can obtain the instantaneous attitude of the asymmetric, rigid spacecraft.

Example 13.2. A rigid spacecraft with principal moments of inertia $J_{xx} = 4000$ kg.m^2, $J_{yy} = 7500$ kg.m^2, and $J_{zz} = 8500$ kg.m^2 has initial angular velocity $\omega(0) = (0.1, -0.2, 0.5)^T$ rad/s and an initial attitude $\psi(0) = 0, \theta(0) = \frac{\pi}{2}, \phi(0) = 0$. Simulate the subsequent rotation of the spacecraft.

Since the given initial condition is relatively large, the approximation of small disturbance is invalid, and both precession and nutation must be properly simulated. We carry out the three-degree-of-freedom simulation by solving the nonlinear, torque-free Euler equations, and the kinematic equations,

dictionary in this regard is very helpful: "precession" is derived from the Latin word *praecedere*, which means the *act of preceding* and is directly relevant to the motion of a spinning, axisymmetric, prolate body, wherein the rotation of spin axis, $\dot{\psi}$, precedes the spinning motion, $\omega_z = n$, itself. On the other hand, "nutation" is derived from the Latin word *nutare*, which means *to nod* and describes the nodding motion, $\dot{\theta}$, of the spin axis.

[3] Jacobi [41] derived a closed-form solution for Euler's equations of torque-free, asymmetric spacecraft [Eq. (13.32)] in terms of the *Jacobian elliptic functions*. However, due to the complexity in evaluating these functions [2], we shall avoid their use here and carry out numerical integration of Eq. (13.32).

Eq. (13.41), with the use of a fourth-order Runge–Kutta algorithm (Appendix A) encoded in the intrinsic MATLAB function, *ode45.m*. The time derivatives of the motion variables, $\omega_x, \omega_y, \omega_z, \psi, \theta, \phi$, required by *ode45.m* are supplied by the program *spacerotation.m*, which is tabulated in Table 13.1. The simulation is carried out for 40 s by specifying the initial condition in the call for *ode45.m* as follows:

```
>> [t,x]=ode45(@spacerotation,[0 40],[0.1 -0.2 0.5 0 0.5*pi 0]');
>> subplot(121),plot(t,x(:,1:3)*180/pi),hold on,...
   subplot(122),plot(t,x(:,4:6)*180/pi)%time evolution of motion variables
```

Table 13.1. M-file *spacerotation.m* for the Torque-free Equations of Rotational Motion

```
function xdot=spacerotation(t,x)
%program for torque-free rotational dynamics and Euler 3-1-3 kinematics
%of rigid spacecraft
%x(1)=omega_x, x(2)=omega_y, x(3)=omega_z (angular velocity in rad/s)
%x(4)=psi, x(5)=theta, x(6)=phi (rad)
%(c) 2006 Ashish Tewari
J1=4000; J2=7500; J3=8500; %principal moments of inertia (kg.m^2)
xdot(1,1)=x(2)*x(3)*(J2-J3)/J1;
xdot(2,1)=x(1)*x(3)*(J3-J1)/J2;
xdot(3,1)=x(1)*x(2)*(J1-J2)/J3;
xdot(4,1)=(sin(x(6))*x(1)+cos(x(6))*x(2))/sin(x(5));
xdot(5,1)=cos(x(6))*x(1)-sin(x(6))*x(2);
xdot(6,1)=x(3)-(sin(x(6))*cos(x(5))*x(1)+cos(x(6))*cos(x(5))*x(2))/sin(x(5));
```

The resulting time-history plots of the motion variables are shown in Fig. 13.3. The precession is evident in the oscillation of $\omega_x, \omega_y, \psi, \phi$, while the nutation is observed in the the oscillation of ω_z, θ. Such a complex motion would be completely missed in a simulation with the approximate, linearized equations, Eq. (13.45), whereby an erroneous result of $\omega_z = 0.5$ rad/s $(= 28.65^\circ/s)$ would be obtained.

13.6 Spacecraft with Attitude Thrusters

Spin stabilization of torque-free spacecraft is a cheap (fuel-free) and simple procedure, compared to stabilization with externally applied torques. However, controlling the motion of a spinning body for carrying out the necessary attitude maneuvers is a complex task. Generally, all spacecraft have a *reaction control system* (RCS) that employs a pair of rocket thrusters—called *attitude thrusters*—about each principal axis for performing attitude maneuvers. When torques about each principal axis are applied for stability and control, the spacecraft is said to be *three-axis stabilized*, as opposed to spin-stabilized.

The attitude thrusters of an RCS are operated in pairs with equal and opposite thrust, such that the net external force remains unaffected. The

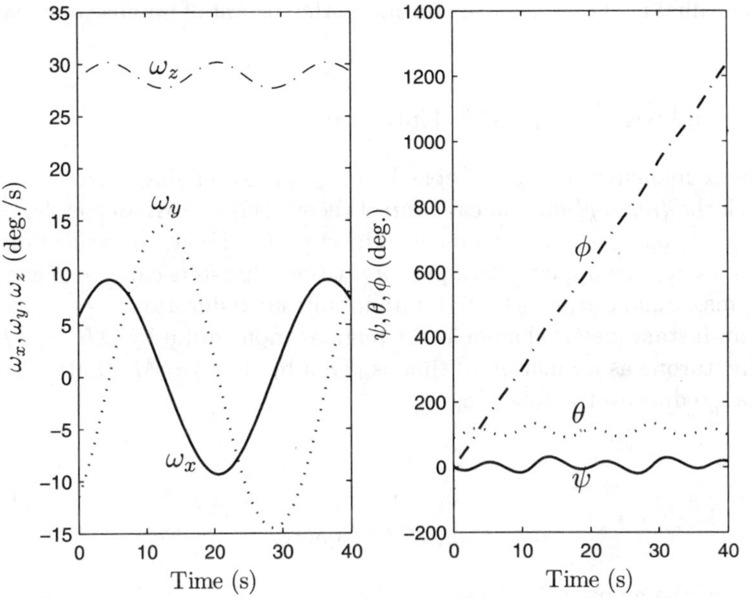

Fig. 13.3. Simulated precession and nutation of an asymmetric spacecraft.

firing of thrusters is limited to short bursts, which can be approximated by *torque impulses*. A torque impulse is defined as a torque of infinite magnitude acting for an infinitesimal duration, thereby causing an instantaneous change in the angular momentum of the spacecraft about the axis of application. The concept of the torque impulse is very useful in analyzing the single-axis rotation of spacecraft, as it allows us to utilize the well-known linear system theory [43], wherein the governing linear differential equation is solved in a closed form with the use of the *unit impulse function*, $\delta(t)$, which denotes an impulse of unit magnitude.[4] The change in angular momentum caused by an impulsive torque, $\mathbf{M}(t) = \mathbf{M}(0)\delta(t)$, can be obtained as the total area under the torque vs. time graph, given by

$$\Delta\mathbf{H} \doteq \int_{-\infty}^{\infty} \mathbf{M}(t)\mathrm{d}t = \int_{-\infty}^{\infty} \mathbf{M}(0)\delta(t)\mathrm{d}t = \mathbf{M}(0) \, . \tag{13.53}$$

[4] The unit impulse function (also known as the *Dirac delta function*), $\delta(t - t_0)$, denoting a unit impulse applied at time $t = t_0$, has the useful property,

$$\int_{-\infty}^{\infty} f(t)\delta(t - t_0)\mathrm{d}t = f(t_0) \, ,$$

where $f(t)$ is a single-valued function.

Thus, the torque impulse causes an instantaneous change in the angular momentum, equal to the value of the torque at the instant of impulse application, $t = 0$.

13.6.1 Single-Axis Impulsive Rotation

A complex maneuver can be designed as a sequence of single-axis rotations, for which the *time-optimal*, linear control theory [42] is most amenable.[5] Consider a rigid spacecraft with moment of inertia, J_{zz}, about the axis of desired rotation, oz, and equipped with a pair of attitude thrusters capable of exerting a large, maximum torque, $M_z(0)$, for an infinitesimal duration, $\Delta t \to 0$, which causes an instantaneous change in the angular momentum by $\Delta H_z = M_z(0)$. Since the torque as a function of time is given by $M_z(t) = M_z(0)\delta(t)$, Euler's equations reduce to the following:

$$\begin{aligned} \dot{\omega}_x &= 0, \\ \dot{\omega}_y &= 0, \\ J_{zz}\dot{\omega}_z &= M_z(0)\delta(t). \end{aligned} \tag{13.54}$$

In terms of the angular displacement about oz, θ, the last of Eq. (13.54) can be written as

$$\ddot{\theta} = \frac{M_z(0)}{J_{zz}}\delta(t) \, , \tag{13.55}$$

whose solution is easily obtained by successive integration using Laplace transform [4] to be

$$\begin{aligned} \omega_z(t) = \dot{\theta} &= \omega_z(0) + \frac{M_z(0)}{J_{zz}} u_s(t), \\ \theta(t) &= \theta(0) + \omega_z(0)t + \frac{M_z(0)}{J_{zz}} r(t), \end{aligned} \tag{13.56}$$

where $\theta(0), \omega_z(0)$ refer to the initial condition immediately before torque application, $u_s(t) \doteq \int \delta(t)dt$ is the *unit step function* applied at $t = 0$, defined by

$$u_s(t - t_0) = \begin{cases} 0, & t < t_0 \, , \\ 1, & t \geq t_0, \end{cases} \tag{13.57}$$

---

[5] Time-optimal control, as the name implies, is a special branch of optimal control theory, which deals with the problem of optimizing time in a general dynamical system. When the applied inputs are limited in magnitude (such as in the case of rocket thrusters), and the system is governed by linear differential equations, the maximum principle of *Pontryagin* dictates that the inputs of the maximum possible magnitude should be applied in order to minimize the total time of a given displacement of the system. Pontryagin's principle is directly applicable to single-axis maneuvers of rigid spacecraft by attitude thrusters.

and $r(t) \doteq \int u_s(t)dt$ is the *unit ramp function* applied at $t = 0$, defined by

$$r(t - t_0) = \begin{cases} 0, & t < t_0\,, \\ t - t_0, & t \geq t_0\,. \end{cases} \tag{13.58}$$

In a practical application, the thruster torque, $M_z(0)$, is not infinite, and the time interval, Δt, over which the torque acts, tends to zero. However, since Δt is much smaller than the period of the maneuver, it is a good approximation (and a valuable one) to assume an impulsive thruster torque, and to employ Eq. (13.56) as the approximate solution. Equation (13.56) implies that the response to a single impulse is a linearly increasing displacement and a step change in the speed. Therefore, if the maneuvering requirement is for a step change in angular velocity (called a *spin-up maneuver*), a single impulse is sufficient. However, if a given single-axis displacement is desired—called a *rest-to-rest maneuver*—one has to apply another impulse of opposite direction, $-M_z(0)\delta(t-\tau)$, in order to stop the rotation at time $t = \tau$, when the desired displacement has been reached. Since the governing differential equation, Eq. (13.55), is linear, its solution obeys the *principle of linear superposition* [43], which allows a weighted addition of the responses to individual impulses to yield the total displacement caused by multiple impulses. Therefore, the net response to two equal and opposite impulses applied after an interval $t = \tau$ is given by

$$\begin{aligned} \omega_z(t) &= \frac{M_z(0)}{J_{zz}}[u_s(t) - u_s(t-\tau)]\,, \\ \theta(t) &= \frac{M_z(0)}{J_{zz}}[r(t) - r(t-\tau)] + \omega_z\tau = \theta_d\,. \end{aligned} \tag{13.59}$$

Hence, the angular velocity becomes zero, and a desired constant displacement, $\theta(t) = \theta_d$, is reached at $t = \tau$. The magnitude of θ_d can be controlled by varying the time τ at which the second impulse is applied (Fig. 13.4). The application of two equal and opposite impulses of maximum magnitude for achieving a time-optimal displacement is called *bang-bang* control. This is an *open-loop* control, requiring only the desired displacement, as opposed to *closed-loop* control [43], for which the knowledge of instantaneous displacement, $\theta(t)$, is also required. The bang-bang, time-optimal, open-loop control is exactly applicable to any linear system without resistive and dissipative external forces. However, even when a small damping force is present, one can approximately apply this approach to control linear systems.

13.6.2 Attitude Maneuvers of Spin-Stabilized Spacecraft

Attitude thrusters can be used for controlling the attitude of a spin-stabilized, axisymmetric spacecraft, which involves multi-axis rotation (precession). If the spin rate is constant ($\omega_z = n$), the governing differential equations describing

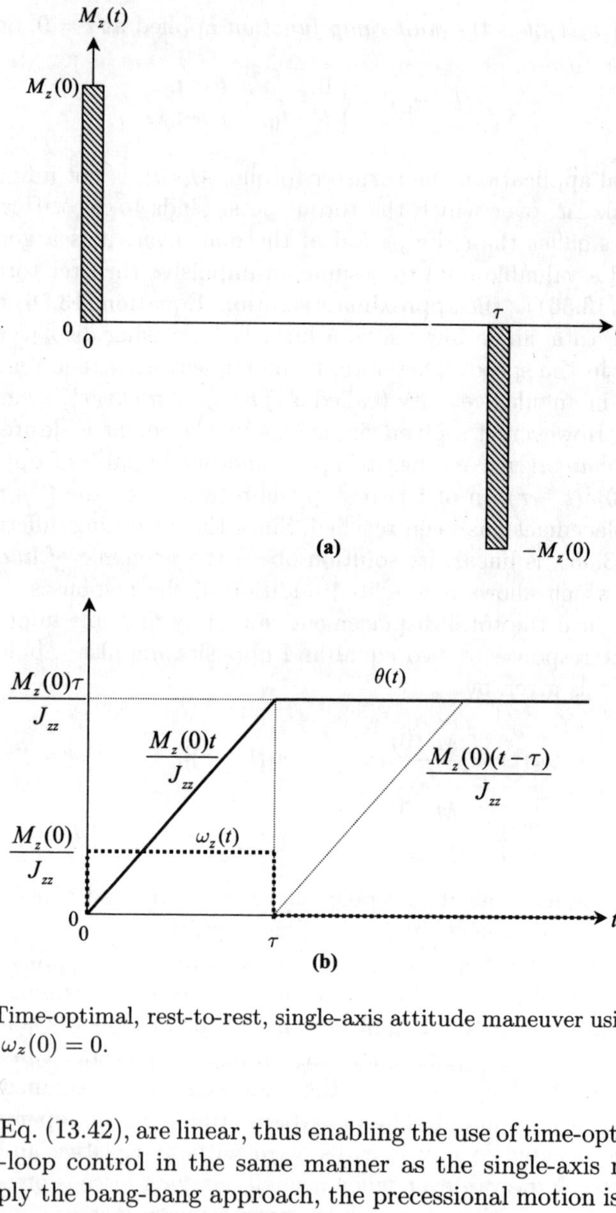

Fig. 13.4. Time-optimal, rest-to-rest, single-axis attitude maneuver using thrusters with $\theta(0) = \omega_z(0) = 0$.

precession, Eq. (13.42), are linear, thus enabling the use of time-optimal, bang-bang, open-loop control in the same manner as the single-axis rotation. In order to apply the bang-bang approach, the precessional motion is excited by applying a torque normal to the spin axis and then exerting another equal and opposite torque to stop the precession when the desired spin-axis orientation has been reached. However, contrary to single-axis rotation, the principal axes of a precessing body are not fixed in space. Hence, the directions of the two torque impulses are referred to the inertial axes.

Let a change of the spin axis be desired through application of thruster torque impulses, as shown in Fig. 13.5. After the application of the first impulse, $\Delta\mathbf{H_1}$, the angular momentum changes instantaneously from $\mathbf{H_0} = J_{zz}n\mathbf{k}$ to its new value $\mathbf{H_1} = \mathbf{H_0} + \Delta\mathbf{H_1}$, such that a nutation angle of $\beta = \frac{\theta_d}{2}$ is obtained. We select the orientation of the inertial frame such that oZ is along the intermediate angular momentum vector, $\mathbf{H_1}$, and oX coincides with the principal axis ox at time $t = 0$. Therefore, we have $\psi(0) = 0, \theta(t) = \frac{\theta_d}{2}, \phi(0) = 0$ in terms of the 3-1-3 Euler angles. It is clear from Fig. 13.5 that the first torque impulse applied normal to the spin axis at $t = 0$ is equal to

$$\Delta\mathbf{H_1} = J_{zz}n\tan\frac{\theta_d}{2}\left(\cos\frac{\theta_d}{2}\mathbf{J} + \sin\frac{\theta_d}{2}\mathbf{K}\right) = J_{zz}n\tan\frac{\theta_d}{2}\mathbf{j} \tag{13.60}$$

and causes a positive rotation of the angular momentum vector about $-\mathbf{I}$. Since the angular momentum has been deflected from the spin axis, the precessional motion is excited and is allowed to continue for half inertial spin ($\phi = \pi$) until $\mathbf{i} = -\mathbf{I}$. At that precise instant, the second impulse,

$$\Delta\mathbf{H_2} = J_{zz}n\tan\frac{\theta_d}{2}\left(\cos\frac{\theta_d}{2}\mathbf{J} - \sin\frac{\theta_d}{2}\mathbf{K}\right) = J_{zz}n\tan\frac{\theta_d}{2}\mathbf{j}\,, \tag{13.61}$$

is applied in order to stop the precession by causing a positive rotation of the angular momentum vector about $\mathbf{I}$. The angular momenta at the beginning and end of the precession are given in terms of the instantaneous principal axes by

$$\begin{aligned}\mathbf{H_1} &= J_{zz}n\mathbf{k} + J_{zz}n\tan\frac{\theta_d}{2}\mathbf{j},\\ \mathbf{H_2} &= J_{zz}n\mathbf{k}\,.\end{aligned} \tag{13.62}$$

It is important to emphasize that the principal axes used in the expressions for $\mathbf{H_1}$ and $\mathbf{H_2}$ are at different instants, separated in time by half the inertial spin time period. The time taken to undergo half inertial spin is given by Eq. (13.42) to be

$$t_{1/2} = \pi/\dot{\phi} = \frac{J_{xx}\pi}{n \mid J_{xx} - J_{zz} \mid}\,. \tag{13.63}$$

It is clear from Eq. (13.63) that the time it takes to reach the final position is large if the spin rate, n, is small or if the two moments of inertia are close to each other.

Although the two impulses are opposite in direction relative to the inertial frame, they have the same orientation in the the instantaneous body-fixed principal frame. Hence, the same pair of attitude thrusters can be used to both start and stop the precession after multiples of half inertial spin ($\phi = \pm\pi, 2\pi, \ldots$). However, in order to achieve the largest possible deflection of the spin axis—which is equal to θ_d and happens when $\mathbf{H_0}, \mathbf{H_1}, \mathbf{H_2}$ all lie in the same plane—the precession angle, ψ, must have changed exactly by $\pm 180°$ when the precession is stopped, which requires that $\mid \dot{\psi} \mid = \mid \dot{\phi} \mid$. On equating the magnitudes of the inertial spin and precession rates in Eq. (13.42), it is

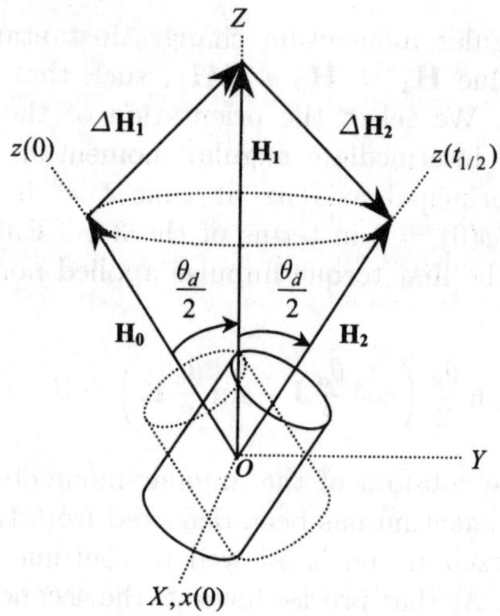

Fig. 13.5. Time-optimal attitude control of a spin-stabilized, axisymmetric spacecraft using thrusters.

clear that the matching of precession with inertial spin is possible if and only if

$$\cos\frac{\theta_d}{2} = \frac{J_{zz}}{\mid J_{xx} - J_{zz} \mid} . \tag{13.64}$$

Because the cosine of an angle cannot exceed unity, this implies that precession and inertial spin can be synchronized only for prolate bodies with $J_{xx} > 2J_{zz}$. Equation (13.64) gives the largest possible angular deflection of the spin axis (θ_d) that can be achieved with a given pair of attitude thrusters and is obtained when $\mid \psi \mid = \mid \phi \mid = \pi$. Since the nutation angle, $\beta = \frac{\theta_d}{2}$, is determined purely by the impulse magnitude, its value can be different from that given by Eq. (13.64), in which case the total angular deviation of the spin axis is less than 2β. From the foregoing discussion, it is clear that for a greater flexibility in performing spin-axis maneuvers, more than one pair of attitude thrusters (or more than two impulses) should be employed.

Since the applied torque magnitude for each impulse, $\mathbf{M_y}$, is proportional to $\tan(\frac{\theta_d}{2})$, it follows that a change of spin axis by $\theta_d = 180°$ would be infinitely expensive. Because impulsive maneuvers are impossible in practice, one must take into account the nonzero time, Δt, of thruster firing, which leads to an average thruster torque requirement $\mathbf{M_y} = \frac{\Delta \mathbf{H_1}}{\Delta t}$. In simulating the spacecraft response due to thruster firing, one must carefully model the actual variation of the thruster torque with time. There are two distinct ways of simulating the bang-bang, impulse response of spin-stabilized, axisymmetric spacecraft: (a) calculating the precessional angular velocity components, ω_x, ω_y, due to the applied impulses, and using them as an initial condition to simulate the ensuing torque-free motion, or (b) directly simulating the response to the applied impulses by solving the equations of motion with a nonzero torque. Of these two, the former is an initial response describing precession between the two impulses, while the latter includes the impulse response caused by the impulses themselves. Since $\theta \neq 0$ for the first method, we can use the 3-1-3 Euler angles for a nonsingular attitude simulation. However, the second approach begins with a zero nutation angle before the application of the first impulse; thus, the 3-1-3 Euler angle representation is unsuitable; instead, the 3-2-1 Euler angle representation should be employed in (b). The kinematic equations of motion in terms of the $(\psi)_3, (\theta)_2, (\phi)_1$ Euler angles are easily derived using the methods of Chapter 2 to be

$$\begin{Bmatrix} \dot{\psi} \\ \dot{\theta} \\ \dot{\phi} \end{Bmatrix} = \frac{1}{\cos\theta} \begin{pmatrix} 0 & \sin\phi & \cos\phi \\ 0 & \cos\phi\cos\theta & -\sin\phi\cos\theta \\ 1 & \sin\phi\sin\theta & \cos\phi\sin\theta \end{pmatrix} \begin{Bmatrix} \omega_x \\ \omega_y \\ \omega_z \end{Bmatrix} . \tag{13.65}$$

Here we employ the initial angular momentum vector, $\mathbf{H_0}$, to be the inertial axis, oZ. Therefore, the nutation angle, β, is given by

$$\cos\beta \doteq \mathbf{K}\cdot\mathbf{k} = \cos\theta\cos\phi\,, \tag{13.66}$$

which determines β uniquely, as $\beta \leq \pi$. However, in this case, the nutation angle, β, denotes the *total* deviation of the spin axis from its original position (rather than the deviation from the intermediate angular momentum, $\mathbf{H_1}$, of the 3-1-3 Euler angle representation shown in Fig. 13.5). We have seen above that a 180° deflection of the spin axis requires an infinite impulse magnitude, which is practically impossible. Hence, we are necessarily simulating an impulse response with $\beta < \pi$, for which the 3-2-1 Euler angles are nonsingular. We shall apply both simulation methods, (a) and (b), in the following example.

Example 13.3. Consider an axisymmetric, spin-stabilized, rigid spacecraft with principal moments of inertia $J_{xx} = J_{yy} = 1500$ kg.m^2 and $J_{zz} = 500$ kg.m^2 and spin rate $\omega_z = 1$ rad/s. A pair of attitude thrusters mounted normal to the spin axis produces a constant torque at each one-hundredth second firing. Simulate the bang-bang response to two thruster firings spaced half a precession period apart, in order to achieve the maximum spin-axis deflection.

We begin by using the approach of simulating the response to initial conditions by writing a program called *spacesymmthrust.m*, which is tabulated in Table 13.2. This program calculates the necessary impulse magnitudes for achieving the maximum spin-axis deviation possible with a synchronization of the inertial spin with precession, such that $\psi = -\phi = \pi$ at the end of the second impulse. The resulting impulse magnitudes are translated into the initial conditions for the angular velocity and nutation angle, and response to the initial conditions following the impulses is simulated by solving the torque-free equations—encoded as *spacesymm.m* (Table 13.3)—by the MATLAB Runge–Kutta solver, *ode45.m*. The MATLAB statement for invoking the program, and its effects, is given as follows:

```
>> spacesymmthrust
   thd2 = 1.0472
   Ts = 4.7124
```

which implies $\frac{\theta_d}{2} = 1.0472$ rad (60°) and $t_{1/2} = 4.7124$ s. The impulsive thruster torque required for this maneuver is calculated as follows:

$$M_y = \frac{J_{zz} n \tan\frac{\theta_d}{2}}{\Delta t} = \frac{(500)(1)\tan 60°}{0.01} = 86,602.54 \text{ s},$$

which is a rather large magnitude, considering the size of the spacecraft (e.g., a pair of thrusters symmetrically placed 2 m away from oy must produce a thrust of 21,650.64 N for 0.01 s). The resulting plots of $\omega_x(t), \omega_y(t), \omega_{xy}(t)$ and $\psi(t), \phi(t)$ are shown in Figs. 13.6 and 13.7, respectively. It is clear from these plots that the inertial spin and precession are synchronous, with both ψ and ϕ reaching 180° simultaneously at the end of the second applied impulse $(t = t_{1/2})$. The effect of the two impulses is to instantaneously increase the angular velocity component, ω_y, thereby starting and stopping precession. Since ϕ and ψ are synchronized, the single pair of attitude thrusters firing about oy achieves the maximum possible deflection of the spin axis by 120°.

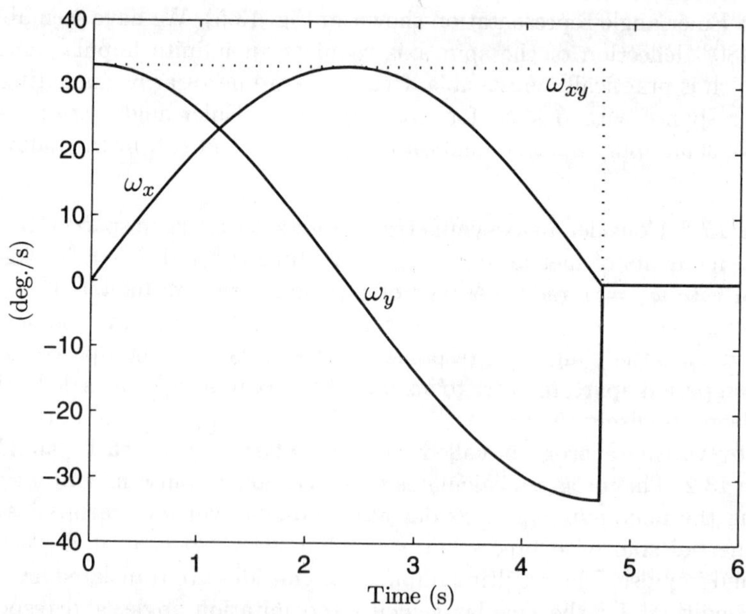

Fig. 13.6. Angular velocity response of a prolate, spin-stabilized spacecraft undergoing impulsive attitude maneuver (simulation by initial response).

Table 13.2. M-file *spacesymmthrust.m* for the Simulation of Impulsive Attitude Maneuver of a Spin-stabilized Spacecraft

```
%program for rotational dynamics and Euler 3-1-3 kinematics
%of rigid, axisymmetric, spin-stabilized spacecraft
%due to torque pulses about 'oy' principal axis
%x(1)=omega_x, x(2)=omega_y (angular velocity in rad/s)
%x(3)=psi, x(4)=phi (rad)
%u = impulsive torque about 'oy' axis (N-m)
%(c) 2006 Ashish Tewari
J1=1500; J3=500; %principal moments of inertia (kg.m^2)
thd2=acos(J3/(J1-J3))
T=0.01;
n=1; %rad/s
%thd2=atan(umax*T/(n*J3))
Ts=pi/abs(n*(1-J3/J1))
x=[];
x(1,1)=0;
x(2,1)=J3*n*tan(thd2)/J1;
x(3,1)=0;x(4,1)=0;
[t1,x1]=ode45(@spacesymm,[0 Ts],x);
N=size(t1,1);
x(1,1)=0;
x(2,1)=0;
x(3,1)=x1(N,3);x(4,1)=x1(N,4);
[t2,x2]=ode45(@spacesymm,[Ts+T Ts+T+1.5],x);
t=[t1;t2];x=[x1;x2];
dtr=pi/180;
plot(t,x(:,1:2)/dtr,t,sqrt(x(:,1).*x(:,1)+x(:,2).*x(:,2))/dtr),...
    xlabel('Time (s)'),ylabel('Precession angular velocity (deg./s)')
figure
plot(t,x(:,3)/dtr,t,x(:,4)/dtr),xlabel('Time (s)'),...
    ylabel('Precession angle, \psi, inertial spin angle, \phi (deg.)')
```

Table 13.3. M-file *spacesymm.m* for the Torque-free Equations for a Rigid, Axisymmetric, Spinning Spacecraft

```
function xdot=spacesymm(t,x)
%program for rotational dynamics and Euler 3-1-3 kinematics
%of rigid, axisymmetric, spin-stabilized spacecraft
%x(1)=omega_x, x(2)=omega_y (angular velocity in rad/s)
%x(3)=psi, x(4)=phi (rad)
%(c) 2006 Ashish Tewari
J1=1500; J3=500; %principal moments of inertia (kg.m^2)
n=1; %rad/s
%umax=1000;%torque magnitude
%T=0.01;%impulse duration
%thd2=atan(umax*T/(n*J3));%nutation angle
thd2=acos(J3/(J1-J3));
xdot(1,1)=x(2)*n*(J1-J3)/J1;%Euler's eqn.(1)
xdot(2,1)=x(1)*n*(J3-J1)/J1;%Euler's eqn.(2)
xdot(3,1)=(sin(x(4))*x(1)+cos(x(4))*x(2))/sin(thd2); %precession rate
xdot(4,1)=n*(1-J3/J1); %inertial spin rate
```

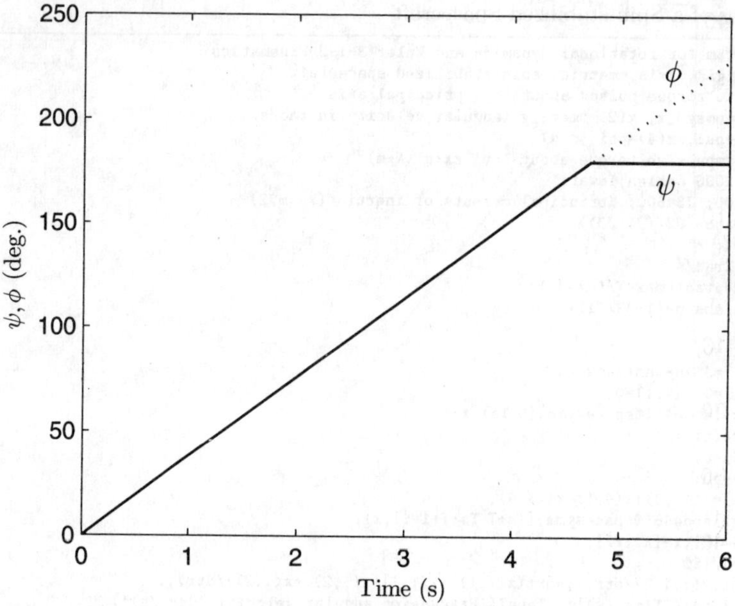

Fig. 13.7. Precession and inertial spin angles of a prolate, spin-stabilized spacecraft undergoing impulsive attitude maneuver (simulation by initial response).

Next we consider direct simulation using numerical integration of the spacecraft Euler's equations and 3-2-1 Euler kinematics with bang-bang torque impulses. For this purpose, a program called *spaceimpulse.m* (Table 13.4) provides the differential equations of motion to the MATLAB Runge–Kutta solver, *ode45.m*. The numerical integration requires a smaller maximum time step and relative tolerance than the default values of *ode45.m* because of the necessity of modeling impulsive torque. The statements for the execution of the program are given below, and the resulting plots of the state variables are shown in Figs. 13.8 and 13.9. The angular velocity response (Fig. 13.8) is identical to Fig. 13.6, whereas the 3-2-1 Euler angles produce a spin-axis deviation of $\beta = 120°$, as expected at the end of the impulse sequence. It is again emphasized that this extremely large and rapid maneuver is atypical of the actual spacecraft.

```
>> options=odeset('MaxStep',0.001,'RelTol',1e-5);
>> [t,x]=ode45(@spaceimpulse,[0 6],[0 0 0 0 0]',options);
```

13.6.3 Asymmetric Spacecraft Maneuvers by Attitude Thrusters

Unfortunately, the foregoing discussion of time-optimal, bang-bang control cannot be extended to a simultaneous, arbitrary rotation of an asymmetric

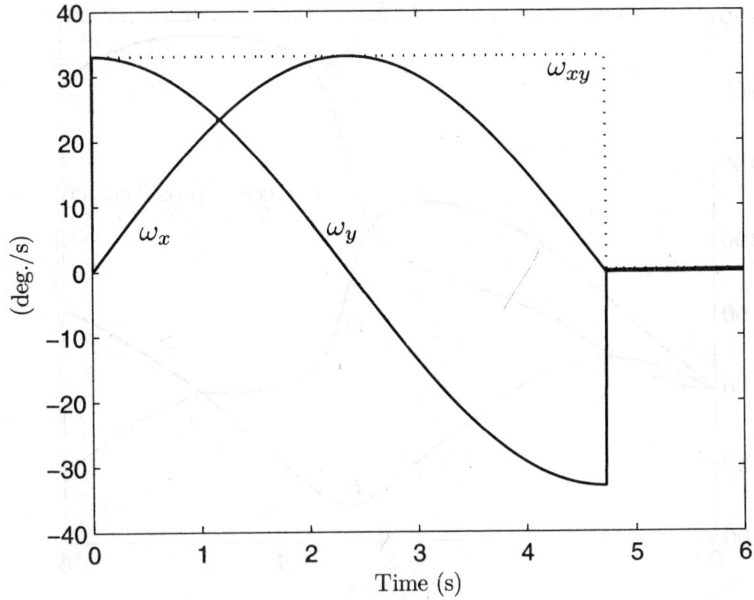

Fig. 13.8. Angular velocity response of a prolate, spin-stabilized spacecraft undergoing impulsive attitude maneuver (direct simulation with torque impulses).

Table 13.4. M-file *spaceimpulse.m* for State Equations of a Spin-stabilized Spacecraft with Bang-Bang Torque Impulses

```
function xdot=spaceimpulse(t,x)
%program for rotational dynamics and Euler (psi)_3 (theta)_2 (phi)_1
%kinematics of a rigid, axisymmetric spacecraft under the
%application of two torque impulses about 'oy' axis, spaced
%half-precession period apart
%x(1)=omega_x, x(2)=omega_y  (angular velocity in rad/s)
%x(3)=psi, x(4)=theta, x(5)=phi (rad)
%(c) 2006 Ashish Tewari
J1=1500; J3=500; %principal moments of inertia (kg.m^2)
thd2=acos(J3/(J1-J3));
n=1; %spin rate (rad/s)
T=0.01; %duration of impulse (s)
umax=J3*n*tan(thd2)/T; %maximum torque of impulse (N-m)
Ts=T+pi/abs(n*(1-J3/J1)); %time of application of second impulse (s)
if t>=0 && t<=T
    u=umax;
elseif t>Ts && t<=Ts+T
    u=umax;
else
    u=0;
end
xdot(1,1)=x(2)*n*(J1-J3)/J1;
xdot(2,1)=x(1)*n*(J3-J1)/J1+u/J1;
xdot(3,1)=(sin(x(5))*x(2)+cos(x(5))*n)/cos(x(4));
xdot(4,1)=cos(x(5))*x(2)-sin(x(5))*n;
xdot(5,1)=x(1)+(sin(x(5))*x(2)+cos(x(5))*n)*tan(x(4));
```

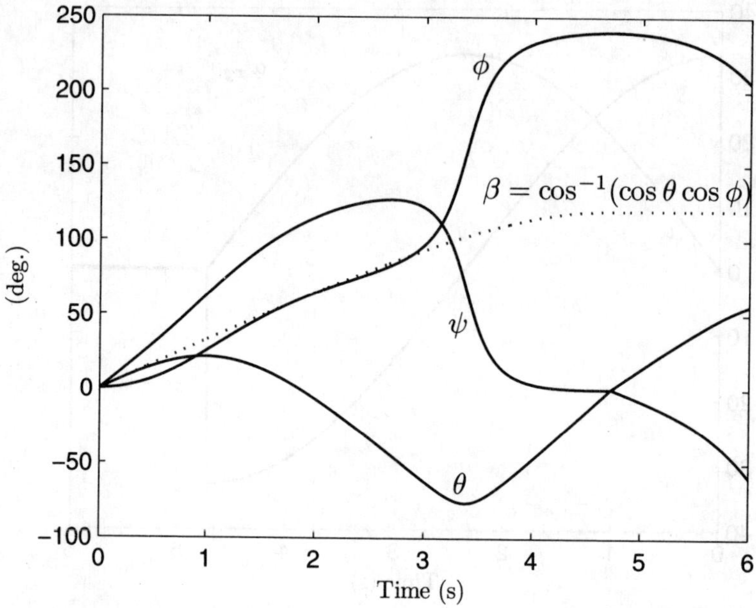

Fig. 13.9. 3-2-1 Euler angles and nutation angle of a prolate, spin-stabilized spacecraft undergoing impulsive attitude maneuver (direct simulation with torque impulses).

spacecraft about two or three axes. This is due to the nonlinear nature of asymmetric Euler's equations when more than one angular velocity components is nonzero, in which case the linear superposition of solutions does not hold, and the time-optimal control is not possible in a closed form. However, if the rotations are small, Euler's equations are rendered linear by approximation, and the bang-bang approach is valid. A practical method of dealing with large, multi-axis, rest-to-rest rotations is to apply them in a sequential manner. For such an approach, attitude thrusters about any two principal axes are capable of producing an arbitrary orientation (such as the 3-1-3 Euler angle attitude representations). Of course, one may choose to fix attitude thrusters about the minor and major axes, thereby precluding the unstable intermediate axis rotation. We have already covered single-axis rotations; thus, modeling of multiple, sequential, single-axis rotations requires no further discussion.

There are advanced closed-loop control algorithms [44] for deriving thruster torques for a large and rapid maneuver of asymmetric spacecraft. Simulating the attitude response of a spacecraft to such torques with simultaneous, large, multi-axis rotations is therefore essential. Numerical integration of nonlinear, coupled Euler's equations with applied torque and kinematic differential equations is feasible through Runge–Kutta and other iterative methods

(Appendix A). Let us simulate a general impulsive maneuver with the standard Runge–Kutta solver of MATLAB, *ode45.m.*

Example 13.4. A rigid spacecraft with principal moments of inertia $J_{xx} = 400$ kg.m^2, $J_{yy} = 750$ kg.m^2, and $J_{zz} = 850$ kg.m^2 has three pairs of thrusters, each capable of generating a torque with adjustable magnitude and duration about a principal axis. The spacecraft is initially at rest, with initial attitude in terms of the 3-1-3 Euler angles given by $\psi(0) = 0, \theta(0) = \frac{\pi}{2}, \phi(0) = 0$. Simulate the attitude response of the spacecraft for 10 s to the following torque profile:

$$\mathbf{M} = \begin{cases} 1000\mathbf{i} - 1000\mathbf{k} \text{ N.m}, & 0 \leq t \leq 1 \text{ s}, \\ -1000\mathbf{i} - 750\mathbf{j} + 750\mathbf{k} \text{ N.m}, & 5 < t \leq 5.97 \text{ s}, \\ \mathbf{0}, & t > 5.97 \text{ s}. \end{cases}$$

We begin by writing a program called *spacethruster.m* (Table 13.5) to provide the governing differential equations of motion with the specified torque to the MATLAB Runge–Kutta solver, *ode45.m.* The numerical integration is carried out with a smaller relative tolerance (10^{-5}) than the default value used in *ode45.m* because of the step changes in the torque. The statements for the execution of the program are given below, and the resulting plots of the state variables are shown in Figs. 13.10 and 13.11. There is a large change of attitude and angular velocity during the maneuver. At the end of the maneuver, the angular velocity becomes a near-zero constant, resulting in an almost constant attitude. It is possible to reduce the residual angular velocity to exactly zero by either using bang-bang thruster impulses as explained above, or using momentum wheels described in the next section.

```
>> options=odeset('RelTol',1e-5);
>> [t,x]=ode45(@spacethruster,[0 10],[0 0 0 0 pi/2 0]',options);
```

13.7 Spacecraft with Rotors

As the frequent use of the attitude thruster reaction control system (RCS) for stabilization and control entails a large fuel expenditure, most three-axis stabilized spacecraft additionally employ *momentum exchange devices* (MED), which consist of spinning rotors capable of exerting an internal torque on the spacecraft about each principal axis. As the MED are rotated by electric motors that derive their power from solar arrays of the spacecraft, they provide a fuel-free means of attitude control in the normal operation of the spacecraft. We shall consider here how a spacecraft with MED can be modeled and simulated accurately.

Consider a spacecraft with principal inertia tensor J and angular velocity resolved in the principal axes $\boldsymbol{\omega} = (\omega_x, \omega_y, \omega_z)^T$. Now consider a rotor with inertia tensor, J_r, about the spacecraft's principal axes, rotating with an angular velocity relative to the spacecraft, $\boldsymbol{\omega}_\mathbf{r} = (\omega_{rx}, \omega_{ry}, \omega_{rz})^T$, also resolved in

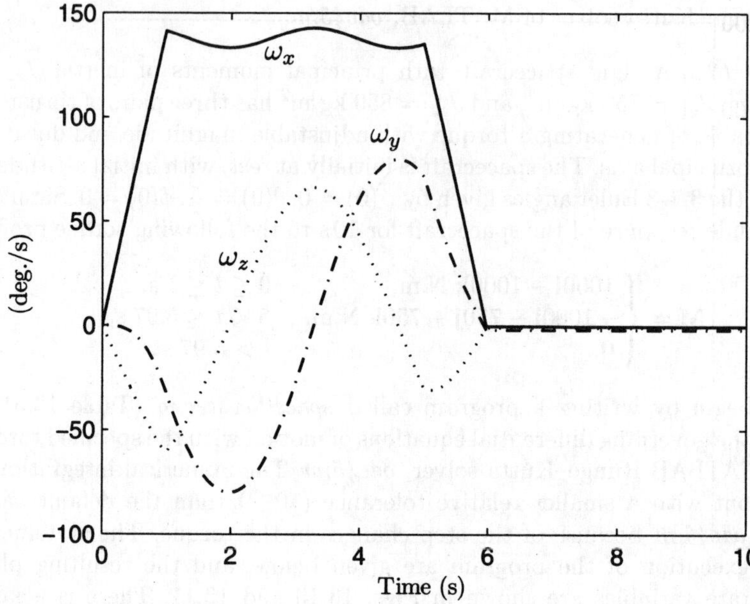

Fig. 13.10. Angular velocity response of an asymmetric spacecraft to the prescribed torque profile.

Table 13.5. M-file *spacethruster.m* for State Equations of an Asymmetric Spacecraft with Specified Torque Profile

```
function xdot=spacethruster(t,x)
%program for rotational dynamics and Euler 3-1-3 kinematics
%of rigid spacecraft with arbitrary torque profile
%x(1)=omega_x, x(2)=omega_y, x(3)=omega_z (angular velocity in rad/s)
%x(4)=psi, x(5)=theta, x(6)=phi (rad)
%(c) 2006 Ashish Tewari
J1=400; J2=750; J3=850; %principal moments of inertia (kg.m^2)
if t>=0 && t<=1
    u=[1000;0;-1000];
elseif t>5 && t<=5.97
    u=[-1000;-750;750];
else
    u=[0;0;0];
end
xdot(1,1)=x(2)*x(3)*(J2-J3)/J1+u(1)/J1;
xdot(2,1)=x(1)*x(3)*(J3-J1)/J2+u(2)/J2;
xdot(3,1)=x(1)*x(2)*(J1-J2)/J3+u(3)/J3;
xdot(4,1)=(sin(x(6))*x(1)+cos(x(6))*x(2))/sin(x(5));
xdot(5,1)=cos(x(6))*x(1)-sin(x(6))*x(2);
xdot(6,1)=x(3)-(sin(x(6))*cos(x(5))*x(1)+cos(x(6))*cos(x(5))*x(2))/sin(x(5));
```

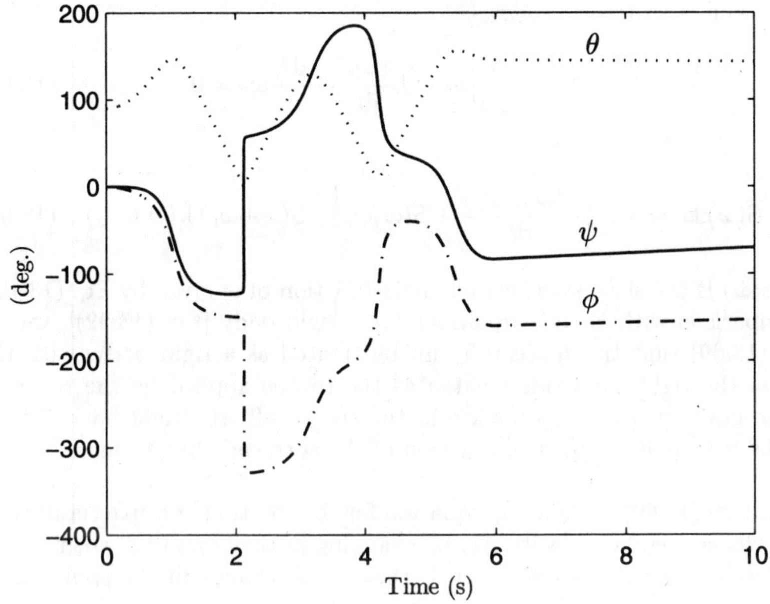

Fig. 13.11. Attitude response of an asymmetric spacecraft to the prescribed torque profile.

the spacecraft's principal frame.[6] The net angular momentum of the system (spacecraft and rotor) is the following:

$$\mathbf{H} = \mathsf{J}\boldsymbol{\omega} + \mathsf{J}_r(\boldsymbol{\omega} + \boldsymbol{\omega}_\mathbf{r}) \,, \tag{13.67}$$

[6] A transformation of the inertia tensor in the rotor's principal frame to that in the spacecraft's principal frame can be easily performed through the *parallel axes theorem*. The theorem states that the inertia tensor of a mass, m, about a parallelly displaced body frame, J, can be derived from that in the original body frame, J', by the following expression:

$$\mathsf{J} = \mathsf{J}' + m \begin{pmatrix} \Delta y^2 + \Delta z^2 & -\Delta x \Delta y & -\Delta x \Delta z \\ -\Delta x \Delta y & \Delta x^2 + \Delta z^2 & -\Delta y \Delta z \\ -\Delta x \Delta z & -\Delta y \Delta z & \Delta x^2 + \Delta y^2 \end{pmatrix} ,$$

where $\Delta x, \Delta y, \Delta z$ are the components of the parallel displacement of the body frame. After translating the principal frame of the rotor to the spacecraft's center of mass by the parallel displacement, a rotation is performed to align the rotor's principal axes with that of the spacecraft. If this rotation is represented by the coordinate transformation of Eq. (13.27), the inertia tensor transformed through the rotation is given by Eq. (13.31). The parallel axis theorem is also useful in deriving the inertia tensor of a complex shaped body composed of several smaller bodies with known inertia tensors.

the time derivative of which is zero (because no external torque acts on the system), and is written as follows:

$$\frac{d\mathbf{H}}{dt} = (\mathsf{J} + \mathsf{J}_r)\frac{d\boldsymbol{\omega}}{dt} + \frac{d\mathsf{J}}{dt}\boldsymbol{\omega} + \mathsf{J}_r\frac{d\boldsymbol{\omega}_r}{dt} + \frac{d\mathsf{J}_r}{dt}\boldsymbol{\omega}_r = \mathbf{0}\,, \tag{13.68}$$

or,

$$\mathsf{J}\frac{\partial\boldsymbol{\omega}}{\partial t} + \mathsf{S}(\boldsymbol{\omega})\mathsf{J}\boldsymbol{\omega} = -\mathsf{J}_r\left[\frac{\partial(\boldsymbol{\omega}+\boldsymbol{\omega}_r)}{\partial t} + \mathsf{S}(\boldsymbol{\omega})\boldsymbol{\omega}_r\right] - \mathsf{S}(\boldsymbol{\omega}+\boldsymbol{\omega}_r)\mathsf{J}_r(\boldsymbol{\omega}+\boldsymbol{\omega}_r)\,, \tag{13.69}$$

where $\mathsf{S}(\boldsymbol{\omega})$ is the skew-symmetric matrix function of $\boldsymbol{\omega}$ given by Eq. (13.19). On comparison with Euler's equations for a rigid body [Eq. (13.32)], we see in Eq. (13.69) that the spacecraft can be treated as a rigid body, with the terms on the right-hand side treated as the torque applied by the rotor on the spacecraft. If several rotors are in the spacecraft, the right-hand side of Eq. (13.69) is replaced by a summation of the corresponding terms of all the rotors.

Equation (13.69) is a general equation for the rotation of a spacecraft with a rotor whose angular velocity can be changing in time due to a varying spin rate as well as a varying spin axis. If there is no change in the spin axis of the rotor relative to the spacecraft, the rotor's angular momentum about a given principal axis is directly exchanged with that of the spacecraft by merely changing the rotor's spin rate. Such a rotor with its axis fixed relative to the spacecraft is called a *reaction wheel* when used in a nonspin-stabilized spacecraft. When a large rotor is used to control a spin-stabilized, axisymmetric spacecraft, with its axis aligned with the spacecraft's spin axis, the configuration is called a *dual-spin spacecraft*. Alternatively, if the rotor's angular velocity relative to the spacecraft is fixed, but its axis is capable of tilting with respect to the spacecraft, thereby applying a *gyroscopic* torque arising out of the last term on right-hand side of Eq. (13.69) the rotor can be used to control the attitude of a nonspinning, asymmetric spacecraft. Such a rotor with a variable spin axis is called a *control moment gyroscope* (CMG). In some advanced spacecraft, the rotor can have a variable spin rate as well a variable axis and is called a *variable-speed control moment gyroscope* (VSCMG). Therefore, a VSCMG is the most general momentum exchange device, and the models for a reaction wheel and a CMG can be easily derived from it by simply neglecting some specific terms on the right-hand side of Eq. (13.69). We will briefly consider how a VSCMG and a dual-spin spacecraft can be modeled appropriately.

13.7.1 Variable-Speed Control Moment Gyroscope

Consider an axisymmetric rotor with a variable spin rate, mounted at a rigid spacecraft's center of mass in such a way that its spin axis is free to rotate in all directions (Fig. 13.12). Such a rotor is termed a *fully gimbaled gyroscope*, and the arrangement that allows it to rotate freely about the spacecraft is called *gimbaling*. Gimbaling can be carried out either using mechanical rotor supports hinged about the three principal axes of the spacecraft (called *gimbals*) or using a magnetic suspension. Of these, the former is more commonly employed. A motor is used to apply the necessary torque on the VSCMG rotor relative to the spacecraft about each principal axis, in order to move the rotor in a desired manner, thereby controlling the motion of the spacecraft. Let $\mathbf{M_r}$ be the torque applied on the rotor. Then we can write the equations of motion of the rotor relative to the spacecraft as follows:

$$\mathbf{M_r} = \mathsf{J_r}\frac{\partial \boldsymbol{\omega}_\mathbf{r}}{\partial t} + \mathsf{S}(\boldsymbol{\omega}_\mathbf{r})\mathsf{J_r}\boldsymbol{\omega}_\mathbf{r} \,, \tag{13.70}$$

where $\mathsf{S}(\boldsymbol{\omega}_\mathbf{r})$ is the skew-symmetric matrix form of $\boldsymbol{\omega}_\mathbf{r}$ given by Eq. (13.19). The motion of the spacecraft is described by the dynamic equations, Eq. (13.69), and the kinematic equations representing the attitude. Since the instantaneous attitude of the spacecraft's principal axes can be arbitrary, we will employ the nonsingular quaternion representation, $\mathbf{q}, q_4$ (Chapter 2). The attitude kinematics of the spacecraft in terms of the quaternion are given by (Chapter 2)

$$\frac{\mathrm{d}\{\mathbf{q}, q_4\}^T}{\mathrm{d}t} = \frac{1}{2}\Omega\{\mathbf{q}(t), q_4(t)\}^T \,, \tag{13.71}$$

where Ω is the following skew-symmetric matrix of the angular velocity components:

$$\Omega = \begin{pmatrix} 0 & \omega_z & -\omega_y & \omega_x \\ -\omega_z & 0 & \omega_x & \omega_y \\ \omega_y & -\omega_x & 0 & \omega_z \\ -\omega_x & -\omega_y & -\omega_z & 0 \end{pmatrix} . \tag{13.72}$$

For the general simulation of an attitude maneuver, Eqs. (13.69), (13.70), and (13.71) must be integrated in time, with given initial conditions, $\boldsymbol{\omega}(0), \boldsymbol{\omega}_\mathbf{r}(0)$, and $\mathbf{q}(0), q_4(0)$, and a prescribed motor torque profile, $\mathbf{M}(t)$. In addition, the rotor's inertia tensor, $\mathsf{J_r}$, which depends on the orientation of the rotor relative to the spacecraft, must be known at the beginning of the maneuver.

Example 13.5. For the spacecraft with the inertia tensor and initial condition given in Example 13.2, consider the a rotor, initially at rest relative to the spacecraft, with the following inertia tensor in the spacecraft's principal frame (not included in J):

$$\mathsf{J_r} = \begin{pmatrix} 50 & -10 & 0 \\ -10 & 100 & 15 \\ 0 & 15 & 250 \end{pmatrix} \text{ kg.m}^2 \,.$$

A three-axis motion of the rotor is initiated by the application of the following motor torque profile beginning at $t = 0$:

$$\mathbf{M_r} = \begin{cases} 7\mathbf{i} - 10\mathbf{j} - 200\mathbf{k} \text{ N.m}, & 0 \geq t < 5 \text{ s}\,, \\ -7\mathbf{i} + 10\mathbf{j} \text{ N.m}, & 5 < t < 10 \text{ s}\,, \\ \mathbf{0}, & t \geq 10 \text{ s}, \end{cases}$$

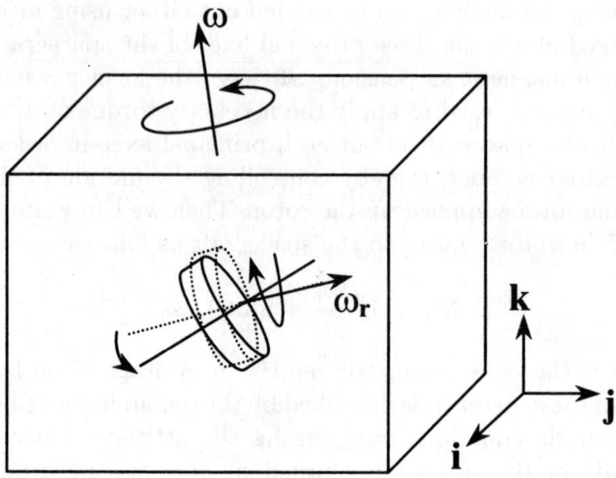

Fig. 13.12. A spacecraft with a VSCMG.

Simulate the response of the spacecraft for $0 \leq t \leq 40$ s.

Assuming the spacecraft and the VSCMG rotor to be rigid bodies. Neglecting friction in the rotor gimbals, we can model the system with Eqs. (13.69), (13.70), and (13.71), which are integrated in time using the Runge–Kutta algorithm of MATLAB, *ode45.m*. The time derivatives of the state variables, $\boldsymbol{\omega}, \boldsymbol{\omega}_\mathbf{r}, \mathbf{q}, q_4$, are obtained from the equations of motion and are programmed in the M-file *spacevscmg.m* (Table 13.6), along with the given motor torque profile. Another program, called *skew.m* (Table 13.7), is written for evaluation the skew-symmetric form of a vector according to Eq. (13.19) within *spacevscmg.m*. The following MATLAB statements are used to specify the initial condition (through *rot313.m* and *quaternion.m* of Chapter 2) and integrate the equations of motion:

```
>> C=rot313(0.5*pi,0,0) %rotation matrix for the initial s/c attitude

C = 1.0000  0        0
    0       0.0000   1.0000
    0      -1.0000   0.0000

>> q0=quaternion(C) %initial quaternion of s/c

q0 =    0.707106781186547       0       0       0.707106781186547

>> [t,x]=ode45(@spacevscmg,[0 40],[0.1 -0.2 0.5 0 0 0 q0]');
```

The rotor's relative angular velocity response and the angular velocity and attitude response of the spacecraft to the VSCMG motion are plotted in Figs. 13.13–13.15. Note that the VSCMG attains an almost constant relative speed about the principal axis *oz* after 5 s, with small amplitude oscillation about a mean value of $-225°$/s (-3.93 rad/s). The relative angular velocity components, ω_{rx}, ω_{ry}, however, display much larger amplitude oscillations

Table 13.6. M-file *spacevscmg.m* for the Equations of Motion of a Rigid Spacecraft with a VSCMG

```
function xdot=spacevscmg(t,x)
%program for torque-free rotational dynamics and quaternion kinematics
%of rigid spacecraft with a VSCMG
%x(1)=omega_rx, x(2)=omega_ry, x(3)=omega_rz (rotor relative ang. vel. (rad/s))
%x(4)=omega_x, x(5)=omega_y, x(6)=omega_z (spacecraft ang. vel. (rad/s))
%x(7)=q1, x(8)=q2, x(9)=q3, x(10)=q(4) (quaternion)
%this function needs the m-file "skew.m"
%(c) 2006 Ashish Tewari
J=diag([4000;7500;8500]); %principal inertia tensor (kg.m^2)
Jr=[50 -10 0;-10 100 15;0 15 250]; %rotor's inertia tensor (kg.m^2)
if t>=0 && t<5
Mr=[7;-10;-200];
elseif t>5 && t<10
   Mr=[-7;10;0];
else
   Mr=[0;0;0];
end
wr=[x(1);x(2);x(3)];
w=[x(4);x(5);x(6)];
q=[x(7);x(8);x(9);x(10)];
dwr=inv(Jr)*(Mr-skew(wr)*Jr*wr);
dw=-inv(J+Jr)*(skew(w)*J*w+Jr*(dwr+skew(w)*wr)+skew(w+wr)*Jr*(w+wr));
S=[0 w(3,1) -w(2,1) w(1,1);
   -w(3,1) 0 w(1,1) w(2,1);
   w(2,1) -w(1,1) 0 w(3,1);
   -w(1,1) -w(2,1) -w(3,1) 0];
dq=0.5*S*q;
xdot(1,1)=dwr(1,1);
xdot(2,1)=dwr(2,1);
xdot(3,1)=dwr(3,1);
xdot(4,1)=dw(1,1);
xdot(5,1)=dw(2,1);
xdot(6,1)=dw(3,1);
xdot(7,1)=dq(1,1);
xdot(8,1)=dq(2,1);
xdot(9,1)=dq(3,1);
xdot(10,1)=dq(4,1);
```

Table 13.7. M-file *skew.m* for the Evaluation of a Skew-symmetric Matrix

```
function S=skew(v)
%skew-symmetric matrix, S (3x3), form of vector v (3x1)
S=[0 -v(3) v(2);v(3) 0 -v(1);-v(2) v(1) 0];
```

about mean values of 0 and $-22.5°$/s (-0.393 rad/s), respectively. Due to this motion of the rotor, the spacecraft displays a much smoother angular velocity response in the given duration, when compared to the response of the same spacecraft to the specific initial condition without the VSCMG. This implies that a part of the spacecraft's angular momentum is absorbed by the rotor. However, the response has by no means reached a steady state, and the angular velocity components ω_x, ω_y show a divergent (unstable) behavior (Fig. 13.13). In order to study the long-term response, we plot the angular speed, $\mid \boldsymbol{\omega} \mid$, and principal angle, Φ, of the spacecraft in Fig. 13.16. It is evident that the spacecraft's rotation keeps on increasing almost steadily with

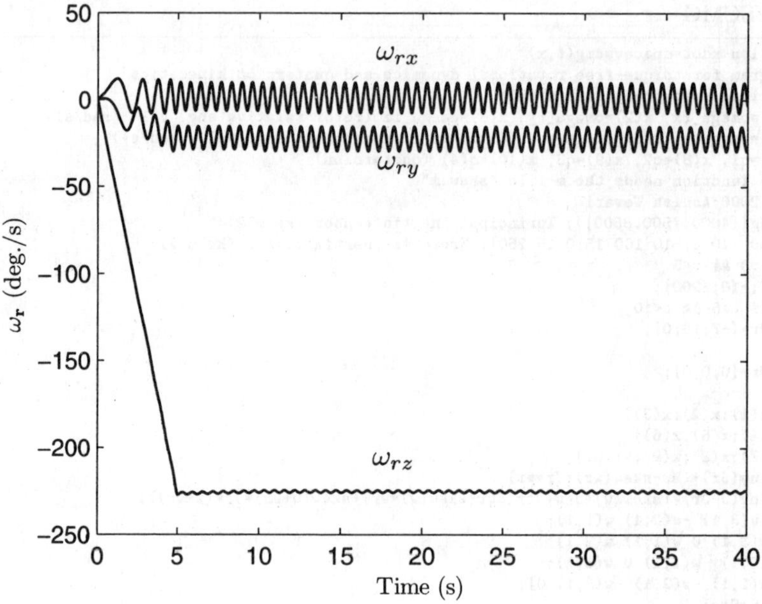

Fig. 13.13. Simulated VSCMG rotor angular velocity relative to the spacecraft.

time, and the principal angle has a randomly oscillatory tendency due to the transfer of kinetic energy from the rotor. Such an unstable response is caused by the undamped motion of the VSCMG, after the motors have ceased applying a torque. In an realistic case, bearing friction would eventually bring the rotor to rest relative to the spacecraft, thereby damping spacecraft's motion. In a practical application, the motor torque is carefully controlled in order to achieve a desired spacecraft orientation and velocity. This generally requires a *feedback loop* (Chapter 14) for measuring spacecraft's attitude and angular velocity, and applying it as an input to a *controller* in order to generate a *control* torque in real time.

13.7.2 Dual-Spin Spacecraft

Often, spacecraft are required to be prolate in shape. This is because a prolate spacecraft fits neatly into the long, aerodynamically efficient payload bays of the launch vehicles. As we have seen earlier, spin about the minor axis is unstable because of internal energy dissipation. However, by using a rotor in a dual-spin configuration, the prolate spacecraft can be spin stabilized about its (minor) axis of symmetry. Such an approach is commonly employed in spin stabilizing communications satellites. Consider a prolate spacecraft with

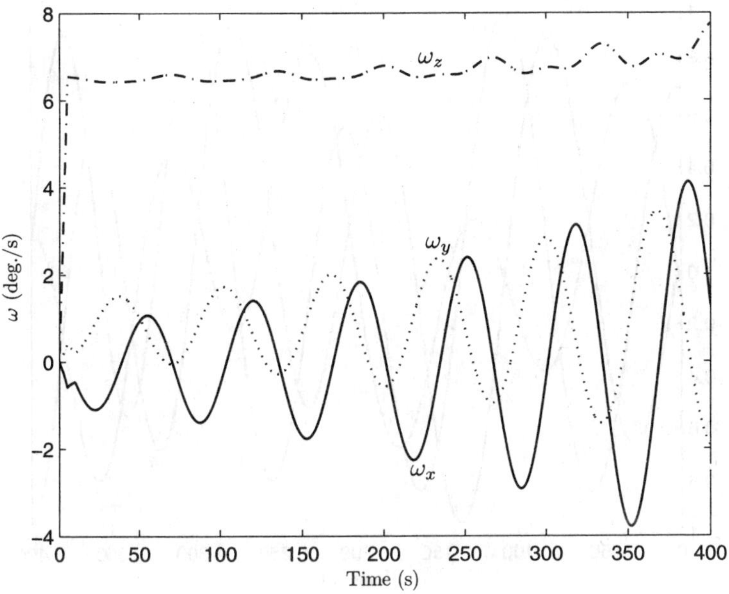

Fig. 13.14. Simulated angular velocity response of the spacecraft with a VSCMG rotor.

a large rotor about its axis of symmetry, and a platform on which a communications payload is mounted (Fig. 13.17). It is required that the platform must be spinning at a very small rate (generally the rate of rotation of the planet relative to the orbit), ω_p, such that the communications antennae are always pointed toward the receiving station. The net angular momentum of the dual-spin configuration in the presence of a lateral disturbance, ω_x, ω_y, is obtained from Eq. (13.67) to be

$$\mathbf{H} = [J_p\omega_p + J_r(\omega_p + \omega_r)]\mathbf{k} + J_{xy}(\omega_x\mathbf{i} + \omega_y\mathbf{j}) , \tag{13.73}$$

where J_p is the moment of inertia of the platform about the spin axis, J_r is the moment of inertia of the rotor about the spin axis, and J_{xy} is the moment of inertia of the total system (platform and rotor) about the lateral (major) axis. The rotational kinetic energy of the system can be expressed as

$$T = \frac{1}{2}(J_p + J_r)\omega_p^2 + \frac{1}{2}J_r\omega_r^2 + J_r\omega_r\omega_p + \frac{1}{2}J_{xy}\omega_{xy}^2 , \tag{13.74}$$

where $\omega_{xy}^2 = \omega_x^2 + \omega_y^2$. Although the net angular momentum is conserved, the rotational kinetic energy is *not* conserved due to internal energy dissipation caused by friction between the platform and the rotor, and sloshing of the propellants in the RCS mounted on the rotor. The internal dissipation of

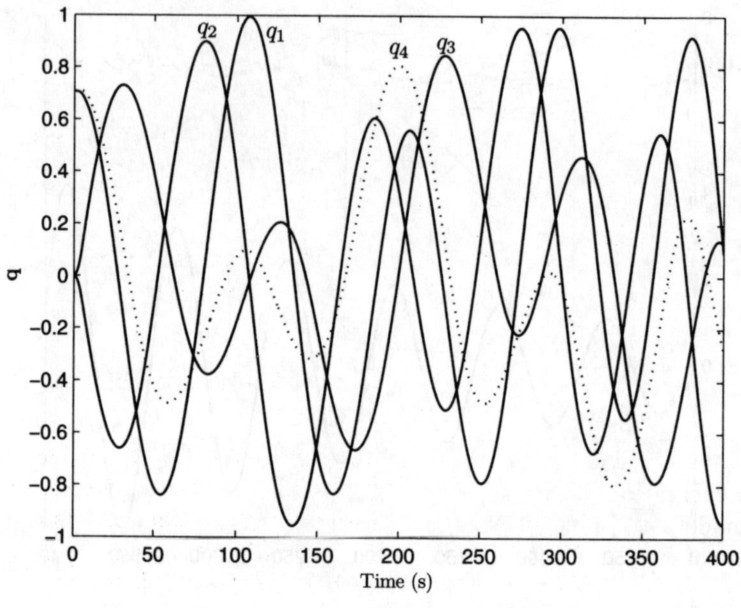

Fig. 13.15. Simulated attitude response of the spacecraft with a VSCMG rotor.

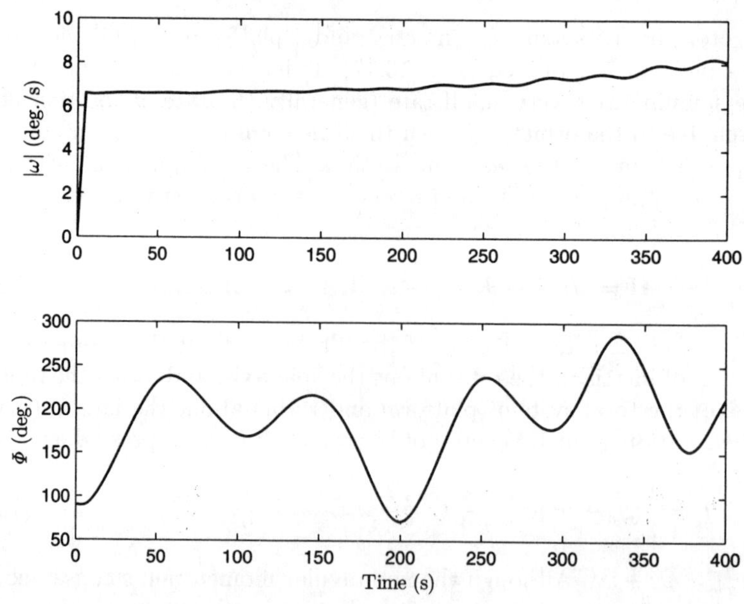

Fig. 13.16. Spacecraft's angular speed and principal rotation caused by VSCMG.

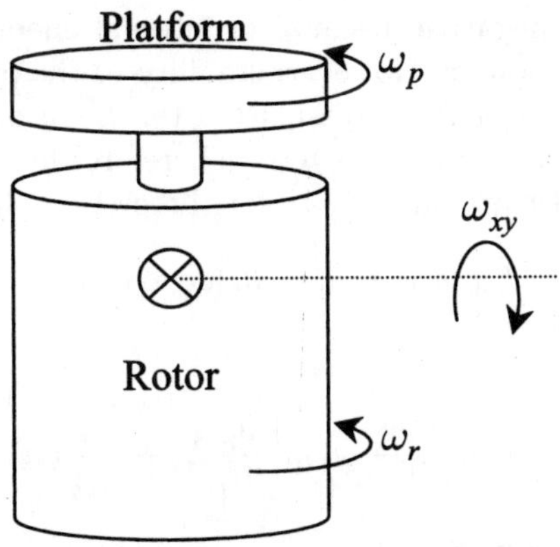

Fig. 13.17. A dual-spin spacecraft.

kinetic energy for the platform is different from that of the rotor, and one must model each as a separate rigid body with different frictional torques. The rate of change of total rotational kinetic energy is given by

$$\dot{T} = J_p \omega_p \dot{\omega}_p + J_r(\omega_p + \omega_r)(\dot{\omega}_p + \dot{\omega}_r) + J_{xy}\omega_{xy}\dot{\omega}_{xy} \ . \tag{13.75}$$

Noting that the rate of change of angular momentum magnitude is zero, we have the following from Eq. (13.73):

$$H\dot{H} = [J_p\omega_p + J_r(\omega_p + \omega_r)][J_p\dot{\omega}_p + J_r(\dot{\omega}_p + \dot{\omega}_r)] + J_{xy}^2\omega_{xy}\dot{\omega}_{xy} = 0 \ , \tag{13.76}$$

from which the term pertaining to the rate of change of kinetic energy by precession can be calculated as

$$J_{xy}\omega_{xy}\dot{\omega}_{xy} = -\frac{1}{J_{xy}}[J_p\omega_p + J_r(\omega_p + \omega_r)][J_p\dot{\omega}_p + J_r(\dot{\omega}_p + \dot{\omega}_r)] \ . \tag{13.77}$$

By substituting Eq. (13.77) into Eq. (13.75) we have

$$\dot{T} = \dot{T}_p + \dot{T}_r \ , \tag{13.78}$$

where $\dot{T}_p$ and $\dot{T}_r$ represent the rate of change of kinetic energy of the platform and rotor, respectively, given by

$$\dot{T}_p = J_p[\omega_p - \frac{1}{J_{xy}}\{J_p\omega_p + J_r(\omega_p + \omega_r)\}]\dot{\omega}_p \ , \tag{13.79}$$

and

$$\dot{T}_r = J_r[\omega_r + \omega_p - \frac{1}{J_{xy}}\{J_p\omega_p + J_r(\omega_p + \omega_r)\}](\dot{\omega}_p + \dot{\omega}_r) \ . \tag{13.80}$$

Both $\dot{T}_p$ and $\dot{T}_r$ are negative, because of internal energy dissipation due to friction and sloshing liquids. However, stability of the motion depends upon the relative magnitude of these dissipation terms, in order that the kinetic energy of precession is reduced to zero. Therefore, for stability it is crucial that the rotor provides an *energy sink* for the precessional motion, i.e.,

$$J_{xy}\omega_{xy}\dot{\omega}_{xy} = (\dot{T}_p - J_p\omega_p\dot{\omega}_p) + [\dot{T}_r - J_r(\omega_p + \omega_r)(\dot{\omega}_p + \dot{\omega}_r)] < 0 \,, \quad (13.81)$$

or,

$$-J_{xy}\omega_{xy}\dot{\omega}_{xy} = [J_p\omega_p + J_r(\omega_p + \omega_r)]\left[\frac{J_p}{J_{xy}}\dot{\omega}_p + \frac{J_r}{J_{xy}}(\dot{\omega}_p + \dot{\omega}_r)\right] > 0 \,, \quad (13.82)$$

which leads to the requirement

$$J_p\dot{\omega}_p + J_r(\dot{\omega}_p + \dot{\omega}_r) > 0 \,, \quad (13.83)$$

because $\omega_p > 0$ and $\omega_r > 0$. Since ω_p is small, we can neglect second-order terms involving it and its time derivative, leading to the approximations

$$\dot{T}_p \approx -\frac{J_pJ_r}{J_{xy}}(\omega_p + \omega_r)\dot{\omega}_p,$$

$$\dot{T}_r \approx J_r\left(1 - \frac{J_r}{J_{xy}}\right)(\omega_p + \omega_r)(\dot{\omega}_p + \dot{\omega}_r) \,. \quad (13.84)$$

It is to be noted that both the energy dissipation terms are negative. Therefore, if the rotor is oblate ($J_{xy} < J_r$), it follows from Eq. (13.84) that $\dot{\omega}_p > 0$ and $\dot{\omega}_r > 0$. For a prolate rotor ($J_{xy} > J_r$), and $\dot{\omega}_r < 0$. Hence, the platform and an oblate rotor speed up, while a prolate rotor slows down in the presence of the lateral disturbance, ω_{xy}. Thus, the stability requirement of Eq. (13.83) is unconditionally met by an oblate rotor. However, in a practical case the rotor is usually prolate, for which stability requires that

$$(J_p + J_r)\dot{\omega}_p > -J_r\dot{\omega}_r \,. \quad (13.85)$$

In terms of the energy dissipation terms, the stability requirement for a prolate rotor is obtained by eliminating $\dot{\omega}_p$ and $\dot{\omega}_r$ from Eqs. (13.84) and (13.85), and making the assumption $\omega_p \ll \omega_r$:

$$-\dot{T}_p > -\dot{T}_r\frac{J_r}{J_{xy} - J_r} \,. \quad (13.86)$$

Hence, for a stable configuration of a prolate spacecraft with a small spin rate coupled with a prolate rotor, the platform must lose kinetic energy at a greater rate than the rotor. Due to friction between the rotor and the platform, the rotor's spin rate decreases, and the platform speeds up, even in the absence of a lateral disturbance. If uncorrected, both rotor and platform will be eventually spinning at the same rate, which leads to an unstable configuration. In order to prevent this, a motor is used to continually apply a small torque to the rotor bearing. Most communications satellites employ a dual-spin configuration. A recent interesting application of the dual-spin stabilization was in the *Galileo* interplanetary spacecraft of NASA. This spacecraft had an inertial (nonspinning) platform for carrying out communications with the earth during its six-year-long voyage to Jupiter, while its rotor, on which several navigational and scientific sensors were mounted, rotated at three revolutions per minute.

In summary, a prolate spacecraft is unconditionally stabilized about its minor spin axis by an oblate rotor. However, if a prolate rotor is to be used for the same purpose, the spacecraft must lose its kinetic energy at a greater rate than that of the rotor. In order to model the dynamics of a dual-spin spacecraft by differential equations, one has to apply the conservation of angular momentum [Eq. (13.68)] to the system, as well as derive Euler's equations for the rotor alone, taking into account the internal energy dissipation by friction and sloshing.

13.7.3 Gravity Gradient Spacecraft

A spacecraft in a low-altitude orbit can generate an appreciable torque due to the variation of the gravity force along its dimensions, called the *gravity gradient* torque. Such a torque is considered negligible in atmospheric flight, because of the much larger aerodynamic moments. However, in space, the gravity gradient torque is large enough to exert a stabilizing (or de-stabilizing) influence over a spacecraft. The magnitude of gravity gradient can be increased by employing a long boom in the desired direction. For a large spacecraft (such as the *space station*) in low orbit, the gravity gradient torque is capable of overwhelming the attitude control system over time if not properly compensated for. This was an important reason why the *Skylab* mission came to a premature end in the 1970s. We shall model the gravity gradient dynamics and carry out a linear stability analysis for determining stable spacecraft attitudes. Consider a spacecraft in a low, circular orbit. The gravity gradient torque experienced by the craft can be written as follows:

$$\mathbf{M_g} = \int \boldsymbol{\rho} \times \mathbf{g} dm , \tag{13.87}$$

where $\boldsymbol{\rho}$ locates an elemental mass, dm, relative to the spacecraft's center of mass (Fig. 13.18). The acceleration due to gravity, $\mathbf{g}$, is appoximated by Newton's law of gravitation for a spherical planet,[7] and can be expanded using the binomial theorem as follows:

---

[7] The oblateness effects have a negligible influence on the gravity gradient torque and are ignored in a linear stability analysis.

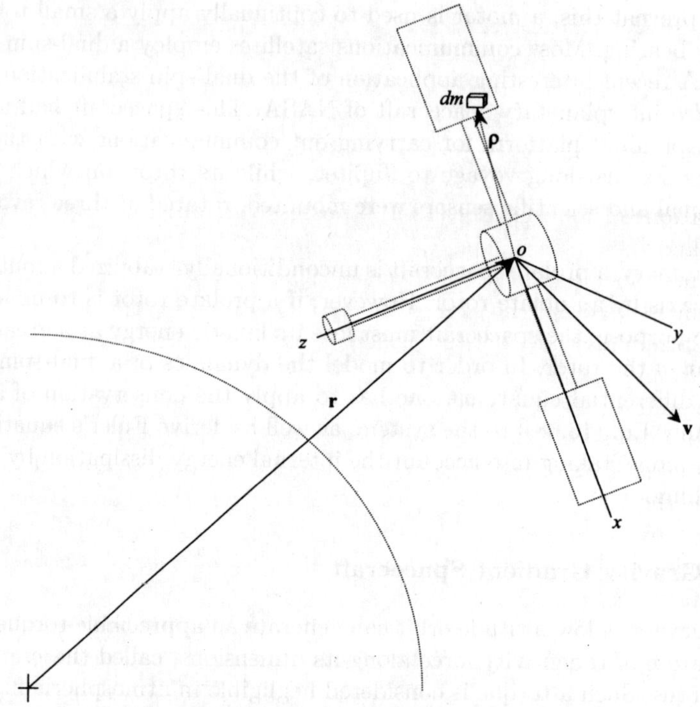

Fig. 13.18. A gravity gradient spacecraft with principal axes *oxyz*.

$$
\begin{aligned}
\mathbf{g} &= -GM\frac{\mathbf{r}+\boldsymbol{\rho}}{|\,\mathbf{r}+\boldsymbol{\rho}\,|^3} \\
&= \frac{GM(\mathbf{r}+\boldsymbol{\rho})}{r^3}\left(1-3\frac{\mathbf{r}\cdot\boldsymbol{\rho}}{r^2}+\ldots\right) , \qquad (13.88)
\end{aligned}
$$

where M denotes the planetary mass. Ignoring the second- and higher-order terms in Eq. (13.88), and carrying out the integral of Eq. (13.87) in terms of the body-referenced components of $\mathbf{r} = X\mathbf{i} + Y\mathbf{j} + Z\mathbf{k}$ and $\boldsymbol{\rho} = x\mathbf{i} + y\mathbf{j} + z\mathbf{k}$ (where $\mathbf{i}, \mathbf{j}, \mathbf{k}$ are the spacecraft's principal body axes), we have

$$
\mathbf{M_g} = M_{gx}\mathbf{i} + M_{gy}\mathbf{j} + M_{gz}\mathbf{k} \, , \qquad (13.89)
$$

where

$$
\begin{aligned}
M_{gx} &= \frac{3GM}{r^5}YZ(J_{zz}-J_{yy}), \\
M_{gy} &= \frac{3GM}{r^5}XZ(J_{xx}-J_{zz}), \qquad (13.90) \\
M_{gz} &= \frac{3GM}{r^5}XY(J_{yy}-J_{xx}).
\end{aligned}
$$

Substituting the gravity gradient torque components into Euler's equations, Eq. (13.19), we have

$$\begin{aligned}
J_{xx}\dot{\omega}_x + \omega_y\omega_z(J_{zz} - J_{yy}) &= \frac{3GM}{r^5}YZ(J_{zz} - J_{yy}), \\
J_{yy}\dot{\omega}_y + \omega_x\omega_z(J_{xx} - J_{zz}) &= \frac{3GM}{r^5}XZ(J_{xx} - J_{zz}), \\
J_{zz}\dot{\omega}_z + \omega_x\omega_y(J_{yy} - J_{xx}) &= \frac{3GM}{r^5}XY(J_{yy} - J_{xx}).
\end{aligned} \tag{13.91}$$

The equations of motion, Eq. (13.91), possess three distinct equilibrium attitudes (and their mirror images) for which any two of the angular velocity components vanish, and the third equals the orbital frequency, n. Hence, one of the principal axes of the spacecraft must be normal to the orbital plane in the equilibrium attitude. Let the principal axis normal to the orbit plane be $\mathbf{j}$. In order to investigate the stability of the equilibrium points, we consider the general equilibrium attitude where the remaining two principal axes are along the velocity direction ($\mathbf{i}$) and toward the planet's ceter ($\mathbf{k}$), respectively. The relative magnitudes of the principal moments of inertia, J_{xx}, J_{yy}, J_{zz}, would determine the stability of the equilibrium points. We shall consider small perturbations from the general equilibrium attitude, represented by the 3-2-1 Euler angles ψ (*yaw*), θ (*pitch*), and ϕ (*roll*), respectively. Such an attitude representation is common in aircraft applications.

Let the equilibrium attitude of the spacecraft be given by the undisturbed body axes, $\mathbf{i^e}, \mathbf{j^e}, \mathbf{k^e}$. The inertial angular velocity of the undisturbed triad, $\mathbf{i^e}, \mathbf{j^e}, \mathbf{k^e}$, resolved in the instantaneous body axes, $\mathbf{i}, \mathbf{j}, \mathbf{k}$, after a small attitude perturbation, ϕ, θ, ψ, is $n\mathbf{j^e} = n\psi\mathbf{i} + n\mathbf{j} - n\phi\mathbf{k}$, while the angular velocity disturbance from the equilibrium attitude is given by $\dot{\phi}\mathbf{i} + \dot{\theta}\mathbf{j} + \dot{\psi}\mathbf{k}$. Therefore, the inertial angular velocity of the spacecraft becomes

$$\boldsymbol{\omega} = (\dot{\phi} + n\psi)\mathbf{i} + (n + \dot{\theta})\mathbf{j} + (\dot{\psi} - n\phi)\mathbf{k}\,. \tag{13.92}$$

The position vector resolved in the body axes is

$$\mathbf{r} = r(-\sin\theta\mathbf{i} + \sin\phi\cos\theta\mathbf{j} + \cos\phi\cos\theta\mathbf{k})\,, \tag{13.93}$$

which leads to $X \approx -r\theta$, $Y \approx r\phi$, and $Z \approx r$ for the small perturbation, which, substituted into the Euler's equations, Eq. (13.91), along with the angular velocity, Eq. (13.92), yield the following linearized equations of rotational motion:

$$\ddot{\phi} = \frac{(J_{xx} - J_{yy} + J_{zz})n}{J_{xx}}\dot{\psi} - \frac{4n^2(J_{yy} - J_{zz})}{J_{xx}}\phi \tag{13.94}$$

$$\ddot{\theta} = -\frac{3n^2(J_{xx} - J_{zz})}{J_{yy}}\theta, \tag{13.95}$$

$$\ddot{\psi} = -\frac{(J_{xx} - J_{yy} + J_{zz})n}{J_{zz}}\dot{\phi} - \frac{n^2(J_{yy} - J_{xx})}{J_{zz}}\psi. \tag{13.96}$$

Clearly, the small-disturbance, linear pitching motion is decoupled from the roll-yaw dynamics and can be solved in a closed form. If $J_{xx} > J_{zz}$, the pitching motion is a stable oscillation of constant amplitude given by

$$\theta(t) = \theta(0) \cos n\sqrt{\frac{3(J_{xx} - J_{zz})}{J_{yy}}} t \,. \tag{13.97}$$

This undamped pitching oscillation is called *libration* and requires an active damping mechanism, such as through a reaction wheel (Chapter 14). The coupled roll-yaw dynamics, Eqs. (13.94) and (13.96)—also called *nutation*—is seen to have the following characteristic equation:

$$s^4 + n^2(1 + 3j_x + j_x j_z)s^2 + 4n^4 j_x j_z = 0 \,, \tag{13.98}$$

where

$$j_x \doteq \frac{J_{yy} - J_{zz}}{J_{xx}} \,,$$
$$j_z \doteq \frac{J_{yy} - J_{xx}}{J_{zz}} \,. \tag{13.99}$$

For stability, all roots, s, of the characteristic equation should have non-positive real parts (Chapter 14), which implies real and negative values of both the quadratic solutions, s^2, and leads to the following necessary and sufficient stability conditions:

$$1 + 3j_x + j_x j_z \geq 4\sqrt{j_x j_z} \,,$$
$$j_x j_z > 0. \tag{13.100}$$

It can be shown [2] that for a spacecraft with internal energy dissipation, the only stable gravity gradient attitude is the one with $J_{yy} > J_{xx} > J_{zz}$, since it results in the lowest kinetic energy, apart from satisfying the stability criteria, Eq. (13.100). Thus, the minor axis should point toward (or away from) the planet's center, while the major axis should lie along the orbit normal. Such an attitude is adopted for most asymmetric spacecraft in low orbits and is also the common attitude of the moons in our solar system. For small—or nearly axisymmetric—satellites, a long boom with an end mass can provide an effective gravity gradient stabilization.

Example 13.6. Consider the International Space Station (ISS) with the following inertia tensor [47]:

$$\mathsf{J} = \begin{pmatrix} 127908568 & 3141229 & 7709108 \\ 3141229 & 107362480 & 1345279 \\ 7709108 & 1345279 & 200432320 \end{pmatrix} \text{ kg.m}^2.$$

Simulate the gravity gradient motion of the ISS in a stable attitude at 93-min circular earth orbit, in response to an initial yaw-rate disturbance of 10^{-5} rad.

Table 13.8. M-file *gravitygrad.m* for Gravity Gradient Spacecraft's Nonlinear State Equations

```
function xdot=gravitygrad(t,x)
%program for gravity gradient rotational dynamics and Euler 3-2-1 kinematics
%of rigid spacecraft
%x(1)=omega_x, x(2)=omega_y, x(3)=omega_z (angular velocity in rad/s)
%x(4)=phi, x(5)=theta, x(6)=psi (rad.)
%(c) 2006 Ashish Tewari
mu = 3.986004e14;
Jzz=106892554.975429;
Jxx=127538483.852694;
Jyy=201272329.171876;
n=2*pi/(60*93);
r=(mu/n^2)^(1/3);
X=-r*sin(x(5));
Y=r*sin(x(4))*cos(x(5));
Z=r*cos(x(4))*cos(x(5));
pdot=-(Jzz-Jyy)*(x(2)*x(3)-3*mu*Y*Z/r^5)/Jxx;
qdot=-(Jxx-Jzz)*(x(1)*x(3)-3*mu*X*Z/r^5)/Jyy;
rdot=-(Jyy-Jxx)*(x(1)*x(2)-3*mu*X*Y/r^5)/Jzz;
phidot=x(1)+(x(2)*sin(x(4))+x(3)*cos(x(4)))/cos(x(5));
thetadot=x(2)*cos(x(4))-x(3)*sin(x(4));
psidot=(x(2)*sin(x(4))+x(3)*cos(x(4)))/cos(x(5));
xdot=[pdot;qdot;rdot;phidot;thetadot;psidot];
```

We begin by computing the principal inertia tensor as follows:

```
>> J=[127908568      3141229      7709108;
      3141229    107362480    1345279;
      7709108     1345279     200432320];

>> [V,D]=eig(J)

V=   0.1471     0.9835     0.1052
    -0.9891     0.1460     0.0178
     0.0021    -0.1067     0.9943

D = 106892554.98      0                0
               0     127538483.85     0
               0      0              201272329.17
```

For a stable gravity gradient attitude, we require $J_{xx} = 127,538,483.85$ kg.m^2, $J_{yy} = 201,272,329.17$ kg.m^2, and $J_{zz} = 106,892,554.98$ kg.m^2. The coordinate transformation matrix to the principal body axes is given by V computed above. We choose to employ the complete set of nonlinear Euler equations, Eq. (13.91), along with the 3-2-1 Euler kinematics (Chapter 2) for a faithful simulation of the coupled motion. The simulation is carried out for two complete orbits using the stiff Runge–Kutta solver of MATLAB, *ode23s*. The equations of motion are encoded in the M-file *gravitygrad.m*, which is tabulated in Table 13.8. The response of the spacecraft is plotted in Figs. 13.19 and 13.20. The stability of the equilibrium attitude is evident, with the yaw response being of the largest angle, while roll response has the highest rate. The weak coupling between roll-yaw (nutation) and pitch (libration) motions is clear in this example. The frequency of roll oscillation is observed to be approximately 0.0016755 rad/s, which falls between the linear roll-yaw frequencies, 0.000919 rad/s and 0.00197 rad/s. The pitch and yaw oscillations are nonharmonic due to the nonlinear coupling effects, which are significant even for the small yaw disturbance considered here.

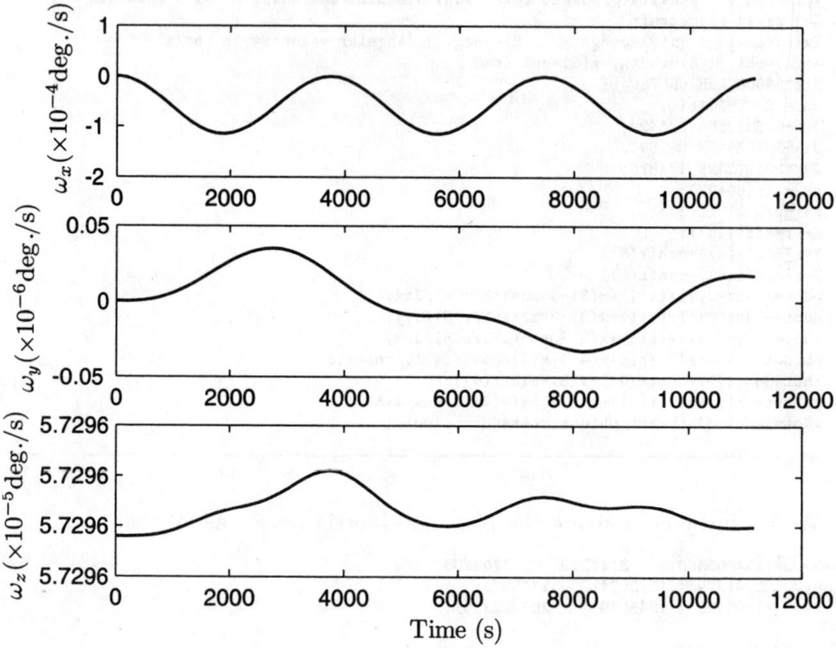

Fig. 13.19. Angular rate response of the gravity gradient ISS to an initial yaw-rate disturbance.

13.8 Attitude Motion in Atmospheric Flight

The trajectory of an atmospheric flight vehicle is very sensitive to aerodynamic force, which are strong functions of the vehicle's attitude relative to the flight path. Thus, rotational motion about the center of mass is crucial for atmospheric flight stability and control. When considering the rotational dynamics of aerospace vehicles within the atmosphere, one can still employ Euler's equations, Eq. (13.18), with the assumption of a rigid vehicle, and taking into account the aerodynamic torque generated by the rotation of the vehicle, as well as a control torque applied either by the pilot, or by an automatic control system. Since the torque generated by gravity is always negligible in comparison with the aerodynamic torque, the vector **M** in Eq. (13.18) is almost entirely a sum of the aerodynamic torque and the control torque. The aerodynamic torque can be a nonlinear function of the vehicle's attitude and angular velocity relative to the atmosphere, and can be obtained through experimental, semi-empirical, or computational fluid dynamics data. The control

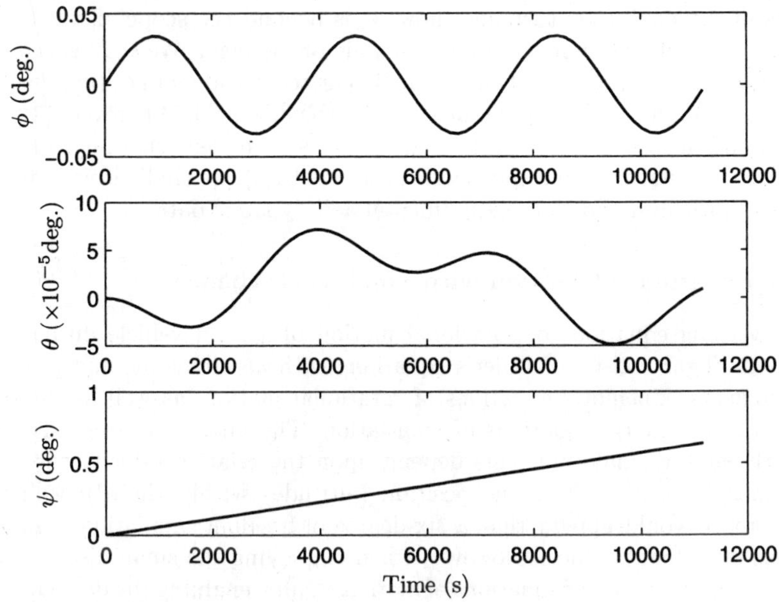

Fig. 13.20. Euler angle response of the gravity gradient ISS to an initial yaw-rate disturbance.

torque can be generated either by aerodynamic means through the deflection of control surfaces, or by propulsive means through thrust deflection. Most atmospheric flight vehicles employ aerodynamic control torques of one kind or another, due to the ease by which such torques can be created. However, there are certain flight situations where an aerodynamic control torque is infeasible, such as the vertical take-off of airplanes and launch vehicles, and the initial phase of atmospheric entry, wherein the dynamic pressure is not large enough to create a sufficient control torque. Moreover, in certain highly agile missiles and fighter airplanes, the vehicle's design precludes the generation of required control torque purely by aerodynamic means. In all such cases, thrust vectoring is employed by rotating the thrust vector relative to the body axes, in order to create the required control torque.

Since most atmospheric flight vehicles are designed to operate efficiently with a low drag, their attitude maneuvers do not create large flow disturbances in normal operation. Therefore, the assumptions of small-disturbance aerodynamics (Chapter 10) remain valid during a general attitude maneuver within the atmosphere. However, there are special circumstances where the small-disturbance approximation is invalid, namely the separated flowfield of a stalled flight, strong normal shock waves during transonic flight, and strong viscous interactions and entropy gradients in hypersonic flight. In such cases, the aerodynamic forces and moments must be derived through wind-tunnel tests, flight tests, or by advanced computational fluid dynamic models of the nonlinear, turbulent flow. It is beyond the scope of this book to discuss modeling of nonlinear aerodynamic phenomena. We shall generally follow the common practice of employing linearized aerodynamics that results from the assumption of small disturbances in the flow field. Wherever such an approximation cannot be applied (such as post-stall meneuvers of fighter aircraft, rolling missiles, and atmospheric entry vehicles), we shall either employ simple empirical methods, or experimental aerodynamic data.

13.8.1 Equations of Motion with Small Disturbance

The governing equations of rotational motion of a rigid vehicle during atmospheric flight consist of Euler's equations with aerodynamic and propulsive moments, kinematic equations of rotational motion, as well as the dynamic and kinematic equations of translation. The latter are necessary because the aerodynamic moments depend upon the relative velocity through the atmosphere, as well as the position (altitude) within the atmosphere. Therefore, it would appear that a six-degree-of-freedom simulation is indispensible for a flight vehicle. However, when employing the small-disturbance theory, a simplification of equations of motion results, enabling the de-coupling of the degrees of freedom, as seen below.

Let us begin with the vehicle initially in a steady, flight dynamic *equilibrium*, with planet-centered position, r^e, δ^e, l^e, and relative velocity in the local horizon frame, v^e, ϕ^e, A^e. This equilibrium condition is chosen such that the velocity of the center of mass relative to the atmosphere is a constant, and the angular velocity components of the vehicle about the center of mass, referred to a body-fixed frame, are time-invariant. Such an equilibrium condition could be an unaccelerated, rectilinear flight, or a steady, curved flight (steady coordinated turn, steady roll, entry trajectory, etc.). In this regard, our treatment of small-disturbance rotational motion is more general than the rectilinear flight equilibrium commonly found in textbooks on flight stability and control [45], [46]. The equilibrium condition generates a reference trajectory about which the vehicle's rotation is to be studied, after a small flow disturbance is applied to the vehicle at some time, taken to be $t = 0$. The equilibrium prevailing immediately before the disturbance is called the *equilibrium point*. The aerodynamic force and moment vectors (and their components), as well as the state variables, at the equilibrium point are denoted by the superscript e, whereas the quantities immediately following the application of the disturbance, are denoted by prime. The disturbances themselves are indicated by normal symbols. A disturbed quantity, such as the relative velocity, $\mathbf{v}'$, is thus written as

$$\mathbf{v}' = \mathbf{v}^e + \Delta \mathbf{v} \,. \qquad (13.101)$$

The flow-field disturbance applied at $t = 0$ causes an instantaneous deflection of the relative velocity vector and serves as the initial condition for the vehicle's motion. In order to study the stability of the equilibrium point, it is sufficient to study the vehicle's response to a small disturbance, which, as pointed out above, is easier to model than that of a large flow disturbance. The primary objective of the rotational stability analysis is, thus, to model the small-disturbance attitude motion caused by an instantaneous change in the relative velocity. The attitude motion, in turn, causes a change in the external force and moment. It must be clear that instead of considering the response of the flight vehicle to the application of an external force and moment, we are interested in the changes in the external force and moment caused by a small disturbance in the vehicle's velocity, which results in a rotational motion of the vehicle. If the ensuing motion beginning from a given equilibrium point is such that the flow-disturbance increases with time, we have an unstable equilibrium point. On the other hand, if the changes in the external force and moment caused by the rotational motion tend to alleviate the disturbance, the equilibrium point is said to be stable. Consider a body-fixed frame $(oxyz)$

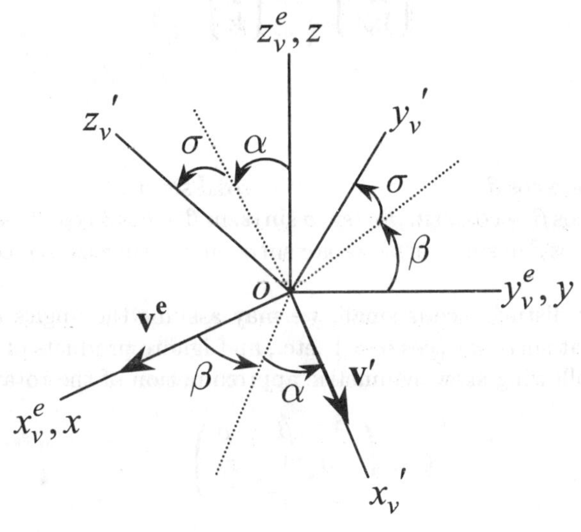

Fig. 13.21. The stability axes, $(oxyz)$, and the disturbed wind axes, $(ox'_v y'_v z'_v)$.

with origin at the vehicle's center of mass such that the axis ox along the instantaneous relative velocity vector at the equilibrium point. The axes of $oxyz$ can thus be chosen to be parallel to the wind axes, $(Sx_v y_v z_v)$ (Chapter 12), at the equilibrium point, $t = 0$. Such a coordinate system, depicted in Fig. 13.21, is referred to as the *stability axes*, and is quite useful in representing aerodynamic force and moment, as well as in analyzing the stability of the rotational motion. The instantaneous rotation of the velocity vector caused by the applied flow disturbance leads to the displaced wind axes, whose orientation can be described relative to the stability axes using the 3-2-1 Euler angles, $\mathsf{C} = \mathsf{C}_1(\sigma)\mathsf{C}_2(\alpha)\mathsf{C}_3(\beta)$, as shown in Fig. 13.21, where σ, α, β denote the changes in the aerodynamic bank angle, the angle of attack, and the sideslip angle, respectively. Therefore, the instantaneous changes in the flight-path angle and the velocity azimuth are $\phi = \alpha$ and $A = \beta$, and the velocity vector, immediately after the flow disturbance at $t = 0$, is given by

$$\begin{aligned} v' &= v^e + v, \\ \phi' &= \phi^e + \alpha, \\ A' &= A^e + \beta. \end{aligned} \tag{13.102}$$

The quantities v, α, β are to be regarded as the instantaneous flow disturbance, to which the rotational response is desired. The instantaneously displaced wind axes brought to the center of mass, $ox'_v y'_v z'_v$, are depicted in Fig. 13.21. The coordinate transformation between the stability and wind axes is given by

$$\begin{Bmatrix} \mathbf{i_v}' \\ \mathbf{j_v}' \\ \mathbf{k_v}' \end{Bmatrix} = \mathsf{C} \begin{Bmatrix} \mathbf{i} \\ \mathbf{j} \\ \mathbf{k} \end{Bmatrix}, \tag{13.103}$$

where

$$\mathsf{C} =$$

$$\begin{pmatrix} \cos\alpha\cos\beta & \cos\alpha\sin\beta & -\sin\alpha \\ (\sin\sigma\sin\alpha\cos\beta - \cos\sigma\sin\beta) & (\sin\sigma\sin\alpha\sin\beta + \cos\sigma\cos\beta) & \sin\sigma\cos\alpha \\ (\cos\sigma\sin\alpha\cos\beta + \sin\sigma\sin\beta) & (\cos\sigma\sin\alpha\sin\beta - \sin\sigma\cos\beta) & \cos\sigma\cos\alpha \end{pmatrix}. \tag{13.104}$$

Since the flow disturbance is small, we may assume the angles σ, α, β to be small, such that $\sin\alpha \approx \alpha, \cos\alpha \approx 1$, etc., and ignore products of angles. This leads to the following skew-symmetric approximation of the rotation matrix:

$$\mathsf{C} \approx \begin{pmatrix} 1 & \beta & -\alpha \\ -\beta & 1 & \sigma \\ \alpha & -\sigma & 1 \end{pmatrix}. \tag{13.105}$$

The kinematic relationship between the disturbance caused in the direction of the velocity vector (given by ϕ, A), and the flow-disturbance angles, α, β, is then derived as follows:

$$\begin{aligned} \mathbf{v}' &\approx v^e\mathbf{i} + v\mathbf{i} + v^e A\mathbf{j} - v^e\phi\mathbf{k} \\ &= (v^e + v)\mathbf{i_v}' = (v^e + v)(\mathbf{i} + \beta\mathbf{j} - \alpha\mathbf{k}) \\ &\approx (v^e + v)\mathbf{i} + v^e\beta\mathbf{j} - v^e\alpha\mathbf{k}, \end{aligned} \tag{13.106}$$

from which it follows that $\phi \approx \alpha$ and $A \approx \beta$.

The net translational acceleration of the center of mass relative to the wind axes was derived in Chapter 12, whose equilibrium and disturbed values are denoted here by $\mathbf{a_v}^e$ and $\mathbf{a_v}'$, respectively, such that

$$\mathbf{a_v}' = \mathbf{a_v}^e + \mathbf{a_v} \,. \tag{13.107}$$

However, it is desired to express all motion variables in the stability axes, which is a body-fixed frame. Let $\boldsymbol{\omega}^e$ be the angular velocity of the stability axes relative to the wind axes at the equilibrium point. Following the usual aeronautical nomenclature of *roll rate*, $P^e \doteq \omega_x^e$, *pitch rate*, $Q^e \doteq \omega_y^e$, and *yaw rate*, $R^e \doteq \omega_z^e$, we can express the disturbed angular velocity of the vehicle about its center of mass referred to the stability axes as

$$\begin{aligned}\boldsymbol{\omega}' &= P'\mathbf{i} + Q'\mathbf{j} + R'\mathbf{k} \\ &= (P^e + P)\mathbf{i} + (Q^e + Q)\mathbf{j} + (R^e + R)\mathbf{k} \\ &= P^e\mathbf{i} + Q^e\mathbf{j} + R^e\mathbf{k} + P\mathbf{i} + Q\mathbf{j} + R\mathbf{k} \\ &= \boldsymbol{\omega}^e + \boldsymbol{\omega} \,,\end{aligned} \tag{13.108}$$

where P, Q, R are the angular rate disturbances. The translational acceleration at equilibrium point, referred to the stability axes is then obtained as follows:

$$\begin{aligned}\mathbf{a}^e &= \mathbf{a_v}^e - \boldsymbol{\omega}^e \times \mathbf{v}^e \\ &= \mathbf{a_v}^e - (P^e\mathbf{i} + Q^e\mathbf{j} + R^e\mathbf{k}) \times (v^e\mathbf{i}) \\ &= \mathbf{a_v}^e - v^e(R^e\mathbf{j} - Q^e\mathbf{k}) \,.\end{aligned} \tag{13.109}$$

Similarly, the disturbed translational acceleration referred to the stability axes is given by

$$\begin{aligned}\mathbf{a}' &= \mathbf{a_v}' - \omega' \times \mathbf{v}' \\ &= \mathbf{a_v}' - (P'\mathbf{i} + Q'\mathbf{j} + R'\mathbf{k}) \times (v'\mathbf{i_v}') \\ &= \mathbf{a_v}' - v^e[-(\alpha Q^e + \beta R^e)\mathbf{i} + (\alpha P^e + R^e + R)\mathbf{j} \\ &\quad +(\beta P^e - Q^e - Q)\mathbf{k}] + v(Q^e\mathbf{k} - R^e\mathbf{j}) \,,\end{aligned} \tag{13.110}$$

where the small-disturbance assumption has been made. Finally, the disturbance translational acceleration is obtained by subtracting Eq. (13.109) from Eq. (13.110) as

$$\begin{aligned}\mathbf{a} &= \mathbf{a}' - \mathbf{a}^e \\ &= \mathbf{a_v} - v^e[-(\alpha Q^e + \beta R^e)\mathbf{i} + (\alpha P^e + R)\mathbf{j} + (\beta P^e - Q)\mathbf{k}] \\ &\quad +v(Q^e\mathbf{k} - R^e\mathbf{j}) \,.\end{aligned} \tag{13.111}$$

Another kinematic relationship is possible by considering the angular velocity of the stability axes relative to the instantaneous wind axes. This difference in the angular velocities of the two frames can be written as $\omega = P\mathbf{i} + \bar{Q}\mathbf{j} + \bar{R}\mathbf{k}$, where $\bar{Q}, \bar{R}$ are the differential pitch and yaw rates due to the relative rotation. We can differentiate Eq. (13.103) to obtain (Chapter 2)

$$\frac{d\mathsf{C}}{dt} = -\mathsf{CS}(\omega) \,, \tag{13.112}$$

where

$$S(\omega) = \begin{pmatrix} 0 & -\bar{R} & \bar{Q} \\ \bar{R} & 0 & -P \\ -\bar{Q} & P & 0 \end{pmatrix} , \quad (13.113)$$

resulting in

$$\dot{\sigma} \approx P . \quad (13.114)$$

Here, we have chosen not to express the time derivatives of α, β in terms of the unknown variables $\bar{Q}, \bar{R}$, which have to be obtained from the solution of the combined translation and rotational equations of motion. Instead, we can derive these derivatives in the following manner.

It is our objective to derive the time derivatives of the velocity components from the disturbance translational dynamic equation of motion, expressed in the stability axes as follows:

$$\mathbf{f} = m\mathbf{a} , \quad (13.115)$$

where $\mathbf{f} = \mathbf{f}' - \mathbf{f}^e$ is the net disturbance force resolved is the stability axes. From Chapter 12, it is clear that the net external force is a vector sum of the gravity, aerodynamic, and thrust forces. It can be generally assumed that the changes in the position, r, δ, l, are negligible during the small-disturbance motion. Thus, we have

$$r' \approx r^e, \quad \delta' \approx \delta^e, \quad l' \approx l^e,$$
$$g'_c \approx g^e_c, \quad g'_\delta \approx g^e_\delta . \quad (13.116)$$

These assumptions make the magnitude of the gravity force essentially unchanged by the small disturbance. However, its components resolved in the stability axes are functions of the disturbances. It is also to be noted that the gravitational components depend upon the instantaneous vehicle attitude relative to the local horizon, and are independent of the translatory motion represented by α, β. Therefore, it is necessary to model the gravity disturbance in terms of the stability axes rotation, such as through the 3-2-1 Euler angles, Ψ (yaw angle), Θ (pitch angle), and Φ (roll angle), representing the change in the body attitude relative to a *north, east, down* (NED) triad, $\mathbf{I}, \mathbf{J}, \mathbf{K}$. Since Φ, Θ, Ψ are small, the Euler angle singularity (Chapter 2) is avoided. The coordinate transformation between the stability axes and the NED local horizon frame is given by

$$\begin{Bmatrix} \mathbf{i} \\ \mathbf{j} \\ \mathbf{k} \end{Bmatrix} = C_1(\Phi)C_2(\Theta)C_3(\Psi) \begin{Bmatrix} \mathbf{I} \\ \mathbf{J} \\ \mathbf{K} \end{Bmatrix} . \quad (13.117)$$

The acceleration due to gravity in the NED frame is the following:

$$\frac{\mathbf{f_g}^e}{m} \doteq \mathbf{g}^e = g^e_c \mathbf{K} + g^e_\delta \mathbf{I}. \quad (13.118)$$

Using the small-disturbance approximation, the gravity disturbance can be resolved in the stability axes as follows:

$$\begin{aligned}\mathbf{g} &= \mathbf{g}' - \mathbf{g}^e \\ &= g_c^e[-\Theta\cos\Theta^e\mathbf{i} + (\Phi\cos\Phi^e\cos\Theta^e - \Theta\sin\Phi^e\sin\Theta^e)\mathbf{j} \\ &\quad -(\Theta\cos\Phi^e\sin\Theta^e + \Phi\sin\Phi^e\cos\Theta^e)\mathbf{k}] \\ &\quad +g_\delta^e[-(\Theta\sin\Theta^e\cos\Psi^e + \Psi\cos\Theta^e\sin\Psi^e)\mathbf{i} \\ &\quad +(\Phi\cos\Phi^e\sin\Theta^e\cos\Psi^e + \Theta\cos\Theta^e\sin\Phi^e\cos\Psi^e - \Psi\sin\Phi^e\sin\Theta^e\sin\Psi^e \\ &\quad +\Phi\sin\Phi^e\sin\Psi^e - \Psi\cos\Phi^e\cos\Psi^e)\mathbf{j} \\ &\quad +(-\Phi\sin\Phi^e\sin\Theta^e\cos\Psi^e + \Theta\cos\Phi^e\cos\Theta^e\cos\Psi^e - \Psi\cos\Phi^e\sin\Theta^e\sin\Psi^e \\ &\quad +\Phi\cos\Phi^e\sin\Psi^e + \Psi\sin\Phi^e\cos\Psi^e)\mathbf{k}]\,.\end{aligned} \tag{13.119}$$

In the derivation of Eq. (13.119)—and in other following derivations—we have used the approximations in the trigonometric terms involving small-disturbance angles, such as

$$\begin{aligned}\sin(\Phi^e + \Phi) &= \sin\Phi^e\cos\Phi + \cos\Phi^e\sin\Phi \approx \sin\Phi^e + \Phi\cos\Phi^e, \\ \cos(\Phi^e + \Phi) &= \cos\Phi^e\cos\Phi - \sin\Phi^e\sin\Phi \approx \cos\Phi^e - \Phi\sin\Phi^e.\end{aligned} \tag{13.120}$$

It is to be noted that the equilibrium attitude of the vehicle is that of the undisturbed stability axes and is given by $\Theta^e = \phi^e, \Psi^e = A^e$.

The sum of disturbed aerodynamic and propulsive force vectors, resolved in the displaced wind axes, is the following:

$$\begin{aligned}\mathbf{f_a}' + \mathbf{f_T}' = (f_T'\cos\epsilon^e\cos\mu^e - D')\mathbf{i_v}' \\ +(f_Y' + f_T'\sin\mu^e)\mathbf{j_v}' \\ -(f_T'\sin\epsilon^e\cos\mu^e + L')\mathbf{k_v}'\,,\end{aligned} \tag{13.121}$$

where we have assumed that the equilibrium thrust angles, ϵ^e, μ^e, are unchanged by the flow disturbance. This is true for most well-designed vehicles, where the thrust is generated either by aerodynamic means or by a rocket engine having a freely swiveling nozzle that always maintains a fixed orientation relative to the wind axes. The equilibrium sum of the aerodynamic and propulsive forces is

$$\begin{aligned}\mathbf{f_a}^e + \mathbf{f_T}^e = (f_T^e\cos\epsilon^e\cos\mu^e - D^e)\mathbf{i} \\ +(f_Y^e + f_T^e\sin\mu^e)\mathbf{j} \\ -(f_T^e\sin\epsilon^e\cos\mu^e + L^e)\mathbf{k}\,.\end{aligned} \tag{13.122}$$

Employing the small-disturbance approximation, the disturbance force arising out of aerodynamics and propulsion is written as:

$$\begin{aligned}\mathbf{f_a} + \mathbf{f_T} = &[f_T \cos\epsilon^e \cos\mu^e - D - \beta(f_Y^e + f_T^e \sin\mu^e)\\ &-\alpha(f_T^e \sin\epsilon^e \cos\mu^e + L^e)]\mathbf{i}\\ &+[f_Y + f_T \sin\mu^e + \beta(f_T^e \cos\epsilon^e \cos\mu^e - D^e)\\ &+\sigma(f_T^e \sin\epsilon^e \cos\mu^e + L^e)]\mathbf{j} \qquad (13.123)\\ &+[-f_T \sin\epsilon^e \cos\mu^e - L + \sigma(f_Y^e + f_T^e \sin\mu^e)\\ &-\alpha(f_T^e \cos\epsilon^e \cos\mu^e - D^e)]\mathbf{k}\,.\end{aligned}$$

We remind ourselves that the aerodynamic and propulsive disturbance terms, L, D, f_Y, f_T, depend upon the flow disturbances, α, β, σ. We shall express the linearized relationships of these disturbance terms a little later.

It now remains to obtain an expression for the disturbance translational acceleration of the center of mass, $\mathbf{a_v}$, resolved in the stability axes. In order to do so, we shall first write the translational acceleration at equilibrium by substituting $\dot{v}^e = \dot{\phi}^e = \dot{A}^e = 0$ into the acceleration derived in Chapter 12, leading to

$$\begin{aligned}\mathbf{a_v}^e &= a_{xv}^e \mathbf{i} + a_{yv}^e \mathbf{j} + a_{zv}^e \mathbf{k}\\ &= v^e\Big[-\frac{v^e}{r^e}\cos^2\phi^e \sin A^e \tan\delta^e\\ &+2\Omega(\sin\phi^e \cos A^e \cos\delta^e - \cos\phi^e \sin\delta^e)\Big]\mathbf{j} \qquad (13.124)\\ &+\Big(\frac{v^e}{r^e}\cos\phi^e + 2\Omega \sin A^e \cos\delta^e\Big)\,,\end{aligned}$$

where the centripetal acceleration terms due to planetary rotational velocity, Ω, are neglected, as they are several orders of magnitude smaller than the other terms. However, we shall (for the time being) retain the Coriolis acceleration terms due to planetary rotation. These are generally negligible for most atmospheric vehicles, except an atmospheric entry vehicle.[8] The disturbance acceleration, $\mathbf{a_v}$, is obtained as follows by subtracting Eq. (13.65) from the disturbed acceleration, $\mathbf{a_v}'$, resolved in stability axes, and applying the small-disturbance approximation:

$$\begin{aligned}\mathbf{a_v} = &(\dot{v} - \beta a_{yv}^e + \alpha a_{zv}^e)\mathbf{i}\\ &+\Big[v^e\dot{\beta} + \frac{v^e}{r^e}\cos\phi^e \tan\delta^e(v^e\alpha \sin\phi^e \sin A^e\\ &-v^e\beta \cos\phi^e \cos A^e - 2v\cos\phi^e \sin A^e)\end{aligned}$$

---

[8] For a typical entry from a low earth orbit, $v^e = 8$ km/s and $r^e = 6500$ km. This yields the maximum centripetal acceleration, $\Omega^2 r^e \approx 0.03$ m/s^2, maximum Coriolis acceleration, $2\Omega v^e \approx 1$ m/s^2, and maximum acceleration due to planetary curvature, $\frac{v^{e2}}{r^e} \approx 10$ m/s^2. Thus, curvature and Coriolis acceleration cannot be ignored, as they are of the same order, and one tenth, respectively, of the magnitude of acceleration due to gravity. The centripetal acceleration terms vanish below first order, when multiplied with a small disturbance, and are thus neglected in a stability analysis.

$$
\begin{aligned}
&+ 2\Omega\{v(\cos\delta^e \sin\phi^e \cos A^e - \sin\delta^e \cos\phi^e) \\
&+ v^e\alpha(\cos\delta^e \cos\phi^e \cos A^e + \sin\delta^e \sin\phi^e) \\
&- v^e\beta \cos\delta^e \sin\phi^e \sin A^e\} + \beta a^e_{xv} - \sigma a^e_{zv}\Big]\mathbf{j} \\
&+ \Big[- v^e\dot{\alpha} + \frac{v^e}{r^e}(2v\cos\phi^e - v^e\alpha\sin\phi^e) \\
&+ 2\Omega\cos\delta^e(v\sin A^e + v^e\beta\cos A^e) \\
&- \alpha a^e_{xv} + \sigma a^e_{yv}\Big]\mathbf{k}\,.
\end{aligned}
\tag{13.125}
$$

Collecting all the terms from Eqs. (13.125), (13.119), (13.123), and (13.111), and substituting them into Eq. (13.115), we have the disturbance force equations:

$$
\begin{aligned}
&f_T\cos\epsilon^e\cos\mu^e - D - \beta(f^e_Y + f^e_T\sin\mu^e) - \alpha(f^e_T\sin\epsilon^e\cos\mu^e + L^e) \\
&\quad -mg^e_c\Theta\cos\Theta^e - mg^e_\delta(\Theta\sin\Theta^e\cos\Psi^e + \Psi\cos\Theta^e\sin\Psi^e) \\
&= m[\dot{v} - \beta a^e_{yv} + \alpha a^e_{zv} + v^e(\alpha Q^e + \beta R^e)]\,.
\end{aligned}
\tag{13.126}
$$

$$
\begin{aligned}
&f_Y + f_T\sin\mu^e + \beta(f^e_T\cos\epsilon^e\cos\mu^e - D^e) \\
&+\sigma(f^e_T\sin\epsilon^e\cos\mu^e + L^e) \\
&+mg^e_c(\Phi\cos\Phi^e\cos\Theta^e - \Theta\sin\Phi^e\sin\Theta^e) \\
&+mg^e_\delta(\Phi\cos\Phi^e\sin\Theta^e\cos\Psi^e + \Theta\cos\Theta^e\sin\Phi^e\cos\Psi^e \\
&-\Psi\sin\Phi^e\sin\Theta^e\sin\Psi^e + \Phi\sin\Phi^e\sin\Psi^e - \Psi\cos\Phi^e\cos\Psi^e) \\
&= m\Big[v^e\dot{\beta} + \frac{v^e}{r^e}\cos\phi^e\tan\delta^e(v^e\alpha\sin\phi^e\sin A^e \\
&-v^e\beta\cos\phi^e\cos A^e - 2v\cos\phi^e\sin A^e) \\
&+2\Omega\{v(\cos\delta^e\sin\phi^e\cos A^e - \sin\delta^e\cos\phi^e) \\
&+v^e\alpha(\cos\delta^e\cos\phi^e\cos A^e + \sin\delta^e\sin\phi^e) \\
&-v^e\beta\cos\delta^e\sin\phi^e\sin A^e\} + \beta a^e_{xv} \\
&-\sigma a^e_{zv} - v^e(\alpha P^e + R) - vR^e\Big]\,.
\end{aligned}
\tag{13.127}
$$

$$
\begin{aligned}
&-f_T\sin\epsilon^e\cos\mu^e - L + \sigma(f^e_Y + f^e_T\sin\mu^e) \\
&-\alpha(f^e_T\cos\epsilon^e\cos\mu^e - D^e) \\
&-mg^e_c(\Theta\cos\Phi^e\sin\Theta^e + \Phi\sin\Phi^e\cos\Theta^e) \\
&+mg^e_\delta(-\Phi\sin\Phi^e\sin\Theta^e\cos\Psi^e \\
&+\Theta\cos\Phi^e\cos\Theta^e\cos\Psi^e - \Psi\cos\Phi^e\sin\Theta^e\sin\Psi^e \\
&+\Phi\cos\Phi^e\sin\Psi^e + \Psi\sin\Phi^e\cos\Psi^e) \\
&= m\Big[- v^e\dot{\alpha} + \frac{v^e}{r^e}(2v\cos\phi^e - v^e\alpha\sin\phi^e) \\
&+2\Omega\cos\delta^e(v\sin A^e + v^e\beta\cos A^e) \\
&-\alpha a^e_{xv} + \sigma a^e_{yv} - v^e(\beta P^e - Q) + vQ^e\Big]\,.
\end{aligned}
\tag{13.128}
$$

These equations will be further expanded when we take into account the linear variation of the aerodynamic and thrust forces with the flow disturbances.

At equilibrium, the vehicle is rotating with a constant, body-referenced angular velocity, $\boldsymbol{\omega}^e = P^e\mathbf{i} + Q^e\mathbf{j} + R^e\mathbf{k}$, and equilibrium torque, $\mathbf{M}^e = \mathcal{L}^e\mathbf{i}+\mathcal{M}^e\mathbf{j}+\mathcal{N}^e\mathbf{k}$, where $\mathcal{L}^e$ is called the *rolling moment*, $\mathcal{M}^e$ the *pitching moment*, and $\mathcal{N}^e$ the *yawing moment*—in standard aeronautical nomenclature—at the equilibrium point. Therefore, Euler's equations of rotational motion Eq. (13.18), expressed in the stability axes at equilibrium, yield the following equations for the torque components at equilibrium:

$$\begin{aligned}
\mathcal{L}^e &= Q^e[R^e(J_{zz} - J_{yy}) - P^e J_{xz} - Q^e J_{yz}] + R^e(P^e J_{xy} + R^e J_{yz}),\\
\mathcal{M}^e &= P^e[R^e(J_{xx} - J_{zz}) + Q^e J_{yz} + P^e J_{xz}] - R^e(Q^e J_{xy} + R^e J_{xz}),\\
\mathcal{N}^e &= Q^e[P^e(J_{yy} - J_{xx}) + Q^e J_{xy} + R^e J_{xz}] - P^e(R^e J_{yz} + P^e J_{xy})
\end{aligned} \tag{13.129}$$

where we note the presence of products of inertia, which are nonzero because $(oxyz)$ is not the principal frame. Most atmospheric flight vehicles possess a plane of symmetry, and often the equilibrium flight condition is such that the velocity vector lies in the plane of symmetry. Such an assumption would greatly simplify Eq. (13.129). However, we shall reserve this assumption for later, because a general flight path may not obey this restriction (e.g., the steady sideslip maneuver of aircraft). In a manner similar to the disturbance force, we can derive the disturbance torque components by subtracting the equilibrium torque from the disturbed torque, resulting in

$$\begin{aligned}
\mathcal{L} = {}& J_{xx}\dot{P} - J_{xy}\dot{Q} - J_{xz}\dot{R} - J_{xz}(QP^e + PQ^e) + 2J_{yz}(RR^e - QQ^e)\\
&+J_{xy}(RP^e + PR^e) + (J_{zz} - J_{yy})(QR^e + RQ^e)\,.
\end{aligned} \tag{13.130}$$

$$\begin{aligned}
\mathcal{M} = {}& J_{yy}\dot{Q} - J_{xy}\dot{P} - J_{yz}\dot{R} - J_{xy}(RQ^e + QR^e) + 2J_{xz}(PP^e - RR^e)\\
&+J_{yz}(QP^e + PQ^e) + (J_{xx} - J_{zz})(RP^e + PR^e)\,.
\end{aligned} \tag{13.131}$$

$$\begin{aligned}
\mathcal{N} = {}& J_{zz}\dot{R} - J_{xz}\dot{P} - J_{yz}\dot{Q} - J_{yz}(RP^e + PR^e) + 2J_{xy}(QQ^e - PP^e)\\
&+J_{xz}(RQ^e + QR^e) + (J_{yy} - J_{xx})(QP^e + PQ^e)\,.
\end{aligned} \tag{13.132}$$

The rate of rotation of the stability axes is affected by the disturbance torque, which in turn, is changed by the flow disturbance. However, due to the rotary inertia of the vehicle, the change in the vehicle's attitude is not instantaneous, but occurs over a period of time. Since the attitude of the vehicle is described by the orientation of the body-fixed stability axes, an appropriate representation can be used for the instantaneous orientation of *oxyz*. If the vehicle is initially at rest, the 3-2-1 Euler angle representation would be non-singular during the rotational motion caused by the small disturbance. For this reason, the 3-2-1 body attitude representation is most popular in aircraft applications. In such a case, the vehicle's attitude at $t = 0$ is given by the equilibrium attitude, Ψ^e, Θ^e, Φ^e, while the perturbation from this attitude is given by the disturbance angles, Ψ, Θ, Φ. However, when the equilibrium state is a general rotary motion, a more appropriate attitude description is via the quaternion (Chapter 2), $\mathbf{q}, q_4$, whose kinematic equations are written as follows:

$$\frac{\mathrm{d}\{\mathbf{q}, q_4\}^T}{\mathrm{d}t} = \frac{1}{2}\Omega\{\mathbf{q}(t), q_4(t)\}^T , \tag{13.133}$$

where Ω is the following skew-symmetric matrix:

$$\Omega = \begin{pmatrix} 0 & (R^e + R) & -(Q^e + Q) & (P^e + P) \\ -(R^e + R) & 0 & (P^e + P) & (Q^e + Q) \\ (Q^e + Q) & -(P^e + P) & 0 & (R^e + R) \\ -(P^e + P) & -(Q^e + Q) & -(R^e + R) & 0 \end{pmatrix} . \tag{13.134}$$

The kinematic equation is to be integrated with the initial condition specified by the equilibrium point attitude, $\mathbf{q}^e, q_4^e$.

The equations of a small-disturbance, rotational motion of an atmospheric flight vehicle, therefore, consist of the coupled set of differential equations, Eqs. (13.114), (13.126)–(13.128), (13.131)–(13.134). The additional coupling terms due to aerodynamics and propulsion in these equations are derived by the linearized stability derivatives.

13.8.2 Stability Derivatives and De-coupled Dynamics

The stability analysis of an atmospheric vehicle requires the functional dependence of the aerodynamic and propulsive force and moment on the disturbance variables. In a general unsteady motion of a flight vehicle, such relationships are non-existent in a closed form, due to the complex effects of turbulence, compressibility, flow separation, and non continuity. Even when simplifying assumptions are made, rarely do we have a closed-form description of the flow field (Chapter 10), and an approximate, numerical solution of partial differential equations is the norm. However, one can employ the small-disturbance approximation to render all aerodynamic relationships essentially linear, irrespective of the flow regime in which the equilibrium point is located. The hallmark of linear dependence of the aero-propulsive force and moment on the disturbance variables is a Taylor series expansion, truncated to first-order terms, such as the following expression for the disturbed pitching moment, $\mathcal{M}'$, of an airplane:

$$\mathcal{M}' = \mathcal{M}^e + \frac{\partial \mathcal{M}}{\partial v} v + \frac{\partial \mathcal{M}}{\partial \alpha} \alpha + \frac{\partial \mathcal{M}}{\partial \dot{\alpha}} \dot{\alpha} + \frac{\partial \mathcal{M}}{\partial Q} Q , \tag{13.135}$$

where the partial derivatives are evaluated at the equilibrium point and are referred to as *stability derivatives*. It is useful to express the stability derivatives in a nondimensional form, which allows us to analyze the characteristics of a particular configuration without having to consider the effects of size, equilibrium speed, and altitude. This is accomplished by dividing the forces by qS, moments by qSl_c, and speed by v^e, where q is the dynamic pressure, S is the reference wing planform area, and l_c is a characteristic length. The angular rates are traditionally expressed in a nondimensional time, $\hat{t} \doteq t\frac{2v^e}{l_c}$, which results in the corresponding nondimensional stability derivatives being multiplied by the factor, $\frac{l_c}{2v^e}$, in the equations of motion. For example, the nondimensionalized pitching moment disturbance can be expressed using Eq. (13.135) in the standard NACA nomenclature as follows:

$$C_m \doteq \frac{\mathcal{M}}{qS\bar{c}} = C_{m_u}u + C_{m_\alpha}\alpha + \frac{\bar{c}}{2v^e}C_{m_{\dot{\alpha}}}\dot{\alpha} + \frac{\bar{c}}{2v^e}C_{m_q}Q\,, \tag{13.136}$$

where

$$\begin{aligned} C_{m_u} &\doteq \frac{v^e}{qS\bar{c}}\frac{\partial\mathcal{M}}{\partial v}, \\ C_{m_\alpha} &\doteq \frac{1}{qS\bar{c}}\frac{\partial\mathcal{M}}{\partial\alpha}, \\ C_{m_{\dot{\alpha}}} &\doteq \frac{1}{qS\bar{c}}\frac{2v^e}{\bar{c}}\frac{\partial\mathcal{M}}{\partial\dot{\alpha}}, \\ C_{m_q} &\doteq \frac{1}{qS\bar{c}}\frac{2v^e}{\bar{c}}\frac{\partial\mathcal{M}}{\partial Q}, \end{aligned} \tag{13.137}$$

$$u \doteq \frac{v}{v^e}, \tag{13.138}$$

and $\bar{c}$ is the wing's mean-aerodynamic chord, representing the characteristic length. The characteristic length (and thus the nondimensional time) indicates the time scale of motion and may be different for the various stability axes. Furthermore, the force and moment relative to each stability axis may depend upon a different set of disturbance quantities. Before pursuing the concept of stability derivatives any further, it is important to define the set of motion variables particular to each stability axis.

All atmospheric flight vehicles possess some form of symmetry, which enables them to achieve a stable equilibrium in normal operation. The least symmetric atmospheric flight vehicle is a lifting configuration—such as the airplane—having only one plane of symmetry, oxz, while a thrust-controlled missile is an axisymmetric vehicle with infinitely many planes of symmetry (and thus non-unique stability axes). In between these two extremes lie most launch vehicles and missiles, with a varying number of symmetry planes. A vehicle with more than one plane of symmetry enjoys inter-exchangeability of two (or more) stability axes, for which the equations of motion are identical. Therefore, the airplane is taken to be the reference vehicle for defining the dependent motion variables for each stability axis, as it results in the most general description of aerodynamic motion. Assuming oxz to be a plane of symmetry, we have $J_{xy} = J_{yz} = 0$. Furthermore, it follows that the motion in the plane of symmetry, called *longitudinal dynamics*, is fundamentally different—and separable from—that outside the plane, which we will refer to as *lateral dynamics*. Hence, longitudinal and lateral dynamics should have distinct sets of motion variables. Clearly, the lateral dynamics involves changes in the "unsymmetrical" variables β, Φ, Ψ, P, R, whereas longitudinal motion involves the remaining variables, namely u, α, Θ, Q. With these assumptions, we can de-couple the longitudinal and lateral dynamics and separate the stability derivatives into the two categories.

13.8.3 Longitudinal Dynamics

The longitudinal dynamic equations—Eqs. (13.126), (13.128), and (13.131)—involve a three-degree-of-freedom motion (translation along ox, oz, and rotation about oy). For a flight in the plane of symmetry, $\beta = \Phi = \Psi = \Phi^e = \mu^e = P^e = R^e = 0$. Hence, the longitudinal equations of motion are written in the following de-coupled form:

$$
\begin{aligned}
m[v^e\dot{u} + \alpha(a^e_{zv} + v^e Q^e)] & \\
= qS[C_{x_u}u + C_{x_\alpha} - \Theta & \left(\cos\Theta^e \frac{mg^e_c}{qS}\right. \\
& \left. + \sin\Theta^e \cos\Psi^e \frac{mg^e_\delta}{qS}\right) + \frac{\bar{c}}{2v^e}(C_{x_{\dot{\alpha}}}\dot{\alpha} + C_{x_q}Q)] \,.
\end{aligned} \tag{13.139}
$$

$$
\begin{aligned}
m[v^e\dot{\alpha} - & \frac{v^e}{r^e}(2v\cos\phi^e - v^e\alpha\sin\phi^e) - 2uv^e\Omega\cos\delta^e\sin A^e \\
& + \alpha a^e_{xv} - uv^e Q^e - Qv^e)] \\
& = qS[C_{z_u}u + C_{z_\alpha}\alpha - \Theta\left(\sin\Theta^e \frac{mg^e_c}{qS} - \cos\Theta^e\cos\Psi^e \frac{mg^e_\delta}{qS}\right) \\
& + \frac{\bar{c}}{2v^e}(C_{z_{\dot{\alpha}}}\dot{\alpha} + C_{z_q}Q)].
\end{aligned} \tag{13.140}
$$

$$
J_{yy}\dot{Q} = qS\bar{c}[C_{m_u}u + C_{m_\alpha}\alpha + \frac{\bar{c}}{2v^e}(C_{m_{\dot{\alpha}}}\dot{\alpha} + C_{m_q}Q)] \,. \tag{13.141}
$$

Upon comparison with Eqs. (13.126) and (13.128), some of the longitudinal force derivatives are directly obtained to be the following:

$$
C_{x_u} = 2(C_T\cos\epsilon^e - C_D) - u\frac{\partial C_D}{\partial u} + u\cos\epsilon^e\frac{\partial C_T}{\partial u}, \tag{13.142}
$$

$$
C_{x_\alpha} = -C_T\sin\epsilon^e - C_L - \frac{\partial C_D}{\partial \alpha} + \cos\epsilon^e\frac{\partial C_T}{\partial \alpha}, \tag{13.143}
$$

$$
C_{z_u} = -2(C_T\sin\epsilon^e + C_L) - u\frac{\partial C_L}{\partial u} - u\sin\epsilon^e\frac{\partial C_T}{\partial u}, \tag{13.144}
$$

$$
C_{z_\alpha} = C_T\cos\epsilon^e - C_D - \frac{\partial C_L}{\partial \alpha} - \sin\epsilon^e\frac{\partial C_T}{\partial \alpha}. \tag{13.145}
$$

Here, C_L, C_D, C_T refer to the lift, drag, and thrust coefficients (Chapters 10 and 11). Except for propeller-engined airplanes, the variation of C_T with speed and angle of attack is negligible. The variation of drag coefficient with speed occurs due to compressibility effects and is especially important in the transonic regime (Chapter 10). Hence, we commonly calculate such derivatives using the Mach number, M, as

$$u\frac{\partial C_D}{\partial u} = M\frac{\partial C_D}{\partial M} .$$

Generally, a well-designed airplane has $C_{m_u} \approx 0$.

C_{m_α} is the most important longitudinal stability derivative and represents the *static longitudinal stability* of the vehicle. It is directly proportional to the distance, Δx, by which the *aerodynamic center*[9] of the vehicle lies aft of the center of mass. For this reason, the said distance is called the *longitudinal static margin*. The longitudinal static margin is affected by the pitching moment contributions of the various components of the vehicle, such as wing, tail (or canard), fuselage, and nacelles. The primary contribution comes from the tail (or canard), where a change in the downwash (or upwash) is caused by the wing due to a change in the angle of attack. In addition, there can be significant changes in the static margin caused by the slipstream of a propeller. Clearly, for longitudinal static stability the vehicle must pitch in the negative direction ("downward" in pilot's viewpoint), whenever the angle of attack increases ($C_{m_\alpha} < 0$). A large majority of atmospheric vehicles have $C_{m_\alpha} < 0$, although some airplanes have been designed to be statically unstable from maneuverability considerations and require either exceptional piloting skills (*Wright 1903 Flyer*) or a closed-loop pitch stabilization system (the *F-16* fighter) for maintaining equilibrium. The static stability also translates into the requirement that the vehicle must produce a positive pitching moment at $\alpha = 0$ for an ability to maintain equilibrium ($C_m = 0$) at positive values of angle of attack (which is the normal situation for airplanes). Since an airplane's wing normally produces a negative pitching moment in order to generate lift at $\alpha = 0$, a stable airplane needs a horizontal stabilizing surface (either a tail, or a canard) to provide the positive C_m at $\alpha = 0$.

The derivatives C_{x_α} and C_{z_α} largely represent the variation of drag and lift coefficients with angle of attack. Of these, C_{z_α}, approximately equaling the negative of lift–curve–slope (C_{L_α}), is the more important and typically falls in the range of 4–6. The lifting effectiveness of the vehicle is measured by the magnitude of C_{z_α}.

The changes in aerodynamic force and moment do not occur instantaneously with the change in the angle of attack, but generally involve a time-lag due to the essentially circulatory flow over the lifting surfaces (Chapter 15).

[9] As defined for a lifting surface in Chapter 10, the aerodynamic center is the unique point about which the pitching moment is independent of the angle of attack. The concept of aerodynamic center can be extended for the whole vehicle, which may have several lifting surfaces. Usually, a vehicle's aerodynamic center is called the *neutral point*, as it indicates the center of mass location for zero longitudinal static margin.

This aerodynamic time lag is referred to *aerodynamic inertia* and is represented by the $\dot{\alpha}$ derivatives. Generally, $C_{x_{\dot{\alpha}}} \approx 0$, as thrust and drag are essentially noncirculatory in nature. On the other hand, the lag in lift and pitching moment can be large for a conventional airplane equipped with a horizontal stabilizer (tail) and have typical values of $C_{z_{\dot{\alpha}}} \approx -1$ and $C_{m_{\dot{\alpha}}} \approx$ --3, respectively. We can derive $C_{z_{\dot{\alpha}}}$ from $C_{m_{\dot{\alpha}}}$ using the *tail arm*, l_t, as

$$C_{z_{\dot{\alpha}}} = \frac{\bar{c}}{l_t} C_{m_{\dot{\alpha}}} \, .$$

At hypersonic speeds encountered by re-entry vehicles, the aerodynamic lag is negligible, which results in all aerodynamic inertia derivatives approximated by zeros.

Finally, the stability derivatives C_{x_q}, C_{z_q} and C_{m_q} represent the effects of the pitch rate on lift and pitching moment. They are caused largely by the change in the angle of attack experienced by the lifting surfaces due to the curvature in the flight path. The derivative C_{m_q} greatly influences the damping in the natural pitching oscillations and is thus known as *damping in pitch*. Generally, $C_{x_q} \approx 0$, while C_{z_q} can be obtained by dividing C_{m_q} by the nondimensional tail (or canard) arm, $\frac{l_t}{\bar{c}}$.

The kinematic attitude relations for the longitudinal motion are expressed as follows:

$$\dot{\Theta} = Q \, , \tag{13.146}$$

where Θ, the disturbance in the pitch angle, is related to angle of attack and disturbance in the flight-path angle by

$$\alpha = \Theta - \phi \, . \tag{13.147}$$

From the last equation, it follows that $\Theta^e = \phi^e$, because we have employed the stability axes.

Example 13.7. A tail-less, delta-winged fighter airplane with $m = 10,455$ kg, $J_{yy} = 121,567$ kg.m^2, $\bar{c} = 6.95$ m, and $S = 60.5$ m^2, is undergoing a pitch-up manuever with $A^e = 0$ and a constant Mach number, $M = 0.94$. The stability derivatives at the given Mach number are the following:

$$\begin{aligned}
C_{m_\alpha} &= -0.31/\text{rad}, \\
C_{m_q} &= -1.44/\text{rad}, \\
C_{m_{\dot{\alpha}}} &= -1/\text{rad}, \\
C_{z_\alpha} &= -2.85/\text{rad}, \\
C_{z_q} &= -2/\text{rad}, \\
C_{z_{\dot{\alpha}}} &= -1.39/\text{rad}, \\
C_{z_u} &= -0.37, \\
C_{x_\alpha} &= -0.144/\text{rad}, \\
C_{x_u} &= -0.048, \\
C_L &= 0.146.
\end{aligned}$$

Simulate the ensuing motion of the airplane after reaching $h^e = 2000$ m and $\delta^e = 45°$, where an angle of attack disturbance is encountered with $v = 0$ and initial condition

(a) $\phi^e = 0$, $Q^e = 0$, $\alpha = 0.01$ rad.
(b) $\phi^e = 0.1$ rad, $Q^e = 0.15$ rad/s, $\alpha = -0.1$ rad.

Table 13.9. M-file *pitchup.m* for Airplane's Longitudinal State Equations

```
function xdot = pitchup(t,x)
%(c) 2006 Ashish Tewari
global S; global c; global m; global Jyy; global rm; global omega;
global v0; global phi0; global A0; global Q0; global Cma; global Cmad;
global Cmq; global Cxu; global Cxa; global Czu; global Cza; global Czad;
global Czq;
%acceleration due to gravity (oblate earth):
delta=x(6)
alt = x(1)
[g,gn]=gravity(alt+rm,delta);
%atmospheric properties:
v = v0*(1+x(2));
atmosp = atmosphere(alt, v, c);
rho = atmosp(2);%density
q= 0.5*rho*v^2;%dynamic pressure
mach = atmosp(3);
CL=m*g/(q*S);
[t alt v mach]
phi=x(4)-x(3); %flight-path angle
%longitudinal dynamics:
hdot=v*sin(phi);
udot = -Q0+q*S*(Cxu*x(2)+Cxa*x(3)...
     +phi*(-cos(phi0)*CL+sin(phi0)*sin(A0)*m*gn/(q*S)))/(m*v0);
alphadot = (2*x(2)*omega*cos(delta)*sin(A0)...
     +2*x(2)*v0/(rm+alt)+x(2)*Q0+x(5)+q*S*(Czu*x(2)+Cza*x(3)...
     +phi*(sin(phi0)*CL-cos(phi0)*cos(A0)*m*gn/(q*S))...
     +c*Czq*x(5)/(2*v0))/(m*v0))/(1-q*S*c*Czad/(2*m*v0^2));
thetadot = x(5);
Qdot = q*S*c*(Cma*x(3)+c*(Cmad*alphadot+Cmq*x(5))/(2*v0))/Jyy;
deltadot=v*cos(phi)*cos(A0)/(rm+alt);
xdot = [hdot; udot; alphadot; thetadot; Qdot; deltadot];
```

The simulation requires a numerical solution to the longitudinal dynamic and kinematic equations, with the prescribed initial condition. The aerodynamic force and moment are allowed to vary with a changing dynamic pressure in this simulation. The necessary computation is performed by the M-file *pitchup.m* tabulated in Table 13.9, which integrates the nonlinear differential equations of motion with the intrinsic MATLAB Runge–Kutta solver *ode45.m*. The results are plotted in Figs. 13.22–13.24. The departure from straight and level equilibrium condition [Case (a)] displays two distinct time scales of the airplane's motion: a rapid and well-damped oscillation in the variables α, Θ, Q with settling time[10] about 1 s and an insignificant change in speed and alti-

[10] In Chapter 14, *settling time* is defined as the time required for the response to decay to within ±2% of the steady state.

tude (Fig. 13.22), and a slower, less damped oscillation in altitude, speed, and pitch angle with a settling time of about 150 s with no appreciable variation in the angle of attack (Fig. 13.23). The clearly defined short- and long-period oscillations form the basis of the approximate longitudinal modes, as discussed ahead. The departure from a steady pitch-up maneuver [Case (b)] is plotted in Fig. 13.24. The steadily increasing altitude and a declining speed with a long-period oscillation in pitch are combined with a short-period pitching motion with variation in the angle of attack. If allowed uncorrected, the motion would quickly lead to the flight speed becoming zero, and then negative (called a *tail slide*).[11] Note that the angle of attack remains small, thus the linear aerodynamic model remains valid, even though the speed falls to zero.

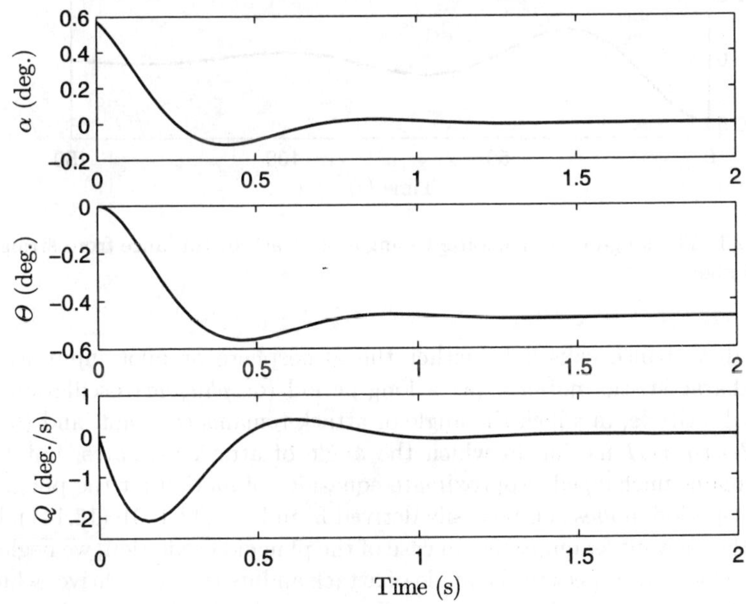

Fig. 13.22. The short-period response to angle of attack disturbance from straight and level flight.

13.8.4 Airplane Longitudinal Modes

The most common equilibrium condition encountered in an airplane is that of straight and level flight ($\Theta^e = Q^e = a^e_{xv} = a^e_{zv} = 0$). In such a condition,

[11] A tail slide is normally avoided, as it causes destruction of trailing-edge control surfaces.

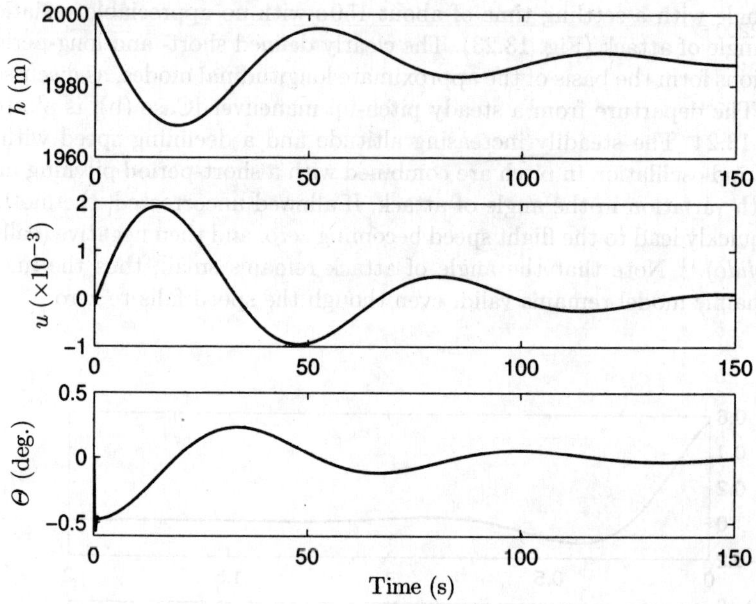

Fig. 13.23. The long-period response to angle of attack disturbance from straight and level flight.

a small disturbance caused by either the atmosphere or pilot input leads to two characteristic motions: (a) a long-period (or *phugoid*) oscillation in speed and altitude, in which the angle of attack remains constant, and (b) a rapid, *short-period* motion in which the angle of attack oscillates, but the speed remains unchanged. Approximate equations of motion for the phugoid and short-period modes can be easily derived from Eqs. (13.139)–(13.141), by making the relevant assumptions. In case of the phugoid oscillation, we neglect all variations with respect to the angle of attack and its time derivative, which amounts to disregarding the pitching motion caused by the change in the angle of attack, and taking $\alpha \approx 0$ in the remaining equations. Therefore, $\Theta \approx \phi$, and the resulting equations for phugoid approximation are the following, after neglecting the terms involving planetary rotation and curvature:

$$\frac{mv^e}{qS}\dot{u} = C_{x_u}u - \Theta\frac{mg_c^e}{qS}. \tag{13.148}$$

$$-\frac{mv^e}{qS}Q = C_{z_u}u + \Theta\cos\Psi^e\frac{mg_\delta^e}{qS} + \frac{\bar{c}}{2v^e}C_{z_q}Q. \tag{13.149}$$

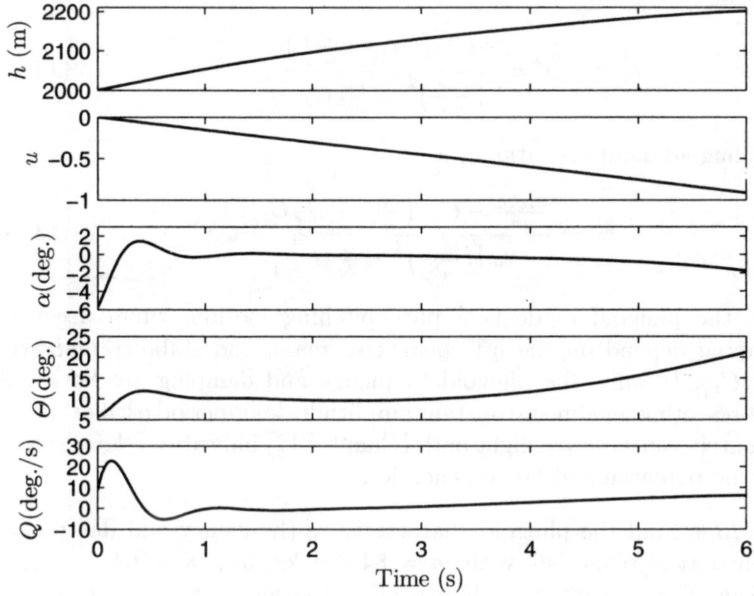

Fig. 13.24. Response to angle of attack disturbance from a steady pitch-up maneuever.

On substituting Eqs. (13.146) and (13.149) into (13.148), we have

$$\left[\left(\frac{mv^e}{qS}\right)^2 \frac{1}{C_{z_u}} + \frac{m\bar{c}}{2qS}\frac{C_{z_q}}{C_{z_u}}\right]\ddot{\Theta} + \left[\frac{mv^e a}{qSC_{z_u}} - \frac{C_{x_u}}{C_{z_u}}\left(\frac{mv^e}{qS} + \frac{\bar{c}}{2v^e}C_{z_q}\right)\right]\dot{\Theta} - qS\left(C_L + a\frac{C_{x_u}}{C_{z_u}}\right)\Theta = 0, \tag{13.150}$$

$$u = -\left(\frac{mv^e}{qS} + \frac{\bar{c}}{2v^e}C_{z_q}\right)\frac{\dot{\Theta}}{C_{z_u}} - \frac{a}{C_{z_u}}\Theta\,, \tag{13.151}$$

where

$$C_L \doteq \frac{mg_c^e}{qS} \tag{13.152}$$

is the equilibrium lift coefficient and

$$a \doteq \cos\Psi^e \frac{mg_\delta^e}{qS}\,. \tag{13.153}$$

Taking the Laplace transform of Eq. (13.150), we can write the characteristic equation for the second-order system (Chapter 14) as follows:

$$s^2 + 2\zeta\omega s + \omega^2 = 0\,, \tag{13.154}$$

where the natural frequency, ω, of the phugoid mode is given by

$$\omega^2 = \frac{-C_{z_u}\left(C_L + a\frac{C_{x_u}}{C_{z_u}}\right)}{\left(\frac{mv^e}{qS}\right)^2 + \frac{m\bar{c}}{2qS}C_{z_q}}, \tag{13.155}$$

and the phugoid damping ratio, ζ, is

$$\zeta = \frac{\frac{mv^e a}{qS} - C_{x_u}\left(\frac{mv^e}{qS} + \frac{\bar{c}}{2v^e}C_{z_q}\right)}{2\omega\left[\left(\frac{mv^e}{qS}\right)^2 + \frac{m\bar{c}}{2qS}C_{z_q}\right]}. \tag{13.156}$$

Thus, the phugoid mode is a pure pitching motion, whose frequency and damping depend on the lift coefficient, mass, and stability derivatives $C_{x_u}, C_{z_u}, C_{z_q}$. Usually, the phugoid frequency and damping are both quite small, representing an almost-constant amplitude, long-period oscillation. The approximately *conservative* flight path (Chapter 12) indicates a slow exchange between the potential and kinetic energies.

Example 13.8. Find the phugoid characteristics (frequency and damping) of a jet transport airplane [46] with $m = 84,891$ kg, $J_{yy} = 3,564,403$ kg.m^2, $\bar{c} = 6.16$ m, $S = 223$ m^2, $\bar{l}_t = 17.8$ m, flying straight and level at 12.2 km altitude, $\Psi^e = 30°$, and $M = 0.62$, where $v^e = 182$ m/s, $q = 5036.79$ N/m^2, and

$$\begin{aligned}
C_{m_\alpha} &= -0.619/\text{rad}, \\
C_{m_q} &= -11.4/\text{rad}, \\
C_{m_{\dot\alpha}} &= -3.27/\text{rad}, \\
C_{z_\alpha} &= -4.46/\text{rad}, \\
C_{z_q} &= C_{m_q}\frac{\bar{c}}{l_t} = -3.94/\text{rad}, \\
C_{z_{\dot\alpha}} &= C_{m_{\dot\alpha}}\frac{\bar{c}}{l_t} = -1.13/\text{rad}, \\
C_{z_u} &= -1.48, \\
C_{x_\alpha} &= 0.392/\text{rad}, \\
C_{x_u} &= -0.088, \\
C_L &= 0.74.
\end{aligned}$$

Substituting these values into Eqs. (13.155) and (13.156), we get the phugoid frequency and damping ratio to be the following:

$$\omega = 0.07627 \text{ rad/s},$$
$$\zeta = 0.04215\,,$$

which results in the time period of $T = \frac{2\pi}{\omega} = 82.38$ s and a settling time (Chapter 14) of $t_s = \frac{4}{\zeta\omega} = 1244$ s.

The short-period longitudinal mode for an airplane is represented by neglecting the variation in the forward speed ($u = \dot{u} = 0$) from a straight and level equilibrium flight, as well as the effects of planetary rotation and curvature. Furthermore, for an airplane the term involving oblate gravitation, g_δ^e, is usually ignored, resulting in a gravity-free dynamical model, given by

$$\frac{mv^e}{qS}(\dot{\alpha} - Q) = C_{z_\alpha}\alpha + \frac{\bar{c}}{2v^e}(C_{z_{\dot{\alpha}}}\dot{\alpha} + C_{z_q}Q), \tag{13.157}$$

$$\frac{J_{yy}}{qS\bar{c}}\dot{Q} = C_{m_\alpha}\alpha + \frac{\bar{c}}{2v^e}(C_{m_{\dot{\alpha}}}\dot{\alpha} + C_{m_q}\dot{\Theta}) \,. \tag{13.158}$$

Here, the equation for forward translation has been discarded. The short-period mode is thus a two-degree-of-freedom motion involving pitch and vertical translation (plunge). The predominant stability derivatives in the short-period mode are $C_{m_\alpha}, C_{z_\alpha}, C_{m_q}, C_{z_q}$. A state-space representation (Chapter 14) of the short-period dynamics is written as follows:

$$\begin{Bmatrix} \dot{\alpha} \\ \dot{\Theta} \\ \dot{Q} \end{Bmatrix} =$$

$$\begin{bmatrix} \frac{C_{z_\alpha}}{\Delta} & 0 & \frac{\frac{mv^e}{qS} + \frac{\bar{c}C_{z_q}}{2v^e}}{\Delta} \\ 0 & 0 & 1 \\ \frac{qS\bar{c}}{J_{yy}}\left(C_{m_\alpha} + \frac{\frac{\bar{c}}{2v^e}C_{m_{\dot{\alpha}}}C_{z_\alpha}}{\Delta}\right) & 0 & \frac{qS\bar{c}^2}{2v^e J_{yy}}\left(C_{m_q} + C_{m_{\dot{\alpha}}}\frac{\frac{mv^e}{qS} + \frac{\bar{c}C_{z_q}}{2v^e}}{\Delta}\right) \end{bmatrix}$$

$$\cdot \begin{Bmatrix} \alpha \\ \Theta \\ Q \end{Bmatrix}, \tag{13.159}$$

where

$$\Delta = \frac{mv^e}{qS} - \frac{\bar{c}C_{z_{\dot{\alpha}}}}{2v^e} \,.$$

Example 13.9. Find the longitudinal short-period characteristics (frequency and damping) of the jet transport airplane of Example 13.8.

The necessary computations are performed with the use of Eq. (13.159) and the following MATLAB statements (assuming all the necessary constants have been defined in the workspace):

```
>> D=m*v/(q*S)-c*Czad/(2*v);
>> A=[Cza/D 0 (m*v/(q*S)+c*Czq/(2*v))/D;
   0 0 1;
   (q*S*c/Jyy)*(Cma+c*Cmad*Cza/(2*v*D)) 0...
   (q*S*c^2/(2*v*Jyy))*(Cmq+Cmad*(m*v/(q*S)+c*Czq/(2*v))/D)]

A =      -0.3238 0    0.9938
               0  0         1.0000
         -1.1668 0        -0.4812

>> damp(A)
```

```
      Eigenvalue                   Damping       FrEq.~(rad/s)

 0.00e+000                       -1.00e+000      0.00e+000
-4.03e-001 + 1.07e+000i           3.51e-001      1.15e+000
-4.03e-001 - 1.07e+000i           3.51e-001      1.15e+000
```

Thus, the short-period natural frequency and damping-ratio are given by

$$\omega = 1.15 \text{ rad/s},$$
$$\zeta = 0.351 \ ,$$

which results in the time period of $T = \frac{2\pi}{\omega} = 5.46$ s and a settling time of $t_s = \frac{4}{\zeta\omega} = 9.91$ s.

The analysis presented above is that of the *controls-fixed* case, where no activation of the aerodynamic control takes place. This condition is rarely met in practice as the control surfaces are not rigidly attached to the vehicle, and thus undergo some deflection whenever the relative flow changes due to the vehicle's motion. The aerodynamic control surface for longitudinal dynamics is the *elevator*. The elevator can take various forms, such as the conventional trailing-edge device on the horizontal tail, the all-moving stabilizer, and the *elevons* of a tail-less aircraft. The vertical force and pitching moment contributions of the elevator are assumed linear and are modeled by the stability derivatives $C_{z_\delta}, C_{m_\delta}$. Often it is necessary to model the actuating mechanism of the elevator using a second-order dynamical system called the *actuator*. In addition, it is also necessary to model the airplane's motion with a *free* elevator, where the surface is completely free to assume an equilibrium position depending upon the angle of attack it experiences. Needless to say, the ideal controls-free condition is rarely met in practise due to inertia, friction, and stiffness of the acuating mechanism. Since the controls-free condition is also the case of zero control force (or moment) exerted by the actuating mechanism, it is desirable to control the free-elevator deflection, often using a smaller trailing-edge surface called a *trim tab*. A pilot uses a trim tab whenever small pitch adjustments are required, and is very useful in relieving the control force. Hence, a trim tab is somewhat like the momentum wheel in a spacecraft, where small attitude changes can be made without direct external force of the rocket thrusters. A variation of the trim tab is the *servo-tab*, wherein the deflection of an all-moving stabilizing surface is controlled by a gearing mechanism.

13.8.5 Lateral Dynamics

The lateral dynamic equations—Eqs. (13.127), (13.130), and (13.132)—involve three degrees of freedom, namely translation along oy (*sideslip*), and rotation about ox (*roll*) and oz (*yaw*). In a nondimensional form, with no longitudinal coupling ($\phi = \alpha = \epsilon = \Theta = Q = v = 0$), these equations can be written as follows:

$$
\begin{aligned}
m(v^e\dot{\beta} &- \frac{v^{e2}}{r^e}\cos^2\Theta^e \tan\delta^e \cos\Psi^e\beta \\
&-2\Omega v^e\beta\cos\delta^e\sin\Theta^e\sin\Psi^e + v^e R) \\
&= mg_c^e\Phi\cos\Phi^e\cos\Theta^e + mg_\delta^e(\Phi\cos\Phi^e\sin\Theta^e\cos\Psi^e \qquad (13.160) \\
&-\Psi\sin\Phi^e\sin\Theta^e\sin\Psi^e + \Phi\sin\Phi^e\sin\Psi^e - \Psi\cos\Phi^e\cos\Psi^e) \\
&+qS[C_{y_\beta}\beta + \frac{b}{2v^e}(C_{y_{\dot\beta}}\dot\beta + C_{y_p}P + C_{y_r}R)].
\end{aligned}
$$

$$
\begin{aligned}
J_{xx}\dot{P} &- J_{xz}\dot{R} + (J_{zz} - J_{yy})RQ^e \\
&= qSb[C_{l_\beta}\beta + \frac{b}{2v^e}(C_{l_{\dot\beta}}\dot\beta + C_{l_p}P + C_{l_r}R)]\,. \qquad (13.161)
\end{aligned}
$$

$$
\begin{aligned}
J_{zz}\dot{R} &- J_{xz}\dot{P} + J_{xz}RQ^e + (J_{yy} - J_{xx})PQ^e \\
&= qSb[C_{n_\beta}\beta + \frac{b}{2v^e}(C_{n_{\dot\beta}}\dot\beta + C_{n_p}P + C_{n_r}R)]\,. \qquad (13.162)
\end{aligned}
$$

The lateral motion thus consists of a translational (sideslip) and two rotational (roll and yaw) degrees of freedom. The lateral stability derivatives are defined in the same manner as that of the longitudinal derivatives, except that the characteristic length is the wing span, b. The derivatives C_{y_β} (side-force due to steady sideslip), C_{n_β} (static directional stability), C_{l_β} (dihedral effect), C_{n_r} (damping in yaw), and C_{l_p} (damping in roll) are the predominant lateral stability derivatives. While the sideslip and yaw rate derivatives are influenced by the fuselage and the vertical tail (fin), the roll-rate derivatives are mainly due to the wing. The static directional stability, C_{n_β}, determines the ability of the aircraft to regain its equilibrium heading, Ψ^e, once displaced by a sideslip. Also known as *weathercock stability*, $C_{n_\beta} > 0$ is required for directional stability and increases with the nondimensional product (called *fin volume ratio*) of the fin arm, l_v, and fin area, S_v. While the predominant contribution to directional stability comes from the fuselage, fin, and nacelles, a swept wing can have a significant stabilizing influence. The dihedral effect, C_{l_β}, is caused by the effective dihedral angle of the wing, as well as the lift produced by the fin. Due to the positive dihedral angle, a negative value of C_{l_β} is created, which turns the aircraft by banking it in the direction opposite to the sideslip. The damping in yaw is caused by the change in the fin's angle of attack due to a yaw rate, which tends to apply an opposite yawing moment ($C_{n_r} < 0$). The damping in roll is due to the change in the wing angle of attack due to a roll rate and is such that an opposing rolling moment is created ($C_{l_p} < 0$). The derivative C_{n_p} is due to the same effect as $C_{l_p}0$, which causes a differential drag on the two wings, thereby generating an *adverse yawing moment* ($C_{n_p} < 0$). The other roll-yaw coupling derivative is C_{l_r}, which is due to a differential lift on the two wings due to a yaw rate. The other rate derivatives ($C_{y_r}, C_{y_p}, C_{y_{\dot\beta}}, C_{n_{\dot\beta}}$) are usually negligible and are ignored in a stability analysis.

The *rudder* is the aerodynamic control surface for yaw and sideslip, while the rolling motion is controlled by the *ailerons*. The rudder is a trailing-edge control surface on the fin and acts like the elevator, while the ailerons are mounted on the wing trailing edges and are differentially deployed to create a rolling moment. The control derivatives, $C_{n_{\delta_a}}, C_{l_{\delta_a}}, C_{n_{\delta_r}}, C_{l_{\delta_r}}, C_{y_{\delta_r}}$, model the linear effects of deploying the aileron and rudder by the respective deflection angles, δ_a and δ_r.

All lateral stability derivatives are strong functions of the Mach number, and some of them can change sign while crossing through the transonic regime. This is especially true for the yawing moment derivatives. Due to reduced static directional stability and yaw damping at supersonic speeds, most supersonic aircraft require a yaw stability augmentation system with a closed-loop activation of rudder.

The kinematic attitude relations for the small-disturbance lateral motion can be expressed in terms of the 3-2-1 Euler angles as follows (Chapter 2):

$$\dot{\Phi} = P + R \tan \Theta^e, \tag{13.163}$$

$$\dot{\Psi} = R \sec \Theta^e. \tag{13.164}$$

Clearly (as pointed out in Chapter 2), $\Theta^e = \pm 90°$ is a point of singularity for this attitude representation.

Example 13.10. Simulate the lateral response of an axisymmetric, ballistic reentry vehicle with the following parameters to be an initial sideslip and roll disturbance of $\beta = 0.01$ rad and $P = -0.001$ rad/s: $m = 92$ kg, $J_{xx} = 0.972\,\text{kg.m}^2$, $J_{yy} = J_{zz} = 9.32\,\text{kg.m}^2$, base radius, $b = 0.22$ m, and base area, $S = 0.152\ \text{m}^2$. The stability derivatives of the vehicle based upon the base area and base radius at the equilibrium flight condition of $v^e = 5000$ m/s, $h^e = 25$ km, $\delta^e = 45°$, $\Psi^e = 190°$, and $\Theta^e = -85°$ are the following:

$$\begin{aligned}
C_{m_\alpha} &= -C_{n_\beta} = -0.52/\text{rad}, \\
C_{m_q} &= C_{n_r} = -8/\text{rad}, \\
C_{z_\alpha} &= C_{y_\beta} = -2.15/\text{rad}, \\
C_{z_q} &= C_{y_r} = -0.35/\text{rad}, \\
C_{x_u} &= -2C_D = -0.2, \\
C_{l_p} &= -0.002/\text{rad}, \\
C_D &= 0.1.
\end{aligned}$$

The other stability derivatives vanish due to the absence of lifting surfaces and the hypersonic speed.

In order for us to perform the simulation, a program named *lateralentry.m*, which is tabulated in Table 13.10, is written to integrate the equations of lateral motion with the intrinsic MATLAB Runge–Kutta solver *ode45.m.* The results are plotted in Figs. 13.25 and 13.26. The sideslip, roll, and yaw angles, as well as the yaw rate, are seen to oscillate in Fig. 13.25 with a decreasing amplitude for the 2 s of simulation, while the roll rate remains nearly constant in the given duration. The decay of roll rate is much slower than the yaw

Table 13.10. M-file *lateralentry.m* for Lateral State Equations of a Re-entry Vehicle

```
function xdot= lateralentry(t,x)
%(c) 2006 Ashish Tewari
global dtr; global mu; global S; global b; global m; global rm; global omega;
global Jxx; global Jyy; global Cnb; global Cyb; global Cnr; global Cyr; global Clp;
global CD; global v0; global Th0; global Psi0; global lat0;
%acceleration due to gravity (oblate earth)
[g,gn]=gravity(x(1),x(2));
chi = Psi0;
cfpa=cos(Th0);sfpa=sin(Th0);
cchi = cos(chi); schi = sin(chi);
cla=cos(lat0);sla=sin(lat0);
%%%atmospheric properties and flow parameters
if x(1)<rm
    x(1)=rm;
end
alt = x(1) - rm;
v  = x(2);
atmosp = atmosphere(alt, v, b);
rho = atmosp(2);
q = 0.5*rho*v^2;
mach = atmosp(3);
[t alt v mach]
Xfo=-q*S*CD;
%state-equations:
raddot = v*sfpa;
veldot=-g*sfpa +gn*cchi*cfpa + Xfo/m...
            +omega*omega*x(1)*cla*(sfpa*cla-cfpa*cchi*sla);
betadot=-x(7)+v*cfpa^2*cchi*tan(lat0)*x(3)/x(1)...
            +2*omega*cla*sfpa*schi*x(3)...
            +g*x(4)*cfpa/v+gn*(x(4)*sfpa*cchi-x(6)*cchi)...
            +q*S*(Cyb*x(3)+b*Cyr*x(7)/(2*v))/(m*v);
phidot=x(5)+x(7)*tan(Th0);
Pdot=q*S*b^2*Clp*x(5)/(2*Jxx*v);
psidot=x(7)/cfpa;
Rdot=q*S*b*(Cnb*x(3)+b*Cnr*x(7)/(2*v))/Jyy;
xdot=[raddot; veldot; betadot; phidot; Pdot; psidot; Rdot];
```

rate due to a much smaller damping factor $\left(\frac{|C_{l_p}|}{J_{xx}} \ll \frac{|C_{n_r}|}{J_{zz}}\right)$. In the 2 s of simulated flight, the altitude drops by nearly 10 km while the speed decays to approximately 4650 m/s. The Mach number stays in the hypersonic range, which implies that the assumption of constant-stability derivatives in the given time is reasonable.

Clearly, the axisymmetric re-entry vehicle of Example 13.10 displays two distinct modes: a short-period dynamics involving β, Φ, Ψ, R, and a long-period pure-rolling mode. The timescales of the two motions depend on the respective moments of inertia and damping derivatives. For an airplane, the presence of wings and stabilizing surfaces increases the roll damping, and gives rise to other stability derivatives (such as the dihedral effect), which result in distinct lateral modes unique to an airplane.

13.8.6 Airplane Lateral Modes

As in the longitudinal dynamics, the lateral motion of an airplane can be represented by a combination of distinct modes, each of which is obtained as

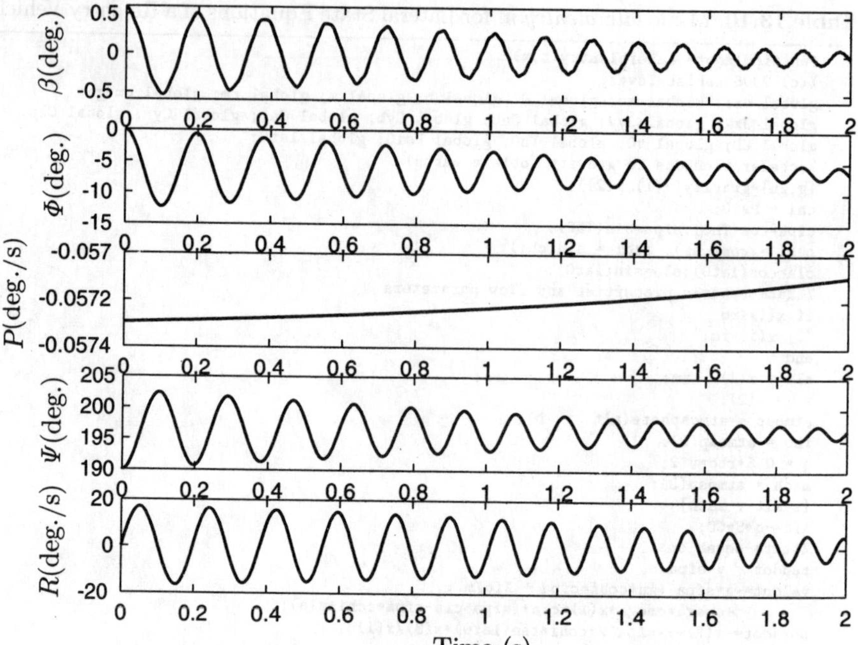

Fig. 13.25. Lateral dynamic response of a re-entry vehicle.

an approximation of the actual equations of motion. The common airplane equilibrium of straight and level flight ($\Theta^e = Q^e = a^e_{xv} = a^e_{yv} = a^e_{zv} = \Phi^e = P^e = R^e = 0$) is the starting point for the lateral modes, the simplest of which is the pure rolling motion created by an aileron input, called the *roll-subsidence mode*, and described by the following rolling moment equation:

$$\frac{J_{xx}}{qSb}\dot{P} = \frac{b}{2v^e}C_{l_p}P \, , \tag{13.165}$$

where $P = \dot{\Phi}$. Clearly, the mode is a first-order dynamical system with an exponentially decaying response,

$$P(t) = P(0)e^{\frac{qSb}{J_{xx}}\frac{b}{2v^e}C_{l_p}t} \, . \tag{13.166}$$

For a given speed and altitude, the rapidity with which the roll rate goes to zero is primarily dependent upon the ratio $\frac{Sb^2C_{l_p}}{J_{xx}}$. However, since J_{xx} is roughly proportional to the square of the wing span, the rate of decay of roll rate is primarily determined by the magnitude of C_{l_p}. The bank angle can be obtained by integrating Eq. (13.166) as follows:

$$\Phi(t) = \int P(t)\mathrm{d}t = \Phi(0) + P(0)\frac{2v^e J_{xx}}{qSb^2C_{l_p}}e^{\frac{qSb}{J_{xx}}\frac{b}{2v^e}C_{l_p}t} \, . \tag{13.167}$$

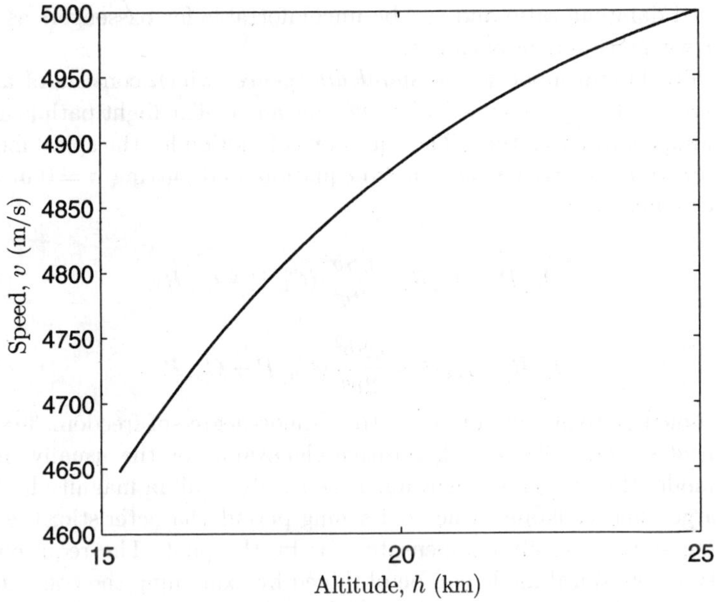

Fig. 13.26. Speed vs. altitude of a re-entry vehicle undergoing lateral motion.

Another lateral approximation is the short-period *Dutch-roll* mode, where the sideslip and yaw are coupled. Such a motion is normally generated by a rudder input. By assuming a negligible rolling motion, the airplane's attitude remains nearly wings' level, and we have $\beta \approx -\Psi$. Hence, the rolling moment and sideslip equations are discarded, and the Dutch-roll dynamics is given by the following yawing moment equation:

$$\frac{J_{zz}}{qSb}\ddot{\beta} - \frac{b}{2v^e} C_{n_r}\dot{\beta} + C_{n_\beta}\beta = 0. \tag{13.168}$$

Clearly, the frequency and damping of the Dutch-roll mode are the following:

$$\omega = \sqrt{\frac{qSbC_{n_\beta}}{J_{zz}}}, \tag{13.169}$$

$$\zeta = -C_{n_r}\sqrt{\frac{qSb^3}{32J_{zz}C_{n_\beta}}}\,. \tag{13.170}$$

Thus, the damping in the Dutch-roll is directly proportional to C_{n_r}, while its frequency is determined by the ratio $\frac{C_{n_\beta}}{J_{zz}}$. The assumption of negligible roll in the traditional Dutch-roll approximation is valid only if the dihedral effect, C_{l_β}, is reasonably small in magnitude. For an airplane with a significantly large magnitude of C_{l_β}, the Dutch-roll mode includes a distinct rolling motion with a reduced damping ratio and can be uncomfortable for passengers as well as bad for weapons-aiming purposes.

A third lateral mode is the *spiral divergence*, which consists of an ever-increasing bank angle, coupled with the yaw angle. The flight path is a slowly steepening coordinated turn. The equations of motion for the spiral mode can be obtained by neglecting the sideslip equation, and putting $\beta = 0$ in the roll and yaw equations:

$$J_{xx}\dot{P} - J_{xz}\dot{R} = \frac{qSb^2}{2v^e}(C_{l_p}P + C_{l_r}R), \tag{13.171}$$

$$J_{zz}\dot{R} - J_{xz}\dot{P} = \frac{qSb^2}{2v^e}(C_{n_p}P + C_{n_r}R) . \tag{13.172}$$

The coupled roll-yaw motion leads to a single-degree-of-freedom, first-order dynamical system with a real, positive eigenvalue for the usually unstable spiral mode. However, the eigenvalue is generally small in magnitude, leading to a large time constant. Due to its long-period characteristic, the slowly diverging spiral is easily compensated for by the pilot. The requirement of stability in the spiral mode can be obtained by examining the constant term in the lateral characteristic equation [45],

$$C_{l_\beta}C_{n_r} - C_{n_\beta}C_{l_r} , \tag{13.173}$$

which must be positive for the real root to lie in the left-half Laplace plane (stable spiral mode). Hence, for spiral stability we require $C_{l_\beta}C_{n_r} > C_{n_\beta}C_{l_r}$. The yawing moment derivatives, C_{n_r}, C_{n_β} are similarly affected by the size of the fin, hence increasing one also results in the increase of the other. The cross-derivative C_{l_r} is primarily dependent upon the lift coefficient, and thus cannot be arbitrarily selected at a given speed–altitude combination. This leaves only the dihedral effect, C_{l_β}, as the design parameter, which can be selected through a proper wing dihedral angle. However, increasing the magnitude of the dihedral effect—while leading to a greater spiral stability—causes the rolling moment in the Dutch-roll motion to become large, resulting in a reduced damping in the coupled roll-yaw-sideslip dynamics. Since damping of the short-period Dutch-roll motion must remain adequate in most airplanes, a small amount of long-period spiral instability is accepted as a compromise in the design.

13.8.7 Rotational Motion of a Launch Vehicle

The atmospheric trajectory of a launch vehicle (or a ballistic missile) may involve an appreciable aerodynamic torque due to the presence of stabilizing fins. As discussed in Chapter 12, it is crucial for a launch vehicle to be maintained at a zero angle of attack and sideslip due to aerodynamic load considerations. The attitude control of launch vehicles primarily involves control torque, $\mathbf{M_c} = M_{cx}\mathbf{i} + M_{cy}\mathbf{j} + M_{cz}\mathbf{k}$, produced by thrust vectoring. Euler's equations of motion for a launch vehicle with pitch-yaw symmetry ($J_{yy} = J_{zz}$) can thus be written as follows:

$$
\begin{aligned}
J_{xx}\dot{P} &= M_{cx} + qSb\frac{b}{2v}C_{l_p}, \qquad (13.174)\\
J_{yy}\dot{Q} + PR(J_{xx} - J_{yy}) &= M_{cy} + qSb[\frac{b}{2v}(C_{m_q}Q + C_{m_{\dot{\alpha}}}\dot{\alpha}) + C_{m_\alpha}\alpha],\\
J_{yy}\dot{R} + PQ(J_{yy} - J_{xx}) &= M_{cz} + qSb[\frac{b}{2v}(C_{n_r}R + C_{n_{\dot{\beta}}}\dot{\beta}) + C_{n_\beta}\beta]\ ,
\end{aligned}
$$

where b refers to the maximum fin span, or the maximum body diameter in case of a vehicle without fins. Due to axisymmetry, the stability derivatives due to pitch and yaw are indistinguishable from one another. Thus, we have $C_{z_\alpha} = C_{y_\beta}$, $C_{m_q} = C_{n_r}$, $C_{m_\alpha} = -C_{n_\beta}$, and $C_{m_{\dot{\alpha}}} = -C_{n_{\dot{\beta}}}$. For most launch vehicles and ballistic missiles, the stability derivatives representing aerodynamic lag in lift and side force are negligible, because of the small size of the lifting surfaces. Thus, we can assume $C_{z_{\dot{\alpha}}} = C_{y_{\dot{\beta}}} \approx 0$. Generally, the roll rate, P, is quite small; hence, no appreciable force is created by the *Magnus effect* [22]. The control torque components M_{cy} and M_{cz} are created by the small thrust deflection angles, ϵ and μ (Fig. 13.27),

$$
\begin{aligned}
M_{cy} &= l_x f_T \epsilon,\\
M_{cz} &= -l_x f_T \mu\ , \qquad (13.175)
\end{aligned}
$$

where l_x is the longitudinal distance of the nozzle from the center of mass, and f_T denotes the thrust. The rolling control torque, M_{cx}, is generated aerodynamically through control surface deflection, δ, and is given by the linear relationship

$$M_{cx} = C_{l_\delta}\delta\ . \qquad (13.176)$$

The control by thrust vectoring produces undesirable lift and side force if there are no opposing aerodynamic force components generated by the body and fins. This is the case during initial lift-off when the airspeed is too small for the opposing aerodynamic force to be created. In such a situation, the lateral support is provided to the vehicle either by the launch tower or by rocket thrusters near the nose of the vehicle. We shall make the assumption that lateral translation of the vehicle is prevented by such a mechanism; therefore, it is not necessary to consider the translational motion when the aerodynamic force and moment are negligible. As the speed increases within a few seconds after launch, the aerodynamic force and moment become sufficiently large for the vehicle to be treated in a manner similar to an aircraft, but with additional pitch-yaw symmetry. It is to be noted that due to the continuously active attitude stabilization system, the angle of attack and sideslip are kept small; thus, the assumption of small disturbances is more valid for the launch

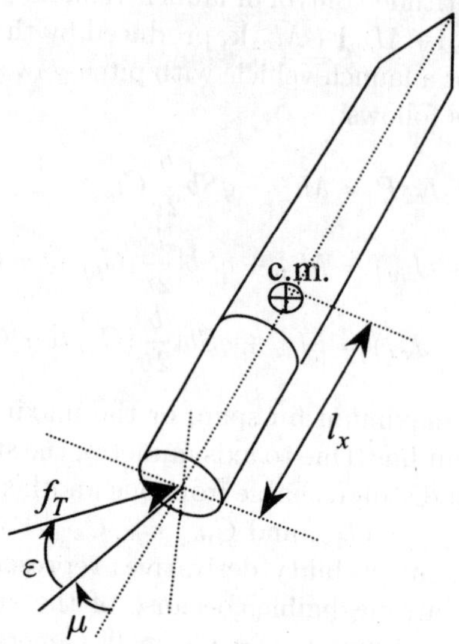

Fig. 13.27. Thrust deflection angles, ϵ and μ.

vehicle than the airplane. Hence, we can confidently utilize the results of the linearized longitudinal translation model presented earlier and write the vehicle's angle of attack (and sideslip) dynamical equations by the short-period approximation relative to a spherical gravity model as follows:

$$\frac{mv}{qS}\dot{\alpha} = \frac{mv}{qS}Q - \Theta\frac{mg}{qS}\sin\Theta^e + C_{z_\alpha}\alpha + \frac{f_T\epsilon}{qS} \tag{13.177}$$

and

$$\frac{mv}{qS}\dot{\beta} = -\frac{mv}{qS}R + \frac{mg}{qS}\Psi\sin\Theta^e + C_{y_\beta}\beta + \frac{f_T\mu}{qS}\,. \tag{13.178}$$

Note that we have adopted the Euler angle representation, $(\Psi)_3, (\Theta)_2, (\Phi)_1$, for the vehicle's attitude relative to the local horizon, which is nonsingular as long as $\Theta^e \neq \pm 90°$. This is acceptable, because the vertical pitch angle occurs only at lift-off, which lies outside our domain of analysis (due to zero aerodynamic force and moment at that point). Furthermore, the angles of attack and sideslip are assumed negligible in comparison with the pitch and yaw angles in the gravity terms.

Example 13.11. Consider the *Vanguard* ballistic missile with the following parameters [46] at flight condition of maximum dynamic pressure, $q = 28,035\ \mathrm{N/m^2}$, which occurs 75 s after launch at relative speed, $v = 392$ m/s, standard altitude 11 km, and mass $m = 6513.2$ kg:

$$\begin{aligned}
b &= 1.1433 \text{ m}, \\
S &= 1.0262 \text{ m}^2, \\
l_x &= 8.2317 \text{ m}, \\
C_{l_p} &= C_{m_{\dot\alpha}} = C_{l_\delta} = 0, \\
\Theta^e &= 68.5^\circ, \\
C_{z_\alpha} &= -3.13 \text{ /rad}, \\
C_{m_\alpha} &= 11.27 \text{ /rad}, \\
f_T &= 133,202.86 \text{ N}, \\
J_{yy} &= 156,452.8 \text{ kg.m}^2, \\
\frac{b}{2v} C_{m_q} &= -0.321 \text{ s}.
\end{aligned} \tag{13.179}$$

Since the missile is not equipped with fins and aerodynamic control surfaces, we do not require roll control; therefore, the first Euler equation yields a missile spinning at a constant rate, which we can take to be zero without any loss of generality.[12] For $P = 0$, the dynamic and kinematic equations of motion of the missile can be represented by the following set of linear differential equations (*state equation* of Chapter 14) at the given flight point:

$$\begin{Bmatrix} \dot Q \\ \dot R \\ \dot\Theta \\ \dot\Psi \\ \dot\alpha \\ \dot\beta \end{Bmatrix} =$$

$$\begin{pmatrix}
\frac{qSb^2}{2vJ_{yy}} C_{m_q} & 0 & 0 & 0 & \frac{qSb}{J_{yy}} C_{m_\alpha} & 0 \\
0 & \frac{qSb^2}{2vJ_{yy}} C_{m_q} & 0 & 0 & 0 & -\frac{qSb}{J_{yy}} C_{m_\alpha} \\
1 & 0 & 0 & 0 & 0 & 0 \\
0 & \frac{1}{\cos\Theta^e} & 0 & 0 & 0 & 0 \\
1 & 0 & -\frac{g\sin\Theta^e}{v} & 0 & \frac{qS}{mv} C_{z_\alpha} & 0 \\
0 & -1 & 0 & \frac{g\sin\Theta^e}{v} & 0 & \frac{qS}{mv} C_{z_\alpha}
\end{pmatrix}
\begin{Bmatrix} Q \\ R \\ \theta \\ \psi \\ \alpha \\ \beta \end{Bmatrix}$$

$$+ \begin{pmatrix}
\frac{l_x f_T}{J_{yy}} & 0 \\
0 & -\frac{l_x f_T}{J_{yy}} \\
0 & 0 \\
0 & 0 \\
\frac{f_T}{mv} & 0 \\
0 & \frac{f_T}{mv}
\end{pmatrix}
\begin{Bmatrix} \epsilon \\ \mu \end{Bmatrix}.$$

[12] The axisymmetric missile without fins would remain nonrolling after being launched at a zero roll rate. Conversely, a rifle bullet continues rolling at a fixed rate after leaving the barrel.

Let us simulate the vehicle's response for 0.1 s due to a step change in both ϵ and μ by 1° at the given flight point using the Simulink block diagram shown in Fig. 13.28. The exponentially increasing response, Fig. 13.29, indicates an unstable equilibrium, which requires an automatic attitude stabilization system. The instability is caused by a positive value of $C_{m_\alpha}(= -C_{n_\beta})$, as well as the effect of gravity on the angle of attack (and sideslip) due to $\theta^e \neq 0$. If large fins are added near the aft part of the vehicle, it could be made statically stable like an airplane and some surface-to-air missiles. The response shown in Fig. 13.29 would differ appreciably from that of the actual vehicle, because we have not modeled structural flexibility and fuel-slosh dynamics.

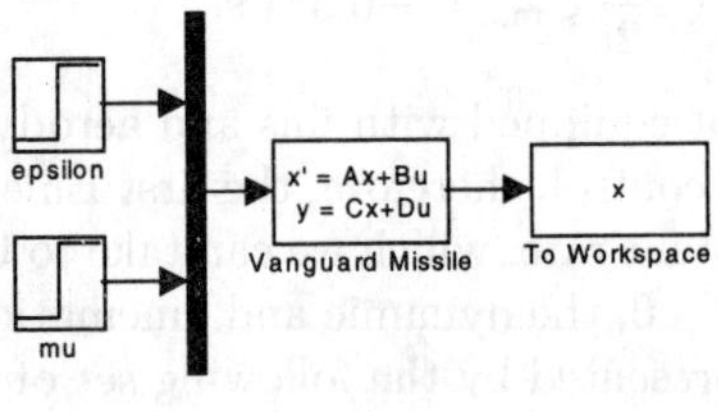

Fig. 13.28. Simulink diagram for step response of the Vanguard missile without attitude stabilization.

13.8.8 Inertia Coupled Dynamics

The separation of atmospheric rotational dynamics into longitudinal and lateral motions involves the assumption of small angular rate disturbances. For some vehicles, such as modern fighter airplanes and missiles, the concentration of mass in the fuselage results in a large difference between the rolling moment of inertia and the pitching (and yawing) inertia. Consequently, the nonlinear coupling terms—such as $(J_{xx} - J_{zz})PR$ in the pitching moment equation—become significant, thereby causing an interaction between the longitudinal and lateral dynamics for even moderate body rates. Inertial coupling has caused some fighter airplanes (e.g., the *Lockheed F-104*) to become unstable when rolling at high rates. While an approximate linear stability analysis is possible by considering the coupled longitudinal short-period and Dutch-roll dynamics [46], it is often necessary to simulate the complete six-degree-of-freedom dynamics of an inertia coupled vehicle, as discussed in Chapter 15. Controlling the inertia coupled dynamics usually requires a multivariable control system (Chapter 14). The case of aerodynamic missiles and artillery shells rolling at high rates is further aggravated by the aerodynamic coupling caused by the *Magnus effect* [22] and generally results in additional stability derivatives in a linearized stability analysis [45]. The aerodynamic behavior of a fighter aircraft rolling at a large angle of attack is essentially

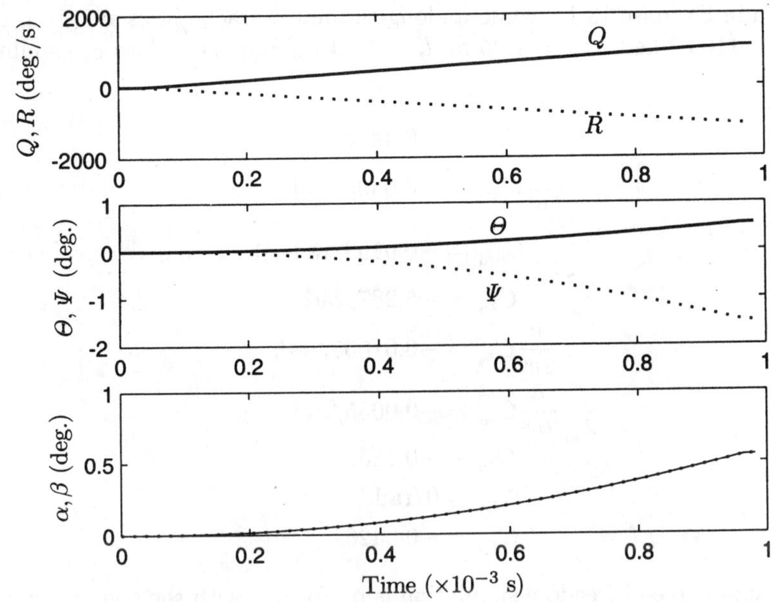

Fig. 13.29. Step response of the Vanguard missile without attitude stabilization.

nonlinear and leads to a complex motion (Chapter 15). Here, we neglect the aerodynamic coupling effects and confine our attention to the inertia coupled dynamics. Euler's equations of rotational dynamics of an inertia coupled vehicle, Eq. (13.19), can be expressed as follows:

$$\dot{P} = \frac{J_{xz}(J_{zz} + J_{xx} - J_{yy})PQ - [J_{xz}^2 + J_{zz}(J_{zz} - J_{yy})]QR + J_{xz}\mathcal{N} + J_{zz}\mathcal{L}}{J_{xx}J_{zz} - J_{xz}^2},$$

$$\dot{Q} = \frac{J_{xz}(R^2 - P^2) + (J_{zz} - J_{xx})PR + \mathcal{M}}{J_{yy}}, \tag{13.180}$$

$$\dot{R} = \frac{J_{xz}(\dot{P} - QR) + (J_{xx} - J_{yy})PQ + \mathcal{N}}{J_{zz}},$$

where $\mathcal{L}, \mathcal{M}, \mathcal{N}$ are the external torque components acting on the vehicle. In addition to the rotational dynamic equations, the linear equations of small-disturbance aerodynamic translation, u, α, β [Eqs. (13.126)–(13.128)], are required for propagating the aerodynamic force and moment in time. Finally, the attitude kinematics are represented by a suitable representation, such as 3-2-1 Euler angles, or the quaternion.

Example 13.12. Simulate the inertia-coupled response of the fighter airplane to an initial roll-rate disturbance of 0.5 rad/s when flying straight and level at standard sea level and $\delta = 45°$, with $A^e = 45°$ and Mach number, $M = 0.797$.

The lateral dynamic data of the airplane at the prescribed Mach number are given in Exercise 13.14, while its longitudinal characteristics are given by $J_{yy} = 36,110.67$ kg.m^2, $\bar{c} = 1.95$ m, $l_t = 5.64$ m, and the following stability derivatives:

$$
\begin{aligned}
C_{m_\alpha} &= -0.44/\text{rad}, \\
\frac{\bar{c}}{2v^e} C_{m_q} &= -0.0305/\text{rad}, \\
\frac{\bar{c}}{2v^e} C_{m_{\dot{\alpha}}} &= -0.0159/\text{rad}, \\
C_{z_\alpha} &= -5.287/\text{rad}, \\
\frac{\bar{c}}{2v^e} C_{z_q} &= -0.01055/\text{rad}, \\
\frac{\bar{c}}{2v^e} C_{z_{\dot{\alpha}}} &= -0.0055/\text{rad}, \\
C_{z_u} &= -0.185, \\
C_{x_\alpha} &= 0/\text{rad}, \\
C_{x_u} &= -0.0426.
\end{aligned}
$$

The six-degree-of-freedom simulation is performed with the coupled equations of motion encoded in *aircoupled.m* (Table 13.11). The response of the aircraft is plotted in Figs. 13.30–13.32. The observed response to the large roll rate can be broken into three distinct phases:

(a) $0 < t \leq 10$ s, during which the roll rate, yaw rate, and sideslip angle undergo a rapidly decaying oscillation. During this time, the speed, pitch angle, pitch rate, and angle of attack are unchanged, while the bank and yaw angles increase slowly with time.
(b) $10 < t \leq 170$ s, in which all variables except the roll rate undergo an unstable long-period oscillation and reach their maximum values near the end of the interval. This interval represents a diving attitude, with ever-increasing peak speed and increasingly negative angle of attack, which leads to a supersonic Mach number and more than double the equilibrium speed at $t = 110$ s. The attitude angles also build up in this phase, with the pitch angle reaching a maximum magnitude of $\Theta = -220°$ at $t = 108$ s, which is a case of inverted flight. By this time, a steep bank and dive are established. The main reason for the increase in the pitch rate during this interval is the small damping in pitch, C_{m_q}, at the given Mach number.
(c) $t > 170$ s, which sees the airplane trying to recover from the unusual pitch attitude in stable pitching oscillations that converge to a steady pitch angle of $\Theta = -90°$. In this phase, rolling and yawing motions increase exponentially, leading to an ever-steepening, downward spiral. This is the classical spiral divergence discussed above. There is a negligible variation in the speed and angle of attack in this phase, while the sideslip angle shows a steep rise in magnitude.

Table 13.11. M-file *aircoupled.m* for Six Degree-of-Freedom, Nonlinear State Equations of Aircraft

```
function xdot=aircoupled(t,x)
%program for inertia-coupled rotational dynamics and Euler 3-2-1 kinematics
%of aircraft and missiles
%x(1)=P, x(2)=Q, x(3)=R (angular velocity in rad/s)
%x(4)=phi, x(5)=theta, x(6)=psi (rad), x(7)=alpha (rad),
%x(8)=beta (rad), x(9)=u
%(c) 2006 Ashish Tewari
global dtr; global S; global c; global b; global lt; global m; global rm;
global omega; global Jxx; global Jyy; global Jzz; global Jxz; global v0;
global phi0; global A0; global Q0; global delta; global h0; global Cma;
global c2vCmad; global c2vCmq; global Cxu; global Cxa; global Czu;
global Cza; global c2vCzad; global c2vCzq; global Cyb; global Cnb;
global Clb; global Cnr; global Cnp; global Clp; global Clr; global Cyr;
%acceleration due to gravity (oblate earth):
[g,gn]=gravity(h0+rm,delta);
%atmospheric properties:
v  = v0*(1+x(9));
atmosp = atmo_sre(h0, v, c);
rho = atmosp(2);%density
q = 0.5*rho*v^2;%dynamic pressure
mach = atmosp(3);
CL=m*g/(q*S);
[t v mach]
%coupled attitude dynamics
udot = -Q0+q*S*(Cxu*x(9)+Cxa*x(7)...
     +x(5)*(-cos(phi0)*CL+sin(phi0)*sin(A0)*m*gn/(q*S)))/(m*v0);
alphadot = (2*x(9)*omega*cos(delta)*sin(A0)...
     +2*x(9)*v0/(rm+h0)+x(9)*Q0+x(2)+q*S*(Czu*x(9)+Cza*x(7)...
     +x(5)*(sin(phi0)*CL-cos(phi0)*cos(A0)*m*gn/(q*S))...
     +c2vCzq*x(2))/(m*v0))/(1-q*S*c2vCzad/(m*v0));
L=q*S*b*(Clb*x(8)+0.5*b*(Clp*x(1)+Clr*x(3))/v);
M=q*S*c*(Cma*x(7)+c2vCmq*x(2)+c2vCmad*alphadot);
N=q*S*b*(Cnb*x(8)+0.5*b*(Cnp*x(1)+Cnr*x(3))/v);
jxz=Jxx*Jzz-Jxz^2;
P=x(1);Q=x(2);R=x(3);
pdot=(Jxz*(Jzz+Jxx-Jyy)*P*Q-(Jxz^2+Jzz*(Jzz-Jyy))*Q*R+Jxz*N+Jzz*L)/jxz;
qdot=(Jxz*(R^2-P^2)+(Jzz-Jxx)*P*R+M)/Jyy;
rdot=(Jxz*(pdot-Q*R)+(Jxx-Jyy)*P*Q+N)/Jzz;
phidot=x(1)+(x(2)*sin(x(4))+x(3)*cos(x(4)))/cos(x(5));
thetadot=x(2)*cos(x(4))-x(3)*sin(x(4));
psidot=(x(2)*sin(x(4))+x(3)*cos(x(4)))/cos(x(5));
betadot=-x(3)+v*cos(phi0)^2*cos(A0)*tan(delta)*x(8)/(rm+h0)...
        +2*omega*cos(delta)*sin(phi0)*sin(A0)*x(8)...
        +g*x(4)*cos(phi0)/v+gn*(x(4)*sin(phi0)*cos(A0)-x(6)*cos(A0))...
        +q*S*(Cyb*x(8)+b*Cyr*x(3)/(2*v))/(m*v);
xdot=[pdot;qdot;rdot;phidot;thetadot;psidot;alphadot;betadot;udot];
```

It is important to note that the flow angles α, β remain small during this nonlinear simulation, thereby validating the assumption of linearized aerodynamics. Hence, the large body rates and angles are not caused by large aerodynamic disturbances, but rather by inertia coupling. The increase in the speed during the steep dive resulting from the high roll-rate disturbance leads to the airplane's entering the supersonic regime. We have not accounted for the variation of the stability derivatives at transonic and supersonic Mach numbers in this simulation, which may cause the aerodynamic torque to be appreciably modified in the last phase. A common condition in transonic

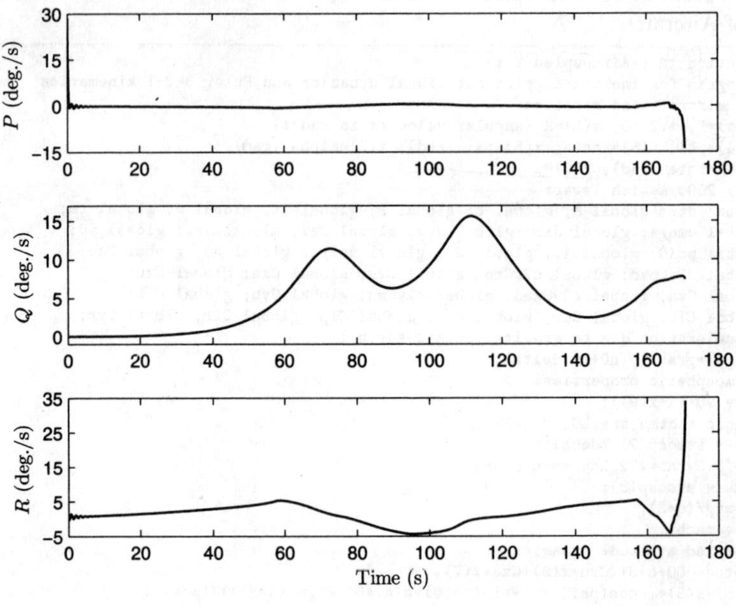

Fig. 13.30. Body-rate response of a fighter aircraft to a large initial roll-rate disturbance.

flight is the *tuck-under* phenomenon, wherein the aerodynamic center moves aft, and the damping derivatives diminish in magnitude as the Mach number increases.

Another situation where inertia coupling becomes important is when the angular momentum of rotors in aircraft engines is taken into account. In such situations, the modeling of inertia coupling is carried out in the same manner as that given above for spacecraft with rotors, by adding the constant angular momenta of the spinning rotors to Euler's equations.

13.9 Summary

Euler's equations of rotational motion govern the rotational dynamics of rigid bodies, and their solution gives the angular velocity at a given instant. Along with the kinematic equations, Euler's equations completely describe the changing attitude of a rigid body under the influence of a time-varying torque vector. When expressed in a body-fixed frame, Euler's equations involve constant moments and products of inertia. In a principal body-fixed frame, the products of inertia vanish, yielding a diagonal inertia tensor. Torque-free motion of rigid spacecraft is an example of conservative rotational maneuvers,

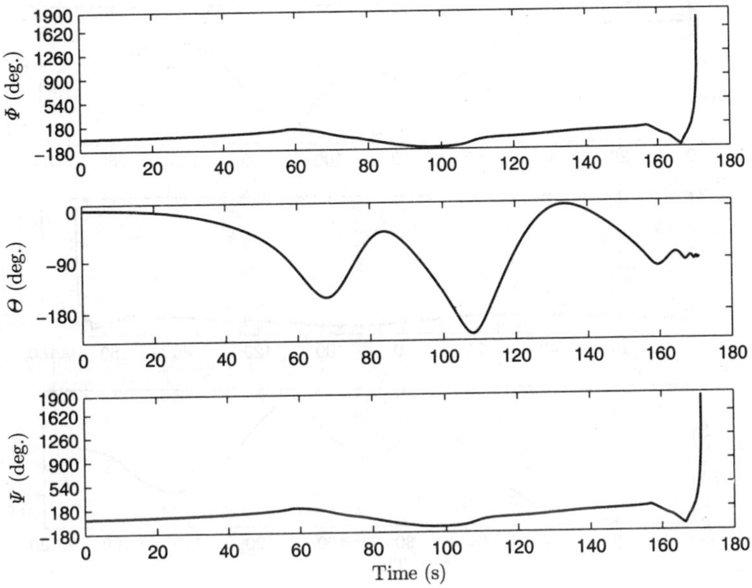

Fig. 13.31. Attitude response of a fighter aircraft to a large initial roll-rate disturbance.

wherein both angular momentum and rotational kinetic energy are conserved. While a rigid spacecraft's rotation about either the minor or the major axis is unconditionally stable, a semirigid spacecraft always tends toward the state of equilibrium with the lowest rotational kinetic energy—a pure spin about the major axis. Time-optimal maneuvers are an important open-loop method of controlling the spin and attitude of spacecraft and consist of at least a pair of suitably timed, equal and opposite torque impulses (bang-bang control). Other methods of stabilizing and controlling spacecraft's attitude motion are the use of rotors (dual-spin, reaction/momentum wheels, and control moment gyroscope), gravity gradient and magnetic torques. When considering the rotational dynamics within the atmosphere, Euler's equations are employed with the assumption of a rigid vehicle, and taking into account the aerodynamic torque generated by the rotation of the vehicle, as well as the control torque applied either by the pilot, or by an automatic control system. The airplane is the generic atmospheric flight vehicle for attitude motion models. The Coriolis acceleration terms due to planetary rotation and flight-path curvature are generally negligible in a rotational model, except for that of an atmospheric entry vehicle. The linearized aerodynamic model employed for airplane stability and control applications is based upon small flow perturbations from an equilibrium point. The small-disturbance approximation also results in the concept of linear stability derivatives, irrespective of the flow regime in which

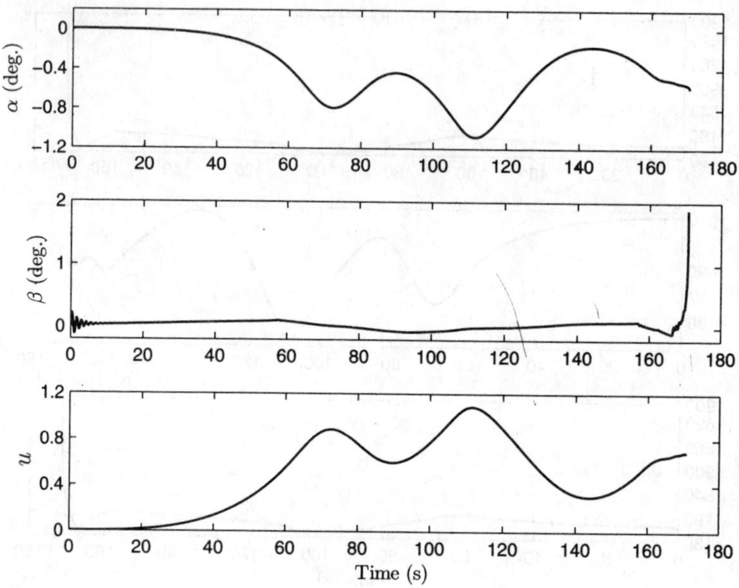

Fig. 13.32. Flow angle and speed response of a fighter aircraft to a large initial roll-rate disturbance.

the equilibrium point is located. Using the plane of symmetry existing in all atmospheric vehicles, one can separate the rotational motion in the plane of symmetry, called *longitudinal dynamics*, from that outside the plane referred to as *lateral dynamics.* A further assumption of de-coupled longitudinal and lateral modes enables a linearized stability analysis commonly applied to airplanes. The attitude control of non-aerodynamic missiles and launch vehicles primarily involves a control torque produced by thrust vectoring, generally leading to a statically unstable configuration. For modern fighter airplanes and missiles, the concentration of mass in the fuselage results in a large difference between the rolling moment of inertia and the pitching (and yawing) inertia. Consequently, the nonlinear, inertial coupling terms in Euler's equations become significant, thereby causing an interaction between the longitudinal and lateral dynamics for even moderate body rates. Hence, a complete six-degree-of-freedom modeling and simulation become necessary for stability and control analysis of inertia-coupled vehicles.

Exercises

13.1. Calculate the principal inertia tensor, $\mathsf{J_p}$, and the principal rotation matrix, $\mathsf{C_p}$, for a spacecraft with the following inertia tensor:

$$\mathsf{J} = \begin{pmatrix} 12500 & -1000 & 3500 \\ -1000 & 62500 & -500 \\ 3500 & -500 & 32000 \end{pmatrix} \text{ kg.m}^2.$$

Use the result to find the angular velocity in the principal frame if the angular velocity in the current body frame is $\boldsymbol{\omega} = (0.15, -0.25, -0.8)^T$ rad/s.

13.2. Using the kinematic equations, Eq. (13.40), show that the precession of an axisymmetric spacecraft obeys the relationship $\tan\phi = \frac{\omega_x}{\omega_y}$.

13.3. Write a program to carry out direct numerical simulation of a bang-bang, impulsive attitude maneuver of the spacecraft with the same moments of inertia as in Example 13.3, but with a spin rate of $n = 0.1$ rad/s, and realistic thruster torque impulses of magnitude 1000 N.m, each applied for $\Delta t = 0.01$ s. What is the time between the impulses and the final deviation of the spin axis?

13.4. Repeat Exercise 13.3 using a reaction wheel spinning about oy instead of the attitude thrusters. The moments of inertia of the wheel in the spacecraft principal axes are $J_{xx} = J_{zz} = 50$ kg.m^2 and $J_{yy} = 150$ kg.m^2. Consider the reaction wheel to be initially at rest relative to the spacecraft. What is the final spin rate of the wheel at the end of the maneuver?

13.5. It is desired to exactly null the final angular velocity of the spacecraft in Example 13.4 by using thruster torque impulses. Design a maneuver that achieves this using linearized Euler's equations with small angular velocity approximation. Determine the smallest number and magnitude of the torque impulses if thruster firing is limited to 0.01 s.

13.6. Carry out the attitude simulation of the VSCMG-equipped spacecraft in Example 13.5 using the modified Rodrigues' parameters (MRP) defined in Chapter 2, and compare the principal rotation angle with that plotted in Fig. 13.16.

13.7. Derive the governing equations of motion for a rigid, asymmetric spacecraft equipped with two reaction wheels, having their spin axes along the major and minor axes of the spacecraft, respectively. Modify the program *spacevscmg.m* to simulate the response of the spacecraft in Example 13.2 with the two reaction wheels of equal moment of inertia of 10 kg.m^2 about their spin axes. Assume that the wheels are initially at rest relative to the spacecraft, whose initial attitude and angular velocity are specified in Example 13.2. At time $t = 0$, a torque of 10 N.m begins acting on each wheel about the spin axis, and remains constant for a period of 10 s, after which it instantaneously drops to zero. Neglect friction in the reaction wheel bearings.

13.8. Write a program to simulate the response of an axisymmetric spacecraft platform with an oblate rotor in dual-spin configuration. The spacecraft has a moment of inertia of 1000 kg.m^2 about its spin axis, and 2000 kg.m^2 about a lateral principal axis. The rotor's moment of inertia about its spin axis is 250 kg.m^2 and 100 kg.m^2 about a lateral principal axis. The centers of mass of the platform and the rotor are offset from the center of mass of the dual-spin configuration by 0.5 m and 2.5 m, respectively. Initially, both the platform and the rotor are spinning in the same direction with angular speeds of 7.27×10^{-5} rad/s and 5 rad/s, respectively, when a lateral angular velocity disturbance of $\omega_{xy} = 0.01$ rad/s, is encountered. Neglect the friction in the rotor bearing.

13.9. Estimate the natural frequencies of gravity gradient motion of the *Seasat* spacecraft with the following characteristics:

$$J_{xx} = J_{yy} = 25,100 \text{ kg.m}^2,$$
$$J_{zz} = 3000 \text{ kg.m}^2,$$
$$n = 0.00105 \text{ rad/s}.$$

Simulate the coupled nonlinear response of the spacecraft to an initial yaw-rate disturbance of 10^{-5} rad.

13.10. Find the phugoid and short-period characteristics of the delta-winged fighter of Example 13.7 if the equilibrium flight path is the straight and level flight of Case (a). How do the settling times compare with those observed in the simulated response of Example 13.7?

13.11. Find the phugoid and short-period characteristics of the jet transport of Example 13.8 if the equilibrium flight path is a quasi-steady climb at standard sea level with $\delta^e = 45°$, $v^e = 150$ m/s, and $\phi^e = 30°$.

13.12. Simulate the longitudinal response of the re-entry vehicle of Example 13.10 to an initial angle of attack disturbance, $\alpha = 0.002$ rad. What are the phugoid and short-period characteristics of the vehicle?

13.13. Model the combined lateral and longitudinal dynamics of the re-entry vehicle of Example 13.10 using the quaternion instead of Euler angles. Use the model to repeat the simulation of Example 13.10 with the given initial condition, except that the initial pitch angle is $\Theta^e = -90°$. Will there be any change in the pitch angle during the resulting motion? Why?

13.14. The aerodynamic data for the *F-94* fighter [46] with $m = 6178.15$ kg, $J_{xx} = 15,004.5$ kg.m^2, $J_{xz} = 455.75$ kg.m^2, $J_{zz} = 50,066.26$ kg.m^2, $b = 11.37$ m, $S = 22.22$ m^2, $A^e = 0$, at Mach number $M = 0.797$ and standard sea level are the following:

$$C_{n_\beta} = 0.1/\text{rad},$$
$$C_{n_r} = -0.134/\text{rad},$$

$$\begin{aligned}
C_{y_\beta} &= -0.546/\text{rad}, \\
C_{y_r} &= 0.287/\text{rad}, \\
C_{l_p} &= -0.39, \\
C_{l_\beta} &= -0.0654/\text{rad}, \\
C_{l_r} &= 0.043, \\
C_L &= 0.0605 \,.
\end{aligned}$$

Simulate the lateral motion of the fighter to an initial sideslip of 0.01 rad from straight and level flight, and identify its lateral modes.

13.15. Repeat the simulation of the *Vanguard* missile dynamics of Example 13.11 using the quaternion for attitude representation.

14

Attitude Control Systems

14.1 Aims and Objectives

- To present modeling and simulation of closed-loop control systems for a large variety of aerospace applications based upon modern control concepts.
- To introduce linear systems theory.
- To provide examples of multivariable control systems applied to aircraft, spacecraft, and rockets.

14.2 Introduction

In the previous chapter we saw how the attitude of an atmospheric flight vehicle can influence its flight path. Maintaining a specific orientation, or changing the orientation with time in a specific manner, is also crucial for mission effectiveness of most spacecraft. Therefore, it is necessary to exercise some form of attitude control in all flight vehicles. In an unmanned vehicle, such control is possible only by an automatic mechanism. Even piloted vehicles routinely employ an automatic attitude control system in order to reduce the pilot work load. In the present chapter, we shall briefly discuss the modeling and simulation of attitude control systems.

Control is the name given to the task of achieving a desired result. The object to be controlled (a flight vehicle) is referred to as *plant*, while the process that exercises the control is called the *controller*. Both plant and controller are *systems*, defined as a self-contained set of physical processes under study. One could graphically depict a system by a box [Fig. 14.1(a)] connected by two arrows, one leading to, and the other away from, the box, called the *input* and *output*, respectively. Such a representation of the system is called a *block diagram*. The input and output each consists of several scalar variables and are mathematically represented as vectors. Figure 14.1(a) shows a system with m scalar input variables and p output variables. In modeling a system, one must account for the relationship between the input and output. This relationship generally takes the form of a differential equation in time if the system is governed by known physical laws, such as the laws of motion (Chapter 4). Such a system is said to be *deterministic*, whereas a system with unknown (or partially known) physical laws is called *nondeterministic*, or *stochastic*. All flight vehicles are designed as deterministic systems, even though some of their input variables may be governed by stochastic processes (such as wind gusts). Similarly, the flight vehicles' control systems are generally deterministic and thus can be modeled by differential equations in time. The condition, or *state*, of a system at a given time is defined by another set of scalar variables, called *state variables*. For example, a flight vehicle's attitude can be described by the quaternion, $\mathbf{q}(t), q_4(t)$, and the body-referenced angular velocity components, $\mathbf{omega}(t)$. Thus, the *state vector* of a flight vehicle's attitude dynamics is $\mathbf{x}(t) = \{\mathbf{q}(t), q_4(t), \mathbf{omega}(t)\}^T$. The vector space formed by the state vector, $\mathbf{x}(t)$, is called a *state space*. Clearly, the state space is not unique (as seen in Chapter 2 for the attitude kinematics), and any system can be described by an infinite number of different state-space representations. A system consisting of

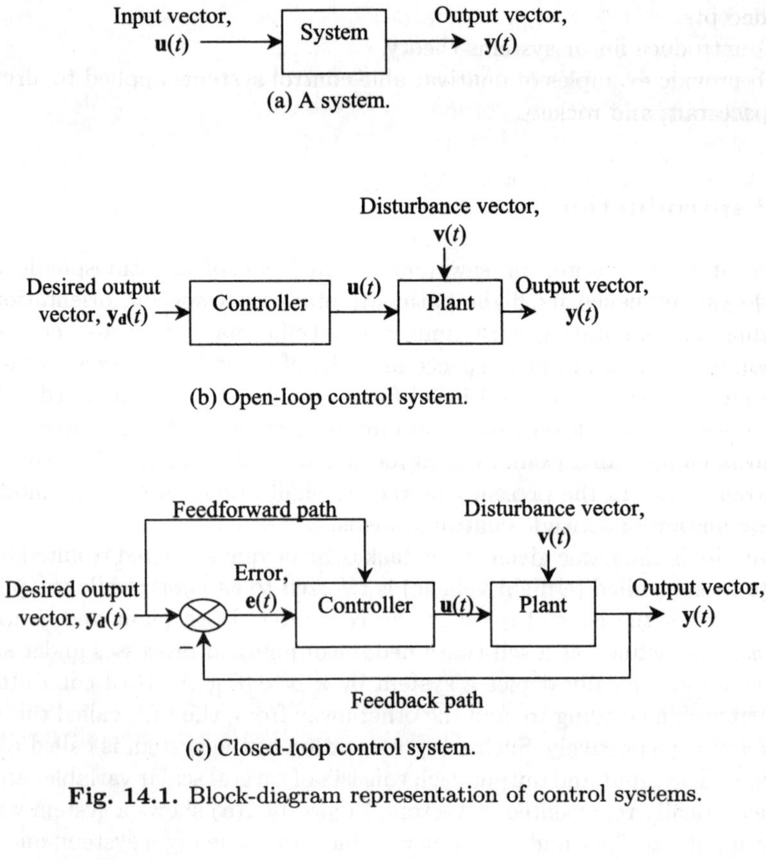

Fig. 14.1. Block-diagram representation of control systems.

the plant and the controller is called a *control system*. The controller exercises control over the plant through the *control input*, $\mathbf{u}(t)$, which is an input to the plant, but an output of the controller. In physical terms, this output can take the form of either a force or a torque (or both) applied to a flight vehicle. It is often necessary to model the process by which such a force, or torque, is generated, called an *actuator*. Generally, there are as many actuators as the number of scalar control inputs. The most common task of a control system is bringing the plant to a desired state, $\mathbf{x_d}(t)$, in the presence of unwanted input variables, $\mathbf{v}(t)$, called *disturbances*. This task can generally be achieved by two kinds of control systems: (a) *open loop*, or (b) *closed loop* [Fig. 14.1(b), (c)]. In an open-loop control system, the controller has no knowledge of the actual state of the plant, $\mathbf{x}(t)$, at a given time, and the control is exercised based upon a model of the plant dynamics as well as an estimate of its initial condition, $\mathbf{x}(0)$. Obviously, such a "blind" application of control would be successful in driving the plant to a desired state, only if the plant model is exact and the disturbances are properly accounted for. It is seldom possible to meet these requirements, and there are always some discrepancies in our estimates of the plant dynamics, initial condition, and the disturbances. Therefore, often a closed-loop control system is utilized, in which the actual state of the plant is provided to the controller through a *feedback* mechanism, so that the control input, $\mathbf{u}(t)$, can be appropriately adjusted. Practically, this feedback consists of either direct measurements of all state variables, which is rarely possible, or an indirect reconstruction of the plant's state $\mathbf{x}(t)$, through a measurement of the plant's output, $\mathbf{y}(t)$, as shown in Fig. 14.1(c). When the plant's output is fed back, the control system is called an *output feedback* system. When the state vector is fed back (either through direct measurement, or by indirect reconstruction through the output), the control system is called a *state feedback* system. Whenever a measurement of a variable is involved, it is necessary to model the dynamics of the measurement process, called the *sensor*. Generally, there are as many sensors as the number of measured scalar variables. In an output feedback system, the plant's output is continually compared with a *desired output*, $\mathbf{y_d}(t)$, as shown in Fig. 14.1(c). The round symbol with a cross in Fig. 14.1(c) represents a *summing junction*, where the various incoming variables are added with appropriate signs (a negative sign indicated near the arrow of a variable implies that the variable is to be subtracted from the other incoming variables), and the result of the arithmetic sum is the variable shown by the outgoing arrow. For example, the input to the controller from the summing junction in Fig. 14.1(c) is the *error* vector,[1]

$$\mathbf{e}(t) = \mathbf{y_d}(t) - \mathbf{y}(t) \,. \tag{14.1}$$

The closed-loop control system works by attempting to drive the error $\mathbf{e}(t)$ to zero over a reasonable time interval through the application of the control

[1] In a state feedback system, the error is defined as the difference between the desired state, $\mathbf{x_d}(t)$, and the plant's state, $\mathbf{x}(t)$.

input, $\mathbf{u}(t)$. Of course, $\mathbf{u}(t)$ should be made dependent upon $\mathbf{e}(t)$. Furthermore, $\mathbf{u}(t)$ may also depend explicitly upon the desired output (or desired state). Thus, it is also necessary to provide the controller with the desired output (or state) through a *feedforward path*, shown in Fig. 14.1(c).

In a modern control system, the actuators are operated through electrical signals as inputs, while the sensors produce electrical signals as their outputs. A controller must manipulate these electrical signals through either *analog* or *digital* processes, which basically regard the signals as continuous or discrete variables of time [43]. One has the option of clubbing the sensors and actuators with either the controller or the plant. Sometimes it may be convenient to model actuators and sensors as separate systems. In Fig. 14.1(b) and (c), we have adopted the convention of including sensors and actuators into the model for the plant. In a general control system, the controller represents mathematical relationships between the plant's output, $\mathbf{y}(t)$, the desired state, $\mathbf{x_d}(t)$, the control input, $\mathbf{u}(t)$, and time, t. These relationships are referred to as *control laws*, and their derivation is the main objective of *control theory*. We shall not attempt to explore control theory, but refer the reader to related textbooks, such as [43]. We shall assume that a control law has been appropriately derived to achieve the desired control system dynamics, and then model the control system accordingly.

The performance of a control system is assessed primarily by the time the plant's output takes (which is also the output of the control system) to come close to (within a specific percentage) the desired state, its deviations from the desired state (called *overshoots*) in the meantime, and the magnitudes of the control input required in the process. Thus, we analyze the behavior of the system at a large time, $t \to \infty$, called the *steady-state response*, as well as that at small values of time when large overshoots from the desired state could occur. A successful control system is the one in which the maximum overshoot is small and the time taken to reach within a small percent of the desired state is also reasonably small. Therefore, a control system is usually approximated by a linear differential equation, resulting from the assumption of small deviations from the desired state (which is regarded as the equilibrium of the linear system, as discussed in Chapter 13). A great simplification occurs by making such an assumption, because we can apply the principle of *linear superposition* (Chapter 13) to a linearized system, which allows a weighted addition of the system output vectors to individual input vectors, in order to yield the total output due to a linear combination of several input vectors. Linear superposition enables us to utilize operational calculus and linear algebraic methods for analysis. Most of the nonlinear systems can be linearized about equilibrium points, and a linear stability analysis would reveal the tendency of either remaining close to, or departing from, a given equilibrium point. We have already carried out linear stability analyses of some equilibrium points in Chapter 13. The same approach can now be extended to linear control systems.

14.3 Linear Systems

Linear system theory is a well-established art [43] and refers to the analysis of a system linearized about a particular equilibrium point. A system is said to be *linear* if its output to the input, $\mathbf{u}(t) = c_1\mathbf{u_1}(t) + c_2\mathbf{u_2}(t)$, is given by $\mathbf{y}(t) = c_1\mathbf{y_1}(t) + c_2\mathbf{y_2}(t)$, where $\mathbf{y_1}(t)$ and $\mathbf{y_2}(t)$ are the outputs of the system to $\mathbf{u_1}(t)$ and $\mathbf{u_2}(t)$, respectively, and c_1, c_2 are arbitrary scalar constants. By inspecting the governing differential equations of a system, it is possible to determine whether it is linear. If the governing differential equations do not contain products and transcendental functions of the state and input variables and their time derivatives, the system is linear. Consider a linear system described by a set of first-order, ordinary differential equations, written in the following matrix form:

$$\dot{\mathbf{x}}(t) = \mathsf{A}(t)\mathbf{x}(t) + \mathsf{B}(t)\mathbf{u}(t) , \tag{14.2}$$

with initial condition

$$\mathbf{x}(0) = \mathbf{x_0} . \tag{14.3}$$

Here, $\mathbf{x}(t)$ is a state vector of the linear system, while $\mathbf{u}(t)$ is the input vector. For the time being, we are ignoring the disturbance inputs to the system, which can be easily included through an additional term on the right-hand side. The set of first-order, ordinary differential equations, Eq. (14.2), is called the *state equation* of the system.[2] The dimension of the state vector is referred to as the *order of the system*. Any set of ordinary differential equations governing the system's behavior can be converted into a state equation by suitably defining the state vector. As pointed out above, a state equation and its initial condition, although nonunique, completely describe the behavior of a system at all time instants, $t \geq 0$. The coefficient matrices, $\mathsf{A}(t)$ and $\mathsf{B}(t)$, can be generally time-varying, and their dimensions depend upon the order of the system as well as the size of the input vector, $\mathbf{u}(t)$.

The general solution of Eq. (14.2), subject to the initial condition of Eq. (14.3), can be written as follows [43]:

$$\mathbf{x}(t) = \Phi(t, 0)\mathbf{x_0} + \int_0^t \Phi(t, \tau)\mathsf{B}(\tau)\mathbf{u}(\tau)\mathrm{d}\tau , \tag{14.4}$$

where $\Phi(t, t_0)$ is called the *state-transition matrix*. The state-transition matrix has the property of transforming the state at time t_0 to another time t if the *applied input is zero* ($\mathbf{u}(t) = \mathbf{0}$), according to

$$\mathbf{x}(t) = \Phi(t, t_0)\mathbf{x}(t_0) . \tag{14.5}$$

For $\mathbf{u}(t) = \mathbf{0}$, the state equation becomes the following *homogeneous* equation:

$$\dot{\mathbf{x}}(t) = \mathsf{A}(t)\mathbf{x}(t) . \tag{14.6}$$

---

2 We can describe nonlinear systems also by a state equation.

Hence, $\Phi(t,0)$ gives the solution to the homogeneous state equation, with the initial condition of Eq. (14.3). Some important properties of the state-transition matrix are the following:

$$\Phi(t,t_0) = -\Phi(t_0,t), \tag{14.7}$$

$$\Phi(t,t_0) = \Phi(t,t_1)\Phi(t_1,t_0), \tag{14.8}$$

$$\frac{\mathrm{d}\Phi(t,t_0)}{\mathrm{d}t} = \mathsf{A}(t)\Phi(t,t_0). \tag{14.9}$$

Equations (14.7)–(14.9) are said to represent the inverse, associative, and derivative properties of the state-transition matrix.

The derivation of the state-transition matrix is a formidable task for a general linear system with time-varying coefficient matrices (called a *time-varying, linear system*). Only in some special cases can the exact, closed-form expressions for $\Phi(t,t_0)$ be derived. Whenever $\Phi(t,t_0)$ cannot be obtained in a closed form, it is necessary to apply approximate numerical techniques for the solution of the state equation, as discussed in Appendix A.

14.3.1 Time-invariant, Linear Systems

Generally, the timescale of rotational dynamics is so small that most flight vehicles can be assumed to have nearly constant coefficient matrices of attitude dynamics at a given equilibrium point. In such a case, the system is approximated to be a *time-invariant, linear system*, with A, B treated as constant matrices. A time-invariant case is easily handled by writing

$$\Phi(t,t_0) = e^{\mathsf{A}(t-t_0)}, \tag{14.10}$$

where e^{M} is called the *matrix exponential* of a square matrix, M, and is defined by the following infinite series (similar to a scalar exponential):

$$e^{\mathsf{M}} \doteq \mathsf{I} + \mathsf{M} + \frac{1}{2}\mathsf{M}^2 + \ldots + \frac{1}{n!}\mathsf{M}^n + \ldots . \tag{14.11}$$

While the evaluation of $e^{\mathsf{A}(t-t_0)}$ by this infinite series would be impossible in practice, a suitable numerical approximation of Eq. (14.11) with only a finite number of terms can be utilized by breaking the time interval, $t - t_0$, into a number of small intervals. This forms the basis of numerical evaluation of the state-transition matrix for time-invariant, linear systems. However, for a small-order system, an exact, closed-form expression for $e^{\mathsf{A}(t-t_0)}$ can be obtained with the use of Laplace's transform.

Consider the Laplace transform of the state equation, Eq. (14.2), for a time-invariant system, subject to the initial condition of Eq. (14.3):

$$s\mathbf{X}(s) - \mathbf{x_0} = \mathsf{A}\mathbf{X}(s) + \mathsf{B}\mathbf{U}(s)\,, \tag{14.12}$$

where $\mathbf{X}(s)$ and $\mathbf{U}(s)$ are the Laplace transforms of $\mathbf{x}(t)$ and $\mathbf{u}(t)$, respectively.[3] In order to derive the state-transition matrix, we make the input vanish, which leads to

$$(s\mathsf{I} - \mathsf{A})\mathbf{X}(s) = \mathbf{x_0}\,, \tag{14.13}$$

or,

$$\mathbf{X}(s) = (s\mathsf{I} - \mathsf{A})^{-1}\mathbf{x_0}\,. \tag{14.14}$$

By taking the inverse Laplace transform of Eq. (14.14), we have

$$\mathbf{x}(t) = \mathcal{L}^{-1}(s\mathsf{I} - \mathsf{A})^{-1}\mathbf{x_0}\,, \tag{14.15}$$

which, upon comparison with Eq. (14.5), yields

$$e^{\mathsf{A}t} = \mathcal{L}^{-1}(s\mathsf{I} - \mathsf{A})^{-1}\,. \tag{14.16}$$

We have already utilized this expression of the state-transition matrix while analyzing the rotational dynamics of spacecraft in Chapter 13.

By substituting Eq. (14.10) into Eq. (14.4), we can write the following general expression for the state of a linear, time-invariant system in the presence of an arbitrary, Laplace transformable input, which begins to act at time $t = 0$ when the system's state was $\mathbf{x}(0) = \mathbf{x_0}$:

$$\mathbf{x}(t) = e^{\mathsf{A}t}\mathbf{x_0} + \int_0^t e^{\mathsf{A}(t-\tau)}\mathsf{B}(\tau)\mathbf{u}(\tau)\mathrm{d}\tau\,. \tag{14.17}$$

The first term on the right-hand side of Eq. (14.17) is called the *initial response* (or transient response), which decays to zero for an asymptotically stable system in the limit $t \to \infty$. However, in the same limit, the integral term may either converge to a finite value (called the *steady state*), or assume the same functional form as that of the input (called the *forced response*). The solution of the linear state equation given by Eq. (14.17) can be obtained by *discretization* of time into small intervals, and evaluating the initial response as well as the integral term by appropriate approximations, as discussed in [43]. Such an approach is programmed in the intrinsic MATLAB function *lsim*, which can also be invoked by a Simulink block diagram (as we will see in the examples discussed ahead).

14.3.2 Linear Stability Criteria

Apart from determining the state-transition matrix, the Laplace transform approach is also useful in a linear stability analysis. Equation (14.13) represents an *eigenvalue problem*, whose solution yields the *eigenvalues*, s, and *eigenvectors*, $\mathbf{X}(s)$. The eigenvalues of the linear system are obtained by the

---

[3] We assume that both $\mathbf{x}(t)$ and $\mathbf{u}(t)$ satisfy the conditions for the existence of a Laplace transform, i.e., they are piecewise-continuous functions bounded by an exponential.

following *characteristic equation*, which results from a nontrivial solution of Eq. (14.13):

$$\mid s\mathsf{I} - \mathsf{A} \mid = 0 \,. \tag{14.18}$$

Hence, the eigenvalues of the constant-coefficient matrix, A, are the roots of the characteristic equation, Eq. (14.18). Generally, the characteristic equation of a system of order n is expressed as a polynomial equation as follows:

$$\mid s\mathsf{I} - \mathsf{A} \mid = s^n + a_{n-1}s^{n-1} + a_{n-2}s^{n-2} + \ldots + a_1 s + a_0 = 0 \,, \tag{14.19}$$

where the *characteristic coefficients*, a_i, $i = 0 \ldots (n-1)$, are invariant with the choice of the state variables (i.e., they are unique parameters of the system). The n complex roots of the characteristic equation (eigenvalues of A) signify an important system property, called *stability*. There are many definitions of stability, but we will regard stability as the system's quality wherein the output to any bounded input is bounded, and the linear system has a tendency to remain close to its equilibrium point after a small, arbitrary initial deviation from it. From the expression of the system's state, Eq. (14.15), it can be deduced that variation of the system's state with time is given by terms containing e^{st}.[4] When an eigenvalue, s_k, is repeated p times, its contribution to the system's state involves terms of the form $t^i e^{s_k t}$, $i = 0 \ldots (p-1)$. Considering that an eigenvalue is generally complex, its imaginary part denotes the frequency of oscillation of the characteristic vector about the equilibrium point, and the real part signifies the growth (or decay) of its amplitude with time. We can now state the following criteria for the stability of a linear, time-invariant system:

(a) If all eigenvalues have negative real parts, the linear system is stable about its equilibrium point. Such a system is called *asymptotically stable*. A system having an eigenvalue with zero real parts (and the remaining eigenvalues with negative real parts) displays oscillatory behavior of a constant amplitude and is said to be *stable* (but not asymptotically stable). If at least one eigenvalue has a positive real part, its contribution to the system's state is an exponentially growing amplitude, and the system is said to be *unstable*.

(b) If a multiple eignevalue of multiplicity p has a zero real part, its contribution to the system's state has terms containing the factors t^i, $i = 0 \ldots (p-1)$, which signify an unbounded behavior with time. In such a case, the linear system would be unstable about its equilibrium. Since complex eigenvalues

[4] It is easily seen that a *characteristic vector*, $\mathbf{x}(t) = \mathbf{X_k} e^{s_k t}$, satisfies the homogeneous state equation, where $\mathbf{X_k}$ is the eigenvector corresponding to the eigenvalue, s_k. The general solution, which must also satisfy the initial condition, $\mathbf{x}(0) = \mathbf{x_0}$, is a combination of all n characteristic vectors. The decomposition of a system's state into the characteristic vectors (or *modes*) is an alternative way of computing the state-transition matrix [43].

occur in conjugate pairs, a repeated eigenvalue with zero real part must also have a zero imaginary part, i.e., $s = 0$.

The stability criteria can be applied to numerically determined eigenvalues from the coefficient matrix, A, or indirectly by inspection of the characteristic polynomial (*Routh–Hurwitz stability criteria* [48]). Since we would normally have a numerical procedure available to us via MATLAB, we can adopt the direct evaluation of the eigenvalues. Much of linear systems theory is traditionally devoted to stability analysis by analytical and graphical methods such as *Root–Locus* and *Nyquist* diagrams [43]. However, these classical methods, while imparting significant insight into the system's behavior, are largely limited to *single-input, single-output* (SISO) systems.

14.3.3 Transfer Matrix and Second-Order Systems

When evaluating a linear, time-invariant SISO system's response to an input, subject to zero initial condition, the concept of a *transfer function* is a valuable system representation. The transfer function, $G(s)$, is defined as the ratio of the scalar output's Laplace transform, $Y(s)$, to that of the scalar input, $U(s)$, subject to a *zero initial condition*,

$$G(s) \doteq \frac{Y(s)}{U(s)} . \tag{14.20}$$

We shall extend the transfer function concept to a multivariable system by defining a *transfer matrix*, $\mathsf{G}(s)$, as follows:

$$\mathbf{Y}(s) = \mathsf{G}(s)\mathbf{U}(s) , \tag{14.21}$$

which is also derived from the system's state equation with a zero initial condition. In order to obtain the transfer matrix, we need an *output equation* relating the system's output vector, $\mathbf{y}(t)$, to the state and input vectors as follows:

$$\mathbf{y}(t) = \mathsf{C}\mathbf{x}(t) + \mathsf{D}\mathbf{u}(t) , \tag{14.22}$$

where C, D are constant-coefficient matrices. If $\mathsf{D} = 0$, there is no direct relationship between the input and the output, and the system is said to be *strictly proper*. By taking the Laplace transform of Eq. (14.22) and substituting Eq. (14.12) with $\mathbf{x_0} = \mathbf{0}$, we have

$$\mathbf{Y}(s) = [\mathsf{C}(s\mathsf{I} - \mathsf{A})^{-1}\mathsf{B} + \mathsf{D}]\mathbf{U}(s) , \tag{14.23}$$

which yields the transfer matrix as

$$\mathsf{G}(s) = [\mathsf{C}(s\mathsf{I} - \mathsf{A})^{-1}\mathsf{B} + \mathsf{D}] . \tag{14.24}$$

From Eq. (14.24) it is clear that the characteristic polynomial is the common denominator $\mid s\mathsf{I} - \mathsf{A} \mid$ of the elements of $\mathsf{G}(s)$, which are rational functions in the Laplace variable, s. For this reason, the characteristic roots (eigenvalues) are also called the system's *poles*.

Apart from the transfer function (or matrix), a linear, time-invariant system is also represented by its outputs to some specialized inputs, subject to zero initial condition. Examples of such inputs are the *singularity functions* (unit *step*, *impulse*, and *ramp* functions), as well as the *purely oscillatory* (sine or cosine) inputs. The system's response to these inputs utilizes well-established linear system theory and enables a convenient classification of its behavior [43]. For example, the response to a unit step function (called *step*, or *indicial* response) is valuable in studying a stable linear control system's *performance* to a step change in the desired output. The time derivative of the step response is the *impulse response*, whose Laplace transform for the SISO system yields the transfer function. The time integral of the step response is the *ramp response*, which is useful in such applications as tracking an object moving with a constant speed. Apart from the singularity functions, some smooth test functions are also applied as inputs for analyzing an unknown system. The response to a simple harmonic input at a particular frequency (called *frequency-response*) reveals important control system properties such as the natural frequencies, and the behavior in the presence of high-frequency, unmodeled disturbances (*noise*). For a detailed information about the responses to special inputs, and their estimation using MATLAB/Simulink, please refer to [43].

Most sensors, actuators, and some SISO plants can be represented as linear, time-invariant, second-order systems by the generic differential equation

$$m\ddot{y}(t) + c\dot{y}(t) + ky(t) = u(t) \,, \tag{14.25}$$

where the constants m,c,k are referred to as *inertia*, *damping*, and *stiffness* parameters, respectively. The transfer function of the second-order system can be expressed as follows:

$$\frac{Y(s)}{U(s)} = \frac{1}{ms^2 + cs + k} \,, \tag{14.26}$$

or in the traditional form,

$$\frac{Y(s)}{U(s)} = \frac{1}{m(s^2 + 2\zeta\omega_n s + \omega_n^2)} \,, \tag{14.27}$$

where $\omega_n \doteq \sqrt{\frac{k}{m}}$ is called the *natural frequency*, and $\zeta \doteq \frac{c}{2m\omega_n}$ is the *damping ratio*. It is clear that the stability of the second-order system depends upon the roots of the characteristic equation $s^2 + 2\zeta\omega_n s + \omega_n^2 = 0$, which are written as

$$s_{1,2} = -\zeta\omega_n \pm \omega_n\sqrt{\zeta - 1} \,. \tag{14.28}$$

Applying the linear stability criteria, it is evident that the system is stable for $\zeta \geq 0$, and unstable for $\zeta < 0$. Furthermore, if $0 \geq \zeta < 1$, the eigenvalues are complex conjugates, and the unforced system's initial response displays an oscillatory behavior with time, while $\zeta > 1$ is the case of real, negative eigenvalues, representing a purely exponential behavior. Most second-order control systems are designed with $0 \geq \zeta < 1$, since the overdamped case of $\zeta \geq 1$ generally requires large control input magnitudes. The performance of a stable second-order control system is often analyzed by its indicial response [response to a unit step function, $U(s) = \frac{1}{s}$, applied at $t = 0$, with zero initial condition], given by

$$y(t) = \frac{1}{m\omega_n^2}\left[1 - e^{-\zeta\omega_n t}\left(\cos\omega_d t + \frac{\zeta}{\sqrt{1-\zeta^2}}\sin\omega_d t\right)\right] \quad (t > 0), \tag{14.29}$$

where $\omega_d \doteq \omega_n\sqrt{1-\zeta^2}$ is the *damped natural frequency* of the system. Figure 14.2 plots $m\omega_n^2 y(t)$ of Eq. (14.29) against the nondimensional time, $\frac{\omega_n t}{2\pi}$, for different values of ζ in the stable range. It is clear that as ζ is increased, the system takes a longer time to reach the desired value of unity for the first time, called the *rise time*. Another indicator of the speed of response is the *settling time*, defined as the time the amplitude takes to settle within $\pm 2\%$ of the steady-state value. The settling time and rise time are nearly the same in the nonoscillatory response of system with $\zeta \geq 1$. On the other hand, for $\zeta < 1$, the system does not stop at the desired output when it reaches it for the first time, but overshoots it (by larger and larger amount as ζ is reduced) and crosses it repeatedly in an expectedly oscillatory response with a decaying amplitude. The rise time, maximum overshoot, and settling time are important performance parameters of the system's indicial response. It is evident in Fig. 14.2 that the increase in speed and decrease of maximum overshoot are contradictory design requirements. A trade-off between the two can be obtained by selecting a value of ζ that is neither too large (causing a sluggish response) nor too small (leading to a large overshoot). In most design applications, $\zeta = \frac{1}{\sqrt{2}} = 0.707$ (shown as a dotted line in Fig. 14.2) is considered ideal in striking the best compromise between speed and maximum overshoot. The concept of second-order system design and analysis can be extended to higher-order systems, which have a dominant pair of complex-conjugate poles near the imaginary axis.[5] In such a case, the system can be approximated by the dominant second-order system. This modeling approach is often valid in even multivariable control systems. For this reason, the concept of a second-order system is a valuable analysis tool, and terms such as "damping ratio" and "natural frequency" are commonly applied to each pair of complex poles in a large-order system.

[5] Poles (eigenvalues) of a system that have the smallest real part magnitude dominate a system's transient response and are thus called *dominant poles*.

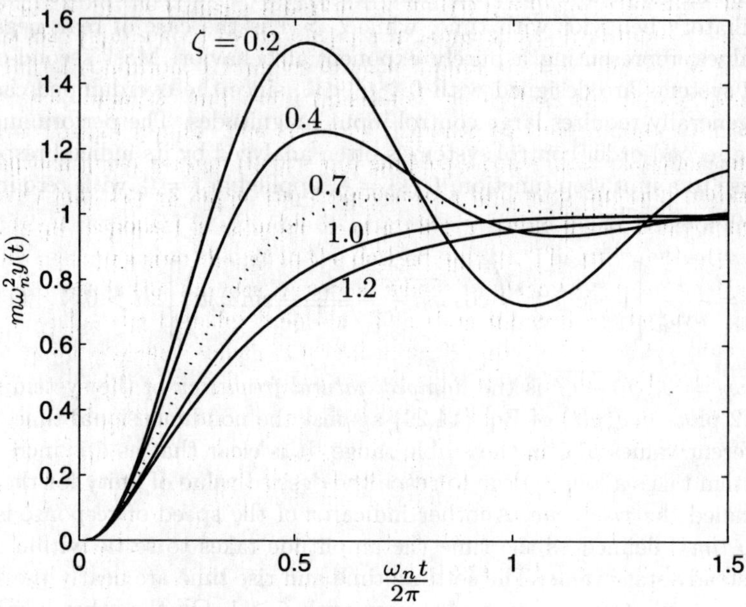

Fig. 14.2. Indicial response of a second-order, SISO system.

14.4 Basic Closed-Loop Systems

We shall now discuss the common SISO control systems. In each case, the principle can be extended to multivariable systems by considering additional feedback loops and feedforward paths. For simplicity, we shall only consider analog control systems, keeping in mind that their modern applications commonly involve implementation through digital electronic circuits [43]. The most common closed-loop control system is a *switching relay* [Fig. 14.3(a)] in which the control input is provided by closing a switch whenever the plant's output falls below the desired output, and opening the same switch as soon as the desired output is reached. Such nonlinear control systems are commonly employed for temperature control through a thermostat. However, in faster plant dynamics (such as atmospheric flight vehicles), switching systems have an unacceptable transient response. Also, the limitation of a unidirectional control input is inadequate in an attitude control application. A modification of the on–off switching relay system is to have two identical actuators (instead of only one) that apply equal but mutually opposite inputs, and a three-position switch capable of choosing among positive, negative, or zero control inputs. Such a control system (called *bang-bang* control), depicted in Fig. 14.3(b), is capable of controlling the attitude of flight vehicles. However, since the maximum possible control input magnitude is always applied, bang-bang control is unsuitable in atmospheric flight vehicles, wherein control surfaces are em-

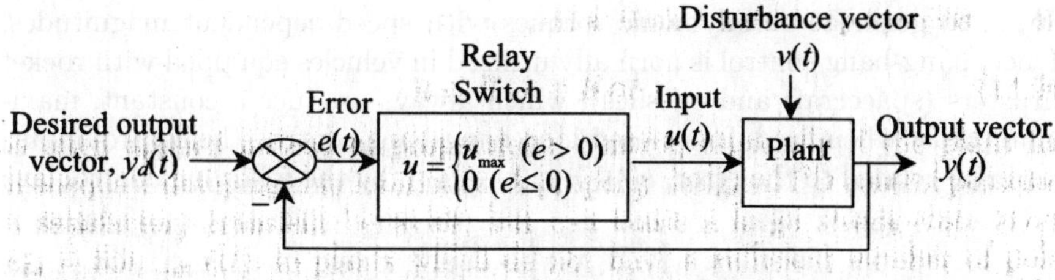

(a) A simple switching control system.

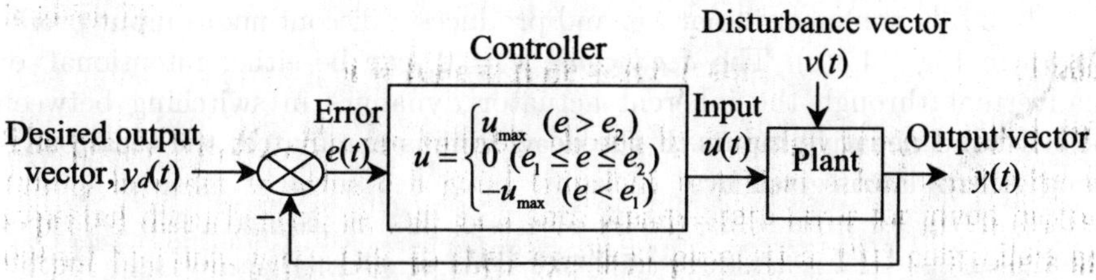

(b) A bang-bang, closed-loop control system.

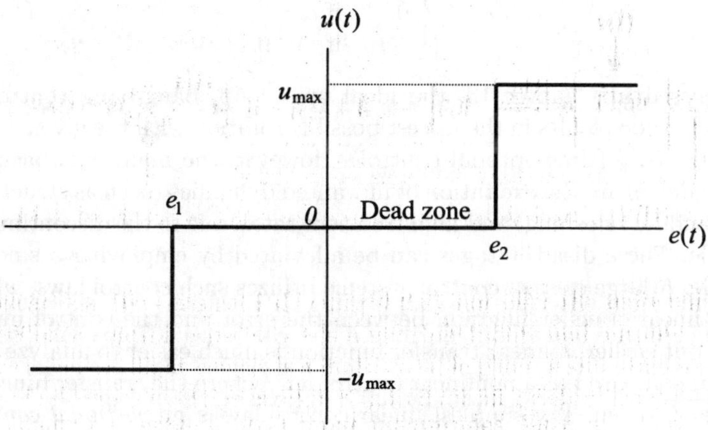

(c) Output of a bang-bang, switching controller.

Fig. 14.3. Switching control systems.

ployed to generate aerodynamic torques with speed-dependent magnitudes. Hence, bang-bang control is normally utilized in vehicles equipped with rocket thrusters (spacecraft and missiles), which always produce a constant, maximum thrust when fired. In a closed-loop bang-bang control, the control input, u, can be related to the error, $e(t) = y_d(t) - y(t)$, by the following control law:

$$u = \begin{cases} u_{\max} & \text{if } e > e_2, \\ 0 & \text{if } e_1 \le e \le e_2, \\ -1 & \text{if } e < e_1, \end{cases} \tag{14.30}$$

where $u_{\max}$ is the maximum control input magnitude. The control law of Eq. (14.30) is nonlinear in nature, and produces a discontinuous input profile shown in Fig. 14.3(c). The *dead zone*, $u = 0$, can be either intentional, or inadvertent through the inherent actuator dynamics in switching between positive and negative inputs. If the dead zone is eliminated, the ideal bang-bang control law is given by

$$u = u_{\max}\text{sgn}(e), \tag{14.31}$$

where sgn(e) is the following signum function of the error, e:

$$\text{sgn}(e) = \begin{cases} 1 & \text{if } e > 0, \\ 0 & \text{if } e = 0, \\ -1 & \text{if } e < 0. \end{cases} \tag{14.32}$$

As discussed in Chapter 13, the ideal open-loop, bang-bang control of spacecraft attitude results in the fastest possible rotation with the given maximum torque, $u_{\max}$ (time-optimal control). However, the main drawbacks of bang-bang control are the excitation of unwanted dynamics (such as structural vibrations and fuel sloshing) and high transient rates, due to the discontinuous input profile. These disadvantages can be alleviated by employing a smooth input profile. A large class of control systems utilizes such control laws, which result in a linear transfer function between the error and the control input. As pointed out earlier, a linear transfer function is much easier to analyze and design when compared to a nonlinear control law, where the transfer function concept breaks down. The simplest linear control law is *proportional control*,

$$u = Ke, \tag{14.33}$$

where K is a positive constant, called the *gain*. With a suitable choice of gain, an acceptable closed-loop performance may be obtained. However, this is not always guaranteed, as the proportional feedback can amplify even small errors (perhaps due to measurement noise), leading to a high sensitivity to disturbances. Thus, the proportional controller often acts analogous to a spring. In order to damp out the oscillations caused by the proportional gain, either the plant must possess satisfactory damping, or an artificial damping mechanism must be provided by the controller. Therefore, a *proportional-derivative* (PD) control,

$$u = K_P e + K_D \dot{e} \, , \tag{14.34}$$

is often applied instead of proportional control, especially if the plant has inadequate damping characteristics. A properly designed PD control possesses a satisfactory transient behavior, but can cause a large *steady-state error*, $e_{ss} \doteq \lim_{t\to\infty} e(t)$, in plants which do not have a sufficient number of poles at the origin ($s = 0$) [43]. In such cases, it becomes imperative to add a controller pole at the origin, which translates into an integral action in time. The resulting control law is termed *proportional-integral-derivative* (PID) and is given by

$$u = K_P e + K_D \dot{e} + K_I \int \mathrm{edt} \, . \tag{14.35}$$

The gains, K_P, K_D, K_I, are suitably chosen by a design process called *PID tuning* in order to achieve a good transient response, a low sensitivity to expected disturbances, as well as a zero steady-state error for given desired output function, $y_d(t)$. Due to their excellent properties, PID controllers are the most commonly used closed-loop devices, especially in SISO plants. A block diagram of PID controller is shown in Fig. 14.4. In some flight-control

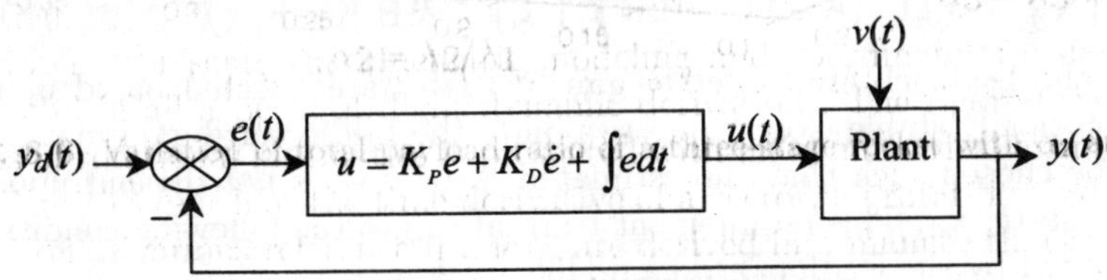

Fig. 14.4. Block diagram of a proportional-integral-derivative (PID) control system.

applications, the classical PID control may not offer the most efficient choice of feedback control, especially when multiple inputs and outputs are involved. Since a generic flight vehicle is a multivariable plant, a simultaneous minimization of all transient errors is often required for an acceptable performance with the smallest possible control input magnitudes. Such an approach is called *optimal control* [43]. Derivation of an optimal control law for a multivariable plant is feasible only through the state-space representation (rather than the classical transfer-function approach). Without considering the optimization problem to be solved, we shall confine ourselves to the modeling of the optimal control laws. The most general representation of a multivariable, linear control law is as follows:

$$\mathbf{u} = \mathsf{K}\hat{\mathbf{e}} + \mathsf{K_d}\mathbf{x_d} \, , \tag{14.36}$$

where $\hat{\mathbf{e}}$ is the *estimated error*,

$$\hat{\mathbf{e}} = \mathbf{x_d} - \hat{\mathbf{x}} \, , \tag{14.37}$$

and K and $\mathsf{K_d}$ are the feedback and feedforward *gain matrices*, respectively. In Eq. (14.37), $\hat{\mathbf{x}}$ is the *estimated state*, whose estimation (or *observation*) is based upon the measured output, $\mathbf{y}$. The process by which the state is estimated is known as an *observer*, whose linear, asymptotically stable dynamics can be represented by [43]

$$\dot{\hat{\mathbf{x}}} = (\mathsf{A} - \mathsf{LC})\hat{\mathbf{x}} + (\mathsf{B} - \mathsf{LD})\mathbf{u} + \mathsf{L}\mathbf{y} \,, \tag{14.38}$$

where L is the *observer gain matrix*.[6] The linear control system design procedure consists of separately designing the linear controller [Eq. (14.36)] and the linear observer [Eq. (14.38)], such that a satisfactory closed-loop performance is achieved [43]. However, a linear observer can be designed for only those plants, *observable* with respect to the measured output, while a linear feedback controller can succeed in only those plants that are *controllable* with respect to the given control input. The properties of controllability and observability are thus crucial in designing a control system [43] for a given plant.

Upon substituting Eqs. (14.36)–(14.38) in the linear, time-invariant plant's state and output equations, we arrive at the following state equation for the closed-loop dynamics:

$$\begin{Bmatrix} \dot{\mathbf{x}} \\ \dot{\hat{\mathbf{x}}} \end{Bmatrix} = \begin{pmatrix} \mathsf{A} & -\mathsf{BK} \\ \mathsf{LC} & (\mathsf{A} - \mathsf{LC} - \mathsf{BK}) \end{pmatrix} \begin{Bmatrix} \mathbf{x} \\ \hat{\mathbf{x}} \end{Bmatrix} + \begin{pmatrix} \mathsf{B}(\mathsf{K} - \mathsf{K_d}) \\ \mathsf{B}(\mathsf{K} - \mathsf{K_d}) \end{pmatrix} \mathbf{x_d} \,. \tag{14.39}$$

While the feedback and observer gain matrices are selected based upon an asymptotically stable closed-loop response, the feedforward gain matrix, $\mathsf{K_d}$, must be chosen such that the estimated error $\hat{\mathbf{e}}$ does not depend upon the desired state, $\mathbf{x_d}$. This requirement translates into the following condition to be satisfied by the feedforward gain [43]:

$$(\mathsf{A_d} - \mathsf{A} + \mathsf{BK_d})\mathbf{x_d} = \mathbf{0} \,, \tag{14.40}$$

where $\mathsf{A_d}$ is the coefficient matrix for the linear, desired state dynamics,

$$\dot{\mathbf{x}}_\mathbf{d} = \mathsf{A_d}\mathbf{x_d} \,. \tag{14.41}$$

It may not always be possible to satisfy Eq. (14.41), which means that a given plant can be made to track only some desired state dynamics with closed-loop asymptotic stability [43].

We are now prepared to discuss some important flight-control systems.

14.5 Implementation of Control System Elements

A classical PID SISO controller, and a linear, multivariable, observer-based tracking system considered above, involve common mathematical operations,

[6] The observer given by Eq. (14.38) is called a *full-order observer*, because it reconstructs the entire state vector from the measured output. However, some state variables can often be directly obtained from the output. Hence, a *reduced-order observer*, which only estimates those state variables that cannot be directly measured, leads to a more efficient model [43].

namely addition (or subtraction), multiplication, differentiation, and integration in time. These operations are performed upon the error and desired state (or output) variables, in order to produce the control inputs. For practical implementation, we must be able to construct such controllers through physical processes that mimic these common operations. This is the basis of analog devices, which act as mechanical, electrical, or electromechanical analogs of elementary mathematical operations.[7] The principal analog devices employed in flight-control applications are *gyroscopic sensors*, *accelerometers*, mechanical, hydraulic, or electromechanical actuators of aerodynamic control surfaces and nozzle gimbals, rocket thrusters, and various electric networks for controller elements (summing junction, gain amplifiers, integrators, etc.) representing mathematical operations. Most linear actuators and mechanical sensors are generally modeled as second-order systems with equivalent inertia, stiffness, and viscous damping. Primary exceptions are the rocket thrusters employed in *reaction control systems* (RCS) of spacecraft, which have a discontinuous, piecewise-linear dependence upon the actuating signal.

Analog electrical networks representing controller elements can have *passive* elements (resistors, inductors, and capacitors), or *active* elements, such as *transformers and operational amplifiers.* For example, the inexpensive *lead-lag* passive network can provide an approximate form of PID control [43], while an operational amplifier can act as a gain, summing device, integrator, or even a nonlinear limiter in more expensive, active networks. In the period 1945–1970, analog network-based flight controllers were in wide use until overtaken by the more versatile and robust digital controllers. This was also the era of the analog computers, which could occupy whole buildings and be prohibitively expensive and cantankerous in performing tasks similar to those of a modern pocket calculator.

The analog electrical networks have been largely replaced by equivalent digital, integrated electronic circuits, which are easier to program and offer more accurate as well as robust operation. Similarly, the traditional mechanical and hydraulic linkages between the controller and the actuators are replaced by electrical wires (*fly-by-wire*) in modern atmospheric vehicles, since electric motors are now the norm as actuating mechanisms. Furthermore, the mechanical gyroscopes and accelerometers are now available in more compact and rugged electronic replacements, such as *ring-laser gyros* and *piezo-electric accelerometers.* We shall not attempt to model the electronic and electro-optic dynamical features of the modern sensors, but assume that their suitable transfer functions are known. It is more apt for our purposes to study the

[7] Although the term *analog devices* originally referred to the physical equivalents of mathematical processes, this term is now applied solely to describe systems whose inputs and outputs are continuous functions of time. In contrast, a *digital device* is a system that has discrete (or discontinuous) inputs and outputs in the time domain.

modeling of mechanical gyroscopes, which continue to find wide application in aerospace vehicles due to their versatile and inexpensive nature.

14.5.1 Gyroscopic Sensors

The first practical analog feedback device employed in a closed-loop flight control system was a *gyroscope* (also called *gyro* in short), wherein a spinning rotor mounted on a restrained *gimbal* could act as either a multiplier (gain) or an integrator of the error signal (an angular rate). For illustration, let us consider a gyroscope used to sense a single-axis rotation of the flight vehicle. As depicted in Fig. 14.5, the single-degree-of-freedom gyroscopic sensor consists of a rotor spun about the vehicle's body axis *ox* relative to the gimbal at a constant angular momentum, H_r, by the use of a *servo-motor*.[8] The gimbal axis, *oy*, can turn by a small angle relative to the flight vehicle through a restraining mechanism. Finally, the flight vehicle is assumed to rotate about the body axis, *oz*, by an inertial angle, ψ. The gimbal axis is usually restrained by a torsional spring of stiffness k, which generates a resisting torque proportional to the small angular displacement, θ, of the gimbal relative to the vehicle. Due to friction in the gimbal axis, some damping is invariably present, which can be enhanced by the addition of a viscous damper. Hence, the net viscous damping constant, c, is assumed to provide a resisting torque proportional to the small gimbal rate, $\dot{\theta}$. Since the angular speed of the rotor relative to gimbal is constant, and the direction of its angular momentum, $\mathbf{H_r}$, does not change relative to the gimbal (due to a rigid construction), there is no change in the rotor's angular momentum relative to the gimbal. Thus, the rate of change of $\mathbf{H_r}$ in the inertial space due to the combined rotation of the vehicle and gimbal is expressed as (Chapter 13)

$$\begin{aligned}\dot{\mathbf{H}}_\mathbf{r} &= \boldsymbol{\omega}_\mathbf{r} \times \mathbf{H_r} = (\dot{\theta}\mathbf{j} + \dot{\psi}\mathbf{k}) \times H_r(\cos\theta\mathbf{i} - \sin\theta\mathbf{k}) \\ &= H_r[-\dot{\theta}(\cos\theta\mathbf{k} + \sin\theta\mathbf{i}) + \dot{\psi}\cos\theta\mathbf{j}] \,. \end{aligned} \tag{14.42}$$

This change of angular momentum results in a torque about the gimbal axis, *oy*, as well as small transverse torques about *ox* and *oz*. While the axial torque causes gimbal rotation, the transverse torques are absorbed by the rotor-gimbal bearing. Since both gimbal angle, θ, and gimbal rate, $\dot{\theta}$, are kept small by the restraining spring and damper, we can approximate Eq. (14.42) by the linear relationship

$$\dot{\mathbf{H}}_\mathbf{r} \approx H_r\dot{\psi}\mathbf{j} \,. \tag{14.43}$$

By Newton's second law for rotational dynamics (Chapter 13), the rate of change of the rotor's angular momentum is equal to the net torque experienced by the rotor, which, by Newton's third law, is equal and opposite to the

[8] A servo-motor is a feedback control system for maintaining a constant angular speed through a tachometer (or an angle encoder) as a sensor and a direct-current (DC) motor.

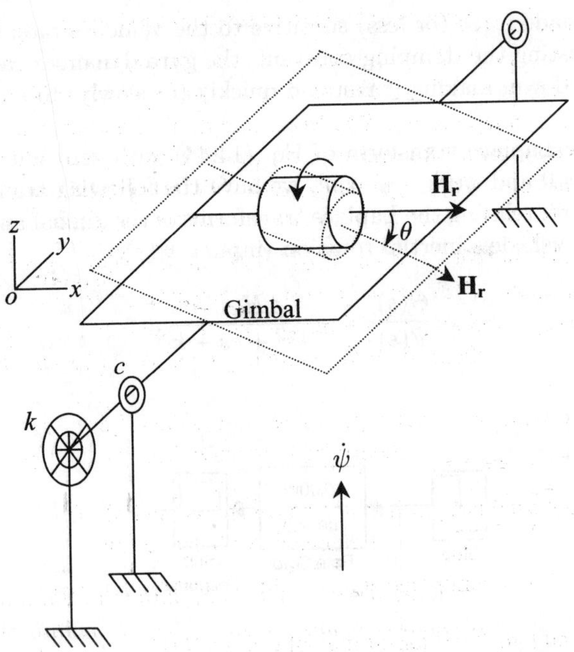

Fig. 14.5. A single-degree-of-freedom rate gyroscope.

torque applied by the rotor on the gimbal. Therefore, the gimbal's linearized dynamical equation can be written as follows:

$$J\ddot{\theta}(t) + c\dot{\theta}(t) + k\theta(t) = -H_r\dot{\psi}(t) \,, \tag{14.44}$$

where J is the moment of inertia of the gimbal and rotor assembly about the axis oy. Equation (14.44) has the following equilibrium solution, $\theta(t) = \theta_e$, in the steady state ($t \to \infty$), obtained by letting $\theta = \dot{\theta} = 0$:

$$\theta_e = -\frac{H_r\dot{\psi}}{k} \,, \tag{14.45}$$

which implies a gimbal angle proportional to the vehicle's rotation rate. For this reason, the gyroscope of Fig. 14.5 is called a *rate gyro*, as it can be calibrated to measure a vehicle's steady rate about the input axis. Two (or more) rate gyros mounted on mutually perpendicular body axis can provide information about a vehicle rotating steadily about multiple body axes. A sudden change in the vehicle's rate, however, will take some time to be registered as the equilibrium gimbal angle output. The time taken to reach the steady state for a given change in the vehicle's rate depends upon the damping constant, c, as well as the moment of inertia, J, while the equilibrium value of the gimbal angle, Eq. (14.45), depends only upon the ratio of the rotor's angular momentum, H_r with the spring stiffness, k. By adjusting this latter ratio, the rate gyro can be made more (or less) sensitive to the vehicle's rate. On the other hand, by adjusting the damping constant, the gyro dynamics can be speeded up, or slowed down, making it respond quickly (or slowly) to a change in the vehicle's rate.

Taking the Laplace transform of Eq. (14.44) with zero initial conditions $[\theta(0) = \dot{\theta}(0) = 0$ and $\psi(0) = \dot{\psi} = 0]$, we have the following transfer function for the rate gyro, relating the Laplace transforms of the gimbal angle (output), $\Theta(s)$, and the vehicle's inertial rotation (input), $\Psi(s)$:

$$\frac{\Theta(s)}{\Psi(s)} = -\frac{H_r s}{Js^2 + cs + k} . \tag{14.46}$$

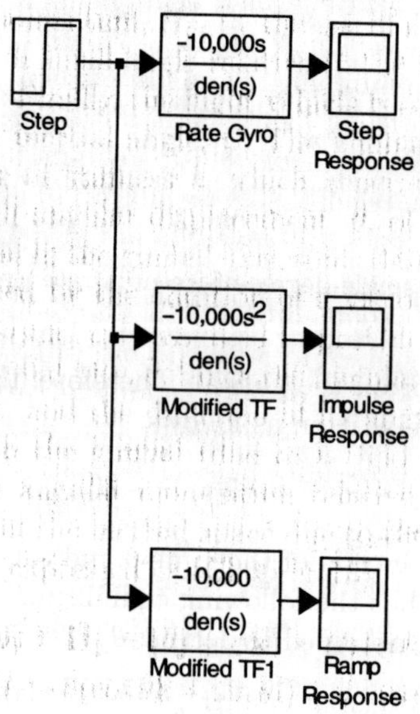

Fig. 14.6. Simulink block diagram for the step, impulse, and ramp response of a rate gyroscope.

Example 14.1. Consider the *MIT* [46] rate gyro with the following characteristics: $H_r = 10^4$ g – cm^2/s, $J = 34$ g – cm^2, $k = 3.03 \times 10^5$ g – cm^2/s^2, and $c = 5000$ g – cm^2/s. Evaluate the gimbal angle response for

(a) an impulsive angular displacement of the vehicle by 0.1 rad.
(b) a step change in the vehicle's angular orientation by 0.1 rad.
(c) a step change in the vehicle's angular rate by 0.1 rad/s.

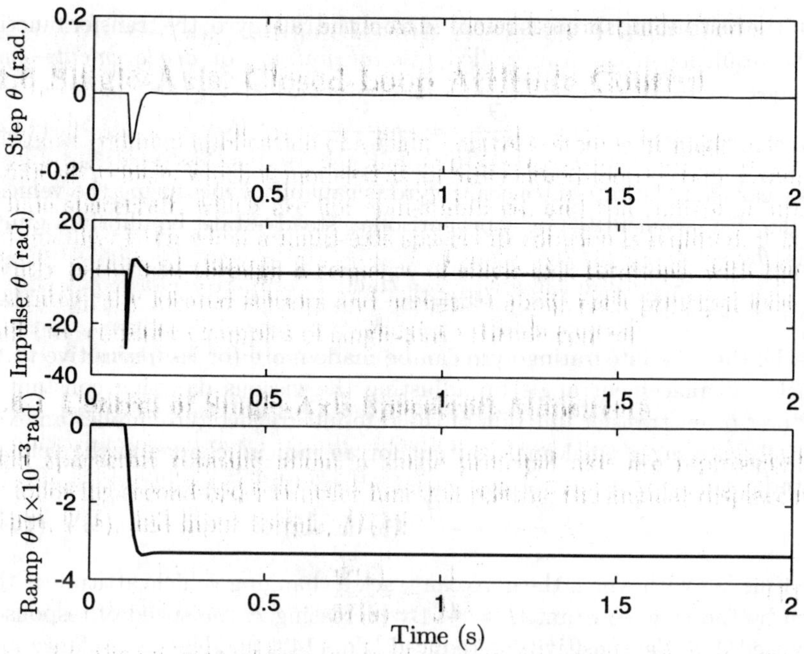

Fig. 14.7. Step, impulse, and ramp gimbal angle response of a rate gyroscope.

The *final value theorem* of the Laplace transform [4] applied to Eq. (14.46) leads to the following steady-state value of the equilibrium gimbal angle to a change in the vehicle's angular orientation:

$$\theta_e = \lim_{s\to 0} s\Theta(s) = \lim_{s\to 0} -\frac{H_r s^2 \Psi(s)}{Js^2 + cs + k} \, .$$

Clearly, the rate gyro would not respond to either an impulsive or a step change in the angular displacement of the vehicle [$\Psi(s) = 0.1$, or $\Psi(s) = \frac{0.1}{s}$], leading to $\theta_e = 0$ in both (a) and (b). However, in case (c), the step change in $\dot{\psi}$ [$\Psi(s) = \frac{0.1}{s^2}$] leads to

$$\theta_e = -\frac{(0.1)(10^4)}{3.03 \times 10^5} = -0.0033 \text{ rad } (-0.1891°) \, .$$

This steady-state output is the same as the one resulting from a constant vehicle rate of 0.1 rad/s in Eq. (14.45). The response to a step change in the angular rate is the same as the response to a ramp input of the given slope in the angular displacement.

The time history of the gimbal angle in the three cases is calculated by the linear Simulink block diagram shown in Fig. 14.6, whose outputs are plotted in Fig. 14.7. Our expectations for the step, impulse, and ramp responses are borne out in this simulation, which is performed with a relative tolerance of 10^{-3} with a variable-step Runge–Kutta solver of the fourth order.

If the restraining spring is removed from the rate gyro, the transfer function of the resulting mechanism (called *rate-integrating—*or *displacement—gyro*) becomes

$$\frac{\Theta(s)}{\Psi(s)} = -\frac{H_r}{Js + c} . \tag{14.47}$$

It is clear that the rate-integrating gyro is capable of measuring the vehicle's angular displacement, ψ, by a proportional, steady-state, equilibrium gimbal angle, θ_e,

$$\theta_e = -\frac{H_r \psi}{c} . \tag{14.48}$$

Clearly, the rate-integrating gyro can be made more (or less) sensitive to the vehicle's displacement angle by adjusting the viscous damping constant, c. The first-order transfer function of the rate-integrating gyro implies an exponentially varying gimbal angle output for an indicial change in the vehicle's attitude,

$$\theta(t) = -\frac{H_r}{c}\left(1 - e^{-\frac{c}{J}t}\right) . \tag{14.49}$$

The time lag with which the gyro can track a changing vehicle attitude is thus given by the time constant, $T = \frac{J}{c}$. By increasing c, the speed of response is increased, but the sensitivity is reduced [Eq. (14.48)]. Hence, a balance must be struck between the speed and sensitivity of a rate-integrating gyro.

The integral action provided by the rate-integrating gyro is a valuable feature in reducing the steady-state error in a closed-loop system and enables PID controller implementation. A combination of rate and rate-integrating gyroscopes can handle most practical attitude control tasks. The *Sperry* autopilot of 1909 was based on such a combination and was tasked principally with maintaining a wings'-level (unbanked) attitude during cruise of the early aircraft. Its utility was demonstrated in the long-distance flights by many aviation pioneers, often in bad weather, or at night (such as the solo, transatlantic flight undertaken by Lindbergh in 1927). In the 1930s and 1940s, the gyroscopic flight-control systems advanced to such applications as unmanned aerial vehicles (*V-1* "flying bomb") and ballistic missiles (*V-2* rocket), and have continued to be useful in the present age in the form of *inertial navigation systems* for airliners, long-range missiles, and spacecraft. When gyroscopes are used as analog devices, their output is an angular rate (or displacement), which can be converted into an electrical signal by a tachometer (or an angle encoder). The resulting current can be amplified and used to drive an electric motor, which, in turn, generates a torque for the movement of a control surface (or a rocket nozzle). Therefore, a gyroscope can act as both a sensor and a controller in attitude-control applications. Alternatively, the sensed electrical signal from a gyroscope (or another analog sensor) can be processed by special electrical networks that act as analog controllers.

14.6 Single-Axis, Closed-Loop Attitude Control

The most common application of a flight-control system is in single-axis rotation of the vehicle, which is modeled as an SISO closed-loop system. Examples include spacecraft, which are not spin-stabilized, and roll control of aircraft and missiles. Even when a multi-axis spacecraft rotation is required, it is frequently performed through a sequence of single-axis rotations, with the use of strategically located sensors and actuators about each principal axis. We shall now consider examples of single-axis attitude control.

14.6.1 Control of Single-Axis Spacecraft Maneuvers

Rigid spacecraft rotating about a single principal axis are represented by the following second-order transfer function relating the angular displacement output, $\Psi(s)$, and input torque, $M(s)$:

$$\frac{\Psi(s)}{M(s)} = \frac{1}{Js^2}, \tag{14.50}$$

where J is the moment of inertia about the concerned axis. Since the transfer function has a double pole at $s = 0$, the spacecraft is unstable when the output is the angular displacement, $\psi(t)$. However, the transfer function between the angular rate, $\dot{\psi}(t)$, and the input torque has a single pole at $s = 0$, indicating a stable plant in terms of angular rate. In Chapter 13, we saw how an open-loop control of rigid, nonspinning spacecraft's single-axis rotations can be performed using a well-designed sequence of impulses provided by a pair of rocket thrusters. Our interest here is in doing the same with the use of closed-loop control. There are two kinds of spacecraft maneuvers: (a) spin maneuver and (b) rest-to-rest slew. In a spin maneuver, the objective is to achieve a given angular velocity in the steady state, while the rest-to-rest slew refers to bringing the spacecraft to a desired rest attitude from another rest attitude. Clearly, the spin maneuver can be easily controlled using a rate gyro and a pair of attitude thrusters in a closed loop. It is more interesting to consider control of a single-axis angular displacement. From our previous remarks and Example 14.1, it is clear that controlling angular displacement requires a rate-integrating gyro (rather than a rate gyro).

We will consider RCS simulation examples with Simulink block diagrams. The plant and gyro sensors can be modeled using the continuous-system Simulink blocks with appropriate transfer functions [49]. The modeling of the controller and actuator dynamics comprising attitude thrusters and a switching control law is through the nonlinear relationship of Eq. (14.30), which can be easily incorporated in a Simulink model by the *sum*, *dead zone*, *sign*, and constant *gain* blocks [49]. In addition, a random disturbance input, arising out of solar radiation and gravity gradient torques, can be added by using a *white-noise* [43] Simulink block generating normally distributed random numbers with a specified intensity, in order to make the model more realistic.

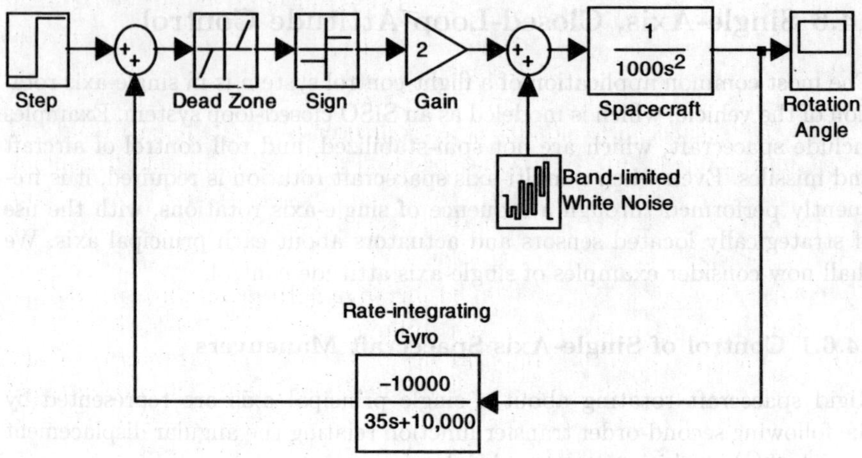

Fig. 14.8. Simulink block diagram for step response of spacecraft control system with rate-integrating gyro.

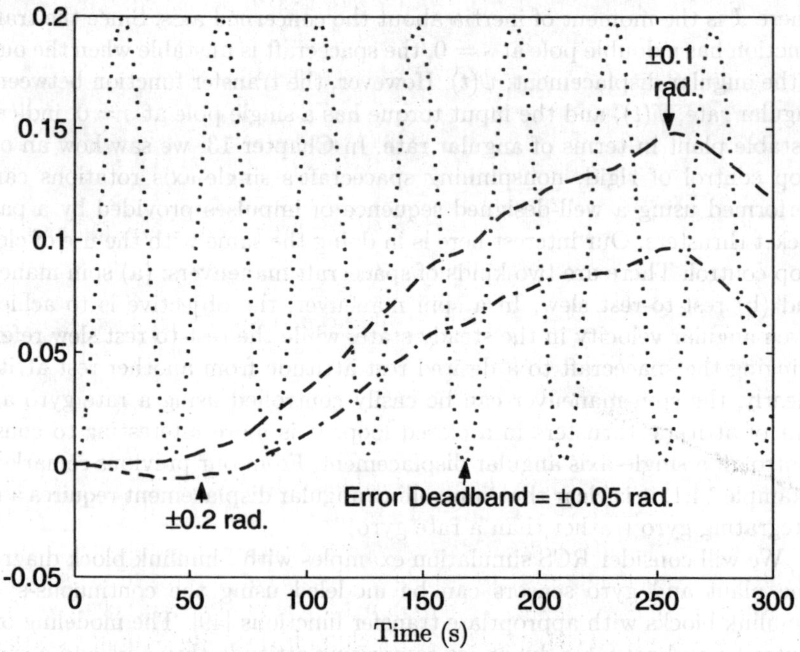

Fig. 14.9. Step response of spacecraft control system with rate-integrating gyro for various dead zones.

Example 14.2. Consider the control of a spacecraft displacement about a principal axis with moment of inertia, $J = 1000$ kg–m^2, using a pair of attitude thrusters that can exert constant torque of 2 N–m about the concerned axis. A rate-integrating gyro with properties given in Problem 14.1 is used as a feedback sensor. It is required to track a step angular displacement of the spacecraft by 0.1 rad. The dead zone of the controller/actuator combination can be fixed at specific values of angle error, $e_2 = -e_1$, called the *deadband*, in reference to Eq. (14.30). We shall explore the following deadbands:

(a) $e_2 = -e_1 = 0.05$ rad.
(b) $e_2 = -e_1 = 0.1$ rad.
(c) $e_2 = -e_1 = 0.2$ rad.

The Simulink block diagram encoded for the required simulation is shown in Fig. 14.8, and the simulated spacecraft angular response for each of the three values of the dead zone is plotted in Fig. 14.9. A white-noise torque disturbance of intensity (power spectral density [43]) 0.01 N.m is specified in the simulation with the *band-limited white-noise* Simulink block. This leads to a disturbing torque amplitude of 4 N.m. The simulation is performed for 300 s using the variable-step *ode45* Runge–Kutta algorithm with a relative tolerance of 10^{-3}. It is clear that the deadband of ± 0.05 rad produces a rapidly oscillating response (40 s wavelength and 0.1 rad. amplitude) about the desired steady state of $\psi_e \doteq 0.1$ rad, while the larger error deadbands yield a much slower and irregular spacecraft rotation with time. In none of the cases is the desired angular displacement achieved, which indicates that the rate-integrating gyro by itself is inadequate in meeting the attitude control requirement.

The push-pull nature of the discontinuous switching actuator torque can excite large spacecraft oscillations for small deadband values if the control is based only on angular displacement. For larger deadbands, the sensitivity of control input to varying displacement error is reduced, thereby lessening the frequency of oscillation, but produces an irregular angle response with large deviations. Hence, pure angular position feedback is unacceptable. The plant transfer function is quite capable of converging to a zero steady-state error for a step change in displacement (due to its double pole at $s = 0$ [43]), provided a damping mechanism (which is absent in the closed-loop system with a rate-integrating gyro). Hence, the PD control, possible by combining the feedbacks through a rate-integrating gyro and a rate gyro, can achieve an acceptable closed-loop performance.

Example 14.3. Repeat the simulation of Example 14.2 by adding the rate gyro of Example 14.1 in parallel feedback with the rate-integrating gyro. The dead zone is almost entirely eliminated with the actuator deadband of ± 0.001 rad. The output of the rate gyro is multiplied by a scaling gain, K_D, such that the desired level of closed-loop damping is achieved. Try the following values of K_D:

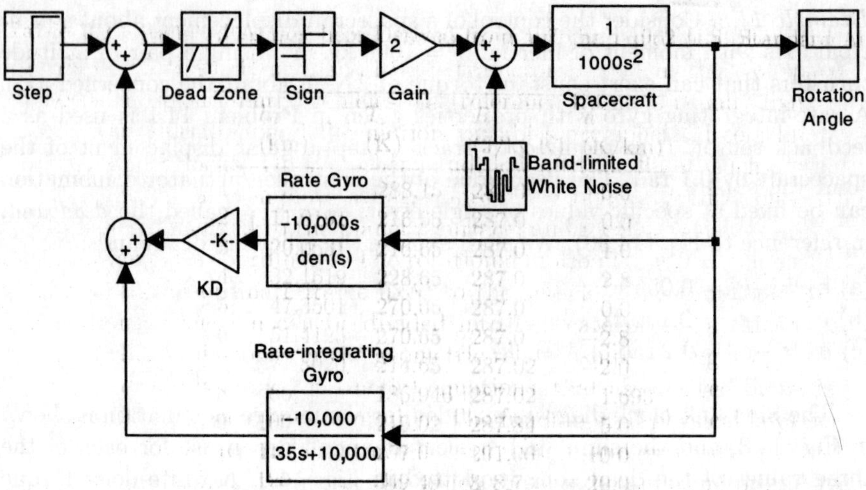

Fig. 14.10. Simulink block diagram for step response of spacecraft RCS with rate and rate-integrating gyros.

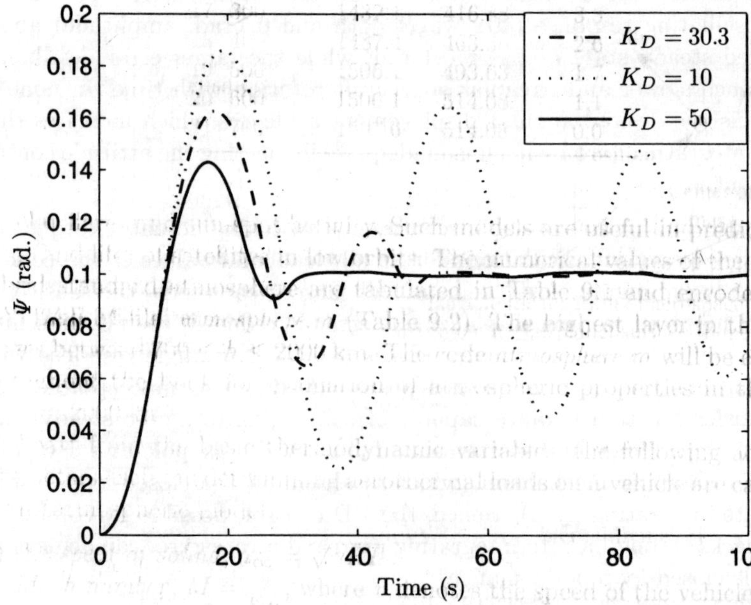

Fig. 14.11. Step response of spacecraft RCS with rate and rate-integrating gyros for various rate gyro gains.

(a) $K_D = 10$.
(b) $K_D = 30.3$.
(c) $K_D = 50$.

The Simulink block diagram for the modified RCS is shown in Fig. 14.10, and the simulated spacecraft angular response for each of the three values of the rate gyro gain is plotted in Fig. 14.11. It is evident that the use of a rate gyro enables the desired angle of 0.1 rad to be achieved quickly and without too many oscillations, in a manner quite similar to a linear, second-order system (even though the RCS is a nonlinear system). The maximum overshoot and settling time are modified by changing the rate gyro gain, K_D, whose increase causes an increase in the "damping." Thus, while $K_D = 10$ leads to the smallest overshoot and the largest settling time, $K_D = 50$ brings the system to the desired state in the smallest time, but with a larger maximum overshoot. A compromise between the conflicting requirements can be achieved by adopting $K_D = 30.3$.

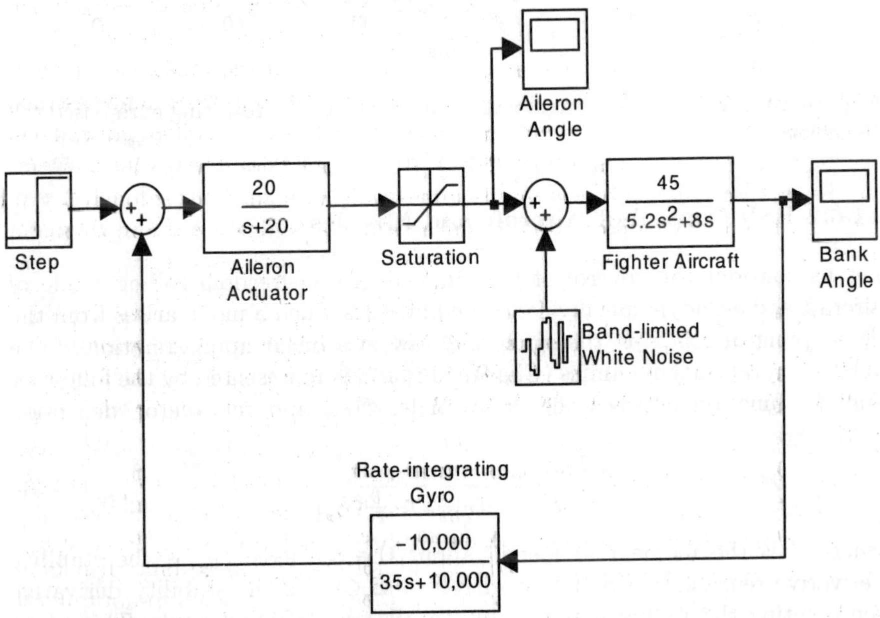

Fig. 14.12. Simulink block diagram for step response of roll autopilot for a fighter airplane with a rate-integrating gyro.

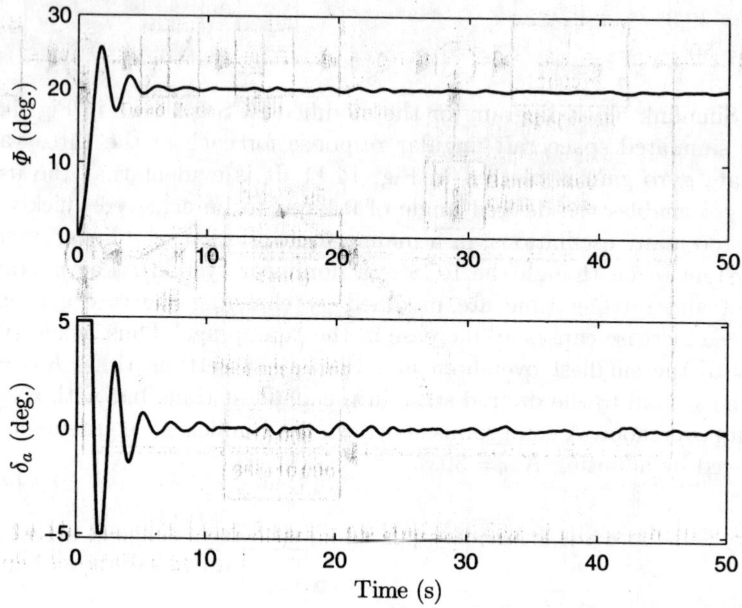

Fig. 14.13. Step response of fighter airplane with a rate-integrating gyro-based roll autopilot.

14.6.2 Roll Control of Aircraft and Missiles

Let us consider the control of the single-degree-of-freedom rolling mode of aircraft and aerodynamically controlled missiles. Such a mode arises from the de-coupling of roll from the pitch and yaw in a linear approximation of the vehicle's rotational dynamics (Chapter 13) and is represented by the following transfer function between the bank angle, $\Phi(s)$, and the aileron deflection angle, δ_a:

$$\frac{\Phi(s)}{\delta_a(s)} = \frac{C_{l_{\delta_a}}}{s(\frac{J}{qSb}s - \frac{b}{2v}C_{l_p})}, \tag{14.51}$$

where J is the moment of inertia about the roll axis, C_{l_p} is the stability derivative representing damping in roll, and $C_{l_{\delta_a}}$ is the stability derivative representing the rolling moment due to aileron deflection angle. This plant has a first-order time constant, $T = -\frac{2vJ}{qSb^2C_{l_p}}$, in addition to a pole at $s = 0$. The aileron actuator can be assumed to be a linear, second-order transfer function with nonlinear saturation limits, $| \delta_a | \leq \delta_{\max}$. Since the plant has a pole at origin, it can produce a desired step change in bank angle in a closed-loop, proportional feedback control system, such as the one approximately provided by a rate-integrating gyro.

Example 14.4. Consider a fighter airplane with wing span, $b = 14$ m, planform area, $S = 56.5$ m^2, moment of inertia about roll axis, $J = 34,700$ kg–m^2, flying straight and level at constant speed, $v = 236.2$ m/s at standard altitude 12.195 km where the dynamic pressure is $q = 8439.4$ N/m^2, $C_{l_p} = -0.27$/rad, and $C_{l_{\delta_a}} = 0.045$/rad. The airplane is equipped with an aileron actuator with the following first-order transfer function between the commanded aileron deflection angle, $\delta_{ac}(s)$, and the actual aileron deflection angle, δ_a:

$$\frac{\delta_a(s)}{\delta_{ac}(s)} = \frac{20}{s+20} \,.$$

Structural limitations restrict the maximum aileron deflection at the given speed to $|\ \delta_a\ | \leq 10°$. A *roll autopilot* is designed with the rate-integrating gyro of Problem 14.1 as the feedback device. Simulate the closed-loop response to a desired step change in bank angle by 20° in the presence of white-noise disturbance of amplitude 0.04 N–m caused by atmospheric turbulence and structural flexibility.

The simulation is carried out by the Simulink block diagram shown in Fig. 14.12, and the simulated aircraft response, $\phi(t), \delta_a(t)$, is plotted in Fig. 14.13. It is seen that the autopilot successfully attains the desired bank angle, within the accuracy of ±0.4°. This accuracy can be improved by employing a *filter* (a linear system with suitable high-frequency response) for the disturbance input [43]. However, another simple method of reducing sensitivity of a bank angle to rolling moment disturbance is to employ a rate gyro in parallel feedback with the rate-integrating gyro, as seen in the next example.

Example 14.5. Let us simulate the modified roll autopilot for the fighter of Example 14.4 by adding the rate gyro of Example 14.1 with scaling gain $K_D = 30.3$ in the feedback loop. The modified Simulink block diagram is shown in Fig. 14.14, with the simulated bank and aileron angle responses plotted in Fig. 14.15. Note the dramatic improvement in both transient and steady-state bank angle response, which has no overshoot or oscillation, and an accuracy of about ±0.1° in achieving the desired angle. However, the aileron angle is seen to oscillate more rapidly (albeit with a small amplitude) in response to the disturbance input. Such an autopilot is ideally suited in those applications where the flight vehicle must provide a stable platform for aiming and firing weapons. Obviously, the aileron in such a control system will always be active for absorbing the rolling moment disturbances.

As seen in Examples 14.3 and 14.5, a stable platform can be provided using a rate gyro coupled in parallel feedback with a rate-integrating gyro. This is the basis of an inertially stabilized platform about multiple axes using a pair of rate and rate-integrating gyros about each axis, and forms the core of an *inertial navigation system* (INS) employed in long-range aircraft, missiles, and spacecraft.

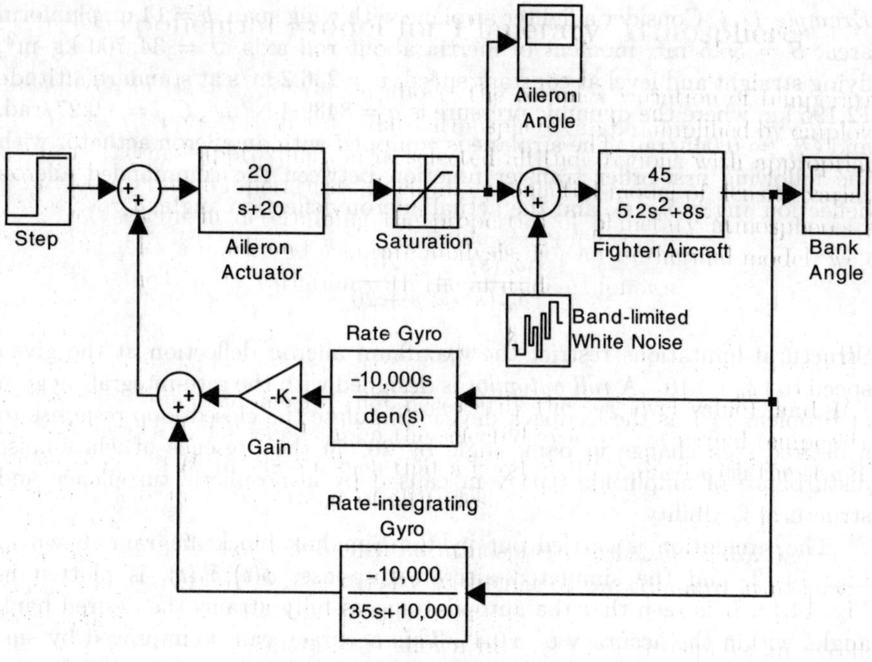

Fig. 14.14. Simulink block diagram for step response of roll autopilot with rate and rate-integrating gyros.

14.7 Multi-Axis Closed-Loop Attitude Control

Most flight-control applications involve multi-axis rotation, as indicated in Chapter 13. The design and analysis of multivariable control systems require a state-space vector-matrix representation of subsystems and often incorporate the linear, time-invariant approximation. We will consider the modeling and simulation examples of some interesting multivariable flight-control systems, and leave it to the reader to extend the methods to other similar applications.

14.7.1 Attitude Stabilization of a Launch Vehicle

As pointed out earlier, all launch vehicles and ballistic missiles are inherently unstable and require an attitude stabilization system for successful operation. At lift-off, a launch vehicle is moving too slowly for aerodynamic stabilization through fins. A similar situation prevails once the vehicle leaves the sensible atmosphere above an altitude of about 100 km. The use of multi-axis rate gyroscopes to sense and feed back the vehicle's departure from the desired equilibrium is a traditional method of attitude stabilization, which was first operationally incorporated in the German *V-2* rocket during World War II. In this vehicle, a simple clockwork mechanism was employed in conjunction with gyroscopic sensors and movable exhaust vanes to navigate toward a fixed

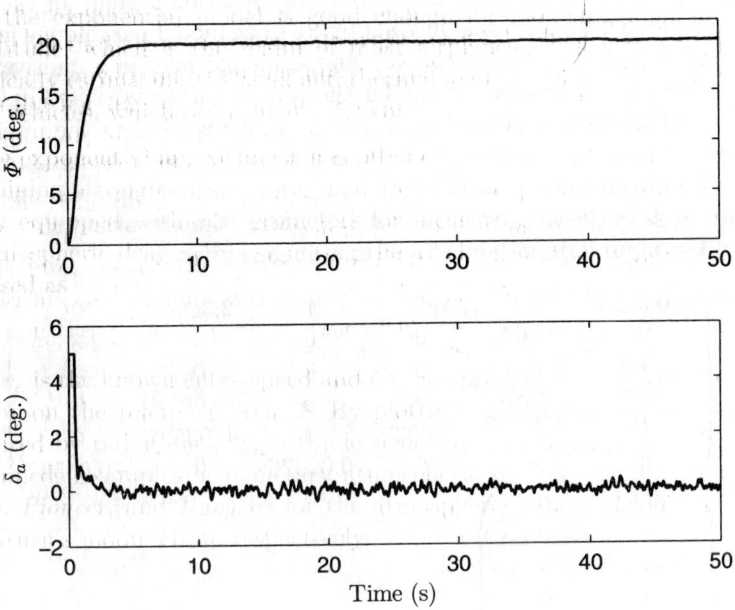

Fig. 14.15. Step response of fighter airplane with rate and rate-integrating gyro-based roll autopilot.

burn-out point. The guidance techniques have since evolved into more accurate techniques, and the necessity of maintaining the vehicle in a precisely controlled attitude has become even more stringent.

Apart from the rate gyro, an accurate sensor of the vehicle's rotational motion is an *accelerometer*, which converts the acceleration sensed normal to a particular axis into electrical voltage. Since acceleration involves the time derivative of a rotational rate, it provides a more sensitive measure of departure from the desired attitude than a rate gyro. The use of multiple accelerometers is also valuable in sensing structural vibration. Consider a single-axis accelerometer calibrated to measure normal acceleration along oz and mounted at distance x from the center of mass, o, normal to the axis of rotation, oy. The normal acceleration measured by the accelerometer is due to the combined effects of pitching and z-translation and can be expressed as

$$a_z = -x\dot{Q} + v\dot{\alpha} \,. \tag{14.52}$$

Since a normal acceleration measurement provides information about a linear and a rotational acceleration, an observer based upon such an output can be used to estimate the concerned state variables. For this reason, and due to their smaller size and cost, accelerometers are replacing rate gyros as primary motion sensors.

Example 14.6. Let us simulate the attitude stabilization system for the *Vanguard* ballistic missile discussed in Chapter 13. The model parameters for this vehicle are provided by Blakelock [46], and extended here in a multivariable control system. A state-space model for the vehicle was constructed in Chapter 13, based on state variables $Q, R, \theta, \psi, \alpha, \beta$. We will select the pitch and yaw rates, Q,R, as well as the normal acceleration, a_y, a_z, measured at location $x = 5$ m from the center of mass, as the four output variables on which the attitude stabilization system is based. A linear, time-invariant state-space model of the nonrolling missile 75 s after launch is given by

$$\mathsf{A} = \begin{pmatrix} -0.0675 & 0 & 0 & 0 & 2.3694 & 0 \\ 0 & -0.0675 & 0 & 0 & 0 & -2.3694 \\ 1 & 0 & 0 & 0 & 0 & 0 \\ 0 & 2.7285 & 0 & 0 & 0 & 0 \\ 1 & 0 & -0.02326 & 0 & -0.03527 & 0 \\ 0 & -1 & 0 & 0.02326 & 0 & -0.03527 \end{pmatrix},$$

$$\mathsf{B} = \begin{pmatrix} 7.0084 & 0 \\ 0 & -7.0084 \\ 0 & 0 \\ 0 & 0 \\ 0.05217 & 0 \\ 0 & 0.05217 \end{pmatrix},$$

$$\mathsf{C} = \begin{pmatrix} 1 & 0 & 0 & 0 & 0 & 0 \\ 0 & 1 & 0 & 0 & 0 & 0 \\ 392.3374 & 0 & -9.1181 & 0 & -11.8469 & 0 \\ 0 & -392.3374 & 0 & 9.1181 & 0 & -11.8469 \end{pmatrix},$$

and

$$\mathsf{D} = \begin{pmatrix} 0 & 0 \\ 0 & 0 \\ 20.4512 & 0 \\ 0 & 20.4512 \end{pmatrix}.$$

The unstable plant has the following eigenvalues of A:

$$s_1 = 1.4758,$$
$$s_2 = -1.6019,$$
$$s_3 = 0.0233,$$
$$s_4 = -1.6206,$$
$$s_5 = 1.4540,$$
$$s_6 = 0.0638.$$

Using the methods of linear optimal control [43], a feedback gain matrix, K, and a full-order observer gain matrix, L, have been obtained in order to stabilize the plant, and are given as follows:

$$\mathsf{K} = \begin{pmatrix} 0.42438 & 0 & 0.20235 & 0 & 0.46026 & 0 \\ 0 & -0.43085 & 0 & -0.13572 & 0 & 0.3087 \end{pmatrix},$$

$$\mathsf{L} = \begin{pmatrix} 93.39015 & 0 & -0.11292 & 0 \\ 0 & 118.76447 & 0 & 0.17771 \\ 27.24025 & 0 & -0.06689 & 0 \\ 0 & 74.18608 & 0 & 0.18239 \\ 26.5547 & 0 & -0.06421 & 0 \\ 0 & -26.86064 & 0 & -0.06507 \end{pmatrix}.$$

The closed-loop system, constructed according to Eq. (14.39), with $\mathbf{x_d} = \mathbf{0}$, has the following eigenvalues:

$$\begin{aligned} s_1 &= -1.4758, \\ s_2 &= -1.6019, \\ s_3 &= -0.0233, \\ s_4 &= -1.6206, \\ s_5 &= -1.4540, \\ s_6 &= -0.0638, \\ s_7 &= -49.1483, \\ s_8 &= -49.1068, \\ s_9 &= -1.4037, \\ s_{10} &= -2.4586, \\ s_{11} &= -0.0099, \\ s_{12} &= -0.01541. \end{aligned}$$

Clearly, the closed-loop system is asymptotically stable, with the dominant poles being s_3, s_6, s_{11}, s_{12}.

In order to build a realistic model, we assume the following second-order transfer function for the two gimbal actuators (called *servos*) [46]:

$$G_{\text{servo}} = \frac{2750}{s^2 + 84s + 2750}\,. \tag{14.53}$$

This transfer function relates the commanded and actual values of the gimbal angles, ϵ, μ. It is further assumed that the gimbal angles are limited to ± 0.1 rad. In addition, the pitch and yaw rate feedback channels are assumed to carry a white noise of nominal amplitude 3×10^{-5} rad/s, which is later increased to an off-nominal value of 10^{-3} rad/s representing especially noisy (bad) gyros. A simulation of response to an initial flow disturbance, $\alpha(0) = \beta(0) = 0.01$ rad, is carried out by the Simulink block diagram shown in Fig. 14.16, using the variable step *ode23s stiff equations* Runge–Kutta solver.[9] The simulated response, $Q(t), R(t), a_z(t), a_y(t), \epsilon(t), \mu(t)$, for the nominal and off-nominal cases is plotted in Figs. 14.17 and 14.18, respectively.

[9] A set of first-order, ordinary differential equations with very different time scales is said to be *stiff*, as it requires a special algorithm for an efficient solution.

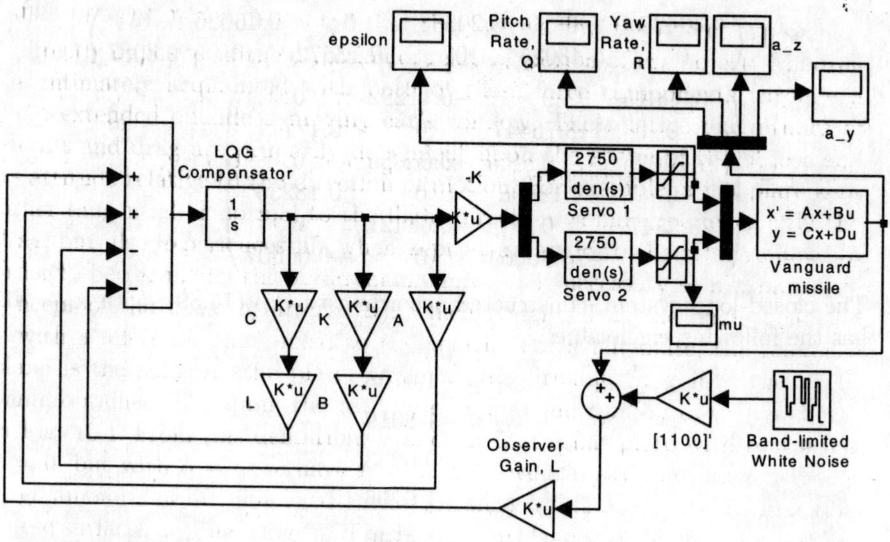

Fig. 14.16. Simulink block diagram for initial response of *Vanguard* missile stabilization system.

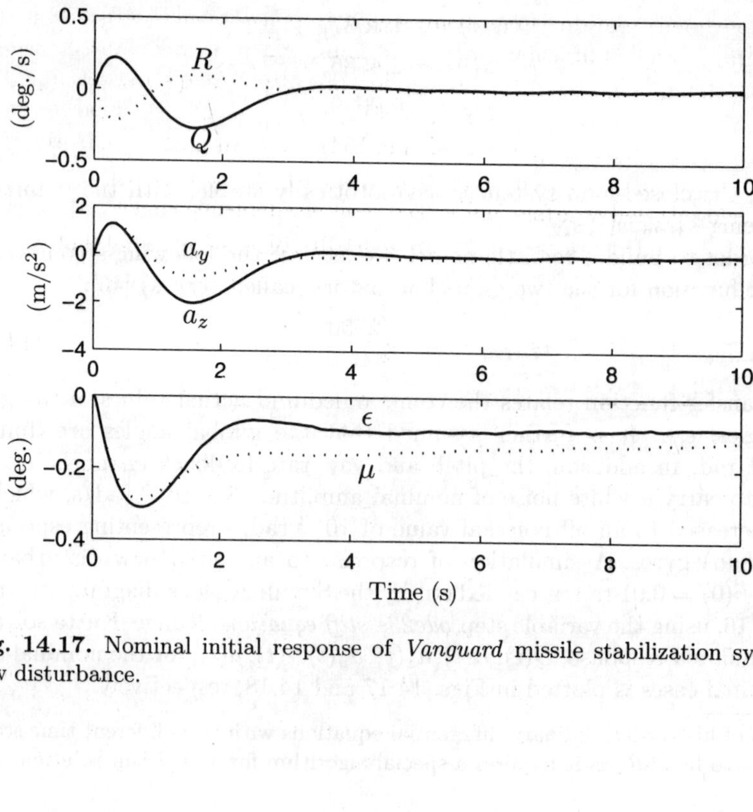

Fig. 14.17. Nominal initial response of *Vanguard* missile stabilization system to flow disturbance.

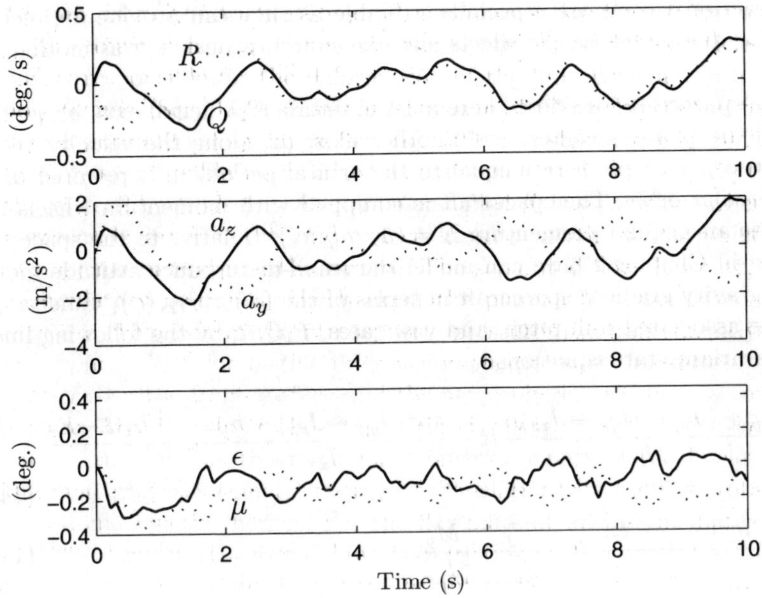

Fig. 14.18. Off-nominal (noisy gyro) initial response of *Vanguard* missile stabilization system to flow disturbance.

It is clear that the response in both the cases remains below 0.3°/s in rate and 2.5 m/s^2 in normal acceleration, while the gimbal angles do not exceed 0.5° magnitude. Although there is a distinct deterioration in the response due to increased measurement noise in the rate channels, the control system is seen to be effective in maintaining equilibrium. The simulation is carried out for 10 s as a longer time would involve changes in the model parameters due to the time-varying nature of the actual plant. A practical method of control system implementation is to vary the gain matrices with the time of flight in a scheduled manner, while employing a time-invariant plant model for obtaining the gains at a given time. This approach is referred to as *gain scheduling* and is commonly employed in high-performance aircraft and missiles.

14.7.2 Reaction Wheel and Magnetic Denutation of Gravity Gradient Spacecraft

Satellites in low- to medium-altitude orbits encounter an appreciable gravity gradient torque—as discussed and modeled in Chapter 13—the presence of which may not always be stabilizing. Therefore, an attitude control system is required for keeping the vehicle pointed in the desired direction, despite environmental torque disturbances. Since the use of RCS thrusters for attitude stabilization is expensive and reduces satellite life, reaction wheels and magnetic coils are commonly employed as actuators for attitude stabilization. The reaction wheels are especially valuable as they can provide appreciable damping due to gyroscopic effects and can simultaneously act as motion sensors.

The spacecraft considered here must maintain a principal axis, oz, pointed toward the planet's center, and another axis, ox, along the velocity vector. Thus, a constant pitch rate equal to the orbital period, n, is required in the near-circular orbit. The spacecraft is equipped with momentum wheels that generate an angular momentum $\mathbf{h} = (h_x, h_y, h_z)^T$ relative to the spacecraft. As seen in Chapter 13, we can model the small-disturbance attitude motion of the gravity gradient spacecraft in terms of the $(\psi)_3, (\theta)_2, (\phi)_1$ Euler angles and the associated roll, pitch, and yaw rates, P, Q, R, by the following linear, time-invariant state equations:

$$\dot{P} = \frac{h_y + (J_{xx} - J_{yy} + J_{zz})n}{J_{xx}} R - \frac{4n^2(J_{yy} - J_{zz}) - h_y n}{J_{xx}} \phi + \frac{h_z n - \dot{h}_x}{J_{xx}} + \frac{M_x}{J_{xx}}, \tag{14.54}$$

$$\dot{Q} = -\frac{3n^2(J_{xx} - J_{zz})}{J_{yy}} \theta - \frac{\dot{h}_y}{J_{yy}} + \frac{M_y}{J_{yy}}, \tag{14.55}$$

$$\dot{R} = -\frac{h_y + (J_{xx} - J_{yy} + J_{zz})n}{J_{zz}} P - \frac{n^2(J_{yy} - J_{xx}) - h_y n}{J_{zz}} \psi - \frac{h_x n + \dot{h}_z}{J_{zz}} + \frac{M_z}{J_{zz}}, \tag{14.56}$$

$$\begin{aligned} \dot{\phi} &= P, \\ \dot{\theta} &= Q, \\ \dot{\psi} &= R\,. \end{aligned} \tag{14.57}$$

Here, the external torque, $\mathbf{M} = (M_x, M_y, M_z)^T$, is the sum of external control torques supplied by RCS thrusters and magnetic torquers, as well as torques due to environmental disturbances such as solar radiation pressure and atmospheric drag. In our analysis, we will assume the disturbances to be modeled by a white noise of suitable intensity.

It is clear from the state equation that the small-disturbance pitch motion is uncoupled from the roll-yaw motion. The roll-yaw coupling is provided by the *bias* angular momentum component, h_y, as well as the orbital rate, n. Two distinct control systems are commonly utilized, namely those utilizing magnetic torquers and reaction wheels, respectively.

Example 14.7. Consider the *Seasat* earth-referenced satellite [2] with roll-yaw reaction wheels, and the following control law:

$$\begin{aligned} \dot{h}_x &= K_\phi \phi + K_P P + n h_z, \\ \dot{h}_z &= -K_y \phi - n h_x, \end{aligned}$$

where K_ϕ, K_P, K_y are constant gains to be determined by design. The feedback is based on the roll error signal measured by a *horizon scanner*. The roll control law has proportional and derivative terms, while yaw control has cross-coupling with roll. The importance of suppressing roll-yaw oscillations is thus the primary objective. The additional terms depending on orbital frequency cancel like terms in the state equations, and the spacecraft is made gravity gradient stable in pitch by having $J_{xx} > J_{zz}$. A wheel about the pitch axis acts as a gyro with a constant angular speed that provides a momentum bias, h_y, for strengthening roll-yaw coupling. This increases the controllability of the plant in roll-yaw motion. The spacecraft is axisymmetric about oz and has the following parameters:

$$\begin{aligned} e &= 0, \\ i &= 108^\circ, \\ J_{xx} = J_{yy} &= 25{,}100 \text{ kg.m}^2, \\ J_{zz} &= 3000 \text{ kg.m}^2, \\ n &= 0.00105 \text{ rad/s}, \\ h_y &= -24.4 \text{ N.m.s.} \end{aligned}$$

A control system design [2] reveals the following suitable gain values:

$$\begin{aligned} K_\phi &= 0.39 \text{ N.m}, \\ K_P &= 116 \text{ N.m.s}, \\ K_y &= 0.08 \text{ N.m.} \end{aligned}$$

We will simulate the response of the control system to a white-noise disturbance of amplitude 0.005 N.m in the roll-yaw torque components, which can be taken to be due to the combined effects of solar radiation, fuel sloshing, and structural vibration. The state-space coefficient matrices of the model based on the state vector, $\mathbf{x} = (P, Q, R, \phi, \theta, \psi)^T$, are given as follows:

$$\mathsf{A} = \begin{pmatrix} 0 & 0 & -0.0008466 & -4.9\times10^{-6} & 0 & 0 \\ 0 & 0 & 0 & 0 & -2.91\times10^{-6} & 0 \\ 0.0070833 & 0 & 0 & 0 & 0 & -8.54\times10^{-6} \\ 1 & 0 & 0 & 0 & 0 & 0 \\ 0 & 1 & 0 & 0 & 0 & 0 \\ 0 & 0 & 1 & 0 & 0 & 0 \end{pmatrix},$$

$$\mathsf{B} = \begin{pmatrix} 3.98406\times10^{-5} & 0 \\ 0 & 0 \\ 0 & 0.0003333 \\ 0 & 0 \\ 0 & 0 \\ 0 & 0 \end{pmatrix}.$$

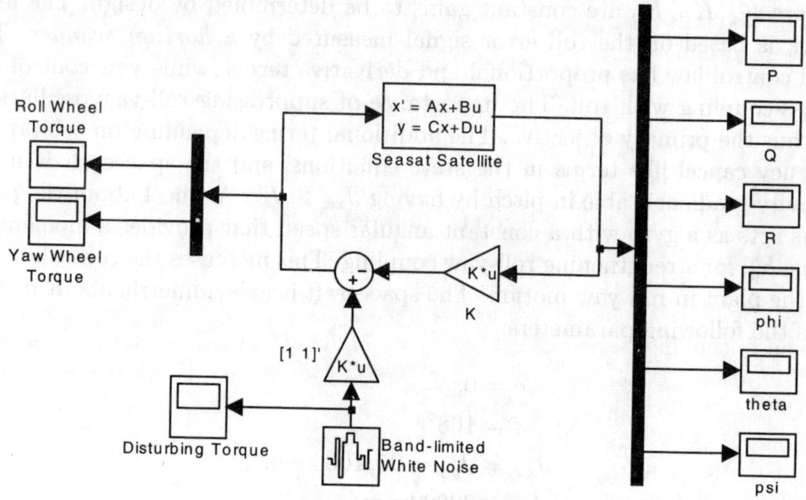

Fig. 14.19. Simulink block diagram for initial response of reaction wheel stabilization system for *Seasat* spacecraft.

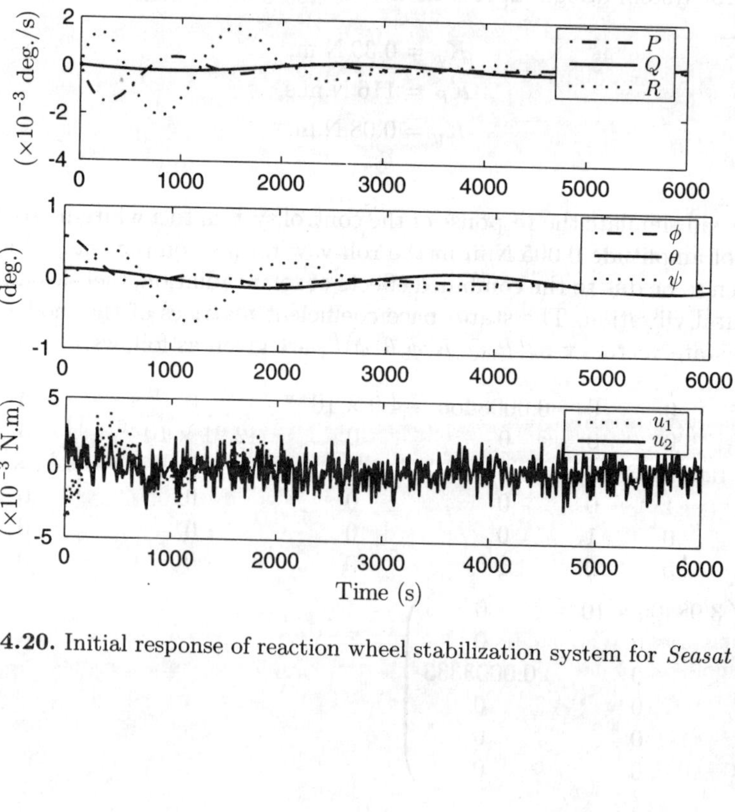

Fig. 14.20. Initial response of reaction wheel stabilization system for *Seasat* spacecraft.

The two control inputs are the torque components generated by the roll and yaw reaction wheels,

$$\mathbf{u} = (h_z n - \dot{h}_x)\mathbf{i} - (h_x n + \dot{h}_z)\mathbf{k} \ ,$$

and are given by the linear control law, $\mathbf{u} - \mathsf{K}\mathbf{x}$, where

$$\mathsf{K} = \begin{pmatrix} 116 & 0 & 0 & 0.39 & 0 & 0 \\ 0 & 0 & 0 & -0.08 & 0 & 0 \end{pmatrix} .$$

The Simulink block diagram for simulating the response for one orbital period to an initial angle error of $\theta = 0.002, \phi = 0.01$ rad in the presence of the given white-noise disturbance is shown in Fig. 14.19 and the resulting state-variable time history is plotted in Fig. 14.20. It is clear that while the pitch response is simple harmonic due to gravity gradient stabilization, the reaction wheels successfully damp out the initial roll error and absorb the disturbing torques with a steady-state accuracy of $\pm 0.1°$ in roll and yaw angles. If the pitch wheel gyro is converted into a reaction wheel and a pitch angle sensor is added to the spacecraft, the single-degree-of-freedom pitch oscillation can be damped out in the same manner as roll-yaw dynamics.

Example 14.8. Consider the magnetic torque stabilization of the *GEOS-3* satellite in a near-polar, circular earth orbit of period 100 min. The axisymmetric ($J_{xx} = J_{yy}$) spacecraft is equipped with a pitch reaction wheel for roll-yaw momentum biasing, an extensible boom with end mass for large gravity gradient torque, and a *magnetic eddy current damper*[10] that generates the following control torque [2]:

$$\mathbf{M} = k_D \mathbf{B} \times \dot{\mathbf{B}} \ , \tag{14.58}$$

where k_D is the damping constant and $\mathbf{B}$ is the strength of the earth's magnetic field, $\mathbf{B_0}$, transformed to the body frame by

$$\mathbf{B} = \mathsf{C}\mathbf{B_0} \ , \tag{14.59}$$

C being the rotation matrix between the geocentric and the body-fixed frames (Chapter 13), approximated for small 3-2-1 Euler angle deflections, ϕ, θ, ψ by

$$\mathsf{C} \approx \begin{pmatrix} 1 & \psi & -\theta \\ -\psi & 1 & \phi \\ \theta & -\phi & 1 \end{pmatrix} . \tag{14.60}$$

[10] A magnetic eddy current damper consists of a conductor plate moving between the poles of an electromagnet and provides a damping force proportional to the relative velocity between the plate and the magnet. The magnet is mounted on a floating platform, such that it can align itself instantaneously with the planet's magnetic field. For details, please refer to [50].

The geomagnetic field in a geocentric frame for a spacecraft in a polar orbit is given in Teslas by

$$\mathbf{B_0} = \frac{7.96 \times 10^6}{r^3} \begin{Bmatrix} \cos\delta \\ 0 \\ \sin\delta \end{Bmatrix}, \tag{14.61}$$

where r is the radial distance of the spacecraft in kilometers, and δ is the latitude, which varies in a circular, polar orbit as $\dot{\delta} = n$. Substituting Eqs. (14.59)–(14.61) into Eq. (14.58), we have

$$\mathbf{M} = -k_D \mathbf{e}, \tag{14.62}$$

where the error $\mathbf{e}$ is given by

$$\mathbf{e} = \begin{Bmatrix} \dfrac{4P\sin^2\delta - R\sin 2\delta + 2n\psi}{\cos^2\delta + 4\sin^2\delta} \\ Q + \dfrac{2n}{\cos^2\delta + 4\sin^2\delta} \\ \dfrac{-P\sin 2\delta + R\cos^2\delta - 2n\phi}{\cos^2\delta + 4\sin^2\delta} \end{Bmatrix}. \tag{14.63}$$

We can employ the spacecraft dynamic equations, Eqs. (14.54)–(14.57) with $h_x = h_z = 0$ and the above-derived magnetic control torque for modeling a general spacecraft in a polar, circular orbit. Due to the dependence of the control torque on latitude, the controller is essentially a linear, time-varying system. For the *GEOS-3* satellite, the parameters are given in [2] as follows:

$$\begin{aligned} e &= 0.0054, \\ i &= 115°, \\ J_{xx} &= J_{yy} = 2157 \text{ kg.m}^2, \\ J_{zz} &= 35.5 \text{ kg.m}^2, \\ n &= 0.00103 \text{ rad/s}, \\ h_y &= -2.41 \text{ N.m.s.} \end{aligned}$$

A nominal controller constant of $k_D = 0.012$ N.m.s is selected for obtaining an acceptable steady-state accuracy (at the cost of transient performance). Let us simulate the response of the spacecraft to an initial angle error of $\theta = 0.002, \phi = 0.01$ rad with the roll-pitch-yaw random disturbing torque of amplitude 6×10^{-5} N.m. The simulation begins when the spacecraft is over the south pole ($\delta = 0$) and continues for two complete orbits. The plant's state-space parameters based on the state vector, $\mathbf{x} = (P, Q, R, \phi, \theta, \psi)^T$, and input magnetic torque, $\mathbf{u} = \mathbf{M}$, are given by

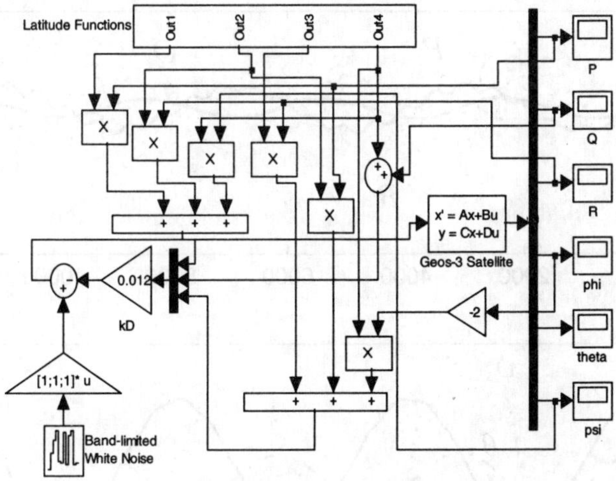

Subsystem Block "Latitude Functions"

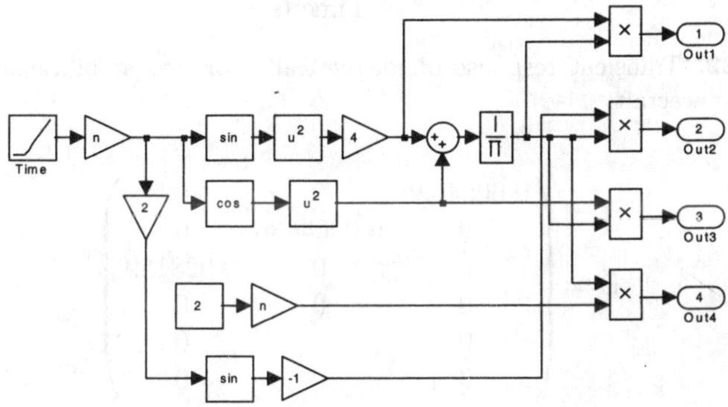

Fig. 14.21. Simulink block diagram for initial response of magnetically torqued stabilization system for *GEOS-3* spacecraft.

$$
\mathsf{A} = \begin{pmatrix}
0 & 0 & -0.0011 & -5.325\times10^{-6} & 0 & 0 \\
0 & 0 & 0 & 0 & -3.13\times10^{-6} & 0 \\
0.066857 & 0 & 0 & 0 & 0 & -6.99\times10^{-5} \\
1 & 0 & 0 & 0 & 0 & 0 \\
0 & 1 & 0 & 0 & 0 & 0 \\
0 & 0 & 1 & 0 & 0 & 0
\end{pmatrix},
$$

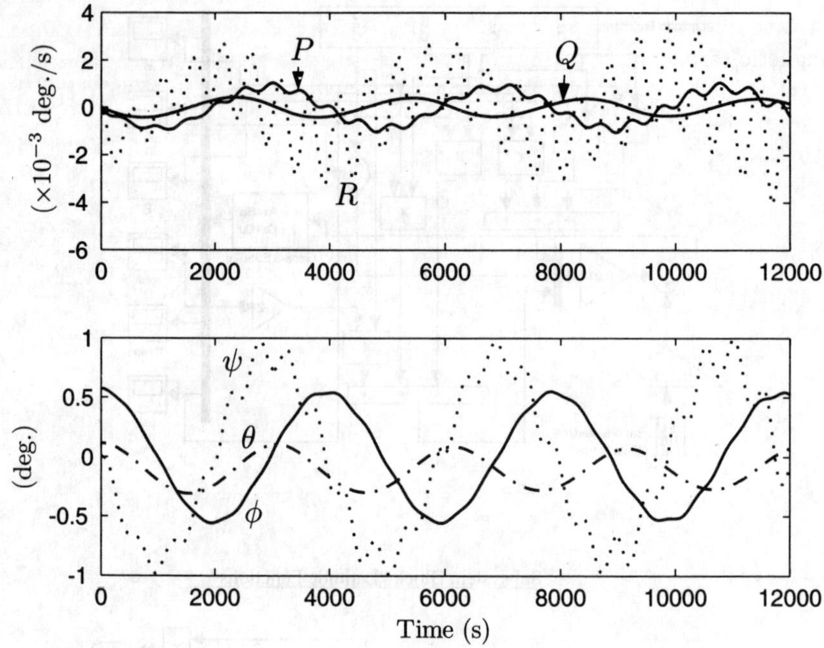

Fig. 14.22. Transient response of magnetically torqued stabilization system for *GEOS-3* spacecraft.

$$\mathsf{B} = \begin{pmatrix} 0.0004636 & 0 & 0 \\ 0 & 0.0004636 & 0 \\ 0 & 0 & 0.028169 \\ 0 & 0 & 0 \\ 0 & 0 & 0 \\ 0 & 0 & 0 \end{pmatrix}.$$

The Simulink block diagram for simulating the initial response for two orbital periods in the presence of the given white-noise disturbance is shown in Fig. 14.21 and the resulting state variable time history is plotted in Fig. 14.22. The main modeling effort occurs in deriving the controller torque components through a separate *Latitude Functions* subsystem block, and providing them as inputs to the spacecraft's state-space model. The transient response shows oscillations in all variables—the largest being in yaw angle and rate—which shows no tendency of subsiding in the span of the first two orbits. In fact, the magnetically damped system has a very large settling time of about 4.2 days, after which the root-mean-square error in Euler angles stays below $\pm 0.5°$. The long-period dynamics can be studied by extending the simulation to 5 days and plotting the response with a larger sampling interval. This is carried out in Fig. 14.23, which has a sampling interval of 10^4 s. The damping of transient response is evident in this plot, and an accuracy of $\pm 0.5°$ in all the attitude angles is observed after about 72 hours. The inadequate damping of magnetic torquers can be improved either by adding roll-yaw reaction wheels or by reducing the pitch-roll inertia for a smaller gravity gradient restoring torque (and thus smaller-amplitude libration). The latter is easily achieved by reducing the deployed boom length at a suitable time.

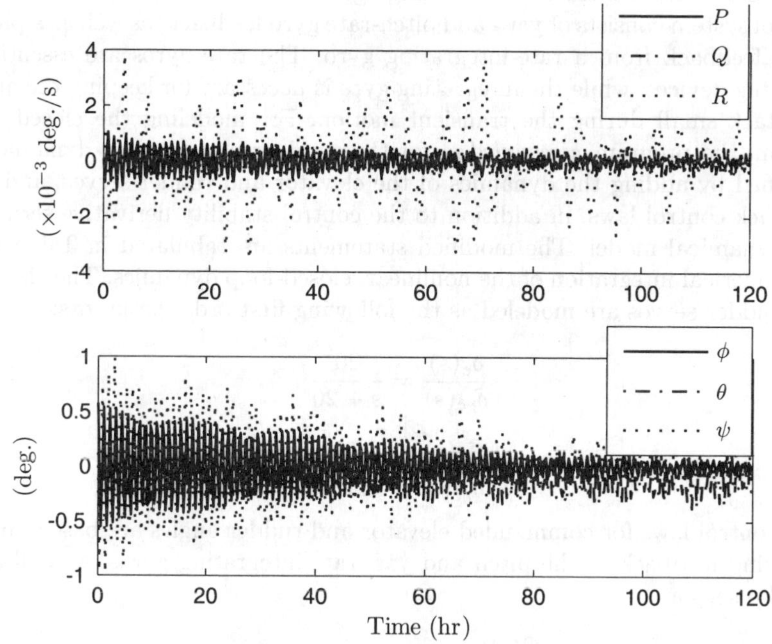

Fig. 14.23. Long-term response of magnetically torqued stabilization system for *GEOS-3* spacecraft.

14.7.3 Control of Aircraft and Missiles with Inertia Coupling

We saw in Chapter 13 how inertia coupling could result in an unintended departure from equilibrium about an axis (pitch) not involved in maneuvering. The effects of inertia coupling are largely confined to fighter airplanes and aerodynamic missiles rolling at high rates. Since the motion is a multi-axis rotation, along with plunge and sideslip, it requires a six-degree-of-freedom, nonlinear model (Chapter 13). From the controls' perspective, inertia coupling offers the advantage of a greater controllability with a given control input. For example, a rudder input is capable of affecting the pitch rate in an inertia-coupled vehicle. This indicates that a single control loop, say about the yaw axis, may be adequate for damping out the coupled dynamic response about the other axes. However, for a better closed-loop performance, as well as increased damping, a multi-axis control system is beneficial. We shall see here how such a control system can be modeled for a fighter aircraft. The application to aerodynamics is similar, if less complex, due to multiple planes of symmetry ($J_{xz} = 0, J_{yy} = J_{zz}$).[11]

Example 14.9. Let us model a flight-control system for damping the inertia-coupled dynamics of the initially rolling fighter aircraft of Example 14.12. The control system consists of yaw- and pitch-rate gyro feedback, as well as a pitch-angle feedback from a rate-integrating gyro. The rate gyros are essentially damping devices, while the integrating gyro is necessary for keeping the angle of attack small during the transient motion. For modeling the closed-loop system, the program *aircoupled.m* for the six-degree-of-freedom dynamics is modified by adding the dynamics of the elevator and rudder servos, and the feedback control laws, in addition to the control stability derivative terms in the dynamical model. The modified statements are tabulated in Table 14.1 for numerical integration of the nonlinear, closed-loop dynamics. The elevator and rudder servos are modeled as the following first-order actuators:

$$\frac{\delta_e(s)}{\delta_{ec}(s)} = \frac{20}{s+20},$$

$$\frac{\delta_r(s)}{\delta_{rc}(s)} = \frac{20}{s+20}.$$

The control laws for commanded elevator and rudder angles are based on the following feedback of the pitch and yaw rate-integrating gyros, as well as a sideslip sensor:

$$\delta_{ec}(s) = 0.1\frac{Q(s)}{s} \quad (|\, Q \,| \geq 0.001 \text{ rad}),$$
$$\delta_{rc}(s) = 0.025\frac{R(s)}{s} - 0.05\beta(s) \quad (|\, R \,| \geq 0.001 \text{ rad}).$$

These transfer functions are converted into differential equations in time before putting them in the MATLAB program. The rate feedbacks are deliberately avoided due to the possibility of self-induced, growing oscillations in the coupled nonlinear dynamics. Furthermore, the control is not activated immediately, but is delayed until the pitch and yaw rates grow to 0.001 rad, in order that the longitudinal short-period and Dutch-roll open-loop dynamics are unhindered during the first few seconds, after which inertia coupling manifests itself in the rapid growth of pitch and yaw rates. If a feedback control is applied in this incipient stage, there is a possibility that the inertia coupling might become amplified, leading to large transient errors, and even instability.

[11] It can be easily seen from Chapter 13 that a fighter aircraft can encounter inertia coupling even for a zero roll rate, while a missile with $P = 0$ has de-coupled lateral and longitudinal dynamics.

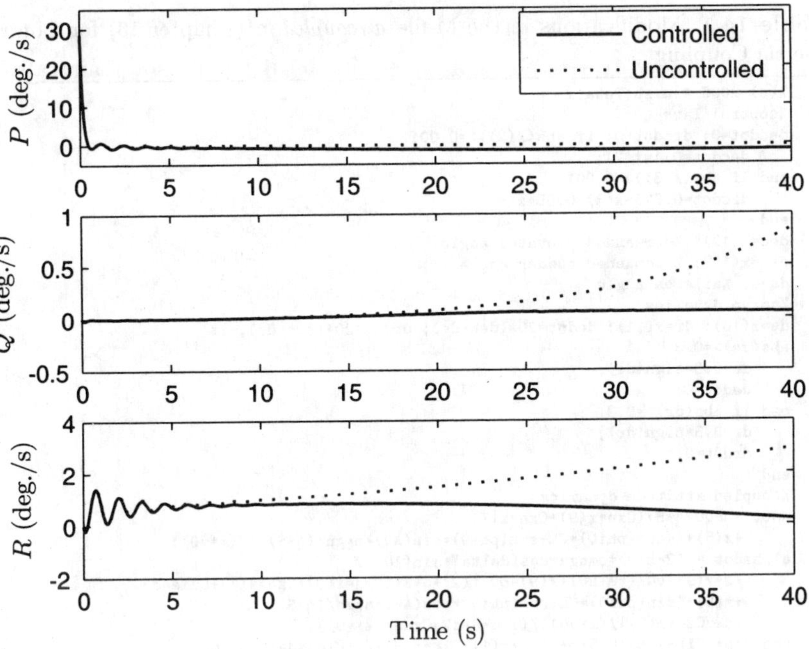

Fig. 14.24. Body rates for the controlled and uncontrolled inertia-coupled dynamics of a fighter aircraft.

The resulting closed-loop simulation plots for the first 40 s are compared with the corresponding plots for uncontrolled inertia-coupled dynamics in Figs. 14.24–14.26, while the closed-loop elevator and rudder deflections are plotted in Fig. 14.27. It can be seen that the roll rate is brought down to a small magnitude without letting the pitch and yaw rates diverge for the controlled case. The pitch and yaw angle deviations are also kept small for the closed-loop system, which is also successful in maintaining small angles of attack and sideslip, with a negligible change in the forward speed compared to that of the open-loop system. For the selected values of the feedback gains, the elevator and rudder angles stay well below their allowable maximum magnitude of 0.3 rad. The aircraft continues turning at a small yaw rate after 40 s. The unstable, long-period spiral mode can be easily controlled by the pilot after the effects of inertia coupling have been removed by the control system.

14.8 Summary

It is necessary to have an attitude control system in all flight vehicles, the most useful design and analysis of which is by the linear systems theory. While a single-input, single-output plant is modeled by a transfer function, multivariable plants are more easily handled by a state-space approach consisting of

Table 14.1. Modifications in the M-file *aircoupled.m* (Chapter 13) for Control of Inertia Coupling

```
%(c) 2006 Ashish Tewari
%control laws
decdot=0; drcdot=0; if abs(x(2))>=0.001
    decdot=0.1*x(2);
end if abs(x(3))>=0.001
    drcdot=0.025*x(3)-0.05*x(8);
end
dec=x(12); %commanded elevator angle
drc=x(13); %commanded rudder angle
da=0; %aileron angle
%servo dynamics
de=x(10); dr=x(11); dedot=20*(dec-de); drdot=20*(drc-dr); if
abs(de)>=0.3
    de=0.3*sign(de);
    dedot=0;
end if abs(dr)>=0.3
    dr=0.3*sign(dr);
    drdot=0;
end
%coupled attitude dynamics
udot = -Q0+q*S*(Cxu*x(9)+Cxa*x(7)...
     +x(5)*(-cos(phi0)*CL+sin(phi0)*sin(A0)*m*gn/(q*S)))/(m*v0);
alphadot = (2*x(9)*omega*cos(delta)*sin(A0)...
     +2*x(9)*v0/(rm+h0)+x(9)*Q0+x(2)+q*S*(Czu*x(9)+Cza*x(7)+Czde*de...
     +x(5)*(sin(phi0)*CL-cos(phi0)*cos(A0)*m*gn/(q*S))...
     +c2vCzq*x(2))/(m*v0))/(1-q*S*c2vCzad/(m*v0));
L=q*S*b*(Clb*x(8)+0.5*b*(Clp*x(1)+Clr*x(3))/v+Clda*da+Cldr*dr);
M=q*S*c*(Cma*x(7)+c2vCmq*x(2)+c2vCmad*alphadot+Cmde*de);
N=q*S*b*(Cnb*x(8)+0.5*b*(Cnp*x(1)+Cnr*x(3))/v+Cndr*dr+Cndrd*drdot+Cnda*da);
jxz=Jxx*Jzz-Jxz^2; P=x(1);Q=x(2);R=x(3);
pdot=(Jxz*(Jzz+Jxx-Jyy)*P*Q-(Jxz^2+Jzz*(Jzz-Jyy))*Q*R+Jxz*N+Jzz*L)/jxz;
qdot=(Jxz*(R^2-P^2)+(Jzz-Jxx)*P*R+M)/Jyy;
rdot=(Jxz*(pdot-Q*R)+(Jxx-Jyy)*P*Q+N)/Jzz;
phidot=x(1)+(x(2)*sin(x(4))+x(3)*cos(x(4)))/cos(x(5));
thetadot=x(2)*cos(x(4))-x(3)*sin(x(4));
psidot=(x(2)*sin(x(4))+x(3)*cos(x(4)))/cos(x(5));
betadot=-x(3)+v*cos(phi0)^2*cos(A0)*tan(delta)*x(8)/(rm+h0)...
       +2*omega*cos(delta)*sin(phi0)*sin(A0)*x(8)...
       +g*x(4)*cos(phi0)/v+gn*(x(4)*sin(phi0)*cos(A0)-x(6)*cos(A0))...
       +q*S*(Cyb*x(8)+Cydr*dr+b*Cyr*x(3)/(2*v))/(m*v);
xdot=[pdot;qdot;rdot;phidot;thetadot;psidot;alphadot;betadot;udot;
      dedot;drdot;decdot;drcdot];
```

a set of first-order, ordinary differential equations in time. Time-invariant, linear systems are commonly used to model a flight-control system with well-established stability and performance criteria. While the controller design by the transfer function approach is based upon the concept of second-order system—extended to higher-order systems with a dominant pair of complex conjugate poles—and a general proportional-integral-derivative (PID) controller, the state-space design is generally based upon the optimal control theory. The basic attitude control system in practically all flight vehicles is based upon gyroscopes as feedback mechanisms, which can provide either proportional (rate gyro) or integral action (rate-integrating gyro) for a PID—or

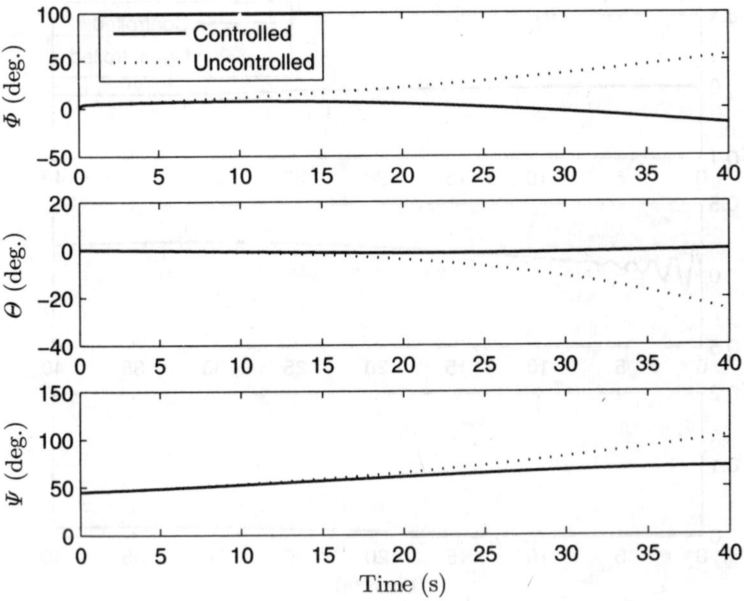

Fig. 14.25. Euler angles for the controlled and uncontrolled inertia-coupled dynamics of a fighter aircraft.

a multivariable—controller setup. Examples of gyro-based control systems include autopilots for airplanes, missiles, and rotating spacecraft. Apart from the rate gyro, a sensitive and accurate sensor of the vehicle's rotational motion is an accelerometer, which has the additional advantage of compact and rugged design. Accelerometer output can provide necessary information for reconstruction of state vector in an observer based control system—such as that employed in thrust vector-controlled, nonrolling ballistic missiles and launch vehicles. Since the use of attitude thrusters for attitude stability and control is expensive and reduces satellite life, reaction/momentum wheels and magnetic coils are commonly employed as actuators in spacecraft. The control of inertia-coupled dynamics of aircraft and missiles requires a multi-axis rotational, six-degree-of-freedom, nonlinear model, which offers the advantage of greater controllability with a given control input, but is difficult to model and analyze. Although a single control loop is adequate for damping out the coupled dynamic response about the other axes, for a better closed-loop performance, as well as increased damping, a multi-axis control system is beneficial.

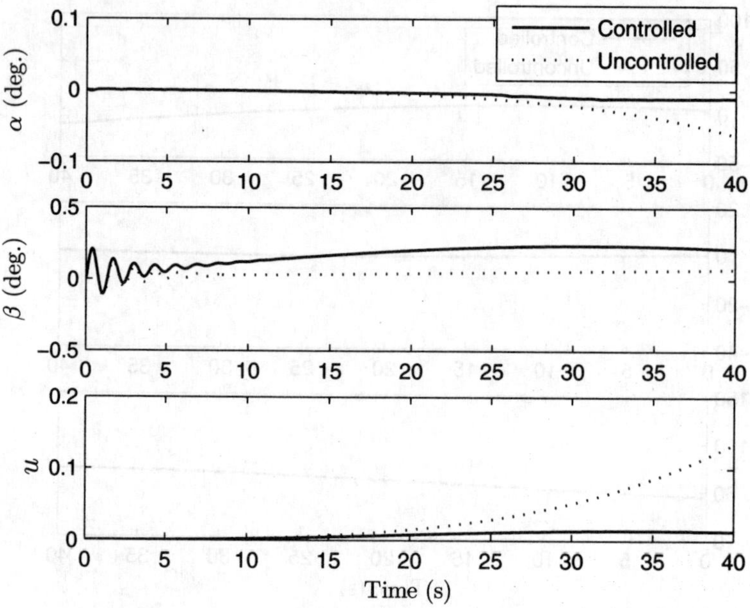

Fig. 14.26. Flow angles and forward speed ratio for the controlled and uncontrolled inertia-coupled dynamics of a fighter aircraft.

Exercises

14.1. Repeat Example 14.1 for the *Honeywell HIG-4* [46] rate-integrating gyro with the following characteristics [46]: $H_r = 10^4$ g–cm^2/s, $J = 35$ g–cm^2, and $c = 10^4$ g–cm^2/s. What differences do you observe compared to the rate gyro considered in the example?

14.2. Simulate a spin maneuver of the spacecraft of Example 14.2 controlled by the given attitude thrusters and the rate gyro of Example 14.1. The maneuver involves a step change in the angular rate by 0.1 rad. Try the following gains multiplying the rate gyro output:
(a) $K_D = 10$.
(b) $K_D = 30.3$.
(c) $K_D = 50$.

14.3. Redesign the multivariable control system for the *Vanguard* missile (Example 14.6) using the *MIT* rate gyros for pitch and yaw rate feedback, and a PD feedback of the normal accelerations, by adjusting the gains for each channel until a satisfactory closed-loop response is obtained for a simultaneous desired step change of 0.01 rad/s in the pitch and yaw rates.

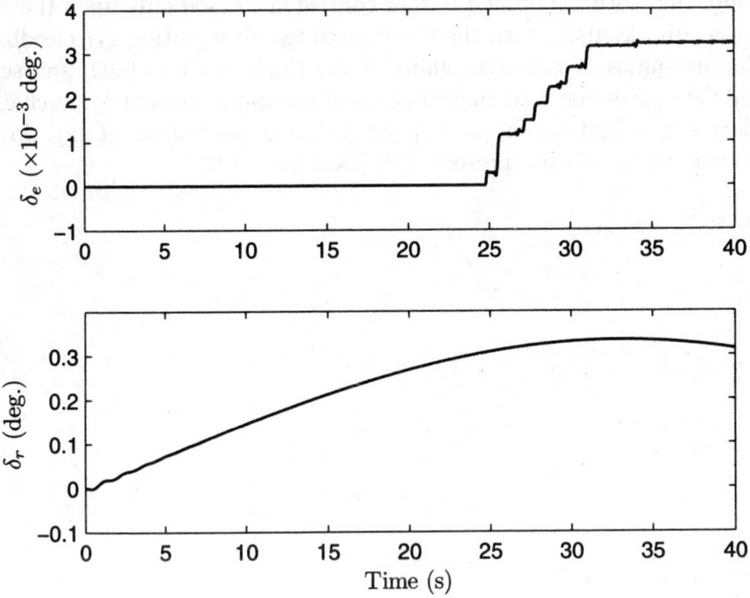

Fig. 14.27. Closed-loop elevator and rudder deflections for the controlled inertia-coupled dynamics of a fighter aircraft.

14.4. Convert the pitch gyro to a pitch reaction wheel, and add a pitch sensor to damp out the gravity gradient oscillation of the *Seasat* satellite of Example 14.7. Then, by using a PD control law similar to the roll reaction wheel, compare the response of the modified control system to the same initial angular errors and disturbing torques with that given in the example.

14.5. Consider a spacecraft in a geosynchronous orbit ($n = 0.000073$ rad/s), which makes the gravity gradient torques negligible. The spacecraft has the same wheel configuration and controller gains as in *Seasat*, but the moments of inertia are $J_{xx} = J_{zz} = 1000$ kg.m^2 and $J_{yy} = 200$ kg.m^2. In addition to the roll-yaw reaction wheels of Example 14.7, a pitch wheel control system is added, such as in Exercise 14.4, to provide three-axis stabilization. Simulate the response of the spacecraft to an initial error of $\theta = \phi = \psi = 0.01$ rad in the presence of a random torque disturbance of maximum magnitude 10^{-4} N.m.

14.6. Repeat the simulation of the *GEOS-3* magnetically damped control system of Example 14.8 with a shorter gravity gradient boom, which reduces the pitch-roll moment of inertia to $J_{xx} = J_{yy} = 300$ kg.m^2 without affecting the other parameters. How does the new spacecraft respond to the same initial error and disturbance as given in Example 14.8?

14.7. Simulate the closed-loop, inertia-coupled dynamics of the fighter aircraft of Example 14.9 with a modified rudder control law based only upon the yaw-rate gyro feedback, along with the given pitch rate-integrating gyro feedback. Try different values of rate-gyro gains in the range −0.1 to 0.02, and select the value that gives the best inertia-coupled damping without large changes in the flow angles and the forward speed. What improvement, if any, do you observe over the simulation presented in Example 14.9?

15

Advanced Modeling and Simulation Concepts

15.1 Aims and Objectives

- To introduce advanced modeling and simulation concepts useful in flight dynamics.
- To provide additional six-degree-of-freedom simulation examples.
- To address modeling of flexible vehicle dynamics with Lagrangian approach.
- To discuss the importance and modeling of unsteady aerodynamics, aeroelasticity, and propellant slosh dynamics in flight dynamic applications.

We have considered above how the translational and rotational motions of an aerospace vehicle can be modeled appropriately, either with or without an automatic control system. In many cases, we have applied relevant assumptions that led to a decoupling of the degrees of freedom, such as small-disturbance atmospheric flight, and modeled additional dynamics as well as disturbances as white-noise inputs. The accuracy of a simulation can be improved by modeling additional degrees of freedom of the vehicle, such as structural flexibility and propellant slosh dynamics, as well as by taking unsteady aerodynamic effects into consideration. Furthermore, external disturbances—such as atmospheric gusts—can be modeled using statistical methods. Finally, control system dynamic models can be enhanced by including nonlinear sensor and actuator dynamics, as well as advanced flight-control laws. In this chapter, we will briefly discuss the modeling and simulation of additional vehicle dynamics.

15.2 Six-Degree-of-Freedom Simulation

An atmospheric flight vehicle has inherently coupled rotational and translational dynamics due to the nature of aerodynamic force and moment. Such a coupling is, however, absent in a spacecraft, where the rotational and translational motions can be studied using separate three-degree-of-freedom models. In most airplanes and lifting entry vehicles, the small-disturbance aerodynamics lead to a decoupling of the motion into linearized longitudinal and lateral dynamics, each of which is modeled as separate three-degree-of-freedom motions. However, there are several instances where the lateral and longitudinal dynamics cannot be decoupled, such as the inertial and aerodynamic coupling caused by a rapidly rolling fighter airplane, missile, or a ballistic entry capsule. In Chapter 13 we considered the inertia-coupled dynamics of a fighter airplane—which can be extended to a rolling missile—and the same was controlled using a multi-axis feedback control system in Example 14.9. In such a case, the six-degree-of-freedom dynamics is modeled by nonlinear state equations, in which the aerodynamic force and moment are governed by linear models. A rapidly maneuvering airplane at a large angle of attack is, however, modeled using nonlinear aerodynamic models, due to the near-stall, or post-stall separated flow on the vehicle. In this section, we shall consider two examples of six-degree-of-freedom simulation, namely a rapidly rolling fighter airplane flying at a large angle of attack and a ballistic entry capsule.

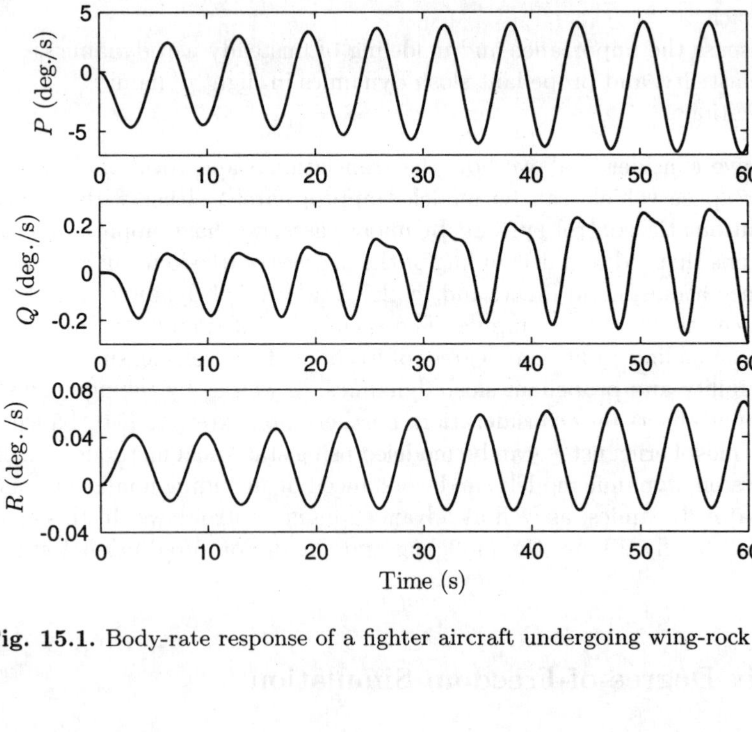

Fig. 15.1. Body-rate response of a fighter aircraft undergoing wing-rock motion.

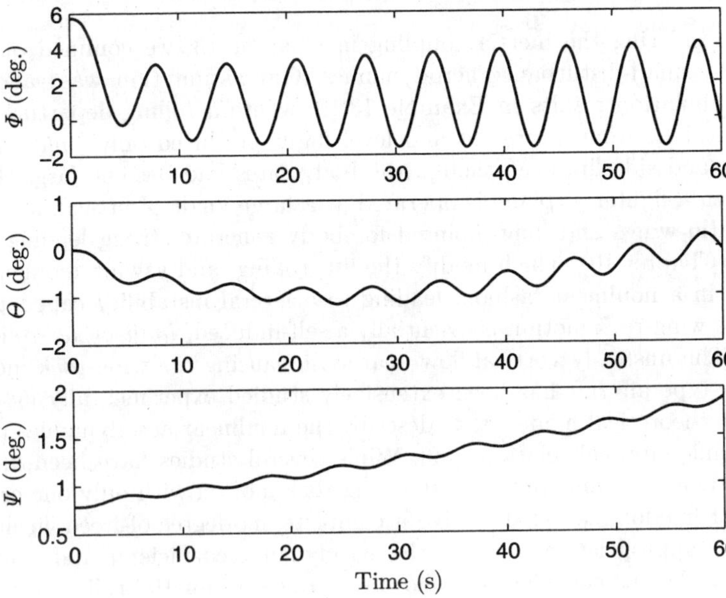

Fig. 15.2. Attitude time history of a fighter aircraft undergoing wing-rock motion.

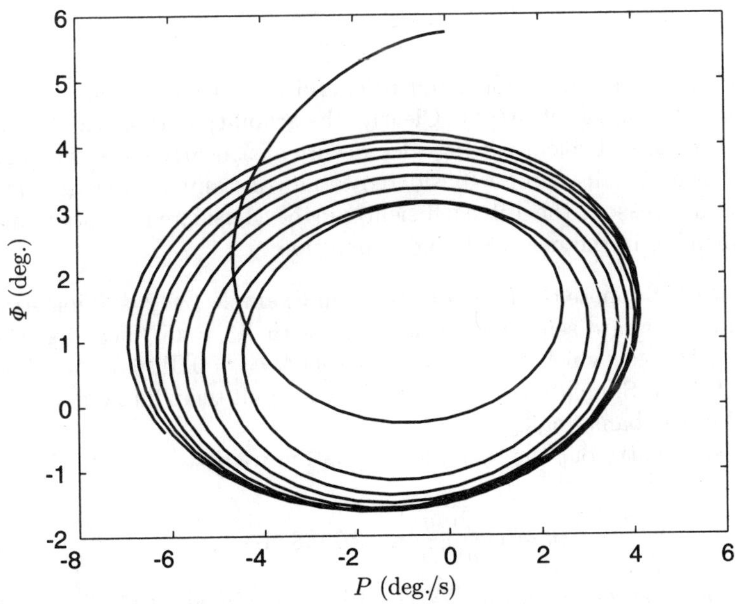

Fig. 15.3. Phase-plane plot of bank angle vs. roll rate during wing-rock motion.

15.2.1 Wing-Rock Motion of a Fighter Airplane

While considering the inertia coupling in Chapter 13, we confined our attention to small-disturbance aerodynamics. Our assumption was validated by the simulation results in Example 13.12, where a rolling departure of a fighter airplane from a straight and level flight produced only small angles of attack and sideslip, even though the body rates had become large. However, when a fighter airplane is operated at a large angle of attack, its small aspect-ratio wings and long, pointed forebody generate strong leading-edge vortices (Chapter 10), which modify the lift, rolling- and yawing-moment coefficients in a nonlinear fashion, leading to a lateral instability called *wing-rock*. The wing-rock motion is essentially a self-induced, *limit-cycle*[1] roll-yaw motion. The unsteady vortical flow near stall, causing the wing-rock motion in fighter-type aircraft has been extensively studied experimentally [58], resulting in theoretical models that describe the nonlinear aerodynamics using simple semi-empirical relations [59]. While several studies have been undertaken of dynamics and control of the wing-rock motion with only one or two degrees of freedom, we shall derive a complete six-degree-of-freedom model by making appropriate changes in the aerodynamic coefficients and stability derivatives. Let us consider the following expression for the rolling-moment coefficient of a fighter airplane at a large angle of attack:

$$C_l = f(\Phi, \bar{P}) + C_{l_\beta}\beta + C_{l_{\delta_a}}\delta_a \, , \tag{15.1}$$

where

$$f(\Phi, P) = a_1\bar{P} + a_2\bar{P}^3 + a_3\Phi^2\bar{P} + a_4\Phi\bar{P}^2 \, , \tag{15.2}$$

$\bar{P} \doteq \frac{b}{2v}P$ is the nondimensional roll rate, and the coefficients $a_0, \ldots, a_4$ are functions of the angle of attack. Clearly, the stability derivative, C_{l_p}, is no longer a constant. Typically, wing-rock occurs at a nearly constant angle of attack, thus $a_0, \ldots, a_4$ can be considered to be constants during the motion. For the same reason, the lift coefficient can be considered constant at the maximum value in the near-stall flight condition.

Example 15.1. Suppose the fighter airplane in Example 13.12 is flying straight and level at standard sea level and $\delta = 45°$, with $A^e = 45°$, near its stalling angle of attack, for which $C_L = 1.6$, $a_1 = -0.04$, $a_2 = 0.0126$, $a_3 = -0.1273$, and $a_4 = 0.5197$. Simulate the response of the aircraft to an initial disturbance of 0.1 rad in the bank angle.

In order to carry out the required simulation, we first calculate the initial speed by

$$v = \sqrt{\frac{2mg}{\rho S C_L}} = 52.7488 \text{ m/s}.$$

Next, the program *aircoupled.m* (Table 13.11) is modified for the present task by changing the rolling-moment equation to the following statement:

[1] Limit cycle is a sustained, constant-amplitude oscillation at a fixed frequency.

```
L=q*S*b*(Clb*x(8)+0.5*b*(Clr*x(3)-0.04*x(1))/v+0.0126*(0.5*x(1)*b/v)^3...
  -0.1273*(0.5*x(1)*b/v)*x(4)^2+0.5197*(0.5*x(1)*b/v)^2*x(4));
```

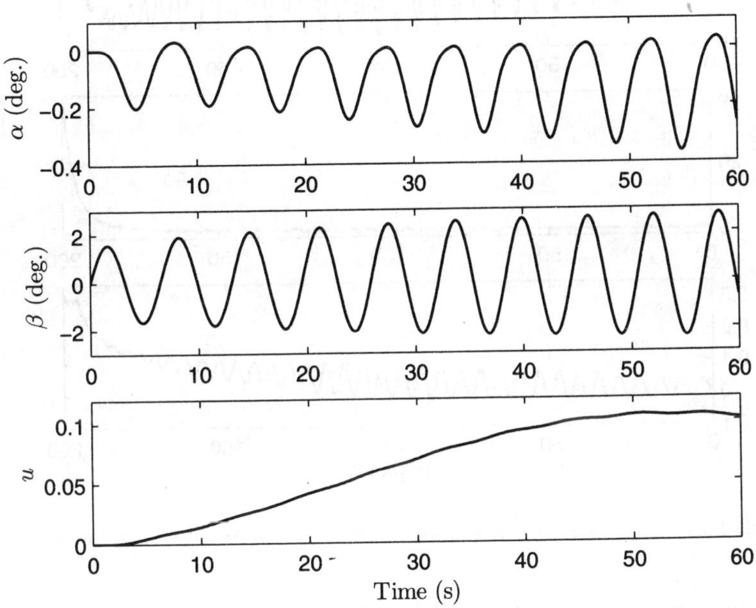

Fig. 15.4. Flow angles and speed disturbance of a fighter aircraft undergoing wing-rock motion.

The simulated state variables are plotted in Figs. 15.1–15.4 for the first 60 s. The bank angle, Φ, and the roll rate, P, display an increasing amplitude rolling oscillation that reaches a limit cycle in about 60 s. The associated oscillation in the heading angle, Ψ, is superimposed on a steady increase with time, while the yaw rate, R, reaches a nearly constant amplitude almost simultaneously with Φ, P. The longitudinal coupling with the lateral motion is evident in an oscillating pitch angle, Θ, and pitch rate, Q. Due to the inherently nonlinear nature of the inertia coupling, the pitch oscillation is not simple harmonic and shows an increasing tendency. The plot of Φ against P (Fig. 15.3)—called a *phase-plane* plot—reveals the limit-cycle behavior of the rolling motion. A nonlinear stability analysis usually requires phase-plane plots, wherein the tendency to remain near an equilibrium point can be easily investigated. Finally, we analyze the disturbances caused in the translational motion by studying the time history of u, α, β, plotted in Fig. 15.4. It is interesting to observe that while the sideslip angle, β, has a simple-harmonic limit-cycle response about $\beta = 0$, that of the angle of attack, α, is non-harmonic and takes place around an equilibrium value of $\alpha = -0.2°$. The amplitude of

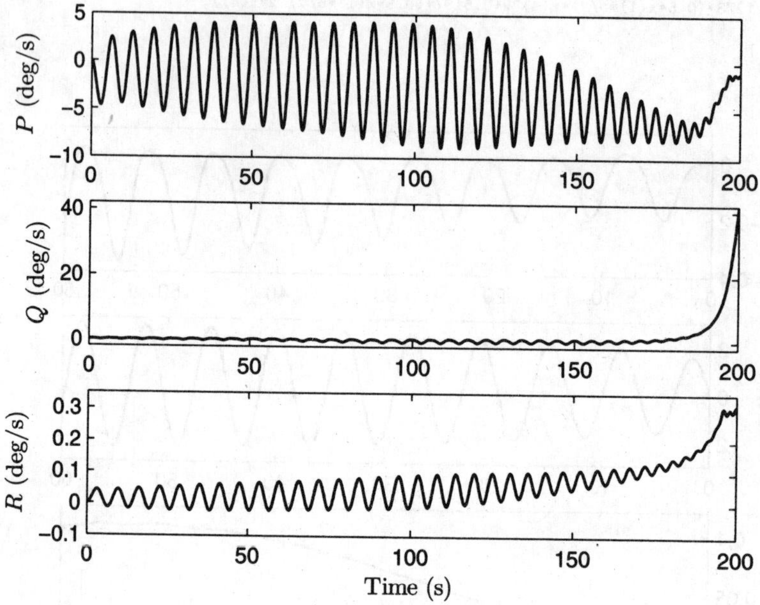

Fig. 15.5. Long period time-history of body-rates in wing-rock motion.

α is about one tenth that of β and is sufficiently small (less than 0.4°) for the constant α assumption in the aerodynamic model to be valid. Due to the negative α, there is a tendency for the speed to increase slightly ($u \leq 0.12$), reach a maximum around 60 s, and then to begin a shallow decline, thereby indicating a long-period oscillation. Therefore, the airplane does not stall and continues its wing-rock motion. However, as the simulation is increased to a longer time (200 s), the inertia coupling ultimately leads to a divergence in the pitch rate, as shown in Fig. 15.5, causing a steep nose-down attitude, and a dive (Fig. 15.6) in which the wing-rock cycle is broken and the roll rate, sideslip, and yaw rate return to small, nearly constant values. The control of wing-rock motion essentially involves breaking the limit cycle by additional damping provided by a closed-loop aileron (or rudder) input based on roll-rate feedback (Chapter 14).

15.2.2 Trajectory and Attitude of a Ballistic Entry Vehicle

Atmospheric entry vehicles experience increasing aerodynamic force and moment as the flight progresses. The ever-changing flow from rarefied to continuum subsonic causes large changes in the force and moment, while the inertia coupling leads to the simultaneous excitation of all the degrees of freedom, especially if the vehicle is rotating during the exo-atmospheric phase. Consequently, six-degree-of-freedom simulations are necessary for entry vehicles. In

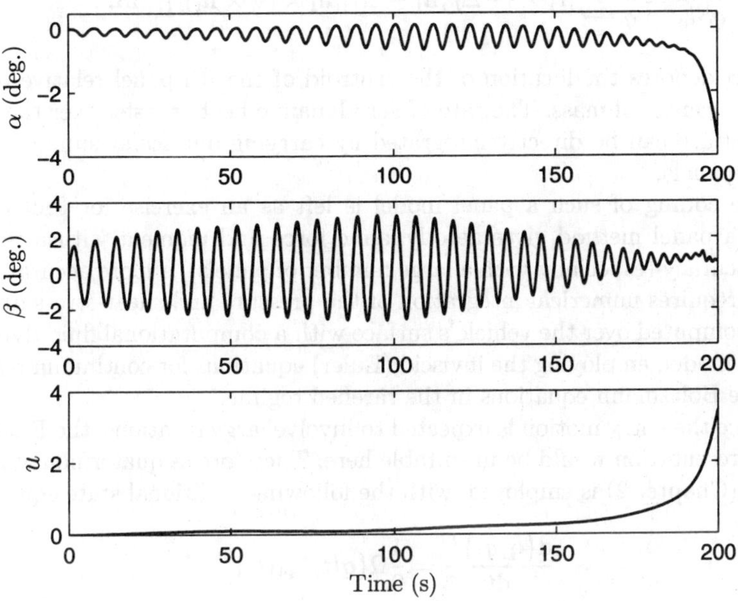

Fig. 15.6. Long-period time history of flow angles and speed disturbance in wing-rock motion.

Chapter 10 we considered the modeling of aerothermal loads in rarefied and continuum regimes, while in Chapter 12 the translational dynamics of both ballistic and lifting entry vehicles were simulated. Here, we shall additionally take into account the aerodynamic moments generated during entry. In order to do so, a panel approximation is employed wherein the vehicle's external surface is approximated by a number, N, of flat panels. On each panel, we carry out the pressure and shear-stress calculation using the free-molecular flow [Eqs. (10.36)–(10.37)], the continuum hypersonic flow by Newtonian approximation of Eq. (10.48), and a suitable boundary-layer model, or the rarefied transition regime by exponential *bridging relations* in the Knudsen number for interpolation between the quantities in the continuum and free-molecular limits. The pressures and shear stresses of all the panels are then summed vectorially for the aerodynamic force over the vehicle as follows:

$$\frac{\mathbf{f_a}}{qS} = \frac{1}{q}\sum_{i=1}^{N}\{-\Delta p_i\mathbf{n_i} + \Delta\tau_i[\mathbf{n_i}\times(\mathbf{v}\times\mathbf{n_i})]\}\,\Delta S_i\,, \tag{15.3}$$

where v, q denote the free-stream speed and dynamic pressure, and $\Delta S_i, \mathbf{n_i}$ are the area and unit normal of the ith panel. The aerodynamic moment is similarly computed by taking the weighted vector summation of the elemental pressure and shear stress by multiplying with appropriate moment arms:

$$\frac{\mathbf{M_a}}{qSl_c} = \frac{1}{q}\sum_{i=1}^{N}[\mathbf{r_i} \times \{-\Delta p_i \mathbf{n_i} + \Delta\tau_i[\mathbf{n_i} \times (\mathbf{v} \times \mathbf{n_i})]\}]\,\Delta S_i \,. \tag{15.4}$$

Here, $\mathbf{r_i}$ denotes the location of the centroid of the ith panel relative to the vehicle's center of mass. The rate of aerodynamic heat transfer over the vehicle's surface can be directly integrated by carrying out scalar summation of all the panels.

The coding of such a panel model is left as an exercise for the reader. While a panel method gives aerodynamic force and moment with a reasonable accuracy (especially with a large number of panels), a more accurate flow model requires numerical integration of the pressure and shear-stress distribution computed over the vehicle's surface with a computational fluid dynamic (CFD) model, employing the inviscid (Euler) equations for continuum regime, and the Boltzmann equations in the rarefied regime.

Since the entry motion is expected to involve large rotations, the Euler angle representation would be unsuitable here. Therefore, a quaternion attitude model (Chapter 2) is employed, with the following additional state equations:

$$\frac{\mathrm{d}\{\mathbf{q}, q_4\}^T}{\mathrm{d}t} = \frac{1}{2}\Omega\{\mathbf{q}(t), q_4(t)\}^T \,, \tag{15.5}$$

where

$$\Omega = \begin{pmatrix} 0 & R & -Q & P \\ -R & 0 & P & Q \\ Q & -P & 0 & R \\ -P & -Q & -R & 0 \end{pmatrix} . \tag{15.6}$$

The flow angles α, β, μ are calculated from the attitude of the body axes relative to the wind axes (Chapter 13).

Example 15.2. Consider the ballistic entry vehicle, with entry conditions described in Example 12.6. Instead of the approximate drag coefficient, we shall employ the flow model based upon the above-described panel approximation. The configuration is a flared sphere cone, with base area $S = 3.24$ m^2, mass 350 kg, and length 1.63 m. The panel configuration of the capsule with $N = 21$ is graphically depicted in Fig. 15.7, with all axes in meters, while its principal moments of inertia about the body axes are $J_{xx} = 80$, $J_{yy} = 120$, $J_{zz} = 130$ kg/m^2.

We shall simulate the coupled translational and rotational motions of the vehicle from the same nominal geocentric position and velocity as given in Example 12.6, and with additional initial condition of $P = 1$, $Q = 0.1$, $R = -0.1$ rad/s, $\alpha = \beta = \mu = 0$. The plots resulting from the simulation are given in Figs. 15.8–15.12. Figures 15.8–15.10 are the plots describing the trajectory, and are quite similar to those obtained in Example 12.6. The latitude and longitude of the impact point (near which the flight path becomes vertical) are obtained as $\delta = 85.48°$, $\lambda = 10.072°$. The rotational motion of the vehicle is completely described by the time history of the body rates (Fig. 15.11), and

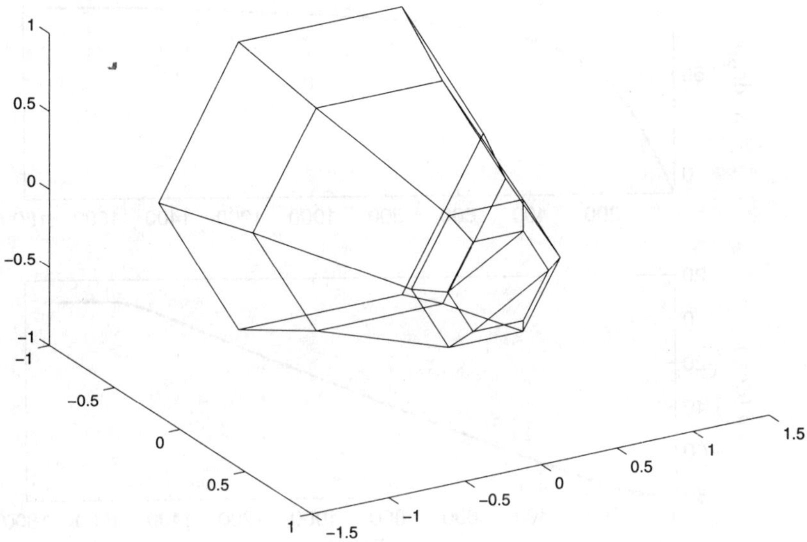

Fig. 15.7. Panel approximation of the ballistic entry vehicle.

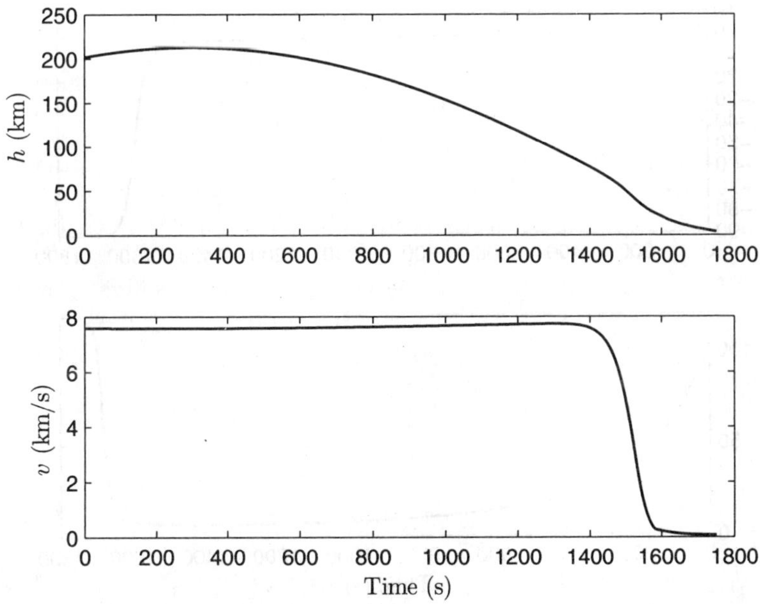

Fig. 15.8. Relative speed and altitude time history of a ballistic entry vehicle.

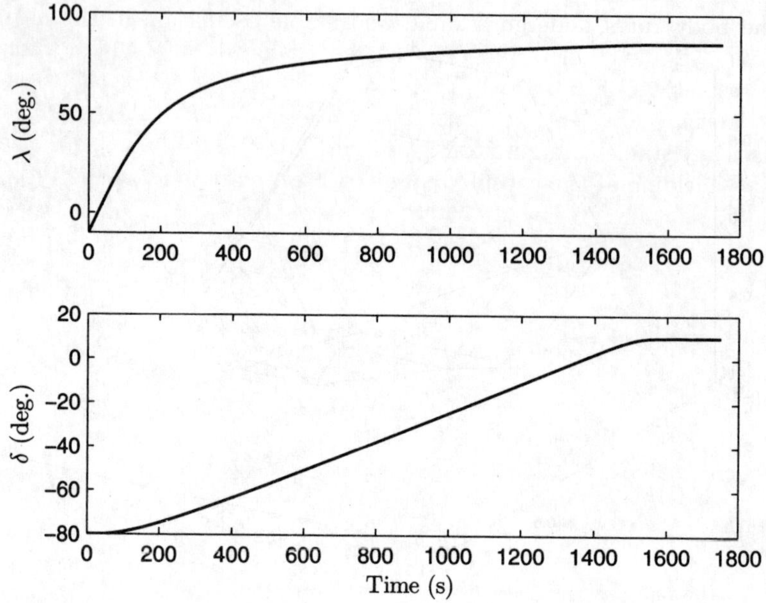

Fig. 15.9. Latitude and longitude time history of a ballistic entry vehicle.

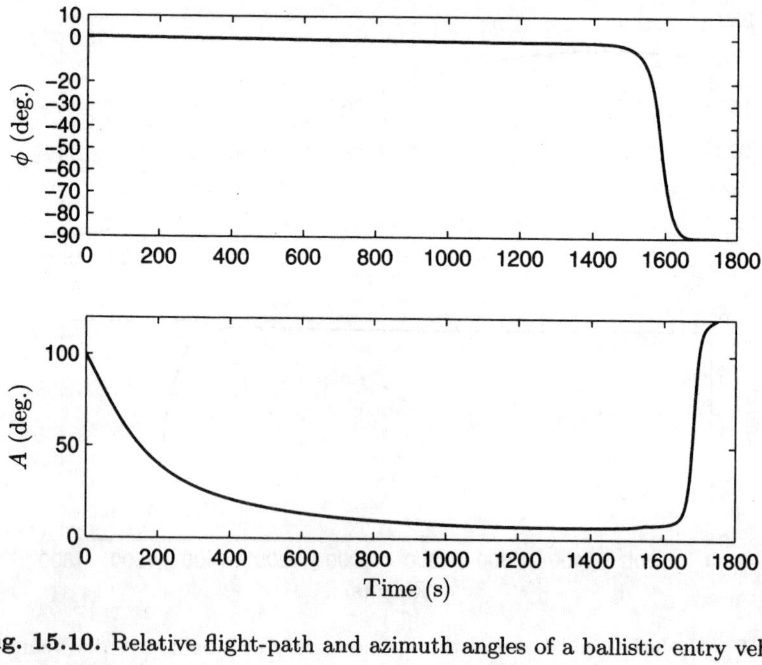

Fig. 15.10. Relative flight-path and azimuth angles of a ballistic entry vehicle.

the attitude parameters (quaternion) (Fig. 15.12). It is interesting to note that the body rates undergo a rapid and stable oscillation in the altitude range $50 < h < 100$ km, and decay to nearly zero below 40-km altitude. Such a behavior indicates the inherent static stability ($C_{m_\alpha} = -C_{n_\beta}$) of the capsule in the rarefied entry regime, somewhat like a badminton shuttle-cock. The statically stable configuration is crucial for mission success, as the selective heat shielding of the capsule requires a precise attitude during the peak thermal loads regime below an altitude of 50 km (Chapter 12).

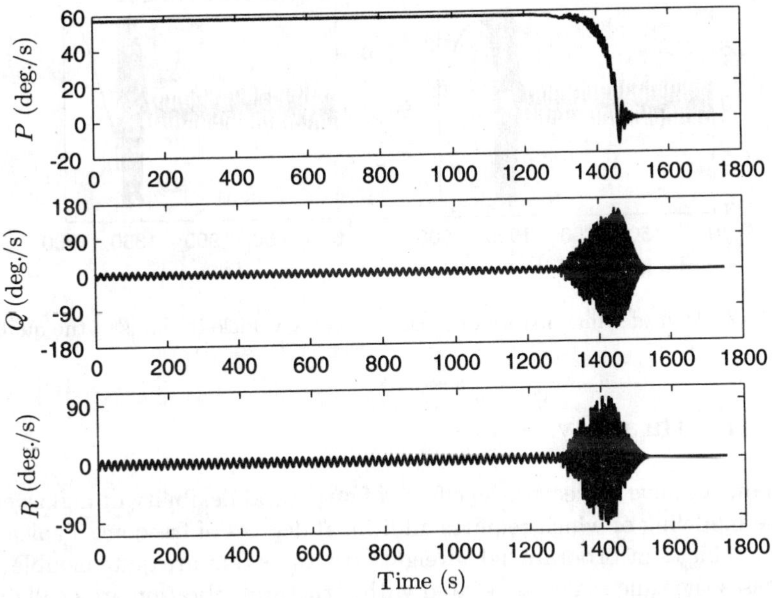

Fig. 15.11. Time history of the body rates of a ballistic entry vehicle.

Other examples of six-degree-of-freedom simulation include missiles, launch vehicles, and lifting entry vehicles, some of which are left to the reader as exercises. The *Aerospace Block-Set* of MATLAB/Simulink (Rel.14.1) software contains modular building blocks for six-degree-of-freedom simulation of aircraft with a nonrotating, flat-earth, and linearized aerodynamics (subsonic/supersonic), along with various wind turbulence models and linear control-system blocks. Examples of six-degree-of-freedom simulation contained in the Aerospace Block-Set are NASA's *HL-10* lifting-body vehicle at low speed and high angle of attack, with nonlinear, subsonic aerodynamics, and an aerodynamically controlled, supersonic air-to-air missile with quaternion for attitude modeling.

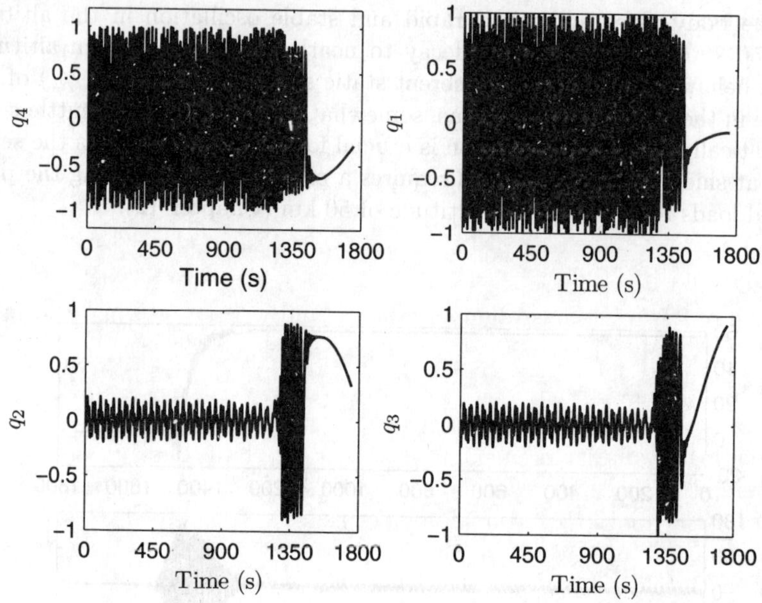

Fig. 15.12. Attitude time history of a ballistic entry vehicle in terms of the quaternion.

15.3 Structural Dynamics

Up to now we have neglected the effects of structural flexibility of flight vehicles, the modeling of which requires additional degrees of freedom. Typically, all flight vehicles incorporate light-weight structures that are quite flexible. In many cases the time scales associated with structural vibration are negligible in comparison with the rotational dynamics of the vehicle, which, in turn, involves a smaller time period compared to the translational motion. In such cases, the coupling between the six-degree-of-freedom motion of the vehicle and the structural dynamics can be ignored. Examples of such vehicles are missiles, launch vehicles, entry vehicles, and some fighter-type aircraft with small aspect ratio, "stiff" wings. However, the rotational motion of spacecraft with large appendages, as well as that of aircraft with moderately large aspect-ratio wings and/or long fueselages, is inherently coupled with that of the structural vibration of the appendages, wings, or fuselages. For example, certain airline transports have the primary wing-bending frequency roughly of the same order as that of the longitudinal short-period mode, while a slender, supersonic bomber can have a fuselage bending mode very close to both short-period and Dutch-roll modes. In such cases, complex interactions can take place between the rigid-body motion, and the structural dynamic response, leading to a significant modification of the stability derivatives, as well as structural instabilities.

The modeling for structural dynamics is based upon the analogy with the single-degree-of-freedom spring-mass-damper system, for which a small spring deflection from a static equilibrium position, x, is given by the differential equation

$$m\ddot{x} + c\dot{x} + kx = f \, , \tag{15.7}$$

where m, c, k, and f are the mass, damping constant, spring stiffness, and applied force, respectively. For a continuous structure with infinite degrees of freedom, an analogous system of equations can be derived using a generalization of the sping-mass-damper system, and written in a matrix form as follows:

$$\mathsf{M}\ddot{\mathbf{z}} + \mathsf{C}\dot{\mathbf{z}} + \mathsf{K}\mathbf{z} = \mathbf{f} \, , \tag{15.8}$$

where $\mathbf{z}$ is a vector of discrete variables called the *generalized coordinates* representing small displacement from the static equilibrium position of the structure, $\mathbf{f}$ is a vector of *generalized forces*, and $\mathsf{M}, \mathsf{K}, \mathsf{C}$ are called the *generalized mass, stiffness*, and *damping* matrices, respectively. The generalized displacements do not necessarily have a physical significance, and arise solely out of mathematical convenience. Note that for an exact representation, $\mathbf{z}$ must have an infinite dimension. However, one finds it sufficient to approximate the motion by only finite degrees of freedom, N, the number of which depends upon the range of natural frequencies (*bandwidth*) of interest. Identifying approximate discrete degrees of freedom for a continuous structure is the realm of elasticity, and the methods employed include the *classical Rayleigh–Ritz* and *finite-element* methods, the *weighted-residual* methods consisting of the *Galerkin* and *collocation* methods, and the *assumed-modes* method. While a detailed discussion of these structural dynamic models is beyond our approach, the reader may refer to introductory texts on elasticity, such as Meirovitch [61].

In the derivation of the discrete set of equations of motion, Eq. (15.8), the energy approach (also called *Lagrangian*, or *variational* method) is often utilized, elements of which were discussed in Chapter 4. In this approach, the work done by an external, nonconservative generalized force on a holonomic system (Chapter 4) described by the generalized coordinates, $\mathbf{z}$, results in a variation of the kinetic energy, T, and potential energy, $-V$, and leads to the following set of equations, called *Lagrange's equations*:

$$\frac{\mathrm{d}}{\mathrm{d}t}\left(\frac{\partial L}{\partial \dot{z}_i}\right) - \frac{\partial L}{\partial z_i} = f_i \;\; (i = 1 \ldots N) \, , \tag{15.9}$$

where $L \doteq T + V$ is the *Lagrangian* of the system. The Lagrangian approach offers a convenient alternative to Newton's laws in deriving equations of motion of a complex dynamical system, because it is based upon scalar variables (generalized coordinates, generalized forces, and energy) rather than the vector variables required in the application of Newton's laws.

The generalized mass and stiffness matrices are obtained using conservative strain energy considerations, which are fairly easily modeled. However, a modeling of the generalized structural damping matrix involves complex considerations of energy dissipation mechanisms within the structure and is generally ignored in a first-order analysis. Assuming one has derived the linear structural dynamic model, Eq. (15.8), in the form of M, K, and identified the generalized force vector, $\mathbf{f}(t)$, as a function of time, a solution $\mathbf{z}(t)$ can be computed using the linear algebraic methods introduced in Chapter 14. The linear equations of motion of an undamped structure in generalized coordinates lead to the following decoupled *modal form*

$$m_i \ddot{z}_i + m_i \omega_i^2 z_i = f_i \quad (i = 1, \ldots, N)\,, \tag{15.10}$$

where ω_i is the natural frequency of the ith mode of the structure obtained by solving the eigenvalue problem associated with the unforced system,

$$(\mathsf{M}\omega_i^2 + \mathsf{K})\mathbf{z}_{0i} \quad (i = 1, \ldots, N)\,. \tag{15.11}$$

The *mode shape* of deformation is indicated by the eigenvector, $\mathbf{z}_{0i}$. Generally, higher modes have a diminishing effect on the structural response. Therefore, only the first few modes are considered important and retained in a model. However, when several modes have nearly the same natural frequency, all of them must be retained for accuracy. For a spacecraft, the above-given approach is commonly employed, and the coupled rotational and structural dynamics are easily modeled. In such a case, $\mathbf{f}(t)$ consists of control force and external disturbance due to gravity gradient, solar radiation, or atmospheric drag. For atmospheric vehicles, however, the generalized force arises due to aerodynamic effects and depends upon the relative velocity of the structure determined by $\dot{\mathbf{z}}$. Furthermore, the forced vibration frequencies differ significantly from the unforced (or *in-vacuo*) natural frequencies of the structure. In order to obtain a solution, this coupling between the aerodynamics and structural dynamics, called *aeroelasticity*, must be carefully modeled, which is the subject of the next section.

Example 15.3. Consider the single-axis rotation of a spacecraft with a rigid hub and two symmetrically located radial, flexible appendages mounted with tip masses. If we use a finite-element approach [62], the first eight structural modal frequencies of the spacecraft (apart from the rigid-body mode, $\omega_0 = 0$) are obtained and tabulated in Table 15.1. A closed-loop vibration suppression system consisting of hub torque input and a normal acceleration output, a_N, at a selected point on one of the appendages, is designed based upon a novel technique [63], wherein an optimal feedback controller penalizes $\dot{a}_N$ and follows a desired hub displacement profile, generated by a feedforward, multi-mode, time-optimal *input-shaping* method. One can choose *a priori* how many structural modes are to be suppressed, while rotating the spacecraft by a desired angle, θ_d. Figure 15.13 shows the closed-loop response of the normal acceleration, hub angle and rate, and the torque input for moving the spacecraft by 1 rad in a rest-to-rest maneuver, while suppressing the first 2 to 8 modes of the structure. It is clear that as the number of modes to be suppressed increases to six, there is a significant reduction in the peak magnitudes of a_N, $\dot{\theta}$, as well as the input torque (which is based upon a feedback of a_N and θ). However, taking into account higher modes does not offer any significant advantage, but only leads to an increase in the size of the control system, which is undesirable.

Table 15.1. The Natural Frequencies of a Flexible Spacecraft

i	ω_i (rad/s)
1	0.763435
2	2.223379
3	5.559157
4	10.77793
5	17.93344
6	27.00576
7	41.35808
8	56.56352

15.4 Unsteady Aerodynamics and Aeroelasticity

The assumption of steady flow becomes invalid when we consider a time-dependent relative flow velocity, thereby necessitating *unsteady aerodynamic* modeling. The unsteady aerodynamics arise when a flight vehicle is suddenly started from rest, encounters a wind gust, or experiences oscillatory motion. The modeling and analysis of unsteady aerodynamics are a complex task, generally requiring the solution of unsteady Euler (or Navier–Stokes) equations (Chapter 10). However, over the past 75 years, a vast compendium of analytical techniques and semi-empirical methods has accumulated for modeling unsteady aerodynamics. The largest contribution of unsteady aerodynamic effects is due to lifting surfaces, which experience a time-dependent lift and pitching moment due to a change in the flow direction. As a result, the lift and moment do not adjust themselves instantaneously to changes in the speed and direction of the flow. The most common model applied to a lifting surface is that of small flow disturbances, for which the inviscid flow assumption remains valid, and approximate linearized analysis in terms of stability derivatives can be applied. However, for nonlifting bodies, and surfaces encountering large flow disturbances, the inviscid flow assumption breaks down due to substantial flow separation.

For an illustration of unsteady aerodynamics, we shall consider the small-disturbance motion of a thin airfoil in incompressible flow. This is the earliest and the most analyzed unsteady aerodynamics problem, attacked first by

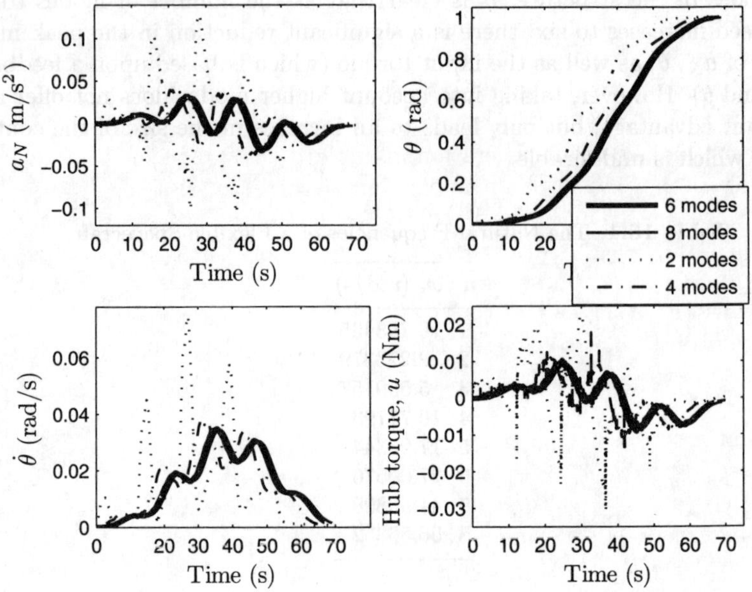

Fig. 15.13. Closed-loop response of a flexible spacecraft with an active vibration suppression system.

researchers in the 1920s and 1930s (Birnbaum, Wagner, Glauert, Theodorsen, Küssner), and is still in wide use for predicting the aeroelastic behavior of large aspect-ratio wings. The assumption of inviscid, incompressible (potential) flow offers the advantage of analytical methods in complex variables and enables closed-form expressions for unsteady lift and pitching moment. We begin by presenting the steady potential flow condition over a thin airfoil at a small angle of attack, where the lift generation involves a *circulation*[2] of the flow over the airfoil. The linear relationship between the strength of the circulation and lift is given by the *Kutta–Joukowski* theorem of potential flow, which requires that the flow at the trailing edge be tangential to the airfoil's chord line (*Kutta condition*) [22]. The Kutta condition is a physical phenomenon and remains approximately valid even in the small-disturbance, unsteady flow. For an inviscid flow, the circulation remains constant around a closed curve containing the same fluid elements, due to *Kelvin's theorem* [22]. This implies that whenever there is a change of circulation around an airfoil, there must be an equal and opposite circulation of the fluid moving past the airfoil in

[2] Circulation, Γ, is defined as the line integral of flow velocity over a closed curve, C,

$$\Gamma \doteq \oint_C \mathbf{v} \cdot d\mathbf{s} \, .$$

its wake. Such a circulation in the wake causes a rotation (vorticity) of the fluid elements in the wake region and is called *vortex shedding*. Hence, an unsteady airfoil motion causes an unsteady wake, which, in turn, induces a vertical velocity component on the airfoil, representing a further change in the effective angle of attack experienced by the airfoil. This complex interaction with wake vortex shedding is a hallmark of unsteady flow. However, there is another mechanism of lift generation in unsteady flow that does not involve a circulation of the flow. Such a phenomenon is easily visualized by considering a flat plate accelerated in the normal direction, which generates lift by displacing the fluid normal to its surface. In a noncirculatory flow, the fluid exerts a normal force on the solid surface which is proportional to the mass of the fluid being displaced and is thus called the *aerodynamic inertia effect*. A flapping airfoil can thus generate a noncirculatory lift even when the forward flow speed is zero. With these preliminaries, we now consider an airfoil of chord $c = 2b$ capable of executing both vertical translation (plunge), h, and a rotation about a fixed axis (pitch), α. The airfoil can be regarded as a typical section of a large aspect-ratio wing and encounters plunge and pitch due to the bending and torsion, respectively, of the wing. Consequently, in an aeroelastic model we restrain the airfoil's motion by a linear and torsional spring representing the local bending and torsional wing stiffnesses. Such a model is in keeping with the approach of generalized displacement presented above, where h, α can be derived from the generalized displacement vector, $\mathbf{z}$, governing the structural motion. Theodorsen (1935) devised a circulatory flow function, $C(k)$, which can be used to represent the unsteady lift and pitching moment of the airfoil oscillating at a frequency ω as follows [64]:

$$L = \pi\rho b^2[\ddot{h} + v\dot{\alpha} - ab\ddot{\alpha}] + 2\pi\rho v b C(k)\left[\dot{h} + v\alpha + b\left(\frac{1}{2} - a\right)\dot{\alpha}\right] \tag{15.12}$$

$$\begin{aligned}\mathcal{M} = {} & \pi\rho b^2\left[ab\ddot{h} - vb\left(\frac{1}{2} - a\right)\dot{\alpha} - b^2\left(\frac{1}{8} + a^2\right)\ddot{\alpha}\right] \\ & + 2\pi\rho v b^2\left(a + \frac{1}{2}\right)C(k)\left[\dot{h} + v\alpha + b\left(\frac{1}{2} - a\right)\dot{\alpha}\right] .\end{aligned} \tag{15.13}$$

Here, $a = \frac{x}{b}$ is the nondimensional location of the pitch axis (Fig. 15.14), $k \doteq \frac{\omega b}{v}$ is a nondimensional frequency of oscillation, called the *reduced frequency*, and

$$C(k) \doteq \frac{\int_1^\infty \frac{y}{\sqrt{y^2-1}} e^{-iky} dy}{\int_1^\infty \sqrt{\frac{y+1}{y-1}} e^{-iky} dy} . \tag{15.14}$$

The unsteady lift and pitching moment are seen to contain noncirculatory terms [those that are independent of $C(k)$] as well as the circulatory terms involving $C(k)$. The Theodorsen function, $C(k)$, is thus a complex function of the reduced frequency, k. For a simple harmonic motion, k is real, while a slightly divergent oscillation is represented by a complex k, with a negative

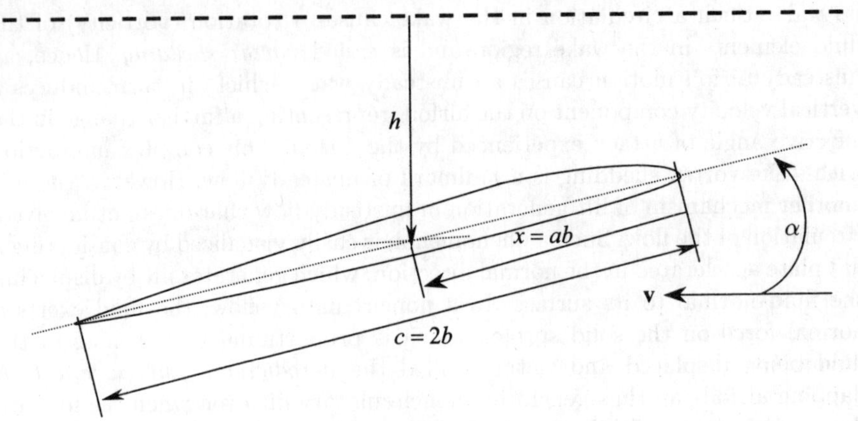

Fig. 15.14. A typical wing airfoil section.

imaginary part. In both of these cases, it is not necessary to evaluate the improper integrals in Eq. (15.14) (which are not always convergent), and $C(k)$ can be computed using the *Hankel functions* of the second kind [5], $H_n^{(2)}$, as follows:

$$C(k) = F(k) + iG(k) = \frac{H_1^{(2)}}{H_1^{(2)} + iH_0^{(2)}} . \tag{15.15}$$

Unfortunately, Hankel functions (also called *Bessel functions* of the third kind) are unavailable in a closed form and require a numerical approximation [5], such as that by the intrinsic MATLAB function *besselh.* In a general application, it is more advantageous to transform $C(k)$ to the Laplace domain. For this purpose, it may be beneficial to consider a related model called *Wagner's function*, $\phi(\bar{t})$, which is the inverse Fourier transform of $\frac{C(k)}{ik}$:

$$\phi(\bar{t}) \doteq \frac{1}{2\pi i} \int_{-\infty}^{\infty} \frac{C(k)}{k} e^{ik\bar{t}} \mathrm{d}k , \tag{15.16}$$

where $\bar{t} \doteq \frac{vt}{b}$ is a nondimensional time. Wagner's function is especially useful in calculating the *indicial* lift (and pitching moment) that arises due to a sudden change in the angle of attack (such as a starting motion, or a sudden vertical gust). A rough approximation of $\phi(\bar{t})$ was given by Garrick [65]:

$$\phi(\bar{t}) \approx 1 - \frac{2}{\bar{t} + 4} . \tag{15.17}$$

A wide range of unsteady problems can be solved using Theodorsen's function, such as the dynamic aeroelastic phenomena of *flutter* and *control surface reversal* [64], the arbitrary motion of the airfoil in the presence of a gust, and estimation of unsteady stability derivatives (such as $C_{z_{\dot{\alpha}}}, C_{m_{\dot{\alpha}}}$). For wings of finite span oscillating in subsonic (compressible) as well as supersonic speeds, a modification of the above approach in the form of a lifting surface, panel methods can yield accurate unsteady aerodynamic models. Such panel methods include the subsonic Doublet–Lattice method [66], the supersonic Potential–Gradient method [67], and the subsonic [68] and supersonic [69] Doublet–Point methods. However, a linearized unsteady aerodynamic modeling is nearly impossible in the transonic regime due to the inherently nonlinear, small-disturbance governing equations of motion, as well as the presence of oscillating shock waves and the attendant unsteady flow separation.

For a practical application, the generalized unsteady aerodynamic force, **f**, must be available in either the time domain or the Laplace domain. In order to develop such a model, a method of *analytic continuation* from the imaginary axis of the Laplace plane is employed in terms of an aerodynamic transfer matrix, $\mathsf{Q}(s)$ defined by

$$\mathbf{F}(s) = \mathsf{Q}(s)\mathbf{Z}(s) , \tag{15.18}$$

where $\mathbf{F}(s), \mathbf{Z}(s)$ are the Laplace transforms of $\mathbf{f}(t), \mathbf{z}(t)$, respectively. The elements of the transfer matrix, $\mathsf{Q}(s)$, are approximated by linear, time-invariant, finite-state, rational functions in s. The numerator coefficients of the rational transfer functions, $Q_{ij}(s)$, are determined by a curve fit with the simple harmonic data obtained from a panel method in the limit $s \to i\omega$. The determination of the denominator constants (called *lag parameters*) of the rational functions, $Q_{ij}(s)$, require additional conditions, such as the minimization of the total least-squares curve fit error summed over a range of frequencies [70]. In the process of modeling $\mathsf{Q}(s)$, attention has to be given to the number of additional state variables (Chapter 14) arising out of the lag parameters. If the additional lag-state variables are large, the modeling (and associated closed-loop design) becomes unwieldy. It is observed [70] that employing repeated pole transfer functions for $Q_{ij}(s)$ reduces the size of the model for a given fit accuracy.

Aeroelasticity refers to the interaction between structural dynamics and unsteady aerodynamics. Using the generalized displacements, the equations for aeroelastic motion can be written in the Laplace domain as follows:

$$[s^2\mathsf{M} + s\mathsf{C} + \mathsf{K}]\mathbf{Z}(s) = \mathsf{Q}(s)\mathbf{Z}(s) , \tag{15.19}$$

where the generalized aerodynamics coefficient matrix is given by an appropriate rational-function approximation. In order to conduct an aeroelastic stability analysis, one must determine the roots of the following characteristic equation

$$| \, s^2\mathsf{M} + s\mathsf{C} + \mathsf{K} - \mathsf{Q}(s) \, | = 0 , \tag{15.20}$$

which can be accomplished numerically by linear algebraic methods. This is the basis of aeroelastic stability analysis. When some characteristic roots (eigenvalues) cross into the right-half s-plane, we have an aeroelastic instability. The aeroelastic instabilities can be either static (*divergence and control surface reversal*) or dynamic (*flutter and control-surface buzz*) [64]. Among these, divergence and flutter are the most dangerous, because they cause catastrophic structural failures. Flutter has received special attention by the airplane designers, since it is rather difficult to predict and is influenced by mass distribution as well as stiffness. At the characteristic flutter speed, the harmonic wing response achieves a near resonance with the unsteady aerodynamics, leading to large amplitude oscillation. A large aspect-ratio wing generally experiences the classical bending-torsion flutter, involving only the first two (or three) structural modes, and a simplified aerodynamic model based upon typical wing sections (*strip method*). On the other hand, a small aspect-ratio wing of a fighter-type airplane may require the inclusion of higher-frequency modes and sophisticated aerodynamic panel methods in a flutter analysis. Flutter can be controlled either passively through wing redesign or actively by employing a control surface in a closed loop with normal wing acceleration. When additional degrees of freedom of control surface deflections—driven by a feedback control law, or vibrating freely—are included in the aeroelastic model, the resulting analysis is termed *aeroservoelasticity*. Examples of aeroservoelasticity include active flutter suppression, active gust-load alleviation, and inadvertent structural modal excitation by an active flight-control system. The last has been a common experience in modern fighter designs, which invariably employ some form of automatic flight stability augmentation. If the adverse aeroservoelastic interactions are not carefully modeled and designed out of the closed-loop system, serious accidents may occur, leading to a complete loss of the aircraft.[3]

Example 15.4. Consider the active flutter suppression of a typical wing section, by a trailing-edge control-surface deflection, δ. Using the following rational-function approximation of Theodorsen's function [71]:

$$C(k) = 0.9962 - \frac{0.1667(ik)}{ik + 0.0553} - \frac{0.3119(ik)}{ik + 0.2861},$$

and a normal acceleration, a_N, feedback, we arrive at a linear, time-invariant state-space representation of a 12th-order aeroservoelastic plant. We test two active control laws for suppressing flutter: (a) the traditional linear, quadratic regulator (LQR) optimal controller, and (b) a novel optimal method based upon the minimization of the output rate (ORW) [72]. The closed-loop responses of the two controllers at the open-loop flutter speed in terms of normal acceleration output, and control-surface input, are plotted in Fig. 15.15. Note the smoother acceleration of a smaller magnitude, as well as a much smaller control input of the ORW controller. However, ORW control requires the control surface to be oscillated more rapidly than the LQR method.

Certain aeroelastic phenomena require nonlinear modeling techniques. One such behavior is the airframe buffet encountered when maneuvering near stall,

[3] Adverse aeroservoelasticity has been encountered by the Lockheed *F-22* and *F-117* fighters, as well as Taiwan's *IDF* fighter prototype.

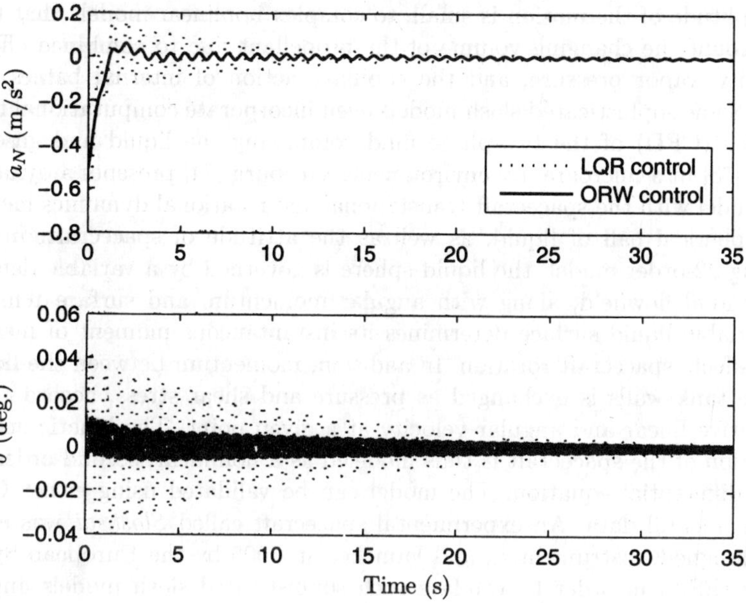

Fig. 15.15. Active flutter suppression of a typical wing airfoil section by two control techniques.

or in the transonic regime, due to unsteady separated flow. The mechanism of buffet is complex and requires an unsteady, turbulent flow model. Other nonlinear aeroelastic effects arise out of large amplitudes or large reduced frequencies and sometimes involve nonlinear structural dynamics.

15.5 Propellant Slosh Dynamics

Many aerospace vehicles contain large liquid propellant masses, whose motion relative to the vehicle—called *slosh*—can excite unwanted rigid and/or structural modes. The vehicles for which slosh becomes a critical issue are launch vehicles and spacecraft with liquid propellants. As a launch vehicle ascends, the propellant is rapidly consumed, leading to partially filled tanks, which aggravates the slosh problem (somewhat like a half-empty soda can). An extreme case of slosh interacting with the vehicle's dynamics and propulsion system is the longitudinal vibration of the entire vehicle, called *pogo oscillation*. An example of pogo was encountered in the *Titan-II* rocket, with a natural frequency of about 10 Hz and a peak acceleration of ± 5 g. The slosh dynamics are much more critical in a spacecraft, where the absence of a strong gravity field increases the amplitude of relative motion. As discussed in Chapter 13, slosh can render the rotational motion about the minor axis unstable.

The modeling of slosh ranges from the second-order linear dynamics when the amplitude of the motion is small, to complex nonlinear models that take into account the changing volume of the propellant due to combined effects of gravity, vapor pressure, and the capillary action of internal baffles and guides. Some sophisticated slosh models even incorporate computational fluid dynamics (CFD) of the two-phase fluid comprising the liquid and gaseous vapour [73] in a microgravity environment. Vreeburg [74] presents a dynamic slosh model with the spacecraft translational and rotational dynamics including a spherical ball of liquid, as well as the attitude of spacecraft. In the resulting 22-order model, the liquid sphere is governed by a variable density and a radial flowfield, along with angular momentum and surface tension. The variable liquid surface determines its instantaneous moment of inertia, which affects spacecraft rotation. In addition, momentum between the liquid and the tank walls is exchanged as pressure and shear stress, according to the relative linear and angular velocity at a given point. The kinetic energy dissipation of the spacecraft is thus modeled as a nonlinear, second-order ordinary differential equation. The model can be validated using either CFD or experimental data. An experimental spacecraft called *Sloshsat* was especially designed, instrumented, and launched in 2005 by the European Space Agency (ESA) in order to validate such sophisticated slosh models and to gather valuable experimental data.

15.6 Summary

The accuracy of a simulation can be improved by modeling additional degrees of freedom of the vehicle, including structural flexibility, propellant slosh dynamics, unsteady aerodynamic, and aeroelasticity, as well as realistic disturbance effects. A complete six-degree-of-freedom model and simulation are necessary in all cases where the lateral and longitudinal rotational dynamics cannot be decoupled from the translational motion, such as in a rapidly rolling fighter airplane, missile, and a ballistic entry vehicle. Modeling of structural dynamics by Lagrangian methods is necessary whenever the coupling between the six-degree-of-freedom motion of the vehicle and the structural dynamics cannot be ignored. When applied to atmospheric vehicles, an accurate model of unsteady aerodynamics and aeroelasticity is required in all cases where structural flexibility is important. Unsteady aerodynamic models are also necessary for rigid vehicles whenever a time-dependent relative flow occurs, such as in a rapidly rotating or suddenly started winged vehicle. Interaction among structural dynamics, unsteady aerodynamics, and control systems (aeroservoelasticity) must be carefully modeled and analyzed for possible instabilities. Modeling of slosh dynamics and its coupling with the vehicle's motion—by semi-empirical or sophisticated CFD models—is a critical issue for launch vehicles and spacecraft with liquid propellants.

Exercises

15.1. Carry out a closed-loop simulation of the fighter aircraft of Example 15.1 with the control system given in Example 14.9. Is the wing-rock suppressed by this control system? Can we suppress the wing-rock, as well as maintain flight equilibrium using only the roll-rate feedback control?

15.2. Build a closed-loop, six-degree-of-freedom simulation for the missile of Example 13.11 using a code that incorporates *rocket.m* (Table 12.14) for the translational motion of a rocket's ascent through the atmosphere, along with the state equations arising out of the control system given in Example 14.6. Use the same stage ratios as given in Example 12.8, but modify the values of mass and thrust at 75 s after launch to be the same as those given for the Example 13.11. Carry out the simulation for flight up to the second-stage separation, and plot the relevant state variables with time.

15.3. Consider a bomber aircraft with a long, slender fuselage. The fuselage is so flexible that the motion of the flight crew in their seats can excite the primary bending mode of natural frequency 6.05 rad/s. The aircraft's aeroelastic motion can be described adequately by a second-order longitudinal short-period mode, a second-order fuselage bending mode, two first-order control-surface (canard and elevator) actuators, and a normal acceleration and a pitch-rate output, resulting in the following state-space model [43]:

$$\mathsf{A} = \begin{pmatrix} 0.4158 & 1.025 & -0.00267 & -0.00011 & -0.08021 & 0 \\ -5.5 & -0.8302 & -0.06549 & -0.0039 & -5.115 & 0.809 \\ 0 & 0 & 0 & 1 & 0 & 0 \\ -1040 & -78.35 & -34.83 & -0.6214 & -865.6 & -631 \\ 0 & 0 & 0 & 0 & -75 & 0 \\ 0 & 0 & 0 & 0 & 0 & -100 \end{pmatrix}$$

$$\mathsf{B}^T = \begin{pmatrix} 0 & 0 & 0 & 0 & 75 & 0 \\ 0 & 0 & 0 & 0 & 0 & 100 \end{pmatrix}$$

$$\mathsf{C} = \begin{pmatrix} -1491 & -146.43 & -40.2 & -0.9412 & -1285 & -564.66 \\ 0 & 1 & 0 & 0 & 0 & 0 \end{pmatrix},$$

$$\mathsf{D} = \begin{pmatrix} 0 & 0 \\ 0 & 0 \end{pmatrix}.$$

Determine the initial response of the airplane to a squirming pilot, resulting in the initial condition $\mathbf{x}(0) = (0, 0, 1, 0, 0, 0)^T$ ft/s. How long does the motion take to damp out, and what are the peak normal acceleration and pitch rate?

A

Numerical Integration of Ordinary Differential Equations

A general set of n ordinary differential equations can be expressed in a vector form as follows:

$$\frac{\mathbf{dy}}{\mathrm{d}t} = \mathbf{f}(t, \mathbf{y}) \ , \tag{A.1}$$

with the initial condition

$$\mathbf{y}(0) = \mathbf{y_0} \ . \tag{A.2}$$

The nonlinear functional, $\mathbf{f}$, possesses continuous time derivatives up to an indefinite order and satisfies the following *Lipschitz condition*:

$$| \mathbf{f}(t, \mathbf{y_1}) - \mathbf{f}(t, \mathbf{y_2}) | \leq c \mid \mathbf{y_1} - \mathbf{y_2} \mid , \tag{A.3}$$

for some constant c. In such a case, the existence of a unique solution to the differential equation is guaranteed [51]. However, such a solution can rarely be found in a closed form and often requires numerical approximation. The approximate numerical solution to Eq. (A.1) involve series expansions of the solution around the given initial condition and can be divided into various categories depending upon the determination of the coefficients of the series.

A.1 Fixed-Step Runge–Kutta Algorithms

A *Runge–Kutta* algorithm of fixed integration time step h, order p, and s stages expresses the solution by the following truncated series:

$$\mathbf{y}(h) = \mathbf{y_0} + h \sum_{k=0}^{s-1} a_k \mathbf{f_k} + \mathcal{O}(h^{p+1}) \ , \tag{A.4}$$

where the neglected part of the series, $\mathcal{O}(h^{p+1})$, is called the *truncation error*, Δ_{TE}. The stages refer to the number of evaluations of the functional required at each time step, apart from that at the initial time ($t = 0$):

$$\mathbf{f_0} = \mathbf{f}(0, \mathbf{y_0}),$$

$$\mathbf{f_k} = \mathbf{f}\left(b_k h, \mathbf{y_0} + h \sum_{i=0}^{k-1} c_{ki}\mathbf{f_i}\right) \quad (k = 1, 2, \ldots, s-1) . \tag{A.5}$$

The coefficients a_k, b_k, c_{ki} are chosen such that the solution is identical to that of a Taylor series approximation of the same order, p, given by:

$$\mathbf{y}(h) = \mathbf{y_0} + h \sum_{k=1}^{p} \frac{h^k}{k!} \frac{\partial^{k-1}\mathbf{f}}{\partial t^{k-1}} |_{t_0, \mathbf{y_0}} . \tag{A.6}$$

The determination of the unknown coefficients through comparison with the equivalent Taylor series involves *constraint equations*, which are typically smaller in number than the number of unknowns. Therefore, some of the unknown coefficients are chosen arbitrarily, called *free parameters*. As the order of the method increases, the number of free parameters also increases. The fixed-step algorithms suffer from a large p for a specific truncation error, thereby requiring a large number of constraint equations per time-step.

A.2 Variable-Step Runge–Kutta Algorithms

In order to improve the efficiency of Runge–Kutta algorithms, the time-step size, h, is made variable in Eq. (A.4), such that the accuracy of the next higher-order algorithm is achieved at each step. The higher-order solution, $\hat{\mathbf{y}}$, of stage r is given by

$$\hat{\mathbf{y}}(h) = \mathbf{y_0} + h \sum_{k=0}^{r-1} \hat{a}_k \mathbf{f_k} + \mathcal{O}(h^{p+2}) , \tag{A.7}$$

where the functional evaluations are now carried out as follows:

$$\mathbf{f_0} = \mathbf{f}(0, \mathbf{y_0}),$$

$$\mathbf{f_k} = \mathbf{f}\left(b_k h, \mathbf{y_0} + h \sum_{i=0}^{k-1} c_{ki}\mathbf{f_i}\right) \quad (k = 1, 2, \ldots, m-1), \tag{A.8}$$

where the number of stages m in the solution is the higher of s and r [$m = \max(s, r)$]. Note that the coefficients b_k, c_{ki} remain the same for the two solutions of adjacent order. The time step is chosen such that the truncation error, which is the difference between the adjacent order solutions,

$$\Delta_{TE} = \mathbf{y}(h) - \hat{\mathbf{y}}(h) = h \sum_{k=0}^{r-1} (a_k - \hat{a}_k)\mathbf{f_k} + \mathcal{O}(h^{p+2}) , \tag{A.9}$$

remains below a specified tolerance, Δ. By expressing Eq. (A.9) as

$$\Delta_{TE} = Kh^{p+1}, \tag{A.10}$$

we can select the time-step size of the next step, h', from the specified tolerance as follows:

$$h' = h\left(\frac{\Delta}{\Delta_{TE}}\right)^{\frac{1}{p+1}} \quad (\Delta_{TE} \le \Delta), \tag{A.11}$$

assuming K remains constant over the next step. In this manner, an accuracy of order $p+1$ is achieved with a method of order p, albeit with an increased number of stages. Therefore, the variable-step Runge–Kutta algorithms are also referred to as Runge–Kutta methods of order $p(p+1)$.

Example A.1. The Runge–Kutta 4(5) algorithm [52], implemented in the MATLAB's intrinsic function *ode45.m* and employed extensively in this book, has coefficients tabulated in Table A.1 and results in the following truncation error:

$$\Delta_{TE} = h\left(-\frac{1}{360}\mathbf{f}_0 + \frac{128}{4275}\mathbf{f}_2 + \frac{2197}{75240}\mathbf{f}_3 - \frac{1}{50}\mathbf{f}_4 - \frac{2}{55}\mathbf{f}_5\right). \tag{A.12}$$

Of course, the coefficients in Table A.1 are by no means unique, since they depend upon the choice of the free parameter values. Fehlberg [52] gives two sets of values for these coefficients. Expressing the coefficients in fractional form makes them independent of the machine round-off errors.

Table A.1. The Coefficients of Runge–Kutta 4(5) Algorithm

k:	0	1	2	3	4	5
a_k:	$\frac{25}{216}$	0	$\frac{1408}{2565}$	$\frac{2197}{4104}$	$-\frac{1}{5}$	-
$\hat{a}_k$:	$\frac{16}{135}$	0	$\frac{6656}{12825}$	$\frac{28561}{56430}$	$-\frac{9}{50}$	$\frac{2}{55}$
b_k:	0	$\frac{1}{4}$	$\frac{3}{8}$	$\frac{12}{13}$	1	$\frac{1}{2}$
c_{k0}:	0	$\frac{1}{4}$	$\frac{3}{32}$	$\frac{1932}{2197}$	$\frac{439}{216}$	$-\frac{8}{27}$
c_{k1}:	-	-	$\frac{9}{32}$	$-\frac{7200}{2197}$	-8	2
c_{k2}:	-	-	-	$\frac{7296}{2197}$	$\frac{3680}{513}$	$-\frac{3544}{2565}$
c_{k3}:	-	-	-	-	$-\frac{845}{4104}$	$\frac{1859}{4104}$
c_{k4}:	-	-	-	-	-	$-\frac{11}{40}$

Apart from the Runge–Kutta methods, other choices are available for low-order integration algorithms, some of which are options in a Simulink [49] simulation:

(a) The *finite-difference* (or *Euler's*) methods, where approximate values are prescribed for the solution at a number of grid points, at which the solution is propagated in time using the values at previous times. These methods, while simple to implement, are computationally inefficient.
(b) The *multistep* explicit and implicit algorithms, such as the *Adams*, *Adams–Bashforth*, and *Adams–Moulton* algorithms [51].
(c) The *predictor-corrector* methods, such as those by *Milne* and *Shampine–Gordon* [51].

The multistep and predictor-corrector methods require sophisticated programming, as their dependence on starting estimates and step sizes may cause convergence and numerical stability problems. In comparison, Runge–Kutta methods are much simpler to program, mainly because their solution begins from a known initial condition, and their truncation error is easily controlled in a straightforward manner by a variable step size. To solve a set of differential equations with a large difference in the time scales (called *stiff equations*), certain implicit multistep algorithms have been especially adapted, such as the *ode23tb* and *ode23s* algorithms of MATLAB [53].

A.3 Runge–Kutta–Nyström Algorithms

When solving a certain class of astronautical problems, such as those involving Cowell's and Encke's formulations for lunar and interplanetary travel (Chapter 6), the low-order time-integration methods given above prove unsuitable, as they result in an accumulation of truncation error over the long times of flight. For such problems, the equations of motion can be written in the following form:

$$\frac{d\mathbf{x}}{dt} = \mathbf{y}(t),$$
$$\frac{d\mathbf{y}}{dt} = \mathbf{f}(t, \mathbf{x}), \tag{A.13}$$

with the initial condition

$$\mathbf{x}(0) = \mathbf{x_0}; \quad \mathbf{y}(0) = \mathbf{y_0} \,. \tag{A.14}$$

The *Runge–Kutta–Nyström* (RKN) method is suitable for integrating the above set of implicit differential equations with a high order (thus small truncation error), and a relatively smaller number of stages compared with the traditional Runge–Kutta algorithm of the same order. The RKN solution of order p and stages s is expressed as follows:

$$\mathbf{x}(h) = \mathbf{x_0} + h\mathbf{y_0} + h^2 \sum_{k=0}^{s-1} a_k \mathbf{z_k} + \mathcal{O}(h^{p+1}),$$
$$\mathbf{y}(h) = \mathbf{y_0} + h \sum_{k=0}^{s-1} b_k \mathbf{z_k} + \mathcal{O}(h^{p+1}), \tag{A.15}$$

where $\mathbf{z_k}$ refers to the following functional evaluations:

$$\mathbf{z_k} = \mathbf{f}\left(\alpha_k h, \mathbf{x_0} + \alpha_k h \mathbf{y_0} + h^2 \sum_{i=0}^{k-1} c_{ki}\mathbf{z_i}\right) (k = 1, 2, \ldots, m-1) . \quad \text{(A.16)}$$

Although the number of stages for a given order is reduced in the RKN algorithms, the determination of coefficients $a_k, b_k, \alpha_k, c_{ki}$ requires the solution of nonlinear constraint equations. Battin [11] presents the procedure for solving the constraint equations for RKN algorithms up to the eighth order. For $p = 8$, we have $s = 8$, but 36 constraint equations and 5 free parameters. Consequently, a significant effort is necessary for evaluating the coefficients, which are then stored and utilized in the solution.

It is possible to achieve a still-higher accuracy with the use of time-step control in a higher-order RKN algorithm. In such a case, time-step control may consider minimizing the truncation error in either $\mathbf{x}$ only [55] or both $\mathbf{x}, \mathbf{y}$ simultaneously [54]. In such applications with adjacent order solutions, the higher-order solution is used to control the step size, h, while the free parameters of the lower-order solution are retained for efficiency.

Example A.2. Battin [11] presents the following elegant formulation for a sixth-order RKN algorithm with only five stages:

$$\begin{aligned} \mathbf{x}(h) &= \mathbf{x_0} + h\mathbf{y_0} + \frac{1}{24}h^2[2\mathbf{z_0} + (5+\sqrt{5})\mathbf{z_2} + (5-\sqrt{5})\mathbf{z_3}] + \mathcal{O}(h^7), \\ \mathbf{y}(h) &= \mathbf{y_0} + \frac{1}{12}h(\mathbf{z_0} + 5\mathbf{z_2} + 5\mathbf{z_3} + \mathbf{z_4}) + \mathcal{O}(h^7) , \end{aligned} \quad \text{(A.17)}$$

where

$$\begin{aligned} \mathbf{z_0} &= \mathbf{f}(0, \mathbf{x_0}), \\ \mathbf{z_1} &= \mathbf{f}\left[\frac{1}{20}(5-\sqrt{5})h, \mathbf{x_0} + \frac{1}{20}(5-\sqrt{5})h\mathbf{y_0} + \frac{1}{80}(3-\sqrt{5})h^2\mathbf{z_0}\right], \\ \mathbf{z_2} &= \mathbf{f}\left\{\frac{1}{10}(5-\sqrt{5})h, \mathbf{x_0} + \frac{1}{10}(5-\sqrt{5})h\mathbf{y_0}\right. \\ &\quad \left. + \frac{1}{60}h^2[(3-\sqrt{5})\mathbf{z_0} + (6-2\sqrt{5})\mathbf{z_1}]\right\}, \\ \mathbf{z_3} &= \mathbf{f}\left\{\frac{1}{10}(5+\sqrt{5})h, \mathbf{x_0} + \frac{1}{10}(5+\sqrt{5})h\mathbf{y_0}\right. \\ &\quad \left. + \frac{1}{60}h^2[(6+2\sqrt{5})\mathbf{z_0} - (8+4\sqrt{5})\mathbf{z_1} + (11+5\sqrt{5})\mathbf{z_2}]\right\}, \\ \mathbf{z_4} &= \mathbf{f}\left\{h, \mathbf{x_0} + h\mathbf{y_0} - \frac{1}{12}h^2[(3+\sqrt{5})\mathbf{z_0}\right. \\ &\quad \left. - (2+6\sqrt{5})\mathbf{z_1} + (2+2\sqrt{5})\mathbf{z_2} - (9-3\sqrt{5})\mathbf{z_3}]\right\} . \end{aligned} \quad \text{(A.18)}$$

This algorithm involves only four free parameters.

Answers to Selected Exercises

Chapter 2

2.2 $\theta = \cos^{-1} c_{33}$, $\psi = \tan^{-1} \frac{-c_{31}}{c_{32}}$, $\phi = \tan^{-1} \frac{c_{13}}{c_{23}}$. Singularities: $\theta = 0, \pm\pi$.

2.3 For a symmetric set, (ψ, θ, ϕ): $\cos \frac{\Phi}{2} = \cos \frac{\theta}{2} \cos \frac{\psi+\phi}{2}$,

$$\mathbf{e} = \frac{1}{2 \sin \Phi} \begin{pmatrix} \sin\theta(\cos\psi + \cos\phi) \\ \sin\theta(\sin\psi - \sin\phi) \\ (1 + \cos\theta)\sin(\psi + \phi) \end{pmatrix}.$$

No.

2.7 (a) $q_1 = \frac{c_{12}+c_{21}}{4q_2}$, $q_3 = \frac{c_{23}+c_{32}}{4q_2}$, $q_4 = \frac{c_{31}-c_{13}}{4q_2}$, $q_2 = \pm\frac{1}{2}\sqrt{1 - c_{11} + c_{22} - c_{33}}$.

2.12

$$\mathbf{p} = \frac{\tan \frac{\Phi}{4}}{2 \sin \Phi} \begin{pmatrix} c_{23} - c_{32} \\ c_{31} - c_{13} \\ c_{12} - c_{21} \end{pmatrix},$$

where $\Phi = \cos^{-1}\{\frac{1}{2}(\mathrm{trace}\mathsf{C} - 1)\}$.

2.14 $\frac{\mathrm{d}\mathsf{C}}{\mathrm{d}t} = -\mathsf{C}\mathsf{S}(\boldsymbol{\omega}_I)$.

2.15

$$\begin{Bmatrix} \dot{\phi} \\ \dot{\theta} \\ \dot{\psi} \end{Bmatrix} = \begin{pmatrix} 1 & \sin\phi\tan\theta & \cos\phi\tan\theta \\ 0 & \cos\phi & -\sin\phi \\ 0 & \sin\phi\sec\theta & \cos\phi\sec\theta \end{pmatrix} \begin{Bmatrix} \omega_x \\ \omega_y \\ \omega_z \end{Bmatrix}.$$

Chapter 3

3.1

$$\mathbf{g} = \begin{Bmatrix} g_\phi \cos\phi\cos\lambda + g_r \sin\phi\cos\lambda \\ g_\phi \cos\phi\sin\lambda + g_r \sin\phi\sin\lambda \\ -g_\phi \sin\phi + g_r \cos\phi \end{Bmatrix}.$$

3.3 $\mathsf{G_g} = -\frac{GM}{r^3}[\mathsf{I} - 3\mathbf{i_r}\mathbf{i_r}^T]$, where $\mathbf{i_r} = \mathbf{r}/r$.

Chapter 4

4.1 5.29515 km/s, 67.621 m/s^2, -4.06455×10^{-4} rad/s^2.

4.2 $\sin\theta = \frac{r}{R}$, $v_0^2 = a_0^2 \frac{R}{r}\sqrt{R^2 - r^2}$.

4.4 $\mathbf{g_e} = (g - \omega^2 R_0 \cos^2\delta)\mathbf{i_d} - \omega^2 R_0 \cos\delta \sin\delta \mathbf{i_n}$.

4.5 $\sqrt{v^2 + \frac{1}{2m} + \frac{3}{5m}}$.

4.7 $-2m\Omega \sin\delta \dot{r} \mathbf{i_\theta}$.

4.9 $r = 87,402.0875$ km, $\theta = 187.6474°$, $v = 1.0353996$ km/s, $\phi = -42.361169°$.

4.13 (a) 10.2774 km/s, (b) 48,499.955 km and 11.048 km/s, (c) 4141.709 s.

4.15 117.2187°.

4.16 March 14, 2062. Position on July 6, 2006: $r = 30.32896$ a.u., $\theta = 173.95°$.

Chapter 5

5.2 $r_a = 6503.555$ km, $\phi = -66.985°$.

5.3 7313.401 km.

5.5 $\mathbf{r} = (-1759.71083, 3563.80532, 5152.49334)^T$,
$\mathbf{v} = (-7.1299085, 0.1594425, -2.9848405)^T$,
$\delta = 52.35377°, \lambda = 116.2789°$,
$\phi = -2.5799376°$, $A = 125.053136°$.

5.7 (a) 3.66024 km/s, (b) 2.9083 km/s.

5.8 $a = 6969.325$ km, $e = 0.73824595$.

5.9 0.10002 km/s. Both impulses are opposite to instantaneous flight direction.

5.10 442.57 mean solar days, 6.4367 km/s.

5.14 776.3544 km.

Chapter 6

6.1 92.6495°.

6.2 $a = 26561.76$ km, $e = 0.74105$, $i = 63.435°$, $\omega = 270°$.

6.6

$$t = \frac{H}{\rho_0 \sqrt{\mu R_e} \frac{C_D A}{m}} (e^{\frac{h_0}{H}} - 1) .$$

Chapter 7

7.1 10.8562 km/s.

7.3 0.0805 and 0.997.

7.6 $t = 2.88T$.

7.8 $e = 0.9944$.

Chapter 8

8.1 1151.72 kg.

8.2 30124.75 kg.

Chapter 9

9.3 18.3642 days.

9.4 $\rho_0 = 65$ kg/m^3, $H = 15.9$ km.

Chapter 10

10.1 63.9 m/s (indicated); 74.1 m/s (true); $M = 0.226$.

10.7 3.3163×10^{-7} m and 0.6538 m.

10.10

$$C_D = \sin^2\theta + \frac{1}{s}\sqrt{\pi \frac{T_w}{T}} \sin\theta + 2\cos^2\theta.$$

10.11

$$C_D = 2 + \frac{2}{3s}\sqrt{\pi \frac{T_w}{T}} .$$

Chapter 11

11.2 $\dot{m} = 840.45$ kg/s.

11.3 $f_T = 22,861.45$ N.

11.4 $\dot{m}_f = 0.4536$ kg/s.

11.5 5.3.

Chapter 12

12.2

$$\begin{aligned}\dot{r} &= v \sin\phi, \\ \dot{\delta} &= \frac{v}{r}\cos\phi\cos A, \\ \dot{\lambda} &= \frac{v\cos\phi\sin A}{r\cos\delta}, \\ m\dot{v} &= f_T\cos\epsilon - D - mg\sin\phi, \\ \dot{A} &= \frac{v}{r}\cos\phi\sin A\tan\delta, \\ mv\dot{\phi} &= m\frac{v^2}{r}\cos\phi + f_T\sin\epsilon + L - mg\cos\phi.\end{aligned}$$

12.3 Flight is confined to a vertical plane ($A =$ const.):

$$\begin{aligned}\dot{h} &= v\sin\phi, \\ \dot{x} &= v\cos\phi\cos A, \\ \dot{y} &= v\cos\phi\sin A, \\ m\dot{v} &= f_T\cos\epsilon - D - mg\sin\phi, \\ \dot{A} &= 0, \\ mv\dot{\phi} &= f_T\sin\epsilon + L - mg\cos\phi.\end{aligned}$$

12.4 (a) $C_L = \sqrt{\frac{C_{Do}}{K}}$. (b) $C_L = \sqrt{\frac{C_{Do}}{3K}}$. (c) $C_L = \sqrt{\frac{3C_{Do}}{K}}$.

12.5 30 min, 12 s.

12.6 (a) $(L/D)_{\max} = 0.5/\sqrt{KC_{D0}} = 17.69$. (b) $s_{\max} = 4181.55$ km; $(L/D) = 15.32$.

12.10 $A = 288.2°$; 6.2 hr.

12.15 (a) $\dot{h}_{\max} = \frac{f_T - \frac{1}{2}\rho v_j^2 S(C_{D0}+KC_L^2)}{mg}$,
where

$$v = v_x\frac{\sqrt{x+\sqrt{x^2+3}}}{\sqrt{3}}, \qquad C_L = \frac{2mg}{\rho S v^2},$$

with x, v_x given in Exercise 12.7.

(b) $C_L = \sqrt{\frac{3C_{D0}}{K}}$; $\dot{h}_{\max} = \frac{P_{esh} - 2\rho v_p^3 SC_{D0}}{mg}$,
where

$$v_p = \sqrt{\frac{2mg\sqrt{K}}{\rho S\sqrt{3C_{D0}}}} .$$

12.16 $\dot{h}_{\max} = 35.49$ m/s; $v = 220.28$ m/s; $\phi \approx 9.3°$ (small).

12.18

$$\dot{s} = v,$$
$$m\dot{v} = f_T - D - \mu_r(mg - L),$$
$$s_G \approx \frac{v_{TO}^2}{2\dot{v}|_{v=\frac{v_{TO}}{\sqrt{2}}}},$$

where $v_{TO} = 1.2v_s$.

12.19

$$\text{(a)} \left|\frac{n - \cos\phi}{n - \cos\phi_0}\right| = e^{-\frac{g}{v_0^2}(h-h_0)} .$$

$$\text{(b)}\ v = v_0 \left|\frac{n - \cos\phi_0}{n - \cos\phi}\right| .$$

12.20 $v_0 = 219.09$ m/s; $h = 1199.5$ m.

12.21 $n = 1.816$.

12.22 (a) $\dot{A} = g\sqrt{n^2 - 1}/v = 26.23°$/s; $n = 9$; $v = 191.663$ m/s.
(b) $\dot{A} = 18.78°$/s; $n = 9$; $v = 267.643$ m/s.

12.23 (b) Subsonic: $\dot{A} = 3.08°$/s; $n = 1.767$; $v = 266$ m/s.
Supersonic: $\dot{A} = 3.23°$/s; $n = 2.86$; $v = 466.57$ m/s.

12.25

$$\frac{ds}{dh} = \cot\phi,$$
$$\frac{dv^2}{dh} = -2\left(\frac{D}{m\sin\phi} + g\right),$$
$$\frac{d\phi}{dh} = -\frac{g}{v^2}\cot\phi.$$

12.26

$$v = v_i e^{-Be^{-}h/H} ,$$

where $B \doteq -\frac{\rho_0 SHC_D}{2m\sin\phi_i}$ is called the *ballistic parameter*.
For $B > 0.5$: maximum deceleration,

$$-\dot{v} = -\frac{v_i^2 \sin\phi_i}{2He},$$

occurs at $h = H\ln 2B$.

For $B \leq 0.5$: the largest deceleration,

$$-\dot{v} = -\frac{v_i^2 Be^{-2B} \sin \phi_i}{H},$$

occurs at $h = 0$.
For $B > 1/3$:

$$\dot{Q}_{\max} = -\frac{mv_i^3 \sin \phi_i}{60He}$$

occurs at $h = H \ln 3B$.
For $B \leq 1/3$: the largest heating rate,

$$\dot{Q} = -\frac{mBe^{-3B} v_i^3 \sin \phi_i}{20H},$$

occurs at $h = 0$.

12.29

$$\begin{aligned}
\dot{h} &= v, \\
m\dot{v} &= -\dot{m}v_e - D - mg, \\
h_f &= -g\frac{(m_f - m_0)^2}{\dot{m}^2} + \frac{v_e}{\dot{m}}\left(m_0 - m_f + m_f \ln \frac{m_f}{m_0}\right), \\
v_f &= g\frac{m_f - m_0}{\dot{m}} - v_e \ln \frac{m_f}{m_0}.
\end{aligned}$$

Chapter 13

13.10 $\omega_p = 0.269$ rad/s, $\zeta_p = 0.102$, $t_{s_p} = 145.7$ s.
$\omega_s = 8.71$ rad/s, $\zeta_s = 0.511$, $t_{s_s} = 0.9$ s.

Chapter 15

15.1 Yes; no.

15.3 12 s, -40 ft/s^2, 0.012 rad/s.

References

[1] Whittaker, E.T.: *Analytical Dynamics of Particles and Rigid Bodies.* Cambridge University Press, Cambridge (1965).

[2] Wertz, J.R. (ed.): *Spacecraft Attitude Determination and Control.* Kluwer Academic Publishers, Dordrecht (1978).

[3] Shuster, M.D.: A survey of attitude representations. *J. Astronautical Sci.*, **41**, 439–517 (1993).

[4] Kreyszig, E.: *Advanced Engineering Mathematics.* Wiley, New York (2001).

[5] Abramowitz, M., and Stegun, I.A.: *Handbook of Mathematical Functions.* Dover Publications, New York (1974).

[6] Lemoine, F.G., Smith, D.E., Rowlands, D.D., Zuber, M.T., Neumann, G.A., and Chinn, D.S.: An improved solution of the gravity field of Mars (GMM-2B) from Mars Global Surveyor. *J. Geophysical Res.*, **106(E10)**, 23359–23376 (2001).

[7] Britting, K.R.: *Inertial Navigation Systems Analysis.* Wiley Interscience, Somerset, NJ (1971).

[8] Vallado, D.A.: *Fundamentals of Astrodynamics and Applications.* Kluwer Academic Press, Dordrecht (2001).

[9] Chobotov, V.A.: *Orbital Mechanics.* American Institute of Aeronautics and Astronautics (AIAA), Education Series, Washington, DC (1996).

[10] Battin, R.H.: *Astronautical Guidance.* McGraw-Hill, New York (1964).

[11] Battin, R.H.: *An Introduction to the Mathematics and Methods of Astrodynamics.* AIAA Education Series, Reston, VA (1999).

[12] Dunham, D.W., and Davis, S.A.: Optimization of a multiple Lunar–Swingby trajectory sequence. *J. Astron. Sci.*, **33**, 275–288 (1985).

[13] Well, K.H.: Use of the method of particular solutions in determining periodic orbits of the earth-moon system. *J. Astronautical Sci.*, **19**, 286–296 (1972).

[14] Byrnes, D.V.: Application of the pseudostate theory to the three-body Lambert problem. *J. Astronautical Sci.*, **37**, 221–232 (1989).

[15] Szebehely, V.: *Theory of Orbits.* Academic Press, New York (1967).

[16] Hill, T.: *An Introduction to Statistical Thermodynamics.* Addison-Wesley, Boston, MA (1960).

[17] Committee on Extension to the Standard Atmosphere (COESA): U.S. Standard Atmosphere 1976. U.S. Government Printing Office, Washington, DC (1976).

[18] COESA: U.S. Standard Atmosphere 1962. U.S. Government Printing Office, Washington, DC (1962).

[19] Regan, F.J., and Anandakrishnan, S.M.: Dynamics of Atmospheric Re-entry. AIAA Education Series, Washington, DC (1993).

[20] Houghton, J.T.: *The Physics of Atmospheres.* Cambridge University Press, Chester, NY (1977).

[21] Schlichting, H.: *Boundary Layer Theory.* McGraw-Hill, New York (1962).

[22] Anderson, J.D.: *Fundamentals of Aerodynamics.* McGraw-Hill, New York (1991).

[23] Anderson, D.A., Tannehill, J.C., and Pletcher, R.H.: *Computational Fluid Mechanics and Heat Transfer.* Hemisphere, Washington, DC (1984).

[24] Schaaf, S.A., and Chambre, P.L.: *Flow of Rarefied Gases. High-Speed Aerodynamics and Jet Propulsion*, III, *Fundamentals of Gas Dynamics*, Princeton University Press, Princeton, NJ (1958).

[25] Chapman, S., and Cowling, T.G.: *The Mathematical Theory of Non-Uniform Gases.* Cambridge University Press, Cambridge, U.K. (1991).

[26] Polhamus, E.C.: Predictions of vortex-lift characteristics by a leading-edge suction analogy. *J. Aircraft*, **8**, 193–199 (1971).

[27] Abbot, I., and Doenhoff, A.: *Theory of Wing Sections.* McGraw-Hill, New York (1949).

[28] Hoak, D.E., et al.: USAF Stability and Control Datcom. Air Force Flight Dynamics Laboratory, Wright-Patterson AFB, OH (1978).

[29] Hayes, W.D., and Probstein, R.F.: *Hypersonic Flow Theory.* Academic Press, New York (1959).

[30] Kerrebrock, J.L.: *Airbreathing Engines.* MIT Press, Cambridge, MA (1974).

[31] Hill, P.G., and Peterson, C.R.: *Mechanics and Thermodynamics of Propulsion.* Addison-Wesley, Reading, MA (1964).

[32] Mattingly, J.D., Heiser, W., and Daley, D.H.: *Aircraft Engine Design.* AIAA Education Series, New York (1987).

[33] Raymer, D.P.: *Aircraft Design: A Conceptual Approach.* AIAA Education Series, Washington, DC (1999).

[34] Miele, A.: *Flight Mechanics.* Addison-Wesley, Reading, MA (1962).

[35] Wing, D.J., et al.: Static internal investigation of a multiaxis thrust vectoring nozzle with variable internal contouring ability. NASA Technical Paper, TP-3628 (1997).

[36] Hoerner, S., and Borst, H.: *Fluid Dynamic Lift.* Hoerner Fluid Dynamics, Bricktown, NJ (1975).

[37] Stoliker, P.C., and Bosworth, J.T.: Evaluation of high-angle-of-attack handling qualities for the *X-31A* using standard evaluation maneuvers. NASA TM-104322 (1996).

[38] Opik, E.J.: *Physics of Meteor Flight in the Atmosphere.* Wiley Interscience, Somerset, NJ (1958).

[39] Loh, W.H.T.: *Dynamics and Thermodynamics of Planetary Entry.* Prentice-Hall, Englewood Cliffs, NJ (1963).

[40] Vinh, N.X.: *Hypersonic and Planetary Entry Flight Mechanics.* University of Michigan Press (1980).

[41] Jacobi, C.G.J.: *Journal fur Math.*, **39**, 293 (1849).

[42] Junkins, J.L., and Turner, J.D.: *Optimal Spacecraft Rotational Maneuvers.* Elsevier, Amsterdam (1986).

[43] Tewari, A.: *Modern Control Design with MATLAB and Simulink*. Wiley, Chichester (2002).

[44] Tewari, A.: Optimal nonlinear spacecraft attitude control through Hamilton–Jacobi formulation. *J. Astronautical Sci.*, **50**, 99–112 (2002).

[45] Etkin, B., and Reid, L.D.: *Dynamics of Flight: Stability and Control*. Wiley, New York (1995).

[46] Blakelock, J.H.: *Automatic Control of Aircraft and Missiles*. Wiley-Interscience, New York (1991).

[47] Kim, J.W., Crassidis, J.L., Vadali, S.R., and Dershowitz, A.L.: International Space Station leak localization using vent torque estimation. *Proc. Intl. Astronautical Congress*, Paper IAC-04-A.4.10 (2004).

[48] D'Azzo, J.J., and Houpis, C.H.: *Linear Control System Analysis and Design: Conventional and Modern*. McGraw-Hill, New York (1988).

[49] Using Simulink®–Simulation and Model-based Design (Version 6). MathWorks Inc., Natick, MA (2004).

[50] Haines, G.A., and Leondes, C.T.: Eddy current nutation dampers for dual-spin satellites. *J. Astronautical Sci.*, **21**, 1 (1973).

[51] Atkinson, K.E.: *An Introduction to Numerical Analysis*. Wiley, New York (2001).

[52] Fehlberg, E.: Low-order classical Runge–Kutta formulas with stepsize control and their application to some heat transfer problems. NASA TR R-315 (1969).

[53] MATLAB®Mathematics (Version 7). MathWorks Inc., Natick, MA (2004).

[54] Bettis, D.: Runge–Kutta–Nyström algorithms. *Celestial Mechanics*, **8**, 229–233 (1973).

[55] Fehlberg, E.: Classical eighth- and lower-order Runge–Kutta–Nyström formulas with stepsize control for special second-order differential equations. NASA TR R-381 (1972).

[56] Stiefel, E.L., and Scheifele, G.: *Linear and Regular Celestial Mechanics*. Springer-Verlag, New York (1971).

[57] Mrinal, K., and Tewari, A.: Trajectory and attitude simulation for aerocapture and aerobraking. *J. Spacecraft and Rockets*, **42**, 684–693 (2005).

[58] Ericsson, L.E., Mendenhall, M.R., and Perkins, S.C.: Review of forebody induced wing rock. *J. Aircraft*, **33**, 253–259 (1996).

[59] Hsu, C.H., and Lan, C.E.: Theory of wing rock. *J. Aircraft*, **22**, 920–924 (1985).

[60] Smail, L.L.: *Trigonometry: Plane and Spherical*. McGraw-Hill, New York (1952).

[61] Meirovitch, L.: *Elements of Vibration Analysis*. McGraw-Hill, New York (1986).

[62] Tewari, A.: Robust model reduction for a flexible spacecraft. *J. Guidance, Control, and Dynamics*, **21**, 809–812 (1998).

[63] Tewari, A.: Active vibration suppression of spacecraft with rate-weighted optimal control and multi-mode command shaping. AIAA-2004-5296, AIAA/AAS Astrodynamics Specialist Conference, Providence, RI (2004).

[64] Bisplinghoff, R.L., Ashley, H., and Halfman, R.L.: *Aeroelasticity*. Addison-Wesley, Cambridge, MA (1955).

[65] Garrick, I.E.: On some reciprocal relations in the theory of nonstationary flows. NACA TR 629 (1938).

[66] Albano, E., and Rodden, W.P.: A doublet-lattice method for calculating lift distributions on oscillating lifting surfaces in subsonic flows. *AIAA Journal*, **7**, 279–285 (1969).

[67] Appa, K.: Constant pressure panel method for supersonic unsteady airload analysis. *J. Aircraft*, **24**, 696–702 (1987).
[68] Ueda, T., and Dowell, E.H.: A new solution method for lifting surfaces in subsonic flow. *AIAA Journal*, **22**, 348–355 (1984).
[69] Tewari, A.: Doublet-point method for supersonic unsteady aerodynamics of nonplanar lifting-surfaces. *J. Aircraft*, **31**, 745–752 (1994).
[70] Eversman, W., and Tewari, A.: Consistent rational function approximation for unsteady aerodynamics. *J. Aircraft*, **28**, 545–552 (1991).
[71] Eversman, W., and Tewari, A.: Modified exponential series approximation for Theodorsen function. *J. Aircraft*, **28**, 553–557 (1991).
[72] Tewari, A.: Output-rate weighted optimal control of aeroelastic systems. *J. Guidance, Control, and Dynamics*, **24**, 409–411 (2001).
[73] Vreeburg, J.P.B., and Veldman, A.E.P.: Transient and sloshing motions in an unsupported container. *Physics of Fluids in Microgravity*, 293–321 (2001).
[74] Vreeburg, J.P.B.: Dynamics and control of a spacecraft with a moving, pulsating ball in a spherical cavity. *Acta Astronautica*, **40**, 257–274 (1997).

Index